STRAPDOWN ANALYTICS

PART 2

STRAPDOWN ANALYTICS

PART 2

Paul G. Savage

Published By:

Strapdown Associates, Inc.
Maple Plain, Minnesota

ISBN: 0-9717786-0-4
(Vol. 2 of 2 Vol. Set)

Foreword

This two volume text provides a detailed comprehensive discourse on the analytics of strapdown inertial navigation systems (INS's), the basic technology used on modern day commercial and military aircraft, guided missiles, surface ships and underwater vehicles. Based on his first-hand experience in this field, the author has provided a unique service to the aerospace industry in preparing this technical dissertation on the algorithms implemented in the strapdown system computer, and the analytics (and software) associated with system software validation, system test, simulation, performance analysis, and the analytical design methodology used in deriving the strapdown equations. Included is an in-depth chapter dealing with Kalman filter theory and its application to the aiding of a strapdown INS.

Strapdown Analytics has been prepared for the reader who may not have had experience in navigation or Kalman filtering. The analytical material presented is derived from scratch, showing the developmental steps in rigorous detail, without relying on reference material for supporting analytics. The book is complicated, yet complete and understandable by analytically inclined graduate students and practicing engineers. The book can be viewed as the text for an advanced course one might take following the introductory course taught by the author, *Introduction To Strapdown Inertial Navigation Systems.*

Contents For Parts 1 And 2

Part 1

Part 2

---

13 Analytical Solutions To The Strapdown Navigation Error Equations

13.0 OVERVIEW

Chapter 12 developed various forms of differential equations describing the dynamic error characteristics of strapdown inertial navigation error terms in various coordinate frames. The complete solution to these equations generally requires a numerical integration process on a digital computer for a particular set of initial conditions, noise characteristics, vertical control loop gains and trajectory profile. Unfortunately, although numerically correct, this procedure does not lend itself to providing an understanding of the various mechanisms that produce navigation error in general. In this chapter, we analyze the Chapter 12 equations under simplifying conditions that lend themselves to closed form analytical solution. The analytical results thus obtained provide the desired insight into the general behavior of the navigation error terms in response to initialization error effects and inertial sensor input error.

The chapter begins with a presentation of generalized vector relations that are used throughout the development of the analytic solutions. We then provide an analytical overview of the general characteristics of the strapdown inertial navigation error response in the vertical and horizontal channels for up to two hours and for long term system operation. Included is a discussion of the "open-loop" versus "closed-loop" vertical channel response and the effect of strapdown inertial sensor scale-factor/misalignment error on attitude/acceleration transformation accuracy. The remaining sections develop analytical solutions to the error differential equations for classical constraint conditions:

- Response to constant sensor errors at constant attitude for up to two hours.
- Solutions for constant sensor errors at rotating attitudes for up to two hours.
- Average long term horizontal solution for constant sensor errors at constant attitude.
- Navigation error response to inertial sensor random output noise.

In general, the coordinate frames we will be using in this chapter are the B, N, E, and I Frames defined in Section 2.2, with the I Frame as defined with more specificity in Section 12.0. An exception is Section 13.5 in which the E, N and I Frames have particular definitions selected for the problem being analyzed.

13.1 USEFUL VECTOR RELATIONSHIPS

The analytical developments in subsections to follow are laborious at times, and expedited by making use of some generalized vector relationships presented here.

Considering a general vector in the N Frame, we can write:

$$\underline{A}^N = \underline{A}_H^N + A_{ZN}\, \underline{u}_{ZN}^N \tag{13.1-1}$$

where

$\underline{A}^N$ = Arbitrary vector with components projected on N Frame coordinate axes.

$\underline{u}_{ZN}^N$ = Unit vector upward along the N Frame Z axis as projected onto N Frame axes. From the definition of the N Frame in Section 2.2, $\underline{u}_{ZN}^N$ lies along the upward local geodetic vertical.

$\underline{A}_H^N$ = Horizontal component of $\underline{A}^N$ defined as the component perpendicular to $\underline{u}_{ZN}^N$.

H = Designation for the horizontal component of the associated vector (i.e., the portion that is perpendicular to $\underline{u}_{ZN}^N$).

A_{ZN} = Vertical component of $\underline{A}^N$ defined as the component along $\underline{u}_{ZN}^N$.

By definition we can write:

$$A_{ZN} = \underline{u}_{ZN}^N \cdot \underline{A}^N \tag{13.1-2}$$

and from (13.1-1):

$$\underline{A}_H^N = \underline{A}^N - \left(\underline{u}_{ZN}^N \cdot \underline{A}^N\right) \underline{u}_{ZN}^N \tag{13.1-3}$$

From (13.1-1) we see that:

$$\underline{u}_{ZN}^N \times \underline{A}^N = \underline{u}_{ZN}^N \times \underline{A}_H^N \tag{13.1-4}$$

Using the Equation (3.1.1-16) general vector triple cross-product rule, the cross product of $\underline{u}_{ZN}^N$ with (13.1-4) is given by:

$$\underline{u}_{ZN}^N \times \left(\underline{u}_{ZN}^N \times \underline{A}^N\right) = -\,\underline{A}_H^N \tag{13.1-5}$$

or with the definition of the cross-product operator in Equation (3.1.1-13):

$$\left(\underline{u}_{ZN}^{N}\times\right)\left(\underline{u}_{ZN}^{N}\times\right)\underline{A}^{N} = -\underline{A}_{H}^{N} \tag{13.1-6}$$

We can also define $\underline{A}_{H}^{N}$ as:

$$\underline{A}_{H}^{N} = I_H\,\underline{A}^{N} \tag{13.1-7}$$

where

I_H = Horizontal identity matrix defined as a diagonal matrix with the 1,1 and 2,2 elements equal to unity and the 3,3 element equal to zero.

From (13.1-6) and (13.1-7) we then see that:

$$\left(\underline{u}_{ZN}^{N}\times\right)\left(\underline{u}_{ZN}^{N}\times\right) = -I_H \tag{13.1-8}$$

Equation (13.1-8) applies for a particular unit vector $\underline{u}_{ZN}^{N}$ defined to lie along the N Frame Z axis. For the more general case of an arbitrary unit vector, the equivalent to (13.1-5) and (13.1-6) is obtained with the Equation (3.1.1-16) vector triple cross-product rule:

$$\underline{u}_{Gen}^{N}\times\left(\underline{u}_{Gen}^{N}\times\underline{A}^{N}\right) = \left(\underline{u}_{Gen}^{N}\times\right)\left(\underline{u}_{Gen}^{N}\times\right)\underline{A}^{N} = \underline{u}_{Gen}^{N}\left(\underline{u}_{Gen}^{N}\cdot\underline{A}^{N}\right) - \underline{A}^{N} \tag{13.1-9}$$

where

$\underline{u}_{Gen}^{N}$ = General unit vector projected on N Frame axes.

Using the Equation (3.1.1-12) vector dot-product operator in the last term in (13.1-9), we get:

$$\left(\underline{u}_{Gen}^{N}\times\right)\left(\underline{u}_{Gen}^{N}\times\right)\underline{A}^{N} = \underline{u}_{Gen}^{N}\left(\underline{u}_{Gen}^{N}\right)^{T}\underline{A}^{N} - \underline{A}^{N} = \left[\underline{u}_{Gen}^{N}\left(\underline{u}_{Gen}^{N}\right)^{T} - I\right]\underline{A}^{N} \tag{13.1-10}$$

Because $\underline{A}^{N}$ has been defined as an arbitrary vector, (13.1-10) then assumes the general form:

$$\left(\underline{u}_{Gen}^{N}\times\right)\left(\underline{u}_{Gen}^{N}\times\right) = \underline{u}_{Gen}^{N}\left(\underline{u}_{Gen}^{N}\right)^{T} - I \tag{13.1-11}$$

An expression for the triple $\left(\underline{u}_{Gen}^{N}\times\right)$ product can also be derived from (13.1-9) as:

$$\begin{aligned}\underline{u}_{Gen}^{N}\times\left[\underline{u}_{Gen}^{N}\times\left(\underline{u}_{Gen}^{N}\times\underline{A}^{N}\right)\right] &= \left(\underline{u}_{Gen}^{N}\times\right)\left(\underline{u}_{Gen}^{N}\times\right)\left(\underline{u}_{Gen}^{N}\times\right)\underline{A}^{N}\\ &= \left(\underline{u}_{Gen}^{N}\times\right)\left[\underline{u}_{Gen}^{N}\left(\underline{u}_{Gen}^{N}\cdot\underline{A}^{N}\right) - \underline{A}^{N}\right] = -\left(\underline{u}_{Gen}^{N}\times\right)\underline{A}^{N}\end{aligned} \tag{13.1-12}$$

In (13.1-12) use was made of the fact that the cross product of a vector with itself is zero. Because $\underline{A}^{N}$ has been defined as an arbitrary vector, (13.1-12) then assumes the general form:

$$\left(\underline{u}_{Gen}^{N}\times\right)\left(\underline{u}_{Gen}^{N}\times\right)\left(\underline{u}_{Gen}^{N}\times\right) = -\left(\underline{u}_{Gen}^{N}\times\right) \qquad (13.1\text{-}13)$$

13.2 GENERAL NAVIGATION ERROR EQUATION CHARACTERISTICS

The general characteristic response of the strapdown inertial navigation error equations can be analyzed using Equations (12.5.4-1). These equations can be revised by incorporating Equation (12.1.2-4) and the N Frame version of (12.3.6.1-29), repeated below:

$$\underline{\omega}_{IN}^{N} = \underline{\omega}_{IE}^{N} + \underline{\omega}_{EN}^{N} \qquad (13.2\text{-}1)$$

$$\underline{\omega}_{EN}^{N} \approx \rho_{ZN}\, \underline{u}_{ZN}^{N} + \frac{1}{R}\left(\underline{u}_{ZN}^{N} \times \underline{v}_{H}^{N}\right) \qquad (13.2\text{-}2)$$

We can also write for the $\underline{\omega}_{IE}^{N}$ earth rate vector using generalized Equations (13.1-1) and (13.1-2):

$$\underline{\omega}_{IE}^{N} = \underline{\omega}_{IE_H}^{N} + \left(\underline{\omega}_{IE}^{N} \cdot \underline{u}_{ZN}^{N}\right) \underline{u}_{ZN}^{N} \qquad (13.2\text{-}3)$$

where

$\underline{\omega}_{IE_H}^{N}$ = Horizontal component of $\underline{\omega}_{IE}^{N}$.

It will prove expeditious in this section to select the N Frame as being of the "Free Azimuth" type which (as defined in Section 4.5) sets ρ_{ZN} equal to the negative of the vertical earth rate component:

$$\rho_{ZN} = -\,\underline{\omega}_{IE}^{N} \cdot \underline{u}_{ZN}^{N} \qquad (13.2\text{-}4)$$

With (13.2-2) - (13.2-4), Equation (13.2-1) becomes:

$$\underline{\omega}_{IN}^{N} = \frac{1}{R}\left(\underline{u}_{ZN}^{N} \times \underline{v}_{H}^{N}\right) + \underline{\omega}_{IE_H}^{N} = \underline{\omega}_{IN_H}^{N} \qquad (13.2\text{-}5)$$

where

$\underline{\omega}_{IN_H}^{N}$ = Horizontal component of $\underline{\omega}_{IN}^{N}$.

It is advantageous for this development to define the $\delta\underline{\upsilon}^{N}$, $\delta\underline{R}^{N}$ velocity/position error terms in Equations (12.5.4-1) as the sum of their horizontal and vertical components:

$$\delta\underline{\upsilon}^{N} = \delta\underline{\upsilon}_{H}^{N} + \delta\upsilon_{R}\, \underline{u}_{ZN}^{N} \qquad (13.2\text{-}6)$$

$$\delta \underline{R}^N = \delta \underline{R}_H^N + \delta R \, \underline{u}_{ZN}^N \tag{13.2-7}$$

where

$\delta \underline{\upsilon}_H^N, \delta \underline{R}_H^N$ = Horizontal components of $\delta \underline{\upsilon}^N$, $\delta \underline{R}^N$.

$\delta \upsilon_R, \delta R$ = Vertical components of $\delta \underline{\upsilon}^N$, $\delta \underline{R}^N$.

With (13.2-5) - (13.2-7) and general Equation (13.1-5), particular cross-product terms in Equations (12.5.4-1) become:

$$\begin{aligned} \underline{\omega}_{IN}^N \times \delta \underline{\upsilon}^N &= \underline{\omega}_{IN\,H}^N \times \delta \underline{\upsilon}_H^N + \left(\underline{\omega}_{IN\,H}^N \times \underline{u}_{ZN}^N \right) \delta \upsilon_R \\ &= \underline{\omega}_{IN\,H}^N \times \delta \underline{\upsilon}_H^N - \underline{u}_{ZN}^N \times \left(\frac{1}{R} \, \underline{u}_{ZN}^N \times \underline{v}_H^N + \underline{\omega}_{IE\,H}^N \right) \delta \upsilon_R \\ &= \underline{\omega}_{IN\,H}^N \times \delta \underline{\upsilon}_H^N + \left(\underline{\omega}_{IE\,H}^N \times \underline{u}_{ZN}^N + \frac{1}{R} \, \underline{v}_H^N \right) \delta \upsilon_R \end{aligned} \tag{13.2-8}$$

and similarly:

$$\underline{\omega}_{IN}^N \times \delta \underline{R}^N = \underline{\omega}_{IN\,H}^N \times \delta \underline{R}_H^N + \left(\underline{\omega}_{IE\,H}^N \times \underline{u}_{ZN}^N + \frac{1}{R} \, \underline{v}_H^N \right) \delta R \tag{13.2-9}$$

Let us now substitute (13.2-6) - (13.2-9) into the Equations (12.5.4-1) velocity/position error rate expressions to obtain the equivalent Free Azimuth N Frame forms:

$$\begin{aligned} \delta \dot{\underline{R}}_H^N + \delta \dot{R} \, \underline{u}_{ZN}^N = {} & \delta \underline{\upsilon}_H^N + \delta \upsilon_R \, \underline{u}_{ZN}^N - \underline{\omega}_{IN\,H}^N \times \delta \underline{R}_H^N - \left(\underline{\omega}_{IE\,H}^N \times \underline{u}_{ZN}^N + \frac{1}{R} \, \underline{v}_H^N \right) \delta R \\ & - C_3 \left(\delta R - \delta h_{Prsr} \right) \underline{u}_{ZN}^N + C_B^N \, \delta \underline{\upsilon}_{Quant} \end{aligned} \tag{13.2-10}$$

$$\begin{aligned} \delta \dot{\underline{\upsilon}}_H^N + \delta \dot{\upsilon}_R \, \underline{u}_{ZN}^N = {} & C_B^N \, \delta \underline{a}_{SF}^B + \underline{a}_{SF}^N \times \underline{\psi}^N - \frac{g}{R} \, \delta \underline{R}_H^N - \underline{\omega}_{IN_H}^N \times \delta \underline{\upsilon}_H^N - \left(\underline{\omega}_{IE_H}^N \times \underline{u}_{ZN}^N + \frac{1}{R} \, \underline{v}_H^N \right) \delta \upsilon_R \\ & + \delta \underline{g}_{Mdl}^N + \left[\left(F(h) \, \frac{g}{R} - C_2 \right) \delta R - \delta e_{vc_3} + C_2 \, \delta h_{Prsr} \right] \underline{u}_{ZN}^N \\ & - \left(\underline{a}_{SF}^N \times \right) C_B^N \, \delta \underline{\alpha}_{Quant} - \left[\left(C_B^N \, \underline{\omega}_{IB}^B \right) \times \right] C_B^N \, \delta \underline{\upsilon}_{Quant} \end{aligned} \tag{13.2-11}$$

$F(h) = 2 \quad \text{For } h \geq 0 \qquad\qquad F(h) = -1 \quad \text{For } h < 0$

The $\underline{\psi}^N$ attitude error and δe_{vc_3} integral control terms in Equations (13.2-11) are provided by the Equations (12.5.4-1) expressions:

$$\dot{\underline{\psi}}^N = - C_B^N \, \delta\underline{\omega}_{IB}^B - \underline{\omega}_{IN}^N \times \underline{\psi}^N + C_B^N \left(\underline{\omega}_{IB}^B \times \delta\underline{\alpha}_{Quant} \right) \tag{13.2-12}$$

$$\delta\dot{e}_{vc_3} = C_1 \left(\delta R - \delta h_{Prsr} \right) \tag{13.2-13}$$

An important characteristic of Equations (13.2-10) - (13.2-13) is the dependence of the $\underline{\psi}^N$ attitude error only on angular rate sensor errors ($\delta\underline{\omega}_{IB}^B$, $\delta\underline{\alpha}_{Quant}$ in Equation (13.2-12)). Recall from Section 12.2.1 Equations (12.2.1-4) and (12.2.1-7) that $\underline{\psi}^N$ represents the rotation angle error vector in the C_B^E matrix projected on N Frame axes. On the other hand, we see from Equations (13.2-10) - (13.2-11) that the position and velocity errors ($\delta\underline{R}^N$, $\delta\underline{\upsilon}^N$) are functions of accelerometer error ($\delta\underline{a}_{SF}^B$, $\delta\underline{\upsilon}_{Quant}$) as well as (through the $\underline{\psi}^N$ coupling) angular rate sensor error ($\delta\underline{\omega}_{IB}^B$, $\delta\underline{\alpha}_{Quant}$). Subsections 13.2.1 through 13.2.4 to follow describe general characteristic solutions to position/velocity error Equations (13.2-10) - (13.2-11) including a discussion of the general effects of strapdown inertial sensor scale-factor/misalignment error on attitude/acceleration-transformation accuracy.

13.2.1 VERTICAL CHANNEL RESPONSE

In order to understand the vertical channel response characteristics of Equations (13.2-10) - (13.2-11), let us first analyze the case in which the vertical controls are eliminated (i.e., the C_1, C_2, C_3 gains are set to zero). The vertical component of (13.2-10) and (13.2-11) with zero vertical channel gains is obtained from the dot product with $\underline{u}_{ZN}^N$, recognizing from (13.2-13) that $\delta e_{vc_3} = 0$ for $C_1 = 0$:

$$\delta\dot{R} = \delta\upsilon_R - \left(\underline{\omega}_{IN_H}^N \times \delta\underline{R}_H^N \right) \cdot \underline{u}_{ZN}^N + \delta\dot{R}_{Force} \tag{13.2.1-1}$$

$$\delta\dot{\upsilon}_R = F(h) \, \frac{g}{R} \, \delta R - \left(\underline{\omega}_{IN_H}^N \times \delta\underline{\upsilon}_H^N \right) \cdot \underline{u}_{ZN}^N + \delta\dot{\upsilon}_{Force}$$

$$F(h) = 2 \quad \text{For} \;\; h \geq 0 \qquad\qquad F(h) = -1 \quad \text{For} \;\; h < 0 \tag{13.2.1-2}$$

with

$$\delta\dot{R}_{Force} = \left(C_B^N \, \delta\underline{\upsilon}_{Quant} \right) \cdot \underline{u}_{ZN}^N$$

$$\delta\dot{\upsilon}_{Force} = \Big\{ C_B^N \, \delta\underline{a}_{SF}^B + \underline{a}_{SF}^N \times \underline{\psi}^N + \delta\underline{g}_{Mdl}^N - \left(\underline{a}_{SF}^N \times \right) C_B^N \, \delta\underline{\alpha}_{Quant} - \left[\left(C_B^N \, \underline{\omega}_{IB}^B \right) \times \right] C_B^N \, \delta\underline{\upsilon}_{Quant} \Big\} \cdot \underline{u}_{ZN}^N \tag{13.2.1-3}$$

where

$\delta\dot{R}_{Force}$ = Equation (13.2.1-1) forcing term.

$\delta\dot{\upsilon}_{Force}$ = Equation (13.2.1-2) forcing term.

Equations (13.2.1-1) and (13.2.1-2) can be combined by taking the time derivative of (13.2.1-1) and substituting (13.2.1-2) for $\delta\dot{\upsilon}_R$. The result after rearrangement is an equation for open-loop vertical channel response that is suitable for characteristics analysis:

$$\delta\ddot{R} - F(h)\,\frac{g}{R}\,\delta R = -\left(\underline{\dot{\omega}}^N_{IN_H} \times \delta\underline{R}^N_H + \underline{\omega}^N_{IN_H} \times \delta\underline{\dot{R}}^N_H\right) \cdot \underline{u}^N_{ZN}$$
$$- \left(\underline{\omega}^N_{IN_H} \times \delta\underline{\upsilon}^N_H\right) \cdot \underline{u}^N_{ZN} + \frac{d}{dt}\,\delta\dot{R}_{Force} + \delta\dot{\upsilon}_{Force} \qquad (13.2.1\text{-}4)$$

$F(h) = 2 \quad \text{For} \quad h \geq 0 \qquad\qquad F(h) = -1 \quad \text{For} \quad h < 0$

Equation (13.2.1-4) with (13.2.1-3) shows that the vertical channel position response (δR) is driven by accelerometer error ($\delta\underline{a}^B_{SF}$), attitude error ($\underline{\psi}^N$), accelerometer and angular rate sensor quantization noise ($\delta\upsilon_{Quant}$ and $\delta\alpha_{Quant}$) and cross-coupling effects between the horizontal and vertical channels ($\delta\underline{R}^N_H$, $\delta\underline{\dot{R}}^N_H$, and $\delta\underline{\upsilon}^N_H$). The cross-coupling effects are generally small and can be ignored for this analysis. Neglecting the cross-coupling allows Equation (13.2.1-4) to be solved independently as a function of initial conditions and input forcing function. In classical fashion, we can then determine the characteristic roots of Equation (13.2.1-4) from the homogeneous portion:

$$\delta\ddot{R} - F(h)\,\frac{g}{R}\,\delta R = 0 \qquad (13.2.1\text{-}5)$$

The roots of (13.2.1-5) are determined from the general trial homogeneous solution:

$$\delta R_{Hmg} = B\, e^{\lambda t} \qquad (13.2.1\text{-}6)$$

where

δR_{Hmg} = Homogeneous trial solution to (13.2.1-5).

λ = Characteristic root of (13.2.1-5).

B = Constant.

Substituting (13.2.1-6) for δR in (13.2.1-5) then yields the characteristic root equation:

$$\lambda^2 - F(h)\,\frac{g}{R} = 0 \tag{13.2.1-7}$$

whose solution is:

$$\lambda = \pm\sqrt{F(h)\,\frac{g}{R}} \tag{13.2.1-8}$$

The complete solution to (13.2.1-5) is a sum of terms of the form (13.2.1-6) for each (13.2.1-8) root with the B coefficients selected to match the initial conditions for δR and $\delta \dot{R}$. Let's look at this solution for the usual case when altitude h (i.e., distance above earth's reference ellipsoid surface) is greater than zero for which (from (13.2.1-4)) $F(h) = 2$. Then:

For $h \geq 0$:

$$\delta R_{HmgTot} = \delta R_0 \cosh\left(\sqrt{\frac{2\,g}{R}}\;t\right) + \delta \dot{R}_0 \sqrt{\frac{R}{2\,g}}\;\sinh\left(\sqrt{\frac{2\,g}{R}}\;t\right) \tag{13.2.1-9}$$

with

$$\cosh x \equiv \frac{1}{2}\left(e^{x} + e^{-x}\right) = 1 + \frac{1}{2\,!}\,x^2 + \frac{1}{4\,!}\,x^4 + \cdots \tag{13.2.1-10}$$

$$\sinh x \equiv \frac{1}{2}\left(e^{x} - e^{-x}\right) = x + \frac{1}{3\,!}\,x^3 + \frac{1}{5\,!}\,x^5 + \cdots \tag{13.2.1-11}$$

where

δR_{HmgTot} = Total homogeneous solution.

δR_0, $\delta \dot{R}_0$ = Initial values for δR, $\delta \dot{R}$ at time $t = 0$.

cosh, sinh = Hyperbolic cosine, sine functions.

It is easily verified by substituting (13.2.1-10) and (13.2.1-11) into (13.2.1-9) that δR_{HmgTot} and its derivative equal δR_0, $\delta \dot{R}_0$ at $t = 0$ as required.

Continuing to neglect the horizontal cross-coupling terms in (13.2.1-4), the total solution for δR is then obtained in classical fashion as the sum of (13.2.1-9) plus the integrated effect of the Equations (13.2.1-1) and (13.2.1-2) forcing terms operating through the (13.2.1-9) δR_0 and $\delta \dot{R}_0$ weighting coefficients at each time instant:

For $h \geq 0$:

$$\delta R = \delta R_0 \cosh\left(\sqrt{\frac{2\,g}{R}}\ t\right) + \delta\dot{R}_0 \sqrt{\frac{R}{2\,g}}\ \sinh\left(\sqrt{\frac{2\,g}{R}}\ t\right)$$
$$+ \int_0^t \delta\dot{R}_{Force}(\tau) \cosh\left(\sqrt{\frac{2\,g}{R}}\ (t - \tau)\right) d\tau \qquad (13.2.1\text{-}12)$$
$$+ \int_0^t \delta\dot{\upsilon}_{Force}(\tau) \sqrt{\frac{R}{2\,g}}\ \sinh\left(\sqrt{\frac{2\,g}{R}}\ (t - \tau)\right) d\tau$$

where

τ = Past time parameter.

t = Current time.

Equation (13.2.1-12) clearly illustrates the exponentially unstable nature of the open-loop vertical channel for positive altitude and for time t larger than the characteristic root time constant $\sqrt{\frac{R}{2\,g}}$. At the surface of the earth on the equator, $\sqrt{\frac{R}{2\,g}} = 9.49$ minutes. For small t compared to $\sqrt{\frac{R}{2\,g}}$, the (13.2.1-12) cosh and sinh functions can be approximated by 1 and $\sqrt{\frac{2\,g}{R}}\ t$ respectively which is a reasonable solution before the exponential enlargement characteristic begins to develop. For t = 9.49 minutes the $\sqrt{\frac{2\,g}{R}}\ t$ argument for the cosh and sinh functions in (13.2.1-12) is unity for which cosh (1) = 1.54 and sinh (1) = 1.18 (compared to the previous small t approximation of 1 for each of these terms). For larger values of t, the magnitudes of the cosh, sinh functions increase exponentially. We can conclude that for positive altitude and navigation times greater than 10 minutes, the unstable nature of the open-loop vertical channel may generate unacceptably large vertical navigation errors. Thus, for extended navigation time applications at positive altitude, the need for vertical channel control becomes a requirement.

The source of the vertical channel instability discussed above can be traced to gravity error along the local vertical created by vertical position error. Because gravity magnitude decreases as the square of distance from the center of the earth (for positive altitude), an upward vertical position error produces a smaller gravity magnitude, hence, a negative gravity magnitude error. The gravity magnitude error so created equals $-\frac{2\,g}{R}\,\delta R$ which decreases downward acceleration, hence, increases upward acceleration (the “2” arises from the inverse “square” gravity magnitude law). The net result is that the upward position acceleration $\delta\ddot{R}$ has a positive addition of $\frac{2\,g}{R}\,\delta R$ which is the source of the $\delta\ddot{R} - 2\,\frac{g}{R}\,\delta R$ term in Equation (13.2.1-4) (when $F(h) = 2$ for positive altitude).

Now, let's look at the vertical channel solution for the case when altitude h is negative and, from (13.2.1-4), $F(h) = -1$. Then, the characteristic λ roots in (13.2.1-8) are $\pm j\sqrt{\frac{g}{R}}$ (in which $j = \sqrt{-1}$), and the classical solution is similar to (13.2.1-12), but with the sinh, cosh functions replaced by sine, cosine functions at the λ root magnitude frequency:

For $h < 0$:

$$
\begin{aligned}
\delta R \; = \; & \delta R_0 \cos\left(\sqrt{\frac{g}{R}}\; t\right) + \delta\dot{R}_0 \sqrt{\frac{R}{g}} \sin\left(\sqrt{\frac{g}{R}}\; t\right) \\
& + \int_0^t \delta\dot{R}_{Force}(\tau) \cos\left(\sqrt{\frac{g}{R}}\;(t - \tau)\right) d\tau \\
& + \int_0^t \delta\dot{\upsilon}_{Force}(\tau) \sqrt{\frac{R}{g}} \sin\left(\sqrt{\frac{g}{R}}\;(t - \tau)\right) d\tau
\end{aligned}
\qquad (13.2.1\text{-}13)
$$

Equation (13.2.1-13) illustrates that for negative altitude, the characteristic response of the open-loop vertical channel is to exhibit undamped oscillations at frequency $\sqrt{\frac{g}{R}}$. At the surface of the earth on the equator, $\sqrt{\frac{g}{R}} = 0.00124$ radians per second which corresponds to an oscillation period of 84 minutes. Interestingly, the horizontal channel response (discussed in the next section) has this same 84 minute undamped oscillation characteristic.

The source of the vertical channel oscillation at negative altitude can also be traced to gravity error along the local vertical created by vertical position error. Because gravity magnitude increases with distance from the center of the earth (for negative altitude), an upward (i.e., away from earth's center) vertical position error produces a larger gravity magnitude, hence, a positive gravity magnitude error. The gravity magnitude error so created equals $\frac{g}{R}\,\delta R$ which increases downward (toward earth center) acceleration, hence, decreases upward acceleration. The net result is that the upward position acceleration $\delta\ddot{R}$ has a negative addition of $\frac{g}{R}\,\delta R$ which is the source of the $\delta\ddot{R} + \frac{g}{R}\delta R$ term in Equation (13.2.1-4) (when F(h) = - 1 for negative altitude).

Although the open-loop vertical channel response for negative altitude is not unstable, the undamped oscillatory response exhibited by Equation (13.2.1-13) is still generally unacceptable for extended navigation times, due to the large vertical position errors created. For example, a 0.1 fps initial velocity error $\delta\dot{R}_0$ would generate an 84 minute period (i.e., 0.00124 rad/sec frequency) oscillatory altitude error with an amplitude of $\frac{0.1}{0.00124} = 80.6$ ft. Thus, to limit the altitude error, we find that vertical channel control is generally also applied for negative as well as positive altitudes.

With vertical channel control applied for both positive and negative altitudes, the governing vertical loop characteristic response equation is developed in the same manner as (13.2.1-4), but with the (13.2-10) - (13.2-11) vertical loop control gains included and $\delta\dot{e}_{vc_3}$ substituted from (13.2-13):

$$
\begin{aligned}
&\delta\ddot{R} + C_3\,\delta\dot{R} + \left(C_2 - F(h)\,\frac{g}{R}\right)\delta R + C_1\int \delta R\,dt \\
&\quad = -\left(\dot{\underline{\omega}}^N_{IN_H} \times \delta\underline{R}^N_H + \underline{\omega}^N_{IN_H} \times \delta\dot{\underline{R}}^N_H\right)\cdot \underline{u}^N_{ZN} \\
&\qquad -\left(\underline{\omega}^N_{IN_H} \times \delta\underline{\upsilon}^N_H\right)\cdot \underline{u}^N_{ZN} + \frac{d}{dt}\,\delta\dot{R}_{Force} + \delta\dot{\upsilon}_{Force} \qquad (13.2.1\text{-}14) \\
&\qquad + C_3\,\delta\dot{h}_{Prsr} + C_2\,\delta h_{Prsr} + C_1\int \delta h_{Prsr}\,dt
\end{aligned}
$$

$F(h) = 2 \quad \text{For } h \geq 0 \qquad\qquad F(h) = -1 \quad \text{For } h < 0$

The vertical velocity error associated with the solution to (13.2.1-14) is obtained from the rearranged form of the vertical component of (13.2-10) including the (13.2.1-3) definition for $\delta\dot{R}_{Force}$:

$$
\delta\upsilon_R = \delta\dot{R} + C_3\left(\delta R - \delta h_{Prsr}\right) + \left(\underline{\omega}^N_{IN_H} \times \delta\underline{R}^N_H\right)\cdot \underline{u}^N_{ZN} - \delta\dot{R}_{Force} \qquad (13.2.1\text{-}15)
$$

The characteristic response of δR can be ascertained from the derivative of (3.2.1-14), using (13.2.1-3) for $\delta\dot{\upsilon}_{Force}$, and subtracting $\frac{d^3}{dt^3}\delta h_{Prsr}$ from both sides of the equal sign:

$$\frac{d^3}{dt^3}\left(\delta R - \delta h_{Prsr}\right) + C_3 \frac{d^2}{dt^2}\left(\delta R - \delta h_{Prsr}\right) + C_2 \frac{d}{dt}\left(\delta R - \delta h_{Prsr}\right) + C_1\left(\delta R - \delta h_{Prsr}\right) = \delta\dot{a}_{SF_{ZN}} - \frac{d^3}{dt^3}\delta h_{Prsr} + \cdots \tag{13.2.1-16}$$

where

$\delta a_{SF_{ZN}}$ = The vertical (N Frame Z axis) component of the $\delta \underline{a}_{SF}$ specific force acceleration error vector (e.g., from accelerometers).

For stable dynamic response control gains, Equation (13.2.1-16) shows that δR will reach a stable bounded solution equal to δh_{Prsr} plus small additions caused by time variations in $\delta a_{SF_{ZN}}$ (due principally to angular maneuvering which changes the coupling of accelerometer errors into the vertical), third derivative changes in δh_{Prsr}, and residual terms in the (13.2.1-14) forcing functions. In practice, the δR position error will approximately equal δh_{Prsr} and $\delta \dot{R}$ will be small. From Equation (13.2.1-15) (and (13.2.1-3) for $\delta \dot{R}_{Force}$), we see that this also corresponds to a small bounded value for $\delta \upsilon_R$. This forms the rationale (for analysis purposes) to neglect the vertical loop coupling into the horizontal position/velocity error equations under closed-loop vertical control conditions.

13.2.2 HORIZONTAL CHANNEL RESPONSE

To analyze the characteristics of the horizontal navigation error channel, we return to Equations (13.2-10) - (13.2-13) and investigate the response under closed-loop vertical channel control conditions. As discussed in Section 13.2.1, this corresponds to δR approximately equal to δh_{Prsr}, and $\delta \upsilon_R$ small. Because δh_{Prsr} is small, δR will also be small, hence, we can neglect both δR and $\delta \upsilon_R$ in Equations (13.2-10) and (13.2-11). We also note that the $\underline{\omega}_{IN\,H}^{N} \times \delta \underline{R}_H^N$ and $\underline{\omega}_{IN\,H}^{N} \times \delta \underline{\upsilon}_H^N$ terms in (13.2-10) - (13.2-11) are cross-products between horizontal vectors which, therefore, are vertical. Based on the previous observations, we see then that the horizontal components of (13.2-10) - (13.2-11) under closed-loop vertical control conditions can be nicely approximated as:

$$\delta \dot{\underline{R}}_H^N \approx \delta \underline{\upsilon}_H^N + \delta \dot{\underline{R}}_{H\,Force}^N \tag{13.2.2-1}$$

$$\delta \dot{\underline{\upsilon}}_H^N \approx -\frac{g}{R}\,\delta \underline{R}_H^N + \delta \dot{\underline{\upsilon}}_{H\,Force}^N \tag{13.2.2-2}$$

with

$$\delta \dot{\underline{R}}^N_{H_{Force}} = \left(C^N_B \, \delta \underline{\upsilon}_{Quant} \right)_H$$

$$\delta \dot{\underline{\upsilon}}^N_{H_{Force}} = \Big\{ C^N_B \, \delta \underline{a}^B_{SF} + \underline{a}^N_{SF} \times \underline{\psi}^N + \delta \underline{g}^N_{Mdl} - \left(\underline{a}^N_{SF} \times \right) C^N_B \, \delta \underline{\alpha}_{Quant} - \left[\left(C^N_B \, \underline{\omega}^B_{IB} \right) \times \right] C^N_B \, \delta \underline{\upsilon}_{Quant} \Big\}_H \tag{13.2.2-3}$$

where

H = Designation for horizontal component of the associated vector.

$\delta \dot{\underline{R}}^N_{H_{Force}}$ = Equation (13.2.2-1) forcing term.

$\delta \dot{\underline{\upsilon}}^N_{H_{Force}}$ = Equation (13.2.2-2) forcing term.

The general characteristics of the horizontal navigation errors are obtained by combining Equations (13.2.2-1) and (13.2.2-2). Differentiating (13.2.2-1), substituting (13.2.2-2) for $\delta \dot{\underline{\upsilon}}^N_H$ and rearrangement yields the characteristic horizontal error response equation:

$$\delta \ddot{\underline{R}}^N_H + \frac{g}{R} \, \delta \underline{R}^N_H = \frac{d}{dt} \, \delta \dot{\underline{R}}^N_{H_{Force}} + \delta \dot{\underline{\upsilon}}^N_{H_{Force}} \tag{13.2.2-4}$$

The homogeneous portion of (13.2.2-4) is:

$$\delta \ddot{\underline{R}}^N_H + \frac{g}{R} \, \delta \underline{R}^N_H = 0 \tag{13.2.2-5}$$

whose characteristic solution is:

$$\delta \underline{R}^N_{H_{Hmg}} = \delta \underline{R}^N_{H_0} \cos \omega_S t + \delta \dot{\underline{R}}^N_{H_0} \frac{\sin \omega_S t}{\omega_S} \tag{13.2.2-6}$$

with

$$\omega_S = \sqrt{\frac{g}{R}} \tag{13.2.2-7}$$

where

$\delta \underline{R}^N_{H_{Hmg}}$ = Solution to homogeneous Equation (13.2.2-5).

$\delta \underline{R}^N_{H_0}, \ \delta \dot{\underline{R}}^N_{H_0}$ = Initial values for $\delta \underline{R}^N_H, \ \delta \dot{\underline{R}}^N_H$.

The solution to (13.2.2-4) is then obtained in classical fashion as the sum of (13.2.2-6) plus the integral of the Equation (13.2.2-1) and (13.2.2-2) forcing functions multiplied by the Equation (13.2.2-6) $\delta \underline{R}_{H_0}^N$, $\delta \dot{\underline{R}}_{H_0}^N$ weighting coefficients:

$$\delta \underline{R}_H^N = \delta \underline{R}_{H_0}^N \cos \omega_S t + \delta \dot{\underline{R}}_{H_0}^N \frac{\sin \omega_S t}{\omega_S}$$
$$+ \int_0^t \left[\delta \dot{\underline{R}}_{H_{Force}}^N(\tau) \cos\big(\omega_S (t - \tau)\big) + \delta \dot{\underline{\upsilon}}_{H_{Force}}^N(\tau) \frac{\sin\big(\omega_S (t - \tau)\big)}{\omega_S} \right] d\tau \qquad (13.2.2\text{-}8)$$

Equation (13.2.2-8) shows that in the Free Azimuth N Frame, the horizontal position error (and its rate of change) contains undamped sinusoidal oscillations at frequency ω_S. Once generated, these oscillation patterns continue to repeat in the Free Azimuth coordinate frame. At the surface of the earth on the equator, $\omega_S = 0.00124$ radians per second which corresponds to an oscillation period of approximately 84 minutes. The horizontal velocity error $\delta \underline{\upsilon}_H^N$ is similarly affected as can be seen from the rearranged form of (13.2.2-1):

$$\delta \underline{\upsilon}_H^N = \delta \dot{\underline{R}}_H^N - \delta \dot{\underline{R}}_{H_{Force}}^N \qquad (13.2.2\text{-}9)$$

It is also instructive to ascertain the response characteristic of the C_B^N matrix horizontal tilt error $\underline{\gamma}_H^N$, which can be determined from the horizontal component of the Equation (12.2.1-17) equivalency relation:

$$\underline{\gamma}_H^N = \underline{\psi}_H^N + \underline{\varepsilon}_H^N \qquad (13.2.2\text{-}10)$$

Anticipating the results of our analysis, we double differentiate (13.2.2-10) and combine it with the product of (13.2.2-10) and ω_S^2 to obtain:

$$\ddot{\underline{\gamma}}_H^N + \omega_S^2 \underline{\gamma}_H^N = \ddot{\underline{\psi}}_H^N + \omega_S^2 \underline{\psi}_H^N + \ddot{\underline{\varepsilon}}_H^N + \omega_S^2 \underline{\varepsilon}_H^N \qquad (13.2.2\text{-}11)$$

The $\left(\ddot{\underline{\varepsilon}}_H^N + \omega_S^2 \underline{\varepsilon}_H^N\right)$ portion of (13.2.2-11) can be defined from the horizontal component of equivalency Equation (12.2.3-19) with (13.1-4):

$$\underline{\varepsilon}_H^N = \frac{1}{R}\left(\underline{u}_{ZN}^N \times \delta \underline{R}_H^N\right) \qquad (13.2.2\text{-}12)$$

Taking the cross-product of Equation (13.2.2-4) with $\underline{u}_{ZN}^{N}$, substituting (13.2.2-12) and its second derivative in the result, and incorporating the ω_S definition from Equation (13.2.2-7) shows that:

$$\ddot{\underline{\varepsilon}}_H^N + \omega_S^2 \, \underline{\varepsilon}_H^N = \frac{1}{R} \, \underline{u}_{ZN}^N \times \left(\frac{d}{dt} \, \delta\dot{\underline{R}}_{H_{Force}}^N + \delta\dot{\underline{\upsilon}}_{H_{Force}}^N \right) \qquad (13.2.2\text{-}13)$$

We also note with (13.2.2-7) and general Equation (13.1-6) that:

$$\omega_S^2 \, \underline{\psi}_H^N = - \frac{g}{R} \, \underline{u}_{ZN}^N \times \left(\underline{u}_{ZN}^N \times \underline{\psi}^N \right) \qquad (13.2.2\text{-}14)$$

Substituting (13.2.2-13) and (13.2.2-14) into (13.2.2-11) then provides:

$$\ddot{\underline{\gamma}}_H^N + \omega_S^2 \, \underline{\gamma}_H^N = \ddot{\underline{\psi}}_H^N + \frac{1}{R} \, \underline{u}_{ZN}^N \times \left(\frac{d}{dt} \, \delta\dot{\underline{R}}_{H_{Force}}^N + \delta\dot{\underline{\upsilon}}_{H_{Force}}^N - g \, \underline{u}_{ZN}^N \times \underline{\psi}^N \right) \qquad (13.2.2\text{-}15)$$

or in more compact notation:

$$\ddot{\underline{\gamma}}_H^N + \omega_S^2 \, \underline{\gamma}_H^N = \ddot{\underline{\gamma}}_{H_{Force}}^N \qquad (13.2.2\text{-}16)$$

with

$$\ddot{\underline{\gamma}}_{H_{Force}}^N \equiv \ddot{\underline{\psi}}_H^N + \frac{1}{R} \, \underline{u}_{ZN}^N \times \left(\frac{d}{dt} \, \delta\dot{\underline{R}}_{H_{Force}}^N + \delta\dot{\underline{\upsilon}}_{H_{Force}}^N - g \, \underline{u}_{ZN}^N \times \underline{\psi}^N \right) \qquad (13.2.2\text{-}17)$$

where

$\ddot{\underline{\gamma}}_{H_{Force}}^N$ = Forcing function for horizontal tilt dynamic Equation (13.2.2-16).

The $\delta\dot{\underline{R}}_{H_{Force}}^N$, $\delta\dot{\underline{\upsilon}}_{H_{Force}}^N$ and $\ddot{\underline{\psi}}_H^N$ terms in (13.2.2-17) are provided by Equations (13.2.2-3) and the horizontal time derivative of (13.2-12). It is also analytically advantageous to define the specific force acceleration $\underline{a}_{SF}^N$ in the Equation (13.2.2-3) $\delta\dot{\underline{\upsilon}}_{H_{Force}}^N$ expression as the sum of a term to balance plumb-bob gravity plus a maneuvering acceleration component:

$$\underline{a}_{SF}^N = g \, \underline{u}_{ZN}^N + \Delta\underline{a}_{SF}^N \qquad (13.2.2\text{-}18)$$

where

$\Delta\underline{a}_{SF}^N$ = Portion of $\underline{a}_{SF}^N$ for maneuvering.

Substituting (13.2.2-3) with (13.2.2-18) and the horizontal time derivative of (13.2-12) into (13.2.2-17) then obtains:

$$\ddot{\underline{\gamma}}_{H\,Force}^{N} = \frac{d}{dt}\left[- C_B^N\, \delta\underline{\omega}_{IB}^{B} - \underline{\omega}_{IN}^{N} \times \underline{\psi}^{N} + C_B^N \left(\underline{\omega}_{IB}^{B} \times \delta\underline{\alpha}_{Quant}\right)\right]_H$$
$$+ \frac{1}{R}\,\underline{u}_{ZN}^{N} \times \left[\frac{d}{dt}\left(C_B^N\, \delta\underline{\upsilon}_{Quant}\right)_H + \left\{ C_B^N\, \delta\underline{a}_{SF}^{B} + \Delta\underline{a}_{SF}^{N} \times \underline{\psi}^{N} + \delta\underline{g}_{Mdl}^{N} \right.\right. \qquad (13.2.2\text{-}19)$$
$$\left.\left. - \left(\underline{a}_{SF}^{N}\times\right) C_B^N\, \delta\underline{\alpha}_{Quant} - \left[\left(C_B^N\, \underline{\omega}_{IB}^{B}\right)\times\right] C_B^N\, \delta\underline{\upsilon}_{Quant}\right\}_H\right]$$

The $\frac{d}{dt}\left(\underline{\omega}_{IN}^{N} \times \underline{\psi}^{N}\right)$ term in (13.2.2-19) can be expanded using (13.2-12) as:

$$\frac{d}{dt}\left(\underline{\omega}_{IN}^{N} \times \underline{\psi}^{N}\right) = \dot{\underline{\omega}}_{IN}^{N} \times \underline{\psi}^{N} + \underline{\omega}_{IN}^{N} \times \dot{\underline{\psi}}^{N}$$
$$= \dot{\underline{\omega}}_{IN}^{N} \times \underline{\psi}^{N} - \underline{\omega}_{IN}^{N} \times \left(\underline{\omega}_{IN}^{N} \times \underline{\psi}^{N}\right) - \underline{\omega}_{IN}^{N} \times \left(C_B^N\, \delta\underline{\omega}_{IB}^{B}\right) \qquad (13.2.2\text{-}20)$$
$$+ \underline{\omega}_{IN}^{N} \times \left[C_B^N \left(\underline{\omega}_{IB}^{B} \times \delta\underline{\alpha}_{Quant}\right)\right]$$

Finally, (13.2.2-20) is applied in (13.2.2-19) to obtain, after rearrangement:

$$\ddot{\underline{\gamma}}_{H\,Force}^{N} = \frac{1}{R}\,\underline{u}_{ZN}^{N} \times \left(C_B^N\, \delta\underline{a}_{SF}^{B} + \Delta\underline{a}_{SF}^{N} \times \underline{\psi}^{N} + \delta\underline{g}_{Mdl}^{N}\right)$$
$$+ \left\{ - \frac{d}{dt}\left(C_B^N\, \delta\underline{\omega}_{IB}^{B}\right) - \left[\left(\dot{\underline{\omega}}_{IN}^{N}\times\right) - \left(\underline{\omega}_{IN}^{N}\times\right)^2\right] \underline{\psi}^{N} + \underline{\omega}_{IN}^{N} \times \left(C_B^N\, \delta\underline{\omega}_{IB}^{B}\right)\right\}_H$$
$$+ \left\{ \frac{d}{dt}\, C_B^N \left(\underline{\omega}_{IB}^{B} \times \delta\underline{\alpha}_{Quant}\right) - \underline{\omega}_{IN}^{N} \times \left[C_B^N \left(\underline{\omega}_{IB}^{B} \times \delta\underline{\alpha}_{Quant}\right)\right]\right\}_H \qquad (13.2.2\text{-}21)$$
$$+ \frac{1}{R}\,\underline{u}_{ZN}^{N} \times \left\{ \frac{d}{dt}\left(C_B^N\, \delta\underline{\upsilon}_{Quant}\right) - \left(\underline{a}_{SF}^{N}\times\right) C_B^N\, \delta\underline{\alpha}_{Quant} - \left[\left(C_B^N\, \underline{\omega}_{IB}^{B}\right)\times\right] C_B^N\, \delta\underline{\upsilon}_{Quant}\right\}_H$$

Equation (13.2.2-16) with (13.2.2-21) shows as expected, that the horizontal tilt error is also characterized by undamped sinusoidal oscillations at frequency ω_S. It is important to also note, however, that the Equation (13.2.2-21) $\ddot{\underline{\gamma}}_{H\,Force}^{N}$ forcing function for (13.2.2-16) contains no significant terms that increase systematically with time. A minor exception is the $\left[\left(\dot{\underline{\omega}}_{IN}^{N}\times\right) - \left(\underline{\omega}_{IN}^{N}\times\right)^2\right] \underline{\psi}^{N}$ term which can have increasing $\underline{\psi}^{N}$ from Equation (13.2-12), however, the average value for $\left[\left(\dot{\underline{\omega}}_{IN}^{N}\times\right) - \left(\underline{\omega}_{IN}^{N}\times\right)^2\right]$ is small. The $\Delta\underline{a}_{SF}^{N} \times \underline{\psi}^{N}$ term in (13.2.2-21) has no prevailing build-up rate because $\Delta\underline{a}_{SF}^{N}$ is generally transitory and of relative short duration. We conclude from Equations (13.2.2-16) and (13.2.2-21), therefore, that the characteristic response of the C_B^N matrix horizontal tilt $\underline{\gamma}_H^N$ is to exhibit bounded sinusoidal oscillation patterns that are stationary in the Free Azimuth N Frame. The response is similar to that of a Foucault pendulum; i.e., a pendulum suspended from a point which is free to oscillate in any plane (such as the pendulum created by a suspended plumb-bob). For this reason, the Free Azimuth coordinate frame has sometimes been denoted as the "Foucault" frame. A Foucault pendulum whose pivot (or suspension) point is fixed relative to the earth will, if disturbed, generate an oscillation pattern that is stationary in a locally level Free Azimuth coordinate frame. Because the Free Azimuth frame has zero angular rotation rate about the vertical, an observer on the earth (who is rotating with the earth) will view it as rotating about the vertical in the direction opposite to earth's rotation, at a rate equal to the vertical component of earth's rate. Thus, the oscillation plane of a Foucault pendulum to such an observer will appear to also rotate about the vertical at the negative of earth's vertical rotation rate.

To make the inertial navigation system tilt error pendulum analogy complete, the pendulum would require an 84 minute period. Such an 84 minute pendulum has been denoted as a "Schuler" pendulum in honor of Dr. Maximilian Schuler, a German scientist who was active during the World War I time period. Dr. Schuler reasoned that an 84 minute pendulum would remain vertical under accelerations of the suspension point. Thus, with such a pendulum, an accurate vertical reference could be established in a dynamic environment such as on a moving ship. His rationale for selecting an 84 minute pendulum period was based on the horizontal acceleration of the pendulum pivot point over the earth surface producing a pendulum angular acceleration (through its pendulocity) that would match the equivalent angular acceleration of the pivot point over the earth's curved surface. Dr. Schuler demonstrated analytically that the pendulum frequency to make this match equals $\sqrt{\frac{g}{R}}$, i.e., the ω_S oscillation frequency for the inertial navigation system horizontal tilt (and velocity/position) errors. Without disturbances, the Schuler pendulum remains vertical under pivot accelerations. If disturbed (e.g., by physically touching the pendulum), the Schuler pendulum oscillates at ω_S, now known as the "Schuler frequency". Similarly, without error, the inertial navigation system maintains a horizontal reference (the N Frame) that remains horizontal under system acceleration. If the navigation system contains errors, it causes the N Frame to exhibit Schuler oscillations about the vertical. Thus, by designing an inertial navigation system, we have in effect, constructed an artificial Schuler pendulum that exhibits the same behavior as a "real" Schuler pendulum.

The source of inertial navigation system Schuler oscillations can be traced to the error in the system computer gravity model created by position error across the local vertical. The gravity model defines gravity to be downward along the local vertical. A horizontal position error creates a tilt error in the apparent local vertical (i.e., the local horizontal appears rotated from true horizontal by the angular movement over the earth's surface associated with the horizontal position error). The apparent local vertical tilt error, thereby, produces a small horizontal gravity error component in opposition to the horizontal position error. From the previous discussion, it can be easily demonstrated that the resulting gravity error is $-\frac{g}{R}\,\delta\underline{R}_H^N$ which is the Schuler effect that has appeared in our error equations. The gravity error, in turn, adds $-\frac{g}{R}\,\delta\underline{R}_H^N$ to the horizontal acceleration error $\delta\ddot{\underline{R}}_H^N$, which is the source of the $\delta\ddot{\underline{R}}_H^N + \frac{g}{R}\,\delta\underline{R}_H^N$ term in Equation (13.2.2-4).

13.2.3 LONG TERM APPROXIMATE POSITION ERROR SOLUTION

Let us begin this subsection with the relationship between the horizontal position error $\delta\underline{R}_H^N$ and the C_N^E matrix error $\underline{\varepsilon}^N$ using the rearranged horizontal component of equivalency Equation (12.2.3-16) with general Equation (13.1-4):

$$\delta\underline{R}_H^N = -R\left(\underline{u}_{ZN}^N \times \underline{\varepsilon}_H^N\right) \tag{13.2.3-1}$$

The horizontal component of equivalency relation (12.2.1-17) rearranged is:

$$\underline{\varepsilon}_H^N = -\underline{\psi}_H^N + \underline{\gamma}_H^N \tag{13.2.3-2}$$

Substituting (13.2.3-2) into (13.2.3-1) then finds:

$$\delta\underline{R}_H^N = R\,\underline{u}_{ZN}^N \times \underline{\psi}_H^N - R\,\underline{u}_{ZN}^N \times \underline{\gamma}_H^N \tag{13.2.3-3}$$

Section 13.2.2 (Equations (13.2.2-16) with (13.2.2-21)) showed that the solution for $\underline{\gamma}_H^N$ in Equation (13.2.3-3) is characterized by undamped Schuler oscillations with no significant terms that increase systematically with time. In contrast, the $\underline{\psi}_H^N$ term in (13.2.3-3) can grow systematically with time as shown in Equation (13.2-12) due principally to angular rate sensor error. Thus, we conclude that for long term navigation position error analysis, all but the $\underline{\psi}_H^N$ term in (13.2.3-3) can be safely ignored, yielding the following simplified form for assessing $\delta\underline{R}_H^N$ long term behavior:

$$\delta\underline{R}_{H_{LngTrm}}^N \approx R\left(\underline{u}_{ZN}^N \times \underline{\psi}_H^N\right) \tag{13.2.3-4}$$

where

$\delta\underline{R}_{H_{LngTrm}}^N$ = Approximate value for $\delta\underline{R}_H^N$ for long term inertial navigation system error analysis.

The $\underline{\psi}_H^N$ term in (13.2.3-4) is the solution to Equation (13.2-12).

We should also recognize that although Equation (13.2.3-4) was derived in a Free Azimuth N Frame, the result can now be transformed to any coordinate frame in which it is convenient to define $\underline{\psi}_H$. Section 13.5 elaborates on this point in more detail in developing a solution for $\underline{\psi}_H^N$ under analytically defined trajectory conditions for long term navigation error application of Equation (13.2.3-4).

13.2.4 STRAPDOWN INERTIAL SENSOR SCALE-FACTOR/MISALIGNMENT ERROR EFFECTS

Strapdown inertial sensor misalignment (i.e., the angular rate sensor triad or the accelerometer triad) can be defined as the sum of two effects; misalignment of an inertial sensor triad (angular rate sensor or accelerometer) relative to the inertial sensor reference axes (i.e., the B Frame), and orthogonality error between input axes of two sensors in a particular sensor triad. Figure 13.2.4-1 illustrates each effect for the X - Y axes of the angular rate sensor triad:

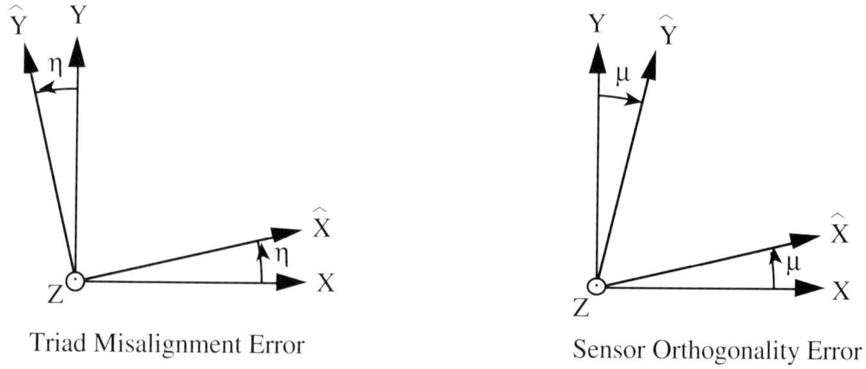

Figure 13.2.4-1 Angular Rate Sensor Triad Misalignment And Sensor Orthogonality Error

In Figure 13.2.4-1, axes $\hat{X}$, $\hat{Y}$ are the compensated input axis directions for the X, Y angular rate sensors, and axes X, Y are the nominal angular rate sensor input axis directions (i.e., the X, Y axes of the B Frame). Figure 13.2.4-1 shows the angular rate sensor misalignment errors for a "Triad Misalignment Error" situation when the angular rate sensor triad is misaligned as a whole from the nominal axes (with the $\hat{X}$, $\hat{Y}$ axes orthogonal), and for a "Sensor Orthogonality Error" situation when the misalignment of the X, Y angular rate sensors makes the $\hat{X}$, $\hat{Y}$ axes non-orthogonal. For the Triad Misalignment Error illustration, $\hat{X}$ is misaligned from X by η (which couples Y axis rate into $\hat{X}$) and $\hat{Y}$ is misaligned from Y by η (which couples X axis rate into $\hat{Y}$). For the Sensor Orthogonality Error illustration, $\hat{X}$, $\hat{Y}$ are misaligned from X, Y by μ. The Figure 13.2.4-1 situation can be related to terms in the angular rate sensor triad sensor scale-factor/misalignment error matrix using the Equation (12.4-13) definition:

$$\delta K_{Scal/Mis} = \begin{bmatrix} \delta K_{XX} & \delta K_{XY} & \delta K_{XZ} \\ \delta K_{YX} & \delta K_{YY} & \delta K_{YZ} \\ \delta K_{ZX} & \delta K_{ZY} & \delta K_{ZZ} \end{bmatrix} \tag{13.2.4-1}$$

where

$\delta K_{Scal/Mis}$ = Angular rate sensor triad scale-factor/misalignment residual error matrix remaining after application of compensation corrections to sensor outputs.

δK_{ii} = Scale factor error for the i axis angular rate sensor.

δK_{ij} = Misalignment error for the i axis angular rate sensor coupling B Frame j axis angular rate into the sensor input (for $i \neq j$).

Returning to Figure 13.2.4-1, for the triad misalignment case, the angle between $\hat{Y}$ and X is greater than 90 degrees (for positive η), hence, η couples X rate negatively into $\hat{Y}$. In contrast, the angle between $\hat{X}$ and Y is less than 90 degrees (for positive η), hence, η couples Y rate positively into $\hat{X}$. For both the $\hat{X}$ and $\hat{Y}$ sensor axes, the coupling gain is equal to η. This corresponds to the situation in Equation (13.2.4-1) when $\delta K_{XY} = - \delta K_{YX}$, which represents a skew symmetric misalignment error characteristic in $\delta K_{Scal/Mis}$ (about the diagonal). For the Figure 13.2.4-1 orthogonality error case, the angles between $\hat{Y}$ and X and between $\hat{X}$ and Y are both less than 90 degrees (for positive μ), hence, μ couples rate positively from X, Y into $\hat{Y}$, $\hat{X}$ with a coupling gain of μ. This corresponds to the situation in Equation (13.2.4-1) when $\delta K_{XY} = \delta K_{YX}$, which represents a symmetric misalignment error characteristic in $\delta K_{Scal/Mis}$ (about the diagonal). In general, both the triad misalignment and orthogonality error situations in Figure 13.2.4-1 exist simultaneously in $\delta K_{Scal/Mis}$ for the angular rate sensor triad between the X/Y, Y/Z and Z/X axes, and in $\delta L_{Scal/Mis}$ for the accelerometer triad, where

$\delta L_{Scal/Mis}$ = Accelerometer triad scale-factor/misalignment residual error matrix remaining after application of compensation corrections to sensor outputs.

We can generalize the previous observations by stating that misalignment errors for a sensor triad as a whole (angular rate sensor or accelerometer triad) are characterized by skew symmetric off-diagonal terms in $\delta K_{Scal/Mis}$, $\delta L_{Scal/Mis}$ while sensor non-orthogonality error misalignments in a sensor triad are characterized by symmetric off-diagonal terms in $\delta K_{Scal/Mis}$, $\delta L_{Scal/Mis}$. We also note from Equation (13.2.4-1) that inertial sensor scale factor errors are characterized by terms along the diagonal of $\delta K_{Scal/Mis}$, $\delta L_{Scal/Mis}$. Based on this generalization, we define $\delta K_{Scal/Mis}$, $\delta L_{Scal/Mis}$ as being composed of the sum of triad misalignment and sensor scale-factor/orthogonality components:

$$\delta K_{Scal/Mis} = \delta K_{Scal/Orth} + \delta K_{TMis} \qquad \delta L_{Scal/Mis} = \delta L_{Scal/Orth} + \delta L_{TMis} \tag{13.2.4-2}$$

where

$\delta K_{Scal/Orth}$, $\delta L_{Scal/Orth}$ = Portion of $\delta K_{Scal/Mis}$, $\delta L_{Scal/Mis}$ containing the scale factor and orthogonality error terms. Based on the previous discussion, $\delta K_{Scal/Orth}$ and $\delta L_{Scal/Orth}$ are symmetric matrices about the diagonal with the diagonal elements generally non-zero.

δK_{TMis}, δL_{TMis} = Portion of $\delta K_{Scal/Mis}$, $\delta L_{Scal/Mis}$ containing the triad misalignment error terms. Based on the previous discussion, δK_{TMis} and δL_{TMis} are skew-symmetric matrices about the diagonal, with zero for the diagonal elements.

From their definition, a symmetric matrix equals its transpose and a skew-symmetric matrix equals the negative of its transpose. Using this rule we can combine Equations (13.2.4-2) with their transpose to find after rearrangement:

$$\begin{aligned}
\delta K_{Scal/Orth} &= \frac{1}{2}\left(\delta K_{Scal/Mis} + \delta K_{Scal/Mis}^{T}\right) \\
\delta K_{TMis} &= \frac{1}{2}\left(\delta K_{Scal/Mis} - \delta K_{Scal/Mis}^{T}\right) \\
\delta L_{Scal/Orth} &= \frac{1}{2}\left(\delta L_{Scal/Mis} + \delta L_{Scal/Mis}^{T}\right) \\
\delta L_{TMis} &= \frac{1}{2}\left(\delta L_{Scal/Mis} - \delta L_{Scal/Mis}^{T}\right)
\end{aligned} \tag{13.2.4-3}$$

Because δK_{TMis}, δL_{TMis} are skew-symmetric, they can be expressed in the equivalent form:

$$\delta K_{TMis} = \left(\delta \underline{K}_{TMis}\times\right) \qquad \delta L_{TMis} = \left(\delta \underline{L}_{TMis}\times\right) \tag{13.2.4-4}$$

where

$\delta \underline{K}_{TMis}, \delta \underline{L}_{TMis}$ = Angular rate sensor triad and accelerometer triad misalignment error vectors.

$\left(\delta \underline{K}_{TMis} \times\right), \left(\delta \underline{L}_{TMis} \times\right)$ = Skew symmetric forms of $\delta \underline{K}_{TMis}, \delta \underline{L}_{TMis}$ as defined by generalized Equation (3.1.1-14).

We can also decompose $\delta K_{Scal/Orth}$ and $\delta L_{Scal/Orth}$ into their constituents:

$$\delta L_{Scal/Orth} = \delta L_{Scal} + \delta L_{Orth} \qquad \delta K_{Scal/Orth} = \delta K_{Scal} + \delta K_{Orth} \tag{13.2.4-5}$$

where

$\delta K_{Scal}, \delta L_{Scal}$ = Scale factor error portions of $\delta K_{Scal/Orth}, \delta L_{Scal/Orth}$ which are the diagonal elements.

$\delta K_{Orth}, \delta L_{Orth}$ = Orthogonality misalignment error portions of $\delta K_{Scal/Orth}$ and $\delta L_{Scal/Orth}$ consisting of the off-diagonal elements. The orthogonality errors represent misalignment between the sensors of the angular rate sensor triad or the accelerometer triad.

In order to characterize the effect of inertial sensor scale-factor/misalignment error on attitude and acceleration transformation accuracy, let us now define the inertial sensor error associated with these effects from Equations (12.4-12) and (12.4-14) as:

$$\delta \underline{\omega}_{IB_{Scal/Mis}}^{B} \equiv \delta K_{Scal/Mis} \, \underline{\omega}_{IB}^{B} \qquad \delta \underline{a}_{SF_{Scal/Mis}}^{B} \equiv \delta L_{Scal/Mis} \, \underline{a}_{SF}^{B} \tag{13.2.4-6}$$

where

$\delta \underline{\omega}_{IB_{Scal/Mis}}^{B}, \delta \underline{a}_{SF_{Scal/Mis}}^{B}$ = Angular rate sensor, accelerometer error vector components produced by inertial sensor scale-factor/misalignment errors.

With (13.2.4-2), (13.2.4-4) and (13.2.4-5), Equations (13.2.4-6) are given by:

$$\begin{aligned} \delta \underline{\omega}_{IB_{Scal/Mis}}^{B} &= \delta K_{Scal} \, \underline{\omega}_{IB}^{B} + \delta \underline{K}_{TMis} \times \underline{\omega}_{IB}^{B} + \delta K_{Orth} \, \underline{\omega}_{IB}^{B} \\ \delta \underline{a}_{SF_{Scal/Mis}}^{B} &= \delta L_{Scal} \, \underline{a}_{SF}^{B} + \delta \underline{L}_{TMis} \times \underline{a}_{SF}^{B} + \delta L_{Orth} \, \underline{a}_{SF}^{B} \end{aligned} \tag{13.2.4-7}$$

Equations (13.2.4-7) are now in a convenient form for investigating their impact on attitude/acceleration-transformation accuracy.

We first analyze the effect of $\delta\underline{\omega}_{IB_{Scal/Mis}}^{B}$ on attitude error build-up by analyzing the impact on $\dot{\underline{\psi}}^{I}$ from Equation (12.3.7.1-4):

$$\dot{\underline{\psi}}^{I} = - C_B^I \, \delta\underline{\omega}_{IB}^{B} \tag{13.2.4-8}$$

where

$\underline{\psi}^{I}$ = Angular error vector associated with C_B^I, the direction cosine matrix that transforms vectors from B to I Frame axes.

From Equation (13.2.4-8) we see that $\underline{\psi}^{I}$ error build-up is directly affected by the magnitude of $\delta\underline{\omega}_{IB}^{B}$. The magnitude squared of $\delta\underline{\omega}_{IB}^{B}$ is derived from the magnitude of $\underline{\omega}_{IB}^{B}$ using general Equation (3.1.1-2):

$$\omega_{IB}^{2} = \underline{\omega}_{IB}^{B} \cdot \underline{\omega}_{IB}^{B} \tag{13.2.4-9}$$

where

ω_{IB} = Magnitude of $\underline{\omega}_{IB}^{B}$.

Taking the differential of (13.2.4-9) then obtains for the $\underline{\omega}_{IB}^{B}$ magnitude error:

$$\delta\omega_{IB} = \frac{1}{\omega_{IB}} \underline{\omega}_{IB}^{B} \cdot \delta\underline{\omega}_{IB}^{B} \tag{13.2.4-10}$$

where

$\delta\omega_{IB}$ = Error in ω_{IB}.

The effect of angular rate sensor scale-factor/misalignment error on $\delta\omega_{IB}$ can now be evaluated by substituting $\delta\underline{\omega}_{IB_{Scal/Mis}}^{B}$ from (13.2.4-7) for $\delta\underline{\omega}_{IB}^{B}$ in (13.2.4-10):

$$\delta\omega_{IB_{Scal/Mis}} = \frac{1}{\omega_{IB}} \underline{\omega}_{IB}^{B} \cdot \left[\left(\delta K_{Scal} + \delta K_{Orth} \right) \underline{\omega}_{IB}^{B} + \delta\underline{K}_{TMis} \times \underline{\omega}_{IB}^{B} \right] \tag{13.2.4-11}$$

where

$\delta\omega_{IB_{Scal/Mis}}$ = Portion of $\delta\omega_{IB}$ produced by angular rate sensor scale-factor/misalignment error.

Because $\delta \underline{K}_{TMis} \times \underline{\omega}_{IB}^{B}$ in (13.2.4-11) is perpendicular to $\underline{\omega}_{IB}^{B}$, its dot product with $\underline{\omega}_{IB}^{B}$ is zero, hence, (13.2.4-11) reduces to:

$$\delta\omega_{IB_{Scal/Mis}} = \frac{1}{\omega_{IB}} \, \underline{\omega}_{IB}^{B} \cdot \left[\left(\delta K_{Scal} + \delta K_{Orth} \right) \underline{\omega}_{IB}^{B} \right] \tag{13.2.4-12}$$

Equation (13.2.4-12) shows that the error in the magnitude of inertially sensed angular rate is affected by angular rate sensor scale factor and orthogonality error but not by angular rate sensor triad misalignment error.

The effect of angular rate sensor triad misalignment error on $\underline{\psi}^{I}$ can be analyzed with Equation (13.2.4-8) using the $\delta \underline{K}_{TMis}$ term in (13.2.4-7) for $\delta\underline{\omega}_{IB}^{B}$:

$$\dot{\underline{\psi}}_{TMis}^{I} = - C_{B}^{I} \left(\delta \underline{K}_{TMis} \times \underline{\omega}_{IB}^{B} \right) = C_{B}^{I} \left(\underline{\omega}_{IB}^{B} \times \delta \underline{K}_{TMis} \right) = C_{B}^{I} \left(\underline{\omega}_{IB}^{B} \times \right) \delta \underline{K}_{TMis} \tag{13.2.4-13}$$

where

$\underline{\psi}_{TMis}^{I}$ = Portion of $\underline{\psi}^{I}$ produced by angular rate sensor triad misalignment.

Equation (13.2.4-13) can be simplified by noting from Equations (12.1.4-18) that:

$$\dot{C}_{B}^{I} = C_{B}^{I} \left(\underline{\omega}_{IB}^{B} \times \right) \tag{13.2.4-14}$$

With (13.2.4-14), (13.2.4-13) reduces to:

$$\dot{\underline{\psi}}_{TMis}^{I} = \dot{C}_{B}^{I} \, \delta \underline{K}_{TMis} \tag{13.2.4-15}$$

whose integral is simply:

$$\Delta \underline{\psi}_{TMis}^{I} = \left(C_{B}^{I} - C_{B_0}^{I} \right) \delta \underline{K}_{TMis} \tag{13.2.4-16}$$

where

B_0 = B Frame orientation in inertial space at time $t = 0$.

$\Delta \underline{\psi}_{TMis}^{I}$ = Change in $\underline{\psi}^{I}$ since time $t = 0$ due to angular rate sensor triad misalignment.

Equation (13.2.4-16) shows that the effect of angular rate sensor triad misalignment on attitude error is to add an attitude error equal to the change in the I Frame projection of the misalignment angle error vector (from the B Frame) since time $t = 0$. The same effect holds for any coordinate frame in which $\underline{\psi}$ is being evaluated, not only the I Frame, as should be readily apparent by transformation of (13.2.4-16) to the coordinate frame of choice.

To analyze the effect of inertial sensor misalignment on acceleration transformation accuracy we begin with the transform relation:

$$\underline{a}_{SF_{XForm}}^{A} = C_B^A \, \underline{a}_{SF}^B = C_I^A \, C_B^I \, \underline{a}_{SF}^B \tag{13.2.4-17}$$

where

A = Arbitrary coordinate frame in which velocity is to be calculated by acceleration integration. The A Frame might be the N Frame, the I Frame or some other frame, depending on the application.

$\underline{a}_{SF_{XForm}}^{A}$ = Specific force acceleration projected on the A Frame.

From its definition, $\underline{a}_{SF_{XForm}}^{A}$ will be integrated (with other terms) to obtain system velocity in the A Frame. To understand the effect of inertial sensor misalignment on $\underline{a}_{SF_{XForm}}^{A}$, hence velocity error build-up, we take the differential of (13.2.4-17), apply (12.3.7.1-10) for δC_B^I, and expand:

$$\begin{aligned}
\delta\underline{a}_{SF_{XForm}}^{A} &= C_I^A \left(\delta C_B^I \, \underline{a}_{SF}^B + C_B^I \, \delta\underline{a}_{SF}^B \right) + \delta C_I^A \, C_B^I \, \underline{a}_{SF}^B \\
&= C_I^A \left(C_B^I \, \delta\underline{a}_{SF}^B - \left(\underline{\psi}^I \times\right) C_B^I \, \underline{a}_{SF}^B \right) + \cdots \\
&= C_I^A \left(C_B^I \, \delta\underline{a}_{SF}^B - C_B^I \, C_I^B \left(\underline{\psi}^I \times\right) C_B^I \, \underline{a}_{SF}^B \right) + \cdots
\end{aligned} \tag{13.2.4-18}$$

In Equation (13.2.4-18), only the terms directly affected by inertial sensor error are identified (i.e., $\underline{\psi}^I$ from angular rate sensor error in (13.2.4-8) and $\delta\underline{a}_{SF}^B$ accelerometer error). The remaining terms are identified by the $\cdots$ notation (e.g., the δC_I^A term which is typically a function of velocity and position produced from the integral of sensor error). Applying generalized Equation (3.1.1-40) to (13.2.4-18), combining matrix products, and factorization then yields the compressed form:

$$\delta\underline{a}_{SF_{XForm}}^{A} = C_B^A \left(\delta\underline{a}_{SF}^B - \underline{\psi}^B \times \underline{a}_{SF}^B \right) + \cdots \tag{13.2.4-19}$$

It is instructive to analyze the particular effects of initial attitude error and inertial sensor misalignment by expanding $\underline{\psi}^B$ in (13.2.4-19) as:

$$\underline{\psi}^B = C_A^B \, \underline{\psi}_0^A + \Delta\underline{\psi}_{TMis}^B + \Delta\underline{\psi}_{Orth}^B + \Delta\underline{\psi}_{Other}^B \qquad (13.2.4\text{-}20)$$

where

$\underline{\psi}_0^A$ = Initial value of the attitude error in the A Frame.

$\Delta\underline{\psi}_{Orth}^B$ = Change in $\underline{\psi}^B$ since time t = 0 due to angular rate sensor orthogonality error (between angular rate sensors).

$\Delta\underline{\psi}_{Other}^B$ = Contributions to $\underline{\psi}^B$ from all angular rate sensor error sources other than angular rate sensor misalignment.

The $\Delta\underline{\psi}_{TMis}^B$ term in (13.2.4-20) is from (13.2.4-16):

$$\Delta\underline{\psi}_{TMis}^B = C_I^B \left(C_B^I - C_{B_0}^I \right) \delta\underline{K}_{TMis} = \left(I - C_{B_0}^B \right) \delta\underline{K}_{TMis} \qquad (13.2.4\text{-}21)$$

The $\Delta\underline{\psi}_{Orth}^B$ term in (13.2.4-20) can be derived from (13.2.4-8) using the δK_{Orth} term in (13.2.4-7) for $\delta\underline{\omega}_{IB}^B$:

$$\dot{\underline{\psi}}_{Orth}^I = - C_B^I \left(\delta K_{Orth} \, \underline{\omega}_{IB}^B \right) \qquad (13.2.4\text{-}22)$$

Integrating (13.2.4-22) and transforming to the B Frame finds for $\Delta\underline{\psi}_{Orth}^B$:

$$\Delta\underline{\psi}_{Orth}^B = - C_I^B \int_0^t C_B^I \left(\delta K_{Orth} \, \underline{\omega}_{IB}^B \right) d\tau \qquad (13.2.4\text{-}23)$$

The $\delta\underline{a}_{SF}^B$ term in (13.2.4-19) is similarly expanded with (13.2.4-7) as:

$$\delta\underline{a}_{SF}^B = \delta\underline{L}_{TMis} \times \underline{a}_{SF}^B + \delta L_{Orth} \, \underline{a}_{SF}^B + \delta\underline{a}_{SF_{Other}}^B \qquad (13.2.4\text{-}24)$$

where

$\delta\underline{a}_{SF_{Other}}^B$ = Accelerometer error terms other than accelerometer misalignment.

Substituting (13.2.4-20), (13.2.4-21), (13.2.4-23) and (13.2.4-24) into Equation (13.2.4-19) then yields:

$$
\begin{aligned}
\delta \underline{a}^{A}_{SF_{XForm}} &= C^{A}_{B} \Bigg\{ \delta \underline{L}_{TMis} \times \underline{a}^{B}_{SF} + \delta L_{Orth} \, \underline{a}^{B}_{SF} + \delta \underline{a}^{B}_{SF_{Other}} - \Bigg[C^{B}_{A} \, \underline{\psi}^{A}_{0} \\
&\quad + \left(I - C^{B}_{B_0} \right) \delta \underline{K}_{TMis} - C^{B}_{I} \int_0^t C^{I}_{B} \left(\delta K_{Orth} \, \underline{\omega}^{B}_{IB} \right) d\tau + \Delta \underline{\psi}^{B}_{Other} \Bigg] \times \underline{a}^{B}_{SF} \Bigg\} \\
&= C^{A}_{B} \left[\left(\delta \underline{L}_{TMis} - \delta \underline{K}_{TMis} \right) \times \underline{a}^{B}_{SF} \right] + C^{A}_{B} \left[\left(C^{B}_{B_0} \, \delta \underline{K}_{TMis} \right) \times \underline{a}^{B}_{SF} \right] \\
&\quad - C^{A}_{B} \left[\left(C^{B}_{A} \, \underline{\psi}^{A}_{0} \right) \times \underline{a}^{B}_{SF} \right] + C^{A}_{B} \left[\left(C^{B}_{I} \int_0^t C^{I}_{B} \left(\delta K_{Orth} \, \underline{\omega}^{B}_{IB} \right) d\tau \right) \times \underline{a}^{B}_{SF} \right] \\
&\quad + C^{A}_{B} \left(\delta L_{Orth} \, \underline{a}^{B}_{SF} + \delta \underline{a}^{B}_{SF_{Other}} - \Delta \underline{\psi}^{B}_{Other} \times \underline{a}^{B}_{SF} \right)
\end{aligned}
\tag{13.2.4-25}
$$

Particular terms in (13.2.4-25) can be converted to an alternative form using generalized Equation (3.1.1-38):

$$
\begin{aligned}
C^{A}_{B} \left[\left(C^{B}_{B_0} \, \delta \underline{K}_{TMis} \right) \times \underline{a}^{B}_{SF} \right] &= C^{A}_{B} \, C^{B}_{B_0} \left(\delta \underline{K}_{TMis} \times \right) C^{B_0}_{B} \, C^{B}_{A} \, C^{A}_{B} \, \underline{a}^{B}_{SF} \\
&= C^{A}_{B_0} \left(\delta \underline{K}_{TMis} \times \right) C^{B_0}_{A} \, \underline{a}^{A}_{SF} = \left(C^{A}_{B_0} \, \delta \underline{K}_{TMis} \right) \times \underline{a}^{A}_{SF}
\end{aligned}
\tag{13.2.4-26}
$$

$$
C^{A}_{B} \left[\left(C^{B}_{A} \, \underline{\psi}^{A}_{0} \right) \times \underline{a}^{B}_{SF} \right] = C^{A}_{B} \, C^{B}_{A} \left(\underline{\psi}^{A}_{0} \times \right) C^{A}_{B} \, \underline{a}^{B}_{SF} = \underline{\psi}^{A}_{0} \times \underline{a}^{A}_{SF}
$$

We also note that:

$$
\begin{aligned}
C^{B}_{I} \int_0^t C^{I}_{B} \left(\delta K_{Orth} \, \underline{\omega}^{B}_{IB} \right) d\tau &= C^{B}_{B_0} \, C^{B_0}_{I} \int_0^t C^{I}_{B_0} \, C^{B_0}_{B} \left(\delta K_{Orth} \, \underline{\omega}^{B}_{IB} \right) d\tau \\
&= C^{B}_{B_0} \int_0^t C^{B_0}_{B} \left(\delta K_{Orth} \, \underline{\omega}^{B}_{IB} \right) d\tau
\end{aligned}
\tag{13.2.4-27}
$$

Substituting (13.2.4-26) and (13.2.4-27) into (13.2.4-25) obtains an expression that provides insight into the effect of inertial sensor misalignment on acceleration transformation error, hence, velocity error build-up:

$$\begin{aligned}\delta\underline{a}^{A}_{SF_{XForm}} &= C^{A}_{B}\left[\left(\delta\underline{L}_{TMis} - \delta\underline{K}_{TMis}\right)\times \underline{a}^{B}_{SF}\right] - \left(\underline{\psi}^{A}_{0} - C^{A}_{B_0}\,\delta\underline{K}_{TMis}\right)\times \underline{a}^{A}_{SF} \\ &+ C^{A}_{B}\left[\delta L_{Orth}\;\underline{a}^{B}_{SF} + \left(C^{B}_{B_0}\int_0^t C^{B_0}_{B}\left(\delta K_{Orth}\;\underline{\omega}^{B}_{IB}\right)d\tau\right)\times \underline{a}^{B}_{SF}\right] \\ &+ C^{A}_{B}\left(\delta\underline{a}^{B}_{SF_{Other}} - \Delta\underline{\psi}^{B}_{Other}\times \underline{a}^{B}_{SF}\right)\end{aligned} \qquad (13.2.4\text{-}28)$$

Equation (13.2.4-28) shows that the angular rate sensor triad misalignment $\delta\underline{K}_{TMis}$ acts as an offset on the initial attitude error $\underline{\psi}^{A}_{0}$. This is an analytical statement that initial attitude error relative to actual sensor axes is the governing factor impacting velocity error build-up. In addition, the accelerometer triad misalignment $\delta\underline{L}_{TMis}$ that impacts acceleration transformation accuracy is relative to the angular rate sensor triad misalignment $\delta\underline{K}_{TMis}$. Hence, only the relative misalignment between the angular rate sensor and accelerometer triads affects velocity error build-up. This all makes sense intuitively when one recognizes that the B Frame is defined as the "nominal" sensor frame from which $\underline{\psi}^{A}$, $\delta\underline{L}_{TMis}$, $\delta\underline{K}_{TMis}$ are measured. In other words, the selection of a particular B Frame orientation relative to the inertial sensor triads should have no impact on velocity error growth, provided that the B Frame attitude in the system computer is properly initialized based on the selected B Frame orientation. We also note that the δL_{Orth} orthogonality error term in (13.2.4-28) represents relative accelerometer-to-accelerometer misalignment. In conjunction with $\delta\underline{L}_{TMis}$ being relative to angular rate sensor triad misalignment $\delta\underline{K}_{TMis}$ (as already discussed), we can further generalize that only relative misalignment between any accelerometer and the angular rate sensor triad will affect velocity error build-up.

The previous discussion has formed the basis for selecting the B Frame as the "mean angular rate sensor axis" frame in which the angular rate sensor triad misalignment compensation corrections are "nominally" defined to be zero. This selection is unique for a particular sensor assembly and is established implicitly during sensor assembly calibration when the misalignment calibration coefficients are determined. Using the "mean angular rate sensor axis" frame for the B Frame minimizes the magnitude of the angular rate sensor misalignment compensation coefficients (for orthogonality correction) which minimizes second order error effects if first order approximations are used for misalignment compensation. A similar argument can be made for use of "mean accelerometer axes" for the B Frame. Since the modern day optical type angular rate sensors generally have stable alignment characteristics, the "mean angular rate sensor axis" approach has been favored. An important point to note, however, is that use of "mean angular rate sensor axes" does not set $\delta\underline{K}_{TMis}$ to zero because

$\delta \underline{K}_{TMis}$ represents the error in the compensated angular rate sensor triad alignment including compensation error and physical mounting instabilities.

We finally note in Equation (13.2.4-28) that the δK_{Orth} angular rate sensor orthogonality error (a sensor-to-sensor relative misalignment error term) can impact velocity error growth, depending on the angular rate (and corresponding attitude) history. Interestingly, Equation (13.2.4-28) also shows that for a particular elapsed $C_B^{B_0}$ attitude history, the effect of δK_{Orth} can be completely erased if the inverse attitude history is traced using the corresponding negative angular rate history $\underline{\omega}_{IB}^{B}$. This effect should be understood when structuring strapdown inertial system rotation test profiles.

13.3 NAVIGATION ERRORS FOR CONSTANT ATTITUDE AND CONSTANT SENSOR ERRORS

In this section we analyze the effect of constant inertial sensor errors on navigation inaccuracy for situations when the sensor (body) B Frame is at a fixed attitude relative to the wander azimuth version of the N Frame. From Section 4.5, the wander azimuth frame is defined to be locally level, with the inertial angular rate about the vertical equal to the vertical component of earth's rate (which is equivalent to setting the vertical transport rate term ρ_{ZN} to zero). For a stationary system relative to the earth, the azimuth wander frame will thereby "keep up" with earth rate around the vertical, hence, will appear stationary relative to an observer in the E Frame. For a system traversing a great circle trajectory, an azimuth wander N Frame will remain at a fixed azimuth orientation relative to the velocity vector (Note - A great circle trajectory is defined as a constant altitude flight path parallel to the earth that is in an earth fixed plane passing through earth's center). Since the B Frame typically remains at a fixed orientation relative to the velocity vector on a great circle trajectory, the B Frame will thereby also remain at a fixed orientation relative to the N Frame along a great circle. Thus, the case when the B Frame is fixed relative to the wander azimuth N Frame represents a real situation in some applications.

As an aside, it is easily demonstrated (as stipulated above) that the wander azimuth N Frame remains fixed relative to the velocity vector along a great circle trajectory by considering that the velocity vector along a great circle is constant, horizontal (i.e., perpendicular to a unit vector along the local vertical), and in a fixed plane relative to the earth. Analytically, this is equivalent in the N Frame to:

$$\underline{v}^N = v\, \underline{u}_{GC}^N \times \underline{u}_{ZN}^N = v \left(C_E^N\, \underline{u}_{GC}^E \right) \times \underline{u}_{ZN}^N \qquad (13.3\text{-}1)$$

where

v = Velocity vector $\underline{v}^N$ magnitude.

$\underline{u}_{ZN}^N$ = Unit vector along the N Frame Z axis, which from the Section 2.2 definition, is along the local upward geodetic vertical.

$\underline{u}_{GC}^N$ = Unit vector perpendicular to the great circle plane and considered to be constant in the E Frame. Because $\underline{u}_{ZN}^N$ lies in the great circle plane, $\underline{u}_{GC}^N$ is also perpendicular to $\underline{u}_{ZN}^N$.

The time derivative of (13.3-1) is obtained by recognizing $\underline{u}_{GC}^E$ to be constant by definition and applying the transpose of (4.4.1.1-1) for $\dot{C}_E^N$ (with the transpose of $(\underline{\rho}^N \times)$ equal to its negative):

$$\dot{\underline{v}}^N = v \left(\dot{C}_E^N \, \underline{u}_{GC}^E \right) \times \underline{u}_{ZN}^N = - v \left[\left(\underline{\rho}^N \times \right) C_E^N \, \underline{u}_{GC}^E \right] \times \underline{u}_{ZN}^N = - v \, \underline{u}_{ZN}^N \times \left(\underline{u}_{GC}^N \times \underline{\rho}^N \right) \qquad (13.3\text{-}2)$$

For the wander azimuth N Frame, the vertical component of the $\underline{\rho}^N$ transport rate is zero, hence, from the (13.2-2) ω_{EN}^N expression (for $\underline{\rho}^N$) we can write to first order accuracy (neglecting e terms) using generalized Equation (13.1-4):

$$\underline{\rho}^N \approx \frac{1}{R} \, \underline{u}_{ZN}^N \times \underline{v}^N \qquad \text{For Wander Azimuth Case} \qquad (13.3\text{-}3)$$

Substituting (13.3-3) for $\underline{\rho}^N$ and (13.3-1) for $\underline{v}^N$ in the (13.3-2) $\underline{u}_{GC}^N \times \underline{\rho}^N$ term, applying (13.1-5) and remembering that $\underline{u}_{GC}^N$ has no vertical component obtains:

$$\begin{aligned} \underline{u}_{GC}^N \times \underline{\rho}^N &= \frac{1}{R} \, \underline{u}_{GC}^N \times \left(\underline{u}_{ZN}^N \times \underline{v}^N \right) \\ &= \frac{1}{R} \, \underline{u}_{GC}^N \times \left[\underline{u}_{ZN}^N \times v \left(\underline{u}_{GC}^N \times \underline{u}_{ZN}^N \right) \right] = \frac{1}{R} \, \underline{u}_{GC}^N \times \left(v \, \underline{u}_{GC}^N \right) = 0 \end{aligned} \qquad (13.3\text{-}4)$$

Thus, with (13.3-4) we see that (13.3-2) becomes simply:

$$\dot{\underline{v}}^N = 0 \qquad \begin{array}{l} \text{For a Great Circle Trajectory} \\ \text{And A Wander Azimuth N Frame} \end{array} \qquad (13.3\text{-}5)$$

Because $\underline{v}^N$ is constant in the N Frame by virtue of Equation (13.3-5), the orientation of the N Frame axes relative to $\underline{v}^N$ is constant as originally stipulated.

Returning to the problem at hand, we develop the constant-attitude/constant-sensor-error navigation error solution from Equations (12.5.3-1) with the sensor quantization noise, sensor random output noise, gravity error and δh_{Prsr} terms omitted, with r_l approximated by R (as in (12.3.6.1-26)), with (13.2.2-18) substituted for $\underline{a}_{SF}^N$ in the $\delta\dot{\underline{V}}^N$ expression, with (13.3-3) for $\underline{\omega}_{EN}^N$ in the $\dot{\varepsilon}_{ZN}$ expression, and applying generalized Equation (13.1-4). The result in the wander azimuth frame (ρ_{ZN} and $\delta\rho_{ZN} = 0$) is:

$$\dot{\underline{\gamma}}^N = - C_B^N \, \delta\underline{\omega}_{IB}^B - \underline{\omega}_{IN}^N \times \underline{\gamma}^N + \frac{1}{R} \underline{\omega}_{IE}^N \times \left(\underline{u}_{ZN}^N \times \delta\underline{R}^N\right) + \left(\underline{\omega}_{IE}^N \times \underline{u}_{ZN}^N\right) \varepsilon_{ZN} + \delta\underline{\omega}_{EN}^N$$

$$\delta\underline{\omega}_{IB}^B = \delta K_{Scal/Mis} \, \underline{\omega}_{IB}^B + \delta\underline{K}_{Bias}$$

$$\dot{\varepsilon}_{ZN} = - \frac{1}{R^2} \left[\left(\underline{u}_{ZN}^N \times \underline{v}_H^N\right) \cdot \delta\underline{R}_H^N \right]$$

$$\delta\dot{\underline{V}}^N = C_B^N \, \delta\underline{a}_{SF}^B + g \, \underline{u}_{ZN}^N \times \underline{\gamma}_H^N + \Delta\underline{a}_{SF}^N \times \left[\underline{\gamma}^N - \frac{1}{R} \left(\underline{u}_{ZN}^N \times \delta\underline{R}_H^N\right) - \varepsilon_{ZN} \, \underline{u}_{ZN}^N \right]$$
$$- \left(\underline{\omega}_{IE}^N + \underline{\omega}_{IN}^N\right) \times \delta\underline{V}^N + \left[\left(F(h) \frac{g}{R} - C_2 \right) \delta R - \delta e_{vc_3} \right] \underline{u}_{ZN}^N$$

$$F(h) = 2 \quad \text{For } h \geq 0 \qquad\qquad F(h) = -1 \quad \text{For } h < 0$$

$$\delta\underline{a}_{SF}^B = \delta L_{Scal/Mis} \, \underline{a}_{SF}^B + \delta\underline{L}_{Bias} \tag{13.3-6}$$

$$\delta\dot{\underline{R}}^N = \delta\underline{V}^N - \underline{\omega}_{EN}^N \times \delta\underline{R}^N - C_3 \, \delta R \, \underline{u}_{ZN}^N$$

$$\delta\dot{e}_{vc_3} = C_1 \, \delta R$$

$$\delta\underline{\omega}_{EN}^N = \frac{1}{R} \left(\underline{u}_{ZN}^N \times \delta\underline{V}^N\right) - v_{ZN} \frac{1}{R^2} \left(\underline{u}_{ZN}^N \times \delta\underline{R}^N\right)$$
$$+ \frac{1}{R} \underline{v}_H^N \, \varepsilon_{ZN} - \frac{1}{R^2} \left(\underline{u}_{ZN}^N \times \underline{v}^N\right) \delta R$$

$$\delta\underline{R}_H^N = \delta\underline{R}^N - \delta R \, \underline{u}_{ZN}^N$$

$$\delta R = \underline{u}_{ZN}^N \cdot \delta\underline{R}^N$$

$$\Delta\underline{a}_{SF}^N = \underline{a}_{SF}^N - g \, \underline{u}_{ZN}^N$$

where

$\underline{\gamma}_H^N$ = Horizontal component of $\underline{\gamma}^N$.

The $\underline{\omega}_{IN}^{N}$ term in (13.3-6) is from (13.2-1) and (13.2-2) with ρ_{ZN} set to zero for the Wander Azimuth N Frame:

$$\underline{\omega}_{IN}^{N} = \underline{\omega}_{IE}^{N} + \frac{1}{R}\,\underline{u}_{ZN}^{N} \times \underline{v}_{H}^{N} \tag{13.3-7}$$

The horizontal component of $\underline{\gamma}^{N}$ is generally small compared to the vertical component which justifies neglecting the $\underline{\gamma}^{N}$ horizontal component in $\underline{\omega}_{IN}^{N} \times \underline{\gamma}^{N}$ of the Equations (13.3-6) $\dot{\underline{\gamma}}^{N}$ expression. Hence, with (13.3-7) and generalized Equations (13.1-4) and (13.1-5) we can approximate:

$$\underline{\omega}_{IN}^{N} \times \underline{\gamma}^{N} \approx \underline{\omega}_{IN}^{N} \times \gamma_{ZN}\,\underline{u}_{ZN}^{N} = \gamma_{ZN}\left(\underline{\omega}_{IE_H}^{N} \times \underline{u}_{ZN}^{N} + \frac{1}{R}\,\underline{v}_{H}^{N}\right) \tag{13.3-8}$$

where

γ_{ZN} = Vertical component of $\underline{\gamma}^{N}$.

$\underline{\omega}_{IE_H}^{N}$ = Horizontal component of $\underline{\omega}_{IE}^{N}$.

For navigation times up to 2 hours, the position error $\delta\underline{R}^{N}$ does not generally grow to a large enough value to be significant in the Equations (13.3-6) $\dot{\underline{\gamma}}^{N}$ expression (through the $\underline{\omega}_{IE}^{N}$ earth rate), in the $\delta\dot{\underline{R}}^{N}$ expression (through $\underline{\omega}_{EN}^{N}$ coupling), or in the $\delta\underline{\omega}_{EN}^{N}$ expression (through v_{ZN} and $\underline{u}_{ZN}^{N} \times \underline{v}^{N}$). Based on the heading initialization discussion in Sections 14.1 and 14.2, the initial value of ε_{ZN} in (13.3-6) can be arbitrarily set to zero (if initial heading error is identified with $\hat{C}_{B}^{N}$ rather then $\hat{C}_{N}^{E}$ heading misalignment) or equivalently, can be considered explicitly part of $\hat{C}_{N}^{E}$ initial heading error with the initial γ_{ZN} heading error in $\hat{C}_{B}^{N}$ set to zero. We choose the former for error analysis purposes (i.e., setting initial ε_{ZN} to zero) with the understanding that initial γ_{ZN} contains the full initial heading error. Additionally, for navigation times of less than 2 hours, ε_{ZN} does not have sufficient time to grow large enough (from $\delta\underline{R}^{N}$ - See the (13.3-6) $\dot{\varepsilon}_{ZN}$ equation) to have significant impact in the $\dot{\underline{\gamma}}^{N}$, $\delta\dot{\underline{V}}^{N}$ or $\delta\underline{\omega}_{EN}^{N}$ expressions. Thus, ε_{ZN} in total has minor impact on Equations (13.3-6). Finally, $\delta\underline{V}^{N}$ feedback in the $\delta\dot{\underline{V}}^{N}$ equation (through $\underline{\omega}_{IE}^{N} + \underline{\omega}_{IN}^{N}$) is comparatively small compared to other terms for up to 2 hour navigation times. Dropping the small effects as negligible, substitution of $\delta\underline{\omega}_{EN}^{N}$ from (13.3-6) in the $\delta\dot{\underline{V}}^N$ expression, substitution of (13.3-8) for $\underline{\omega}_{IN}^N \times \underline{\gamma}^N$ in the $\dot{\underline{\gamma}}^N$ expression, and application of (13.1-4) to particular terms, then results in the following simplified version of the Equations (13.3-6) navigation error expressions:

$$\begin{aligned}
\dot{\underline{\gamma}}^N &\approx - C_B^N \, \delta\underline{\omega}_{IB}^B - \gamma_{ZN} \left(\underline{\omega}_{IE_H}^N \times \underline{u}_{ZN}^N + \frac{1}{R} \underline{v}_H^N \right) + \frac{1}{R} \left(\underline{u}_{ZN}^N \times \delta\underline{V}_H^N \right) \\
\delta\dot{\underline{V}}^N &\approx C_B^N \, \delta\underline{a}_{SF}^B + g \, \underline{u}_{ZN}^N \times \underline{\gamma}_H^N + \Delta\underline{a}_{SF}^N \times \left[\underline{\gamma}^N - \frac{1}{R} \left(\underline{u}_{ZN}^N \times \delta\underline{R}_H^N \right) \right] \\
&\quad + \left[\left(F(h) \frac{g}{R} - C_2 \right) \delta R - \delta e_{vc_3} \right] \underline{u}_{ZN}^N \qquad (13.3\text{-}9) \\
F(h) &= 2 \quad \text{For} \quad h \geq 0 \qquad\qquad F(h) = -1 \quad \text{For} \quad h < 0 \\
\delta\dot{\underline{R}}^N &= \delta\underline{V}^N - C_3 \, \delta R \, \underline{u}_{ZN}^N \\
\delta\dot{e}_{vc_3} &= C_1 \, \delta R
\end{aligned}$$

For the case being investigated of the B Frame at fixed attitude relative to the Wander Azimuth N Frame, the inertial B Frame rate $\underline{\omega}_{IB}^B$ equals the N Frame inertial rate $\underline{\omega}_{IN}^B$ so that:

$$\underline{\omega}_{IB}^B = \underline{\omega}_{IN}^B = \left(C_B^N \right)^T \underline{\omega}_{IN}^N \qquad (13.3\text{-}10)$$

We can also write for the B Frame specific force acceleration using Equation(13.2.2-18) for $\underline{a}_{SF}^N$:

$$\underline{a}_{SF}^B = \left(C_B^N \right)^T \underline{a}_{SF}^N = \left(C_B^N \right)^T \left(g \, \underline{u}_{ZN}^N + \Delta\underline{a}_{SF}^N \right) \qquad (13.3\text{-}11)$$

Using (13.3-10) and (13.3-11), the angular rate sensor and accelerometer error terms in (13.3-6) become:

$$\begin{aligned}
\delta\underline{\omega}_{IB}^B &= \delta K_{Scal/Mis} \left(C_B^N \right)^T \underline{\omega}_{IN}^N + \delta\underline{K}_{Bias} \\
\delta\underline{a}_{SF}^B &= \delta L_{Scal/Mis} \left(C_B^N \right)^T \left(g \, \underline{u}_{ZN}^N + \Delta\underline{a}_{SF}^N \right) + \delta\underline{L}_{Bias}
\end{aligned} \qquad (13.3\text{-}12)$$

Equations (13.3-9) with (13.3-12) can be solved analytically if we make simplifying assumptions for the navigation system trajectory distance and maneuver acceleration. For the trajectory distance, we assume that the navigation time is of short enough duration that the $\underline{\omega}_{IE_H}^N$ horizontal earth rate term can be considered constant. For the maneuver acceleration $\Delta\underline{a}_{SF}^N$, we assume that it occurs horizontally and impulsively at the start of navigation and zero thereafter, imparting a change in velocity at time zero. From Equations (13.3-9) and (13.3-12) we see that the $\delta \dot{\underline{V}}^N$ and $\delta \underline{a}_{SF}^B$ contain $\Delta \underline{a}_{SF}^N$ which directly impacts $\delta \underline{V}^N$ at time zero. The result is that the initial conditions for Equations (13.3-9) for the time instant immediately following application of the $\Delta \underline{a}_{SF}^N$ impulse are:

$$\gamma_{ZN_{0+}} = \gamma_{ZN_0} \qquad \underline{\gamma}_{H_{0+}}^N = \underline{\gamma}_{H_0}^N$$

$$\delta V_{ZN_{0+}} = \delta V_{ZN_0} + \left[C_B^N \, \delta L_{Scal/Mis} \left(C_B^N\right)^T \Delta \underline{v}_{SF_H}^N \right]_{ZN} + \left\{ \Delta \underline{v}_{SF_H}^N \times \left[\underline{\gamma}_{H_0}^N - \frac{1}{R} \left(\underline{u}_{ZN}^N \times \delta \underline{R}_{H_0}^N \right) \right] \right\} \cdot \underline{u}_{ZN}^N \tag{13.3-13}$$

$$\delta \underline{V}_{H_{0+}}^N = \delta \underline{V}_{H_0}^N + \left[C_B^N \, \delta L_{Scal/Mis} \left(C_B^N\right)^T \Delta \underline{v}_{SF_H}^N \right]_H - \gamma_{ZN_0} \, \underline{u}_{ZN}^N \times \Delta \underline{v}_{SF_H}^N$$

$$\delta R_{0+} = \delta R_0 \qquad \delta \underline{R}_{H_{0+}}^N = \delta \underline{R}_{H_0}^N$$

where

0 = Subscript indicating initial value prior to application of the $\Delta \underline{a}_{SF}^N$ impulse.

0+ = Subscript indicating value immediately following application of the $\Delta \underline{a}_{SF}^N$ impulse.

ZN = Subscript indicating Z component in the N Frame (i.e., along $\underline{u}_{ZN}^N$).

H = Subscript indicating horizontal component (i.e., perpendicular to $\underline{u}_{ZN}^N$).

$\Delta \underline{v}_{SF_H}^N$ = Velocity change produced by the $\Delta \underline{a}_{SF}^N$ impulse.

Following application of the $\Delta \underline{a}_{SF}^N$ impulse, $\Delta \underline{a}_{SF}^N$ is zero. We will also assume that the velocity is constant following application of the $\Delta \underline{a}_{SF}^N$ impulse. Thus, the velocity and Equations (13.3-12) following $\Delta \underline{a}_{SF}^N$ application are:

$$\underline{v}_H^N = \underline{v}_{H_0}^N + \Delta \underline{v}_{SF_H}^N$$

$$\delta \underline{\omega}_{IB}^B = \delta K_{Scal/Mis} \left(C_B^N\right)^T \underline{\omega}_{IN}^N + \delta \underline{K}_{Bias} \tag{13.3-14}$$

$$\delta \underline{a}_{SF}^B = g \, \delta L_{Scal/Mis} \left(C_B^N\right)^T \underline{u}_{ZN}^N + \delta \underline{L}_{Bias}$$

Following application of the $\Delta \underline{a}_{SF}^{N}$ impulse, the $\Delta \underline{a}_{SF}^{N}$ term can also be removed from Equations (13.3-9). Expansion of Equations (13.3-9) into horizontal and vertical components at this point is advantageous to expedite analytical solution. Thus, recognizing that cross-products of $\underline{u}_{ZN}^{N}$ with horizontal vectors are horizontal, (13.3-9) becomes:

$$\dot{\gamma}_{ZN} = - \left(C_B^N \, \delta \underline{\omega}_{IB}^{B} \right)_{ZN}$$

$$\dot{\underline{\gamma}}_H^N \approx - \left(C_B^N \, \delta \underline{\omega}_{IB}^{B} \right)_H - \gamma_{ZN} \left(\underline{\omega}_{IE_H}^{N} \times \underline{u}_{ZN}^{N} + \frac{1}{R} \underline{v}_H^N \right) + \frac{1}{R} \left(\underline{u}_{ZN}^{N} \times \delta \underline{V}_H^N \right)$$

$$\delta \dot{V}_{ZN} = \left(C_B^N \, \delta \underline{a}_{SF}^{B} \right)_{ZN} + \left(F(h) \frac{g}{R} - C_2 \right) \delta R - \delta e_{vc_3}$$

$$F(h) = 2 \quad \text{For } h \geq 0 \qquad F(h) = -1 \quad \text{For } h < 0 \qquad (13.3\text{-}15)$$

$$\delta \dot{\underline{V}}_H^N \approx \left(C_B^N \, \delta \underline{a}_{SF}^{B} \right)_H + g \, \underline{u}_{ZN}^{N} \times \underline{\gamma}_H^N$$

$$\delta \dot{R} = \delta V_{ZN} - C_3 \, \delta R$$

$$\delta \dot{\underline{R}}_H^N = \delta \underline{V}_H^N$$

$$\delta \dot{e}_{vc_3} = C_1 \, \delta R$$

Equations (13.3-15) with (13.3-14) for the velocity and sensor error terms, and (13.3-13) for initial conditions (following $\Delta \underline{a}_{SF}^{N}$ impulse application) are now in a form that allows simple analytical solution. The following subsections solve these equations for two situations; without vertical loop control for short navigation times (i.e., less than 10 minutes) and with vertical loop control for navigation times of up to 2 hours.

13.3.1 SHORT TERM SOLUTION AT CONSTANT ATTITUDE WITH FREE INERTIAL VERTICAL CHANNEL

For the short term solution, we seek the solution to Equations (13.3-15) with (13.3-13) and (13.3-14) without vertical loop control (i.e., C_1, C_2, C_3, δe_{vc_3} equal to zero). For short term, we can approximate the $\delta \underline{V}_H^N$ and δR terms in the Equations (13.3-15) $\dot{\underline{\gamma}}_H^N$ and $\delta \dot{V}_{ZN}$ expressions to equal the Equation (13.3-13) initial values. Thus, the form of (13.3-15) we will analyze is:

$$\dot{\gamma}_{ZN} = - \left(C_B^N \, \delta\underline{\omega}_{IB}^B \right)_{ZN}$$

$$\dot{\underline{\gamma}}_H^N \approx - \left(C_B^N \, \delta\underline{\omega}_{IB}^B \right)_H - \gamma_{ZN} \left(\underline{\omega}_{IE_H}^N \times \underline{u}_{ZN}^N + \frac{1}{R} \, \underline{v}_H^N \right) + \frac{1}{R} \left(\underline{u}_{ZN}^N \times \delta\underline{V}_{H_{0+}}^N \right)$$

$$\delta\dot{V}_{ZN} = \left(C_B^N \, \delta\underline{a}_{SF}^B \right)_{ZN} + F(h) \, \frac{g}{R} \, \delta R_{0+}$$

$$F(h) = 2 \quad \text{For } h \geq 0 \qquad\qquad F(h) = -1 \quad \text{For } h < 0 \tag{13.3.1-1}$$

$$\delta\dot{\underline{V}}_H^N \approx \left(C_B^N \, \delta\underline{a}_{SF}^B \right)_H + g \, \underline{u}_{ZN}^N \times \underline{\gamma}_H^N$$

$$\delta\dot{R} = \delta V_{ZN}$$

$$\delta\dot{\underline{R}}_H^N = \delta\underline{V}_H^N$$

To simplify the analysis, we hold the C_B^N attitude and altitude (i.e., R) constant, and approximate $\underline{\omega}_{IE}^N$ as constant over the short navigation time. Using (13.3-7) for $\underline{\omega}_{IN}^N$ and (13.3-14) for $\underline{v}_H^N$, we also assume that the inertial sensor error contributors in (13.3-14) are constant so the sensor error terms in (13.3.1-1) become:

$$C_B^N \, \delta\underline{\omega}_{IB}^B = \delta\underline{\omega}_{IB}^N$$

$$= C_B^N \left\{ \delta K_{Scal/Mis} \left(C_B^N \right)^T \left[\underline{\omega}_{IE}^N + \frac{1}{R} \, \underline{u}_{ZN}^N \times \left(\underline{v}_{H_0}^N + \Delta\underline{v}_{SF_H}^N \right) \right] + \delta\underline{K}_{Bias} \right\} = \text{Constant}$$

$$C_B^N \, \delta\underline{a}_{SF}^B = \delta\underline{a}_{SF}^N = C_B^N \left[g \, \delta L_{Scal/Mis} \left(C_B^N \right)^T \underline{u}_{ZN}^N + \delta\underline{L}_{Bias} \right] = \text{Constant} \tag{13.3.1-2}$$

where

$$\delta\underline{\omega}_{IB}^N, \, \delta\underline{a}_{SF}^N = \text{N Frame components of } \delta\underline{\omega}_{IB}, \, \delta\underline{a}_{SF}.$$

Using (13.3-13) for initial conditions, (13.3.1-2) for sensor errors, (13.3-14) for $\underline{v}_H^N$, and the previous simplifying assumptions, Equations (13.3.1-1) can then be easily integrated in the sequence shown to yield:

$$\gamma_{ZN} = \gamma_{ZN_0} - t\,\delta\omega_{IB_{ZN}}$$

$$\begin{aligned}\underline{\gamma}_H^N = {} & \underline{\gamma}_{H_0}^N - t\,\delta\underline{\omega}_{IB_H}^N - \left(t\,\gamma_{ZN_0} - \frac{1}{2}t^2\,\delta\omega_{IB_{ZN}}\right)\left[\underline{\omega}_{IE_H}^N \times \underline{u}_{ZN}^N + \frac{1}{R}\left(\underline{v}_{H_0}^N + \Delta\underline{v}_{SF_H}^N\right)\right] \\ & + t\,\frac{1}{R}\,\underline{u}_{ZN}^N \times \left\{\delta\underline{V}_{H_0}^N + \left[C_B^N\,\delta L_{Scal/Mis}\left(C_B^N\right)^T \Delta\underline{v}_{SF_H}^N\right]_H - \gamma_{ZN_0}\,\underline{u}_{ZN}^N \times \Delta\underline{v}_{SF_H}^N\right\}\end{aligned}$$

$$\begin{aligned}\delta V_{ZN} = {} & \delta V_{ZN_0} + \left[C_B^N\,\delta L_{Scal/Mis}\left(C_B^N\right)^T \Delta\underline{v}_{SF_H}^N\right]_{ZN} \\ & + \left\{\Delta\underline{v}_{SF_H}^N \times \left[\underline{\gamma}_{H_0}^N - \frac{1}{R}\left(\underline{u}_{ZN}^N \times \delta\underline{R}_{H_0}^N\right)\right]\right\} \cdot \underline{u}_{ZN}^N + t\left(\delta a_{SF_{ZN}} + F(h)\,\frac{g}{R}\,\delta R_0\right)\end{aligned}$$

$$F(h) = 2 \quad \text{For} \quad h \geq 0 \qquad\qquad F(h) = -1 \quad \text{For} \quad h < 0$$

$$\begin{aligned}\delta\underline{V}_H^N = {} & \delta\underline{V}_{H_0}^N + \left[C_B^N\,\delta L_{Scal/Mis}\left(C_B^N\right)^T \Delta\underline{v}_{SF_H}^N\right]_H - \gamma_{ZN_0}\left(\underline{u}_{ZN}^N \times \Delta\underline{v}_{SF_H}^N\right) \\ & + t\,\delta\underline{a}_{SF_H}^N + g\,\underline{u}_{ZN}^N \times \Big\langle t\,\underline{\gamma}_{H_0}^N - \frac{1}{2}t^2\,\delta\underline{\omega}_{IB_H}^N \\ & - \left(\frac{1}{2}t^2\,\gamma_{ZN_0} - \frac{1}{6}t^3\,\delta\omega_{IB_{ZN}}\right)\left[\underline{\omega}_{IE_H}^N \times \underline{u}_{ZN}^N + \frac{1}{R}\left(\underline{v}_{H_0}^N + \Delta\underline{v}_{SF_H}^N\right)\right] \\ & + \frac{1}{2}t^2\,\frac{1}{R}\,\underline{u}_{ZN}^N \times \left\{\delta\underline{V}_{H_0}^N + \left[C_B^N\,\delta L_{Scal/Mis}\left(C_B^N\right)^T \Delta\underline{v}_{SF_H}^N\right]_H - \gamma_{ZN_0}\,\underline{u}_{ZN}^N \times \Delta\underline{v}_{SF_H}^N\right\}\Big\rangle\end{aligned} \qquad (13.3.1\text{-}3)$$

$$\begin{aligned}\delta R = {} & \delta R_0 + t\,\delta V_{ZN_0} + t\left[C_B^N\,\delta L_{Scal/Mis}\left(C_B^N\right)^T \Delta\underline{v}_{SF_H}^N\right]_{ZN} \\ & + t\left\{\Delta\underline{v}_{SF_H}^N \times \left[\underline{\gamma}_{H_0}^N - \frac{1}{R}\left(\underline{u}_{ZN}^N \times \delta\underline{R}_{H_0}^N\right)\right]\right\} \cdot \underline{u}_{ZN}^N + \frac{1}{2}t^2\left(\delta a_{SF_{ZN}} + F(h)\,\frac{g}{R}\,\delta R_0\right)\end{aligned}$$

$$\begin{aligned}\delta\underline{R}_H^N = {} & \delta\underline{R}_{H_0}^N + t\,\delta\underline{V}_{H_0}^N + t\left[C_B^N\,\delta L_{Scal/Mis}\left(C_B^N\right)^T \Delta\underline{v}_{SF_H}^N\right]_H - t\,\gamma_{ZN_0}\left(\underline{u}_{ZN}^N \times \Delta\underline{v}_{SF_H}^N\right) \\ & + \frac{1}{2}t^2\,\delta\underline{a}_{SF_H}^N + g\,\underline{u}_{ZN}^N \times \Big\langle \frac{1}{2}t^2\,\underline{\gamma}_{H_0}^N - \frac{1}{6}t^3\,\delta\underline{\omega}_{IB_H}^N \\ & - \left(\frac{1}{6}t^3\,\gamma_{ZN_0} - \frac{1}{24}t^4\,\delta\omega_{IB_{ZN}}\right)\left[\underline{\omega}_{IE_H}^N \times \underline{u}_{ZN}^N + \frac{1}{R}\left(\underline{v}_{H_0}^N + \Delta\underline{v}_{SF_H}^N\right)\right] \\ & + \frac{1}{6}t^3\,\frac{1}{R}\,\underline{u}_{ZN}^N \times \left\{\delta\underline{V}_{H_0}^N + \left[C_B^N\,\delta L_{Scal/Mis}\left(C_B^N\right)^T \Delta\underline{v}_{SF_H}^N\right]_H - \gamma_{ZN_0}\,\underline{u}_{ZN}^N \times \Delta\underline{v}_{SF_H}^N\right\}\Big\rangle\end{aligned}$$

where

$\delta\underline{\omega}_{IB_H}^N$, $\delta\omega_{IB_{ZN}}$ = Horizontal and upward vertical components of $\delta\underline{\omega}_{IB}^N$ in (13.3.1-2).

$\delta\underline{a}_{SF_H}^N$, $\delta a_{SF_{ZN}}$ = Horizontal and upward vertical components of $\delta\underline{a}_{SF}^N$ in (13.3.1-2).

Applying generalized Equation (13.1-5) and grouping common error source terms in (13.3.1-3), we then obtain the final short-term/fixed-attitude solution:

$$
\begin{aligned}
\gamma_{ZN} &= \gamma_{ZN_0} - t\, \delta\omega_{IB_{ZN}} \\
\underline{\gamma}^N_H &= \underline{\gamma}^N_{H_0} - t\, \gamma_{ZN_0} \left(\underline{\omega}^N_{IE_H} \times \underline{u}^N_{ZN} + \frac{1}{R}\, \underline{v}^N_{H_0} \right) + t\, \frac{1}{R}\, \underline{u}^N_{ZN} \times \delta\underline{V}^N_{H_0} \\
&\quad + t\, \frac{1}{R}\, \underline{u}^N_{ZN} \times \left[C^N_B\, \delta L_{Scal/Mis} \left(C^N_B \right)^T \Delta\underline{v}^N_{SF_H} \right]_H - t\, \delta\underline{\omega}^N_{IB_H} \\
&\quad + \frac{1}{2}\, t^2\, \delta\omega_{IB_{ZN}} \left[\underline{\omega}^N_{IE_H} \times \underline{u}^N_{ZN} + \frac{1}{R} \left(\underline{v}^N_{H_0} + \Delta\underline{v}^N_{SF_H} \right) \right]
\end{aligned}
$$

$$
\begin{aligned}
\delta V_{ZN} &= \delta V_{ZN_0} + F(h)\, \frac{g}{R}\, t\, \delta R_0 + \left\{ \Delta\underline{v}^N_{SF_H} \times \left[\underline{\gamma}^N_{H_0} - \frac{1}{R} \left(\underline{u}^N_{ZN} \times \delta\underline{R}^N_{H_0} \right) \right] \right\} \cdot \underline{u}^N_{ZN} \\
&\quad + \left[C^N_B\, \delta L_{Scal/Mis} \left(C^N_B \right)^T \Delta\underline{v}^N_{SF_H} \right]_{ZN} + t\, \delta a_{SF_{ZN}}
\end{aligned}
$$

$$
F(h) = 2 \quad \text{For } h \geq 0 \qquad\qquad F(h) = -1 \quad \text{For } h < 0
$$

$$
\begin{aligned}
\delta\underline{V}^N_H &= \left(1 - \frac{1}{2}\, \frac{g}{R}\, t^2 \right) \delta\underline{V}^N_{H_0} + g\, t\, \underline{u}^N_{ZN} \times \underline{\gamma}^N_{H_0} \\
&\quad - \gamma_{ZN_0} \left[\underline{u}^N_{ZN} \times \Delta\underline{v}^N_{SF_H} + \frac{1}{2}\, \frac{g}{R}\, t^2\, R \left(\underline{\omega}^N_{IE_H} + \frac{1}{R}\, \underline{u}^N_{ZN} \times \underline{v}^N_{H_0} \right) \right] \qquad (13.3.1\text{-}4) \\
&\quad + \left(1 - \frac{1}{2}\, \frac{g}{R}\, t^2 \right) \left\{ C^N_B\, \delta L_{Scal/Mis} \left(C^N_B \right)^T \Delta\underline{v}^N_{SF_H} \right\}_H + t\, \delta\underline{a}^N_{SF_H} \\
&\quad - \frac{1}{2}\, \frac{g}{R}\, t^2\, R\, \underline{u}^N_{ZN} \times \delta\underline{\omega}^N_{IB_H} + \frac{1}{6}\, \frac{g}{R}\, t^3\, R\, \delta\omega_{IB_{ZN}} \left[\underline{\omega}^N_{IE_H} + \frac{1}{R}\, \underline{u}^N_{ZN} \times \left(\underline{v}^N_{H_0} + \Delta\underline{v}^N_{SF_H} \right) \right]
\end{aligned}
$$

$$
\begin{aligned}
\delta R &= \delta R_0 + t\, \delta V_{ZN_0} + \frac{1}{2}\, F(h)\, \frac{g}{R}\, t^2\, \delta R_0 + t \left\{ \Delta\underline{v}^N_{SF_H} \times \left[\underline{\gamma}^N_{H_0} - \frac{1}{R} \left(\underline{u}^N_{ZN} \times \delta\underline{R}^N_{H_0} \right) \right] \right\} \cdot \underline{u}^N_{ZN} \\
&\quad + t \left[C^N_B\, \delta L_{Scal/Mis} \left(C^N_B \right)^T \Delta\underline{v}^N_{SF_H} \right]_{ZN} + \frac{1}{2}\, t^2\, \delta a_{SF_{ZN}}
\end{aligned}
$$

$$
\begin{aligned}
\delta\underline{R}^N_H &= \delta\underline{R}^N_{H_0} + \left(t - \frac{1}{6}\, \frac{g}{R}\, t^3 \right) \delta\underline{V}^N_{H_0} + \frac{1}{2}\, g\, t^2\, \underline{u}^N_{ZN} \times \underline{\gamma}^N_{H_0} \\
&\quad - \gamma_{ZN_0} \left[t\, \underline{u}^N_{ZN} \times \Delta\underline{v}^N_{SF_H} + \frac{1}{6}\, \frac{g}{R}\, t^3\, R \left(\underline{\omega}^N_{IE_H} + \frac{1}{R}\, \underline{u}^N_{ZN} \times \underline{v}^N_{H_0} \right) \right] \\
&\quad + \left(t - \frac{1}{6}\, \frac{g}{R}\, t^3 \right) \left\{ C^N_B\, \delta L_{Scal/Mis} \left(C^N_B \right)^T \Delta\underline{v}^N_{SF_H} \right\}_H + \frac{1}{2}\, t^2\, \delta\underline{a}^N_{SF_H} \\
&\quad - \frac{1}{6}\, \frac{g}{R}\, t^3\, R\, \underline{u}^N_{ZN} \times \delta\underline{\omega}^N_{IB_H} + \frac{1}{24}\, \frac{g}{R}\, t^4\, R\, \delta\omega_{IB_{ZN}} \left[\underline{\omega}^N_{IE_H} + \frac{1}{R}\, \underline{u}^N_{ZN} \times \left(\underline{v}^N_{H_0} + \Delta\underline{v}^N_{SF_H} \right) \right]
\end{aligned}
$$

Equations (13.3.1-4) reveal the following navigation error characteristics in response to the principal sources of navigation error. The heading error γ_{ZN} grows linearly (t) with vertical angular rate sensor error. The horizontal tilt $\underline{\gamma}_H^N$ grows linearly with initial heading error, initial horizontal velocity error, and horizontal angular rate sensor error. The vertical velocity error δV_{ZN} grows linearly with vertical accelerometer error. The horizontal velocity error $\delta \underline{V}_H^N$ grows linearly with initial horizontal tilt and horizontal accelerometer error, and parabolically (t^2) with initial heading error and horizontal angular rate sensor error. The vertical position error δR grows linearly with initial vertical velocity error and parabolically with vertical accelerometer error. The horizontal position error $\delta \underline{R}_H^N$ grows linearly with initial horizontal velocity error, parabolically with initial horizontal tilt and horizontal accelerometer error, and as t^3 with initial heading error and horizontal angular rate sensor error. The effect of the initial horizontal velocity impulse $\Delta \underline{v}_{SF}^N$ is to add initial velocity error in the earth-rate/transport-rate coupling term, to produce an initial horizontal velocity error and a linearly increasing horizontal position error proportional to initial heading error, and to generate an initial velocity error and linearly increasing position error proportional to accelerometer scale-factor/misalignment error.

13.3.2 UP TO TWO HOUR HORIZONTAL SOLUTION AT CONSTANT ATTITUDE WITH CONTROLLED VERTICAL CHANNEL

For the solution to Equations (13.3-15) that is valid for up to 2 hours under controlled vertical channel conditions, we assume that δR and δV_{ZN} will be controlled to zero by the control gains so that only the horizontal velocity/position solutions are of interest. Then Equations (13.3-15) simplify to:

$$\begin{aligned}
\dot{\gamma}_{ZN} &= -\left(C_B^N \, \delta\underline{\omega}_{IB}^B\right)_{ZN} \\
\dot{\underline{\gamma}}_H^N &\approx -\left(C_B^N \, \delta\underline{\omega}_{IB}^B\right)_H - \gamma_{ZN}\left(\underline{\omega}_{IE_H}^N \times \underline{u}_{ZN}^N + \frac{1}{R}\,\underline{v}_H^N\right) + \frac{1}{R}\left(\underline{u}_{ZN}^N \times \delta\underline{V}_H^N\right) \\
\delta\dot{\underline{V}}_H^N &\approx \left(C_B^N \, \delta\underline{a}_{SF}^B\right)_H + g\,\underline{u}_{ZN}^N \times \underline{\gamma}_H^N \\
\delta\dot{\underline{R}}_H^N &= \delta\underline{V}_H^N
\end{aligned} \qquad (13.3.2\text{-}1)$$

Equations (13.3-14) still apply for the velocity and inertial sensor terms in (13.3.2-1), and Equations (13.3-13) still apply for initial conditions immediately following application of the initial $\Delta\underline{a}_{SF}^N$ impulse.

The analytical solution to (13.3.2-1) is obtained by first noting that the $\dot{\underline{\gamma}}^N_H$ and $\delta\dot{\underline{V}}^N_H$ expressions are coupled together by the $\delta\underline{V}^N_H$ and $\underline{\gamma}^N_H$ terms, that the $\dot{\underline{\gamma}}^N_H$, $\delta\dot{\underline{V}}^N_H$ expressions are driven by sensor errors and the $\dot{\gamma}_{ZN}$ solution, and that the $\delta\dot{\underline{R}}^N_H$ solution is driven by the $\delta\dot{\underline{V}}^N_H$ solution. The general solution process is to solve for γ_{ZN} first, then determine the coupled $\dot{\underline{\gamma}}^N_H$, $\delta\dot{\underline{V}}^N_H$ solution, and finally find $\delta\underline{R}^N_H$ from $\delta\underline{V}^N_H$. For our simplified model we also assume (as in Section 13.3.1) that the distance traveled will be short enough that $\underline{\omega}^E_{IN}$ can be approximated as constant, that attitude in the form of C^N_B is constant, and that inertial sensor error contributors in (13.3-14) are constant so Equations (13.3.1-2) apply. Then the solution to the (13.3.2-1) $\dot{\gamma}_{ZN}$ expression for time t following application of the initial $\Delta\underline{a}^N_{SF}$ impulse is simply:

$$\gamma_{ZN} = \gamma_{ZN_{0+}} - t\ \delta\omega_{IB_{ZN}} \tag{13.3.2-2}$$

The $\gamma_{ZN_{0+}}$ initial condition is provided in Equations (13.3-13).

The coupled $\dot{\underline{\gamma}}^N_H$, $\delta\dot{\underline{V}}^N_H$ expressions in (13.3.2-1) are solved by taking the derivative of $\delta\dot{\underline{V}}^N_H$ (subject to the previous simplifying assumptions), substituting $\dot{\underline{\gamma}}^N_H$ from (13.3.2-1) and γ_{ZN} from (13.3.2-2) in the result, using (13.3-14) for $\underline{v}^N_H$, and application of generalized Equation (13.1-6) to obtain:

$$\begin{aligned}
\delta\ddot{\underline{V}}^N_H &= g\ \underline{u}^N_{ZN} \times \dot{\underline{\gamma}}^N_H \\
&= g\ \underline{u}^N_{ZN} \times \Big\{ - \delta\underline{\omega}^N_{IB_H} - \left(\gamma_{ZN_{0+}} - t\ \delta\omega_{IB_{ZN}}\right) \Big[\underline{\omega}^N_{IE_H} \times \underline{u}^N_{ZN} \\
&\qquad + \frac{1}{R}\left(\underline{v}^N_{H_0} + \Delta\underline{v}^N_{SF_H}\right) \Big] + \frac{1}{R}\left(\underline{u}^N_{ZN} \times \delta\underline{V}^N_H\right) \Big\} \\
&= - \frac{g}{R}\ \delta\underline{V}^N_H - \frac{g}{R}\ R \Big\{ \underline{u}^N_{ZN} \times \delta\underline{\omega}^N_{IB_H} \\
&\qquad + \left(\gamma_{ZN_{0+}} - t\ \delta\omega_{IB_{ZN}}\right) \Big[\underline{\omega}^N_{IE_H} + \frac{1}{R}\ \underline{u}^N_{ZN} \times \left(\underline{v}^N_{H_0} + \Delta\underline{v}^N_{SF_H}\right)\Big]\Big\}
\end{aligned} \tag{13.3.2-3}$$

or upon rearrangement:

$$\delta \ddot{\underline{V}}_H^N + \omega_S^2 \, \delta \underline{V}_H^N = - \omega_S^2 \, R \left\{ \underline{u}_{ZN}^N \times \delta \underline{\omega}_{IB_H}^N + \left(\gamma_{ZN_{0+}} - t \, \delta\omega_{IB_{ZN}} \right) \left[\underline{\omega}_{IE_H}^N + \frac{1}{R} \, \underline{u}_{ZN}^N \times \left(\underline{v}_{H_0}^N + \Delta \underline{v}_{SF_H}^N \right) \right] \right\} \tag{13.3.2-4}$$

with

$$\omega_S = \sqrt{\frac{g}{R}} \tag{13.3.2-5}$$

in which ω_S is, of course, our familiar 84 minute period Schuler frequency (See Section 13.2.2).

Equation (13.3.2-4) has the compressed form:

$$\delta \ddot{\underline{V}}_H^N + \omega_S^2 \, \delta \underline{V}_H^N = \underline{D}_1 + \underline{D}_2 \, t \tag{13.3.2-6}$$

with

$$\underline{D}_1 = - \omega_S^2 \, R \left\{ \underline{u}_{ZN}^N \times \delta \underline{\omega}_{IB_H}^N + \gamma_{ZN_{0+}} \left[\underline{\omega}_{IE_H}^N + \frac{1}{R} \, \underline{u}_{ZN}^N \times \left(\underline{v}_{H_0}^N + \Delta \underline{v}_{SF_H}^N \right) \right] \right\}$$

$$\underline{D}_2 = \omega_S^2 \, R \, \delta\omega_{IB_{ZN}} \left[\underline{\omega}_{IE_H}^N + \frac{1}{R} \, \underline{u}_{ZN}^N \times \left(\underline{v}_{H_0}^N + \Delta \underline{v}_{SF_H}^N \right) \right] \tag{13.3.2-7}$$

For the problem at hand, $\underline{D}_1$, $\underline{D}_2$ are treated as constants.

Equation (13.3.2-6) can be solved by first finding the general homogeneous solution that fits $\delta \ddot{\underline{V}}_H^N + \omega_S^2 \, \delta \underline{V}_H^N$ to zero, finding the "Particular" solution that fits $\delta \ddot{\underline{V}}_H^N + \omega_S^2 \, \delta \underline{V}_H^N$ to $\underline{D}_1 + \underline{D}_2 \, t$, then summing the two and setting the homogeneous solution coefficients to match the Equations (13.3-13) initial conditions. As in Section 13.2.1, the homogeneous solution involves solving for the roots of the (13.3.2-6) characteristic equation:

$$\lambda^2 + \omega_S^2 = 0 \tag{13.3.2-8}$$

for which:

$$\lambda = \pm j \, \omega_S$$

where

$$j = \text{Imaginary parameter } \sqrt{-1}.$$

The homogeneous solution is the sum of terms of the form $B \, e^{\lambda t}$ for each root. We also know that:

$$\sin x = \frac{e^{jx} - e^{-jx}}{2j} \qquad \cos x = \frac{e^{jx} + e^{-jx}}{2} \tag{13.3.2-9}$$

so that the homogeneous solution is:

$$\delta \underline{V}_{H_{Hmg}}^{N} = \underline{C}_1 \sin \omega_S t + \underline{C}_2 \cos \omega_S t \tag{13.3.2-10}$$

where

$\delta \underline{V}_{H_{Hmg}}^{N}$ = Homogeneous solution to (13.3.2-6).

$\underline{C}_1, \underline{C}_2$ = Constants to be subsequently determined based on initial condition constraints.

The Particular solution to (13.3.2-6) has the form:

$$\delta \underline{V}_{H_{Prt}}^{N} = \underline{C}_3 + \underline{C}_4 \, t \tag{13.3.2-11}$$

where

$\delta \underline{V}_{H_{Prt}}^{N}$ = Particular solution to (13.3.2-6).

$\underline{C}_3, \underline{C}_4$ = Constants to satisfy (13.3.2-6).

The $\underline{C}_3, \underline{C}_4$ constants are determined by substituting (13.3.2-11) for $\delta \underline{V}_H^N$ in (13.3.2-6) and equating coefficients of like powers of t:

$$\delta \ddot{\underline{V}}_{H_{Prt}}^{N} + \omega_S^2 \, \delta \underline{V}_{H_{Prt}}^{N} = \omega_S^2 \left(\underline{C}_3 + \underline{C}_4 \, t \right) = \underline{D}_1 + \underline{D}_2 \, t \tag{13.3.2-12}$$

for which:

$$\underline{C}_3 = \frac{1}{\omega_S^2} \underline{D}_1 \qquad \underline{C}_4 = \frac{1}{\omega_S^2} \underline{D}_2 \tag{13.3.2-13}$$

The total solution is then constructed as the sum of (13.3.2-10) and (13.3.2-11) with (13.3.2-13):

$$\delta \underline{V}_H^N = \underline{C}_1 \sin \omega_S t + \underline{C}_2 \cos \omega_S t + \frac{1}{\omega_S^2} \left(\underline{D}_1 + \underline{D}_2 \, t \right) \tag{13.3.2-14}$$

The $\underline{C}_1, \underline{C}_2$ coefficients in (13.3.2-14) can now be evaluated by setting t equal to zero in (13.3.2-14) and its derivative, and introducing initial condition constraints:

$$\delta \underline{V}^{N}_{H_{t=0}} = \underline{C}_2 + \frac{1}{\omega_S^2}\,\underline{D}_1 \qquad (13.3.2\text{-}15)$$

$$\delta \dot{\underline{V}}^{N}_{H_{t=0}} = \left(\underline{C}_1\,\omega_S \cos\omega_S t - \underline{C}_2\,\omega_S \sin\omega_S t + \frac{1}{\omega_S^2}\,\underline{D}_2\right)_{t=0} = \underline{C}_1\,\omega_S + \frac{1}{\omega_S^2}\,\underline{D}_2 \qquad (13.3.2\text{-}16)$$

From the Equations (13.3.2-1) $\delta \dot{\underline{V}}^{N}_{H}$ expression with (13.3.1-2), we also know that:

$$\delta \dot{\underline{V}}^{N}_{H_{t=0}} = \delta \underline{a}^{N}_{SF_H} + g\,\underline{u}^{N}_{ZN} \times \underline{\gamma}^{N}_{H_{t=0}} \qquad (13.3.2\text{-}17)$$

Combining (13.3.2-16) and (13.3.2-17) yields:

$$\delta \underline{a}^{N}_{SF_H} + g\,\underline{u}^{N}_{ZN} \times \underline{\gamma}^{N}_{H_{t=0}} = \underline{C}_1\,\omega_S + \frac{1}{\omega_S^2}\,\underline{D}_2 \qquad (13.3.2\text{-}18)$$

The solution for $\underline{C}_1$, $\underline{C}_2$ is obtained from (13.3.2-15) and (13.3.2-18) by rearrangement and equating the t = 0 terms to the Equations (13.3-13) "0+" initial values:

$$\underline{C}_1 = \frac{1}{\omega_S}\left(\delta \underline{a}^{N}_{SF_H} + g\,\underline{u}^{N}_{ZN} \times \underline{\gamma}^{N}_{H_{0+}} - \frac{1}{\omega_S^2}\,\underline{D}_2\right) \qquad \underline{C}_2 = \delta \underline{V}^{N}_{H_{0+}} - \frac{1}{\omega_S^2}\,\underline{D}_1 \qquad (13.3.2\text{-}19)$$

The solution for $\delta \underline{V}^{N}_{H}$ (i.e., the solution to Equations (13.3.2-1)) is then determined by substituting (13.3.2-19) into (13.3.2-14) and rearranging:

$$\begin{aligned}\delta \underline{V}^{N}_{H} = {} & \delta \underline{V}^{N}_{H_{0+}} \cos\omega_S t + \left(\delta \underline{a}^{N}_{SF_H} + g\,\underline{u}^{N}_{ZN} \times \underline{\gamma}^{N}_{H_{0+}}\right)\frac{\sin\omega_S t}{\omega_S} \\ & + \underline{D}_1\,\frac{(1 - \cos\omega_S t)}{\omega_S^2} + \frac{1}{\omega_S^2}\,\underline{D}_2\left(t - \frac{\sin\omega_S t}{\omega_S}\right)\end{aligned} \qquad (13.3.2\text{-}20)$$

with $\delta \underline{V}^{N}_{H_{0+}}$, $\underline{\gamma}^{N}_{H_{0+}}$ provided from Equations (13.3-13) and $\underline{D}_1$, $\underline{D}_2$ from Equations (13.3.2-7).

The $\underline{\gamma}^{N}_{H}$ solution to Equations (13.3.2-1) is found as the integral of the $\dot{\underline{\gamma}}^{N}_{H}$ expression. Using the definitions for $\underline{D}_1$, $\underline{D}_2$, the Equations (13.3-14) velocity expression, the (13.3.2-2) solution for γ_{ZN}, and general Equation (13.1-5), the $\dot{\underline{\gamma}}^{N}_{H}$ expression in (13.3.2-1) is first compacted to the form:

$$\dot{\underline{\gamma}}_H^N = - \frac{1}{\omega_S^2 R} \underline{u}_{ZN}^N \times \left(\underline{D}_1 + \underline{D}_2 \, t \right) + \frac{1}{R} \left(\underline{u}_{ZN}^N \times \delta \underline{V}_H^N \right) \tag{13.3.2-21}$$

Then, substituting (13.3.2-20) for $\delta \underline{V}_H^N$, applying (13.1-5), and using Schuler frequency expression (13.3.2-5), we obtain:

$$\begin{aligned} \dot{\underline{\gamma}}_H^N &= \frac{1}{R} \underline{u}_{ZN}^N \times \delta \underline{V}_{H_{0+}}^N \cos \omega_S t \\ &+ \left(\frac{1}{R} \underline{u}_{ZN}^N \times \delta \underline{a}_{SF_H}^N - \omega_S^2 \, \underline{\gamma}_{H_{0+}}^N \right) \frac{\sin \omega_S t}{\omega_S} - \frac{1}{\omega_S^2 R} \underline{u}_{ZN}^N \times \left(\underline{D}_1 \cos \omega_S t + \underline{D}_2 \frac{\sin \omega_S t}{\omega_S} \right) \end{aligned} \tag{13.3.2-22}$$

The bounded integral of (13.3.2-22) from time = 0 to t with $\underline{\gamma}_{H_{0+}}^N$ as the initial condition provides the $\underline{\gamma}_H^N$ solution:

$$\begin{aligned} \underline{\gamma}_H^N &= \underline{\gamma}_{H_{0+}}^N \cos \omega_S t + \frac{1}{R} \left(\underline{u}_{ZN}^N \times \delta \underline{V}_{H_{0+}}^N \right) \frac{\sin \omega_S t}{\omega_S} \\ &+ \frac{1}{R} \underline{u}_{ZN}^N \times \delta \underline{a}_{SF_H}^N \frac{(1 - \cos \omega_S t)}{\omega_S^2} - \frac{1}{\omega_S^2 R} \underline{u}_{ZN}^N \times \left(\underline{D}_1 \frac{\sin \omega_S t}{\omega_S} + \underline{D}_2 \frac{(1 - \cos \omega_S t)}{\omega_S^2} \right) \end{aligned} \tag{13.3.2-23}$$

The $\delta \underline{R}_H^N$ solution to Equations (13.3.2-1) is easily calculated from the $\delta \dot{\underline{R}}_H^N$ expression by the simple bounded integration of (13.3.2-20) from time = 0 to t :

$$\begin{aligned} \delta \underline{R}_H^N &= \delta \underline{R}_{H_{0+}}^N + \delta \underline{V}_{H_{0+}}^N \frac{\sin \omega_S t}{\omega_S} + \left(\delta \underline{a}_{SF_H}^N + g \, \underline{u}_{ZN}^N \times \underline{\gamma}_{H_{0+}}^N \right) \frac{(1 - \cos \omega_S t)}{\omega_S^2} \\ &\quad + \underline{D}_1 \frac{1}{\omega_S^2} \left(t - \frac{\sin \omega_S t}{\omega_S} \right) + \frac{1}{\omega_S^2} \underline{D}_2 \left(\frac{1}{2} t^2 - \frac{(1 - \cos \omega_S t)}{\omega_S^2} \right) \end{aligned} \tag{13.3.2-24}$$

We now substitute Equations (13.3.2-7) for $\underline{D}_1$, $\underline{D}_2$, the $\delta \underline{\omega}_{IB}^B$, $\delta \underline{a}_{SF}^B$ inertial sensor error expressions from (13.3-14), the (13.3-13) initial condition relations, (13.3-7) for $\underline{\omega}_{IN}^N$, and $\underline{v}_H^N$ from (13.3-14) into (13.3.2-2), (13.3.2-20), (13.3.2-23) and (13.3.2-24). After grouping common error source terms, applying ω_S Equation (13.3.2-5), and applying general Equation (13.1-5), the final result is:

$$\gamma_{ZN} = \gamma_{ZN_0} - t\,\delta\omega_{IB_{ZN}}$$

$$\begin{aligned}
\underline{\gamma}_H^N = {} & \underline{\gamma}_{H_0}^N \cos\omega_S t - \gamma_{ZN_0}\left(\underline{\omega}_{IE_H}^N \times \underline{u}_{ZN}^N + \frac{1}{R}\underline{v}_{H_0}^N\right)\frac{\sin\omega_S t}{\omega_S} + \frac{1}{R}\left(\underline{u}_{ZN}^N \times \delta\underline{V}_{H_0}^N\right)\frac{\sin\omega_S t}{\omega_S} \\
& + \underline{u}_{ZN}^N \times \left[C_B^N\,\delta L_{Scal/Mis}\left(C_B^N\right)^T \Delta\underline{v}_{SF_H}^N \frac{1}{R}\frac{\sin\omega_S t}{\omega_S}\right]_H \\
& + \frac{1}{R}\underline{u}_{ZN}^N \times \delta\underline{a}_{SF_H}^N \frac{(1-\cos\omega_S t)}{\omega_S^2} - \delta\underline{\omega}_{IB_H}^N \frac{\sin\omega_S t}{\omega_S} \\
& + \delta\omega_{IB_{ZN}}\left[\underline{\omega}_{IE_H}^N \times \underline{u}_{ZN}^N + \frac{1}{R}\left(\underline{v}_{H_0}^N + \Delta\underline{v}_{SF_H}^N\right)\right]\frac{(1-\cos\omega_S t)}{\omega_S^2}
\end{aligned}$$

$$\begin{aligned}
\delta\underline{V}_H^N = {} & \delta\underline{V}_{H_0}^N \cos\omega_S t + g\left(\underline{u}_{ZN}^N \times \underline{\gamma}_{H_0}^N\right)\frac{\sin\omega_S t}{\omega_S} \\
& - \gamma_{ZN_0}\left[\underline{u}_{ZN}^N \times \Delta\underline{v}_{SF_H}^N + R\left(\underline{\omega}_{IE_H}^N + \frac{1}{R}\underline{u}_{ZN}^N \times \underline{v}_{H_0}^N\right)(1-\cos\omega_S t)\right] \\
& + \left[C_B^N\,\delta L_{Scal/Mis}\left(C_B^N\right)^T \Delta\underline{v}_{SF_H}^N\right]_H \cos\omega_S t + \delta\underline{a}_{SF_H}^N \frac{\sin\omega_S t}{\omega_S} \\
& - R\,\underline{u}_{ZN}^N \times \delta\underline{\omega}_{IB_H}^N (1-\cos\omega_S t) + R\,\delta\omega_{IB_{ZN}}\left[\underline{\omega}_{IE_H}^N + \frac{1}{R}\underline{u}_{ZN}^N \times \left(\underline{v}_{H_0}^N + \Delta\underline{v}_{SF_H}^N\right)\right]\left(t - \frac{\sin\omega_S t}{\omega_S}\right)
\end{aligned} \qquad (13.3.2\text{-}25)$$

$$\begin{aligned}
\delta\underline{R}_H^N = {} & \delta\underline{R}_{H_0}^N + \delta\underline{V}_{H_0}^N \frac{\sin\omega_S t}{\omega_S} + g\left(\underline{u}_{ZN}^N \times \underline{\gamma}_{H_0}^N\right)\frac{(1-\cos\omega_S t)}{\omega_S^2} \\
& - \gamma_{ZN_0}\left[\left(\underline{u}_{ZN}^N \times \Delta\underline{v}_{SF_H}^N\right)t + R\left(\underline{\omega}_{IE_H}^N + \frac{1}{R}\underline{u}_{ZN}^N \times \underline{v}_{H_0}^N\right)\left(t - \frac{\sin\omega_S t}{\omega_S}\right)\right] \\
& + \left[C_B^N\,\delta L_{Scal/Mis}\left(C_B^N\right)^T \Delta\underline{v}_{SF_H}^N\right]_H \frac{\sin\omega_S t}{\omega_S} + \delta\underline{a}_{SF_H}^N \frac{(1-\cos\omega_S t)}{\omega_S^2} \\
& - R\,\underline{u}_{ZN}^N \times \delta\underline{\omega}_{IB_H}^N\left(t - \frac{\sin\omega_S t}{\omega_S}\right) \\
& + R\,\delta\omega_{IB_{ZN}}\left[\underline{\omega}_{IE_H}^N + \frac{1}{R}\underline{u}_{ZN}^N \times \left(\underline{v}_{H_0}^N + \Delta\underline{v}_{SF_H}^N\right)\right]\left(\frac{1}{2}t^2 - \frac{(1-\cos\omega_S t)}{\omega_S^2}\right)
\end{aligned}$$

The $\delta\underline{\omega}_{IB}^N$ and $\delta\underline{a}_{SF}^N$ sensor error terms in (13.3.2-25) are provided by Equations (13.3.1-2).

If Equations (13.3.2-25) (using (13.3.2-5) for ω_S and with Taylor series expansion for the sine, cosine terms) are compared with Equations (13.3.1-4), it will be found that they agree for short time periods to first order (and to second, third and fourth order for some terms). Equations (13.3.2-25) reveal the following navigation error characteristics in response to the principal sources of navigation error.

The heading error γ_{ZN} grows linearly (t) with vertical angular rate sensor error. The horizontal tilt $\underline{\gamma}_H^N$ responds as a sine wave Schuler oscillation to initial heading error, initial horizontal velocity error and horizontal angular rate sensor error, and as an offset cosine wave Schuler oscillation to horizontal accelerometer error and vertical angular rate sensor error. Initial horizontal tilt becomes a $\underline{\gamma}_H^N$ cosine wave Schuler oscillation response.

The horizontal velocity error $\delta\underline{V}_H^N$ responds as a sine wave Schuler oscillation to initial horizontal tilt and horizontal accelerometer error, as an offset cosine wave Schuler oscillation to initial heading error and horizontal angular rate sensor error, and as an unbounded ramp in time t (plus a sine wave Schuler oscillation) to vertical angular rate sensor error. Initial horizontal velocity error becomes a $\delta\underline{V}_H^N$ cosine wave Schuler oscillation response.

The horizontal position error $\delta\underline{R}_H^N$ responds as a sine wave Schuler oscillation to initial horizontal velocity error, as an offset cosine wave Schuler oscillation to initial horizontal tilt and horizontal accelerometer error, as an unbounded ramp in t (plus a sine wave Schuler oscillation) to initial heading error and horizontal angular rate sensor error, and as a t^2 buildup (plus an offset cosine wave Schuler oscillation) to vertical angular rate sensor error.

The effect of the initial horizontal velocity impulse $\Delta\underline{v}_{SF}^N$ is add to initial velocity error in the transport rate coupling term, to produce a horizontal velocity error cosine wave Schuler oscillation and horizontal position error sine wave Schuler oscillation proportional to accelerometer scale-factor/misalignment error, and to generate an offset horizontal velocity error and linearly ramping horizontal position error proportional to initial heading error.

13.4 NAVIGATION ERRORS FOR ROTATING ATTITUDE AND CONSTANT SENSOR ERRORS

In the previous section we investigated the navigation error equation response to constant sensor errors in a non-rotating environment. In this section we investigate the navigation error response to constant sensor errors in a rotating B Frame for four different trajectory conditions:

- Constant high rate spinning.
- High rotation rate when the rotation rate vector is also rotating (simulating rotating free body angular motion).
- Circular trajectory profile at constant angular rate about the vertical.
- Circular trajectory profile at constant rate about the vertical when the angular rate equals the Schuler frequency.

13.4.1 CONSTANT HIGH RATE SPINNING ABOUT NON-ROTATING AXIS

For the constant rate spinning example we investigate the angular rate error, attitude error build-up and acceleration transformation error produced by constant inertial sensor errors.

13.4.1.1 ANGULAR RATE VECTOR ERROR CHARACTERISTICS

The angular rate vector is considered constant in the B Frame and described by:

$$\underline{\omega}_{IB}^{B} = \omega_{IB}\, \underline{u}_{\omega}^{B} \tag{13.4.1.1-1}$$

where

ω_{IB} = Magnitude of $\underline{\omega}_{IB}^{B}$ (considered constant).

$\underline{u}_{\omega}^{B}$ = B Frame projection of a unit vector along $\underline{\omega}_{IB}^{B}$ (considered constant).

We also define the angular rate sensor derived equivalent to Equation (13.4.1.1-1):

$$\underline{\tilde{\omega}}_{IB}^{B} = \tilde{\omega}_{IB}\, \underline{\tilde{u}}_{\omega}^{B} \tag{13.4.1.1-2}$$

where

$\tilde{\ }$ = Indicator for parameter input to the system computer, hence, containing errors. The input parameter without the $\tilde{\ }$ is defined to be the idealized version of the input parameter, hence, error free. In this section, the $\tilde{\ }$ parameters are derived from compensated inertial sensor inputs which contain error because the compensation has imperfections and/or the sensor error characteristics change following calibration.

Of interest is the error in the $\underline{\tilde{\omega}}_{IB}^{B}$ magnitude ($\tilde{\omega}_{IB}$) and the angular error in the $\underline{\tilde{\omega}}_{IB}^{B}$ direction vector $\underline{\tilde{u}}_{\omega}^{B}$, each of which is defined by:

$$\delta\omega_{IB} \equiv \tilde{\omega}_{IB} - \omega_{IB} \qquad (13.4.1.1\text{-}3)$$

$$\underline{\beta}^B \equiv \underline{u}_\omega^B \times \underline{\tilde{u}}_\omega^B \qquad (13.4.1.1\text{-}4)$$

where

$\delta\omega_{IB}, \underline{\beta}^B$ = Magnitude and angular errors in $\underline{\tilde{\omega}}_{IB}^B$.

The $\delta\omega_{IB}$ magnitude error in $\underline{\tilde{\omega}}_{IB}^B$ can be expressed in terms if the angular rate sensor vector error $\delta\underline{\omega}_{IB}^B$ as follows. We first note from general Equation (3.1.1-2) that:

$$\omega_{IB}^2 = \underline{\omega}_{IB}^B \cdot \underline{\omega}_{IB}^B \qquad (13.4.1.1\text{-}5)$$

The $\delta\omega_{IB}$ error is then obtained from the differential of (13.4.1.1-5):

$$\omega_{IB}\,\delta\omega_{IB} = \underline{\omega}_{IB}^B \cdot \delta\underline{\omega}_{IB}^B \qquad (13.4.1.1\text{-}6)$$

or upon rearrangement and application of (13.4.1.1-1):

$$\delta\omega_{IB} = \underline{u}_\omega^B \cdot \delta\underline{\omega}_{IB}^B \qquad (13.4.1.1\text{-}7)$$

The $\underline{\beta}^B$ angular error in $\underline{\tilde{\omega}}_{IB}^B$ can be expressed in terms of the angular rate sensor vector error $\delta\underline{\omega}_{IB}^B$ by first defining $\underline{\tilde{u}}_\omega^B$ as:

$$\underline{\tilde{u}}_\omega^B = \underline{u}_\omega^B + \delta\underline{u}_\omega^B \qquad (13.4.1.1\text{-}8)$$

where

$\delta\underline{u}_\omega^B$ = Error in $\underline{\tilde{u}}_\omega^B$.

Substituting (13.4.1.1-8) into (13.4.1.1-4) yields the equivalent form:

$$\underline{\beta}^B = \underline{u}_\omega^B \times \delta\underline{u}_\omega^B \qquad (13.4.1.1\text{-}9)$$

The $\delta\underline{u}_\omega^B$ term in (13.4.1.1-9) is then obtained from the differential of (13.4.1.1-1):

$$\delta\underline{\omega}_{IB}^B = \delta\omega_{IB}\,\underline{u}_\omega^B + \omega_{IB}\,\delta\underline{u}_\omega^B \qquad (13.4.1.1\text{-}10)$$

or after rearrangement:

$$\delta \underline{u}_{\omega}^{B} = - \frac{\delta\omega_{IB}}{\omega_{IB}} \underline{u}_{\omega}^{B} + \frac{1}{\omega_{IB}} \delta\underline{\omega}_{IB}^{B} \tag{13.4.1.1-11}$$

Substituting (13.4.1.1-11) into (13.4.1.1-9) then gives:

$$\underline{\beta}^{B} = \frac{1}{\omega_{IB}} \underline{u}_{\omega}^{B} \times \delta\underline{\omega}_{IB}^{B} \tag{13.4.1.1-12}$$

Equations (13.4.1.1-7) and (13.4.1.1-12) define the magnitude error ($\delta\omega_{IB}$) and angular error ($\underline{\beta}^{B}$) in $\tilde{\underline{\omega}}_{IB}^{B}$ as a function of the angular rate sensor error vector $\delta\underline{\omega}_{IB}^{B}$. We now define $\delta\underline{\omega}_{IB}^{B}$ in terms of its contributing elements so that $\delta\omega_{IB}$ and $\underline{\beta}^{B}$ can be expressed in expanded form. For the case being considered of constant sensor errors, $\delta\underline{\omega}_{IB}^{B}$ is from (12.5.6-1) (neglecting $\delta\underline{\omega}_{Rand}$):

$$\delta\underline{\omega}_{IB}^{B} = \delta K_{Scal/Mis} \, \underline{\omega}_{IB}^{B} + \delta\underline{K}_{Bias} \tag{13.4.1.1-13}$$

With (13.4.1.1-13), (13.4.1.1-1) for $\underline{\omega}_{IB}^{B}$, and generalized Equations (3.1.1-12) - (3.1.1-13), Equations (13.4.1.1-7) and (13.4.1.1-12) become:

$$\delta\omega_{IB} = \omega_{IB} \left(\underline{u}_{\omega}^{B}\right)^{T} \delta K_{Scal/Mis} \, \underline{u}_{\omega}^{B} + \left(\underline{u}_{\omega}^{B}\right)^{T} \delta\underline{K}_{Bias} \tag{13.4.1.1-14}$$

$$\underline{\beta}^{B} = \left(\underline{u}_{\omega}^{B} \times\right) \delta K_{Scal/Mis} \, \underline{u}_{\omega}^{B} + \frac{1}{\omega_{IB}} \underline{u}_{\omega}^{B} \times \delta\underline{K}_{Bias} \tag{13.4.1.1-15}$$

Equations (13.4.1.1-14) and (13.4.1.1-15) show that $\tilde{\underline{\omega}}_{IB}^{B}$ is misaligned from $\underline{\omega}_{IB}^{B}$ (by $\underline{\beta}^{B}$) and changed in length from $\underline{\omega}_{IB}^{B}$ (by $\delta\omega_{IB}$) due to angular rate sensor bias, misalignment and scale factor error ($\delta\underline{K}_{Bias}$, $\delta K_{Scal/Mis}$). The $\delta\omega_{IB}$ error is of particular concern because it leads to unbounded angle error buildup around the spin axis when integrated into attitude. It is instructive to expand Equation (13.4.1.1-14) for $\delta\omega_{IB}$ in terms of the components of $\delta\underline{K}_{Bias}$, $\delta K_{Scal/Mis}$ and $\underline{u}_{\omega}^{B}$ which we shall define as:

$$\delta\underline{K}_{Bias} = \begin{bmatrix} \delta K_{BX} \\ \delta K_{BY} \\ \delta K_{BZ} \end{bmatrix} \qquad \delta K_{Scal/Mis} = \begin{bmatrix} \delta K_{XX} & \delta K_{XY} & \delta K_{XZ} \\ \delta K_{YX} & \delta K_{YY} & \delta K_{YZ} \\ \delta K_{ZX} & \delta K_{ZY} & \delta K_{ZZ} \end{bmatrix} \qquad \underline{u}_{\omega}^{B} = \begin{bmatrix} u_{\omega XB} \\ u_{\omega YB} \\ u_{\omega ZB} \end{bmatrix} \tag{13.4.1.1-16}$$

Substituting (13.4.1.1-16) into (13.4.1.1-14) then yields for $\delta\omega_{IB}$:

$$\begin{aligned}\delta\omega_{IB} = \omega_{IB}\Big[& u_{\omega XB}^2\,\delta K_{XX} + u_{\omega YB}^2\,\delta K_{YY} + u_{\omega ZB}^2\,\delta K_{ZZ} \\ & + u_{\omega XB}\,u_{\omega YB}\left(\delta K_{XY} + \delta K_{YX}\right) + u_{\omega YB}\,u_{\omega ZB}\left(\delta K_{YZ} + \delta K_{ZY}\right) \\ & + u_{\omega ZB}\,u_{\omega XB}\left(\delta K_{ZX} + \delta K_{XZ}\right)\Big] + u_{\omega XB}\,\delta K_{BX} + u_{\omega YB}\,\delta K_{BY} + u_{\omega ZB}\,\delta K_{BZ}\end{aligned} \qquad (13.4.1.1\text{-}17)$$

Equation (13.4.1.1-17) shows that angular rate magnitude error is produced by angular rate sensor scale factor and bias error components along the spin axis (as may have been expected), but also from orthogonality error components (i.e., the $\left(\delta K_{ij} + \delta K_{ji}\right)$ terms represent angular rate sensor axis orthogonality error, as discussed in Section 13.2.4 leading to Equations (13.2.4-3)). Note that all orthogonality error contributions become zero if the spin axis lies along any of the B Frame axes (i.e., any of the angular rate sensor axes). Under this condition, only one of the $\underline{u}_{\omega}^{B}$ components in (13.4.1.1-17) is non-zero which sets all $u_{\omega i}\,u_{\omega j}$ to zero. The previous orthogonality error observation is the basis for having the preferred angular rate sensor mounting in a spinning vehicle with the angular rate sensor axes either parallel or perpendicular to the spin axis.

13.4.1.2 ATTITUDE AND ACCELERATION TRANSFORMATION ERROR CHARACTERISTICS

For analytical expediency, we will describe the spinning vehicle attitude and transformation error effects in a non-rotating inertial coordinate frame (i.e., the I Frame). The attitude error we will analyze is $\underline{\psi}^{I}$, the error in the C_B^I matrix defined in Equations (12.3.7.1-2):

$$\left(\underline{\psi}^{I}\times\right) = I - \hat{C}_B^I\left(C_B^I\right)^T \qquad (13.4.1.2\text{-}1)$$

and whose error rate is produced from $\delta\underline{\omega}_{IB}^{B}$ angular rate sensor error as described by Equation (12.3.7.1-4):

$$\dot{\underline{\psi}}^{I} = -\,C_B^I\,\delta\underline{\omega}_{IB}^{B} \qquad (13.4.1.2\text{-}2)$$

The acceleration transformation error we will analyze is the error in $C_B^I\,\underline{a}_{SF}^{B}$ caused by the C_B^I error ($\underline{\psi}^{I}$) and the accelerometer error $\delta\underline{a}_{SF}^{B}$. These effects are represented by the first two terms on the right side of the Equations (12.3.7.1-15) velocity error rate expression, which with (3.1.1-13) is:

$$\delta \dot{\underline{\upsilon}}^{I}_{Snsr} = C^{I}_{B}\, \delta \underline{a}^{B}_{SF} + \left(\underline{a}^{I}_{SF} \times\right) \underline{\psi}^{I} \qquad (13.4.1.2\text{-}3)$$

where

$\delta \dot{\underline{\upsilon}}^{I}_{Snsr}$ = Acceleration transformation error in the I Frame caused by angular rate sensor and accelerometer error.

For analysis purposes we also define:

$$C^{I}_{B} = C^{I}_{B_0}\, C^{B_0}_{B} \qquad (13.4.1.2\text{-}4)$$

where

B_0 = B Frame attitude relative to the I Frame at navigation time $t = 0$. The B_0 Frame is considered fixed in the I Frame, hence, $C^{I}_{B_0}$ is constant.

We consider the B Frame to be rotating at angular rate $\underline{\omega}_{IB}$ which for the problem being analyzed, is considered constant in the B Frame. The B_0 Frame is constant in the I Frame (by definition), hence:

$$\underline{\omega}^{B}_{B_0 B} = \underline{\omega}^{B}_{IB} \qquad (13.4.1.2\text{-}5)$$

where

$\underline{\omega}^{B}_{B_0 B}$ = Angular rate of the B Frame relative to the B_0 Frame in B Frame coordinates (and constant because $\underline{\omega}^{B}_{IB}$ is defined to be constant).

For a constant B Frame angular rate vector (relative to the B_0 Frame), generalized Equation (3.3.5-14) shows that the rotation vector $\underline{\phi}$ associated with the B Frame $C^{B_0}_{B}$ attitude matrix has a rate of change equal to $\underline{\omega}^{B}_{B_0 B}$. This is easily verified by setting $\underline{\phi}$ in (3.3.5-14) to $\underline{\omega}^{B}_{B_0 B}\, t$ (i.e., the integral of $\underline{\omega}^{B}_{B_0 B}$). The cross-product terms in (3.3.5-14) of $\underline{\phi}$ with $\underline{\omega}^{B}_{B_0 B}$ are thereby zero, leaving the $\underline{\phi}$ rate of change equal to $\underline{\omega}^{B}_{B_0 B}$. Thus, with (13.4.1.2-5) we can write:

$$\phi = \omega_{IB}\, t \qquad (13.4.1.2\text{-}6)$$

where

ϕ = Magnitude of the $C^{B_0}_{B}$ attitude rotation vector $\underline{\phi}$.

ω_{IB} = Magnitude of $\underline{\omega}^{B}_{IB}$.

and

$$\underline{\phi} = \phi \, \underline{u}_{\omega}^{B} \tag{13.4.1.2-7}$$

where

$\underline{u}_{\omega}^{B}$ = Unit vector along $\underline{\omega}_{IB}^{B}$ (considered constant).

Generalized Equation (3.2.2.1-4) shows that $C_{B}^{B_0}$ corresponding to Equation (13.4.1.2-7) is given by:

$$C_{B}^{B_0} = I + \sin\phi \left(\underline{u}_{\omega}^{B}\times\right) + (1 - \cos\phi)\left(\underline{u}_{\omega}^{B}\times\right)\left(\underline{u}_{\omega}^{B}\times\right) \tag{13.4.1.2-8}$$

With generalized equation (13.1-11) for $\left(\underline{u}_{\omega}^{B}\times\right)\left(\underline{u}_{\omega}^{B}\times\right)$, (13.4.1.2-8) has the alternate form:

$$C_{B}^{B_0} = \underline{u}_{\omega}^{B}\left(\underline{u}_{\omega}^{B}\right)^{T} + \sin\phi \left(\underline{u}_{\omega}^{B}\times\right) - \cos\phi \left(\underline{u}_{\omega}^{B}\times\right)\left(\underline{u}_{\omega}^{B}\times\right) \tag{13.4.1.2-9}$$

We then substitute (13.4.1.2-9) and (13.4.1.2-4) into (13.4.1.2-2) and (13.4.1.2-3) to obtain:

$$\dot{\underline{\psi}}^{I} = - C_{B_0}^{I}\left[\underline{u}_{\omega}^{B}\left(\underline{u}_{\omega}^{B}\right)^{T} + \sin\phi \left(\underline{u}_{\omega}^{B}\times\right) - \cos\phi \left(\underline{u}_{\omega}^{B}\times\right)\left(\underline{u}_{\omega}^{B}\times\right)\right] \delta\underline{\omega}_{IB}^{B} \tag{13.4.1.2-10}$$

$$\delta\dot{\underline{\upsilon}}_{Snsr}^{I} = \left(\underline{a}_{SF}^{I}\times\right)\underline{\psi}^{I} + C_{B_0}^{I}\left[\underline{u}_{\omega}^{B}\left(\underline{u}_{\omega}^{B}\right)^{T} + \sin\phi \left(\underline{u}_{\omega}^{B}\times\right) - \cos\phi \left(\underline{u}_{\omega}^{B}\times\right)\left(\underline{u}_{\omega}^{B}\times\right)\right] \delta\underline{a}_{SF}^{B} \tag{13.4.1.2-11}$$

Equations (13.4.1.2-10) and (13.4.1.2-11) with (13.4.1.2-6) are now in a form from which we can draw conclusions regarding the impact of constant sensor errors in rotating environments. First we should understand the analytical operations performed by the $\underline{u}_{\omega}^{B}\left(\underline{u}_{\omega}^{B}\right)^{T}$, $\left(\underline{u}_{\omega}^{B}\times\right)$ and $\left(\underline{u}_{\omega}^{B}\times\right)\left(\underline{u}_{\omega}^{B}\times\right)$ operators in these equations. The $\underline{u}_{\omega}^{B}\left(\underline{u}_{\omega}^{B}\right)^{T}$ operator produces an output vector along $\underline{u}_{\omega}^{B}$ with magnitude equal to the component of its input vector (i.e., $\delta\underline{\omega}_{IB}^{B}$ or $\delta\underline{a}_{SF}^{B}$) along $\underline{u}_{\omega}^{B}$. The $\left(\underline{u}_{\omega}^{B}\times\right)$ operator produces an output vector that is perpendicular to $\underline{u}_{\omega}^{B}$, with magnitude equal to the component of the input vector perpendicular to $\underline{u}_{\omega}^{B}$. The $\left(\underline{u}_{\omega}^{B}\times\right)\left(\underline{u}_{\omega}^{B}\times\right)$ operator produces a vector equal in magnitude to the $\left(\underline{u}_{\omega}^{B}\times\right)$ operator output vector, but which is also perpendicular to the $\left(\underline{u}_{\omega}^{B}\times\right)$ operator output vector (and perpendicular to $\underline{u}_{\omega}^{B}$). Hence, the outputs from the three operators are mutually perpendicular. Because the $\left(\underline{u}_{\omega}^{B}\times\right)$ and $\left(\underline{u}_{\omega}^{B}\times\right)\left(\underline{u}_{\omega}^{B}\times\right)$ operators in (13.4.1.2-10) and (13.4.1.2-11) multiply $\sin\phi$ and $\cos\phi$, and because ϕ (from Equation (13.4.1.2-6)) grows linearly with time t at spin rate ω_{IB}, we can conclude that the combined effect of the $\left(\underline{u}_{\omega}^{B}\times\right)$ and $\left(\underline{u}_{\omega}^{B}\times\right)\left(\underline{u}_{\omega}^{B}\times\right)$ operators is to produce an output vector with magnitude equal to the input vector component perpendicular to $\underline{u}_{\omega}^{B}$, but which rotates in the I Frame at ω_{IB} around $\underline{u}_{\omega}^{B}$.

Let us now address the analytical solution to (13.4.1.2-10) and (13.4.1.2-11) under constant inertial sensor error conditions. Ignoring sensor noise effects for the problem at hand, the $\delta\underline{\omega}_{IB}^{B}$ angular rate sensor error vector in (13.4.1.2-10) is given by Equation (13.4.1.1-13). From (13.4.1.1-13) we see that for constant $\underline{\omega}_{IB}^{B}$ (the case being analyzed), $\delta\underline{\omega}_{IB}^{B}$ is constant. Under these conditions we can analytically integrate Equation (13.4.1.2-10) from time = 0 to t with (13.4.1.2-6) for ϕ, to obtain for $\underline{\psi}^{I}$:

$$\begin{aligned}\underline{\psi}^{I} &= \underline{\psi}_{0}^{I} - C_{B_0}^{I}\left[t\,\underline{u}_{\omega}^{B}\left(\underline{u}_{\omega}^{B}\right)^{T} + \frac{(1-\cos\omega_{IB}\,t)}{\omega_{IB}}\left(\underline{u}_{\omega}^{B}\times\right) - \frac{\sin\omega_{IB}\,t}{\omega_{IB}}\left(\underline{u}_{\omega}^{B}\times\right)\left(\underline{u}_{\omega}^{B}\times\right)\right]\delta\underline{\omega}_{IB}^{B} \\ &= \underline{\psi}_{0}^{I} - C_{B_0}^{I}\left\{t\,\underline{u}_{\omega}^{B}\left(\underline{u}_{\omega}^{B}\right)^{T}\delta\underline{\omega}_{IB}^{B} + \left[(1-\cos\omega_{IB}\,t)\,I - \sin\omega_{IB}\,t\left(\underline{u}_{\omega}^{B}\times\right)\right]\frac{1}{\omega_{IB}}\left(\underline{u}_{\omega}^{B}\times\right)\delta\underline{\omega}_{IB}^{B}\right\}\end{aligned} \qquad (13.4.1.2\text{-}12)$$

where

$\underline{\psi}_{0}^{I}$ = Initial value for $\underline{\psi}^{I}$ at time t = 0.

From (13.4.1.2-12), we see that the $\underline{\psi}^{I}$ component along $\underline{u}_{\omega}^{B}$ (from the $t\,\underline{u}_{\omega}^{B}\left(\underline{u}_{\omega}^{B}\right)^{T}$ term) is a ramp in time t proportional to the component of $\delta\underline{\omega}_{IB}^{B}$ along $\underline{u}_{\omega}^{B}$, and the component of $\delta\underline{\omega}_{IB}^{B}$ perpendicular to $\underline{u}_{\omega}^{B}$ (the $\left(\underline{u}_{\omega}^{B}\times\right)\delta\underline{\omega}_{IB}^{B}$ term) has no systematic build-up effect on $\underline{\psi}^{I}$. Thus, the spinning motion effectively compensates for constant angular rate sensor error perpendicular to $\underline{u}_{\omega}^{B}$. This effect forms the basis for the construction of some strapdown inertial navigation systems in which the inertial sensor assembly is mounted on a turn-table that provides forced rotation to the angular rate sensors. Attitude error build-up perpendicular to the turn-table rotation axis is thereby reduced. We also note from Equations (13.4.1.1-7) and (13.4.1.1-12) that the $\left(\underline{u}_{\omega}^{B}\right)^{T}\delta\underline{\omega}_{IB}^{B}$ and $\frac{1}{\omega_{IB}}\,\underline{u}_{\omega}^{B}\times\delta\underline{\omega}_{IB}^{B}$ terms in (13.4.1.2-12) represent the $\underline{\tilde{\omega}}_{IB}^{B}$ magnitude and tilt errors, $\delta\omega_{IB}$ and $\underline{\beta}^{B}$. Thus, the systematic build-up effect on $\underline{\psi}^{I}$ is produced by the $\underline{\tilde{\omega}}_{IB}^{B}$ magnitude error $\delta\omega_{IB}$, while the $\underline{\tilde{\omega}}_{IB}^{B}$ tilt error $\underline{\beta}^{B}$ produces a component in $\underline{\psi}^{I}$ (the $\hat{C}_{B}^{I}$ angular error vector) that rotates around the spin axis.

For constant inertial sensor errors, $\delta \underline{a}_{SF}^{B}$ in (13.4.1.2-11) is from (12.5.6-1) (neglecting $\delta \underline{a}_{Rand}$):

$$\delta \underline{a}_{SF}^{B} = \delta L_{Scal/Mis}\, \underline{a}_{SF}^{B} + \delta \underline{L}_{Bias} \qquad (13.4.1.2\text{-}13)$$

The $\underline{a}_{SF}^{B}$ term in (13.4.1.2-13) can be expressed as the sum of a portion to balance gravity plus a maneuver acceleration, as in the B Frame version of (13.2.2-18):

$$\underline{a}_{SF}^{B} = g\, \underline{u}_{ZN}^{B} + \Delta \underline{a}_{SF}^{B} \qquad (13.4.1.2\text{-}14)$$

where

$\underline{u}_{ZN}^{B}$ = Unit vector along the N Frame Z axis projected on the B Frame, which from the N Frame definition in Section 2.2, is along the upward geodetic vertical.

Similarly, the $\underline{a}_{SF}^{I}$ term in (13.4.1.2-11) can be expressed as:

$$\underline{a}_{SF}^{I} = C_{B_0}^{I}\, C_{B}^{B_0} \left(g\, \underline{u}_{ZN}^{B} + \Delta \underline{a}_{SF}^{B} \right) \qquad (13.4.1.2\text{-}15)$$

With (13.4.1.2-9) for $C_{B}^{B_0}$ in (13.4.1.2-15), $\underline{a}_{SF}^{B}$ from (13.4.1.2-14) in (13.4.1.2-13), and $\underline{a}_{SF}^{I}$ from (13.4.1.2-15) in (13.4.1.2-11), it is difficult to define general conditions that make the components of $\delta \dot{\underline{\upsilon}}_{Snsr}^{I}$ Equation (13.4.1.2-11) constant to enable simple analytical integration. One particular condition that permits simplification is the case when $\Delta \underline{a}_{SF}^{B}$ is zero and the spin axis is vertical so that $\underline{u}_{\omega}^{B} = \underline{u}_{ZN}^{B}$ for which (13.4.1.2-14) - (13.4.1.2-15) reduce to:

$$\underline{a}_{SF}^{B} = g\, \underline{u}_{\omega}^{B} \qquad (13.4.1.2\text{-}16)$$

$$\underline{a}_{SF}^{I} = g\, C_{B_0}^{I}\, \underline{u}_{\omega}^{B_0} \qquad (13.4.1.2\text{-}17)$$

where

$\underline{u}_{\omega}^{B_0}$ = The $\underline{u}_{\omega}$ rotation axis unit vector projected on B_0 Frame axes.

With (13.4.1.2-16) for $\underline{a}_{SF}^{B}$ in (13.4.1.2-13) (and recognizing $\underline{u}_{\omega}^{B}$ as constant from its definition), the $\delta \underline{a}_{SF}^{B}$ accelerometer error term in $\delta \dot{\underline{\upsilon}}_{Snsr}^{I}$ Equation (13.4.1.2-11) is constant. Applying generalized Equation (3.1.1-38) to (13.4.1.2-17) and, since $\underline{u}_{\omega}^{B}$ is constant, using (3.2.2.1-6) (which sets $\underline{u}_{\omega}^{B_0} = \underline{u}_{\omega}^{B}$), the $\left(\underline{a}_{SF}^{I} \times \right)$ term in (13.4.1.2-11) becomes:

$$\left(\underline{a}_{SF}^{I}\times\right) = g\left[\left(C_{B_0}^{I}\,\underline{u}_{\omega}^{B_0}\right)\times\right] = g\left[\left(C_{B_0}^{I}\,\underline{u}_{\omega}^{B}\right)\times\right] = g\,C_{B_0}^{I}\left(\underline{u}_{\omega}^{B}\times\right)\left(C_{B_0}^{I}\right)^{T} \qquad (13.4.1.2\text{-}18)$$

Substituting (13.4.1.2-18) into $\delta\dot{\underline{\upsilon}}_{Snsr}^{I}$ Equation (13.4.1.2-11) using (13.4.1.2-12) for $\underline{\psi}^{I}$, substituting (13.4.1.2-6) for ϕ, applying generalized Equation (13.1-13), and treating $\delta\underline{a}_{SF}^{B}$ as constant then provides an expression that can be readily integrated:

$$\begin{aligned}
\delta\dot{\underline{\upsilon}}_{Snsr}^{I} &= g\left(C_{B_0}^{I}\,\underline{u}_{\omega}^{B}\right)\times\underline{\psi}_0^{I} \\
&\quad - g\,C_{B_0}^{I}\left(\underline{u}_{\omega}^{B}\times\right)\left[t\,\underline{u}_{\omega}^{B}\left(\underline{u}_{\omega}^{B}\right)^{T} + \frac{(1-\cos\omega_{IB}t)}{\omega_{IB}}\left(\underline{u}_{\omega}^{B}\times\right) - \frac{\sin\omega_{IB}t}{\omega_{IB}}\left(\underline{u}_{\omega}^{B}\times\right)\left(\underline{u}_{\omega}^{B}\times\right)\right]\delta\underline{\omega}_{IB}^{B} \\
&\quad + C_{B_0}^{I}\left[\underline{u}_{\omega}^{B}\left(\underline{u}_{\omega}^{B}\right)^{T} + \sin\omega_{IB}t\left(\underline{u}_{\omega}^{B}\times\right) - \cos\omega_{IB}t\left(\underline{u}_{\omega}^{B}\times\right)\left(\underline{u}_{\omega}^{B}\times\right)\right]\delta\underline{a}_{SF}^{B} \qquad (13.4.1.2\text{-}19)\\
&= g\left(C_{B_0}^{I}\,\underline{u}_{\omega}^{B}\right)\times\underline{\psi}_0^{I} - g\,C_{B_0}^{I}\left[\frac{(1-\cos\omega_{IB}t)}{\omega_{IB}}\left(\underline{u}_{\omega}^{B}\times\right)\left(\underline{u}_{\omega}^{B}\times\right) + \frac{\sin\omega_{IB}t}{\omega_{IB}}\left(\underline{u}_{\omega}^{B}\times\right)\right]\delta\underline{\omega}_{IB}^{B} \\
&\quad + C_{B_0}^{I}\left[\underline{u}_{\omega}^{B}\left(\underline{u}_{\omega}^{B}\right)^{T} + \sin\omega_{IB}t\left(\underline{u}_{\omega}^{B}\times\right) - \cos\omega_{IB}t\left(\underline{u}_{\omega}^{B}\times\right)\left(\underline{u}_{\omega}^{B}\times\right)\right]\delta\underline{a}_{SF}^{B}
\end{aligned}$$

The integral of (13.4.1.2-19) from time = 0 to t (for $\delta\underline{a}_{SF}^{B}$ constant) defines the velocity error produced from constant inertial sensor errors in the vertical rotation environment:

$$\begin{aligned}
\delta\underline{\upsilon}_{Snsr}^{I} &= \delta\underline{\upsilon}_{Snsr_0}^{I} + g\,t\left(C_{B_0}^{I}\,\underline{u}_{\omega}^{B}\right)\times\underline{\psi}_0^{I} \\
&\quad - g\,C_{B_0}^{I}\left[\left(t - \frac{\sin\omega_{IB}t}{\omega_{IB}}\right)\left(\underline{u}_{\omega}^{B}\times\right) + \frac{(1-\cos\omega_{IB}t)}{\omega_{IB}}\,I\right]\frac{1}{\omega_{IB}}\left(\underline{u}_{\omega}^{B}\times\right)\delta\underline{\omega}_{IB}^{B} \qquad (13.4.1.2\text{-}20)\\
&\quad + C_{B_0}^{I}\left[t\,\underline{u}_{\omega}^{B}\left(\underline{u}_{\omega}^{B}\right)^{T} + \frac{(1-\cos\omega_{IB}t)}{\omega_{IB}}\left(\underline{u}_{\omega}^{B}\times\right) - \frac{\sin\omega_{IB}t}{\omega_{IB}}\left(\underline{u}_{\omega}^{B}\times\right)\left(\underline{u}_{\omega}^{B}\times\right)\right]\delta\underline{a}_{SF}^{B}
\end{aligned}$$

where

$\delta\underline{\upsilon}_{Snsr_0}^{I}$ = Initial value for $\delta\underline{\upsilon}_{Snsr}^{I}$ at time t = 0.

Similar to the attitude error build-up, Equation (13.4.1.2-20) shows that the angular rotation for the case being investigated converts the constant accelerometer error perpendicular to the rotation axis into an oscillatory (rotating) velocity error at the spin frequency. Thus, strapdown inertial navigation systems using rotation turn-tables for angular rate sensor accuracy enhancement, also reduce velocity error build-up perpendicular to the turn-table rotation axis created by accelerometer error. The component of accelerometer error along the spin axis integrates into a ramping velocity error along the spin axis. Equation (13.4.1.2-20) also shows that the angular rate sensor error produces a ramping velocity error build-up across the spin axis (in addition to oscillating errors at the spin frequency). From the discussion following $\underline{\psi}^{I}$ Equation (13.4.1.2-12), the velocity ramp produced from angular rate sensor error is directly attributed to the $\underline{\beta}^{B}$ tilt in $\tilde{\underline{\omega}}_{IB}^{B}$ (i.e., $\underline{\beta}^{B} = \frac{1}{\omega_{IB}} \left(\underline{u}_{\omega}^{B} \times\right) \delta\underline{\omega}_{IB}^{B}$ in (13.4.1.2-20) multiplying $t \left(\underline{u}_{\omega}^{B} \times\right)$).

The effect of $\underline{\beta}^{B}$ is to produce an average $\underline{\psi}^{I}$ attitude error (see Equation (13.4.1.2-12)) that cross-couples $\underline{a}_{SF}^{I}$ in Equation (13.4.1.2-11) from the vertical in this case, into the horizontal plane and perpendicular to the $\tilde{\underline{\omega}}_{IB}^{B}$ tilt error angle vector $\underline{\beta}^{B}$.

13.4.2 HIGH RATE SPINNING ABOUT ROTATING AXIS

The previous section analyzed inertial sensor generated angular rate, attitude and acceleration transformation error in a rotating environment in which the angular rate vector was constant in magnitude and direction. In this section we analyze the attitude error rate in a rotating environment in which the angular rate vector is also rotating. Such an angular rate environment is produced by rotating rigid body motion in the absence of applied torques which, through the rotational equivalent of Newton's laws of motion, is governed by the body angular momentum in inertial space being constant (Reference 8 - Sections 5-3 and 5-4). The angular momentum of a rigid body is defined as:

$$\underline{H}^{P} = J^{P} \underline{\omega}_{IB}^{P} \tag{13.4.2-1}$$

where

J^{P} = Moment of inertia tensor of the rigid body in body fixed coordinate frame P. The components of J^{P} are a function of the rigid body mass distribution and the orientation of the P frame.

P = Coordinate frame fixed in the rigid body along the "principal" moment of inertia axes. The principal moment of inertia axes are defined to be the axes in which J^{P} is a diagonal matrix. Note that the B Frame which is also fixed in the rigid body, is defined to be nominally aligned to the strapdown inertial sensor axis frame, hence, is not necessarily aligned with the P Frame (even nominally).

$\underline{\omega}_{IB}^{P}$ = Angular rate of the rigid body B Frame relative to inertial space projected on coordinate Frame P axes.

$\underline{H}^{P}$ = Angular momentum of the rigid body in P Frame coordinates.

Using generalized Equation (3.4-6), the time rate of change of $\underline{H}^{P}$ is related to the time rate of change in our non-rotating inertial Frame I through:

$$\underline{\dot{H}}^P = C_I^P \, \underline{\dot{H}}^I + \underline{\omega}_{PI}^P \times \underline{H}^P = C_I^P \, \underline{\dot{H}}^I + \underline{\omega}_{BI}^P \times \underline{H}^P \qquad (13.4.2\text{-}2)$$

where

C_I^P = Direction cosine matrix that transforms vectors from the I Frame to the P Frame.

$\underline{\omega}_{PI}^P$ = Angular rate of the I Frame relative to the P Frame projected on coordinate Frame P axes.

$\underline{\omega}_{BI}^P$ = Angular rate of the I Frame relative to the B Frame projected on coordinate Frame P axes, and equal to $\underline{\omega}_{PI}^P$ because the P and B Frames are defined to be fixed in the rigid body, hence, fixed relative to one another.

In the absence of applied torques (i.e., free spinning body motion), Newton's laws applied to rotating mass state that $\underline{\dot{H}}^I = 0$. Recognizing that $\underline{\omega}_{BI}^P$ is the negative of $\underline{\omega}_{IB}^P$ then allows us to write for (13.4.2-2):

$$\underline{\dot{H}}^P = - \underline{\omega}_{IB}^P \times \underline{H}^P \qquad (13.4.2\text{-}3)$$

Equation (13.4.2-3) with (13.4.2-1) for $\underline{H}^P$ are now expanded in component form. We first define the components of $\underline{\omega}_{IB}^P$ and J^P as:

$$\underline{\omega}_{IB}^P = \begin{bmatrix} \omega_{IB_{XP}} \\ \omega_{IB_{YP}} \\ \omega_{IB_{ZP}} \end{bmatrix} \qquad J^P = \begin{bmatrix} J_{SA} & 0 & 0 \\ 0 & J_{CA} & 0 \\ 0 & 0 & J_{CA} \end{bmatrix} \qquad (13.4.2\text{-}4)$$

where

$\omega_{IB_{XP}}, \omega_{IB_{YP}}, \omega_{IB_{ZP}}$ = P Frame components of $\underline{\omega}_{IB}^P$.

J_{SA} = Moment of inertia of the body around its spin axis. For this analysis, the body is assumed to be nominally spinning about the P Frame X axis.

J_{CA} = Moment of inertia of the body across the spin axis. For this analysis we will assume that the body has mass symmetry around the spin axis so that the moment of inertia around any axis perpendicular to the spin axis is J_{CA}. Hence, the Y and Z components of J^P are equal to J_{CA}.

Substituting (13.4.2-1) into (13.4.2-3) with (13.4.2-4) then yields for the X component:

$$\dot{\omega}_{IB_{XP}} = 0 \tag{13.4.2-5}$$

Thus:

$$\omega_{IB_{XP}} \equiv \omega_{Spin} = \text{Constant} \tag{13.4.2-6}$$

where

ω_{Spin} = Spin rate of the body about the P Frame X axis.

From Equation (13.4.2-1) with (13.4.2-4) and (13.4.2-6) we can also write:

$$H_{XP} = J_{SA}\, \omega_{Spin} = \text{Constant} \tag{13.4.2-7}$$

where

H_{XP} = P Frame X axis component of $\underline{H}^P$.

Now, let's decompose $\underline{H}$ into two parts, a component along the P Frame X axis (i.e., the spin axis) and a component perpendicular to the spin axis. Without specific coordinate frame designation we write:

$$\underline{H} = H_{XP}\, \underline{u}_{Spin} + H_{\perp}\, \underline{u}_{\perp} \tag{13.4.2-8}$$

where

$\underline{H}$ = Angular momentum vector in general without specific coordinate frame designation.

$\underline{u}_{Spin}$ = Unit vector along the P Frame X axis (the body spin axis).

$\underline{u}_{\perp}$ = Unit vector perpendicular to $\underline{u}_{Spin}$ along the component of $\underline{H}$ perpendicular to $\underline{u}_{Spin}$.

$H_{\perp}$ = Magnitude of the $\underline{H}$ component perpendicular to $\underline{u}_{Spin}$.

Figure 13.4.2-1 illustrates Equation (13.4.2-8).

Generalized Equations (3.1.1-2) and (3.1.1-29) (with $\underline{V} = \underline{W}$) show that the magnitude of any vector is the same in any coordinate frame. Thus for the problem at hand, because $\underline{H}$ is constant in inertial space, its magnitude is constant in inertial space and in all coordinate frames. Therefore, the length of $\underline{H}$ in Figure 13.4.2-1 is constant. We also know from Equation (13.4.2-7) that H_{XP} in Figure 13.4.2-1 is constant. Since H_{XP} is the projection of $\underline{H}$ on $\underline{u}_{Spin}$, we conclude that β must also be constant, thus $\underline{u}_{Spin}$ and $\underline{u}_{\perp}$ will maintain the fixed geometry relative to $\underline{H}$ shown in the figure. Based on this observation we define a new coordinate Frame $\perp$ (X, Z axes shown in Figure 13.4.2-1) where:

$\perp$ = Right handed orthogonal coordinate frame with X along $\underline{u}_{Spin}$, Y axis along $\underline{H} \times \underline{u}_{Spin}$ with the Z axis in the $\underline{H}$ - $\underline{u}_{Spin}$ plane and perpendicular to X, Y in the right hand sense (i.e., along $\underline{u}_{\perp}$).

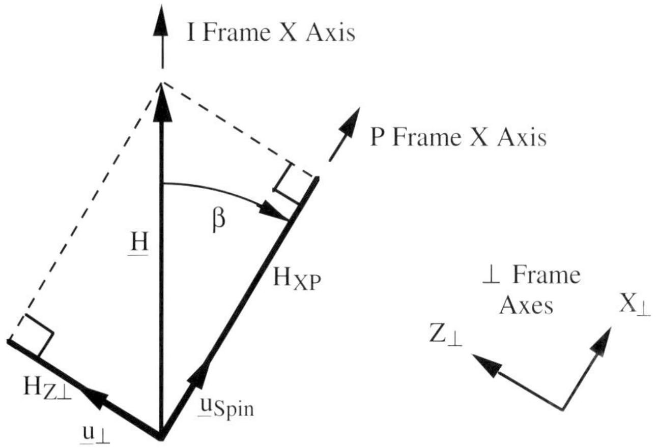

Figure 13.4.2-1 Angular Momentum Vector Components

We also define the I Frame for this problem with the X axis along $\underline{H}$ (See Figure 13.4.2-1), the Y, Z axes perpendicular to $\underline{H}$ such that the Figure 13.4.2-1 $\underline{u}_{Spin}$ - $\underline{u}_{\perp}$ plane is rotated from the I Frame Z - X plane by an angle α measured positive about the I Frame X axis. When $\alpha = 0$, the Y_I and $Y_{\perp}$ axes are the same. From this definition, β (in Figure 13.4.2-1) and α form a two Euler angle set relating the I and $\perp$ Frames. The Method of Least Work diagram in Figure 13.4.2-2 illustrates this geometry as well as the P Frame orientation at Euler angle ϕ around $\underline{u}_{Spin}$ from the $\underline{u}_{Spin}$ - $\underline{u}_{\perp}$ plane and where:

α, β, ϕ = Euler angles relating the P and I Frames. The α, β Euler angles relate the $\perp$ and I Frames.

Also shown in Figure 13.4.2-2 are the α, β, ϕ rates of change and the components of $\underline{\omega}_{IB}$ in Frame $\perp$ and Frame P (as explained in general in Section 3.3.3.3).

From the previous discussion we know that:

$$\beta = \text{Constant} \tag{13.4.2-9}$$

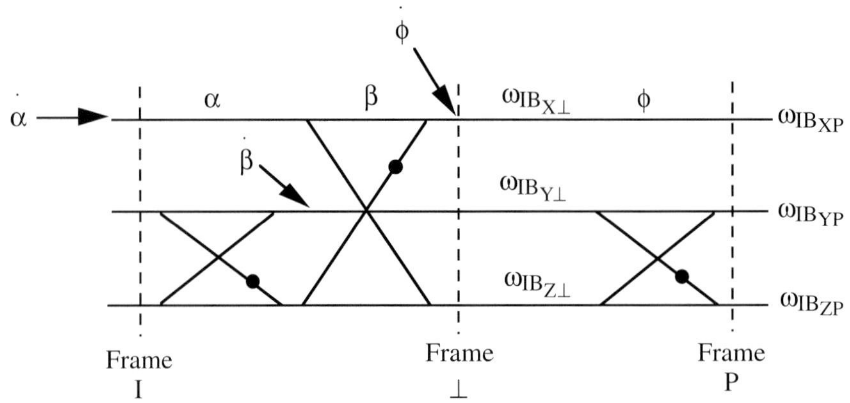

Figure 13.4.2-2 Coordinate Frame Geometry and Angular Rates

Since β is constant, $\dot{\beta} = 0$, and the $\underline{\omega}_{IB}$ angular rate components in Frame $\perp$ are from Figure 13.4.2-2 (with (13.4.2-6):

$$\underline{\omega}_{IB}^{\perp} = \begin{bmatrix} \dot{\alpha}\cos\beta + \dot{\phi} \\ 0 \\ \dot{\alpha}\sin\beta \end{bmatrix} = \begin{bmatrix} \omega_{IB_{XP}} \\ 0 \\ \dot{\alpha}\sin\beta \end{bmatrix} = \begin{bmatrix} \omega_{Spin} \\ 0 \\ \dot{\alpha}\sin\beta \end{bmatrix} \tag{13.4.2-10}$$

The β angle is determined with Figure 13.4.2-1 from the components of $\underline{H}$ in the $\perp$ Frame:

$$\tan\beta = H_{Z\perp} / H_{XP} \tag{13.4.2-11}$$

The H_{XP} component in (13.4.2-11) is given by Equation (13.4.2-7). The $H_{Z\perp}$ component in (13.4.2-11) is found from $\underline{H}$ in the $\perp$ Frame using (13.4.2-1):

$$\underline{H}^{\perp} = C_P^{\perp}\,\underline{H}^P = C_P^{\perp}\,J^P\,\underline{\omega}_{IB}^P = C_P^{\perp}\,J^P\left(C_P^{\perp}\right)^T C_P^{\perp}\,\underline{\omega}_{IB}^P = J^{\perp}\,\underline{\omega}_{IB}^{\perp} \tag{13.4.2-12}$$

with:

$$J^{\perp} = C_P^{\perp}\,J^P\left(C_P^{\perp}\right)^T \tag{13.4.2-13}$$

where

$C_P^{\perp}$ = Direction cosine matrix that transforms vectors from the P Frame to the $\perp$ Frame.

From Figure 13.4.2-2 we see that:

$$C_P^{\perp} = \begin{bmatrix} 1 & 0 & 0 \\ 0 & \cos\phi & -\sin\phi \\ 0 & \sin\phi & \cos\phi \end{bmatrix} \tag{13.4.2-14}$$

Substituting (13.4.2-14) with J^P from (13.4.2-4) into (13.4.2-13), we find that because of the symmetry in J^P around the spin axis:

$$J^{\perp} = J^P \tag{13.4.2-15}$$

Thus, (13.4.2-12) is simply:

$$\underline{H}^{\perp} = J^P \underline{\omega}_{IB}^{\perp} \tag{13.4.2-16}$$

The $\underline{H}^{\perp}$ components can now be obtained from (13.4.2-16) with (13.4.2-10) for $\underline{\omega}_{IB}^{\perp}$ and (13.4.2-4) for J^P:

$$\underline{H}^{\perp} \equiv \begin{bmatrix} H_{X\perp} \\ H_{Y\perp} \\ H_{Z\perp} \end{bmatrix} = \begin{bmatrix} J_{SA}\,\omega_{Spin} \\ 0 \\ J_{CA}\,\dot{\alpha}\sin\beta \end{bmatrix} \tag{13.4.2-17}$$

With (13.4.2-17) for $H_{Z\perp}$ and (13.4.2-7) for H_{XP}, Equation (13.4.2-11) for tan β becomes:

$$\tan\beta = \frac{J_{CA}}{J_{SA}}\,\frac{\dot{\alpha}\sin\beta}{\omega_{Spin}} \tag{13.4.2-18}$$

or solving for $\dot{\alpha}$ and remembering from (13.4.2-9) that β is constant:

$$\dot{\alpha} = \frac{J_{SA}}{J_{CA}}\,\omega_{Spin}\sec\beta = \text{Constant} \tag{13.4.2-19}$$

The X component of Equation (13.4.2-10) is:

$$\dot{\alpha}\cos\beta + \dot{\phi} = \omega_{Spin} \tag{13.4.2-20}$$

or, solving for $\dot{\phi}$, using $\dot{\alpha}$ from (13.4.2-19):

$$\dot{\phi} = \left(1 - \frac{J_{SA}}{J_{CA}}\right) \omega_{Spin} = \text{Constant} \tag{13.4.2-21}$$

Thus, we see that free rotating body motion corresponds to a constant spin rate about the spin axis, in which the spin axis rotates at a constant rate $\dot{\alpha}$ about an inertial axis (the inertial axis being along $\underline{H}$ in Figure 13.4.2-2) and in which the spin axis is inclined by a constant angle β from the inertial axis.

The α, β, ϕ Euler angles can also be expressed from (13.4.2-9), (13.4.2-19) and (13.4.2-21) as:

$$\alpha = \alpha_0 + \dot{\alpha}\, t \qquad \beta = \text{Constant} \qquad \phi = \phi_0 + \dot{\phi}\, t \tag{13.4.2-22}$$

where

α_0, ϕ_0 = Initial values for α, ϕ at time t = 0.

To determine the attitude error rate produced from strapdown angular rate sensors in the hypothesized angular motion environment, we will also need the components of $\underline{\omega}_{IB}^{P}$ which can be obtained from:

$$\underline{\omega}_{IB}^{P} = \left(C_{P}^{\perp}\right)^{T} \underline{\omega}_{IB}^{\perp} \tag{13.4.2-23}$$

Using (13.4.2-14) for $C_{P}^{\perp}$ and (13.4.2-10) for $\underline{\omega}_{IB}^{\perp}$, Equation (13.4.2-23) becomes:

$$\underline{\omega}_{IB}^{P} = \begin{bmatrix} \omega_{IB_{XP}} \\ \omega_{IB_{YP}} \\ \omega_{IB_{ZP}} \end{bmatrix} = \begin{bmatrix} \omega_{Spin} \\ \dot{\alpha} \sin\beta \sin\phi \\ \dot{\alpha} \sin\beta \cos\phi \end{bmatrix} \tag{13.4.2-24}$$

Equations (13.4.2-22) - (13.4.2-24) describe the rotary motion of a free body with moment of inertia symmetry about the spin axis. These relations will now be used to define the attitude error rate generated by strapdown angular rate sensors in such an environment. We begin with the Equation (12.3.7.1-4) I Frame version of the attitude error rate and the Equation (13.4.1.1-13) angular rate sensor error expression for constant sensor errors. Using the (3.2.1-5) chain rule, these equations are:

$$\dot{\underline{\psi}}^{I} = - C_{P}^{I}\, C_{B}^{P}\, \delta\underline{\omega}_{IB}^{B} \tag{13.4.2-25}$$

$$\delta\underline{\omega}_{IB}^{B} = \delta K_{Scal/Mis}^{B}\, \underline{\omega}_{IB}^{B} + \delta\underline{K}_{Bias}^{B} \tag{13.4.2-26}$$

Note that for clarity, the angular rate sensor error coefficients $\delta K^{B}_{Scal/Mis}$ and $\delta \underline{K}^{B}_{Bias}$ have been given a superscript B to identify them as being applied in the B Frame as indicated in Equation (13.4.2-26). We can also write for $\underline{\omega}^{B}_{IB}$:

$$\underline{\omega}^{B}_{IB} = \left(C^{P}_{B}\right)^{T} \underline{\omega}^{P}_{IB} \qquad (13.4.2\text{-}27)$$

Substituting (13.4.2-27) into (13.4.2-26), and the result into (13.4.2-25) finds:

$$\dot{\underline{\psi}}^{I} = - C^{I}_{P}\left[C^{P}_{B}\, \delta K^{B}_{Scal/Mis} \left(C^{P}_{B}\right)^{T} \underline{\omega}^{P}_{IB} + C^{P}_{B}\, \delta \underline{K}^{B}_{Bias}\right] \qquad (13.4.2\text{-}28)$$

or

$$\dot{\underline{\psi}}^{I} = - C^{I}_{P}\left(\delta K^{P}_{Scal/Mis}\, \underline{\omega}^{P}_{IB} + \delta \underline{K}^{P}_{Bias}\right) \qquad (13.4.2\text{-}29)$$

with the P Frame angular rate sensor error coefficients defined by:

$$\delta K^{P}_{Scal/Mis} = C^{P}_{B}\, \delta K^{B}_{Scal/Mis} \left(C^{P}_{B}\right)^{T} \qquad (13.4.2\text{-}30)$$

$$\delta \underline{K}^{P}_{Bias} = C^{P}_{B}\, \delta \underline{K}^{B}_{Bias} \qquad (13.4.2\text{-}31)$$

We identify the components of $\delta K^{P}_{Scal/Mis}$ and $\delta \underline{K}^{P}_{Bias}$ as:

$$\delta K^{P}_{Scal/Mis} = \begin{bmatrix} \delta K_{S/M_{XXP}} & \delta K_{S/M_{XYP}} & \delta K_{S/M_{XZP}} \\ \delta K_{S/M_{YXP}} & \delta K_{S/M_{YYP}} & \delta K_{S/M_{YZP}} \\ \delta K_{S/M_{ZXP}} & \delta K_{S/M_{ZYP}} & \delta K_{S/M_{ZZP}} \end{bmatrix} \qquad \delta \underline{K}^{P}_{Bias} = \begin{bmatrix} \delta K_{Bias_{XP}} \\ \delta K_{Bias_{YP}} \\ \delta K_{Bias_{ZP}} \end{bmatrix} \qquad (13.4.2\text{-}32)$$

Using the (13.4.2-32) and (13.4.2-24) component definitions, Equation (13.4.2-29) can be expressed in component form as:

$$\dot{\psi}_{S/M_{l}} = - \sum_{m,n} C_{PI}(l,m)\, \delta K_{S/M_{mnP}}\, \omega_{IB_{nP}} = - \sum_{m,n} \delta K_{S/M_{mnP}}\, C_{PI}(l,m)\, \omega_{IB_{nP}} \qquad (13.4.2\text{-}33)$$

$$\dot{\psi}_{Bias_{l}} = - \sum_{m} C_{PI}(l,m)\, \delta K_{Bias_{mP}} \qquad (13.4.2\text{-}34)$$

where

$\dot{\psi}_{S/M_{l}}$ = Portion of $\dot{\underline{\psi}}^{I}$ component l generated from angular rate sensor scale-factor/misalignment error (i.e., from $\delta K^{P}_{Scal/Mis}$).

$\dot{\psi}_{Bias_l}$ = Portion of $\underline{\dot{\psi}}^I$ component l generated from angular rate sensor bias error (i.e., from $\delta \underline{K}^P_{Bias}$).

$C_{PI}(l,m)$ = Element in row l, column m of C^I_P.

m, n, l = Indices equal to 1, 2 or 3 representing X, Y, Z components respectively.

The total $\underline{\dot{\psi}}^I$ component l equals the sum of $\dot{\psi}_{S/M_l}$ and $\dot{\psi}_{Bias_l}$.

From Figure 13.4.2-2, the elements of C^I_P are:

$$
\begin{aligned}
C_{PI}(1,1) &= \cos\ \beta \\
C_{PI}(1,2) &= \sin\ \beta \sin \phi \\
C_{PI}(1,3) &= \sin\ \beta \cos \phi \\
\\
C_{PI}(2,1) &= \sin\ \alpha \sin \beta \\
C_{PI}(2,2) &= \cos\ \alpha \cos \phi - \sin \alpha \cos \beta \sin \phi \\
C_{PI}(2,3) &= -\cos\ \alpha \sin \phi - \sin \alpha \cos \beta \cos \phi \\
\\
C_{PI}(3,1) &= -\cos\ \alpha \sin \beta \\
C_{PI}(3,2) &= \sin\ \alpha \cos \phi + \cos \alpha \cos \beta \sin \phi \\
C_{PI}(3,3) &= -\sin\ \alpha \sin \phi + \cos \alpha \cos \beta \cos \phi
\end{aligned}
\qquad (13.4.2\text{-}35)
$$

Using (13.4.2-35) for $C_{PI}(l,m)$, (13.4.2-24) for $\omega_{IB_{nP}}$, and (13.4.2-22) for α, β, ϕ, we make the following observations regarding the $C_{PI}(l,m)\ \omega_{IB_{nP}}$ products in Equation (13.4.2-33):

$$
\begin{aligned}
C_{PI}(1,1)\ \omega_{IB_{XP}} &= \omega_{Spin} \cos \beta \\
C_{PI}(1,2)\ \omega_{IB_{YP}} &= \frac{1}{2} \dot{\alpha} \sin^2\beta + \text{Oscillatory Terms} \\
C_{PI}(1,3)\ \omega_{IB_{ZP}} &= \frac{1}{2} \dot{\alpha} \sin^2\beta + \text{Oscillatory Terms}
\end{aligned}
\qquad (13.4.2\text{-}36)
$$

$$
\text{For } l = 2 \ \text{ or } \ l = 3 \ \text{ or } \ l = 1,\ m \neq n: \quad C_{PI}(l,m)\ \omega_{IB_{nP}} = \text{Oscillatory Terms}
$$

Note from Equations (13.4.2-19) and (13.4.2-22) that $\dot{\alpha}$ and β are constant. We conclude, therefore, from Equation (13.4.2-36), that the only $C_{PI}(l,m)\ \omega_{IB_{nP}}$ terms containing non-oscillatory elements have $l = 1$ and $m = n$. These terms multiply $\delta K_{S/M_{XXP}}$, $\delta K_{S/M_{YYP}}$,

$\delta K_{S/M_{ZZP}}$ in $\dot{\psi}_{S/M_l}$ Equation (13.4.2-33), and are only present in $\dot{\psi}_{S/M_1}$. It is the constant terms in $\dot{\psi}_{S/M_l}$ that affect attitude error build-up in $\underline{\psi}^I$. The oscillatory components create only bounded $\underline{\psi}^I$ oscillations. Hence, if we wish to calculate the average attitude error build-up rate in $\underline{\psi}^I$, we need only use the constant portions of Equations (13.4.2-36) and $\delta K_{S/M_{XXP}}$, $\delta K_{S/M_{YYP}}$, $\delta K_{S/M_{ZZP}}$ in the Equation (13.4.2-33) $\dot{\psi}_{S/M_1}$ expression, with the average attitude error build-up rate in $\dot{\psi}_{S/M_2}$, $\dot{\psi}_{S/M_3}$ recognized to be zero.

A similar but simpler rationale applies to $\dot{\psi}_{Bias_l}$ Equation (13.4.2-34) for which it is recognized from Equations (13.4.2-35) with (13.4.2-22) for α, β, ϕ that:

$$
\begin{aligned}
&C_{PI}(1,1) \;=\; \text{Constant} \\
&\text{For } l,m \neq 1,1: \quad C_{PI}(l,m) \;=\; \text{Oscillatory Terms}
\end{aligned}
\tag{13.4.2-37}
$$

Thus, from Equation (13.4.2-34), $\dot{\psi}_{Bias_1}$ produces an average $\underline{\psi}^I$ build-up rate from $\delta K_{Bias_{XP}}$ operating through $C_{PI}(1,1)$, while the average build-up rate from $\dot{\psi}_{Bias_2}$, $\dot{\psi}_{Bias_3}$ is zero.

Let us now calculate analytical expressions for $\delta K_{S/M_{XXP}}$, $\delta K_{S/M_{YYP}}$, $\delta K_{S/M_{ZZP}}$ and $\delta K_{Bias_{XP}}$ for evaluating the average attitude error build-up rate. First we define:

$$
\delta K^B_{Scal/Mis} = \begin{bmatrix} \delta K_{S/M_{XXB}} & \delta K_{S/M_{XYB}} & \delta K_{S/M_{XZB}} \\ \delta K_{S/M_{YXB}} & \delta K_{S/M_{YYB}} & \delta K_{S/M_{YZB}} \\ \delta K_{S/M_{ZXB}} & \delta K_{S/M_{ZYB}} & \delta K_{S/M_{ZZB}} \end{bmatrix} \qquad \delta K^B_{Bias} = \begin{bmatrix} \delta K_{Bias_{XB}} \\ \delta K_{Bias_{YB}} \\ \delta K_{Bias_{ZB}} \end{bmatrix} \tag{13.4.2-38}
$$

We then substitute (13.4.2-38) into (13.4.2-30) and (13.4.2-31) to obtain the desired analytical scalar forms:

$$
\begin{aligned}
\delta K_{S/M_{XXP}} = \; & C^2_{BP}(1,1)\, \delta K_{S/M_{XXB}} + C^2_{BP}(1,2)\, \delta K_{S/M_{YYB}} + C^2_{BP}(1,3)\, \delta K_{S/M_{ZZB}} \\
& + C_{BP}(1,1)\, C_{BP}(1,2) \left(\delta K_{S/M_{XYB}} + \delta K_{S/M_{YXB}}\right) \\
& + C_{BP}(1,2)\, C_{BP}(1,3) \left(\delta K_{S/M_{YZB}} + \delta K_{S/M_{ZYB}}\right) \\
& + C_{BP}(1,3)\, C_{BP}(1,1) \left(\delta K_{S/M_{ZXB}} + \delta K_{S/M_{XZB}}\right)
\end{aligned}
\tag{13.4.2-39}
$$

(Continued)

$$\begin{aligned}\delta K_{S/M_{YYP}} = {} & C_{BP}^2(2,1)\, \delta K_{S/M_{XXB}} + C_{BP}^2(2,2)\, \delta K_{S/M_{YYB}} + C_{BP}^2(2,3)\, \delta K_{S/M_{ZZB}} \\ & + C_{BP}(2,1)\, C_{BP}(2,2) \left(\delta K_{S/M_{XYB}} + \delta K_{S/M_{YXB}}\right) \\ & + C_{BP}(2,2)\, C_{BP}(2,3) \left(\delta K_{S/M_{YZB}} + \delta K_{S/M_{ZYB}}\right) \\ & + C_{BP}(2,3)\, C_{BP}(2,1) \left(\delta K_{S/M_{ZXB}} + \delta K_{S/M_{XZB}}\right)\end{aligned} \qquad \text{(13.4.2-39) (Continued)}$$

$$\begin{aligned}\delta K_{S/M_{ZZP}} = {} & C_{BP}^2(3,1)\, \delta K_{S/M_{XXB}} + C_{BP}^2(3,2)\, \delta K_{S/M_{YYB}} + C_{BP}^2(3,3)\, \delta K_{S/M_{ZZB}} \\ & + C_{BP}(3,1)\, C_{BP}(3,2) \left(\delta K_{S/M_{XYB}} + \delta K_{S/M_{YXB}}\right) \\ & + C_{BP}(3,2)\, C_{BP}(3,3) \left(\delta K_{S/M_{YZB}} + \delta K_{S/M_{ZYB}}\right) \\ & + C_{BP}(3,3)\, C_{BP}(3,1) \left(\delta K_{S/M_{ZXB}} + \delta K_{S/M_{XZB}}\right)\end{aligned}$$

$$\delta K_{Bias_{XP}} = C_{BP}(1,1)\, \delta K_{Bias_{XB}} + C_{BP}(1,2)\, \delta K_{Bias_{YB}} + C_{BP}(1,3)\, \delta K_{Bias_{ZB}}$$

The final step is to substitute (13.4.2-39) and (13.4.2-36) - (13.4.2-37) (neglecting the oscillatory terms) into the axis 1 component of Equations (13.4.2-33) - (13.4.2-34), and add the results to obtain the average attitude error build-up rate around I Frame axis 1. The average attitude error build-up rate around I Frame axes 2 and 3 are set to zero. The process is expedited by noting from generalized Equations (3.2.1-6) that direction cosine matrix columns represent orthogonal unit vectors, hence, the magnitude of any column is 1 and the dot product between any two columns is zero. Thus:

For $m = 1$ to 3:

$$\begin{aligned} & C_{BP}^2(2,m) + C_{BP}^2(3,m) = 1 - C_{BP}^2(1,m) \\ & C_{BP}(2,m)\, C_{BP}(2,m+1) + C_{BP}(3,m)\, C_{BP}(3,m+1) \\ & \qquad = - C_{BP}(1,m)\, C_{BP}(1,m+1)\end{aligned} \qquad \text{(13.4.2-40)}$$

in which the condition $m + 1$ is defined to be 1 when $m = 3$.

Applying the previous described procedure with Equation (13.4.2-40) then yields the final result:

$$\begin{aligned}\dot{\psi}_{Avg_1} = {} & - \left[C_{BP}^2(1,1)\, \omega_{Spin} \cos\beta + \left(1 - C_{BP}^2(1,1)\right) \frac{1}{2}\, \dot{\alpha} \sin^2\beta \right] \delta K_{S/M_{XXB}} \\ & - \left[C_{BP}^2(1,2)\, \omega_{Spin} \cos\beta + \left(1 - C_{BP}^2(1,2)\right) \frac{1}{2}\, \dot{\alpha} \sin^2\beta \right] \delta K_{S/M_{YYB}}\end{aligned} \qquad \text{(13.4.2-41)}$$

(Continued)

$$
\begin{aligned}
& - \left[C_{BP}^2(1,3)\ \omega_{Spin} \cos\beta + \left(1 - C_{BP}^2(1,3)\right) \frac{1}{2}\ \dot{\alpha} \sin^2\beta \right] \delta K_{S/M_{ZZB}} \\
& - \left[C_{BP}(1,1)\ C_{BP}(1,2) \left(\omega_{Spin} \cos\beta - \frac{1}{2}\ \dot{\alpha} \sin^2\beta\right) \right] \left(\delta K_{S/M_{XYB}} + \delta K_{S/M_{YXB}}\right) \\
& - \left[C_{BP}(1,2)\ C_{BP}(1,3) \left(\omega_{Spin} \cos\beta - \frac{1}{2}\ \dot{\alpha} \sin^2\beta\right) \right] \left(\delta K_{S/M_{YZB}} + \delta K_{S/M_{ZYB}}\right) \\
& - \left[C_{BP}(1,3)\ C_{BP}(1,1) \left(\omega_{Spin} \cos\beta - \frac{1}{2}\ \dot{\alpha} \sin^2\beta\right) \right] \left(\delta K_{S/M_{ZXB}} + \delta K_{S/M_{XZB}}\right) \\
& - \cos\beta \left(C_{BP}(1,1)\ \delta K_{Bias_{XB}} + C_{BP}(1,2)\ \delta K_{Bias_{YB}} + C_{BP}(1,3)\ \delta K_{Bias_{ZB}}\right)
\end{aligned}
\qquad \text{(13.4.2-41) (Continued)}
$$

$$
\dot{\psi}_{Avg_2} = 0 \qquad \dot{\psi}_{Avg_3} = 0
$$

where

$\dot{\psi}_{Avg_i}$ = Average value of the ith component of $\dot{\underline{\psi}}^I$.

13.4.3 SOLUTION FOR HORIZONTAL CIRCULAR TRAJECTORY PROFILE FOR UP TO TWO HOURS WITH CONTROLLED VERTICAL CHANNEL

In this section we analyze navigation errors for an INS with a controlled vertical channel that is navigating along a circular horizontal trajectory at constant velocity (e.g., simulating an aircraft in a stationary holding pattern relative to the earth). To simplify the analysis we restrict the navigation time to up to two hours so that we can ignore some earth rate terms. We also assume that the vehicle velocity is slow enough that transport rate can be neglected compared to earth rate.

With these approximations, the following simplified version of Equations (13.3.2-1) apply:

$$
\begin{aligned}
\dot{\gamma}_{ZN} &= -\left(C_B^N\ \delta\underline{\omega}_{IB}^B\right)_{ZN} \\
\dot{\underline{\gamma}}_H^N &= -\left(C_B^N\ \delta\underline{\omega}_{IB}^B\right)_H - \gamma_{ZN}\left(\underline{\omega}_{IE_H}^N \times \underline{u}_{ZN}^N\right) + \frac{1}{R}\left(\underline{u}_{ZN}^N \times \delta\underline{V}_H^N\right) \\
\delta\dot{\underline{V}}_H^N &\approx \left(C_B^N\ \delta\underline{a}_{SF}^B\right)_H + g\ \underline{u}_{ZN}^N \times \underline{\gamma}_H^N \\
\delta\dot{\underline{R}}_H^N &= \delta\underline{V}_H^N
\end{aligned}
\qquad \text{(13.4.3-1)}
$$

An analytical expression for the C_B^N matrix in (13.4.3-1) can be derived from the sensor assembly B Frame angular rate which we shall define as:

$$\underline{\omega}_{EB}^B = \underline{u}_\omega^B \, \omega_{EB} \tag{13.4.3-2}$$

where

$\underline{\omega}_{EB}^B$ = B Frame angular rate relative to the earth fixed E Frame as projected on B Frame axes. For a steady turning rate at constant velocity along the hypothesized circular trajectory, $\underline{\omega}_{EB}^B$ will be constant.

$\underline{u}_\omega^B$ = Unit vector along $\underline{\omega}_{EB}^B$. For the hypothesized circular trajectory at constant velocity, $\underline{u}_\omega^B$ will be parallel to a vertical located at the center of the trajectory circle.

ω_{EB} = Magnitude of $\underline{\omega}_{EB}^B$.

For the assumed constant $\underline{\omega}_{EB}^B$ angular rate, the B Frame rotates around $\underline{u}_\omega^B$ at ω_{EB}. The relationship between the B Frame current attitude and its initial attitude can then be described by a rotation vector along $\underline{u}_\omega^B$ with magnitude:

$$\phi = \int_0^t \omega_{EB} \, dt = \omega_{EB} \, t \tag{13.4.3-3}$$

where

$\phi, \underline{u}_\omega^B$ = Magnitude and direction of the rotation vector describing the current B Frame attitude relative to its starting attitude (considered fixed to the earth). See Section 3.2.2 for definition of the "rotation vector".

Generalized Equations (3.2.2.1-4) and (3.2.2.1-6) show that the direction cosine matrix associated with the previous rotation vector is given by:

$$C_B^{B_0} = I + \sin\phi \left(\underline{u}_\omega^B \times\right) + (1 - \cos\phi) \left(\underline{u}_\omega^B \times\right)\left(\underline{u}_\omega^B \times\right) \tag{13.4.3-4}$$

where

B_0 = Initial orientation of the B Frame (considered fixed in the earth fixed E Frame).

$C_B^{B_0}$ = Direction cosine matrix that transforms vectors from the B Frame to the B_0 Frame.

We now apply generalized Equation (3.1.1-40) for particular terms in (13.4.3-4):

$$\left(\underline{u}_\omega^B \times\right) = \left(C_B^N\right)^T \left(\underline{u}_\omega^N \times\right) C_B^N$$
$$\left(\underline{u}_\omega^B \times\right)\left(\underline{u}_\omega^B \times\right) = \left(C_B^N\right)^T \left(\underline{u}_\omega^N \times\right)\left(\underline{u}_\omega^N \times\right) C_B^N \qquad (13.4.3\text{-}5)$$

Substituting (13.4.3-5) and (13.4.3-3) into (13.4.3-4), and factoring out the C_B^N, $\left(C_B^N\right)^T$ terms yields:

$$C_B^{B_0} = \left(C_B^N\right)^T \left[I + \sin \omega_{EB}\, t \left(\underline{u}_\omega^N \times\right) + (1 - \cos \omega_{EB}\, t) \left(\underline{u}_\omega^N \times\right)\left(\underline{u}_\omega^N \times\right)\right] C_B^N \qquad (13.4.3\text{-}6)$$

To simplify the analysis, we assume that the trajectory circle radius is small compared to earth's radius, hence, we can approximate $\underline{u}_\omega^N$ as lying along the local vertical and:

$$\underline{u}_\omega^N \approx \underline{u}_{ZN}^N \qquad (13.4.3\text{-}7)$$

From (13.1-8) and (13.1-11) we can also write:

$$\left(\underline{u}_{ZN}^N \times\right)\left(\underline{u}_{ZN}^N \times\right) = - I_H$$
$$I - I_H = I + \left(\underline{u}_{ZN}^N \times\right)\left(\underline{u}_{ZN}^N \times\right) = \underline{u}_{ZN}^N \left(\underline{u}_{ZN}^N\right)^T \qquad (13.4.3\text{-}8)$$

where

I_H = Horizontal diagonal matrix with the 1,1 and 2,2 elements equal to unity and the 3,3 element equal to zero.

With (13.4.3-7) - (13.4.3-8), Equation (13.4.3-6) becomes:

$$C_B^{B_0} = \left(C_B^N\right)^T \left[\underline{u}_{ZN}^N \left(\underline{u}_{ZN}^N\right)^T + \left(\underline{u}_{ZN}^N \times\right) \sin \omega_{EB}\, t + I_H \cos \omega_{EB}\, t\right] C_B^N \qquad (13.4.3\text{-}9)$$

We now obtain C_B^N from (13.4.3-9) through:

$$C_B^N = C_B^N\, C_B^{B_0} \left(C_B^{B_0}\right)^T \qquad (13.4.3\text{-}10)$$

If we substitute Equation (13.4.3-9) for the middle $C_B^{B_0}$ term in (13.4.3-10), the (13.4.3-9) $\left(C_B^N\right)^T$ term will be canceled by the (13.4.3-10) C_B^N term, leaving the (13.4.3-9) square bracketed term with $C_B^N \left(C_B^{B_0}\right)^T$ (or $C_{B_0}^N$) on its right. Thus, application of (13.4.3-9) in (13.4.3-10) obtains:

$$C_B^N = \left[\underline{u}_{ZN}^N \left(\underline{u}_{ZN}^N\right)^T + \left(\underline{u}_{ZN}^N \times\right) \sin \omega_{EB}\, t + I_H \cos \omega_{EB}\, t \right] C_{B_0}^N \qquad (13.4.3\text{-}11)$$

With (13.4.3-11) for C_B^N, we can derive an analytical solution to Equations (13.4.3-1) for the case of constant inertial sensor error components. From Equations (12.5.6-1), for constant sensor error components we omit the noise terms, and the (13.4.3-1) $\delta\underline{\omega}_{IB}^B$, $\delta\underline{a}_{SF}^B$ inertial sensor error vectors become:

$$\delta\underline{\omega}_{IB}^B = \delta K_{Scal/Mis}\, \underline{\omega}_{IB}^B + \delta\underline{K}_{Bias} \qquad \delta\underline{a}_{SF}^B = \delta L_{Scal/Mis}\, \underline{a}_{SF}^B + \delta\underline{L}_{Bias} \qquad (13.4.3\text{-}12)$$

For the constant velocity steady turn trajectory, $\underline{a}_{SF}^B$ in (13.4.3-12) will be constant equal to the negative of plumb-bob gravity in the B Frame. The $\underline{\omega}_{IB}^B$ term in (13.4.3-12) equals the sum of earth's rate plus B Frame angular rate relative to the earth ($\underline{\omega}_{EB}^B$). For the assumed trajectory circle radius much smaller that earth's radius, as the B Frame turns about the vertical, the horizontal earth rate component will appear oscillatory in the B Frame (at frequency ω_{EB}), and the B Frame vertical earth rate component will be approximately constant. If we assume that the trajectory turning rate is large compared to earth's rate, we can neglect the oscillatory horizontal earth rate component in $\underline{\omega}_{IB}^B$. Since $\underline{\omega}_{EB}^B$ is constant, we can thereby approximate $\underline{\omega}_{IB}^B$ in (13.4.3-12) as constant. Thus, for $\underline{a}_{SF}^B$ and $\underline{\omega}_{IB}^B$ approximately constant in (13.4.3-12), for our assumed constant δK, δL coefficients, the $\delta\underline{a}_{SF}^B$ and $\delta\underline{\omega}_{IB}^B$ sensor errors will also be constant.

Understanding the implications of the constant sensor error assumption, we now seek an analytical solution to Equations (13.4.3-1) with (13.4.3-11) using constants for $\delta\underline{\omega}_{IB}^B$, $\delta\underline{a}_{SF}^B$. Because of the relatively short navigation time and limited trajectory range, we will also approximate the $\underline{\omega}_{IE_H}^N$ horizontal earth rate component in (13.4.3-1) as constant. We begin by applying C_B^N from (13.4.3-11) to find the horizontal and vertical sensor error components in Equations (13.4.3-1)

$$C_B^N \, \delta\underline{\omega}_{IB}^B = \left[\underline{u}_{ZN}^N \left(\underline{u}_{ZN}^N\right)^T + \left(\underline{u}_{ZN}^N \times\right) \sin \omega_{EB} \, t + I_H \cos \omega_{EB} \, t \right] C_{B_0}^N \, \delta\underline{\omega}_{IB}^B$$

$$C_B^N \, \delta\underline{a}_{SF}^B = \left[\underline{u}_{ZN}^N \left(\underline{u}_{ZN}^N\right)^T + \left(\underline{u}_{ZN}^N \times\right) \sin \omega_{EB} \, t + I_H \cos \omega_{EB} \, t \right] C_{B_0}^N \, \delta\underline{a}_{SF}^B \qquad (13.4.3\text{-}13)$$

The $\underline{u}_{ZN}^N \left(\underline{u}_{ZN}^N\right)^T$ operator in (13.4.3-13) transmits the vertical component of its input, the I_H operator transmits the horizontal input component, and the $\left(\underline{u}_{ZN}^N \times\right)$ operator generates a resultant that is perpendicular to $\underline{u}_{ZN}^N$, hence, horizontal. Thus from (13.4.3-13) we can write:

$$\left(C_B^N \, \delta\underline{\omega}_{IB}^B\right)_{ZN} = \left(C_{B_0}^N \, \delta\underline{\omega}_{IB}^B\right)_{ZN}$$

$$\left(C_B^N \, \delta\underline{\omega}_{IB}^B\right)_H = \left[\left(\underline{u}_{ZN}^N \times\right) \sin \omega_{EB} \, t + I_H \cos \omega_{EB} \, t \right] \left(C_{B_0}^N \, \delta\underline{\omega}_{IB}^B\right)_H \qquad (13.4.3\text{-}14)$$

$$\left(C_B^N \, \delta\underline{a}_{SF}^B\right)_H = \left[\left(\underline{u}_{ZN}^N \times\right) \sin \omega_{EB} \, t + I_H \cos \omega_{EB} \, t \right] \left(C_{B_0}^N \, \delta\underline{a}_{SF}^B\right)_H$$

where

H = Designation for the horizontal component of the associated vector.

ZN = Designation for the vertical component (along $\underline{u}_{ZN}^N$) of the associated vector.

With (13.4.3-14), the integral of the vertical attitude error rate in (13.4.3-1) is:

$$\gamma_{ZN} = \gamma_{ZN_0} - \left(C_{B_0}^N \, \delta\underline{\omega}_{IB}^B\right)_{ZN} t \qquad (13.4.3\text{-}15)$$

where

γ_{ZN_0} = Initial vertical attitude error (i.e., initial heading error).

The derivative of the (13.4.3-1) horizontal velocity error expression (with constant g because of the horizontal trajectory) is:

$$\delta\ddot{\underline{V}}_H^N = \frac{d}{dt} \left(C_B^N \, \delta\underline{a}_{SF}^B\right)_H + g \, \underline{u}_{ZN}^N \times \dot{\underline{\gamma}}_H^N \qquad (13.4.3\text{-}16)$$

Using (13.4.3-14) for $\left(C_B^N \, \delta\underline{a}_{SF}^B\right)_H$, Equation (13.4.3-1) for $\dot{\underline{\gamma}}_H^N$, (13.4.3-15) for γ_{ZN}, and applying generalized Equation (13.1-5), we find for particular terms in (13.4.3-16):

$$\frac{d}{dt}\left(C_B^N \, \delta\underline{a}_{SF}^B\right)_H = \omega_{EB} \left[\left(\underline{u}_{ZN}^N \times\right) \cos \omega_{EB}\, t - I_H \sin \omega_{EB}\, t\right] \left(C_{B_0}^N \, \delta\underline{a}_{SF}^B\right)_H$$

$$\underline{u}_{ZN}^N \times \dot{\underline{\gamma}}_H^N = - \underline{u}_{ZN}^N \times \left(C_B^N \, \delta\underline{\omega}_{IB}^B\right)_H - \left[\gamma_{ZN_0} - \left(C_{B_0}^N \, \delta\underline{\omega}_{IB}^B\right)_{ZN} t\right] \underline{\omega}_{IE_H}^N - \frac{1}{R} \, \delta\underline{V}_H^N \qquad (13.4.3\text{-}17)$$

with (13.4.3-14) for $\left(C_B^N \, \delta\underline{\omega}_{IB}^B\right)_H$ using (13.1-8) and (13.1-13):

$$\underline{u}_{ZN}^N \times \left(C_B^N \, \delta\underline{\omega}_{IB}^B\right)_H = \left[- I_H \sin \omega_{EB}\, t + \left(\underline{u}_{ZN}^N \times\right) \cos \omega_{EB}\, t\right] \left(C_{B_0}^N \, \delta\underline{\omega}_{IB}^B\right)_H \qquad (13.4.3\text{-}18)$$

Substituting (13.4.3-17) - (13.4.3-18) into (13.4.3-16) then obtains after rearrangement:

$$\begin{aligned} \delta\ddot{\underline{V}}_H^N + \frac{g}{R} \, \delta\underline{V}_H^N = \; & \omega_{EB} \left[\left(\underline{u}_{ZN}^N \times\right) \cos \omega_{EB}\, t - I_H \sin \omega_{EB}\, t\right] \left(C_{B_0}^N \, \delta\underline{a}_{SF}^B\right)_H \\ & - g \left[- I_H \sin \omega_{EB}\, t + \left(\underline{u}_{ZN}^N \times\right) \cos \omega_{EB}\, t\right] \left(C_{B_0}^N \, \delta\underline{\omega}_{IB}^B\right)_H \\ & - g \left[\gamma_{ZN_0} - \left(C_{B_0}^N \, \delta\underline{\omega}_{IB}^B\right)_{ZN} t\right] \underline{\omega}_{IE_H}^N \end{aligned} \qquad (13.4.3\text{-}19)$$

or in more compact notation:

$$\delta\ddot{\underline{V}}_H^N + \omega_S^2 \, \delta\underline{V}_H^N = \underline{D}_{\gamma z} + \underline{D}_c \cos \omega_{EB}\, t + \underline{D}_s \sin \omega_{EB}\, t + \underline{D}_t\, t \qquad (13.4.3\text{-}20)$$

with

$$\begin{aligned} \omega_S &= \sqrt{\frac{g}{R}} = \text{Schuler frequency} \\ \underline{D}_{\gamma z} &= - g \, \gamma_{ZN_0} \, \underline{\omega}_{IE_H}^N \\ \underline{D}_s &= g \left(C_{B_0}^N \, \delta\underline{\omega}_{IB}^B\right)_H - \omega_{EB} \left(C_{B_0}^N \, \delta\underline{a}_{SF}^B\right)_H \\ \underline{D}_c &= \underline{u}_{ZN}^N \times \left[\omega_{EB} \left(C_{B_0}^N \, \delta\underline{a}_{SF}^B\right)_H - g \left(C_{B_0}^N \, \delta\underline{\omega}_{IB}^B\right)_H\right] = - \underline{u}_{ZN}^N \times \underline{D}_s \\ \underline{D}_t &= g \left(C_{B_0}^N \, \delta\underline{\omega}_{IB}^B\right)_{ZN} \underline{\omega}_{IE_H}^N \end{aligned} \qquad (13.4.3\text{-}21)$$

In classical fashion, we construct the "Particular" solution to (13.4.3-20) as:

$$\delta\underline{V}_{H_{Prt}}^N = \underline{C}_0 + \underline{C}_c \cos \omega_{EB}\, t + \underline{C}_s \sin \omega_{EB}\, t + \underline{C}_t\, t \qquad (13.4.3\text{-}22)$$

where

$$\underline{\delta V}^N_{H_{Prt}} = \text{Particular solution for } \underline{\delta V}^N_H \text{ in (13.4.3-20).}$$

The second derivative of (13.4.3-22) is;

$$\underline{\delta \ddot{V}}^N_{H_{Prt}} = -\,\omega_{EB}^2 \left(\underline{C}_c \cos \omega_{EB}\, t + \underline{C}_s \sin \omega_{EB}\, t\right) \tag{13.4.3-23}$$

Substituting (13.4.3-22) for $\underline{\delta V}^N_H$ and (13.4.3-23) for $\underline{\delta \ddot{V}}^N_H$ in (13.4.3-20) with $\underline{D}_c$ from (13.4.3-21) yields:

$$\begin{aligned} &\left(\omega_S^2 - \omega_{EB}^2\right)\left(\underline{C}_c \cos \omega_{EB}\, t + \underline{C}_s \sin \omega_{EB}\, t\right) + \omega_S^2 \left(\underline{C}_0 + \underline{C}_t\, t\right) \\ &\qquad = \underline{D}_{\gamma Z} + \underline{D}_s \sin \omega_{EB}\, t - \left(\underline{u}^N_{ZN} \times \underline{D}_s\right) \cos \omega_{EB}\, t + \underline{D}_t\, t \end{aligned} \tag{13.4.3-24}$$

Equation (13.4.3-24) is solved for the (13.4.3-22) $\underline{\delta V}^N_{H_{Prt}}$ coefficients by equating coefficients of like terms, whence:

$$\begin{aligned} &\underline{C}_0 = \frac{1}{\omega_S^2}\,\underline{D}_{\gamma Z} \qquad \underline{C}_s = \frac{1}{\left(\omega_S^2 - \omega_{EB}^2\right)}\,\underline{D}_s \\ &\underline{C}_c = -\frac{1}{\left(\omega_S^2 - \omega_{EB}^2\right)}\left(\underline{u}^N_{ZN} \times \underline{D}_s\right) = -\,\underline{u}^N_{ZN} \times \underline{C}_s \qquad \underline{C}_t = \frac{1}{\omega_S^2}\,\underline{D}_t \end{aligned} \tag{13.4.3-25}$$

The complete solution for $\underline{\delta V}^N_H$ is now obtained as the sum of the homogeneous part with the (13.4.3-22) and (13.4.3-25) particular part. As in Section 13.2.2, we construct the homogeneous solution as the sum of terms for the characteristic roots of the $\underline{\delta V}^N_H$ differential equation which, for Equation (13.4.3-20), are sinusoids at the Schuler frequency ω_S. Thus, using (13.4.3-22) with the $\underline{C}_c$ expression in (13.4.3-25) for the Particular solution, the overall solution for $\underline{\delta V}^N_H$ is:

$$\begin{aligned} \underline{\delta V}^N_H = &\ \underline{C}_1 \cos \omega_S\, t + \underline{C}_2 \sin \omega_S\, t \\ &+ \underline{C}_0 + \underline{C}_s \sin \omega_{EB}\, t - \left(\underline{u}^N_{ZN} \times \underline{C}_s\right) \cos \omega_{EB}\, t + \underline{C}_t\, t \end{aligned} \tag{13.4.3-26}$$

and for its derivative:

$$\begin{aligned} \underline{\delta \dot{V}}^N_H = &-\underline{C}_1\, \omega_S \sin \omega_S\, t + \omega_S\, \underline{C}_2 \cos \omega_S\, t \\ &+ \omega_{EB}\, \underline{C}_s \cos \omega_{EB}\, t + \omega_{EB} \left(\underline{u}^N_{ZN} \times \underline{C}_s\right) \sin \omega_{EB}\, t + \underline{C}_t \end{aligned} \tag{13.4.3-27}$$

The $\underline{C}_1, \underline{C}_2$ coefficients are set to satisfy Equations (13.4.3-26) - (13.4.3-27) at time t = 0, which with $\delta\dot{\underline{V}}_H^N$ from (13.4.3-1) and $\left(C_B^N \, \delta\underline{a}_{SF}^B\right)_H$ from (13.4.3-14) are:

$$
\begin{aligned}
\delta\underline{V}_{H_0}^N &= \underline{C}_1 + \underline{C}_0 - \underline{u}_{ZN}^N \times \underline{C}_s \\
\delta\dot{\underline{V}}_{H_0}^N &= \omega_S \, \underline{C}_2 + \omega_{EB} \, \underline{C}_s + \underline{C}_t \\
&= \left(C_B^N \, \delta\underline{a}_{SF}^B\right)_{H_0} + g \, \underline{u}_{ZN}^N \times \underline{\gamma}_{H_0}^N = \left(C_{B_0}^N \, \delta\underline{a}_{SF}^B\right)_H + g \, \underline{u}_{ZN}^N \times \underline{\gamma}_{H_0}^N
\end{aligned}
\qquad (13.4.3\text{-}28)
$$

where

$$
\delta\underline{V}_{H_0}^N, \, \delta\dot{\underline{V}}_{H_0}^N, \, \underline{\gamma}_{H_0}^N = \text{Initial values for } \delta\underline{V}_H^N, \, \delta\dot{\underline{V}}_H^N, \, \underline{\gamma}_H^N \text{ (at time t = 0).}
$$

Rearranging (13.4.3-28) then yields:

$$
\begin{aligned}
\underline{C}_1 &= \delta\underline{V}_{H_0}^N - \underline{C}_0 + \underline{u}_{ZN}^N \times \underline{C}_s \\
\underline{C}_2 &= \frac{1}{\omega_S}\left[\left(C_{B_0}^N \, \delta\underline{a}_{SF}^B\right)_H + g \, \underline{u}_{ZN}^N \times \underline{\gamma}_{H_0}^N\right] - \frac{\omega_{EB}}{\omega_S}\,\underline{C}_s - \frac{1}{\omega_S}\,\underline{C}_t
\end{aligned}
\qquad (13.4.3\text{-}29)
$$

The complete solution for $\delta\underline{V}_H^N$ is finally obtained from (13.4.3-26) with (13.4.3-29) for $\underline{C}_1$ and $\underline{C}_2$, Equation (13.4.3-25) for $\underline{C}_0, \underline{C}_s, \underline{C}_t$, Equation (13.4.3-21) for $\underline{D}_{\gamma z}, \underline{D}_s, \underline{D}_t$, and selective substitution of ω_S from (13.4.3-21):

$$
\begin{aligned}
\delta\underline{V}_H^N = {} & \delta\underline{V}_{H_0}^N \cos\omega_S t - R \, \gamma_{ZN_0} \, \underline{\omega}_{IE_H}^N (1 - \cos\omega_S t) + g\left(\underline{u}_{ZN}^N \times \underline{\gamma}_{H_0}^N\right)\frac{\sin\omega_S t}{\omega_S} \\
& - \frac{\omega_{EB}}{\left(\omega_S^2 - \omega_{EB}^2\right)}\left[\left(\sin\omega_{EB} t - \frac{\omega_S}{\omega_{EB}}\sin\omega_S t\right) I - (\cos\omega_{EB} t - \cos\omega_S t)\left(\underline{u}_{ZN}^N \times\right)\right]\left(C_{B_0}^N \, \delta\underline{a}_{SF}^B\right)_H \\
& + R\left(C_{B_0}^N \, \delta\underline{\omega}_{IB}^B\right)_{ZN} \underline{\omega}_{IE_H}^N\left(t - \frac{\sin\omega_S t}{\omega_S}\right) \\
& + R\,\frac{\omega_S^2}{\left(\omega_S^2 - \omega_{EB}^2\right)}\left[\left(\sin\omega_{EB} t - \frac{\omega_{EB}}{\omega_S}\sin\omega_S t\right) I - (\cos\omega_{EB} t - \cos\omega_S t)\left(\underline{u}_{ZN}^N \times\right)\right]\left(C_{B_0}^N \, \delta\underline{\omega}_{IB}^B\right)_H
\end{aligned}
\qquad (13.4.3\text{-}30)
$$

The corresponding horizontal position error solution is determined per Equations (13.4.3-1) as the integral from time = 0 to t of the Equation (13.4.3-30) horizontal velocity error solution:

$$
\begin{aligned}
\delta \underline{R}_H^N = {} & \delta \underline{R}_{H_0}^N + \delta \underline{V}_{H_0}^N \frac{1}{\omega_S} \sin \omega_S t - R\, \gamma_{ZN_0}\, \underline{\omega}_{IE_H}^N \left(t - \frac{1}{\omega_S} \sin \omega_S t \right) + g \left(\underline{u}_{ZN}^N \times \underline{\gamma}_{H_0}^N \right) \frac{(1 - \cos \omega_S t)}{\omega_S^2} \\
& + \frac{1}{\left(\omega_S^2 - \omega_{EB}^2 \right)} \left[(\cos \omega_{EB} t - \cos \omega_S t)\, I + \left(\sin \omega_{EB} t - \frac{\omega_{EB}}{\omega_S} \sin \omega_S t \right) \left(\underline{u}_{ZN}^N \times \right) \right] \left(C_{B_0}^N\, \delta \underline{a}_{SF}^B \right)_H \\
& + R \left(C_{B_0}^N\, \delta \underline{\omega}_{IB}^B \right)_{ZN} \underline{\omega}_{IE_H}^N \left[\frac{1}{2} t^2 - \frac{(1 - \cos \omega_S t)}{\omega_S^2} \right] \qquad (13.4.3\text{-}31) \\
& + R \frac{\omega_S}{\left(\omega_S^2 - \omega_{EB}^2 \right)} \left[\left(\frac{\omega_S}{\omega_{EB}} (1 - \cos \omega_{EB} t) - \frac{\omega_{EB}}{\omega_S} (1 - \cos \omega_S t) \right) I \right. \\
& \qquad \left. - \left(\frac{\omega_S}{\omega_{EB}} \sin \omega_{EB} t - \sin \omega_S t \right) \left(\underline{u}_{ZN}^N \times \right) \right] \left(C_{B_0}^N\, \delta \underline{\omega}_{IB}^B \right)_H
\end{aligned}
$$

where

$\delta \underline{R}_{H_0}^N$ = Initial value for $\delta \underline{R}_H^N$ (at time t = 0).

Equations (13.4.3-30) and (13.4.3-31) show that the horizontal angular rate and accelerometer errors produce position and velocity error oscillations at the Schuler frequency and at the frequency of circular trajectory traversal. In addition, no long term position error build-up is created from horizontal sensor error, and for rapid turning rates (ω_{EB}) compared to the ω_S Schuler frequency, the oscillatory effects on navigation error are attenuated. The previous situation is unique to strapdown (compared to gimbaled) inertial navigation systems, and stems from the rotation of the strapdown sensors which averages their impact on navigation error build-up. This contrasts with the strapdown INS solution for a straight trajectory (Equations (13.3.2-25)) which show a ramping position error build-up due to horizontal angular rate sensor error coupled with Schuler oscillations from horizontal strapdown inertial sensor errors, but without the ω_{EB} turning rate attenuation. The effect of vertical angular rate sensor error and initial condition errors is the same for straight and circular trajectories. For the case when ω_{EB} approaches ω_S, Equations (13.4.3-30) - (13.4.3-31) show that the circular trajectory has the effect of amplifying position/velocity error build-up due to horizontal inertial sensor error (due to the $\omega_S^2 - \omega_{EB}^2$ term). The effect is analogous to the response of a pendulum to driving forces near its natural frequency of oscillation. For the singular case when $\omega_{EB} = \omega_S$, the (13.4.3-30) - (13.4.3-31) solution is indeterminate. The next section develops a solution for the circular trajectory that is valid for the particular situation when $\omega_{EB} = \omega_S$ (known as Schuler pumping).

13.4.4 SOLUTION FOR HORIZONTAL CIRCULAR TRAJECTORY PROFILE AT SCHULER FREQUENCY WITH CONTROLLED VERTICAL CHANNEL

Consider the case of a strapdown INS with a controlled vertical channel that is navigating along a circular trajectory in which the frequency of rotation around the circle matches the Schuler frequency. Subject to the same constraints delineated in Section 13.4.3, the applicable navigation error equations are provided by Equations (13.4.3-20) - (13.4.3-21) with $\omega_{EB} = \omega_S$:

$$\delta \ddot{\underline{V}}_H^N + \omega_S^2 \, \delta \underline{V}_H^N = \underline{D}_{\gamma z} + \underline{D}_s \sin \omega_S t - \left(\underline{u}_{ZN}^N \times \underline{D}_s \right) \cos \omega_S t + \underline{D}_t \, t \qquad (13.4.4\text{-}1)$$

$$\begin{aligned}
\omega_S &= \sqrt{\frac{g}{R}} = \text{Schuler frequency} \\
\underline{D}_{\gamma z} &= - g \, \gamma_{ZN_0} \, \underline{\omega}_{IE_H}^N \\
\underline{D}_s &= g \left(C_{B_0}^N \, \delta \underline{\omega}_{IB}^B \right)_H - \omega_S \left(C_{B_0}^N \, \delta \underline{a}_{SF}^B \right)_H \\
\underline{D}_t &= g \left(C_{B_0}^N \, \delta \underline{\omega}_{IB}^B \right)_{ZN} \underline{\omega}_{IE_H}^N
\end{aligned} \qquad (13.4.4\text{-}2)$$

Following the same procedure as in Section 13.4.3, the solution to Equation (13.4.4-1) is obtained by first constructing a "Particular" solution that we anticipate to fit the (13.4.4-1) forcing function:

$$\delta \underline{V}_{H_{Prt}}^N = \underline{C}_0 + \underline{C}_{ct} \, t \cos \omega_S t + \underline{C}_{st} \, t \sin \omega_S t + \underline{C}_t \, t \qquad (13.4.4\text{-}3)$$

Substituting (13.4.4-3) and its second derivative into (13.4.4-1) for $\delta \underline{V}_H^N$ then gives after combining terms:

$$\begin{aligned}
& - 2 \, \omega_S \, \underline{C}_{ct} \sin \omega_S t + 2 \, \omega_S \, \underline{C}_{st} \cos \omega_S t + \omega_S^2 \left(\underline{C}_0 + \underline{C}_t \, t \right) \\
& \quad = \underline{D}_{\gamma z} + \underline{D}_s \sin \omega_S t - \left(\underline{u}_{ZN}^N \times \underline{D}_s \right) \cos \omega_S t + \underline{D}_t \, t
\end{aligned} \qquad (13.4.4\text{-}4)$$

from which, after equating coefficients of like terms:

$$\begin{aligned}
& \underline{C}_0 = \frac{1}{\omega_S^2} \underline{D}_{\gamma z} \qquad \underline{C}_{ct} = - \frac{1}{2 \, \omega_S} \underline{D}_s \\
& \underline{C}_{st} = - \frac{1}{2 \, \omega_S} \left(\underline{u}_{ZN}^N \times \underline{D}_s \right) = \underline{u}_{ZN}^N \times \underline{C}_{ct} \qquad \underline{C}_t = \frac{1}{\omega_S^2} \underline{D}_t
\end{aligned} \qquad (13.4.4\text{-}5)$$

As in Section 13.4.3, the total solution is then constructed as the sum of the Particular part (from (13.4.4-3) and (13.4.4-5)) and the homogeneous part (based on the characteristic Schuler frequency oscillation roots). Thus:

$$\begin{aligned} \delta\underline{V}_H^N &= \underline{C}_1 \cos\omega_S t + \underline{C}_2 \sin\omega_S t + \underline{C}_0 \\ &\quad + \underline{C}_{ct}\, t \cos\omega_S t + \left(\underline{u}_{ZN}^N \times \underline{C}_{ct}\right) t \sin\omega_S t + \underline{C}_t\, t \end{aligned} \qquad (13.4.4\text{-}6)$$

and for its derivative:

$$\begin{aligned} \delta\dot{\underline{V}}_H^N &= -\,\underline{C}_1\, \omega_S \sin\omega_S t + \omega_S\, \underline{C}_2 \cos\omega_S t + \underline{C}_{ct} \cos\omega_S t + \left(\underline{u}_{ZN}^N \times \underline{C}_{ct}\right) \sin\omega_S t \\ &\quad -\,\underline{C}_{ct}\, \omega_S\, t \sin\omega_S t + \left(\underline{u}_{ZN}^N \times \underline{C}_{ct}\right) \omega_S\, t \cos\omega_S t + \underline{C}_t \end{aligned} \qquad (13.4.4\text{-}7)$$

The $\underline{C}_1$, $\underline{C}_2$ coefficients are set to satisfy Equations (13.4.4-6) and (13.4.4-7) at time $t = 0$ using $\delta\dot{\underline{V}}_H^N$ from Equations (13.4.3-1):

$$\delta\dot{\underline{V}}_H^N = \left(C_B^N\, \delta\underline{a}_{SF}^B\right)_H + g\, \underline{u}_{ZN}^N \times \underline{\gamma}_H^N \qquad (13.4.4\text{-}8)$$

Equations (13.4.4-6) - (13.4.4-8) at $t = 0$ then yields:

$$\begin{aligned} \underline{C}_1 &= \delta\underline{V}_{H_0}^N - \underline{C}_0 \\ \underline{C}_2 &= \frac{1}{\omega_S}\left[\left(C_{B_0}^N\, \delta\underline{a}_{SF}^B\right)_H + g\, \underline{u}_{ZN}^N \times \underline{\gamma}_{H_0}^N - \underline{C}_{ct} - \underline{C}_t\right] \end{aligned} \qquad (13.4.4\text{-}9)$$

The complete solution for $\delta\underline{V}_H^N$ is obtained using (13.4.4-6) with (13.4.4-9) for $\underline{C}_1$ and $\underline{C}_2$, Equation (13.4.4-5) for $\underline{C}_0$, $\underline{C}_{ct}$, $\underline{C}_t$, Equation (13.4.4-2) for $\underline{D}_{\gamma Z}$, $\underline{D}_S$, $\underline{D}_t$, and selective substitution of ω_S from (13.4.4-2):

$$\begin{aligned} \delta\underline{V}_H^N &= \delta\underline{V}_{H_0}^N \cos\omega_S t - R\, \gamma_{ZN_0}\, \underline{\omega}_{IE_H}^N \left(1 - \cos\omega_S t\right) + g\left(\underline{u}_{ZN}^N \times \underline{\gamma}_{H_0}^N\right) \frac{\sin\omega_S t}{\omega_S} \\ &\quad + \frac{1}{2}\left\{ t\left[I \cos\omega_S t + \sin\omega_S t \left(\underline{u}_{ZN}^N \times\right)\right] + I\, \frac{\sin\omega_S t}{\omega_S} \right\} \left(C_{B_0}^N\, \delta\underline{a}_{SF}^B\right)_H \\ &\quad + R\left(C_{B_0}^N\, \delta\underline{\omega}_{IB}^B\right)_{ZN} \underline{\omega}_{IE_H}^N \left(t - \frac{\sin\omega_S t}{\omega_S}\right) \\ &\quad - R\, \frac{\omega_S}{2}\left\{ t\left[I \cos\omega_S t + \sin\omega_S t \left(\underline{u}_{ZN}^N \times\right)\right] - I\, \frac{\sin\omega_S t}{\omega_S} \right\} \left(C_{B_0}^N\, \delta\underline{\omega}_{IB}^B\right)_H \end{aligned} \qquad (13.4.4\text{-}10)$$

The corresponding horizontal position error solution is derived by integrating (13.4.4-10) from time = 0 to t in accordance with (13.4.3-1):

$$
\begin{aligned}
\delta \underline{R}_H^N = \; & \delta \underline{R}_{H_0}^N + \delta \underline{V}_{H_0}^N \frac{1}{\omega_S} \sin \omega_S t - R\, \gamma_{ZN_0}\, \underline{\omega}_{IE_H}^N \left(t - \frac{\sin \omega_S t}{\omega_S} \right) + g \left(\underline{u}_{ZN}^N \times \underline{\gamma}_{H_0}^N \right) \frac{(1 - \cos \omega_S t)}{\omega_S^2} \\
& + \frac{1}{2\, \omega_S} \left\{ t \left[I \sin \omega_S t - \cos \omega_S t \left(\underline{u}_{ZN}^N \times \right) \right] - \frac{1}{\omega_S} \left[(1 - \cos \omega_S t)\, I - \sin \omega_S t \left(\underline{u}_{ZN}^N \times \right) \right] \right. \\
& \qquad \left. + I \frac{(1 - \cos \omega_S t)}{\omega_S} \right\} \left(C_{B_0}^N\, \delta \underline{a}_{SF}^B \right)_H \\
& + R \left(C_{B_0}^N\, \delta \underline{\omega}_{IB}^B \right)_{ZN} \underline{\omega}_{IE_H}^N \left[\frac{1}{2} t^2 - \frac{(1 - \cos \omega_S t)}{\omega_S^2} \right] \\
& - \frac{R}{2} \left\{ t \left[I \sin \omega_S t - \cos \omega_S t \left(\underline{u}_{ZN}^N \times \right) \right] - \frac{1}{\omega_S} \left[(1 - \cos \omega_S t)\, I - \sin \omega_S t \left(\underline{u}_{ZN}^N \times \right) \right] \right. \\
& \qquad \left. - I \frac{(1 - \cos \omega_S t)}{\omega_S} \right\} \left(C_{B_0}^N\, \delta \underline{\omega}_{IB}^B \right)_H
\end{aligned}
\tag{13.4.4-11}
$$

Equations (13.4.4-10) - (13.4.4-11) illustrate the principal effect of Schuler pumping (i.e., trajectory turning rate ω_{EB} equal to Schuler frequency ω_S) on horizontal navigation error; the creation of an oscillating navigation error at Schuler frequency in response to horizontal inertial sensor error, with an unbounded linearly ramping envelope component. As for the Section 13.4.3 general circular trajectory situation, no pure unbounded linear position ramping error exists in response to horizontal angular rate sensor error (as it does along the Section 13.3.2 straight trajectory). However, the slope of the Schuler envelope for the Schuler pumping case is half the magnitude of the pure linear position error ramp slope along the straight trajectory (See Equations (13.3.2-25)), and is created along both N Frame horizontal axes (in quadrature) for each angular rate sensor error component. For the straight trajectory case, a given angular rate sensor error component creates position error build-up along only one direction in the N Frame. A ramping Schuler oscillation envelope is generally considered a serious effect because it is accompanied by a ramping velocity oscillation envelope. In contrast, along the straight trajectory, the same angular rate sensor error source would create a velocity offset with a bounded amplitude Schuler oscillation.

It is also to be noted that the Equations (13.4.4-10) - (13.4.4-11) navigation error response to vertical angular rate sensor and initial condition error along the Schuler circle is identical to the error response in Section 13.4.3 for the general circular trajectory case.

13.5 LONG TERM POSITION ERROR FOR CONSTANT ATTITUDE AND SENSOR ERRORS

Thus far, in Sections 13.3 - 13.4 we have developed analytical solutions to the strapdown inertial navigation error equations for limited navigation times (up to two hours) for which the effect of position error on earth rate components can be neglected. In this section we deal with the long term position error solution in which the Schuler oscillation effects can be ignored. The applicable long term position error equation we will analyze is Equation (13.2.3-4) which has been derived in Section 13.2.3:

$$\delta \underline{R}^N_{H_{LngTrm}} = R \left(\underline{u}^N_{ZN} \times \underline{\psi}^N_H \right) \tag{13.5-1}$$

where

$\delta \underline{R}^N_{H_{LngTrm}}$ = Approximate value for $\delta \underline{R}^N_H$ for long term inertial navigation system error analysis in which Schuler oscillation effects have been neglected.

$\underline{\psi}^N_H$ = Horizontal component of $\underline{\psi}^N$.

The $\underline{\psi}^N$ attitude error in (13.5-1) is from Equation (12.3.7.1-4) with the (3.2.1-5) chain rule:

$$\dot{\underline{\psi}}^I = - C^I_N \, C^N_B \, \delta \underline{\omega}^B_{IB} \tag{13.5-2}$$

$$\underline{\psi}^N = \left(C^I_N \right)^T \underline{\psi}^I \tag{13.5-3}$$

where

C^I_N = Direction cosine matrix that transforms vectors from the N Frame to the I Frame.

The $\delta \underline{\omega}^B_{IB}$ angular rate sensor error vector in (13.5-2) is given in general by the $\delta \underline{\omega}^B_{IB}$ expression in Equations (12.5.6-1):

$$\delta \underline{\omega}^B_{IB} = \delta K_{Scal/Mis} \, \underline{\omega}^B_{IB} + \delta \underline{K}_{Bias} + \delta \underline{\omega}_{Rand} \tag{13.5-4}$$

For long term error analysis, the $\delta \underline{K}_{Bias}$ term in (13.5-4) dominates for typical angular rate sensors used in strapdown systems designed for long term inertial navigation (e.g., ring laser gyros). To simplify the analysis we will, therefore, only include $\delta \underline{K}_{Bias}$ with the further simplification that $\delta \underline{K}_{Bias}$ be considered constant for the navigation period. Thus:

$$\delta\underline{\omega}_{IB}^{B} \approx \delta\underline{K}_{Bias} = \text{Constant} \tag{13.5-5}$$

In order to find an analytical to Equations (13.5-1) - (13.5-3) with (13.5-5) we define a simple but meaningful trajectory for the navigation system as a great circle relative to the earth at constant velocity, with selected analysis coordinate frames connecting the B and I Frames as illustrated in the Figure 13.5-1 "Method of Least Work" diagram (See Section 3.2.3.3 for interpretation):

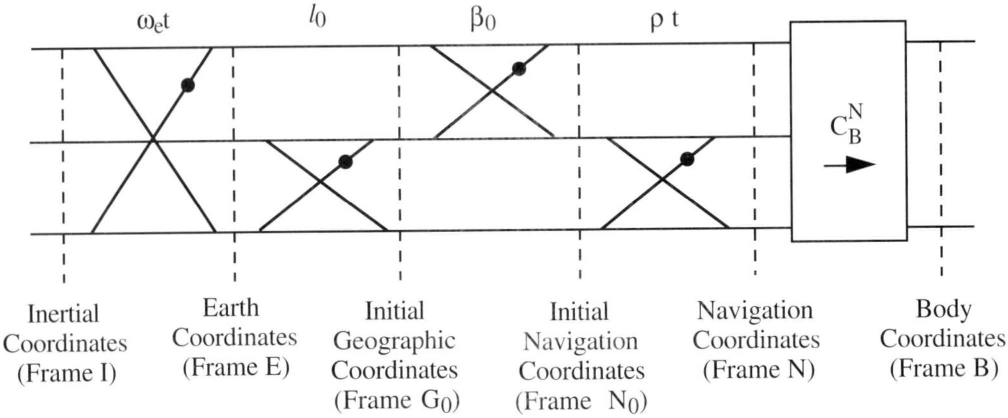

Figure 13.5-1 Analysis Coordinate Frame Relationships

where

t = Time from trajectory start.

E, G_0 and N_0 = Earth fixed coordinate frames. E is the earth reference frame with Y along the earth polar rotation axis and Z in the plane of the starting meridian at trajectory time $t = 0$ (in contrast with the Section 2.2 E Frame that has Z in the Greenwich meridian plane). G_0 is a geographic frame with Y axis north and Z axis along the upward vertical at the $t = 0$ trajectory starting location. The N_0 initial navigation frame has the Z axis vertical at the $t = 0$ starting location with the Y axis in the initial velocity direction.

N = Local navigation coordinate frame with X axis perpendicular to the great circle plane, Y axis along the velocity vector and Z axis up at time t along the trajectory.

I = Non-rotating inertial reference frame. The I Frame is defined in Figure 13.5-1 to be coincident with the E Frame at navigation time $t = 0$.

ω_e = Earth's angular rotation rate. The $\omega_e t$ angle is the angle rotated through by the earth (relative to inertial space) since navigation time $t = 0$.

l_0 = Initial latitude.

β_0 = Angle between the initial north direction and the initial velocity vector measured positive around a downward vertical (also known as the initial "track angle").

ρ = Transport rate representing the angular rate of the INS over the earth's surface (relative to the earth) along the great circle. The $\rho\, t$ angle is the great circle angle traversed by the INS since navigation time $t = 0$. Since the velocity relative to the earth is assumed constant, ρ is also constant.

From Figure 13.5-1 we write for the elements of C_N^I in Equations (13.5-2) - (13.5-3):

$$
\begin{aligned}
C_{NI}(1,1) &= \cos\omega_e t \cos\beta_0 + \sin\omega_e t \sin l_0 \sin\beta_0 \\
C_{NI}(1,2) &= \cos\omega_e t \sin\beta_0 \cos\rho t + \sin\omega_e t \left(- \sin l_0 \cos\beta_0 \cos\rho t - \cos l_0 \sin\rho t\right) \\
C_{NI}(1,3) &= \cos\omega_e t \sin\beta_0 \sin\rho t + \sin\omega_e t \left(- \sin l_0 \cos\beta_0 \sin\rho t + \cos l_0 \cos\rho t\right) \\
\\
C_{NI}(2,1) &= - \cos l_0 \sin\beta_0 \\
C_{NI}(2,2) &= \cos l_0 \cos\beta_0 \cos\rho t - \sin l_0 \sin\rho t \\
C_{NI}(2,3) &= \cos l_0 \cos\beta_0 \sin\rho t + \sin l_0 \cos\rho t \\
\\
C_{NI}(3,1) &= - \sin\omega_e t \cos\beta_0 + \cos\omega_e t \sin l_0 \sin\beta_0 \\
C_{NI}(3,2) &= - \sin\omega_e t \sin\beta_0 \cos\rho t + \cos\omega_e t \left(- \sin l_0 \cos\beta_0 \cos\rho t - \cos l_0 \sin\rho t\right) \\
C_{NI}(3,3) &= - \sin\omega_e t \sin\beta_0 \sin\rho t + \cos\omega_e t \left(- \sin l_0 \cos\beta_0 \sin\rho t + \cos l_0 \cos\rho t\right)
\end{aligned}
\qquad (13.5\text{-}6)
$$

where

$C_{NI}(i,j)$ = Element of C_N^I in row i, column j.

The analytical solution for $\underline{\psi}^I$ in Equation (13.5-3) is the integral of (13.5-2) from time = 0 to t using (13.5-5) - (13.5-6) and, for simplicity, assuming C_B^N to be constant (i.e., constant sensor axis B Frame orientation relative to the N Frame):

$$
\underline{\psi}^I = \underline{\psi}_0^I - \left(\int_0^t C_N^I \, dt \right) C_B^N \, \delta\underline{K}_{Bias} \qquad (13.5\text{-}7)
$$

where

$\underline{\psi}_0^I$ = Initial value for $\underline{\psi}^I$ at time $t = 0$.

The $\underline{\psi}_0^{I}$ initial attitude error in (13.5-7) can be expressed in terms if its N_0 Frame components if we recognize that the I and E Frames have been defined to be coincident at t = 0:

$$\underline{\psi}_0^{I} = \underline{\psi}_0^{E} = C_{N_0}^{E}\,\underline{\psi}_0^{N_0} \tag{13.5-8}$$

With (13.5-7) and (13.5-8), Equation (13.5-3) for $\underline{\psi}^{N}$ becomes:

$$\underline{\psi}^{N} = \left(C_N^I\right)^T \left[C_{N_0}^{E}\,\underline{\psi}_0^{N_0} - \left(\int_0^t C_N^I \, dt \right) C_B^N \, \delta\underline{K}_{Bias} \right] \tag{13.5-9}$$

Components of the $C_{N_0}^{E}$ matrix in (13.5-9) are from Figure 13.5-1 (or Equations (13.5-6) at t = 0):

$$\begin{aligned}
C_{N_0E}(1,1) &= \cos \beta_0 \\
C_{N_0E}(1,2) &= \sin \beta_0 \\
C_{N_0E}(1,3) &= 0 \\
\\
C_{N_0E}(2,1) &= -\cos l_0 \sin\beta_0 \\
C_{N_0E}(2,2) &= \cos l_0 \cos\beta_0 \\
C_{N_0E}(2,3) &= \sin l_0 \\
\\
C_{N_0E}(3,1) &= \sin l_0 \sin\beta_0 \\
C_{N_0E}(3,2) &= -\sin l_0 \cos\beta_0 \\
C_{N_0E}(3,3) &= \cos l_0
\end{aligned} \tag{13.5-10}$$

where

$$C_{N_0E}(i,j) = \text{Element in row i, column j of } C_{N_0}^{E}.$$

Summarizing and collecting results at this point, the long term position error $\delta\underline{R}_{H_{LngTrm}}^{N}$ is provided by Equations (13.5-1) and (13.5-9):

$$\begin{aligned}
\delta\underline{R}_{H_{LngTrm}}^{N} &= R\left(\underline{u}_{ZN}^{N} \times \underline{\psi}_{H}^{N}\right) \\
\underline{\psi}^{N} &= \left(C_N^I\right)^T C_{N_0}^{E}\,\underline{\psi}_0^{N_0} - \left(C_N^I\right)^T IC_N^I \, C_B^N \, \delta\underline{K}_{Bias}
\end{aligned} \tag{13.5-11}$$

with

$$IC_N^I \equiv \int_0^t C_N^I \, dt \tag{13.5-12}$$

where

IC_N^I = Integral of C_N^I from time t = 0.

and with (13.5-6) and (13.5-10) for C_N^I, $C_{N_0}^E$.

Equations (13.5-10) - (13.5-11) for $\delta \underline{R}_{H_{LngTrm}}^N$ are solved once the C_N^I integral (IC_N^I) in (13.5-12) is evaluated using (13.5-6) for C_N^I. Evaluation of (13.5-12) is expedited by first defining the following for particular terms in (13.5-6):

$$\begin{aligned}
g_{c\omega}(t) &\equiv \cos \omega_e t \\
g_{s\omega}(t) &\equiv \sin \omega_e t \\
g_{c\rho}(t) &\equiv \cos \rho t \\
g_{s\rho}(t) &\equiv \sin \rho t \\
g_{cc}(t) &\equiv \cos \omega_e t \cos \rho t = \frac{1}{2}\left(\cos(\omega_e+\rho)t + \cos(\omega_e-\rho)t\right) \\
g_{cs}(t) &\equiv \cos \omega_e t \sin \rho t = \frac{1}{2}\left(\sin(\omega_e+\rho)t - \sin(\omega_e-\rho)t\right) \\
g_{sc}(t) &\equiv \sin \omega_e t \cos \rho t = \frac{1}{2}\left(\sin(\omega_e+\rho)t + \sin(\omega_e-\rho)t\right) \\
g_{ss}(t) &\equiv \sin \omega_e t \sin \rho t = \frac{1}{2}\left(- \cos(\omega_e+\rho)t + \cos(\omega_e-\rho)t\right)
\end{aligned} \tag{13.5-13}$$

With (13.5-13), Equations (13.5-6) become for the C_N^I elements:

$$\begin{aligned}
C_{NI}(1,1) &= \cos \beta_0 \, g_{c\omega}(t) + \sin l_0 \sin \beta_0 \, g_{s\omega}(t) \\
C_{NI}(1,2) &= \sin \beta_0 \, g_{cc}(t) - \sin l_0 \cos \beta_0 \, g_{sc}(t) - \cos l_0 \, g_{ss}(t) \\
C_{NI}(1,3) &= \sin \beta_0 \, g_{cs}(t) - \sin l_0 \cos \beta_0 \, g_{ss}(t) + \cos l_0 \, g_{sc}(t) \\
C_{NI}(2,1) &= - \cos l_0 \sin \beta_0 \\
C_{NI}(2,2) &= \cos l_0 \cos \beta_0 \, g_{c\rho}(t) - \sin l_0 \, g_{s\rho}(t) \\
C_{NI}(2,3) &= \cos l_0 \cos \beta_0 \, g_{s\rho}(t) + \sin l_0 \, g_{c\rho}(t)
\end{aligned} \tag{13.5-14}$$

(Continued)

$$
\begin{aligned}
C_{NI}(3,1) &= -\cos\beta_0\, g_{s\omega}(t) + \sin l_0 \sin\beta_0\, g_{c\omega}(t) \\
C_{NI}(3,2) &= -\sin\beta_0\, g_{sc}(t) - \sin l_0 \cos\beta_0\, g_{cc}(t) - \cos l_0\, g_{cs}(t) \\
C_{NI}(3,3) &= -\sin\beta_0\, g_{ss}(t) - \sin l_0 \cos\beta_0\, g_{cs}(t) + \cos l_0\, g_{cc}(t)
\end{aligned}
\qquad \text{(13.5-14) (Continued)}
$$

The only time varying elements in (13.5-14) are the Equation (13.5-13) terms. Thus, the integral of (13.5-14) for IC_N^I per Equation (13.5-12) is a function of the integrated (13.5-13) terms which we shall define as:

$$
\begin{aligned}
Ig_{c\omega}(t) &\equiv \int_0^t g_{c\omega}(\tau)\, d\tau = \frac{\sin\omega_e t}{\omega_e} \\
Ig_{s\omega}(t) &\equiv \int_0^t g_{s\omega}(\tau)\, d\tau = \frac{\left(1 - \cos\omega_e t\right)}{\omega_e} \\
Ig_{c\rho}(t) &\equiv \int_0^t g_{c\rho}(\tau)\, d\tau = \frac{\sin\rho t}{\rho} \\
Ig_{s\rho}(t) &\equiv \int_0^t g_{s\rho}(\tau)\, d\tau = \frac{\left(1 - \cos\rho t\right)}{\rho} \\
Ig_{cc}(t) &\equiv \int_0^t g_{cc}(\tau)\, d\tau = \frac{1}{2}\left[\frac{\sin(\omega_e+\rho)t}{\omega_e+\rho} + \frac{\sin(\omega_e-\rho)t}{\omega_e-\rho}\right] \\
Ig_{cs}(t) &\equiv \int_0^t g_{cs}(\tau)\, d\tau = \frac{1}{2}\left[\frac{\left(1 - \cos(\omega_e+\rho)t\right)}{\omega_e+\rho} - \frac{\left(1 - \cos(\omega_e-\rho)t\right)}{\omega_e-\rho}\right] \\
Ig_{sc}(t) &\equiv \int_0^t g_{sc}(\tau)\, d\tau = \frac{1}{2}\left[\frac{\left(1 - \cos(\omega_e+\rho)t\right)}{\omega_e+\rho} + \frac{\left(1 - \cos(\omega_e-\rho)t\right)}{\omega_e-\rho}\right] \\
Ig_{ss}(t) &\equiv \int_0^t g_{ss}(\tau)\, d\tau = \frac{1}{2}\left[-\frac{\sin(\omega_e+\rho)t}{\omega_e+\rho} + \frac{\sin(\omega_e-\rho)t}{\omega_e-\rho}\right]
\end{aligned}
\qquad \text{(13.5-15)}
$$

Some of the elements in Equations (13.5-15) become indeterminate when ρ approaches zero, plus ω_e or minus ω_e. Using Taylor series expansion for the appropriate sinusoidal terms in these elements, the particular affected terms can be evaluated to first order under the previous conditions, whence:

For $\rho \approx 0$:

$$Ig_{c\rho}(t) \approx t \qquad Ig_{s\rho}(t) \approx \frac{1}{2}\rho\, t^2 \approx 0$$

For $\rho \approx \omega_e$:

$$\begin{aligned}
Ig_{cc}(t) &\approx \frac{1}{2}\left[t + \frac{\sin(\omega_e+\rho)t}{\omega_e+\rho}\right] & Ig_{cs}(t) &\approx \frac{1}{2}\left[\frac{\left(1 - \cos(\omega_e+\rho)t\right)}{\omega_e+\rho}\right] \\
Ig_{sc}(t) &\approx \frac{1}{2}\left[\frac{\left(1 - \cos(\omega_e+\rho)t\right)}{\omega_e+\rho}\right] & Ig_{ss}(t) &\approx \frac{1}{2}\left[t - \frac{\sin(\omega_e+\rho)t}{\omega_e+\rho}\right]
\end{aligned} \qquad (13.5\text{-}16)$$

For $\rho \approx -\omega_e$:

$$\begin{aligned}
Ig_{cc}(t) &\approx \frac{1}{2}\left[t + \frac{\sin(\omega_e-\rho)t}{\omega_e-\rho}\right] & Ig_{cs}(t) &\approx -\frac{1}{2}\left[\frac{1 - \cos(\omega_e-\rho)t}{\omega_e-\rho}\right] \\
Ig_{sc}(t) &\approx \frac{1}{2}\left[\frac{1 - \cos(\omega_e-\rho)t}{\omega_e-\rho}\right] & Ig_{ss}(t) &\approx \frac{1}{2}\left[-t + \frac{\sin(\omega_e-\rho)t}{\omega_e-\rho}\right]
\end{aligned}$$

Then, using (13.5-15) - (13.5-16), the elements of IC_N^I are obtained per Equation (13.5-12) as the integral of Equations (13.5-14):

$$\begin{aligned}
IC_{NI}(1,1) &= \cos\beta_0\, Ig_{c\omega}(t) + \sin l_0 \sin\beta_0\, Ig_{s\omega}(t) \\
IC_{NI}(1,2) &= \sin\beta_0\, Ig_{cc}(t) - \sin l_0 \cos\beta_0\, Ig_{sc}(t) - \cos l_0\, Ig_{ss}(t) \\
IC_{NI}(1,3) &= \sin\beta_0\, Ig_{cs}(t) - \sin l_0 \cos\beta_0\, Ig_{ss}(t) + \cos l_0\, Ig_{sc}(t) \\
\\
IC_{NI}(2,1) &= -t \cos l_0 \sin\beta_0 \\
IC_{NI}(2,2) &= \cos l_0 \cos\beta_0\, Ig_{c\rho}(t) - \sin l_0\, Ig_{s\rho}(t) \\
IC_{NI}(2,3) &= \cos l_0 \cos\beta_0\, Ig_{s\rho}(t) + \sin l_0\, Ig_{c\rho}(t) \\
\\
IC_{NI}(3,1) &= -\cos\beta_0\, Ig_{s\omega}(t) + \sin l_0 \sin\beta_0\, Ig_{c\omega}(t) \\
IC_{NI}(3,2) &= -\sin\beta_0\, Ig_{sc}(t) - \sin l_0 \cos\beta_0\, Ig_{cc}(t) - \cos l_0\, Ig_{cs}(t) \\
IC_{NI}(3,3) &= -\sin\beta_0\, Ig_{ss}(t) - \sin l_0 \cos\beta_0\, Ig_{cs}(t) + \cos l_0\, Ig_{cc}(t)
\end{aligned} \qquad (13.5\text{-}17)$$

where

$IC_{NI}(i,j)$ = Element of IC_N^I in row i, column j.

The general analytical solution for the long term position error $\delta \underline{R}_{H_{LngTrm}}^{N}$ is given by Equations (13.5-11) using (13.5-10) for $C_{N_0}^{E}$, Equation (13.5-14) with (13.5-13) for C_{N}^{I}, and Equation (13.5-17) with (13.5-15) - (13.5-16) for IC_{N}^{I}.

It is instructive to also generate a $\delta \underline{R}_{H_{LngTrm}}^{N}$ analytic solution under the special case of a slowly moving vehicle for which ρ and $\rho\, t$ can be considered negligible. For simplicity we also set $\beta_0 = 0$ so that the Y axis of the N_0 Frame is north and, since ρ is assumed small, the N and G_0 Frames can be approximated to be coincident. Then, using (13.1-4), Equations (13.5-11) - (13.5-12) simplify to:

$$\delta \underline{R}_{H_{LngTrm}}^{G_0} = R \left(\underline{u}_{ZG_0}^{G_0} \times \underline{\psi}^{G_0} \right)$$
$$\underline{\psi}^{G_0} = \left(C_{G_0}^{I} \right)^T C_{G_0}^{E} \underline{\psi}_0^{G_0} - \left(C_{G_0}^{I} \right)^T IC_{G_0}^{I} C_B^{G_0} \delta \underline{K}_{Bias} \tag{13.5-18}$$

$$IC_{G_0}^{I} \equiv \int_0^t C_{G_0}^{I} \, dt \tag{13.5-19}$$

Equations (13.5-18) - (13.5-19) can also be expressed as:

$$\delta \underline{R}_{H_{LngTrm}}^{G_0} = A_{\psi_0} \underline{\psi}_0^{G_0} + A_{\delta K_{Bias}} \delta \underline{K}_{Bias}^{G_0} \qquad \delta \underline{K}_{Bias}^{G_0} = C_B^{G_0} \delta \underline{K}_{Bias} \tag{13.5-20}$$

with

$$A_{\psi_0} = R \left(\underline{u}_{ZG_0}^{G_0} \times \right) \left(C_{G_0}^{I} \right)^T C_{G_0}^{E} \qquad A_{\delta K_{Bias}} = - R \left(\underline{u}_{ZG_0}^{G_0} \times \right) \left(C_{G_0}^{I} \right)^T IC_{G_0}^{I} \tag{13.5-21}$$

where

A_{ψ_0}, $A_{\delta K_{Bias}}$ = Long term horizontal position error sensitivity matrices to initial attitude error and G_0 Frame angular rate sensor bias error.

The $C_{G_0}^{I}$ matrix terms in (13.5-21) are equivalently:

$$C_{G_0}^{I} = C_E^{I} C_{G_0}^{E} \qquad \left(C_{G_0}^{I} \right)^T = \left(C_{G_0}^{E} \right)^T \left(C_E^{I} \right)^T \tag{13.5-22}$$

Recognizing from Figure 13.5-1 that $C_{G_0}^{E}$ is constant allows us to write for $IC_{G_0}^{I}$ from (13.5-19) with (13.5-22) for $C_{G_0}^{I}$:

$$IC^I_{G_0} = IC^I_E \, C^E_{G_0} \tag{13.5-23}$$

$$IC^I_E \equiv \int_0^t C^I_E \, dt \tag{13.5-24}$$

Using (13.5-23) and $\left(C^I_{G_0}\right)^T$ from (13.5-22), Equation (13.5-21) becomes:

$$\begin{aligned} A_{\psi_0} &= R\left(\underline{u}^{G_0}_{ZG_0}\times\right)\left(C^E_{G_0}\right)^T \left(C^I_E\right)^T C^E_{G_0} \\ A_{\delta K_{Bias}} &= - R\left(\underline{u}^{G_0}_{ZG_0}\times\right)\left(C^E_{G_0}\right)^T \left(C^I_E\right)^T IC^I_E \, C^E_{G_0} \end{aligned} \tag{13.5-25}$$

From Figure 13.5-1 we now write for the individual matrices in (13.5-25):

$$C^E_{G_0} = \begin{bmatrix} 1 & 0 & 0 \\ 0 & \cos l_0 & \sin l_0 \\ 0 & -\sin l_0 & \cos l_0 \end{bmatrix} \qquad C^I_E = \begin{bmatrix} \cos \omega_e t & 0 & \sin \omega_e t \\ 0 & 1 & 0 \\ -\sin \omega_e t & 0 & \cos \omega_e t \end{bmatrix} \tag{13.5-26}$$

Then, using C^I_E from (13.5-26) in (13.5-24):

$$IC^I_E = \begin{bmatrix} \dfrac{\sin \omega_e t}{\omega_e} & 0 & \dfrac{\left(1 - \cos \omega_e t\right)}{\omega_e} \\ 0 & t & 0 \\ -\dfrac{\left(1 - \cos \omega_e t\right)}{\omega_e} & 0 & \dfrac{\sin \omega_e t}{\omega_e} \end{bmatrix} \tag{13.5-27}$$

Finally, we substitute (13.5-26) and (13.5-27) into (13.5-25) to obtain for the position error sensitivity matrices:

$$A_{\psi 0} = \begin{bmatrix} R \sin l_0 \sin \omega_e t & -R\left(1 - \sin^2 l_0 \,(1 - \cos \omega_e t)\right) & -R \sin l_0 \cos l_0 \,(1 - \cos \omega_e t) \\ R \cos \omega_e t & R \sin l_0 \sin \omega_e t & -R \cos l_0 \sin \omega_e t \\ 0 & 0 & 0 \end{bmatrix}$$

$$A_{\delta K_{Bias}} = \begin{bmatrix} -R \sin l_0 \dfrac{(1 - \cos \omega_e t)}{\omega_e} & R\left[t - \sin^2 l_0 \left(t - \dfrac{\sin \omega_e t}{\omega_e}\right)\right] & R \sin l_0 \cos l_0 \left(t - \dfrac{\sin \omega_e t}{\omega_e}\right) \\ -R \dfrac{\sin \omega_e t}{\omega_e} & -R \sin l_0 \dfrac{(1 - \cos \omega_e t)}{\omega_e} & R \cos l_0 \dfrac{(1 - \cos \omega_e t)}{\omega_e} \\ 0 & 0 & 0 \end{bmatrix} \tag{13.5-28}$$

When interpreting the Equations (13.5-28) sensitivities, recall from Equations (13.5-20) that they are for horizontal position error in the G_0 Frame which has X east, Y north and Z vertical. Thus, the third row is zero because we are evaluating horizontal position error components. Note that the sensitivity to initial attitude error is to create only an oscillatory position error response at earth rate frequency (i.e., one revolution per day). Similar horizontal position error oscillations at earth rate frequency are generated from angular rate sensor bias error, however, the north (Y) and vertical (Z) bias errors also produce an unbounded linearly ramping east (X) position error build-up (in the first row).

For short term navigation times (e.g., one to two hours) when $\omega_e t$ is small, with Taylor series expansion for the sinusoidal terms in Equations (13.5-28) and retaining only the dominant term for each input component (i.e., the dominant term in each matrix column), the (13.5-28) matrices become the approximate forms:

$$A_{\psi 0} \approx \begin{bmatrix} 0 & -R & 0 \\ R & 0 & -R \cos l_0 \,\omega_e t \\ 0 & 0 & 0 \end{bmatrix}$$

$$A_{\delta K_{Bias}} \approx \begin{bmatrix} 0 & R\,t & 0 \\ -R\,t & 0 & R \cos l_0 \,\omega_e \dfrac{1}{2} t^2 \\ 0 & 0 & 0 \end{bmatrix} \tag{13.5-29}$$

It is informative to compare the long term position response of Equation (13.5-20) with the Equation (13.3.2-25) up-to-two-hour approximate horizontal position error expression developed in Section 13.3.2. To make the comparison, we use the approximate (13.5-29) forms for $A_{\psi 0}$ and $A_{\delta K_{Bias}}$ which are reformatted based on cross-product operator definition Equation (3.1.1-14):

$$
\begin{aligned}
A_{\psi 0} &\approx R\left(\underline{u}_{ZG_0}^{G_0}\times\right) + R\cos l_0\,\omega_e\,t\left(\underline{u}_{XG_0}^{G_0}\times\right)_H \\
A_{\delta K_{Bias}} &\approx - R\,t\left(\underline{u}_{ZG_0}^{G_0}\times\right) - R\cos l_0\,\omega_e\,\frac{1}{2}\,t^2\left(\underline{u}_{XG_0}^{G_0}\times\right)_H
\end{aligned}
\qquad (13.5\text{-}30)
$$

where

$\underline{u}_{XG_0}^{G_0}$ = Unit vector along the G_0 Frame X axis (i.e., East).

$\left(\underline{u}_{XG_0}^{G_0}\times\right)_H$ = Cross-product operator form of $\underline{u}_{XG_0}^{G_0}$ with the third row set to zero (i.e., only transmitting the X and Y horizontal (H) components).

The $\underline{\psi}_0^{G_0}$ term in (13.5-20) can also be reformatted in terms of the Equation (13.3.2-25) error parameters using (12.2.1-17), (12.2.3-19) and (13.1-4) with $\underline{\gamma}^N$ represented as the sum of its horizontal and vertical components:

$$
\underline{\psi}_0^{G_0} = \underline{\gamma}_{H_0}^{G_0} + \gamma_{ZG_0}\,\underline{u}_{ZG_0}^{G_0} - \frac{1}{R}\left(\underline{u}_{ZG_0}^{G_0} \times \delta\underline{R}_{H_0}^{G_0}\right) - \varepsilon_{ZG_0}\,\underline{u}_{ZG_0}^{G_0} \qquad (13.5\text{-}31)
$$

where

$\underline{\gamma}_{H_0}^{G_0}, \gamma_{ZG_0}, \delta\underline{R}_{H_0}^{G_0}, \varepsilon_{ZG_0}$ = Initial G_0 Frame values for the Section 13.3 $\underline{\gamma}_H^N$, γ_{ZN}, $\delta\underline{R}_H^N$, ε_{ZN} error parameters. For reference, Equations (13.3.2-25) which we will be comparing against (13.5-20), is based on Equations (13.3-6) of Section 13.3.

Substituting (13.5-30) - (13.5-31) into (13.5-20) then obtains the equivalent simplified form for short term navigation times:

$$
\begin{aligned}
\delta\underline{R}_{H_{LngTrm}}^{G_0} = {} & \delta\underline{R}_{H_0}^{G_0} + R\,\underline{u}_{ZG_0}^{G_0} \times \underline{\gamma}_{H_0}^{G_0} - R\cos l_0\,\omega_e\,t\,\underline{u}_{YG_0}^{G_0}\left(\gamma_{ZG_0} - \varepsilon_{ZG_0}\right) \\
& - R\,t\,\underline{u}_{ZG_0}^{G_0} \times \delta\underline{K}_{Bias_H}^{G_0} + R\cos l_0\,\omega_e\,\frac{1}{2}\,t^2\,\delta K_{Bias_{ZG0}}\,\underline{u}_{YG_0}^{G_0}
\end{aligned}
\qquad (13.5\text{-}32)
$$

where

$\delta\underline{K}_{Bias_H}^{G_0}, \delta K_{Bias_{ZG0}}$ = Horizontal and vertical components of $\delta\underline{K}_{Bias}^{G_0}$.

Equation (13.5-32) can now be compared against the $\delta \underline{R}_H^N$ horizontal position error expression in Equations (13.3.2-25) by setting the (13.3.2-25) N Frame equal to the Geo Frame, equating ω_S^2 in (13.3.2-25) to g / R (per Equation (13.2.2-7)), and recognizing that the horizontal earth rate component $\underline{\omega}_{IE_H}$ in (13.3.2-25) lies along the G_0 Frame Y axis and equals $\omega_e \cos l_0$. (The last point is easily seen using Figure 13.5-1 if we inject the full earth rate vector (of magnitude ω_e) along the Y axis of the E Frame (i.e., earth's polar rotation axis) and find its projection on the G_0 Frame horizontal (X, Y) axes.) Based on the previous equalities we see then that, neglecting Schuler oscillation effects, the (13.3.2-25) $\delta \underline{R}_H^N$ response to $\delta \underline{R}_{H0}^{G0}$, $\underline{\gamma}_{H0}^{G0}$, γ_{ZG0}, $\delta \underline{K}_{Bias}^{G0}$ is identical to the (13.5-32) response. The only term in (13.5-32) that has no equivalency in (13.3.2-25) is ε_{ZG0}, the value for the heading error in the initial C_N^E matrix. The heading initialization discussion following Equation (13.3-8) shows why this term has been legitimately dropped from Equations (13.3-15) for which (13.3.2-25) is the short term (1 to 2 hour) solution. The previous referenced initialization discussion also points out that, depending on the initialization technique utilized, ε_{ZG0} can also be considered to be part of γ_{ZG0}.

Finally, it is to be noted in Equations (13.5-28) that the only truly unbounded long term position error response is produced (through $A_{\delta K_{Bias}}$) in the east (X) direction by north (Y) and vertical (Z) angular rate sensor bias error. In the (13.5-29) simplified short term version of (13.5-28), what appears to be a ramping north (Y) position error response from vertical (Z) attitude error (through $A_{\psi 0}$) and east (X) angular rate sensor bias error (through $A_{\delta K_{Bias}}$), is actually the starting leg of a sine wave at earth rate frequency with the first peak occurring at 6 hours. What appears in (13.5-29) to be an unbounded parabolically increasing north (Y) position error response to vertical (Z) angular rate sensor bias error (through $A_{\delta K_{Bias}}$) is actually (in (13.5-28)) the starting leg of an offset cosine wave at earth rate frequency with the first peak occurring at 12 hours. The net result is that many of the errors that were believed to produce unbounded position error drift during short term (up to two hour) navigation periods are actually bounded sinusoids at earth rate frequency. In the more general case studied earlier when transport rate was not treated as negligible (Equations (13.5-10) - (13.5-17)), the transport rate also plays a part in the long term error bounding, depending on whether the transport rate is adding to or subtracting from earth rate, and the orientation of the great circle flight path relative to the earth's polar axis. Thus, overall system performance measured as average position error drift per navigation period (i.e., nautical miles error per hour) is actually improved for longer navigation periods. This basic effect can be directly traced to the strapdown inertial sensors maintaining a stationary attitude relative to the local flight path direction and thereby tracing a circular path. As we have seen in Section 13.4.3, navigating along a circular path attenuates the long term position drift for the inertial components whose input axes are rotated by sensor assembly angular motion following the flight path.

13.6 NAVIGATION ERROR FROM SENSOR OUTPUT RANDOM NOISE DURING NAVIGATION

Thus far we have analyzed the response of the strapdown navigation error equations to constant inertial sensor errors and initial position, velocity, attitude uncertainties during the navigation period. In this section we analyze the navigation error response to inertial component output random noise (i.e., the $\delta\underline{\omega}_{Rand}$, $\delta\underline{a}_{Rand}$ terms in Equations (12.5.6-1)) for the up-to-two-hour and short term cases analyzed in Sections 13.3.2 and 13.3.1. The method we will use to develop analytical solutions to the differential navigation error equations is based on treating the sensor noise terms as a sequence of random impulses, each creating a change in particular navigation error terms at the instant it occurs, which then propagate forward in time as did the initial navigation error terms in Sections 13.3.1 - 13.3.2. The complete navigation error response at any particular time during navigation is then obtained as a convolution of all the random noise impulse effects up to that point in time.

13.6.1 HORIZONTAL SOLUTION TO RANDOM NOISE FOR UP TO TWO HOURS WITH CONTROLLED VERTICAL CHANNEL

As outlined in Section 13.6, the navigation error response to random noise is derived from the response to initial navigation error uncertainties; specifically the response to initial velocity and attitude errors. For the one to two hour case with controlled vertical channel, the response to initial condition uncertainties can be obtained from the Section 13.3.2 analytical navigation error solution as provided by Equations (13.3.2-25). For this section we will analyze navigation errors under constant velocity conditions, therefore, the $\Delta\underline{v}_{SF_H}^{N}$ velocity change in Equations (13.3.2-25) will be equated to zero and the $\underline{v}_{H_0}^{N}$ initial horizontal velocity will be equated simply to the horizontal velocity $\underline{v}_{H}^{N}$ (assumed constant). Retaining only the Equation (13.3.2-25) response to initial velocity/attitude errors and selectively substituting for the Schuler frequency ω_S from (13.3.2-5) then yields for the navigation errors:

$$\gamma_{ZN} = \gamma_{ZN_0}$$

$$\underline{\gamma}_H^N = \underline{\gamma}_{H_0}^N \cos\omega_S t - \gamma_{ZN_0}\left(\underline{\omega}_{IE_H}^N \times \underline{u}_{ZN}^N + \frac{1}{R}\,\underline{v}_H^N\right)\frac{\sin\omega_S t}{\omega_S} + \frac{1}{R}\left(\underline{u}_{ZN}^N \times \delta\underline{V}_{H_0}^N\right)\frac{\sin\omega_S t}{\omega_S}$$

$$\delta\underline{V}_H^N = \delta\underline{V}_{H_0}^N \cos\omega_S t + g\left(\underline{u}_{ZN}^N \times \underline{\gamma}_{H_0}^N\right)\frac{\sin\omega_S t}{\omega_S} - \gamma_{ZN_0} R\left(\underline{\omega}_{IE_H}^N + \frac{1}{R}\,\underline{u}_{ZN}^N \times \underline{v}_H^N\right)(1 - \cos\omega_S t) \tag{13.6.1-1}$$

$$\delta\underline{R}_H^N = \delta\underline{V}_{H_0}^N \frac{\sin\omega_S t}{\omega_S} + R\left(\underline{u}_{ZN}^N \times \underline{\gamma}_{H_0}^N\right)(1 - \cos\omega_S t) - \gamma_{ZN_0} R\left(\underline{\omega}_{IE_H}^N + \frac{1}{R}\,\underline{u}_{ZN}^N \times \underline{v}_H^N\right)\left(t - \frac{1}{\omega_S}\sin\omega_S t\right)$$

We note in passing that Equations (13.3.2-25) from which (13.6.1-1) was derived, were based on the assumption that the B to N Frame attitude (as manifested in C_B^N) is constant. Since C_B^N does not appear in (13.6.1-1), this limitation does not apply, and Equations (13.6.1-1) are also valid for arbitrary time varying C_B^N.

Equations (13.6.1-1) represent the velocity/attitude error initial condition response of navigation error differential Equations (13.3.2-1):

$$\dot{\gamma}_{ZN} = -\left(C_B^N\,\delta\underline{\omega}_{IB}^B\right)_{ZN}$$

$$\dot{\underline{\gamma}}_H^N \approx -\left(C_B^N\,\delta\underline{\omega}_{IB}^B\right)_H - \gamma_{ZN}\left(\underline{\omega}_{IE_H}^N \times \underline{u}_{ZN}^N + \frac{1}{R}\,\underline{v}_H^N\right) + \frac{1}{R}\left(\underline{u}_{ZN}^N \times \delta\underline{V}_H^N\right)$$

$$\delta\dot{\underline{V}}_H^N \approx \left(C_B^N\,\delta\underline{a}_{SF}^B\right)_H + g\,\underline{u}_{ZN}^N \times \underline{\gamma}_H^N \tag{13.6.1-2}$$

$$\delta\dot{\underline{R}}_H^N = \delta\underline{V}_H^N$$

The inertial sensor errors we are analyzing in this section ($\delta\underline{\omega}_{IB}^B$, $\delta\underline{a}_{SF}^B$) are the random components which from (12.5.6-1) are given by:

$$\delta\underline{\omega}_{IB}^B = \delta\underline{\omega}_{Rand} \qquad \delta\underline{a}_{SF}^B = \delta\underline{a}_{Rand} \tag{13.6.1-3}$$

To analyze the response of (13.6.1-2) to the (13.6.1-3) sensor errors, we treat $\underline{\delta\omega}_{Rand}$, $\underline{\delta a}_{Rand}$ as a sequence of random impulses and analyze the response of (13.6.1-2) to each. Each impulse is integrated in Equations (13.6.1-2), producing an instantaneous change in velocity/attitude at the time the impulse occurs. From Equations (13.6.1-2) with (3.1.1-12) we see that the velocity/attitude changes are given by:

$$\begin{aligned}
\Delta\gamma_{ZN_i} &= -\,\underline{u}^N_{ZN} \cdot \left(C^{N_i}_{B_i}\,\underline{\varepsilon}_{\omega Rnd_i}\right) = -\left(\underline{u}^N_{ZN}\right)^T \left(C^{N_i}_{B_i}\,\underline{\varepsilon}_{\omega Rnd_i}\right) \\
\Delta\underline{\gamma}^N_{H_i} &= -\left(C^{N_i}_{B_i}\,\underline{\varepsilon}_{\omega Rnd_i}\right)_H = -\left(C^{N_i}_{B_i}\right)_H \underline{\varepsilon}_{\omega Rnd_i} \\
\Delta\delta\underline{V}^N_{H_i} &= \left(C^{N_i}_{B_i}\,\underline{\varepsilon}_{aRnd_i}\right)_H = \left(C^{N_i}_{B_i}\right)_H \underline{\varepsilon}_{aRnd_i}
\end{aligned} \tag{13.6.1-4}$$

where

$\underline{\varepsilon}_{\omega Rnd_i}$, $\underline{\varepsilon}_{aRnd_i}$ = Random impulse components of $\underline{\delta\omega}_{Rand}$, $\underline{\delta a}_{Rand}$ occurring at time instant i.

$C^{N_i}_{B_i}$ = Value of C^N_B at time instant i.

$\left(C^{N_i}_{B_i}\right)_H$ = $C^{N_i}_{B_i}$ with the third row (representing the vertical) set to zero.

$\Delta\gamma_{ZN_i}$, $\Delta\underline{\gamma}^N_{H_i}$, $\Delta\delta\underline{V}^N_{H_i}$ = Changes in γ_{ZN}, $\underline{\gamma}^N_H$, $\delta\underline{V}^N_H$ produced at time instant i by $\underline{\varepsilon}_{\omega Rnd_i}$ and $\underline{\varepsilon}_{aRnd_i}$.

Let us now analyze the effect of $\Delta\gamma_{ZN_i}$, $\Delta\underline{\gamma}^N_{H_i}$, $\Delta\delta\underline{V}^N_{H_i}$ in Equations (13.6.1-4) on $\delta\underline{V}^N_H$. Clearly, the effect is to create a $\delta\underline{V}^N_H$ corresponding with the Equations (13.6.1-1) $\delta\underline{V}^N_H$ response to γ_{ZN_0}, $\underline{\gamma}^N_{H_0}$, $\delta\underline{V}^N_{H_0}$, but with the time variable t replaced by the time interval from when $\Delta\gamma_{ZN_i}$, $\Delta\underline{\gamma}^N_{H_i}$ and $\Delta\delta\underline{V}^N_{H_i}$ were created. Thus, from the $\delta\underline{V}^N_H$ expression in (13.6.1-1) we have:

$$\begin{aligned}
\Delta_i\delta\underline{V}^N_H(t) = {} & \Delta\delta\underline{V}^N_{H_i} \cos\omega_S\,(t-\tau_i) + g\left(\underline{u}^N_{ZN} \times \Delta\underline{\gamma}^N_{H_i}\right)\frac{\sin\omega_S\,(t-\tau_i)}{\omega_S} \\
& - \Delta\gamma_{ZN_i}\,R\left(\underline{\omega}^N_{IE_H} + \frac{1}{R}\,\underline{u}^N_{ZN} \times \underline{v}^N_H\right)\left(1 - \cos\omega_S\,(t-\tau_i)\right)
\end{aligned} \tag{13.6.1-5}$$

where

τ_i = Time that $\Delta\gamma_{ZN_i}$, $\Delta\underline{\gamma}^N_{H_i}$ and $\Delta\delta\underline{V}^N_{H_i}$ were created due to $\underline{\varepsilon}_{\omega Rnd_i}$, $\underline{\varepsilon}_{aRnd_i}$.

$\Delta_i \delta \underline{V}_H^N(t)$ = Change in $\delta \underline{V}_H^N$ at time t produced by $\Delta \gamma_{ZN_i}$, $\Delta \underline{\gamma}_{H_i}^N$, $\Delta \delta \underline{V}_{H_i}^N$ created at time τ_i.

Equation (13.6.1-5) can also be expressed in the more compact form:

$$\Delta_i \delta \underline{V}_H^N(t) = \Phi_{V_H V_H}(t, \tau_i)\, \Delta \delta \underline{V}_{H_i}^N + \Phi_{V_H \gamma_H}(t, \tau_i)\, \Delta \underline{\gamma}_{H_i}^N + \Phi_{V_H \gamma_Z}(t, \tau_i)\, \Delta \gamma_{ZN_i} \tag{13.6.1-6}$$

in which, using general Equation (3.1.1-13):

$$\begin{aligned} \Phi_{V_H V_H}(t,\tau_i) &= \cos \omega_S (t-\tau_i) \\ \Phi_{V_H \gamma_H}(t,\tau_i) &= g \frac{\sin \omega_S (t-\tau_i)}{\omega_S} \left(\underline{u}_{ZN}^N \times \right) \\ \Phi_{V_H \gamma_Z}(t,\tau_i) &= - R \left(\underline{\omega}_{IE_H}^N + \frac{1}{R} \underline{u}_{ZN}^N \times \underline{v}_H^N \right) \left(1 - \cos \omega_S (t-\tau_i)\right) \end{aligned} \tag{13.6.1-7}$$

where

$\Phi_{V_H k}(t,\tau_i)$ = "Error state transition" element defining the response of $\delta \underline{V}_H^N$ at time t to initial conditions on variable k (i.e., γ_{ZN}, $\underline{\gamma}_H^N$ or $\delta \underline{V}_H^N$) at time τ_i.

The total $\delta \underline{V}_H^N$ response to random noise is the sum at time t of all the $\Delta_i \delta \underline{V}_H^N(t)$ responses:

$$\delta \underline{V}_H^N(t) = \sum_{i=1}^{n} \Delta_i \delta \underline{V}_H^N(t) \tag{13.6.1-8}$$

where

n = Total number of τ_i's since navigation time t = 0.

Equations (13.6.1-6) - (13.6.1-8) with (13.6.1-4) comprise what is known as a "deterministic" set from which $\delta \underline{V}_H^N$ can be evaluated for a particular $\underline{\varepsilon}_{\omega Rnd_i}$, $\underline{\varepsilon}_{aRnd_i}$ sequence. Since $\underline{\varepsilon}_{\omega Rnd_i}$, $\underline{\varepsilon}_{aRnd_i}$ is random by nature, every real navigation "run" will have a different $\underline{\varepsilon}_{\omega Rnd_i}$, $\underline{\varepsilon}_{aRnd_i}$ history, hence, a different $\delta \underline{V}_H^N$ response. In order to characterize the $\delta \underline{V}_H^N$ response for an ensemble of navigation runs, we resort to the statistical error definition:

$$P_{V_H V_H}(t) \equiv \mathcal{E} \left[\delta \underline{V}_H^N(t) \left(\delta \underline{V}_H^N(t) \right)^T \right] \tag{13.6.1-9}$$

where

$\mathcal{E}(\,)$ = Expected value operator representing the ensemble average of the quantity in brackets.

$P_{V_H V_H}(t)$ = Horizontal velocity error covariance matrix at time t.

The component form of (13.6.1-9) is:

$$P_{V_H V_H}(t) = \begin{bmatrix} \mathcal{E}(\delta V_{XN}^2) & \mathcal{E}(\delta V_{XN}\,\delta V_{YN}) & 0 \\ \mathcal{E}(\delta V_{YN}\,\delta V_{XN}) & \mathcal{E}(\delta V_{YN}^2) & 0 \\ 0 & 0 & 0 \end{bmatrix} \tag{13.6.1-10}$$

where

$\delta V_{XN}, \delta V_{YN}$ = N Frame X, Y components of $\delta \underline{V}_H^N$. Note, that the Z (vertical) term is zero because $\delta \underline{V}_H^N$ represents horizontal velocity error components.

From (13.6.1-10) we see that the diagonal elements of $P_{V_H V_H}(t)$ equal the average square of the $\delta \underline{V}_H^N$ components, each known as the component "variance" about the mean, and which for the assumed random noise processes, the mean (or simple linear average) is zero. The square root of each diagonal element is called the "standard deviation".

Let us now expand upon Equation (13.6.1-9) using (13.6.1-8) for $\delta \underline{V}_H^N(t)$:

$$\begin{aligned} P_{V_H V_H}(t) &= \mathcal{E}\left[\left(\sum_{i=1}^{n} \Delta_i \delta \underline{V}_H^N(t)\right)\left(\sum_{i=1}^{n} \Delta_i \delta \underline{V}_H^N(t)\right)^T\right] \\ &= \mathcal{E}\left[\left(\sum_{i=1}^{n} \Delta_i \delta \underline{V}_H^N(t)\right)\left(\sum_{j=1}^{n} \left(\Delta_j \delta \underline{V}_H^N(t)\right)^T\right)\right] \\ &= \sum_{i=1}^{n} \mathcal{E}\left[\Delta_i \delta \underline{V}_H^N(t) \left(\Delta_i \delta \underline{V}_H^N(t)\right)^T\right] + \sum_{\substack{i=1 \\ i \neq j}}^{n} \sum_{j=1}^{n} \mathcal{E}\left[\Delta_i \delta \underline{V}_H^N(t) \left(\Delta_j \delta \underline{V}_H^N(t)\right)^T\right] \end{aligned} \tag{13.6.1-11}$$

We formally characterize the $\underline{\varepsilon}_{\omega Rnd_i}$, $\underline{\varepsilon}_{aRnd_i}$ noise impulses as having a zero ensemble mean and to be statistically independent from one i to any other i. Then the second term in (13.6.1-11) averages to zero and (13.6.1-11) reduces to:

$$P_{V_H V_H}(t) = \sum_{i=1}^{n} \mathcal{E}\left[\Delta_i \delta \underline{V}_H^N(t) \left(\Delta_i \delta \underline{V}_H^N(t) \right)^T \right] \tag{13.6.1-12}$$

Substituting (13.6.1-6) into the (13.6.1-12) argument, expanding and equating the average product of different noise components to zero (because of their assumed independence), and noting the form of the Φ elements defined in (13.6.1-7), obtains:

$$\begin{aligned}
&\mathcal{E}\left[\Delta_i \delta \underline{V}_H^N(t) \left(\Delta_i \delta \underline{V}_H^N(t) \right)^T \right] \\
&= \mathcal{E}\Big[\left(\Phi_{V_H V_H}(t, \tau_i)\, \Delta \delta \underline{V}_{H_i}^N + \Phi_{V_H \gamma_H}(t, \tau_i)\, \Delta \underline{\gamma}_{H_i}^N + \Phi_{V_H \gamma_Z}(t, \tau_i)\, \Delta \gamma_{ZN_i} \right) \\
&\qquad \left(\Phi_{V_H V_H}(t, \tau_i)\, \Delta \delta \underline{V}_{H_i}^N + \Phi_{V_H \gamma_H}(t, \tau_i)\, \Delta \underline{\gamma}_{H_i}^N + \Phi_{V_H \gamma_Z}(t, \tau_i)\, \Delta \gamma_{ZN_i} \right)^T \Big] \\
&= \mathcal{E}\left[\left(\Phi_{V_H V_H}(t, \tau_i)\, \Delta \delta \underline{V}_{H_i}^N \right) \left(\Phi_{V_H V_H}(t, \tau_i)\, \Delta \delta \underline{V}_{H_i}^N \right)^T \right] \\
&\qquad + \mathcal{E}\left[\left(\Phi_{V_H \gamma_H}(t, \tau_i)\, \Delta \underline{\gamma}_{H_i}^N \right) \left(\Phi_{V_H \gamma_H}(t, \tau_i)\, \Delta \underline{\gamma}_{H_i}^N \right)^T \right] \\
&\qquad + \mathcal{E}\left[\left(\Phi_{V_H \gamma_Z}(t, \tau_i)\, \Delta \gamma_{ZN_i} \right) \left(\Phi_{V_H \gamma_Z}(t, \tau_i)\, \Delta \gamma_{ZN_i} \right)^T \right] \\
&= \Phi_{V_H V_H}^2(t, \tau_i)\, \mathcal{E}\left[\Delta \delta \underline{V}_{H_i}^N \left(\Delta \delta \underline{V}_{H_i}^N \right)^T \right] \\
&\qquad + \Phi_{V_H \gamma_H}(t, \tau_i)\, \mathcal{E}\left[\Delta \underline{\gamma}_{H_i}^N \left(\Delta \underline{\gamma}_{H_i}^N \right)^T \right] \left(\Phi_{V_H \gamma_H}(t, \tau_i) \right)^T \\
&\qquad + \mathcal{E}\left(\Delta \gamma_{ZN_i} \right)^2 \Phi_{V_H \gamma_Z}(t, \tau_i) \left(\Phi_{V_H \gamma_Z}(t, \tau_i) \right)^T
\end{aligned} \tag{13.6.1-13}$$

The expected value terms in (13.6.1-13) can be evaluated as functions of sensor noise using (13.6.1-4):

$$\begin{aligned}
&\mathcal{E}\left(\Delta \gamma_{ZN_i} \right)^2 = \left(\underline{u}_{ZN}^N \right)^T C_{B_i}^{N_i}\, \mathcal{E}\left[\underline{\varepsilon}_{\omega Rnd_i} \left(\underline{\varepsilon}_{\omega Rnd_i} \right)^T \right] \left(C_{B_i}^{N_i} \right)^T \underline{u}_{ZN}^N \\
&\mathcal{E}\left[\Delta \underline{\gamma}_{H_i}^N \left(\Delta \underline{\gamma}_{H_i}^N \right)^T \right] = \left(C_{B_i}^{N_i} \right)_H \mathcal{E}\left[\underline{\varepsilon}_{\omega Rnd_i} \left(\underline{\varepsilon}_{\omega Rnd_i} \right)^T \right] \left[\left(C_{B_i}^{N_i} \right)_H \right]^T \\
&\mathcal{E}\left[\Delta \delta \underline{V}_{H_i}^N \left(\Delta \delta \underline{V}_{H_i}^N \right)^T \right] = \left(C_{B_i}^{N_i} \right)_H \mathcal{E}\left[\underline{\varepsilon}_{aRnd_i} \left(\underline{\varepsilon}_{aRnd_i} \right)^T \right] \left[\left(C_{B_i}^{N_i} \right)_H \right]^T
\end{aligned} \tag{13.6.1-14}$$

To simplify the analysis we make further assumptions regarding the $\underline{\varepsilon}_{\omega Rnd_i}$, $\underline{\varepsilon}_{aRnd_i}$ angular rate sensor triad and accelerometer triad random noise vectors. Specifically, we assume each inertial sensor in a sensor triad has the same output noise variance, and that the output noise from a given sensor in a triad is independent from output noise generated by other sensors in the triad. Then the off-diagonal terms in the (13.6.1-14) expected value terms are zero and the diagonal elements are equal to the individual sensor type variance:

$$\mathcal{E}\left[\underline{\varepsilon}_{\omega Rnd_i}\left(\underline{\varepsilon}_{\omega Rnd_i}\right)^T\right] = I\,\mathcal{E}\left(\varepsilon^2_{\omega Rnd_i}\right)$$
$$\mathcal{E}\left[\underline{\varepsilon}_{aRnd_i}\left(\underline{\varepsilon}_{aRnd_i}\right)^T\right] = I\,\mathcal{E}\left(\varepsilon^2_{aRnd_i}\right) \qquad (13.6.1\text{-}15)$$

where

$\mathcal{E}\left(\varepsilon^2_{\omega Rnd_i}\right), \mathcal{E}\left(\varepsilon^2_{aRnd_i}\right)$ = Individual angular rate sensor and accelerometer random output noise variances at time instant i.

I = Identity matrix.

The following identity will also prove useful:

$$\left(C^N_B\right)_H I \left[\left(C^N_B\right)_H\right]^T = \left(C^N_B\right)_H \left[\left(C^N_B\right)_H\right]^T = \left\{\left(C^N_B\right)\left[\left(C^N_B\right)_H\right]^T\right\}_H$$
$$= \left\{\left[\left(C^N_B\right)_H \left(C^N_B\right)^T\right]^T\right\}_H = \left\langle\left\{\left[C^N_B \left(C^N_B\right)^T\right]_H\right\}^T\right\rangle_H = \left\{\left[I_H\right]^T\right\}_H = I_H \qquad (13.6.1\text{-}16)$$

where

I_H = Identity matrix with element 3,3 (the vertical component) set to zero.

Substituting (13.6.1-15) into (13.6.1-14) and applying (13.6.1-16) then gives:

$$\mathcal{E}\left[\Delta\underline{\gamma}^N_{H_i}\left(\Delta\underline{\gamma}^N_{H_i}\right)^T\right] = \left(C^{N_i}_{B_i}\right)_H \left[I\,\mathcal{E}\left(\varepsilon^2_{\omega Rnd_i}\right)\right]\left[\left(C^{N_i}_{B_i}\right)_H\right]^T$$
$$= \left(C^{N_i}_{B_i}\right)_H \left[\left(C^{N_i}_{B_i}\right)_H\right]^T \mathcal{E}\left(\varepsilon^2_{\omega Rnd_i}\right) = I_H\,\mathcal{E}\left(\varepsilon^2_{\omega Rnd_i}\right)$$

$$\mathcal{E}\left(\Delta\gamma_{ZN_i}\right)^2 = \left(\underline{u}^N_{ZN}\right)^T C^{N_i}_{B_i}\left[I\,\mathcal{E}\left(\varepsilon^2_{\omega Rnd_i}\right)\right]\left(C^{N_i}_{B_i}\right)^T \underline{u}^N_{ZN} \qquad (13.6.1\text{-}17)$$
$$= \left(\underline{u}^N_{ZN}\right)^T C^{N_i}_{B_i}\left(C^{N_i}_{B_i}\right)^T \underline{u}^N_{ZN}\,\mathcal{E}\left(\varepsilon^2_{\omega Rnd_i}\right) = \mathcal{E}\left(\varepsilon^2_{\omega Rnd_i}\right)$$

$$\mathcal{E}\left[\Delta\delta\underline{V}^N_{H_i}\left(\Delta\delta\underline{V}^N_{H_i}\right)^T\right] = I_H\,\mathcal{E}\left(\varepsilon^2_{aRnd_i}\right)$$

With (13.6.1-17) and the form of the Φ's from (13.6.1-7), Equation (13.6.1-13) becomes:

$$\mathcal{E}\left[\Delta_i\delta\underline{V}^N_H(t)\left(\Delta_i\delta\underline{V}^N_H(t)\right)^T\right] = \Phi^2_{V_HV_H}(t,\tau_i)\,I_H\,\mathcal{E}\left(\varepsilon^2_{aRnd_i}\right)$$
$$+ \Phi_{V_H\gamma_H}(t,\tau_i)\,I_H\left(\Phi_{V_H\gamma_H}(t,\tau_i)\right)^T \mathcal{E}\left(\varepsilon^2_{\omega Rnd_i}\right) \qquad (13.6.1\text{-}18)$$
$$+ \Phi_{V_H\gamma_Z}(t,\tau_i)\left(\Phi_{V_H\gamma_Z}(t,\tau_i)\right)^T \mathcal{E}\left(\varepsilon^2_{\omega Rnd_i}\right)$$

(Continued)

$$= \Phi_{V_H V_H}^2(t,\tau_i)\, I_H\, \mathcal{E}\left(\varepsilon_{aRnd_i}^2\right) + \Phi_{V_H \gamma_H}(t,\tau_i)\left(\Phi_{V_H \gamma_H}(t,\tau_i)\right)^T \mathcal{E}\left(\varepsilon_{\omega Rnd_i}^2\right) + \Phi_{V_H \gamma_Z}(t,\tau_i)\left(\Phi_{V_H \gamma_Z}(t,\tau_i)\right)^T \mathcal{E}\left(\varepsilon_{\omega Rnd_i}^2\right) \qquad \text{(13.6.1-18) (Continued)}$$

Substituting (13.6.1-18) into (13.6.1-12) with clever insertion of $\Delta\tau_i$, we then find for $P_{V_H V_H}(t)$:

$$P_{V_H V_H}(t) = \sum_{i=1}^{n} \left[\Phi_{V_H V_H}^2(t,\tau_i)\, I_H \frac{1}{\Delta\tau_i} \mathcal{E}\left(\varepsilon_{aRnd_i}^2\right) \Delta\tau_i + \Phi_{V_H \gamma_H}(t,\tau_i)\left(\Phi_{V_H \gamma_H}(t,\tau_i)\right)^T \frac{1}{\Delta\tau_i} \mathcal{E}\left(\varepsilon_{\omega Rnd_i}^2\right) \Delta\tau_i + \Phi_{V_H \gamma_Z}(t,\tau_i)\left(\Phi_{V_H \gamma_Z}(t,\tau_i)\right)^T \frac{1}{\Delta\tau_i} \mathcal{E}\left(\varepsilon_{\omega Rnd_i}^2\right) \Delta\tau_i \right] \qquad \text{(13.6.1-19)}$$

where

$\Delta\tau_i$ = Time interval over which $\underline{\varepsilon}_{\omega Rnd_i}$, $\underline{\varepsilon}_{aRnd_i}$ operate.

Equation (13.6.1-19) is now in classical form for reduction to an integral expression if we introduce the following notation:

$$q_{aRnd}(\tau) \equiv \lim \left[\frac{1}{\Delta\tau_i} \mathcal{E}\left(\varepsilon_{aRnd_i}^2\right) \right]_{\Delta\tau_i \to 0}$$

$$q_{\omega Rnd}(\tau) \equiv \lim \left[\frac{1}{\Delta\tau_i} \mathcal{E}\left(\varepsilon_{\omega Rnd_i}^2\right) \right]_{\Delta\tau_i \to 0} \qquad \text{(13.6.1-20)}$$

$$\int_{\tau=0}^{\tau=t} f(\tau)\, d\tau = \lim \left(\sum_{i=1}^{n} f(\tau_i)\, \Delta\tau_i \right)_{\Delta\tau_i \to 0}$$

where

$q_{\omega Rnd}(\tau)$, $q_{aRnd}(\tau)$ = Angular rate sensor and accelerometer random output noise variance densities.

With (13.6.1-20), Equation (13.6.1-19) becomes in the limit as $\Delta\tau_i \to 0$:

$$P_{V_H V_H}(t) = I_H \int_{\tau=0}^{\tau=t} \Phi^2_{V_H V_H}(t,\tau)\, q_{aRnd}(\tau)\, d\tau$$
$$+ \int_{\tau=0}^{\tau=t} \Phi_{V_H \gamma_H}(t,\tau) \left(\Phi_{V_H \gamma_H}(t,\tau)\right)^T q_{\omega Rnd}(\tau)\, d\tau \qquad (13.6.1\text{-}21)$$
$$+ \int_{\tau=0}^{\tau=t} \Phi_{V_H \gamma_Z}(t,\tau) \left(\Phi_{V_H \gamma_Z}(t,\tau)\right)^T q_{\omega Rnd}(\tau)\, d\tau$$

For constant inertial sensor process noise densities, Equation (13.6.1-21) can be explicitly evaluated by substitution of (13.6.1-7) for the error state transition elements, recognizing that the transpose of a cross-product operator is its negative, and employing generalized Equation (13.1-8):

$$\int_{\tau=0}^{\tau=t} \Phi^2_{V_H V_H}(t,\tau)\, d\tau = \int_{\tau=0}^{\tau=t} \cos^2 \omega_S(t-\tau)\, d\tau = \frac{1}{2}\left(t + \frac{\sin 2\,\omega_S t}{2\,\omega_S}\right)$$

$$\int_{\tau=0}^{\tau=t} \Phi_{V_H \gamma_H}(t,\tau) \left(\Phi_{V_H \gamma_H}(t,\tau)\right)^T d\tau = -\, g^2 \left(\underline{u}^N_{ZN}\times\right)^2 \frac{1}{\omega_S^2} \int_{\tau=0}^{\tau=t} \sin^2 \omega_S (t-\tau)\, d\tau$$
$$= I_H \frac{g^2}{\omega_S^2} \frac{1}{2}\left(t - \frac{\sin 2\,\omega_S t}{2\,\omega_S}\right)$$

$$\int_{\tau=0}^{\tau=t} \Phi_{V_H \gamma_Z}(t,\tau) \left(\Phi_{V_H \gamma_Z}(t,\tau)\right)^T d\tau \qquad (13.6.1\text{-}22)$$

$$= R^2 \left(\underline{\omega}^N_{IE_H} + \frac{1}{R}\, \underline{u}^N_{ZN} \times \underline{v}^N_H\right)\left(\underline{\omega}^N_{IE_H} + \frac{1}{R}\, \underline{u}^N_{ZN} \times \underline{v}^N_H\right)^T \int_{\tau=0}^{\tau=t} \left(1 - \cos \omega_S(t-\tau)\right)^2 d\tau$$

$$= R^2 \left(\underline{\omega}^N_{IE_H} + \frac{1}{R}\, \underline{u}^N_{ZN} \times \underline{v}^N_H\right)\left(\underline{\omega}^N_{IE_H} + \frac{1}{R}\, \underline{u}^N_{ZN} \times \underline{v}^N_H\right)^T \int_{\tau=0}^{\tau=t} \left(1 - 2\cos \omega_S(t-\tau) + \cos^2 \omega_S(t-\tau)\right) d\tau$$
$$= R^2 \left(\underline{\omega}^N_{IE_H} + \frac{1}{R}\, \underline{u}^N_{ZN} \times \underline{v}^N_H\right)\left(\underline{\omega}^N_{IE_H} + \frac{1}{R}\, \underline{u}^N_{ZN} \times \underline{v}^N_H\right)^T \left(\frac{3}{2}\, t - \frac{2 \sin \omega_S t}{\omega_S} + \frac{\sin 2\,\omega_S t}{4\,\omega_S}\right)$$

Finally, we substitute (13.6.1-22) into (13.6.1-21) to obtain the analytical solution for $P_{V_H V_H}(t)$ under constant sensor noise density conditions:

$$
\begin{aligned}
P_{V_H V_H}(t) &= I_H \frac{1}{2}\left(t + \frac{\sin 2\,\omega_S t}{2\,\omega_S}\right) q_{aRnd} \\
&+ \left\{ I_H \frac{g^2}{\omega_S^2}\frac{1}{2}\left(t - \frac{\sin 2\,\omega_S t}{2\,\omega_S}\right) + R^2 \left(\underline{\omega}_{IE_H}^{N} + \frac{1}{R}\,\underline{u}_{ZN}^{N} \times \underline{v}_{H}^{N}\right)\left(\underline{\omega}_{IE_H}^{N} + \frac{1}{R}\,\underline{u}_{ZN}^{N} \times \underline{v}_{H}^{N}\right)^T \left(\frac{3}{2}\,t \right.\right. \\
&\qquad \left.\left. - \frac{2 \sin \omega_S t}{\omega_S} + \frac{\sin 2\,\omega_S t}{4\,\omega_S}\right)\right\} q_{\omega Rnd}
\end{aligned}
\qquad (13.6.1\text{-}23)
$$

The identical process is used to obtain the attitude and horizontal position error covariances from Equations (13.6.1-1) and (13.6.1-4):

$$
P_{\gamma_Z \gamma_Z}(t) = t \; q_{\omega Rnd} \qquad (13.6.1\text{-}24)
$$

$$
\begin{aligned}
P_{\gamma_H \gamma_H}(t) &= I_H \frac{1}{R^2}\frac{1}{\omega_S^2}\frac{1}{2}\left(t - \frac{\sin 2\,\omega_S t}{2\,\omega_S}\right) q_{aRnd} \\
&+ \left\{ I_H \frac{1}{2}\left(t + \frac{\sin 2\,\omega_S t}{2\,\omega_S}\right) + \left(\underline{\omega}_{IE_H}^{N} \times \underline{u}_{ZN}^{N} + \frac{1}{R}\,\underline{v}_{H}^{N}\right)\left(\underline{\omega}_{IE_H}^{N} \times \underline{u}_{ZN}^{N} + \frac{1}{R}\,\underline{v}_{H}^{N}\right)^T \right. \\
&\qquad \left. \frac{1}{\omega_S^2}\frac{1}{2}\left(t - \frac{\sin 2\,\omega_S t}{2\,\omega_S}\right)\right\} q_{\omega Rnd}
\end{aligned}
\qquad (13.6.1\text{-}25)
$$

$$
\begin{aligned}
P_{R_H R_H}(t) &= I_H \frac{1}{\omega_S^2}\frac{1}{2}\left(t - \frac{\sin 2\,\omega_S t}{2\,\omega_S}\right) q_{aRnd} \\
&+ \left\{ I_H R^2 \left(\frac{3}{2}\,t - \frac{2 \sin \omega_S t}{\omega_S} + \frac{\sin 2\,\omega_S t}{4\,\omega_S}\right) + R^2 \left(\underline{\omega}_{IE_H}^{N} + \frac{1}{R}\,\underline{u}_{ZN}^{N} \times \underline{v}_{H}^{N}\right)\left(\underline{\omega}_{IE_H}^{N} \right.\right. \\
&\left.\left. + \frac{1}{R}\,\underline{u}_{ZN}^{N} \times \underline{v}_{H}^{N}\right)^T \left[\frac{1}{3}\,t^3 - \frac{2 \sin \omega_S t}{\omega_S^3} + \frac{2 \cos \omega_S t}{\omega_S^2}\,t + \frac{1}{\omega_S^2}\frac{1}{2}\left(t - \frac{\sin 2\,\omega_S t}{2\,\omega_S}\right)\right]\right\} q_{\omega Rnd}
\end{aligned}
\qquad (13.6.1\text{-}26)
$$

Equations (13.6.1-23) - (13.6.1-26) describe the attitude, velocity and position error response to inertial sensor random noise operating during the navigation period for up to two hour navigation times and for constant sensor noise densities. These equations do not include the effect of inertial sensor random noise operating during the initial alignment period (prior to navigation mode engagement) which impacts initial heading accuracy. The effect of random noise on the initial alignment process is addressed in Section 14.6 of Chapter 14.

It is also instructive to investigate the position error produced from inertial sensor random noise using the long term navigation position error approximation of Equation (13.2.3-4) to illustrate the inadequacy of this equation to properly account for the random noise effect. Equation (13.2.3-4) for the average long term position error $\delta \underline{R}_{H_{LngTrm}}^{N}$ using $\underline{\psi}^{N}$ from attitude error Equation (12.3.7.1-4) with the (3.2.1-5) chain rule is:

$$\delta \underline{R}_{H_{LngTrm}}^{N} = R \left(\underline{u}_{ZN}^{N} \times \underline{\psi}_{H}^{N} \right) \tag{13.6.1-27}$$

$$\dot{\underline{\psi}}^{I} = - C_{N}^{I} \, C_{B}^{N} \, \delta \underline{\omega}_{IB}^{B} \tag{13.6.1-28}$$

$$\underline{\psi}^{N} = \left(C_{N}^{I} \right)^{T} \underline{\psi}^{I} \tag{13.6.1-29}$$

The long term position variance is from (13.6.1-27) and (13.6.1-29):

$$\begin{aligned} P_{R_{HLng} R_{HLng}} &= \mathcal{E} \left[\delta \underline{R}_{H_{LngTrm}}^{N} \left(\delta \underline{R}_{H_{LngTrm}}^{N} \right)^{T} \right] \\ &= R^{2} \, \mathcal{E} \left[\left(\underline{u}_{ZN}^{N} \times \right) \underline{\psi}_{H}^{N} \left(\underline{\psi}_{H}^{N} \right)^{T} \left(\underline{u}_{ZN}^{N} \times \right)^{T} \right] \\ &= - R^{2} \, \mathcal{E} \left[\left(\underline{u}_{ZN}^{N} \times \right) \left(C_{N}^{I} \right)_{H}^{T} \underline{\psi}^{I} \left(\underline{\psi}^{I} \right)^{T} \left(C_{N}^{I} \right)_{H} \left(\underline{u}_{ZN}^{N} \times \right) \right] \\ &= - R^{2} \left(\underline{u}_{ZN}^{N} \times \right) \left(C_{N}^{I} \right)_{H}^{T} \mathcal{E} \left[\underline{\psi}^{I} \left(\underline{\psi}^{I} \right)^{T} \right] \left(C_{N}^{I} \right)_{H} \left(\underline{u}_{ZN}^{N} \times \right) \end{aligned} \tag{13.6.1-30}$$

where

$P_{R_{HLng} R_{HLng}}$ = Covariance matrix for $\delta \underline{R}_{H_{LngTrm}}^{N}$ as defined by Equation (13.6.1-27).

The $\mathcal{E} \left[\underline{\psi}^{I} \left(\underline{\psi}^{I} \right)^{T} \right]$ term in (13.6.1-30) produced by random sensor output noise is evaluated (as in the development of Equation (13.6.1-23)) by considering $\delta \underline{\omega}_{IB}^{B}$ in (13.6.1-28) as a sequence of random impulses, and $\underline{\psi}^{I}$ as the sum of the (13.6.1-28) responses to each ith impulse. As in the development of (13.6.1-23), we then treat each sensor in the angular rate sensor triad as having equal noise statistics, go to the limit to convert the summation to an integral, and set the angular rate sensor noise density constant:

$$\Delta \underline{\psi}_{i}^{I} = - C_{N_i}^{I} \, C_{B_i}^{N_i} \, \underline{\varepsilon}_{\omega Rnd_i} \tag{13.6.1-31}$$

$$\underline{\psi}^{I} = \sum_{i=1}^{n} \Delta \underline{\psi}_{i}^{I} = - \sum_{i=1}^{n} C_{N_i}^{I} \, C_{B_i}^{N_i} \, \underline{\varepsilon}_{\omega Rnd_i} \tag{13.6.1-32}$$

$$
\begin{aligned}
\mathcal{E}\left[\underline{\psi}^{I}\left(\underline{\psi}^{I}\right)^{T}\right] &= \mathcal{E}\left\{\sum_{i=1}^{n} C_{N_i}^{I}\, C_{B_i}^{N_i}\, \underline{\varepsilon}_{\omega Rnd_i} \sum_{j=1}^{n} \left(C_{N_j}^{I}\, C_{B_j}^{N_j}\, \underline{\varepsilon}_{\omega Rnd_j}\right)^{T}\right\} \\
&= \sum_{i=1}^{n} C_{N_i}^{I}\, C_{B_i}^{N_i}\, \mathcal{E}\left[\underline{\varepsilon}_{\omega Rnd_i}\left(\underline{\varepsilon}_{\omega Rnd_i}\right)^{T}\right]\left(C_{B_i}^{N_i}\right)^{T}\left(C_{N_i}^{I}\right)^{T} \\
&= \sum_{i=1}^{n} C_{N_i}^{I}\, C_{B_i}^{N_i}\, I\, \mathcal{E}\left(\varepsilon_{\omega Rnd_i}^{2}\right)\left(C_{B_i}^{N_i}\right)^{T}\left(C_{N_i}^{I}\right)^{T} = \sum_{i=1}^{n} I\, \mathcal{E}\left(\varepsilon_{\omega Rnd_i}^{2}\right) \\
&= I \sum_{i=1}^{n} \frac{1}{\Delta\tau_i}\, \mathcal{E}\left(\varepsilon_{\omega Rnd_i}^{2}\right) \Delta\tau_i = I \int_{\tau=0}^{\tau=t} q_{\omega Rnd}(\tau)\, d\tau = I\, t\, q_{\omega Rnd}
\end{aligned}
\qquad (13.6.1\text{-}33)
$$

With (13.6.1-33), Equation (13.6.1-30) for $P_{R_{HLng}R_{HLng}}$ becomes:

$$
\begin{aligned}
P_{R_{HLng}R_{HLng}} &= -R^{2}\left(\underline{u}_{ZN}^{N}\times\right)\left[\left(C_{N}^{I}\right)_{H}\right]^{T} I\, t\, q_{\omega Rnd}\left(C_{N}^{I}\right)_{H}\left(\underline{u}_{ZN}^{N}\times\right) \\
&= -R^{2}\left(\underline{u}_{ZN}^{N}\times\right) I_{H}\left(\underline{u}_{ZN}^{N}\times\right) t\, q_{\omega Rnd} = -R^{2}\left(\underline{u}_{ZN}^{N}\times\right)\left(\underline{u}_{ZN}^{N}\times\right) t\, q_{\omega Rnd}
\end{aligned}
\qquad (13.6.1\text{-}34)
$$

or, finally, with generalized Equation (13.1-8):

$$
P_{R_{HLng}R_{HLng}} = I_{H}\, R^{2}\, t\, q_{\omega Rnd} \qquad (13.6.1\text{-}35)
$$

If $P_{R_{HLng}R_{HLng}}$ Equation (13.6.1-35) is compared with $P_{R_H R_H}$ Equation (13.6.1-26), it will be seen that (13.6.1-26) contains an $I_H\, R^2 \frac{3}{2} t\, q_{\omega Rnd}$ term versus the comparable $I_H\, R^2\, t\, q_{\omega Rnd}$ term in (13.6.1-35) (i.e., 50% larger), Equation (13.6.1-26) contains additional Schuler terms which become negligible compared to $I_H\, R^2 \frac{3}{2} t\, q_{\omega Rnd}$ for long navigation times, and Equation (13.6.1-26) contains a $\frac{1}{3} t^3\, q_{\omega Rnd}$ term multiplied by earth-rate/transport-rate products, also small compared to $I_H\, R^2 \frac{3}{2} t\, q_{\omega Rnd}$ (at least for up to two hour navigation periods for which (13.6.1-26) is valid). The net result is that Equation (13.6.1-35) provides a solution that is 33% lower than the more accurate (13.6.1-26) result. The reason for the discrepancy is the effect of random Schuler oscillation build-up that is properly accounted for in (13.6.1-26) but which is ignored in the approximate long term position error Equation (13.6.1-27) from which (13.6.1-35) was derived.

13.6.2 SHORT TERM SOLUTION WITH FREE INERTIAL VERTICAL CHANNEL

In order to develop the navigation error response to random noise for short term navigation periods (e.g., less than 10 minutes) with a free inertial vertical channel, we begin (as in Section 13.6.1) with the response to initial velocity/attitude error from solution Equation (13.3.1-4), while setting $\Delta \underline{v}_{SF_H}^N$ to zero (no maneuvering), and equating $\underline{V}_{H_0}^N$ to $\underline{V}_H^N$ (assumed constant):

$$\gamma_{ZN} = \gamma_{ZN_0}$$

$$\underline{\gamma}_H^N \approx \underline{\gamma}_{H_0}^N - t\, \gamma_{ZN_0} \left(\underline{\omega}_{IE_H}^N \times \underline{u}_{ZN}^N + \frac{1}{R} \underline{v}_H^N \right) + t \frac{1}{R} \left(\underline{u}_{ZN}^N \times \delta \underline{V}_{H_0}^N \right)$$

$$\delta V_{ZN} = \delta V_{ZN_0}$$

$$\delta \underline{V}_H^N = \left(1 - \frac{1}{2} \frac{g}{R} t^2 \right) \delta \underline{V}_{H_0}^N + g\, t\, \underline{u}_{ZN}^N \times \underline{\gamma}_{H_0}^N - \gamma_{ZN_0} \frac{1}{2} g\, t^2 \left(\underline{\omega}_{IE_H}^N + \frac{1}{R} \underline{u}_{ZN}^N \times \underline{v}_H^N \right) \qquad (13.6.2\text{-}1)$$

$$\delta R = t\, \delta V_{ZN_0}$$

$$\delta \underline{R}_H^N = \left(t - \frac{1}{6} \frac{g}{R} t^3 \right) \delta \underline{V}_{H_0}^N + \frac{1}{2} g\, t^2 \left(\underline{u}_{ZN}^N \times \underline{\gamma}_{H_0}^N \right) - \gamma_{ZN_0} \frac{1}{6} g\, t^3 \left(\underline{\omega}_{IE_H}^N + \frac{1}{R} \underline{u}_{ZN}^N \times \underline{v}_H^N \right)$$

Equations (13.6.2-1) are the response of navigation error Equations (13.3.1-1) (repeated below) to initial velocity/attitude errors:

$$\dot{\gamma}_{ZN} = - \left(C_B^N \, \delta \underline{\omega}_{IB}^B \right)_{ZN}$$

$$\dot{\underline{\gamma}}_H^N \approx - \left(C_B^N \, \delta \underline{\omega}_{IB}^B \right)_H - \gamma_{ZN} \left(\underline{\omega}_{IE_H}^N \times \underline{u}_{ZN}^N + \frac{1}{R} \underline{v}_H^N \right) + \frac{1}{R} \left(\underline{u}_{ZN}^N \times \delta \underline{V}_{H_0}^N \right)$$

$$\delta \dot{V}_{ZN} = \left(C_B^N \, \delta \underline{a}_{SF}^B \right)_{ZN} + F(h) \frac{g}{R} \delta R_0$$

$$F(h) = 2 \quad \text{For} \quad h \geq 0 \qquad\qquad F(h) = -1 \quad \text{For} \quad h < 0 \qquad (13.6.2\text{-}2)$$

$$\delta \dot{\underline{V}}_H^N \approx \left(C_B^N \, \delta \underline{a}_{SF}^B \right)_H + g\, \underline{u}_{ZN}^N \times \underline{\gamma}_H^N$$

$$\delta \dot{R} = \delta V_{ZN}$$

$$\delta \dot{\underline{R}}_H^N = \delta \underline{V}_H^N$$

The response of Equations (13.6.2-2) to random $\delta \underline{\omega}_{IB}^B$, $\delta \underline{a}_{SF}^B$ inertial sensor errors can be derived following the identical procedure used in Section 13.6.1 (i.e., using (13.6.2-1) to calculate the response of individual sensor random noise impulses, then summing to obtain the total for all noise impulses). The process can be expedited by noting that the short term navigation error Equations (13.6.2-2) are identical to the up-to-two-hour navigation error Equations (13.6.1-2) for the attitude and horizontal velocity/position errors (except for the approximation of using $\delta \underline{V}_{H_0}^{N}$ for $\delta \underline{V}_{H}^{N}$) in the (13.6.2-2) $\dot{\gamma}_{H}^{N}$ expression). Hence, as an alternative derivation procedure for the attitude and horizontal position/velocity error response to sensor random noise, the short term response can be obtained from the up-to-two-hour solution (Equations (13.6.1-23) - (13.6.1-26)) by replacing the Schuler sinusoids with truncated Taylor series expansions and selective Schuler frequency (ω_S) substitution using (13.3.2-5). Using either procedure (both are useful exercises for validity checking one against the other), the following is the result:

$$
\begin{aligned}
P_{\gamma_Z \gamma_Z}(t) &= t\ q_{\omega Rnd} \\
P_{\gamma_H \gamma_H}(t) &= \frac{1}{R^2} I_H \frac{1}{3} t^3\ q_{aRnd} \\
&\quad + \left\{ I_H\, t + \left(\underline{\omega}_{IE_H}^{N} \times \underline{u}_{ZN}^{N} + \frac{1}{R} \underline{v}_H^{N} \right) \left(\underline{\omega}_{IE_H}^{N} \times \underline{u}_{ZN}^{N} + \frac{1}{R} \underline{v}_H^{N} \right)^{T} \frac{1}{3} t^3 \right\} q_{\omega Rnd} \\
P_{V_Z V_Z}(t) &= t\ q_{aRnd} \\
P_{V_H V_H}(t) &= \left(t - \frac{g}{R} \frac{1}{3} t^3 + \frac{g^2}{R^2} \frac{1}{20} t^5 \right) I_H\ q_{aRnd} \\
&\quad + \left\{ I_H \frac{1}{3} g^2 t^3 + \left(\underline{\omega}_{IE_H}^{N} + \frac{1}{R} \underline{u}_{ZN}^{N} \times \underline{v}_H^{N} \right) \left(\underline{\omega}_{IE_H}^{N} + \frac{1}{R} \underline{u}_{ZN}^{N} \times \underline{v}_H^{N} \right)^{T} \frac{1}{20} g^2 t^5 \right\} q_{\omega Rnd} \\
P_{RR}(t) &= \frac{1}{3} t^3\ q_{aRnd} \\
P_{R_H R_H}(t) &= \left(\frac{1}{3} t^3 - \frac{g}{R} \frac{1}{15} t^5 + \frac{g^2}{R^2} \frac{1}{252} t^7 \right) I_H\ q_{aRnd} \\
&\quad + \left\{ I_H \frac{1}{20} g^2 t^5 + \left(\underline{\omega}_{IE_H}^{N} + \frac{1}{R} \underline{u}_{ZN}^{N} \times \underline{v}_H^{N} \right) \left(\underline{\omega}_{IE_H}^{N} + \frac{1}{R} \underline{u}_{ZN}^{N} \times \underline{v}_H^{N} \right)^{T} \frac{1}{252} g^2 t^7 \right\} q_{\omega Rnd}
\end{aligned}
\tag{13.6.2-3}
$$

14 Quasi-Stationary Initialization Error Equations And Solutions

14.0 OVERVIEW

In Chapters 12 and 13 we analyzed the effect of inertial sensor and initial condition errors on strapdown inertial navigation system attitude, velocity and position error, treating the initial condition errors as independent variables. In fact, for most inertial systems, some of the initial condition errors (particularly attitude) are produced by the inertial sensors during the "Initial Alignment" mode of system operation (immediately prior to engaging the "Navigation Mode"). Section 6.1.2 described the final segment of Initial Alignment (identified as "Fine Alignment") under quasi-stationary environmental conditions in which the vehicle carrying the INS is stationary except for random (and bounded) position disturbances due to wind gusts, stores/fuel loading, passenger/crew boarding, etc. Section 15.2.2 and its subsections examine the more general situation in which the initial alignment process is performed on a moving base. In this chapter we analyze the quasi-stationary alignment process to assess the effect of inertial sensor and disturbance errors on initial attitude determination.

The following sections review the quasi-stationary Fine Alignment analytical process and derives a set of Fine Alignment error equations that establish the navigation mode attitude initialization error. Closed-form analytical solutions to the initialization error equations are then developed for three conditions:

- Constant inertial sensor errors.
- Linearly increasing horizontal accelerometer error.
- Random inertial sensor errors in a random quasi-stationary disturbance environment.

Preceding the section on random error sources is a discussion of the correlation between navigation errors caused by initial attitude and inertial sensor error, recognizing that the attitude initialization error is created (in part) by the same inertial sensors used during navigation.

The principal coordinate frames used in this chapter are the B, L, N, E and I Frames defined in Section 2.2.

14.1 FINE ALIGNMENT ANALYTICAL PROCESS EQUATIONS

We begin by restating Equations (6.1.2-2) governing the Fine Alignment process in the N Frame with the addition of a "Measurement" vector and a "Reference Position Divergence" vector:

$$
\begin{aligned}
\dot{C}_B^N &= C_B^N \left(\underline{\omega}_{IB}^B \times\right) - \left(\underline{\omega}_{IN}^N \times\right) C_B^N \\
\underline{\omega}_{IN}^N &= \underline{\omega}_{IE}^N + \underline{\omega}_{Tilt}^N \\
\underline{Z}^N &= \Delta\underline{R}_H^N - \Delta\underline{R}_{Ref_H}^N \\
\Delta\underline{R}_{Ref_H}^N &= 0 \\
\dot{\underline{\omega}}_{IE_H}^N &= K_1 \, \underline{u}_{ZN}^N \times \underline{Z}^N \\
\underline{\omega}_{Tilt}^N &= K_2 \, \underline{u}_{ZN}^N \times \underline{Z}^N \\
\underline{\omega}_{IE}^N &= \underline{\omega}_{IE_H}^N + \underline{u}_{ZN}^N \, \omega_e \sin l \\
\dot{\underline{v}}_H^N &= \left(C_B^N \, \underline{a}_{SF}^B\right)_H - K_3 \, \underline{Z}^N \\
\Delta\dot{\underline{R}}_H^N &= \underline{v}_H^N - K_4 \, \underline{Z}^N
\end{aligned}
\qquad (14.1\text{-}1)
$$

where

$\underline{Z}^N$ = The "measurement" vector used for alignment process feedback through the alignment process gains K_1, K_2, K_3, K_4. The form of Equation (14.1-1) with $\underline{Z}^N$ as defined, is a more general form than in Equations (6.1.2-2), and has been introduced here in anticipation of its use in Section 15.2.1 in which the process gains are developed from Kalman filter theory.

$\Delta\underline{R}_{Ref_H}^N$ = Horizontal reference position divergence representing the quasi-stationary dynamic movement of the INS. The value for $\Delta\underline{R}_{Ref_H}^N$ is approximated as zero (for lack of a measure of true horizontal position motion). The $\Delta\underline{R}_{Ref_H}^N$ concept will be used later when developing the alignment process error equations to account for the effect of dynamic disturbance on initial alignment error.

Equations (14.1-1) generate a C_B^N matrix at the end of Fine Alignment that is aligned to the local plumb-bob vertical, and provides N Frame horizontal earth rate components from which the C_B^N initial heading (orientation relative to true north) is obtained. Sections 6.2.1 and 6.2.2 discuss two methods for accounting for the C_B^N initial heading; 1. Initializing the wander angle in the C_N^E matrix to correspond with the determined C_B^N heading, or 2. Rotating the C_B^L matrix about the vertical at Fine Alignment completion through the negative of the determined heading so that heading in the rotated C_B^L is zero (Recall from Chapter 6 that the L and N Frames are defined to be parallel to one another, hence, adjusting the C_B^L heading to zero has the identical effect on C_B^N. From an analytical standpoint, we can consider the adjustment to be applied directly to C_B^N.) Each heading initialization technique is described analytically below from the standpoint of C_B^N (in contrast with C_B^L) initialization.

An analytical statement of the C_N^E heading initialization method is developed in Section 6.2.1 (Equation (6.2.1-6)) which shows how we can account for the C_B^N wander angle by initializing the C_N^E matrix second row with the components of:

$$\underline{u}_{YE}^N = \frac{\cos l}{\omega_{IE_H}} \underline{\omega}_{IE_H}^N + \underline{u}_{ZN}^N \sin l \qquad (14.1\text{-}2)$$

where

$\underline{u}_{YE}^N$ = Unit vector along the earth E Frame Y axis as projected on N Frame axes. From the definition of the E Frame in Section 2.2, the Y axis is along the earth polar rotation axis, hence, $\underline{u}_{YE}^E$ is along the earth polar axis.

$\underline{\omega}_{IE_H}^N$ = Horizontal earth rate components at completion of Fine Alignment as calculated from the Equations (14.1-1) dynamic process.

l = Initial geodetic latitude.

Equation (14.1-2) forms the basis for initializing the C_N^E matrix as the means to account for C_B^N initial wander angle.

The alternative wander angle initialization process (by which C_B^N is rotated around the vertical to zero out the wander angle) can be described analytically by defining C_B^N from the N Frame version of Equation (6.2.2-1):

$$C_B^{N+} = C_N^{N+} C_B^N \tag{14.1-3}$$

where

$N+$ = N Frame after application of the wander angle heading adjustment. The N+ Frame then becomes the initial N Frame for the navigation mode.

C_N^{N+} = Direction cosine matrix that transforms vectors from the N Frame (at completion of Fine Alignment) to the N+ Frame immediately after C_B^N wander angle adjustment.

C_B^{N+} = C_B^N matrix at the start of the Navigation Mode.

From Equations (6.2.2-3) with (6.2.2-4) - (6.2.2-6) we can write for C_N^{N+} in (14.1-3):

$$\begin{aligned}\left(C_N^{N+}\right)^T &= \begin{bmatrix} \dfrac{1}{\omega_{IE_H}} \underline{\omega}_{IE_H}^N \times \underline{u}_{ZN}^N & \dfrac{1}{\omega_{IE_H}} \underline{\omega}_{IE_H}^N & \underline{u}_{ZN}^N \end{bmatrix} \\ &= \begin{bmatrix} -\dfrac{1}{\omega_{IE_H}} \left(\underline{u}_{ZN}^N \times\right) \underline{\omega}_{IE_H}^N & \dfrac{1}{\omega_{IE_H}} \underline{\omega}_{IE_H}^N & \underline{u}_{ZN}^N \end{bmatrix}\end{aligned} \tag{14.1-4}$$

Equations (14.1-3) and (14.1-4) represent the alternative initialization method to account for wander angle by rotating C_B^N to set the wander angle to zero.

14.2 QUASI-STATIONARY INITIAL ALIGNMENT ERROR EQUATIONS

Error equations for the quasi-stationary initial alignment process are derived from Equations (14.1-1) using (14.1-2) or (14.1-3) - (14.1-4) for wander angle initialization. The derivation of the error equations begins with the differential of the $\underline{Z}^N$ Measurement in Equations (14.1-1) to obtain the error form:

$$\delta \underline{Z}^N = \delta \Delta \underline{R}_H^N - \delta \Delta \underline{R}_{Ref_H}^N \tag{14.2-1}$$

where

$\delta(\ )$ = Error in the bracketed quantity.

$\delta \Delta \underline{R}_{Ref_H}^N$ = Error in $\Delta \underline{R}_{Ref_H}^N$ caused by its approximation in (14.1-1) as zero.

The $\delta\Delta\underline{R}^N_{Ref_H}$ error can be defined as the difference between its true value and the value utilized in the strapdown INS computer software:

$$\delta\Delta\underline{R}^N_{Ref_H} \equiv \Delta\underline{R}^N_{Ref_H} - \Delta\underline{R}^N_{True_H} \qquad (14.2\text{-}2)$$

where

$\Delta\underline{R}^N_{True_H}$ = Actual horizontal position motion during initial alignment.

We attribute $\Delta\underline{R}^N_{True_H}$ to quasi-stationary random vibration type motion, hence:

$$\Delta\underline{R}^N_{True_H} = \Delta\underline{R}^{N\,\blacklozenge}_{Vib_H} \qquad (14.2\text{-}3)$$

where

$\Delta\underline{R}^{N\,\blacklozenge}_{Vib_H}$ = Horizontal random position disturbance during alignment from a nominally stationary position. The ♦ notation has been included to distinguish between this term in the approximate continuous form equations discussed in this chapter and the equivalent term (but without the ♦) in the actual Fine Alignment equations discussed in Sections 15.2.1 and 15.2.1.1. The actual Fine Alignment process operates with (14.1-1) type $\underline{Z}^N$ measurements that are taken and applied at a finite (discrete) update time through a Kalman gain matrix. Section 15.1.5.3.2 discusses analytical equivalencies between the continuous and discrete equation forms.

Then with (14.2-3) and $\Delta\underline{R}^N_{Ref_H}$ approximated as zero, Equation (14.2-2) is:

$$\delta\Delta\underline{R}^N_{Ref_H} = -\,\Delta\underline{R}^{N\,\blacklozenge}_{Vib_H} \qquad (14.2\text{-}4)$$

and (14.2-1) becomes:

$$\delta\underline{Z}^N = \delta\Delta\underline{R}^N_H + \Delta\underline{R}^{N\,\blacklozenge}_{Vib_H} \qquad (14.2\text{-}5)$$

With (14.2-5) we now write a set of quasi-stationary Fine Alignment error equations as the differential of Equations (14.1-1). For the differentials of the $\dot{C}^N_B$ and $\dot{\underline{v}}^N_H$ expressions, we apply the same procedures that led to Equations (12.3.4-39) and finally, to the revised (12.5.2-1) form for $\dot{\underline{\gamma}}^N$ and $\delta\dot{\underline{v}}^N_H$. In the interests of brevity, the process will only be outlined here.

For the differential of the (14.1-1) $\dot{C}_B^N$ equation, substitute (12.2.1-9) to get (12.3.4-10) for $\dot{\underline{\gamma}}^N$:

$$\dot{\underline{\gamma}}^N = - C_B^N \, \delta\underline{\omega}_{IB}^B - \underline{\omega}_{IN}^N \times \underline{\gamma}^N + \delta\underline{\omega}_{IN}^N \tag{14.2-6}$$

Then in (14.2-6), substitute (12.5-7) for $\underline{\gamma}^N$ in the $\underline{\omega}_{IN}^N \times \underline{\gamma}^N$ term, and substitute the derivative of (12.5-7) for $\dot{\underline{\gamma}}^N$ using $\dot{C}_B^N$ from (14.1-1) and $\delta\underline{\omega}_{Quant}$ for $\delta\dot{\underline{\alpha}}_{Quant}$ (as in (12.5-1)). Rearrange to solve for $\dot{\underline{\gamma}}^{N*}$, identifying $\delta\underline{\omega}_{IB}^B - \delta\underline{\omega}_{Quant}$ as $\delta\underline{\omega}_{IB}^{B*}$ (from (12.4-12) and (12.5-6)). Lastly, redefine $\underline{\gamma}^{N*}$ as $\underline{\gamma}^N$ and $\delta\underline{\omega}_{IB}^{B*}$ as $\delta\underline{\omega}_{IB}^B$, to get the result:

$$\dot{\underline{\gamma}}^N = - C_B^N \, \delta\underline{\omega}_{IB}^B - \underline{\omega}_{IN}^N \times \underline{\gamma}^N + \delta\underline{\omega}_{IN}^N + C_B^N \left(\underline{\omega}_{IB}^B \times \delta\underline{\alpha}_{Quant} \right) \tag{14.2-7}$$

For the differential of the (14.1-1) $\underline{v}_H^N$ equation, substitute (12.2.1-9) and (14.2-5) to get:

$$\delta\dot{\underline{v}}_H^N = \left(C_B^N \, \delta\underline{a}_{SF}^B \right)_H + \left(\underline{a}_{SF}^N \times \underline{\gamma}^N \right)_H - K_3 \, \delta\Delta\underline{R}_H^N - K_3 \, \Delta\underline{R}_{Vib_H}^{N \blacklozenge} \tag{14.2-8}$$

Then in (14.2-8), substitute (12.5-7) for $\underline{\gamma}^N$ in the $\underline{a}_{SF}^N \times \underline{\gamma}^N$ term, and substitute the derivative of (12.5-13) for $\delta\dot{\underline{v}}^N$ using $\dot{C}_B^N$ from (14.1-1) and $\delta\underline{a}_{Quant}$ for $\delta\dot{\underline{\upsilon}}_{Quant}$ (as in (12.5-1)). Rearrange to solve for $\delta\dot{\underline{v}}^{N*}$, identifying $\delta\underline{a}_{SF}^B - \delta\underline{a}_{Quant}$ as $\delta\underline{a}_{SF}^{B*}$ (from (12.4-14) and (12.5-12)). Lastly, redefine $\delta\dot{\underline{v}}^{N*}$ as $\delta\dot{\underline{v}}^N$, $\underline{\gamma}^{N*}$ as $\underline{\gamma}^N$ and $\delta\underline{a}_{SF}^{B*}$ as $\delta\underline{a}_{SF}^B$, to get the result:

$$\begin{aligned} \delta\dot{\underline{v}}_H^N = & \left(C_B^N \, \delta\underline{a}_{SF}^B \right)_H + \left(\underline{a}_{SF}^N \times \underline{\gamma}^N \right)_H - K_3 \, \delta\Delta\underline{R}_H^N - K_3 \, \Delta\underline{R}_{Vib_H}^{N \blacklozenge} \\ & - \left\{ \left(\underline{a}_{SF}^N \times \right) C_B^N \, \delta\underline{\alpha}_{Quant} + \left[\left(C_B^N \, \underline{\omega}_{IB}^B - \underline{\omega}_{IN}^N \right) \times \right] C_B^N \, \delta\underline{\upsilon}_{Quant} \right\}_H \end{aligned} \tag{14.2-9}$$

The differential of the remaining terms in (14.1-1) is found straight-forwardly using (14.2-5) and assuming that latitude l is a known input, hence, error free. With (14.2-7) and (14.2-9) the overall result is:

$$
\begin{aligned}
\dot{\underline{\gamma}}^{N} &= - C_B^N\, \delta\underline{\omega}_{IB}^B - \underline{\omega}_{IN}^N \times \underline{\gamma}^N + \delta\underline{\omega}_{IN}^N + C_B^N \left(\underline{\omega}_{IB}^B \times \delta\underline{\alpha}_{Quant}\right) \\
\delta\underline{\omega}_{IB}^B &= \delta K_{Scal/Mis}\, \underline{\omega}_{IB}^B + \delta\underline{K}_{Bias} + \delta\underline{\omega}_{Rand} \\
\delta\underline{\omega}_{IN}^N &= \delta\underline{\omega}_{IE}^N + \delta\underline{\omega}_{Tilt}^N \\
\delta\underline{\omega}_{IE}^N &= \delta\underline{\omega}_{IE_H}^N \\
\delta\dot{\underline{\omega}}_{IE_H}^N &= K_1\, \underline{u}_{ZN}^N \times \delta\Delta\underline{R}_H^N + K_1\, \underline{u}_{ZN}^N \times \Delta\underline{R}_{Vib_H}^{N\,\blacklozenge} \\
\delta\underline{\omega}_{Tilt}^N &= K_2\, \underline{u}_{ZN}^N \times \delta\Delta\underline{R}_H^N + K_2\, \underline{u}_{ZN}^N \times \Delta\underline{R}_{Vib_H}^{N\,\blacklozenge} \\
\delta\dot{\underline{v}}_H^N &= \left(C_B^N\, \delta\underline{a}_{SF}^B\right)_H + \left(\underline{a}_{SF}^N \times \underline{\gamma}^N\right)_H - K_3\, \delta\Delta\underline{R}_H^N - K_3\, \Delta\underline{R}_{Vib_H}^{N\,\blacklozenge} \\
&\quad - \left\{\left(\underline{a}_{SF}^N \times\right) C_B^N\, \delta\underline{\alpha}_{Quant} + \left[\left(C_B^N\, \underline{\omega}_{IB}^B - \underline{\omega}_{IN}^N\right)\times\right] C_B^N\, \delta\underline{\upsilon}_{Quant}\right\}_H \\
\delta\underline{a}_{SF}^B &= \delta L_{Scal/Mis}\, \underline{a}_{SF}^B + \delta\underline{L}_{Bias} + \delta\underline{a}_{Rand} \\
\delta\Delta\dot{\underline{R}}_H^N &= \delta\underline{v}_H^N - K_4\, \delta\Delta\underline{R}_H^N - K_4\, \Delta\underline{R}_{Vib_H}^{N\,\blacklozenge} + \left(C_B^N\, \delta\underline{\upsilon}_{Quant}\right)_H
\end{aligned}
\qquad (14.2\text{-}10)
$$

It is analytically beneficial at this point to divide the C_B^N attitude error rate expression ($\dot{\underline{\gamma}}^N$) in Equations (14.2-10) into horizontal and vertical components based on:

$$\underline{\gamma}^N = \underline{\gamma}_H^N + \gamma_{ZN}\, \underline{u}_{ZN}^N \qquad (14.2\text{-}11)$$

where

$\underline{\gamma}_H^N, \gamma_{ZN}$ = Horizontal, vertical components of the C_B^N attitude error $\underline{\gamma}^N$.

Recognizing from Equations (14.2-10) that $\delta\underline{\omega}_{IN}^N$ is horizontal, we expand the $\dot{\underline{\gamma}}^N$ expression into horizontal and vertical (along $\underline{u}_{ZN}^N$) components:

$$
\begin{aligned}
\dot{\gamma}_{ZN} &= - \underline{u}_{ZN}^N \cdot \left(C_B^N\, \delta\underline{\omega}_{IB}^B\right) - \underline{u}_{ZN}^N \cdot \left(\underline{\omega}_{IN}^N \times \underline{\gamma}^N\right) + \underline{u}_{ZN}^N \cdot \left[C_B^N \left(\underline{\omega}_{IB}^B \times \delta\underline{\alpha}_{Quant}\right)\right] \\
\dot{\underline{\gamma}}_H^N &= - \left(C_B^N\, \delta\underline{\omega}_{IB}^B\right)_H - \left(\underline{\omega}_{IN}^N \times \underline{\gamma}^N\right)_H + \delta\underline{\omega}_{IN}^N + \left[C_B^N \left(\underline{\omega}_{IB}^B \times \delta\underline{\alpha}_{Quant}\right)\right]_H
\end{aligned}
\qquad (14.2\text{-}12)
$$

The $\underline{\omega}_{IN}^N \times \underline{\gamma}^N$ terms in (14.2-12) can be expanded using (14.1-1) for $\underline{\omega}_{IN}^N$ and $\underline{\omega}_{IE}^N$, (14.2-11) for $\underline{\gamma}^N$, and recognizing that the cross-product of $\underline{u}_{ZN}^N$ with a horizontal vector is horizontal:

$$\underline{u}_{ZN}^{N} \cdot \left(\underline{\omega}_{IN}^{N} \times \underline{\gamma}^{N}\right) = \underline{u}_{ZN}^{N} \cdot \left(\underline{\omega}_{IN_H}^{N} \times \underline{\gamma}_{H}^{N}\right) = \underline{u}_{ZN}^{N} \cdot \left[\left(\underline{\omega}_{IE_H}^{N} + \underline{\omega}_{Tilt}^{N}\right) \times \underline{\gamma}_{H}^{N}\right]$$
$$\approx \underline{u}_{ZN}^{N} \cdot \left(\underline{\omega}_{IE_H}^{N} \times \underline{\gamma}_{H}^{N}\right)$$
$$\left(\underline{\omega}_{IN}^{N} \times \underline{\gamma}^{N}\right)_{H} = \left[\gamma_{ZN}\left(\underline{\omega}_{IE}^{N} + \underline{\omega}_{Tilt}^{N}\right) \times \underline{u}_{ZN}^{N}\right]_{H} + \left[\left(\underline{\omega}_{IE}^{N} + \underline{\omega}_{Tilt}^{N}\right) \times \underline{\gamma}_{H}^{N}\right]_{H} \qquad (14.2\text{-}13)$$
$$= \gamma_{ZN}\left(\underline{\omega}_{IE_H}^{N} + \underline{\omega}_{Tilt}^{N}\right) \times \underline{u}_{ZN}^{N} + \left(\underline{u}_{ZN}^{N} \times \underline{\gamma}_{H}^{N}\right) \omega_{e} \sin l$$
$$\approx \gamma_{ZN}\, \underline{\omega}_{IE_H}^{N} \times \underline{u}_{ZN}^{N} + \left(\underline{u}_{ZN}^{N} \times \underline{\gamma}_{H}^{N}\right) \omega_{e} \sin l$$

Equations (14.2-13) employ the approximation that $\underline{\omega}_{Tilt}^{N}$ is small at the end of Fine Alignment because, from Equations (14.1-1), it is proportional to the measurement $\underline{Z}^{N}$ which is driven toward zero by the alignment feedback dynamic process. As such, the product of $\underline{\omega}_{Tilt}^{N}$ with $\underline{\gamma}^{N}$ components can be considered second order, hence, negligible.

We also recognize in (14.2-12) and in the (14.2-10) $\delta\underline{\omega}_{IB}^{B}$ expression that for the quasi-stationary initialization condition, the average B Frame inertial angular rate $\underline{\omega}_{IB}^{B}$ equals earth rate. Hence, allowing for quasi-stationary angular vibration (as in (12.6-6)), we can write:

$$\underline{\omega}_{IB}^{B} = \left(C_{B}^{N}\right)^{T} \underline{\omega}_{IE}^{N} + \underline{\omega}_{Vib}^{B} \qquad (14.2\text{-}14)$$

where

$\underline{\omega}_{Vib}^{B}$ = Quasi-stationary angular vibration in the B Frame.

Substituting (14.2-14) into the (14.2-10) $\delta\underline{\omega}_{IB}^{B}$ expression, and substituting (14.2-13) and (14.2-14) into (14.2-12) using generalized Equation (3.1.1-38) with $\delta\underline{\omega}_{IN}^{N}$, $\delta\underline{\omega}_{IE}^{N}$ and $\delta\underline{\omega}_{Tilt}^{N}$ from (14.2-10), then yields:

$$\dot{\gamma}_{ZN} = - \underline{u}_{ZN}^{N} \cdot \left(C_{B}^{N}\, \delta\underline{\omega}_{IB}^{B}\right) - \underline{u}_{ZN}^{N} \cdot \left(\underline{\omega}_{IE_H}^{N} \times \underline{\gamma}_{H}^{N}\right) + \underline{u}_{ZN}^{N} \cdot \left[C_{B}^{N}\left(\underline{\omega}_{Vib}^{B} \times \delta\underline{\alpha}_{Quant}\right)\right]$$
$$\dot{\underline{\gamma}}_{H}^{N} = - \left(C_{B}^{N}\, \delta\underline{\omega}_{IB}^{B}\right)_{H} - \gamma_{ZN}\, \underline{\omega}_{IE_H}^{N} \times \underline{u}_{ZN}^{N} - \left(\underline{u}_{ZN}^{N} \times \underline{\gamma}_{H}^{N}\right) \omega_{e} \sin l + \delta\underline{\omega}_{IE_H}^{N}$$
$$+ K_{2}\, \underline{u}_{ZN}^{N} \times \delta\Delta\underline{R}_{H}^{N} + \left(C_{B}^{N}\right)_{H}\left(\underline{\omega}_{Vib}^{B} \times \delta\underline{\alpha}_{Quant}\right) + K_{2}\, \underline{u}_{ZN}^{N} \times \Delta\underline{R}_{Vib_H}^{N} \; \blacklozenge \qquad (14.2\text{-}15)$$

$$\delta\underline{\omega}_{IB}^{B} = \delta K_{Scal/Mis}\left(C_{B}^{N}\right)^{T} \underline{\omega}_{IE}^{N} + \delta\underline{K}_{Bias} + \delta\underline{\omega}_{Rand}$$

where

$$\left(C_B^N\right)_H = C_B^N \text{ with third row set to zero.}$$

In (14.2-15) we have neglected the product of $\underline{\omega}_{IE}^N$ with $\delta\underline{\alpha}_{Quant}$ in the $\dot{\underline{\gamma}}^N$ equations and the $\underline{\omega}_{Vib}^B$ product with $\delta K_{Scal/Mis}$ in the $\delta\underline{\omega}_{IB}^B$ equation as being small with zero average value.

The $\delta\dot{\underline{v}}_H^N$ and $\delta\underline{a}_{SF}^B$ expressions in (14.2-10) can be simplified by recognizing that for quasi-stationary conditions, the average specific force acceleration equals the negative of plumb-bob gravity acting along the local geodetic vertical. Hence, allowing for specific force vibration:

$$\begin{aligned} \underline{a}_{SF}^N &= -\underline{g}_P^N + \underline{a}_{Vib}^N = g\,\underline{u}_{ZN}^N + C_B^N\,\underline{a}_{Vib}^B \\ \underline{a}_{SF}^B &= g\left(C_B^N\right)^T \underline{u}_{ZN}^N + \underline{a}_{Vib}^B \end{aligned} \qquad (14.2\text{-}16)$$

where

g = Plumb-bob gravity magnitude.

$\underline{a}_{Vib}^B$ = Quasi-stationary linear vibration in the B Frame.

With (14.2-16), (14.2-14) and (14.2-11), while recognizing that the cross-product of $\underline{u}_{ZN}^N$ with a vector is horizontal, the $\delta\dot{\underline{v}}_H^N$ and $\delta\underline{a}_{SF}^B$ expressions in (14.2-10) become:

$$\begin{aligned} \delta\dot{\underline{v}}_H^N = {} & \left(C_B^N\,\delta\underline{a}_{SF}^B\right)_H + g\left(\underline{u}_{ZN}^N \times \underline{\gamma}_H^N\right) - K_3\,\delta\Delta\underline{R}_H^N - g\left(\underline{u}_{ZN}^N \times\right) C_B^N\,\delta\underline{\alpha}_{Quant} \\ & - \left(C_B^N\right)_H \left(\underline{a}_{Vib}^B \times \delta\underline{\alpha}_{Quant} + \underline{\omega}_{Vib}^B \times \delta\underline{\upsilon}_{Quant}\right) - K_3\,\Delta\underline{R}_{Vib_H}^{N\,\blacklozenge} \\ \delta\underline{a}_{SF}^B = {} & g\,\delta L_{Scal/Mis}\left(C_B^N\right)^T \underline{u}_{ZN}^N + \delta\underline{L}_{Bias} + \delta\underline{a}_{Rand} \end{aligned} \qquad (14.2\text{-}17)$$

In Equations (14.2-17) we have made the approximation that $\underline{\omega}_{Tilt}^N$ is small (as is $\underline{\omega}_{IE}^N$) so that in the (14.2-10) $\delta\dot{\underline{v}}_H^N$ expression, the product of $\underline{\omega}_{IN}^N$ (the sum of $\underline{\omega}_{Tilt}^N$ and $\underline{\omega}_{IE}^N$ - See (14.1-1)) with $C_B^N\,\delta\underline{\upsilon}_{Quant}$ is negligible, and the product of $\underline{\omega}_{IE}^N$ with $C_B^N\,\delta\underline{\upsilon}_{Quant}$ is negligible. Additionally, we have neglected the product of $C_B^N\,\underline{a}_{Vib}^B$ with $\underline{\gamma}^N$ in the $\delta\dot{\underline{v}}_H^N$ expression, and the product of $\underline{a}_{Vib}^B$ with $\delta L_{Scal/Mis}$ in the $\delta\underline{a}_{SF}^B$ expression, as being small with zero average value.

We now summarize Equations (14.2-15), (14.2-17) and the remaining pertinent expressions from (14.2-10) to obtain the error equations for the Fine Alignment dynamic process that will be used in subsequent sections for analysis.

$$
\begin{aligned}
\dot{\gamma}_{ZN} &= -\underline{u}_{ZN}^{N} \cdot \left(C_B^N \, \delta\underline{\omega}_{IB}^{B}\right) - \underline{u}_{ZN}^{N} \cdot \left(\underline{\omega}_{IE_H}^{N} \times \underline{\gamma}_H^N\right) + \underline{u}_{ZN}^{N} \cdot \left[C_B^N \left(\underline{\omega}_{Vib}^{B} \times \delta\underline{\alpha}_{Quant}\right)\right] \\
\delta\dot{\underline{\omega}}_{IE_H}^{N} &= K_1 \, \underline{u}_{ZN}^{N} \times \delta\Delta\underline{R}_H^N + K_1 \, \underline{u}_{ZN}^{N} \times \Delta\underline{R}_{Vib_H}^{N \, \blacklozenge} \\
\dot{\underline{\gamma}}_H^N &= -\left(C_B^N \, \delta\underline{\omega}_{IB}^{B}\right)_H - \gamma_{ZN} \, \underline{\omega}_{IE_H}^{N} \times \underline{u}_{ZN}^{N} - \left(\underline{u}_{ZN}^{N} \times \underline{\gamma}_H^N\right) \omega_e \sin l + \delta\underline{\omega}_{IE_H}^{N} \\
&\quad + K_2 \, \underline{u}_{ZN}^{N} \times \delta\Delta\underline{R}_H^N + \left(C_B^N\right)_H \left(\underline{\omega}_{Vib}^{B} \times \delta\underline{\alpha}_{Quant}\right) + K_2 \, \underline{u}_{ZN}^{N} \times \Delta\underline{R}_{Vib_H}^{N \, \blacklozenge} \\
\delta\underline{\omega}_{IB}^{B} &= \delta K_{Scal/Mis} \left(C_B^N\right)^T \underline{\omega}_{IE}^{N} + \delta\underline{K}_{Bias} + \delta\underline{\omega}_{Rand} \qquad\qquad (14.2\text{-}18) \\
\delta\dot{\underline{v}}_H^N &= \left(C_B^N \, \delta\underline{a}_{SF}^{B}\right)_H + g \left(\underline{u}_{ZN}^{N} \times \underline{\gamma}_H^N\right) - K_3 \, \delta\Delta\underline{R}_H^N - g \left(\underline{u}_{ZN}^{N} \times\right) C_B^N \, \delta\underline{\alpha}_{Quant} \\
&\quad - \left(C_B^N\right)_H \left(\underline{a}_{Vib}^{B} \times \delta\underline{\alpha}_{Quant} + \underline{\omega}_{Vib}^{B} \times \delta\underline{\upsilon}_{Quant}\right) - K_3 \, \Delta\underline{R}_{Vib_H}^{N \, \blacklozenge} \\
\delta\underline{a}_{SF}^{B} &= g \, \delta L_{Scal/Mis} \left(C_B^N\right)^T \underline{u}_{ZN}^{N} + \delta\underline{L}_{Bias} + \delta\underline{a}_{Rand} \\
\delta\Delta\dot{\underline{R}}_H^N &= \delta\underline{v}_H^N - K_4 \, \delta\Delta\underline{R}_H^N - K_4 \, \Delta\underline{R}_{Vib_H}^{N \, \blacklozenge} + \left(C_B^N \, \delta\underline{\upsilon}_{Quant}\right)_H
\end{aligned}
$$

The purpose of the Fine Alignment process is to initialize the inertial sensor B Frame attitude relative to the earth E Frame. As such, the error parameter that properly characterizes the accuracy of the Fine Alignment process is the error in the C_B^E matrix at entry into the navigation mode. As discussed in Section 12.2.1, the error in C_B^E can be represented in the N Frame by the rotation error vector $\underline{\psi}^N$. The solution to Equations (14.2-18) at Fine Alignment completion provides $\underline{\gamma}_H^N$ and γ_{ZN} (the N Frame error in the C_B^N matrix), and $\delta\underline{\omega}_{IE_H}^{N}$, the error in the estimated horizontal earth rate components used to initialize the wander angle. The remainder of this section develops an equation for $\underline{\psi}^N$ at navigation mode entry as a function of $\underline{\gamma}_H^N$, γ_{ZN} and $\delta\underline{\omega}_{IE_H}^{N}$ at Fine Alignment completion. The $\underline{\psi}^N$ expression will be developed for each of the two methods discussed in Section 14.1 for initializing the wander angle (i.e., Method 1. Setting the C_N^E matrix, and Method 2. Rotating the C_B^N matrix). As we shall see, and as should be expected, the identical expression is obtained for $\underline{\psi}^N$ using either method.

We first develop the equation for $\underline{\psi}^N$ based on wander angle initialization Method 1, setting C_N^E to match the wander angle determined from the estimated horizontal earth rate $\underline{\omega}_{IE_H}^N$. The general relationship between $\underline{\psi}^N$ and $\underline{\gamma}^N$ is from Equation (12.2.1-17):

$$\underline{\psi}^N = \underline{\gamma}^N - \underline{\varepsilon}^N \tag{14.2-19}$$

The $\underline{\varepsilon}^N$ term in (14.2-19) represents the angular error in the C_N^E matrix due to wander angle initialization using $\underline{\omega}_{IE_H}^N$ (which is in error). Assuming correct C_N^E initialization for latitude, longitude (See Equations (6.2.1-6) - (6.2.1-9)), Equation (12.2.3-35) shows that:

$$\underline{\varepsilon}^N = \delta\alpha \, \underline{u}_{ZN}^N \tag{14.2-20}$$

where

$\delta\alpha$ = Wander angle initialization error in C_N^E.

With (14.2-11) for $\underline{\gamma}^N$ and (14.2-20), Equation (14.2-19) becomes:

$$\underline{\psi}^N = \underline{\gamma}_H^N + \underline{u}_{ZN}^N \left(\gamma_{ZN} - \delta\alpha \right) \tag{14.2-21}$$

The $\underline{\varepsilon}^N$ rotation error vector can also be expressed in terms of the C_N^E matrix error (δC_N^E) (as in Equation (12.2.1-11)). Taking the transpose of (12.2.1-11), multiplying on the right by C_N^E, and equating the transpose of $\left(\underline{\varepsilon}^N \times\right)$ to its negative, provides a useful relationship between δC_N^E and $\underline{\varepsilon}^N$:

$$\left(\underline{\varepsilon}^N \times\right) = - \left(\delta C_N^E\right)^T C_N^E \tag{14.2-22}$$

From Equation (6.2.1-3), the C_E^N matrix (i.e., the transpose of C_N^E) is:

$$\left(C_N^E\right)^T = \begin{bmatrix} \underline{u}_{XE}^N & \underline{u}_{YE}^N & \underline{u}_{ZE}^N \end{bmatrix} \tag{14.2-23}$$

where

$\underline{u}_{XE}^N, \underline{u}_{YE}^N, \underline{u}_{ZE}^N$ = Unit vectors along the E Frame X, Y, Z axes projected on N Frame axes.

The differential of (14.2-23) provides $\left(\delta C_N^E\right)^T$ for (14.2-22):

$$\left(\delta C_N^E\right)^T = \left[\begin{array}{ccc} \delta \underline{u}_{XE}^N & \delta \underline{u}_{YE}^N & \delta \underline{u}_{ZE}^N \end{array}\right] \tag{14.2-24}$$

Substituting (14.2-24) and the transpose of (14.2-23) into (14.2-22) then gives:

$$\left(\underline{\varepsilon}^N \times\right) = - \delta \underline{u}_{XE}^N \left(\underline{u}_{XE}^N\right)^T - \delta \underline{u}_{YE}^N \left(\underline{u}_{YE}^N\right)^T - \delta \underline{u}_{ZE}^N \left(\underline{u}_{ZE}^N\right)^T \tag{14.2-25}$$

Because $\underline{u}_{XE}^N$, $\underline{u}_{YE}^N$, $\underline{u}_{ZE}^N$ are mutually perpendicular, multiplication of (14.2-25) on the right by $\underline{u}_{YE}^N$ nullifies all but the middle term, whence:

$$\underline{\varepsilon}^N \times \underline{u}_{YE}^N = - \delta \underline{u}_{YE}^N \tag{14.2-26}$$

The $\delta \underline{u}_{YE}^N$ term in (14.2-26) is the differential of (14.1-2), with which (14.2-26) (assuming zero latitude error) is:

$$\underline{\varepsilon}^N \times \underline{u}_{YE}^N = - \frac{\cos l}{\omega_{IE_H}} \delta \underline{\omega}_{IE_H}^N + \frac{\delta \omega_{IE_H} \cos l}{\omega_{IE_H}^2} \underline{\omega}_{IE_H}^N \tag{14.2-27}$$

An alternative expression for $\underline{\varepsilon}^N \times \underline{u}_{YE}^N$ is obtained from the combination of (14.2-20) for $\underline{\varepsilon}^N$ and (14.1-2) for $\underline{u}_{YE}^N$:

$$\underline{\varepsilon}^N \times \underline{u}_{YE}^N = \delta\alpha \, \underline{u}_{ZN}^N \times \underline{u}_{YE}^N = \delta\alpha \frac{\cos l}{\omega_{IE_H}} \underline{u}_{ZN}^N \times \underline{\omega}_{IE_H}^N \tag{14.2-28}$$

Equating (14.2-27) and (14.2-28) yields:

$$\delta\alpha \frac{\cos l}{\omega_{IE_H}} \underline{u}_{ZN}^N \times \underline{\omega}_{IE_H}^N = - \frac{\cos l}{\omega_{IE_H}} \delta \underline{\omega}_{IE_H}^N + \frac{\delta \omega_{IE_H} \cos l}{\omega_{IE_H}^2} \underline{\omega}_{IE_H}^N \tag{14.2-29}$$

The wander angle error $\delta\alpha$ is extracted from (14.2-29) by multiplying by $\frac{\omega_{IE_H}}{\cos l}$, taking the cross-product with $\underline{\omega}_{IE_H}^N$, and applying the triple vector cross-product identity (Equation (3.1.1-16):

$$\delta\alpha\,\underline{\omega}^N_{IE_H} \times \left(\underline{u}^N_{ZN} \times \underline{\omega}^N_{IE_H}\right) = \delta\alpha\,\omega^2_{IE_H}\,\underline{u}^N_{ZN} = -\,\underline{\omega}^N_{IE_H} \times \delta\underline{\omega}^N_{IE_H} \qquad (14.2\text{-}30)$$

or

$$\delta\alpha = -\frac{1}{\omega^2_{IE_H}}\,\underline{u}^N_{ZN} \cdot \left(\underline{\omega}^N_{IE_H} \times \delta\underline{\omega}^N_{IE_H}\right) \qquad (14.2\text{-}31)$$

We finally substitute (14.2-31) into (14.2-21) to obtain the equation for $\underline{\psi}^N$ (the B Frame attitude error relative to the E Frame) after application of the Method 1 wander angle initialization approach, as a function of the Fine Alignment process error terms $\underline{\gamma}^N_H$, γ_{ZN} and $\delta\underline{\omega}^N_{IE_H}$:

$$\underline{\psi}^N = \underline{\gamma}^N_H + \underline{u}^N_{ZN}\left[\gamma_{ZN} + \frac{1}{\omega^2_{IE_H}}\,\underline{u}^N_{ZN} \cdot \left(\underline{\omega}^N_{IE_H} \times \delta\underline{\omega}^N_{IE_H}\right)\right] \qquad (14.2\text{-}32)$$

As an aside, it is instructive at this point to examine a rearranged form of the earth rate term in (14.2-32). Since north is by definition in the horizontal direction toward earth's positive rotation axis, we can also write:

$$\underline{\omega}^N_{IE_H} = \omega_{IE_H}\,\underline{u}^N_{North} \qquad (14.2\text{-}33)$$

where

$\underline{u}^N_{North}$ = Horizontal unit vector in the direction of true north.

Substituting (14.2-33) in (14.2-32), applying the (3.1.1-35) mixed vector dot/cross product identity, and recognizing that the cross product of north and vertical unit vectors lies east (as in Equation (18.1.1-12)), then finds for the (14.2-32) earth rate term:

$$\begin{aligned}
&\underline{u}^N_{ZN}\left[\frac{1}{\omega^2_{IE_H}}\,\underline{u}^N_{ZN} \cdot \left(\underline{\omega}^N_{IE_H} \times \delta\underline{\omega}^N_{IE_H}\right)\right] \\
&\quad = \underline{u}^N_{ZN}\left[\frac{1}{\omega_{IE_H}}\,\delta\underline{\omega}^N_{IE_H} \cdot \left(\underline{u}^N_{ZN} \times \underline{u}^N_{North}\right)\right] = -\,\underline{u}^N_{ZN}\,\frac{1}{\omega_{IE_H}}\,\delta\underline{\omega}^N_{IE_H} \cdot \underline{u}^N_{East}
\end{aligned} \qquad (14.2\text{-}34)$$

where

$\underline{u}^N_{East}$ = Unit vector along the actual horizontal easterly direction in the N Frame.

Equation (14.2-34) substituted in (14.2-32) expresses the well known relationship that INS initial "platform" heading error produced by horizontal earth rate estimation uncertainty equals the east earth rate estimation error divided by horizontal earth rate magnitude.

Continuing the main discussion, we now develop an equation for $\underline{\psi}^{N}$ based on the Method 2 wander angle initialization approach of rotating C_B^N to nullify the wander angle determined from the Fine Alignment process. The Method 2 process initializes C_N^E based on zero wander angle, hence, since we are assuming error free latitude/longitude initialization, the error in C_N^E (i.e., $\underline{\varepsilon}^{N}$) is zero. Thus, we can write for $\underline{\psi}^{N}$ at initialization completion:

$$\underline{\psi}^{N+} = \underline{\gamma}_{+}^{N+} \tag{14.2-35}$$

where

$N+$ = N Frame following C_B^N rotation for zero wander angle setting.

$\underline{\gamma}_{+}^{N+}$ = Rotation error vector associated with C_B^{N+}, the value for C_B^N following wander angle rotation initialization, and the value for C_B^N for Navigation Mode engagement.

An expression for $\underline{\gamma}_{+}^{N+}$ in (14.2-35) is derived beginning with the differential of Equation (14.1-3):

$$\delta C_B^{N+} = \delta C_N^{N+} \, C_B^N + C_N^{N+} \, \delta C_B^N \tag{14.2-36}$$

The δC_B^N, δC_B^{N+} and δC_N^{N+} terms in (14.2-36) can be expressed in terms of their corresponding rotation error vectors as in Equation (12.2.1-9) for δC_B^N:

$$\delta C_B^N = -\left(\underline{\gamma}^{N} \times\right) C_B^N \qquad \delta C_B^{N+} = -\left(\underline{\gamma}_{+}^{N+} \times\right) C_B^{N+} \qquad \delta C_N^{N+} = -\left(\Delta\underline{\gamma}^{N+} \times\right) C_N^{N+} \tag{14.2-37}$$

where

$\underline{\gamma}^{N}$, $\Delta\underline{\gamma}^{N+}$ = Rotation error vectors associated with C_B^N, C_N^{N+}.

Substituting (14.2-37) into (14.2-36) and applying generalized Equation (3.1.1-38) then gives:

$$\begin{aligned}
\left(\underline{\gamma}_{+}^{N+} \times\right) C_B^{N+} &= \left(\Delta\underline{\gamma}^{N+} \times\right) C_N^{N+} \, C_B^N + C_N^{N+} \left(\underline{\gamma}^{N} \times\right) C_B^N \\
&= \left(\Delta\underline{\gamma}^{N+} \times\right) C_B^{N+} + C_N^{N+} \left(\underline{\gamma}^{N} \times\right) \left(C_N^{N+}\right)^T C_B^{N+} \\
&= \left(\Delta\underline{\gamma}^{N+} \times\right) C_B^{N+} + \left[\left(C_N^{N+} \, \underline{\gamma}^{N}\right) \times\right] C_B^{N+}
\end{aligned} \tag{14.2-38}$$

or after multiplication on the right by the transpose of C_B^{N+} and substitution in (14.2-35):

$$\underline{\psi}^{N+} = \Delta\underline{\gamma}^{N+} + C_N^{N+}\,\underline{\gamma}^N \tag{14.2-39}$$

With (14.2-11) for $\underline{\gamma}^N$ and the transpose of (14.1-4) for C_N^{N+}, (14.2-39) becomes:

$$\underline{\psi}^{N+} = C_N^{N+}\,\underline{\gamma}_H^N + \gamma_{ZN}\,\underline{u}_{ZN}^N + \Delta\underline{\gamma}^{N+} \tag{14.2-40}$$

The $\Delta\underline{\gamma}^{N+}$ angle error vector in (14.2-40) is from (14.2-37):

$$\left(\Delta\underline{\gamma}^{N+}\times\right) = -\,\delta C_N^{N+}\left(C_N^{N+}\right)^T \tag{14.2-41}$$

with $\left(C_N^{N+}\right)^T$ from (14.1-4) and δC_N^{N+} as the differential of the transpose of (14.1-4):

$$\delta C_N^{N+} = \begin{bmatrix} \dfrac{1}{\omega_{IE_H}}\left[\left(\delta\underline{\omega}_{IE_H}^N\right)^T - \dfrac{\delta\omega_{IE_H}}{\omega_{IE_H}}\left(\underline{\omega}_{IE_H}^N\right)^T\right]\left(\underline{u}_{ZN}^N\times\right) \\ \dfrac{1}{\omega_{IE_H}}\left[\left(\delta\underline{\omega}_{IE_H}^N\right)^T - \dfrac{\delta\omega_{IE_H}}{\omega_{IE_H}}\left(\underline{\omega}_{IE_H}^N\right)^T\right] \\ 0 \end{bmatrix} \tag{14.2-42}$$

We also note as in (13.4.1.1-5) - (13.4.1.1-6) that with generalized Equation (3.1.1-12):

$$\left(\underline{\omega}_{IE_H}^N\right)^T\underline{\omega}_{IE_H}^N = \omega_{IE_H}^2 \qquad\qquad \left(\delta\underline{\omega}_{IE_H}^N\right)^T\underline{\omega}_{IE_H}^N = \omega_{IE_H}\,\delta\omega_{IE_H} \tag{14.2-43}$$

Substituting (14.2-42) for δC_N^{N+} and (14.1-4) for $\left(C_N^{N+}\right)^T$ into (14.2-41), and applying (14.2-43) with generalized Equations (13.1-8), (3.1.1-12) and (3.1.1-35) obtains for $\Delta\underline{\gamma}^{N+}$:

$$\left(\Delta\underline{\gamma}^{N+}\times\right) = \begin{bmatrix} 0 & -A & 0 \\ A & 0 & 0 \\ 0 & 0 & 0 \end{bmatrix} \tag{14.2-44}$$

with

$$\begin{aligned} A &= \frac{1}{\omega_{IE_H}^2}\left(\delta\underline{\omega}_{IE_H}^N\right)^T \underline{u}_{ZN}^N \times \underline{\omega}_{IE_H}^N = \frac{1}{\omega_{IE_H}^2}\,\delta\underline{\omega}_{IE_H}^N\cdot\left(\underline{u}_{ZN}^N\times\underline{\omega}_{IE_H}^N\right) \\ &= \frac{1}{\omega_{IE_H}^2}\,\underline{u}_{ZN}^N\cdot\left(\underline{\omega}_{IE_H}^N\times\delta\underline{\omega}_{IE_H}^N\right) \end{aligned} \tag{14.2-45}$$

Using the generalized Equation (3.1.1-14) definition for the cross-product operator, we see from (14.2-44) - (14.2-45) that:

$$\Delta\underline{\gamma}^{N+} = A\,\underline{u}_{ZN}^{N} = \frac{1}{\omega_{IE_H}^{2}}\left[\underline{u}_{ZN}^{N}\cdot\left(\underline{\omega}_{IE_H}^{N}\times\delta\underline{\omega}_{IE_H}^{N}\right)\right]\underline{u}_{ZN}^{N} \tag{14.2-46}$$

Finally, we substitute (14.2-46) into (14.2-40) to obtain the equation for Frame B relative to Frame E attitude initialization error $\underline{\psi}^{N+}$ based on the Method 2 wander angle initialization procedure, in terms of the Fine Alignment process error terms:

$$\underline{\psi}^{N+} = C_{N}^{N+}\,\underline{\gamma}_{H}^{N} + \underline{u}_{ZN}^{N}\left[\gamma_{ZN} + \frac{1}{\omega_{IE_H}^{2}}\,\underline{u}_{ZN}^{N}\cdot\left(\underline{\omega}_{IE_H}^{N}\times\delta\underline{\omega}_{IE_H}^{N}\right)\right] \tag{14.2-47}$$

The N Frame version of (14.2-47) (i.e., the N Frame at the orientation preceding the wander angle rotation correction) is found by multiplication with $\left(C_{N}^{N+}\right)^{T}$ while noting that the form of $\left(C_{N}^{N+}\right)^{T}$ in (14.1-4) has the first two columns perpendicular to $\underline{u}_{ZN}^{N}$. The result is identical to Equation (14.2-32) for the Method 1 wander angle initialization process.

In summary, the overall error equations for the Fine Alignment attitude initialization process are given by Equations (14.2-18) with (14.2-32) for the Method 1 wander angle initialization process, or equivalently, Equations (14.2-18) with Equation (14.2-47) for the Method 2 wander angle initialization process.

14.3 INITIAL ALIGNMENT ERRORS PRODUCED BY CONSTANT INERTIAL SENSOR ERRORS

In this section we analyze $\underline{\psi}^{N}$ attitude initialization error response of the quasi-stationary Fine Alignment initialization process to constant inertial sensor errors using Equations (14.2-18) and (14.2-32) with the noise terms set to zero:

$$\begin{aligned} \dot{\gamma}_{ZN} &= -\,\underline{u}_{ZN}^{N}\cdot\left(C_{B}^{N}\,\delta\underline{\omega}_{IB}^{B}\right) - \underline{u}_{ZN}^{N}\cdot\left(\underline{\omega}_{IE_H}^{N}\times\underline{\gamma}_{H}^{N}\right) \\ \delta\dot{\underline{\omega}}_{IE_H}^{N} &= K_{1}\,\underline{u}_{ZN}^{N}\times\delta\Delta\underline{R}_{H}^{N} \end{aligned} \tag{14.3-1}$$

(Continued)

$$\dot{\underline{\gamma}}_H^N = -\left(C_B^N\ \delta\underline{\omega}_{IB}^B\right)_H - \gamma_{ZN}\ \underline{\omega}_{IE_H}^N \times \underline{u}_{ZN}^N - \left(\underline{u}_{ZN}^N \times \underline{\gamma}_H^N\right) \omega_e \sin l$$
$$+ \delta\underline{\omega}_{IE_H}^N + K_2\ \underline{u}_{ZN}^N \times \delta\Delta\underline{R}_H^N$$
$$\delta\dot{\underline{v}}_H^N = \left(C_B^N\ \delta\underline{a}_{SF}^B\right)_H + g\left(\underline{u}_{ZN}^N \times \underline{\gamma}_H^N\right) - K_3\ \delta\Delta\underline{R}_H^N \qquad \text{(14.3-1) (Continued)}$$
$$\delta\Delta\dot{\underline{R}}_H^N = \delta\underline{v}_H^N - K_4\ \delta\Delta\underline{R}_H^N$$

with

$$\underline{\psi}^N = \underline{\gamma}_H^N + \underline{u}_{ZN}^N \left[\gamma_{ZN} + \frac{1}{\omega_{IE_H}^2}\ \underline{u}_{ZN}^N \cdot \left(\underline{\omega}_{IE_H}^N \times \delta\underline{\omega}_{IE_H}^N\right)\right] \qquad \text{(14.3-2)}$$

and

$$\delta\underline{\omega}_{IB}^B = \delta K_{Scal/Mis} \left(C_B^N\right)^T \underline{\omega}_{IE}^N + \delta\underline{K}_{Bias}$$
$$\delta\underline{a}_{SF}^B = g\ \delta L_{Scal/Mis} \left(C_B^N\right)^T \underline{u}_{ZN}^N + \delta\underline{L}_{Bias} \qquad \text{(14.3-3)}$$

In Equations (14.3-3), because of the quasi-stationary nature of the initial alignment problem being analyzed, C_B^N is on average, constant. Thus, for the assumed constant δK, δL inertial sensor error coefficients, the $\delta\underline{\omega}_{IB}^B$, $\delta\underline{a}_{SF}^B$ inertial sensor error vectors can also be considered constant.

The analytical solution to Equations (14.3-1) - (14.3-2) is expedited if we split (14.3-2) into its horizontal and vertical components:

$$\underline{\psi}_H^N = \underline{\gamma}_H^N \qquad \text{(14.3-4)}$$

$$\psi_{ZN} = \gamma_{ZN} + \frac{1}{\omega_{IE_H}^2}\ \underline{u}_{ZN}^N \cdot \left(\underline{\omega}_{IE_H}^N \times \delta\underline{\omega}_{IE_H}^N\right) \qquad \text{(14.3-5)}$$

where

$\underline{\psi}_H^N$, ψ_{ZN} = Horizontal and vertical components of $\underline{\psi}^N$.

The γ_{ZN} term in (14.3-5) can be replaced by an alternative form through the following development. We first note using the (3.1.1-16) triple vector cross-product identity that:

$$\frac{1}{\omega_{IE_H}^2} \underline{u}_{ZN}^N \cdot \left[\underline{\omega}_{IE_H}^N \times \left(\underline{\omega}_{IE_H}^N \times \underline{u}_{ZN}^N \right) \right] = -1 \tag{14.3-6}$$

With (14.3-6), Equation (14.3-5) is equivalently:

$$\psi_{ZN} = \frac{1}{\omega_{IE_H}^2} \underline{u}_{ZN}^N \cdot \left[\underline{\omega}_{IE_H}^N \times \left(\delta\underline{\omega}_{IE_H}^N - \gamma_{ZN}\, \underline{\omega}_{IE_H}^N \times \underline{u}_{ZN}^N \right) \right] \tag{14.3-7}$$

The round bracketed term in (14.3-7) is from the $\dot{\underline{\gamma}}_H^N$ expression in (14.3-1):

$$\begin{aligned} \delta\underline{\omega}_{IE_H}^N - \gamma_{ZN}\, \underline{\omega}_{IE_H}^N \times \underline{u}_{ZN}^N = {} & \dot{\underline{\gamma}}_H^N + \left(C_B^N\, \delta\underline{\omega}_{IB}^B \right)_H \\ & + \left(\underline{u}_{ZN}^N \times \underline{\gamma}_H^N \right) \omega_e \sin l - K_2\, \underline{u}_{ZN}^N \times \delta\Delta\underline{R}_H^N \end{aligned} \tag{14.3-8}$$

Substitution of (14.3-8) into (14.3-7) then yields:

$$\begin{aligned} \psi_{ZN} = \frac{1}{\omega_{IE_H}^2} \underline{u}_{ZN}^N \cdot \Big\{ \underline{\omega}_{IE_H}^N \times \Big[& \dot{\underline{\gamma}}_H^N + \left(C_B^N\, \delta\underline{\omega}_{IB}^B \right)_H \\ & + \left(\underline{u}_{ZN}^N \times \underline{\gamma}_H^N \right) \omega_e \sin l - K_2\, \underline{u}_{ZN}^N \times \delta\Delta\underline{R}_H^N \Big] \Big\} \end{aligned} \tag{14.3-9}$$

or upon expansion and application of the Equation (3.1.1-16) vector triple cross-product rule:

$$\begin{aligned} \psi_{ZN} = {} & \frac{1}{\omega_{IE_H}^2} \underline{u}_{ZN}^N \cdot \left\{ \underline{\omega}_{IE_H}^N \times \left[\dot{\underline{\gamma}}_H^N + \left(C_B^N\, \delta\underline{\omega}_{IB}^B \right)_H \right] \right\} \\ & + \frac{1}{\omega_{IE_H}^2} \left(\underline{\omega}_{IE_H}^N \cdot \underline{\gamma}_H^N \right) \omega_e \sin l - \frac{K_2}{\omega_{IE_H}^2} \left(\underline{\omega}_{IE_H}^N \cdot \delta\Delta\underline{R}_H^N \right) \end{aligned} \tag{14.3-10}$$

In seeking the analytical solution for $\underline{\psi}^N$ at Fine Alignment completion, we will use the solution to (14.3-1) to find the $\underline{\psi}^N$ components calculated from (14.3-4) and (14.3-10). We now seek such a solution under constant inertial sensor error conditions. For analytical expediency, we also stipulate that the Fine Alignment gains (K_1, K_2, K_3, K_4) generate a stable response in Equations (14.3-1) that will reach a steady state condition by Fine Alignment completion in which transient effects produced by homogeneous roots have decayed to zero. Then, we need only seek the "Particular" solution to (14.3-1) to determine the effect on $\underline{\psi}^N$ initialization error.

With the previous assumptions, we now seek the "Particular" solution to Equations (14.3-1) by trying the following generalized analytical form:

$$
\begin{aligned}
\gamma_{ZN} &= C_{\gamma Z_0} + C_{\gamma Z_1}\, t \\
\underline{\gamma}_H^N &= \underline{C}_{\gamma H_0} + \underline{C}_{\gamma H_1}\, t \\
\delta\underline{\omega}_{IE_H}^N &= \underline{C}_{\omega H_0} + \underline{C}_{\omega H_1}\, t \\
\delta\underline{v}_H^N &= \underline{C}_{v H_0} + \underline{C}_{v H_1}\, t \\
\delta\Delta\underline{R}_H^N &= \underline{C}_{R H_0} + \underline{C}_{R H_1}\, t
\end{aligned}
\qquad (14.3\text{-}11)
$$

where

$C_{\gamma Z_0}, C_{\gamma Z_1}, \underline{C}_{(\,)H_0}, \underline{C}_{(\,)H_1}$ = Constants.

t = Time since start of Fine Alignment.

Substituting (14.3-11) into (14.3-1) yields:

$$
\begin{aligned}
C_{\gamma Z_1} &= -\,\underline{u}_{ZN}^N \cdot \left(C_B^N\, \delta\underline{\omega}_{IB}^B\right) - \underline{u}_{ZN}^N \cdot \left[\underline{\omega}_{IE_H}^N \times \left(\underline{C}_{\gamma H_0} + \underline{C}_{\gamma H_1}\, t\right)\right] \\
\underline{C}_{\omega H_1} &= K_1\, \underline{u}_{ZN}^N \times \left(\underline{C}_{RH_0} + \underline{C}_{RH_1}\, t\right) \\
\underline{C}_{\gamma H_1} &= -\left(C_B^N\, \delta\underline{\omega}_{IB}^B\right)_H - \left(C_{\gamma Z_0} + C_{\gamma Z_1}\, t\right) \underline{\omega}_{IE_H}^N \times \underline{u}_{ZN}^N \\
&\quad - \left[\underline{u}_{ZN}^N \times \left(\underline{C}_{\gamma H_0} + \underline{C}_{\gamma H_1}\, t\right)\right] \omega_e \sin l \\
&\quad + \underline{C}_{\omega H_0} + \underline{C}_{\omega H_1}\, t + K_2\, \underline{u}_{ZN}^N \times \left(\underline{C}_{RH_0} + \underline{C}_{RH_1}\, t\right) \\
\underline{C}_{vH_1} &= \left(C_B^N\, \delta\underline{a}_{SF}^B\right)_H + g\left[\underline{u}_{ZN}^N \times \left(\underline{C}_{\gamma H_0} + \underline{C}_{\gamma H_1}\, t\right)\right] - K_3\left(\underline{C}_{RH_0} + \underline{C}_{RH_1}\, t\right) \\
\underline{C}_{RH_1} &= \underline{C}_{vH_0} + \underline{C}_{vH_1}\, t - K_4\left(\underline{C}_{RH_0} + \underline{C}_{RH_1}\, t\right)
\end{aligned}
\qquad (14.3\text{-}12)
$$

Equations (14.3-12) can be solved for the (14.3-11) coefficients if we now make the expedient assumption of constant Fine Alignment process gains (K_1, K_2, K_3, K_4). In practice, the alignment gains are time varying, however, the major portion of the solution for constant gains will apply for time varying gains, and we will be able to infer the effect of the remaining portion based on heuristic reasoning and numerical (simulation) experience. Proceeding with the constant gain assumption then, we equate coefficients of like powers of t in (14.3-12) which gives:

$$C_{\gamma Z_1} = - \underline{u}_{ZN}^{N} \cdot \left(C_B^N \, \delta \underline{\omega}_{IB}^{B} \right) - \underline{u}_{ZN}^{N} \cdot \left(\underline{\omega}_{IE_H}^{N} \times \underline{C}_{\gamma H_0} \right) \tag{14.3-13}$$

$$0 = - \underline{u}_{ZN}^{N} \cdot \left(\underline{\omega}_{IE_H}^{N} \times \underline{C}_{\gamma H_1} \right) \tag{14.3-14}$$

$$\underline{C}_{\omega H_1} = K_1 \, \underline{u}_{ZN}^{N} \times \underline{C}_{RH_0} \tag{14.3-15}$$

$$0 = K_1 \, \underline{u}_{ZN}^{N} \times \underline{C}_{RH_1} \tag{14.3-16}$$

$$\begin{aligned} \underline{C}_{\gamma H_1} = & - \left(C_B^N \, \delta \underline{\omega}_{IB}^{B} \right)_H - C_{\gamma Z_0} \, \underline{\omega}_{IE_H}^{N} \times \underline{u}_{ZN}^{N} - \left(\underline{u}_{ZN}^{N} \times \underline{C}_{\gamma H_0} \right) \omega_e \sin l \\ & + \underline{C}_{\omega H_0} + K_2 \, \underline{u}_{ZN}^{N} \times \underline{C}_{RH_0} \end{aligned} \tag{14.3-17}$$

$$\begin{aligned} 0 = & - C_{\gamma Z_1} \, \underline{\omega}_{IE_H}^{N} \times \underline{u}_{ZN}^{N} - \left(\underline{u}_{ZN}^{N} \times \underline{C}_{\gamma H_1} \right) \omega_e \sin l \\ & + \underline{C}_{\omega H_1} + K_2 \, \underline{u}_{ZN}^{N} \times \underline{C}_{RH_1} \end{aligned} \tag{14.3-18}$$

$$\underline{C}_{vH_1} = \left(C_B^N \, \delta \underline{a}_{SF}^{B} \right)_H + g \left(\underline{u}_{ZN}^{N} \times \underline{C}_{\gamma H_0} \right) - K_3 \, \underline{C}_{RH_0} \tag{14.3-19}$$

$$0 = g \left(\underline{u}_{ZN}^{N} \times \underline{C}_{\gamma H_1} \right) - K_3 \, \underline{C}_{RH_1} \tag{14.3-20}$$

$$\underline{C}_{RH_1} = \underline{C}_{vH_0} - K_4 \, \underline{C}_{RH_0} \tag{14.3-21}$$

$$0 = \underline{C}_{vH_1} - K_4 \, \underline{C}_{RH_1} \tag{14.3-22}$$

Equations (14.3-13) - (14.3-22) constitute a set of ten simultaneous equations for the coefficient terms in (14.3-12). A procedure for solving these equations for the coefficient terms is now outlined. We first see from (14.3-16), (14.3-20) and (14.3-22) that:

$$\underline{C}_{RH_1} = 0 \tag{14.3-23}$$

$$\underline{C}_{\gamma H_1} = 0 \tag{14.3-24}$$

$$\underline{C}_{vH_1} = 0 \tag{14.3-25}$$

Combining Equations (14.3-15), (14.3-18), (14.3-23) and (14.3-24) then yields:

$$\underline{C}_{\omega H_1} = C_{\gamma Z_1} \, \underline{\omega}_{IE_H}^{N} \times \underline{u}_{ZN}^{N} = K_1 \, \underline{u}_{ZN}^{N} \times \underline{C}_{RH_0} \tag{14.3-26}$$

Taking the cross-product of (14.3-26) with $\underline{u}_{ZN}^{N}$, applying generalized Equation (13.1-6), and dividing by K_1 then obtains:

$$\underline{C}_{RH_0} = -\frac{1}{K_1} C_{\gamma Z_1} \underline{\omega}^N_{IE_H} \qquad (14.3\text{-}27)$$

Equation (14.3-19) with (14.3-25) gives:

$$0 = \left(C^N_B \, \delta\underline{a}^B_{SF}\right)_H + g\left(\underline{u}^N_{ZN} \times \underline{C}_{\gamma H_0}\right) - K_3 \, \underline{C}_{RH_0} \qquad (14.3\text{-}28)$$

or after applying the cross-product with $\underline{u}^N_{ZN}$, utilizing Equation (13.1-6), and dividing by g:

$$\underline{C}_{\gamma H_0} = \frac{1}{g}\left[\underline{u}^N_{ZN} \times \left(C^N_B \, \delta\underline{a}^B_{SF}\right)_H - K_3 \, \underline{u}^N_{ZN} \times \underline{C}_{RH_0}\right] \qquad (14.3\text{-}29)$$

The $\underline{\omega}^N_{IE_H} \times \underline{C}_{\gamma H_0}$ term in Equation (14.3-13) is from (14.3-29) and (14.3-27) with (3.1.1-16) and the knowledge that $\underline{u}^N_{ZN}$ is perpendicular to horizontal vectors:

$$\begin{aligned}
\underline{\omega}^N_{IE_H} \times \underline{C}_{\gamma H_0} &= \frac{1}{g}\, \underline{\omega}^N_{IE_H} \times \left[\underline{u}^N_{ZN} \times \left(C^N_B \, \delta\underline{a}^B_{SF}\right)_H - K_3 \, \underline{u}^N_{ZN} \times \underline{C}_{RH_0}\right] \\
&= \frac{1}{g}\left[\underline{\omega}^N_{IE_H} \cdot \left(C^N_B \, \delta\underline{a}^B_{SF}\right)_H - K_3 \, \underline{\omega}^N_{IE_H} \cdot \underline{C}_{RH_0}\right] \underline{u}^N_{ZN} \qquad (14.3\text{-}30) \\
&= \frac{1}{g}\left[\underline{\omega}^N_{IE_H} \cdot \left(C^N_B \, \delta\underline{a}^B_{SF}\right)_H + \frac{K_3}{K_1}\, \omega^2_{IE_H} \, C_{\gamma Z_1}\right] \underline{u}^N_{ZN}
\end{aligned}$$

Substituting (14.3-30) in (14.3-13) then obtains:

$$C_{\gamma Z_1} = -\underline{u}^N_{ZN} \cdot \left(C^N_B \, \delta\underline{\omega}^B_{IB}\right) - \frac{1}{g}\left[\underline{\omega}^N_{IE_H} \cdot \left(C^N_B \, \delta\underline{a}^B_{SF}\right)_H + \frac{K_3}{K_1}\, \omega^2_{IE_H} \, C_{\gamma Z_1}\right] \qquad (14.3\text{-}31)$$

or upon rearrangement:

$$\left(1 + \frac{K_3}{g\,K_1}\, \omega^2_{IE_H}\right) C_{\gamma Z_1} = -\underline{u}^N_{ZN} \cdot \left(C^N_B \, \delta\underline{\omega}^B_{IB}\right) - \frac{1}{g}\left[\underline{\omega}^N_{IE_H} \cdot \left(C^N_B \, \delta\underline{a}^B_{SF}\right)_H\right] \qquad (14.3\text{-}32)$$

which then gives:

$$C_{\gamma Z_1} = -\frac{1}{1 + \dfrac{K_3}{g\,K_1}\, \omega^2_{IE_H}} \left\{\underline{u}^N_{ZN} \cdot \left(C^N_B \, \delta\underline{\omega}^B_{IB}\right) + \frac{1}{g}\left[\underline{\omega}^N_{IE_H} \cdot \left(C^N_B \, \delta\underline{a}^B_{SF}\right)_H\right]\right\} \qquad (14.3\text{-}33)$$

Substituting (14.3-33) into (14.3-27) obtains for $\underline{C}_{RH_0}$:

$$\underline{C}_{RH_0} = \frac{1}{K_1 + \frac{K_3}{g}\,\omega_{IE_H}^2}\left\{\underline{u}_{ZN}^N \cdot \left(C_B^N\,\delta\underline{\omega}_{IB}^B\right) + \frac{1}{g}\left[\underline{\omega}_{IE_H}^N \cdot \left(C_B^N\,\delta\underline{a}_{SF}^B\right)_H\right]\right\}\underline{\omega}_{IE_H}^N \tag{14.3-34}$$

Using (14.3-34), Equation (14.3-29) for $\underline{C}_{\gamma H_0}$ becomes:

$$\begin{aligned}\underline{C}_{\gamma H_0} &= \frac{1}{g}\left[\underline{u}_{ZN}^N \times \left(C_B^N\,\delta\underline{a}_{SF}^B\right)_H\right] \\ &\quad - \frac{K_3}{g\,K_1 + K_3\,\omega_{IE_H}^2}\left\{\underline{u}_{ZN}^N \cdot \left(C_B^N\,\delta\underline{\omega}_{IB}^B\right) + \frac{1}{g}\left[\underline{\omega}_{IE_H}^N \cdot \left(C_B^N\,\delta\underline{a}_{SF}^B\right)_H\right]\right\}\underline{u}_{ZN}^N \times \underline{\omega}_{IE_H}^N\end{aligned} \tag{14.3-35}$$

Equations (14.3-23) - (14.3-25) and (14.3-33) - (14.3-35) provide analytical expressions for all Equation (14.3-11) coefficients with the exception of $\underline{C}_{vH_0}$, $\underline{C}_{\omega H_0}$, $\underline{C}_{\omega H_1}$ and $C_{\gamma Z_0}$. The calculated coefficients are sufficient to evaluate $\underline{\psi}^N$ at Fine Alignment completion using Equations (14.3-4), (14.3-10) and (14.3-11). Nevertheless, it is important to confirm that a reasonable solution exists for the $\underline{C}_{vH_0}$, $\underline{C}_{\omega H_0}$, $\underline{C}_{\omega H_1}$, $C_{\gamma Z_0}$ terms not yet evaluated to confirm that the selected form of the Particular solution in (14.3-11) adequately fits Equations (14.3-1) (or if it doesn't for all terms, a plausible explanation must be provided). Only then can the individual solutions arrived at be accepted. Proceeding, we determine the remaining terms as a function of the already evaluated Equation (14.3-34) $\underline{C}_{RH_0}$ expression; $\underline{C}_{vH_0}$ is found by combining Equations (14.3-21) and (14.3-23), $\underline{C}_{\omega H_1}$ has already been determined in (14.3-15) as a function of $\underline{C}_{RH_0}$, and $\underline{C}_{\omega H_0}$ is found as a function of $C_{\gamma Z_0}$ and $\underline{C}_{RH_0}$ by combining Equations (14.3-17), (14.3-24) and (14.3-29). Using generalized Equation (13.1-5), the solution for the remaining terms is:

$$\begin{aligned}\underline{C}_{vH_0} &= K_4\,\underline{C}_{RH_0} \\ \underline{C}_{\omega H_1} &= K_1\,\underline{u}_{ZN}^N \times \underline{C}_{RH_0} \\ \underline{C}_{\omega H_0} - C_{\gamma Z_0}\,\underline{\omega}_{IE_H}^N \times \underline{u}_{ZN}^N &= \left(C_B^N\,\delta\underline{\omega}_{IB}^B\right)_H - \frac{1}{g}\left(C_B^N\,\delta\underline{a}_{SF}^B\right)_H\,\omega_e \sin l \\ &\quad + \left[\frac{1}{g}\,K_3\,I\,\omega_e \sin l - K_2\left(\underline{u}_{ZN}^N \times\right)\right]\underline{C}_{RH_0}\end{aligned} \tag{14.3-36}$$

The $C_{\gamma Z_0}$ term in (14.3-36) cannot be explicitly evaluated because it represents part of the homogeneous solution to Equations (14.3-1) (i.e., from the $\dot{\gamma}_{ZN}$ expression). Equations (14.3-36) with (14.3-34) for $\underline{C}_{RH_0}$ completes development of the Equation (14.3-11) Particular solution coefficients (subject to the previous explanation of why $C_{\gamma Z_0}$ cannot be explicitly evaluated). Hence, we conclude that the selected Particular solution fits Equations (14.3-1) and is, therefore, valid. Note, however, that the selected Particular solution is only valid if the calculated solution coefficients are constant as stipulated in (14.3-11) and its application in (14.3-12) (in which derivatives of the assumed constants were equated to zero). Since the solution coefficients are a function (in part) of the K_1, K_2, K_3, K_4 gains, they will only be constant if the gains are constant. Thus, so far, the validity of our Particular solution depends on the constant gain assumption.

We now use the above results in $\underline{\psi}^N$ component Equations (14.3-4) and (14.3-10) to find the solution for the attitude initialization error. In performing this operation, we note from (14.3-24) and the derivative of the (14.3-11) $\underline{\gamma}_H^N$ expression (i.e., $\underline{C}_{\gamma H\,1}$), that the $\dot{\underline{\gamma}}_H^N$ term in (14.3-10) is zero. Additionally, we make use of the following based on the Equation (3.1.1-35) mixed vector dot/cross product identity:

$$\underline{\omega}_{IE_H}^N \cdot \left[\underline{u}_{ZN}^N \times \left(C_B^N \, \delta\underline{a}_{SF}^B \right)_H \right] = - \underline{u}_{ZN}^N \cdot \left[\underline{\omega}_{IE_H}^N \times \left(C_B^N \, \delta\underline{a}_{SF}^B \right)_H \right] \tag{14.3-37}$$

Then, substituting Equation (14.3-37), Equations (14.3-23) - (14.3-24) and (14.3-34) - (14.3-35) for $\underline{C}_{\gamma H\,0}$, $\underline{C}_{\gamma H\,1}$, $\underline{C}_{RH\,0}$, $\underline{C}_{RH\,1}$, and Equations (14.3-11) for $\underline{\gamma}_H^N$ and $\delta\Delta\underline{R}_H^N$ into Equations (14.3-4) and (14.3-10), and applying (3.1.1-35), we find for the $\underline{\psi}^N$ attitude initialization components:

$$\begin{aligned}
\underline{\psi}_H^N &= \frac{1}{g} \left[\underline{u}_{ZN}^N \times \left(C_B^N \, \delta\underline{a}_{SF}^B \right)_H \right] \\
&\quad - \frac{K_3}{g\,K_1 + K_3\,\omega_{IE_H}^2} \left\{ \underline{u}_{ZN}^N \cdot \left(C_B^N \, \delta\underline{\omega}_{IB}^B \right) + \frac{1}{g} \left[\underline{\omega}_{IE_H}^N \cdot \left(C_B^N \, \delta\underline{a}_{SF}^B \right)_H \right] \right\} \underline{u}_{ZN}^N \times \underline{\omega}_{IE_H}^N \\
\psi_{ZN} &= \frac{1}{\omega_{IE_H}^2} \, \underline{u}_{ZN}^N \cdot \left\{ \underline{\omega}_{IE_H}^N \times \left[\left(C_B^N \, \delta\underline{\omega}_{IB}^B \right)_H - \frac{1}{g} \left(C_B^N \, \delta\underline{a}_{SF}^B \right)_H \omega_e \sin l \right] \right\} \\
&\quad - \frac{K_2}{K_1 + \frac{K_3}{g}\,\omega_{IE_H}^2} \left\{ \underline{u}_{ZN}^N \cdot \left(C_B^N \, \delta\underline{\omega}_{IB}^B \right) + \frac{1}{g} \left[\underline{\omega}_{IE_H}^N \cdot \left(C_B^N \, \delta\underline{a}_{SF}^B \right)_H \right] \right\}
\end{aligned} \tag{14.3-38}$$

The leading terms in Equations (14.3-38) are the dominant terms. They show that for constant sensor errors, initial horizontal tilt is created from horizontal accelerometer error and initial heading error is created from the component of horizontal angular rate sensor error perpendicular to $\underline{\omega}_{IE_H}^N$. Since $\underline{\omega}_{IE_H}^N$ lies north, the critical angular rate sensor error component affecting initial heading error lies east. We also note that initial heading error is inversely proportional to the magnitude of horizontal earth rate. From the $\underline{\omega}_{IE}^{N}$ earth rate vector expression in (6.1.2-2), since the vertical component of $\underline{\omega}_{IE}^{N}$ is $\omega_e \sin l$ and the $\underline{\omega}_{IE}^{N}$ magnitude is ω_e, the magnitude of the horizontal $\underline{\omega}_{IE}^{N}$ component, is $\omega_e \cos l$. Thus, for high or low latitude alignments (i.e., near the north or south pole) when the magnitude of l approaches ninety degrees, the initial heading error caused by east angular rate sensor error can be very large. To minimize the effect of inertial sensor error on initial heading determination accuracy, the ideal latitude would be zero (i.e., on the equator).

The (14.3-38) solution for both the horizontal and vertical $\underline{\psi}^{N}$ components have two elements, a portion that is independent of the gains and a portion that is proportional to gain ratios. The gain dependent portion is driven by the { } bracketed term containing the vertical component of angular rate sensor error and the component of accelerometer error along the horizontal earth rate component $\underline{\omega}_{IE_H}^{N}$ (which lies north/south by definition). The source of the bracketed term can be traced to $C_{\gamma Z_1}$ in Equation (14.3-33) through the following logic: $C_{\gamma Z_1}$ impacts $\underline{C}_{RH_0}$ and $\underline{C}_{\gamma H_0}$ (through (14.3-27) and (14.3-29)), which sets $\underline{\gamma}_H^N$ and $\delta\Delta\underline{R}_H^N$ (through (14.3-11) with (14.3-23) and (14.3-24)), thereby affecting $\underline{\psi}_H^N$ and ψ_{ZN} through (14.3-4) and (14.3-10). From Equation (14.3-11), we see that $C_{\gamma Z_1}$ is the vertical attitude error rate $\dot{\gamma}_{ZN}$. Thus, in order to eliminate the gain dependent terms in (14.3-38), we must stipulate conditions that will set $\dot{\gamma}_{ZN}$ to zero.

Recall that the Equation (14.3-38) solution was based on the assumption of constant alignment process gains. We now demonstrate that the gain independent solution in (14.3-38) is also valid for the general case of time varying gains when $\dot{\gamma}_{ZN}$ (the source of the gain dependent terms as discussed in the previous paragraph) is zero. If we set $\dot{\gamma}_{ZN}$ to zero in (14.3-1) and repeat the previous Particular solution process, we will find that $\underline{C}_{RH_0}$ in (14.3-34) is zero, and the remaining coefficients in (14.3-11) become independent of the K_1, K_2, K_3, K_4 gains. Thus, the Particular solution coefficients become constant as stipulated in the (14.3-11) - (14.3-12) trial solution, hence valid, but also valid for the general case of time varying gains which is the normal situation. If we substitute the Particular solution coefficients based on zero $\dot{\gamma}_{ZN}$ in Equations (14.3-11), we then find that the solution to (14.3-1) becomes:

$$\gamma_{ZN} = \gamma_{ZN_0} \qquad \text{Assumes condition of } \dot{\gamma}_{ZN} = 0.$$

$$\delta\underline{\omega}_{IE_H}^{N} = \gamma_{ZN_0}\, \underline{\omega}_{IE_H}^{N} \times \underline{u}_{ZN}^{N} + \left(C_B^N\ \delta\underline{\omega}_{IB}^{B}\right)_H - \frac{1}{g}\left(C_B^N\, \delta\underline{a}_{SF}^{B}\right)_H \omega_e \sin l$$

$$\underline{\gamma}_H^N = \frac{1}{g}\, \underline{u}_{ZN}^{N} \times \left(C_B^N\ \delta\underline{a}_{SF}^{B}\right)_H \tag{14.3-39}$$

$$\delta\underline{v}_H^N = 0$$

$$\delta\Delta\underline{R}_H^N = 0$$

where

$$\gamma_{ZN_0} = \text{Value for } \gamma_{ZN} \text{ at the start of Fine Alignment.}$$

It is easily demonstrated by direct substitution in (14.3-1), that Equations (14.3-39) satisfy (14.3-1) without specifying the nature of the gains. Moreover, if Equations (14.3-39) for $\underline{\gamma}_H^N$ and $\delta\Delta\underline{R}_H^N$ are substituted in Equations (14.3-4) and (14.3-10), we find that:

$$\underline{\psi}_H^N = \frac{1}{g}\left[\underline{u}_{ZN}^{N} \times \left(C_B^N\ \delta\underline{a}_{SF}^{B}\right)_H\right]$$

$$\psi_{ZN} = \frac{1}{\omega_{IE_H}^2}\, \underline{u}_{ZN}^{N} \cdot \left\{\underline{\omega}_{IE_H}^{N} \times \left[\left(C_B^N\ \delta\underline{\omega}_{IB}^{B}\right)_H - \frac{1}{g}\left(C_B^N\, \delta\underline{a}_{SF}^{B}\right)_H \omega_e \sin l\right]\right\} \tag{14.3-40}$$

which is identical to the gain independent portion of (14.3-38). Thus, the gain independent portion of (14.3-38) is valid for arbitrary gain profiles, as the gains produce a stable equilibrium at Fine Alignment completion. Finally, if the Equations (14.3-39) $\underline{\gamma}_H^N$ solution is substituted into the Equations (14.3-1) $\dot{\gamma}_{ZN}$ expression (which we have stipulated to be zero), we find with the (3.1.1-16) triple vector cross-product identity:

$$\begin{aligned}\dot{\gamma}_{ZN} &= -\,\underline{u}_{ZN}^{N} \cdot \left(C_B^N\ \delta\underline{\omega}_{IB}^{B}\right) - \frac{1}{g}\, \underline{u}_{ZN}^{N} \cdot \left\{\underline{\omega}_{IE_H}^{N} \times \left[\underline{u}_{ZN}^{N} \times \left(C_B^N\ \delta\underline{a}_{SF}^{B}\right)_H\right]\right\} \\ &= -\left\{\underline{u}_{ZN}^{N} \cdot \left(C_B^N\ \delta\underline{\omega}_{IB}^{B}\right) + \frac{1}{g}\left[\underline{\omega}_{IE_H}^{N} \cdot \left(C_B^N\ \delta\underline{a}_{SF}^{B}\right)_H\right]\right\}\end{aligned} \tag{14.3-41}$$

The (14.3-41) $\dot{\gamma}_{ZN}$ expression is exactly the critical bracketed term in the gain dependent portion of Equations (14.3-38) (derived based on constant gains). Thus, we see that by setting $\dot{\gamma}_{ZN}$ to zero (the basis for (14.3-39) - (14.3-40)), we are also directly nullifying the (14.3-38) gain dependent term, thereby making (14.3-38) and (14.3-40) equivalent.

For the ψ_{ZN} expression in (14.3-38), the gain independent portion sensor error components are orthogonal to the gain dependent portion sensor error components. Thus, the setting of the gain dependent sensor error components to zero has no effect on the ψ_{ZN} gain independent term. The gain independent portion of ψ_{ZN} is created by horizontal angular rate sensor and accelerometer error components perpendicular to $\underline{\omega}_{IE_H}^N$ (Note that since $\underline{\omega}_{IE_H}^N$ is along a north/south line by definition, the horizontal inertial sensor components perpendicular to $\underline{\omega}_{IE_H}^N$ lie east/west). Compared to the Equation (14.3-38) gain independent ψ_{ZN} error produced by angular rate sensor error, the gain independent accelerometer error effect is generally small and arises from the horizontal tilt it creates (See the $\underline{\gamma}_H^N$ solution from Equations (14.3-4) and (14.3-38)) which couples vertical earth rate ($\omega_e \sin l$) into the (14.3-1) $\dot{\underline{\gamma}}_H^N$ expression.

The gain dependent term in the (14.3-38) ψ_{ZN} expression is generally small compared to the gain independent term. For Kalman filter derived gains (See Section 15.2.1 and its subsections) which are time varying, it can be demonstrated numerically (by simulation), that the effect of the sensor error terms multiplying the gain ratio $\dfrac{K_2}{K_1 + \dfrac{K_3}{g}\,\omega_{IE_H}^2}$ in the Equation (14.3-38) ψ_{ZN} expression is equivalent to substituting 0.5 t for the gain ratio. This same effect can also be deduced from Equation (18.3.1.2-8) which, upon rearrangement, shows that Kalman filter estimated true heading $\hat{\psi}_T$ will contain an error component equal to the vertical angular rate sensor error multiplied by one half the time in alignment T_{Align}. From the previous discussion, we can conclude that an accurate solution for ψ_{ZN}, that is valid for time varying gains, is the Equations (14.3-38) result with $\dfrac{K_2}{K_1 + \dfrac{K_3}{g}\,\omega_{IE_H}^2}$ set to 0.5 t.

From empirical experience, the author can also state that the dominant term in the (14.3-38) $\underline{\psi}_H^N$ expression is the gain independent term. The gain dependent term in $\underline{\psi}_H^N$ is produced in the Equations (14.3-1) $\dot{\underline{\gamma}}_H^N$ expression from horizontal earth rate coupling of γ_{ZN} build-up (due to vertical angular rate sensor error), plus vertical earth rate component ($\omega_e \sin l$) coupling of horizontal tilt (produced by horizontal accelerometer error). The author has no empirical experience regarding the particular characteristic of the $\dfrac{K_3}{g\,K_1 + K_3\,\omega_{IE_H}^2}$ gain ratio term in the

(14.3-38) $\underline{\psi}_H^N$ expression under time varying gain conditions, other than the general experience that $\underline{\psi}_H^N$ is dominated by the gain independent term.

14.4 INITIAL ALIGNMENT ERROR CAUSED BY RAMPING ACCELEROMETER ERROR

In the previous section we analyzed the error response of the Fine Alignment process to constant inertial sensor errors. In this section we analyze the response to a particular type of time varying inertial sensor error that can have serious impact on initial heading determination accuracy; the effect of time varying accelerometer error during the alignment process. For analysis purposes, we will investigate the effect of a linearly changing accelerometer error during Fine Alignment.

To simplify the analysis, it is expedient to neglect the earth-rate/horizontal-tilt coupling terms in the $\dot{\underline{\gamma}}^N$ expressions of general Fine Alignment error Equations (14.2-18). It can be verified from the analytical solutions in this and previous sections, that these terms have only a second order effect on results to be obtained in this section. Thus, neglecting the horizontal-tilt/earth-rate coupling in $\dot{\underline{\gamma}}^N$ and only including linearly varying accelerometer error effects, Equations (14.2-18) simplify to:

$$
\begin{aligned}
\dot{\gamma}_{ZN} &= 0 \\
\delta\dot{\underline{\omega}}_{IE_H}^N &= K_1\, \underline{u}_{ZN}^N \times \delta\Delta\underline{R}_H^N \\
\dot{\underline{\gamma}}_H^N &= -\gamma_{ZN}\, \underline{\omega}_{IE_H}^N \times \underline{u}_{ZN}^N + \delta\underline{\omega}_{IE_H}^N + K_2\, \underline{u}_{ZN}^N \times \delta\Delta\underline{R}_H^N \\
\delta\dot{\underline{v}}_H^N &= \left(C_B^N\, \delta\underline{a}_{SF}^B\right)_H + g\left(\underline{u}_{ZN}^N \times \underline{\gamma}_H^N\right) - K_3\, \delta\Delta\underline{R}_H^N \\
\left(C_B^N\, \delta\underline{a}_{SF}^B\right)_H &= \left(C_B^N\, \delta\dot{\underline{a}}_{SF}^B\right)_H t \\
\delta\Delta\dot{\underline{R}}_H^N &= \delta\underline{v}_H^N - K_4\, \delta\Delta\underline{R}_H^N
\end{aligned}
\qquad (14.4\text{-}1)
$$

Equations (14.3-4) and (14.3-10) for $\underline{\psi}^N$ remain valid for the case being investigated of ramping accelerometer errors. Neglecting the horizontal-tilt/earth-rate coupling in these equations and dropping the angular rate sensor error terms as not being investigated obtains for $\underline{\psi}^N$:

$$\underline{\psi}_H^N = \underline{\gamma}_H^N$$

$$\psi_{ZN} = \frac{1}{\omega_{IE_H}^2}\, \underline{u}_{ZN}^N \cdot \left(\underline{\omega}_{IE_H}^N \times \dot{\underline{\gamma}}_H^N\right) - \frac{K_2}{\omega_{IE_H}^2}\left(\underline{\omega}_{IE_H}^N \cdot \delta\Delta\underline{R}_H^N\right) \qquad (14.4\text{-}2)$$

As in the previous section, we assume that the response of Equations (14.4-1) reaches a steady state at Fine Alignment completion when the transients associated with the characteristic dynamic roots have decayed to zero. Thus, we seek the "Particular" solution to (14.4-1) which we try to fit (as in the previous section) with the following generalized form:

$$\gamma_{ZN} = 0$$

$$\delta\underline{\omega}_{IE_H}^N = \underline{C}_{\omega H_0} + \underline{C}_{\omega H_1}\, t$$

$$\underline{\gamma}_H^N = \underline{C}_{\gamma H_0} + \underline{C}_{\gamma H_1}\, t \qquad (14.4\text{-}3)$$

$$\delta\underline{v}_H^N = \underline{C}_{vH_0} + \underline{C}_{vH_1}\, t$$

$$\delta\Delta\underline{R}_H^N = \underline{C}_{RH_0} + \underline{C}_{RH_1}\, t$$

where

$$\gamma_{ZN_0} = \text{Value for } \gamma_{ZN} \text{ at the start of Fine Alignment.}$$

The solution for γ_{ZN} in (14.4-3) is based on γ_{ZN} equal to zero in (14.4-1) with the initial value of γ_{ZN} set to zero (because we are only analyzing the Fine Alignment error response to ramping accelerometer error, not initial condition uncertainties). Substituting (14.4-3) into (14.4-1) yields:

$$\underline{C}_{\omega H_1} = K_1\, \underline{u}_{ZN}^N \times \left(\underline{C}_{RH_0} + \underline{C}_{RH_1}\, t\right)$$

$$\underline{C}_{\gamma H_1} = \underline{C}_{\omega H_0} + \underline{C}_{\omega H_1}\, t + K_2\, \underline{u}_{ZN}^N \times \left(\underline{C}_{RH_0} + \underline{C}_{RH_1}\, t\right)$$

$$\underline{C}_{vH_1} = \left(C_B^N\, \delta\dot{\underline{a}}_{SF}^B\right)_H t + g\left[\underline{u}_{ZN}^N \times \left(\underline{C}_{\gamma H_0} + \underline{C}_{\gamma H_1}\, t\right)\right] - K_3\left(\underline{C}_{RH_0} + \underline{C}_{RH_1}\, t\right) \qquad (14.4\text{-}4)$$

$$\underline{C}_{RH_1} = \underline{C}_{vH_0} + \underline{C}_{vH_1}\, t - K_4\left(\underline{C}_{RH_0} + \underline{C}_{RH_1}\, t\right)$$

Equations (14.4-4) are readily solvable (as in the previous section) by grouping and equating terms with like powers of t for each (14.4-4) expression, and then solving for the (14.4-3) coefficients by combining results through algebraic manipulation. The coefficients so obtained are constant and independent of the K_1, K_2, K_3, K_4 gains, thereby validating the (14.4-3) model for the general case of time varying gains. Substituting the coefficients into (14.4-3) then yields the (14.4-1) solution:

$$\gamma_{ZN} = 0$$

$$\delta\underline{\omega}_{IE_H}^{N} = \frac{1}{g}\,\underline{u}_{ZN}^{N} \times \left(C_B^N\,\delta\dot{\underline{a}}_{SF}^{B}\right)_H$$

$$\underline{\gamma}_H^N = \frac{1}{g}\,\underline{u}_{ZN}^{N} \times \left(C_B^N\,\delta\dot{\underline{a}}_{SF}^{B}\right)_H t \qquad (14.4\text{-}5)$$

$$\delta\underline{v}_H^N = 0$$

$$\delta\Delta\underline{R}_H^N = 0$$

Substitution of the (14.4-5) results in (14.4-1) further verifies that (14.4-5) satisfies (14.4-1) without gain restriction (other than the assumption of stable gains that will converge (14.4-1) to a stable steady state).

We now obtain the solution for $\underline{\psi}^N$ produced by accelerometer error ramping during Fine Alignment, by substituting (14.4-5) into Equations (14.4-2), and applying the (3.1.1-16) vector triple cross-product identity:

$$\underline{\psi}_H^N = \frac{1}{g}\,\underline{u}_{ZN}^{N} \times \left(C_B^N\,\delta\dot{\underline{a}}_{SF}^{B}\right)_H t$$

$$\psi_{ZN} = \frac{1}{g\,\omega_{IE_H}^2}\left[\underline{\omega}_{IE_H}^{N} \cdot \left(C_B^N\,\delta\dot{\underline{a}}_{SF}^{B}\right)_H\right] \qquad (14.4\text{-}6)$$

Equations (14.4-6) show that the effect of a ramping horizontal accelerometer during Fine Alignment is to produce a horizontal tilt proportional to the cumulative accelerometer error $\left(C_B^N\,\delta\dot{\underline{a}}_{SF}^{B}\right)_H t$, and a heading error proportional to the north component of accelerometer error rate. If Equations (14.4-6) are compared with (14.3-38) we see that the north accelerometer rate (divided by g) has the identical effect on initial heading error as east angular rate sensor error. This places a tight turn-on transient requirement for the north accelerometer to minimize its effect on initial heading error.

14.5 CORRELATION BETWEEN SENSOR ERRORS DURING INITIAL ALIGNMENT AND NAVIGATION

The navigation errors generated in an INS are produced by inertial sensor error, initial attitude/heading error, etc. Thus far we have treated the effects of inertial sensor error and initial attitude error (horizontal tilt and heading error) as independent sources of navigation error. In fact, because the initial alignment process utilizes the inertial sensors, the initial attitude errors for navigation are correlated with the inertial sensor errors during navigation. In this section we analyze the effect of inertial-sensor/initial-attitude error correlation on inertial navigation performance.

We begin with the attitude and velocity error rate expressions from Equations (12.5.1-1):

$$
\begin{aligned}
\dot{\underline{\psi}}^N &= - C_B^N \, \delta\underline{\omega}_{IB}^B - \underline{\omega}_{IN}^N \times \underline{\psi}^N + C_B^N \left(\underline{\omega}_{IB}^B \times \delta\underline{\alpha}_{Quant} \right) \\
\delta\dot{\underline{V}}^N &= C_B^N \, \delta\underline{a}_{SF}^B + \underline{a}_{SF}^N \times \underline{\psi}^N - \frac{g}{R} \, \delta\underline{R}_H^N - \left(\underline{\omega}_{IE}^N + \underline{\omega}_{IN}^N \right) \times \delta\underline{V}^N + \delta\underline{g}_{Mdl}^N \\
&\quad + \left[\left(F(h) \, \frac{g}{R} - C_2 \right) \delta R + C_2 \, \delta h_{Prsr} - \delta e_{vc_3} \right] \underline{u}_{ZN}^N \\
&\quad - \left(\underline{a}_{SF}^N \times \right) C_B^N \, \delta\underline{\alpha}_{Quant} - \left[\left(C_B^N \, \underline{\omega}_{IB}^B + \underline{\omega}_{IE}^N \right) \times \right] C_B^N \, \delta\underline{\upsilon}_{Quant}
\end{aligned}
\qquad (14.5\text{-}1)
$$

$$
F(h) = 2 \quad \text{For} \quad h \geq 0 \qquad\qquad F(h) = -1 \quad \text{For} \quad h < 0
$$

As in past sections, we define particular terms in (14.5-1) as the sum of their constituents:

$$
\begin{aligned}
\underline{a}_{SF}^N &= g \, \underline{u}_{ZN}^N + \Delta\underline{a}_{SF}^N \\
\underline{\omega}_{IN}^N &= \underline{\omega}_{IE}^N + \underline{\omega}_{EN}^N \\
\underline{\psi}^N &= \underline{\psi}_H^N + \psi_{ZN} \, \underline{u}_{ZN}^N
\end{aligned}
\qquad (14.5\text{-}2)
$$

where

$\underline{u}_{ZN}^N$ = Unit vector along the N frame Z axis which lies along the upward local geodetic vertical.

g = Plumb-bob gravity magnitude. The $g \, \underline{u}_{ZN}^N$ term in (14.5-2) represents the component of $\underline{a}_{SF}^N$ (specific force acceleration) required to balance the INS against plumb-bob gravity. If $\underline{a}_{SF}^N$ equaled $g \, \underline{u}_{ZN}^N$ at a stationary earth fixed position, the INS would remain stationary.

$\Delta \underline{a}_{SF}^{N}$ = Component of $\underline{a}_{SF}^{N}$ attributable to variations from the portion used to balance against plumb-bob gravity. $\Delta \underline{a}_{SF}^{N}$ is produced principally by maneuver induced acceleration.

$\underline{\psi}_{H}^{N}$, ψ_{ZN} = Horizontal, vertical components of attitude error $\underline{\psi}^{N}$.

$\underline{\omega}_{IE}^{N}$, $\underline{\omega}_{EN}^{N}$, $\underline{\omega}_{IN}^{N}$ = Earth fixed E Frame angular rate relative to inertial I Frame space (i.e., "earth rate"), navigation N Frame angular rate relative to the E Frame (i.e., "transport rate"), N Frame angular rate relative to the I Frame (i.e., "platform rate"), all projected on N Frame axes.

We can also write for $\underline{\omega}_{IE}^{N}$ from Equations (6.1.2-2):

$$\underline{\omega}_{IE}^{N} = \underline{\omega}_{IE_H}^{N} + \underline{u}_{ZN}^{N}\, \omega_e \sin l \tag{14.5-3}$$

where

ω_e = Earth's inertial angular rate magnitude.

l = Geodetic latitude.

Substituting (14.5-2) and (14.5-3) into (14.5-1) yields for the vertical/horizontal components of $\dot{\underline{\psi}}^{N}$ and the horizontal component of $\delta\dot{\underline{V}}_{H}^{N}$:

$$\dot{\psi}_{ZN} = -\left(C_B^N\, \delta\underline{\omega}_{IB}^{B} + \underline{\omega}_{IN}^{N} \times \underline{\psi}^{N}\right) \cdot \underline{u}_{ZN}^{N} + \cdots$$

$$\begin{aligned}
\dot{\underline{\psi}}_{H}^{N} &= \left[- C_B^N\, \delta\underline{\omega}_{IB}^{B} - \left(\underline{\omega}_{IE_H}^{N} + \underline{u}_{ZN}^{N}\, \omega_e \sin l + \underline{\omega}_{EN}^{N}\right) \times \left(\underline{\psi}_{H}^{N} + \psi_{ZN}\, \underline{u}_{ZN}^{N}\right) + \cdots \right]_H \\
&= -\left(C_B^N\, \delta\underline{\omega}_{IB}^{B}\right)_H - \psi_{ZN}\, \underline{\omega}_{IE_H}^{N} \times \underline{u}_{ZN}^{N} - \omega_e \sin l\, \underline{u}_{ZN}^{N} \times \underline{\psi}_{H}^{N} - \left(\underline{\omega}_{EN}^{N} \times \underline{\psi}^{N}\right)_H + \cdots
\end{aligned} \tag{14.5-4}$$

$$\begin{aligned}
\delta\dot{\underline{V}}_{H}^{N} &= \left(C_B^N\, \delta\underline{a}_{SF}^{B}\right)_H + \left[\left(g\, \underline{u}_{ZN}^{N} + \Delta\underline{a}_{SF}^{N}\right) \times \underline{\psi}^{N}\right]_H + \cdots \\
&= \left(C_B^N\, \delta\underline{a}_{SF}^{B}\right)_H + g\, \underline{u}_{ZN}^{N} \times \underline{\psi}_{H}^{N} + \left(\Delta\underline{a}_{SF}^{N} \times \underline{\psi}^{N}\right)_H + \cdots
\end{aligned}$$

where

H = Designator for horizontal component.

Let us now define $\underline{\psi}^{N}$ and its components as being composed of their initial value plus changes since the start of navigation:

$$\underline{\psi}^N = \underline{\psi}_0^N + \Delta\underline{\psi}^N \qquad \underline{\psi}_H^N = \underline{\psi}_{H_0}^N + \Delta\underline{\psi}_H^N \qquad \psi_{ZN} = \psi_{ZN_0} + \Delta\psi_{ZN} \tag{14.5-5}$$

where

$\underline{\psi}_0^N, \underline{\psi}_{H_0}^N, \psi_{ZN_0}$ = Values for $\underline{\psi}^N, \underline{\psi}_H^N, \psi_{ZN}$ at the end of Fine Alignment which is the start of Navigation.

$\Delta\underline{\psi}^N, \Delta\underline{\psi}_H^N, \Delta\psi_{ZN}$ = Change in $\underline{\psi}^N, \underline{\psi}_H^N, \psi_{ZN}$ since the start of Navigation.

Substituting (14.5-5) into (14.5-4) obtains:

$$\Delta\dot{\psi}_{ZN} = -\left[C_B^N\, \delta\underline{\omega}_{IB}^B + \underline{\omega}_{IN}^N \times \left(\underline{\psi}_0^N + \Delta\underline{\psi}^N\right)\right] \cdot \underline{u}_{ZN}^N + \cdots \tag{14.5-6}$$

$$\begin{aligned}\Delta\dot{\underline{\psi}}_H^N = & -\left(C_B^N\, \delta\underline{\omega}_{IB}^B\right)_H - \psi_{ZN_0}\, \underline{\omega}_{IE_H}^N \times \underline{u}_{ZN}^N - \omega_e \sin l\, \underline{u}_{ZN}^N \times \underline{\psi}_{H_0}^N \\ & - \Delta\psi_{ZN}\, \underline{\omega}_{IE_H}^N \times \underline{u}_{ZN}^N - \omega_e \sin l\, \underline{u}_{ZN}^N \times \Delta\underline{\psi}_H^N - \left[\underline{\omega}_{EN}^N \times \left(\underline{\psi}_0^N + \Delta\underline{\psi}^N\right)\right]_H + \cdots\end{aligned} \tag{14.5-7}$$

$$\delta\dot{\underline{V}}_H^N = \left(C_B^N\, \delta\underline{a}_{SF}^B\right)_H + g\, \underline{u}_{ZN}^N \times \underline{\psi}_{H_0}^N + g\, \underline{u}_{ZN}^N \times \Delta\underline{\psi}_H^N + \left[\Delta\underline{a}_{SF}^N \times \left(\underline{\psi}_0^N + \Delta\underline{\psi}^N\right)\right]_H + \cdots \tag{14.5-8}$$

The effect of inertial-sensor/initial-attitude error correlation is produced by the components of sensor error that are the same during the alignment and navigation modes (i.e., the constant sensor error components). From Equations (14.3-38) we know that for constant sensor errors:

$$\underline{\psi}_{H_0}^N = \frac{1}{g}\left[\underline{u}_{ZN}^N \times \left(C_B^N\, \delta\underline{a}_{SF}^B\right)_{H_{Align}}\right] + \text{Negligible} \tag{14.5-9}$$

$$\begin{aligned}\psi_{ZN_0} = \frac{1}{\omega_{IE_{H/Align}}^2}\, \underline{u}_{ZN}^N \cdot \Big\{ & \underline{\omega}_{IE_{H/Align}}^N \times \left[\left(C_B^N\, \delta\underline{\omega}_{IB}^B\right)_{H_{Align}} \right. \\ & \left. - \frac{1}{g}\left(C_B^N\, \delta\underline{a}_{SF}^B\right)_{H_{Align}} \omega_e \sin l_{Align}\right]\Big\} + \text{Negligible}\end{aligned} \tag{14.5-10}$$

where

Align = Subscript denoting the parameter value during the Fine Alignment process.

and in which the $\delta\underline{a}_{SF}^B$ and $\delta\underline{\omega}_{IB}^B$ sensor errors are considered to only contain constant error contributors. This will be our interpretation for the remainder of this section. Remaining time varying sensor errors can be considered as included in the $\cdots$ terms in (14.5-6) - (14.5-8). The "Negligible" terms in (14.5-9) and (14.5-10) are proportional to vertical angular rate sensor error and north accelerometer error which are small contributions compared to the other terms (See Section 14.3 for discussion).

Equation (14.5-10) can be rearranged using the generalized Equation (3.1.1-35) mixed vector dot/cross product identity:

$$\psi_{ZN_0} = -\frac{1}{\omega_{IE_{H/Align}}^{2}} \left[\left(C_B^N\, \delta\underline{\omega}_{IB}^{B}\right)_{H_{Align}} - \frac{1}{g}\left(C_B^N\, \delta\underline{a}_{SF}^{B}\right)_{H_{Align}} \omega_e \sin l_{Align} \right] \cdot \left[\underline{\omega}_{IE_{H/Align}}^{N} \times \underline{u}_{ZN}^{N} \right]$$

$$= -\frac{1}{\omega_{IE_{H/Align}}} \left[\left(C_B^N\, \delta\underline{\omega}_{IB}^{B}\right)_{H_{Align}} - \frac{1}{g}\left(C_B^N\, \delta\underline{a}_{SF}^{B}\right)_{H_{Align}} \omega_e \sin l_{Align} \right] \cdot \underline{u}_{East_{Align}}^{N} \qquad (14.5\text{-}11)$$

with

$$\underline{u}_{East}^{N} = \frac{1}{\omega_{IE_H}}\, \underline{\omega}_{IE_H}^{N} \times \underline{u}_{ZN}^{N} \qquad (14.5\text{-}12)$$

where

$\underline{u}_{East}^{N}$ = Horizontal unit vector pointing east (i.e., along a horizontal perpendicular to north pointing $\underline{\omega}_{IE_H}^{N}$).

Substituting (14.5-9), (14.5-11) and (14.5-12) into (14.5-7) and applying generalized Equation (13.1-5) then yields for $\Delta\dot{\underline{\psi}}_H^N$:

$$\Delta\dot{\underline{\psi}}_H^N = -\left[\left(C_B^N\, \delta\underline{\omega}_{IB}^{B}\right)_{H} - \frac{1}{g}\, \omega_e \sin l \left(C_B^N\, \delta\underline{a}_{SF}^{B}\right)_{H_{Align}} \right]$$

$$+ \frac{\omega_{IE_H}}{\omega_{IE_{H/Align}}} \left\{ \left[\left(C_B^N\, \delta\underline{\omega}_{IB}^{B}\right)_{H_{Align}} - \frac{1}{g}\left(C_B^N\, \delta\underline{a}_{SF}^{B}\right)_{H_{Align}} \omega_e \sin l_{Align} \right] \cdot \underline{u}_{East_{Align}}^{N} \right\} \underline{u}_{East}^{N} \qquad (14.5\text{-}13)$$

$$- \Delta\psi_{ZN}\, \omega_{IE_H}\, \underline{u}_{East}^{N} - \omega_e \sin l\, \underline{u}_{ZN}^{N} \times \Delta\underline{\psi}_H^N - \left[\underline{\omega}_{EN}^{N} \times \left(\underline{\psi}_0^N + \Delta\underline{\psi}^N \right) \right]_H + \cdots$$

From the discussion following Equation (14.3-38) we can write for the horizontal component of earth rate:

$$\omega_{IE_H} = \omega_e \cos l \qquad (14.5\text{-}14)$$

Substituting (14.5-14) into (14.5-13), (14.5-9) into (14.5-8), and applying generalized Equation (13.1-5) yields the final result for $\Delta\dot{\underline{\psi}}_H^N$ and $\delta\dot{\underline{V}}_H^N$ with $\Delta\dot{\psi}_{ZN}$ from (14.5-6):

$$\Delta\dot{\psi}_{ZN} = -\left[C_B^N \, \delta\underline{\omega}_{IB}^B + \underline{\omega}_{IN}^N \times \left(\underline{\psi}_0^N + \Delta\underline{\psi}^N\right)\right] \cdot \underline{u}_{ZN}^N + \cdots$$

$$\Delta\dot{\underline{\psi}}_H^N = -\left[\left(C_B^N \, \delta\underline{\omega}_{IB}^B\right)_H - \frac{1}{g} \, \omega_e \sin l \left(C_B^N \, \delta\underline{a}_{SF}^B\right)_{H_{Align}}\right]$$
$$+ \frac{\cos l}{\cos l_{Align}} \left\{\left[\left(C_B^N \, \delta\underline{\omega}_{IB}^B\right)_{H_{Align}} - \frac{1}{g} \, \omega_e \sin l_{Align} \left(C_B^N \, \delta\underline{a}_{SF}^B\right)_{H_{Align}}\right] \cdot \underline{u}_{East_{Align}}^N\right\} \underline{u}_{East}^N$$
$$- \Delta\psi_{ZN} \, \omega_e \cos l \, \underline{u}_{East}^N - \omega_e \sin l \, \underline{u}_{ZN}^N \times \Delta\underline{\psi}_H^N - \left[\underline{\omega}_{EN}^N \times \left(\underline{\psi}_0^N + \Delta\underline{\psi}^N\right)\right]_H + \cdots \qquad (14.5\text{-}15)$$

$$\delta\dot{\underline{V}}_H^N = \left(C_B^N \, \delta\underline{a}_{SF}^B\right)_H - \left(C_B^N \, \delta\underline{a}_{SF}^B\right)_{H_{Align}}$$
$$+ g \, \underline{u}_{ZN}^N \times \Delta\underline{\psi}_H^N + \left[\Delta\underline{a}_{SF}^N \times \left(\underline{\psi}_0^N + \Delta\underline{\psi}^N\right)\right]_H + \cdots$$

The $\delta\dot{\underline{V}}_H^N$ expression in (14.5-15) shows that the horizontal tilt created during the initial alignment process generates a horizontal acceleration $\left(C_B^N \, \delta\underline{a}_{SF}^B\right)_{H_{Align}}$ that compensates the constant horizontal accelerometer error during navigation $\left(C_B^N \, \delta\underline{a}_{SF}^B\right)_H$, thereby reducing $\delta\underline{V}_H^N$ horizontal velocity error build-up. The compensation only holds, however, if the attitude C_B^N during navigation matches the attitude during alignment, and the accelerometer error $\delta\underline{a}_{SF}^B$ remains the same during alignment and navigation. The ideal compensation situation occurs for constant attitude and non-maneuvering navigation at the alignment attitude (e.g., on a stationary test fixture). The worst case situation occurs for navigation at an attitude that is in the opposite heading direction from the alignment attitude (e.g., executing a 180 degree heading rotation following alignment) which has the effect of reversing the polarity of $\left(C_B^N \, \delta\underline{a}_{SF}^B\right)_H$ so that in $\delta\dot{\underline{V}}_H^N$, it adds to (rather than subtracts from) $\left(C_B^N \, \delta\underline{a}_{SF}^B\right)_{H_{Align}}$. In most applications, the heading during navigation can be at arbitrary orientation relative to the alignment heading, hence, the potential benefit of alignment-tilt/accelerometer-error compensation cannot be counted on. The net result is a tighter long term stability requirement placed on the accelerometer because calibration from initial tilt is generally not viable.

The $\Delta\dot{\underline{\psi}}_H^N$ expression in (14.5-15) shows that the heading error created during the initial alignment process generates an east earth rate coupling term $\frac{\cos l}{\cos l_{Align}} \left[\left(C_B^N \, \delta\underline{\omega}_{IB}^B\right)_{H_{Align}} - \frac{1}{g} \, \omega_e \sin l_{Align} \left(C_B^N \, \delta\underline{a}_{SF}^B\right)_{H_{Align}}\right] \cdot \underline{u}_{East_{Align}}^N$ that compensates the east component of the inertial sensor error term $- \left[\left(C_B^N \, \delta \underline{\omega}_{IB}^B \right)_H - \frac{1}{g} \, \omega_e \sin l \left(C_B^N \, \delta \underline{a}_{SF}^B \right)_{H_{Align}} \right] \cdot \underline{u}_{East}^N$, thereby reducing $\Delta\psi_H^N$ horizontal tilt error build-up and associated coupling into $\delta \underline{V}_H^N$ horizontal velocity error. The compensation holds if the navigation latitude remains near the alignment latitude, the attitude C_B^N during navigation matches the attitude during alignment, and the inertial sensor errors ($\delta \underline{\omega}_{IB}^B$ and $\delta \underline{a}_{SF}^B$) remain the same during alignment and navigation (Note that the dominant error term in $- \left[\left(C_B^N \, \delta \underline{\omega}_{IB}^B \right)_H - \frac{1}{g} \, \omega_e \sin l \left(C_B^N \, \delta \underline{a}_{SF}^B \right)_{H_{Align}} \right] \cdot \underline{u}_{East}^N$ is the east component of $\delta \underline{\omega}_{IB}$). The ideal compensation situation occurs for constant attitude and non-maneuvering navigation at the alignment attitude (e.g., on a stationary test fixture). The worst case situation occurs for navigation at an attitude C_B^N that is in the opposite heading direction from the alignment attitude (e.g., executing a 180 degree heading rotation following alignment). This has the effect of reversing the polarity of the east angular rate sensor bias $- \left(C_B^N \, \delta \underline{\omega}_{IB}^B \right)_H \cdot \underline{u}_{East}^N$ so that in $\Delta \dot{\underline{\psi}}_H^N$, it adds to (rather than subtracts from) $\frac{\cos l}{\cos l_{Align}} \left(C_B^N \, \delta \underline{\omega}_{IB}^B \right)_{H_{Align}} \cdot \underline{u}_{East_{Align}}^N$. In most applications, the heading during navigation can be at arbitrary orientation relative to the alignment heading, hence, the potential benefit of alignment-heading-error/east-angular-rate-sensor (and accelerometer) error compensation cannot be counted on. The net result is a tighter long term stability requirement placed on the angular rate sensors because calibration from initial heading alignment is generally not viable.

It should be noted that during the alignment process, the estimated horizontal earth rate in the north direction compared with the correct value (i.e., $\omega_e \cos l$) can be utilized to calibrate the north angular rate sensor error. This can be achieved in practice in a strapdown INS by transforming the estimated north angular rate sensor error into the sensor frame (B Frame) in which it would be then applied as part of sensor compensation operations. The utility of this approach depends on the uncertainty in the estimated north error component (due to sensor noise and transitory effects) compared to the calibration accuracy required. For stable angular rate sensors, the process can be enhanced by implementing the calibration procedure as a smoothing summation process over successive initial alignments. With such an approach, the angular rate sensor calibration coefficients determined at the end of each alignment (as described previously) would be attenuated by a weighting coefficient, and then added to the previously determined smoothed calibration coefficients (multiplied by one minus the weighting coefficient). The result would then represent the updated smoothed calibration coefficients and would be used for the sensor calibration correction during navigation.

Finally, we also point out that the potential compensation of sensor error by initial-tilt/heading-error also depends on the navigation position remaining not far removed from the alignment position so that in Equations (14.5-15), the integrated effect of $\underline{\omega}_{EN}^{N}$ transport rate coupling is small, and the components of $\underline{\omega}_{IE}^{N}$ and $\underline{u}_{East}^{N}$ remain near the alignment values.

14.6 INITIAL ALIGNMENT ERROR CAUSED BY RANDOM SENSOR ERRORS AND DISTURBANCES

In this section we will examine the error response of the (14.2-18) quasi-stationary Fine Alignment error equations to random inertial sensor output errors and random external disturbances. The feedback control gains in Equations (14.2-18) are generally derived from Kalman filter theory (as is subsequently discussed in detail in Section 15.2.1 and its subsections). The Fine Alignment error response in this section will be determined in terms of the estimated horizontal earth rate error variance calculated within the Fine Alignment Kalman filter covariance matrix. As discussed in Section 14.2, it is the earth rate estimation error that determines initial heading alignment inaccuracy, the critical performance design parameter for the overall initial alignment operation.

The analysis approach used in this section makes extensive use of fundamental background material from Chapter 15. To more easily comprehend the material in this section, the reader is encouraged to first read Chapter 15, Sections 15.0 through 15.1.2.1.1.3, 15.2.1 through 15.2.1.2, and 15.1.5 through 15.1.5.4.1.

The analytical solutions to be developed will be derived by analytically integrating the continuous form Kalman filter error state covariance propagation/reset equation developed in Section 15.1.5.3.1 and whose general solutions are provided in Sections 15.1.5.4 and 15.1.5.4.1. First we will analytically state the problem to be analyzed in terms of a partitioned form of the quasi-stationary Fine Alignment process error state dynamic equation and its associated Kalman filter input measurement equation. Next we will use the Section 15.1.5.4 equations to attempt a complete solution for the earth rate estimation error that includes all noise terms. Faced with the complexity of the task we will simplify the problem to only include measurement and angular rate sensor output noise. A closed-form analytical solution will be obtained, but will prove too unwieldy for practical use in interpreting general error behavior. To further simplify the analysis, we will then attempt to develop individual solutions in response to each noise source applied separately. The response to measurement noise will be found by direct application of the Section 15.1.5.4 equations. The response to process noise (i.e., with zero measurement noise) will be found by application of the Section 15.1.5.4.1 procedure that uses a revised measurement equation having its measurement noise equal to a partition of the process noise vector. To simplify the process noise analysis, we will apply the Section 15.1.5.4 equations successively for individual process noise inputs acting alone. Finally, we will show that the sum of solutions obtained for the individual measurement and process noise sources is a reasonable approximation to the solution with all noise sources applied simultaneously.

To simplify the notation in the following sections, we will drop the functional (t) designation used in Chapter 15, except for cases where it is necessary for clarity.

14.6.1 ANALYTICAL PROBLEM DEFINITION

The (14.2-18) Fine Alignment process error equations have the following general closed-loop analytical form:

$$\dot{\underline{\zeta}} = A\,\underline{\zeta} + G_P\,\underline{n}_P - K^{\blacklozenge}\,\underline{\nu} \qquad \underline{\nu} = H\,\underline{\zeta} + G_M\,\underline{n}_M^{\blacklozenge} \tag{14.6.1-1}$$

where

$\underline{\zeta}$ = Vector of quasi-stationary Fine Alignment closed-loop controlled error states formed from the $\delta\underline{\omega}_{IE_H}^N$, $\underline{\gamma}_H^N$, $\delta\underline{v}_H^N$, $\delta\Delta\underline{R}_H^N$ terms in Equations (14.2-18).

$K^{\blacklozenge}$ = Matrix formed from the K_1 - K_4 feedback gains in (14.2-18).

$\underline{\nu}$ = $\underline{\zeta}$ feedback correction vector formed from the terms containing $K^{\blacklozenge}$ in (14.2-18).

A, $\underline{n}_P$, G_P = The error state dynamic matrix, process noise vector and process noise dynamic coupling matrix formed from the error equations by appropriate grouping of coefficients multiplying $\underline{\zeta}$ and the noise terms exclusive of those containing $K^{\blacklozenge}$.

H, $\underline{n}_M^{\blacklozenge}$, G_M = The measurement matrix, measurement noise vector and measurement noise dynamic coupling matrix formed from the error equations by appropriate grouping of coefficients multiplying $\underline{\zeta}$ and the noise terms exclusive of those containing $K^{\blacklozenge}$.

$(\,)^{\blacklozenge}$ = Notation signifying that ( ) is a parameter in the continuous form Equations (14.2-18) that is equivalent and analytically related to the same parameter ( ) appearing in the discrete recursive Kalman filter closed-loop configuration (as discussed in Section 15.1.5.3.2) that would be implemented in the actual real-time system computer.

Note that γ_{ZN} in (14.2-18) is not included in the above $\underline{\zeta}$ definition (hence, excluded in the (14.6.1-1) $\underline{\zeta}$ error state dynamic equation). Based on experience, the γ_{ZN} term has negligible impact on initial alignment performance (particularly regarding the portion of γ_{ZN} affected by input noise, the general analysis topic of this section).

Equations (14.6.1-1) represent the continuous form of the actual discrete closed-loop Kalman estimation/control error equations developed in Sections 15.2.1 and 15.2.1.1 for the quasi-stationary Fine Alignment process. Equations (14.6.1-1) are also identical in form to Equations (15.1.5.3.2-22) and (15.1.5.3.2-14) of Section 15.1.5.3.2, the general closed-loop continuous estimation/reset equations. The Section 15.1.5.3.2 closed-loop equations are developed as a special case of the more general (15.1.5.3.2-3) and (15.1.5.3.2-13) - (15.1.5.3.2-14) continuous estimation/control equations:

$$\dot{\underline{x}} = A\,\underline{x} + G_P\,\underline{n}_P + \underline{u}_c^{\blacklozenge}$$

$$\dot{\tilde{\underline{x}}} = \left(A - K^{\blacklozenge} H\right)\tilde{\underline{x}} + K^{\blacklozenge}\,\underline{z}^{\blacklozenge} + \underline{u}_c^{\blacklozenge} \qquad \underline{z}^{\blacklozenge} = H\,\underline{x} + G_M\,\underline{n}_M^{\blacklozenge} \tag{14.6.1-2}$$

where

$\underline{x}$ = Vector of uncontrolled error states.

$\tilde{\underline{x}}$ = The estimate for $\underline{x}$.

$\underline{z}^{\blacklozenge}$ = The continuous form measurement vector.

$K^{\blacklozenge}$ = The continous form estimation gain matrix.

$\underline{u}_c^{\blacklozenge}$ = The continuous form control vector.

Equations (14.6.1-1) are derived from (14.6.1-2) by setting $\underline{u}_c^{\blacklozenge}$ to $- K^{\blacklozenge}\,\underline{z}^{\blacklozenge}$ which, as shown in Section 15.1.5.3.2, is the equivalent to setting $\underline{u}_c^{\blacklozenge}$ such that the $\tilde{\underline{x}}$ estimate is continuously controlled to zero. Section 15.1.5.3.2 shows that under these conditions, the controlled error state vector becomes:

$$\underline{\zeta} = -\,\Delta\underline{x} \tag{14.6.1-3}$$

in which

$$\Delta\underline{x} \equiv \tilde{\underline{x}} - \underline{x} \tag{14.6.1-4}$$

where

$\Delta\underline{x}$ = Uncertainty in $\tilde{\underline{x}}$ (or the error in the estimated error state vector $\tilde{\underline{x}}$).

By (14.6.1-3) we see that we can evaluate $\underline{\zeta}$ in (14.6.1-1) by analyzing $\Delta\underline{x}$. For general noise analysis problems, this is more easily achieved by analyzing the covariance of $\Delta\underline{x}$ defined by (15.1.2.1-4) as:

$$P \equiv \mathcal{E}\left(\Delta\underline{x}\;\Delta\underline{x}^{T}\right) \tag{14.6.1-5}$$

where

P = Estimated error state uncertainty covariance matrix.

$\mathcal{E}(\ )$ = The expected value operator (i.e., average statistical value).

Section 15.1.5.3.1 shows that for a Kalman type estimation process (i.e., using the continuous form Kalman gain matrix for $K^{\blacklozenge}$), the rate of change of the covariance matrix is given by (15.1.5.3.1-23):

$$\dot{P} = A\,P + P\,A^T + G_P\,Q_{P_{Dens}}\,G_P^T - P\,H^T \left(G_M\,R^{\blacklozenge}\,G_M{}^T\right)^{-1} H\,P \qquad (14.6.1\text{-}6)$$

where

$Q_{P_{Dens}}$ = White noise density matrix associated with $\underline{n}_P$.

$R^{\blacklozenge}$ = White noise density matrix associated with $\underline{n}_M^{\blacklozenge}$.

The general analytical integral solution to (14.6.1-6) is provided in Section 15.1.5.4 by Equations (15.1.5.4-7), (15.1.5.4-8) and (15.1.5.4-11).

Thus, we can analyze the (14.6.1-1) ζ response to noise by assessing the response of P as the analytical integral of (14.6.1-6). This will be the general analysis approach we will use in the sub-sections to follow.

Section 15.1.5.3.2 shows that Equation (14.6.1-6) is also valid for a pure estimator type Kalman filter in which the $\underline{u}_c^{\blacklozenge}$ control vector is set to zero. Then (14.6.1-2) becomes for the error state dynamic and measurement equations:

$$\dot{\underline{x}} = A\,\underline{x} + G_P\,\underline{n}_P \qquad \underline{z}^{\blacklozenge} = H\,\underline{x} + G_M\,\underline{n}_M^{\blacklozenge} \qquad (14.6.1\text{-}7)$$

Equation (14.6.1-7) is the generic form we will use in this section as the error state dynamic and measurement equations for Kalman filter design, for which (14.6.1-6) is the resulting covariance rate equation.

Finally, to expedite problem analysis, we will study the partitioned form of ζ which divides ζ into two separate four-component vectors, one for each of the horizontal N Frame axes. Section 15.2.1.1 shows that the partitioned error state vectors are completely independent from one another with each having identical error state uncertainty covariance matrices. The general form of each partitioned error state vector with its process noise vector and pertinent matrices for (14.6.1-6) are given by (15.2.1.1-10) - (15.2.1.1-11) with (15.1.5.3.1-24) for $R^{\blacklozenge}$. To simplify the analysis, we will use normalized versions of the (15.2.1.1-11) partitioned error state and process noise vectors as follows:

$$\underline{x}' = \left(\delta\omega_{IE_H}, \gamma_H, \frac{1}{g}\,\delta v_H, \frac{1}{g}\,\delta\Delta R_H\right)^T$$
$$\underline{n}'_P = \left(\delta\omega_{Rand_H}, \delta\alpha_{Quant_H}, \frac{1}{g}\,\delta a_{Rand_H}, \frac{1}{g}\,\delta\upsilon_{Quant_H}\right)^T \tag{14.6.1-8}$$

where

$'$ = Designation for the normalized version of the equivalent (15.2.1.1-10) - (15.2.1.1-11) * partitioned parameter.

$\underline{x}', \underline{n}'_P$ = Normalized versions of the (15.2.1.1-10) partitioned error state and process noise vectors.

$\delta\omega_{IE_H}, \gamma_H, \delta v_H, \delta\Delta R_H$ = Components of $\delta\underline{\omega}^N_{IE_H}, \underline{\gamma}^N_H, \delta\underline{v}^N_H, \delta\Delta\underline{R}^N_H$ in (14.2-18) along one of the N Frame horizontal axes.

$\delta\omega_{Rand_H}, \delta\alpha_{Quant_H}, \delta a_{Rand_H}, \delta\upsilon_{Quant_H}$ = Components of $\delta\underline{\omega}_{Rand}, \delta\underline{\alpha}_{Quant}, \delta\underline{a}_{Rand}, \delta\underline{\upsilon}_{Quant}$ in (14.2-18) along one of the N Frame horizontal axes.

We also note that for this analysis, the $\underline{n}'_P$ process noise vector excludes the vibration-quantization noise products in Equations (14.2-18) which are generally small compared to the individual quantization noise terms.

Equations (14.6.1-8) are equivalent to operating on the (15.2.1.1-11) vectors with the mapping transformation:

$$\underline{x}' = B\,\underline{x}^* \qquad \underline{n}'_P = B\,\underline{n}_{P^*} \qquad B = \begin{bmatrix} 1 & 0 & 0 & 0 \\ 0 & 1 & 0 & 0 \\ 0 & 0 & \frac{1}{g} & 0 \\ 0 & 0 & 0 & \frac{1}{g} \end{bmatrix} \tag{14.6.1-9}$$

where

B = Normalization matrix.

$\underline{x}^*, \underline{n}_{P^*}$ = Partitioned error state and process noise vectors from (15.2.1.1-11).

From (14.6.1-9) we also see that:

$$\underline{x}^* = B^{-1}\,\underline{x}' \qquad \underline{n}_{P^*} = B^{-1}\,\underline{n}'_P \qquad B^{-1} = \begin{bmatrix} 1 & 0 & 0 & 0 \\ 0 & 1 & 0 & 0 \\ 0 & 0 & g & 0 \\ 0 & 0 & 0 & g \end{bmatrix} \qquad (14.6.1\text{-}10)$$

The normalized version of Equations (15.2.1.1-10) - (15.2.1.1-11) with (15.1.5.3.1-24) is obtained by pre-multiplying the (15.2.1.1-10) $\dot{\underline{x}}^*$ equation by B, substituting (14.6.1-10) for $\underline{x}^*$, $\underline{n}_{P^*}$ in the result, identifying the terms multiplying $\underline{x}'$, $\underline{n}'_P$ as revised error state dynamic and process noise dynamic coupling matrices A', $G_{P'}$, substituting $\underline{x}^*$ from (14.6.1-10) in the (15.2.1.1-10) z^*_n expression and identifying the terms multiplying $\underline{x}'_n$ as the revised measurement matrix H'. For simplicity and compatibility with the (14.6.1-7) generic form, we then drop the * notation in the measurement equation and the ′ notation on the A', $G_{P'}$, H' matrices with the understanding that from this point forward, the affected matrices and vectors are compatible with the (14.6.1-8) defined normalized vector components. The result, including the continuous form measurement equation from (14.6.1-7), is:

$$\dot{\underline{x}}' = A\,\underline{x}' + G_P\,\underline{n}'_P \qquad z_n = H\,\underline{x}'_n + G_M\,n_{M_n} \qquad z^{\blacklozenge} = H\,\underline{x}' + G_M\,n_M^{\blacklozenge} \qquad (14.6.1\text{-}11)$$

$$\underline{x}' = \left(\delta\omega_{IE_H},\ \gamma_H,\ \frac{1}{g}\,\delta v_H,\ \frac{1}{g}\,\delta\Delta R_H\right)^T$$

$$\underline{n}'_P = \left(\delta\omega_{Rand_H},\ \delta\alpha_{Quant_H},\ \frac{1}{g}\,\delta a_{Rand_H},\ \frac{1}{g}\,\delta\upsilon_{Quant_H}\right)^T$$

$$n_M = \Delta r_{Vib_H}$$

$$A = \begin{bmatrix} 0 & 0 & 0 & 0 \\ 1 & 0 & 0 & 0 \\ 0 & 1 & 0 & 0 \\ 0 & 0 & 1 & 0 \end{bmatrix} \qquad Q_{P_{Dens}} = \begin{bmatrix} q_{\omega Rand} & 0 & 0 & 0 \\ 0 & q_{\alpha Quant} & 0 & 0 \\ 0 & 0 & \frac{1}{g^2}\,q_{aRand} & 0 \\ 0 & 0 & 0 & \frac{1}{g^2}\,q_{\upsilon Quant} \end{bmatrix} \qquad (14.6.1\text{-}12)$$

$$G_P = \begin{bmatrix} 0 & 0 & 0 & 0 \\ -1 & 0 & 0 & 0 \\ 0 & -1 & 1 & 0 \\ 0 & 0 & 0 & 1 \end{bmatrix}$$

$$H = \begin{bmatrix} 0 & 0 & 0 & g \end{bmatrix} \qquad R^{\blacklozenge} = P_{RVib_H}\,T_n$$

$$G_M = 1 \qquad P_{RVib_H} = \mathcal{E}\left(\Delta r_{Vib_H}^2\right)$$

where

$q_{\omega Rand}$, $q_{\alpha Quant}$, q_{aRand}, $q_{\upsilon Quant}$ = Noise density for each component of $\delta\underline{\omega}_{Rand}$, $\delta\underline{\alpha}_{Quant}$, $\delta\underline{a}_{Rand}$, $\delta\underline{\upsilon}_{Quant}$. Values for the $q_{\alpha Quant}$ and $q_{\upsilon Quant}$ densities can be found using Equations (15.2.1.2-18). Values used for the q_{aRand} and $q_{\omega Rand}$ densities should be based on the characteristics of the particular accelerometers and angular rate sensors being used.

Δr_{Vib_H} = Horizontal per-axis quasi-stationary position random motion.

P_{RVib_H} = Variance of Δr_{Vib_H}.

n = Computation rate index for the actual recursive Kalman alignment filter.

T_n = Discrete n cycle quasi-stationary Fine Alignment process control feedback cycle time.

$z^{\blacklozenge}$ = The input measurement for the continuous form Kalman filter (a scalar in this particular case).

z_n = The input measurement at cycle n for the n cycle Kalman filter (a scalar in this particular case).

n_{M_n} = The measurement noise at cycle n for the n cycle Kalman filter (a scalar in this particular case).

$\overset{\blacklozenge}{n}_M$ = Continuous form white measurement noise defined implicitly as a function of the discrete measurement noise n_{M_n} by Equation (15.1.5.3.1-30) (a scalar in this particular case).

$R^{\blacklozenge}$ = Continuous form measurement noise covariance matrix defined in (15.1.5.3.1-24) in terms of the n_{M_n} variance R_n. At the end of Section 15.1.5.3.1, $R^{\blacklozenge}$ is shown to be the white noise density matrix associated with $\overset{\blacklozenge}{n}_M$.

Equations (14.6.1-11) and (14.6.1-12) constitute the model we will use for the remainder of this chapter to determine the associated covariance matrix from (14.6.1-6) as representative of the response of (14.6.1-11) to process and measurement noise inputs.

Based on the (14.6.1-12) definitions we also see that:

$$G_P \, Q_{P_{Dens}} \, G_P^T = \begin{bmatrix} 0 & 0 & 0 & 0 \\ 0 & q_2 & 0 & 0 \\ 0 & 0 & q_3 & 0 \\ 0 & 0 & 0 & q_4 \end{bmatrix}$$

$$q_2 \equiv q_{\omega Rand} \qquad q_3 \equiv q_{\alpha Quant} + \frac{1}{g^2} q_{aRand} \qquad q_4 \equiv \frac{1}{g^2} q_{\upsilon Quant} \tag{14.6.1-13}$$

$$H^T \left(G_M \, R^{\blacklozenge} \, G_M^T \right)^{-1} H = \begin{bmatrix} 0 & 0 & 0 & 0 \\ 0 & 0 & 0 & 0 \\ 0 & 0 & 0 & 0 \\ 0 & 0 & 0 & \frac{1}{r} \end{bmatrix} \qquad r \equiv \frac{P_{RVib_H} \, T_n}{g^2}$$

where

q_2, q_3, q_4 = Redefined process noise densities for analytic expediency.

r = Redefined normalized continuous coupled measurement noise variance (for analytic expediency).

14.6.2 ANALYTICAL COVARIANCE SOLUTION ATTEMPTS UNDER COMBINED MEASUREMENT AND PROCESS NOISE

The analytical integral solution to the non-linear covariance rate Equation (14.6.1-6) is found in Section 15.1.5.4 by first performing a variable transformation that converts (14.6.1-6) into the Equation (15.1.5.4-7) coupled homogeneous linear differential equation set:

$$\begin{bmatrix} \dot{\underline{y}} \\ \dot{\underline{\lambda}} \end{bmatrix} = \begin{bmatrix} -A^T & H^T \left(G_M \, R^{\blacklozenge} \, G_M^T \right)^{-1} H \\ G_P \, Q_{P_{Dens}} \, G_P^T & A \end{bmatrix} \begin{bmatrix} \underline{y} \\ \underline{\lambda} \end{bmatrix} \tag{14.6.2-1}$$

where

$\underline{y}, \underline{\lambda}$ = Vectors of the same dimension as P defined by (15.1.5.4-1) and (15.1.5.4-2).

The general homogeneous solution to Equation(14.6.2-1) can be written in terms of its state transition matrix (as in (15.1.5.4-8)) which is then used to obtain the covariance matrix P with Equation (15.1.5.4-11).

Let's attempt to execute this procedure by first defining:

$$\underline{y} \equiv \left(y_1, y_2, y_3, y_4 \right)^T \qquad \underline{\lambda} \equiv \left(\lambda_1, \lambda_2, \lambda_3, \lambda_4 \right)^T \tag{14.6.2-2}$$

where

$$y_i, \lambda_i = \text{Elements i of } \underline{y}, \underline{\lambda}.$$

Then using (14.6.2-2), Equations (14.6.1-12) for A, and (14.6.1-13) for $H^T \left(G_M R^{\blacklozenge} G_M^T\right)^{-1} H$ and $G_P Q_{P_{Dens}} G_P^T$, Equation (14.6.2-1) becomes in scalar form:

$$\begin{aligned} \dot{y}_1 &= -y_2 & \dot{\lambda}_1 &= 0 \\ \dot{y}_2 &= -y_3 & \dot{\lambda}_2 &= q_2\, y_2 + \lambda_1 \\ \dot{y}_3 &= -y_4 & \dot{\lambda}_3 &= q_3\, y_3 + \lambda_2 \\ \dot{y}_4 &= \frac{1}{r}\lambda_4 & \dot{\lambda}_4 &= q_4\, y_4 + \lambda_3 \end{aligned} \qquad (14.6.2\text{-}3)$$

Finding an analytic solution to (14.6.2-3) would be a challenging and exceedingly complex task. For simplicity we will attempt to find a solution in which we delete all but the most significant q_i process noise term. From experience we know that the angular rate sensor output noise q_2 is generally the dominant q_i term so we approximate (14.6.2-3) as:

For $q_3 \approx 0$ and $q_4 \approx 0$:

$$\begin{aligned} \dot{y}_1 &= -y_2 & \dot{\lambda}_1 &= 0 \\ \dot{y}_2 &= -y_3 & \dot{\lambda}_2 &= q_2\, y_2 + \lambda_1 \\ \dot{y}_3 &= -y_4 & \dot{\lambda}_3 &= \lambda_2 \\ \dot{y}_4 &= \frac{1}{r}\lambda_4 & \dot{\lambda}_4 &= \lambda_3 \end{aligned} \qquad (14.6.2\text{-}4)$$

An analytical solution to (14.6.2-4) can be found through its Laplace transform:

$$\begin{aligned} S\, Y_1 - y_{1_0} &= -Y_2 & S\, \Lambda_1 - \lambda_{1_0} &= 0 \\ S\, Y_2 - y_{2_0} &= -Y_3 & S\, \Lambda_2 - \lambda_{2_0} &= q_2\, Y_2 + \Lambda_1 \\ S\, Y_3 - y_{3_0} &= -Y_4 & S\, \Lambda_3 - \lambda_{3_0} &= \Lambda_2 \\ S\, Y_4 - y_{4_0} &= \frac{1}{r}\Lambda_4 & S\, \Lambda_4 - \lambda_{4_0} &= \Lambda_3 \end{aligned} \qquad (14.6.2\text{-}5)$$

or with rearrangement:

$$\begin{array}{llll} \text{a)} & Y_1 = \frac{1}{S}\left(-Y_2 + y_{1_0}\right) & \text{e)} & \Lambda_1 = \frac{1}{S}\lambda_{1_0} \\ \text{b)} & Y_2 = \frac{1}{S}\left(-Y_3 + y_{2_0}\right) & \text{f)} & \Lambda_2 = \frac{1}{S}\left(q_2\, Y_2 + \Lambda_1 + \lambda_{2_0}\right) \\ \text{c)} & Y_3 = \frac{1}{S}\left(-Y_4 + y_{3_0}\right) & \text{g)} & \Lambda_3 = \frac{1}{S}\left(\Lambda_2 + \lambda_{3_0}\right) \\ \text{d)} & Y_4 = \frac{1}{S}\left(\frac{1}{r}\Lambda_4 + y_{4_0}\right) & \text{h)} & \Lambda_4 = \frac{1}{S}\left(\Lambda_3 + \lambda_{4_0}\right) \end{array} \qquad (14.6.2\text{-}6)$$

where

a) - h) = Individual equation designators.

S = Laplace transform parameter.

Y_i, Λ_i = Laplace transforms of y_i, λ_i.

y_{i_0}, λ_{i_0} = Initial values for y_i, λ_i at time $t = 0$.

Equations (14.6.2-6) can now be analytically combined to solve for each Y_i, Λ_i. The Λ_4 component is found by substituting g) into h), f) into the result, e) and b) into the result, c) into the result, d) into the result, and rearranging. The result is:

$$\left(S^6 - \frac{q_2}{r}\right)\Lambda_4 = q_2\left(y_{4_0} - S\, y_{3_0} + S^2\, y_{2_0}\right) + S^2\,\lambda_{1_0} + S^3\,\lambda_{2_0} + S^4\,\lambda_{3_0} + S^5\,\lambda_{4_0} \qquad (14.6.2\text{-}7)$$

We then define the characteristic polynomial:

$$F \equiv S^6 - \frac{q_2}{r} \qquad (14.6.2\text{-}8)$$

with which, (14.6.2-7) becomes for Λ_4:

$$\Lambda_4 = \frac{1}{F}\left[q_2\left(y_{4_0} - S\, y_{3_0} + S^2\, y_{2_0}\right) + S^2\,\lambda_{1_0} + S^3\,\lambda_{2_0} + S^4\,\lambda_{3_0} + S^5\,\lambda_{4_0}\right] \qquad (14.6.2\text{-}9)$$

The Λ_2 to Λ_3 and Y_1 to Y_4 solutions are found by substituting (14.6.2-9) in (14.6.2-6) d), d) in c), c) in b), b) in a), b) and e) in f), and f) in g). The results with e) are in matrix form:

$$
\begin{bmatrix} Y_1 \\ Y_2 \\ Y_3 \\ Y_4 \\ \Lambda_1 \\ \Lambda_2 \\ \Lambda_3 \\ \Lambda_4 \end{bmatrix}
= \frac{1}{F}
\begin{bmatrix}
\frac{F}{S} & -S^4 & S^3 & -S^2 & -\frac{1}{r\,S^2} & -\frac{1}{r\,S} & -\frac{1}{r} & -\frac{S}{r} \\
0 & S^5 & -S^4 & S^3 & \frac{1}{r\,S} & \frac{1}{r} & \frac{S}{r} & \frac{S^2}{r} \\
0 & -\frac{q_2}{r} & S^5 & -S^4 & -\frac{1}{r} & -\frac{S}{r} & -\frac{S^2}{r} & -\frac{S^3}{r} \\
0 & \frac{q_2}{r}\,S & -\frac{q_2}{r} & S^5 & \frac{S}{r} & \frac{S^2}{r} & \frac{S^3}{r} & \frac{S^4}{r} \\
0 & 0 & 0 & 0 & \frac{F}{S} & 0 & 0 & 0 \\
0 & q_2\,S^4 & -q_2\,S^3 & q_2\,S^2 & S^4 & S^5 & \frac{q_2}{r} & \frac{q_2}{r}\,S \\
0 & q_2\,S^3 & -q_2\,S^2 & q_2\,S & S^3 & S^4 & S^5 & \frac{q_2}{r} \\
0 & q_2\,S^2 & -q_2\,S & q_2 & S^2 & S^3 & S^4 & S^5
\end{bmatrix}
\begin{bmatrix} y_{1_0} \\ y_{2_0} \\ y_{3_0} \\ y_{4_0} \\ \lambda_{1_0} \\ \lambda_{2_0} \\ \lambda_{3_0} \\ \lambda_{4_0} \end{bmatrix}
\qquad (14.6.2\text{-}10)
$$

The inverse Laplace transform of (14.6.2-10) yields the form required to identify the state transition matrix elements in (15.1.5.4-8) for the desired covariance solution using (15.1.5.4-11). The Laplace transform inversion process is expedited by first defining:

$$
\Theta_0 \equiv \frac{1}{S} \qquad \Theta_1 \equiv \frac{S^5}{F} \qquad \Theta_2 \equiv \frac{S^4}{F} \qquad \Theta_3 \equiv \frac{S^3}{F} \qquad \Theta_4 \equiv \frac{S^2}{F}
$$
$$
\Theta_5 \equiv \frac{S}{F} \qquad \Theta_6 \equiv \frac{1}{F} \qquad \Theta_7 \equiv \frac{1}{F\,S} \qquad \Theta_8 \equiv \frac{1}{F\,S^2} \qquad (14.6.2\text{-}11)
$$

With (14.6.2-11), Equation (14.6.2-10) becomes:

$$
\begin{bmatrix} Y_1 \\ Y_2 \\ Y_3 \\ Y_4 \\ \Lambda_1 \\ \Lambda_2 \\ \Lambda_3 \\ \Lambda_4 \end{bmatrix}
=
\begin{bmatrix}
\Theta_0 & -\Theta_2 & \Theta_3 & -\Theta_4 & -\frac{1}{r}\Theta_8 & -\frac{1}{r}\Theta_7 & -\frac{1}{r}\Theta_6 & -\frac{1}{r}\Theta_5 \\
0 & \Theta_1 & -\Theta_2 & \Theta_3 & \frac{1}{r}\Theta_7 & \frac{1}{r}\Theta_6 & \frac{1}{r}\Theta_5 & \frac{1}{r}\Theta_4 \\
0 & -\frac{q_2}{r}\Theta_6 & \Theta_1 & -\Theta_2 & -\frac{1}{r}\Theta_6 & -\frac{1}{r}\Theta_5 & -\frac{1}{r}\Theta_4 & -\frac{1}{r}\Theta_3 \\
0 & \frac{q_2}{r}\Theta_5 & -\frac{q_2}{r}\Theta_6 & \Theta_1 & \frac{1}{r}\Theta_5 & \frac{1}{r}\Theta_4 & \frac{1}{r}\Theta_3 & \frac{1}{r}\Theta_2 \\
0 & 0 & 0 & 0 & \Theta_0 & 0 & 0 & 0 \\
0 & q_2\,\Theta_2 & -q_2\,\Theta_3 & q_2\,\Theta_4 & \Theta_2 & \Theta_1 & \frac{q_2}{r}\Theta_6 & \frac{q_2}{r}\Theta_5 \\
0 & q_2\,\Theta_3 & -q_2\,\Theta_4 & q_2\,\Theta_5 & \Theta_3 & \Theta_2 & \Theta_1 & \frac{q_2}{r}\Theta_6 \\
0 & q_2\,\Theta_4 & -q_2\,\Theta_5 & q_2\,\Theta_6 & \Theta_4 & \Theta_3 & \Theta_2 & \Theta_1
\end{bmatrix}
\begin{bmatrix} y_{1_0} \\ y_{2_0} \\ y_{3_0} \\ y_{4_0} \\ \lambda_{1_0} \\ \lambda_{2_0} \\ \lambda_{3_0} \\ \lambda_{4_0} \end{bmatrix}
\tag{14.6.2-12}
$$

To further simplify the analysis, let's only deal with the situation when the initial covariance matrix has the earth rate uncertainty term as the non-zero element. Based on the form of $\underline{x}'$ in (14.6.1-12) we then write:

$$
P'_0 = \begin{bmatrix} P_{\Omega_0} & 0 & 0 & 0 \\ 0 & 0 & 0 & 0 \\ 0 & 0 & 0 & 0 \\ 0 & 0 & 0 & 0 \end{bmatrix} \tag{14.6.2-13}
$$

where

P'_0 = Initial value of P' defined below.

P' = The estimated $\underline{x}'$ uncertainty covariance matrix.

P_{Ω_0} = Initial value for the expected value of $\delta\omega_{IE_H}^2$.

From (14.6.2-13), the inverse Laplace transform of (14.6.2-12), and the general (15.1.5.4-8) format, we see that the terms required to find the covariance P' from (15.1.5.4-11) (with P(0) identified as P'_0) are:

$$\Phi_{yy}(t) + \Phi_{y\lambda}(t)\ P'_0 = \begin{bmatrix} \left(\theta_0(t) - \dfrac{P_{\Omega_0}}{r}\,\theta_8(t)\right) & -\,\theta_2(t) & \theta_3(t) & -\,\theta_4(t) \\ \dfrac{P_{\Omega_0}}{r}\,\theta_7(t) & \theta_1(t) & -\,\theta_2(t) & \theta_3(t) \\ -\dfrac{P_{\Omega_0}}{r}\,\theta_6(t) & -\dfrac{q_2}{r}\,\theta_6(t) & \theta_1(t) & -\,\theta_2(t) \\ \dfrac{P_{\Omega_0}}{r}\,\theta_5(t) & \dfrac{q_2}{r}\,\theta_5(t) & -\dfrac{q_2}{r}\,\theta_6(t) & \theta_1(t) \end{bmatrix} \tag{14.6.2-14}$$

$$\Phi_{\lambda y}(t) + \Phi_{\lambda\lambda}(t)\ P'_0 = \begin{bmatrix} P_{\Omega_0}\,\theta_0(t) & 0 & 0 & 0 \\ P_{\Omega_0}\,\theta_2(t) & q_2\,\theta_2(t) & -\,q_2\,\theta_3(t) & q_2\,\theta_4(t) \\ P_{\Omega_0}\,\theta_3(t) & q_2\,\theta_3(t) & -\,q_2\,\theta_4(t) & q_2\,\theta_5(t) \\ P_{\Omega_0}\,\theta_4(t) & q_2\,\theta_4(t) & -\,q_2\,\theta_5(t) & q_2\,\theta_6(t) \end{bmatrix} \tag{14.6.2-15}$$

where

$\theta_i(t)$ = Inverse Laplace transform of Θ_i.

For positive time t, the $\theta_0(t)$ term in (14.6.2-15) (the inverse Laplace transform of Θ_0 in (14.6.2-11)) is 1. From (15.1.5.4-11) and (14.6.2-14) - (14.6.2-15) we then find for the horizontal earth rate variance (the critical term affecting initial alignment accuracy):

$$P_\Omega = \frac{Cftr_{11}(t)}{Dtr(t)}\,P_{\Omega_0} \tag{14.6.2-16}$$

where

$Dtr(t)$ = Determinant of $\Phi_{yy}(t) + \Phi_{y\lambda}(t)\ P'_0$ in (14.6.2-14).

$Cftr_{11}(t)$ = Cofactor for the 1,1 element for $\Phi_{yy}(t) + \Phi_{y\lambda}(t)\ P'_0$ in (14.6.2-14) defined as the determinant of the matrix formed from $\Phi_{yy}(t) + \Phi_{y\lambda}(t)\ P'_0$ with column 1 and row 1 deleted.

P_Ω = The expected value of $\delta\omega^2_{IE_{H'}}$, the square of the (14.6.1-12) earth rate term in $\underline{x}'$, and the 1,1 element of P'.

To obtain a complete closed-form analytical solution for P_Ω from (14.6.2-14) - (14.6.2-16) it remains to find analytical expressions for the $\theta_i(t)$ inverse Laplace transforms of the (14.6.2-11) Θ_i's. The following outlines the general process. First we define:

$$S_0 \equiv \left(\frac{q_2}{r}\right)^{-\frac{1}{6}} \tag{14.6.2-17}$$

Based on (14.6.2-17), F in (14.6.2-8) factors into:

$$F = S^6 - \frac{q_2}{r} = S^6 - S_0^6 = \left(S^2 - S_0^2\right)\left(S^2 + S_0\, S + S_0^2\right)\left(S^2 - S_0\, S + S_0^2\right) \tag{14.6.2-18}$$

Using (14.6.2-18), the $\frac{1}{F}$ term in (14.6.2-11) is by partial fraction expansion:

$$\frac{1}{F} = \frac{1}{6\, S_0^5}\left[\frac{2\, S_0}{\left(S^2 - S_0^2\right)} - \frac{(S + 2\, S_0)}{\left(S^2 + S_0\, S + S_0^2\right)} + \frac{(S - 2\, S_0)}{\left(S^2 - S_0\, S + S_0^2\right)}\right] \tag{14.6.2-19}$$

or in general for the Θ_i terms in (14.6.2-11) for i = 1 to 6:

For i = 1 to 6:

$$\Theta_i = \frac{S^{6-i}}{F} = \frac{(a_i\, S + b_i)}{\left(S^2 - S_0^2\right)} + \frac{(c_i\, S + d_i)}{\left(S^2 + S_0\, S + S_0^2\right)} + \frac{(e_i\, S + f_i)}{\left(S^2 - S_0\, S + S_0^2\right)} \tag{14.6.2-20}$$

where

a_i, b_i, etc. = Constants as in (14.6.2-19) for the i = 6 case.

With (14.6.2-20), the Θ_7 expression in (14.6.2-11) is:

$$\begin{aligned}\Theta_7 = \frac{1}{F\, S} &= \frac{(a_6\, S + b_6)}{S\left(S^2 - S_0^2\right)} + \frac{(c_6\, S + d_6)}{S\left(S^2 + S_0\, S + S_0^2\right)} + \frac{(e_6\, S + f_6)}{S\left(S^2 - S_0\, S + S_0^2\right)} \\ &= \frac{a_6}{\left(S^2 - S_0^2\right)} + \frac{b_6}{S\left(S^2 - S_0^2\right)} + \frac{(c_6\, S + d_6)}{S\left(S^2 + S_0\, S + S_0^2\right)} + \frac{(e_6\, S + f_6)}{S\left(S^2 - S_0\, S + S_0^2\right)}\end{aligned} \tag{14.6.2-21}$$

Using

$$\frac{1}{S\left(S^2 - S_0^2\right)} = \frac{1}{S_0^2}\left[\frac{S}{\left(S^2 - S_0^2\right)} - \frac{1}{S}\right] \tag{14.6.2-22}$$

Equation (14.6.2-21) becomes:

$$\Theta_7 = \frac{a_6}{\left(S^2 - S_0^2\right)} + \frac{b_6}{S_0^2}\left[\frac{S}{\left(S^2 - S_0^2\right)} - \frac{1}{S}\right] + \frac{\left(c_6\, S + d_6\right)}{S\left(S^2 + S_0\, S + S_0^2\right)} + \frac{\left(e_6\, S + f_6\right)}{S\left(S^2 - S_0\, S + S_0^2\right)} \tag{14.6.2-23}$$

Similarly, Θ_8 is:

$$\begin{aligned}
\Theta_8 &= \frac{1}{F\, S^2} = \frac{\left(a_6\, S + b_6\right)}{S^2\left(S^2 - S_0^2\right)} + \frac{\left(c_6\, S + d_6\right)}{S^2\left(S^2 + S_0\, S + S_0^2\right)} + \frac{\left(e_6\, S + f_6\right)}{S^2\left(S^2 - S_0\, S + S_0^2\right)} \\
&= \frac{a_6}{S\left(S^2 - S_0^2\right)} + \frac{b_6}{S^2\left(S^2 - S_0^2\right)} + \frac{\left(c_6\, S + d_6\right)}{S^2\left(S^2 + S_0\, S + S_0^2\right)} + \frac{\left(e_6\, S + f_6\right)}{S^2\left(S^2 - S_0\, S + S_0^2\right)} \\
&= \frac{a_6}{S_0^2}\left[\frac{S}{\left(S^2 - S_0^2\right)} - \frac{1}{S}\right] + \frac{b_6}{S^2\left(S^2 - S_0^2\right)} + \frac{\left(c_6\, S + d_6\right)}{S^2\left(S^2 + S_0\, S + S_0^2\right)} + \frac{\left(e_6\, S + f_6\right)}{S^2\left(S^2 - S_0\, S + S_0^2\right)}
\end{aligned} \tag{14.6.2-24}$$

and for completeness, Θ_0 from (14.6.2-11) is:

$$\Theta_0 = \frac{1}{S} \tag{14.6.2-25}$$

Upon reviewing Equations (14.6.2-20) and (14.6.2-23) - (14.6.2-25) for Θ_0 - Θ_8 it will be found that all are linearly composed of polynomials, each having one of the following forms:

$$\begin{gathered}
\frac{1}{S} \qquad \frac{S}{\left(S^2 - S_0^2\right)} \qquad \frac{1}{\left(S^2 - S_0^2\right)} \qquad \frac{1}{S^2\left(S^2 - S_0^2\right)} \\
\frac{\left(a\, S + b\right)}{\left(S^2 + S_0\, S + S_0^2\right)} \qquad \frac{\left(a\, S + b\right)}{\left(S^2 - S_0\, S + S_0^2\right)} \qquad \frac{\left(a\, S + b\right)}{S\left(S^2 + S_0\, S + S_0^2\right)} \\
\frac{\left(a\, S + b\right)}{S\left(S^2 - S_0\, S + S_0^2\right)} \qquad \frac{\left(a\, S + b\right)}{S^2\left(S^2 + S_0\, S + S_0^2\right)} \qquad \frac{\left(a\, S + b\right)}{S^2\left(S^2 - S_0\, S + S_0^2\right)}
\end{gathered} \tag{14.6.2-26}$$

where

a, b = General coefficients that have different particular values for each (14.6.2-26) term.

Inverse Laplace transforms of each of the (14.6.2-26) expressions are directly available from standard Laplace transform pair tables (e.g., Reference 10 - Laplace Transform Pair Numbers 3, 19, 66, 83, 114, 115 and 116).

Thus, (14.6.2-16) with (14.6.2-14), (14.6.2-20), (14.6.2-23) - (14.6.2-25), and the inverse Laplace transforms of (14.6.2-26) constitute a complete closed-form solution for P_Ω under the assumed simplifying assumptions of q_3 and q_4 being negligible (zero) and P'_0 having P_{Ω_0} as the only non-zero element. The final result is far too complex for any practical application.

In order to simplify the analytics further, the following sections will resort to finding the P' response to individual noise terms acting independently. The solution approach will be as in previous sections based on the analytical integral to the continuous form covariance rate Equation (14.6.1-6) using its equivalent linear form (14.6.2-1) with (15.1.5.4-8) and (15.1.5.4-11) as the general integral solution formulas. The individual noise input solutions can (hopefully) then be summed to determine the solution under combined noise inputs. It should be clear at the outset, however, that the combined solutions so obtained will not be identical to the true integral solution with all noise terms operating simultaneously. This is because Equation (14.6.1-6) is based on the implicit assumption that the associated Kalman filter (and the computed feedback gains thereof) will be designed based only on those noise terms appearing in (14.6.1-6). In the actual Kalman filter, the design is based on all noise terms present. Thus, a solution to (14.6.1-6) based on including only a selected individual noise source, will differ from the response of the actual system to the same noise source. A section is included at the end of this chapter that analyzes the sum of individual noise input solutions to (14.6.1-6) compared to the simultaneous multiple noise input solution, and shows that the former is a reasonable approximation to the latter.

14.6.3 COVARIANCE SOLUTION UNDER ONLY MEASUREMENT NOISE

Consider the solution to (14.6.2-1) for the case when the measurement noise n_M in our (14.6.1-11) error model is the only active noise source (i.e., the $\underline{n}'_P$ process noise is zero). Then Equations (14.6.2-3) (derived from (14.6.2-1)) become:

$$\begin{aligned}
&a)\ \dot{y}_1 = -y_2 && e)\ \dot{\lambda}_1 = 0 \\
&b)\ \dot{y}_2 = -y_3 && f)\ \dot{\lambda}_2 = \lambda_1 \\
&c)\ \dot{y}_3 = -y_4 && g)\ \dot{\lambda}_3 = \lambda_2 \\
&d)\ \dot{y}_4 = \frac{1}{r}\lambda_4 && h)\ \dot{\lambda}_4 = \lambda_3
\end{aligned} \tag{14.6.3-1}$$

Equations (14.6.3-1) are easily analytically integrated in the order e), f), g), h), d), c), b), a) as follows:

$$\begin{aligned}
\lambda_1 &= \lambda_{1_0} \\
\lambda_2 &= \lambda_{2_0} + \lambda_{1_0}\, t \\
\lambda_3 &= \lambda_{3_0} + \lambda_{2_0}\, t + \lambda_{1_0} \frac{1}{2\,!} t^2 \\
\lambda_4 &= \lambda_{4_0} + \lambda_{3_0}\, t + \lambda_{2_0} \frac{1}{2\,!} t^2 + \lambda_{1_0} \frac{1}{3\,!} t^3 \\
y_4 &= y_{4_0} + \frac{1}{r}\left(\lambda_{4_0}\, t + \lambda_{3_0} \frac{1}{2\,!} t^2 + \lambda_{2_0} \frac{1}{3\,!} t^3 + \lambda_{1_0} \frac{1}{4\,!} t^4\right) \\
y_3 &= y_{3_0} - y_{4_0}\, t - \frac{1}{r}\left(\lambda_{4_0} \frac{1}{2\,!} t^2 + \lambda_{3_0} \frac{1}{3\,!} t^3 + \lambda_{2_0} \frac{1}{4\,!} t^4 + \lambda_{1_0} \frac{1}{5\,!} t^5\right) \\
y_2 &= y_{2_0} - y_{3_0}\, t + y_{4_0} \frac{1}{2\,!} t^2 + \frac{1}{r}\left(\lambda_{4_0} \frac{1}{3\,!} t^3 + \lambda_{3_0} \frac{1}{4\,!} t^4 + \lambda_{2_0} \frac{1}{5\,!} t^5 + \lambda_{1_0} \frac{1}{6\,!} t^6\right) \\
y_1 &= y_{1_0} - y_{2_0}\, t + y_{3_0} \frac{1}{2\,!} t^2 - y_{4_0} \frac{1}{3\,!} t^3 - \frac{1}{r}\left(\lambda_{4_0} \frac{1}{4\,!} t^4 + \lambda_{3_0} \frac{1}{5\,!} t^5 + \lambda_{2_0} \frac{1}{6\,!} t^6 + \lambda_{1_0} \frac{1}{7\,!} t^7\right)
\end{aligned} \tag{14.6.3-2}$$

or in matrix form:

$$
\begin{bmatrix} y_1 \\ y_2 \\ y_3 \\ y_4 \\ \lambda_1 \\ \lambda_2 \\ \lambda_3 \\ \lambda_4 \end{bmatrix}
=
\begin{bmatrix}
1 & -t & \frac{1}{2!}t^2 & -\frac{1}{3!}t^3 & -\frac{1}{r}\frac{1}{7!}t^7 & -\frac{1}{r}\frac{1}{6!}t^6 & -\frac{1}{r}\frac{1}{5!}t^5 & -\frac{1}{r}\frac{1}{4!}t^4 \\
0 & 1 & -t & \frac{1}{2!}t^2 & \frac{1}{r}\frac{1}{6!}t^6 & \frac{1}{r}\frac{1}{5!}t^5 & \frac{1}{r}\frac{1}{4!}t^4 & \frac{1}{r}\frac{1}{3!}t^3 \\
0 & 0 & 1 & -t & -\frac{1}{r}\frac{1}{5!}t^5 & -\frac{1}{r}\frac{1}{4!}t^4 & -\frac{1}{r}\frac{1}{3!}t^3 & -\frac{1}{r}\frac{1}{2!}t^2 \\
0 & 0 & 0 & 1 & \frac{1}{r}\frac{1}{4!}t^4 & \frac{1}{r}\frac{1}{3!}t^3 & \frac{1}{r}\frac{1}{2!}t^2 & \frac{1}{r}t \\
0 & 0 & 0 & 0 & 1 & 0 & 0 & 0 \\
0 & 0 & 0 & 0 & t & 1 & 0 & 0 \\
0 & 0 & 0 & 0 & \frac{1}{2!}t^2 & t & 1 & 0 \\
0 & 0 & 0 & 0 & \frac{1}{3!}t^3 & \frac{1}{2!}t^2 & t & 1
\end{bmatrix}
\begin{bmatrix} y_{1_0} \\ y_{2_0} \\ y_{3_0} \\ y_{4_0} \\ \lambda_{1_0} \\ \lambda_{2_0} \\ \lambda_{3_0} \\ \lambda_{4_0} \end{bmatrix}
\tag{14.6.3-3}
$$

Comparing (14.6.3-3) with the (15.1.5.4-8) general form we see that:

$$
\Phi_{yy}(t) = \begin{bmatrix}
1 & -t & \frac{1}{2!}t^2 & -\frac{1}{3!}t^3 \\
0 & 1 & -t & \frac{1}{2!}t^2 \\
0 & 0 & 1 & -t \\
0 & 0 & 0 & 1
\end{bmatrix}
$$

$$
\Phi_{y\lambda}(t) = \begin{bmatrix}
-\frac{1}{r}\frac{1}{7!}t^7 & -\frac{1}{r}\frac{1}{6!}t^6 & -\frac{1}{r}\frac{1}{5!}t^5 & -\frac{1}{r}\frac{1}{4!}t^4 \\
\frac{1}{r}\frac{1}{6!}t^6 & \frac{1}{r}\frac{1}{5!}t^5 & \frac{1}{r}\frac{1}{4!}t^4 & \frac{1}{r}\frac{1}{3!}t^3 \\
-\frac{1}{r}\frac{1}{5!}t^5 & -\frac{1}{r}\frac{1}{4!}t^4 & -\frac{1}{r}\frac{1}{3!}t^3 & -\frac{1}{r}\frac{1}{2!}t^2 \\
\frac{1}{r}\frac{1}{4!}t^4 & \frac{1}{r}\frac{1}{3!}t^3 & \frac{1}{r}\frac{1}{2!}t^2 & \frac{1}{r}t
\end{bmatrix}
\tag{14.6.3-4}
$$

(Continued)

$$\Phi_{\lambda y}(t) = 0 \qquad \Phi_{\lambda\lambda}(t) = \begin{bmatrix} 1 & 0 & 0 & 0 \\ t & 1 & 0 & 0 \\ \frac{1}{2\,!}t^2 & t & 1 & 0 \\ \frac{1}{3\,!}t^3 & \frac{1}{2\,!}t^2 & t & 1 \end{bmatrix} \tag{14.6.3-4 (Continued)}$$

As in the previous section, we simplify the analysis by considering the case when the earth rate uncertainty variance P_{Ω_0} is the only member of initial P′. Then (14.6.2-13) applies which, with (14.6.3-4), yields for terms in general solution Equation (15.1.5.4-11):

$$\Phi_{yy}(t) + \Phi_{y\lambda}(t)\; P'_0 = \begin{bmatrix} \left(1 - \frac{1}{r}\frac{1}{7\,!}t^7\, P_{\Omega_0}\right) & -\,t & \frac{1}{2\,!}t^2 & -\frac{1}{3\,!}t^3 \\ \frac{1}{r}\frac{1}{6\,!}t^6\, P_{\Omega_0} & 1 & -\,t & \frac{1}{2\,!}t^2 \\ -\frac{1}{r}\frac{1}{5\,!}t^5\, P_{\Omega_0} & 0 & 1 & -\,t \\ \frac{1}{r}\frac{1}{4\,!}t^4\, P_{\Omega_0} & 0 & 0 & 1 \end{bmatrix}$$

$$\Phi_{\lambda y}(t) + \Phi_{\lambda\lambda}(t)\; P'_0 = \begin{bmatrix} P_{\Omega_0} & 0 & 0 & 0 \\ t\, P_{\Omega_0} & 0 & 0 & 0 \\ \frac{1}{2\,!}t^2\, P_{\Omega_0} & 0 & 0 & 0 \\ \frac{1}{3\,!}t^3\, P_{\Omega_0} & 0 & 0 & 0 \end{bmatrix} \tag{14.6.3-5}$$

The determinant of the first expression in (14.6.3-5) is:

$$\begin{aligned} Dtr(t) &= 1 - \frac{1}{r}\frac{1}{7\,!}t^7\, P_{\Omega_0} - \frac{1}{r}\frac{1}{6\,!}t^6\, P_{\Omega_0}\,(-\,t) - \frac{1}{r}\frac{1}{5\,!}t^5\, P_{\Omega_0}\left(t^2 - \frac{1}{2\,!}t^2\right) \\ &\quad - \frac{1}{r}\frac{1}{4\,!}t^4\, P_{\Omega_0}\left(-\,t^3 - \frac{1}{3\,!}t^3 + \frac{1}{2\,!}t^3 + \frac{1}{2\,!}t^3\right) \\ &= 1 + \frac{1}{r}t^7\, P_{\Omega_0}\left[-\frac{1}{7\,!} + \frac{1}{6\,!} - \frac{1}{5\,!\,2\,!} + \frac{1}{4\,!\,3\,!}\right] \\ &= 1 + \frac{1}{r}t^7\, P_{\Omega_0}\,\frac{1}{7\,!}\left(-\,1 + 7 - \frac{7\times 6}{2\,!} + \frac{7\times 6\times 5}{3\,!}\right) \\ &= 1 + \frac{1}{r}t^7\, P_{\Omega_0}\left(\frac{20}{7\,!}\right) = 1 + \frac{1}{252\,r}t^7\, P_{\Omega_0} \end{aligned} \tag{14.6.3-6}$$

where

$$\mathrm{Dtr(t)} = \text{Determinant of } \Phi_{yy}(t) + \Phi_{y\lambda}(t)\, P'_0.$$

With (14.6.3-6) we find for the inverse of the first expression in (14.6.3-5):

$$\left(\Phi_{yy}(t) + \Phi_{y\lambda}(t)\, P'_0\right)^{-1} = \left(1 + \frac{1}{252\, r}\, t^7\, P_{\Omega_0}\right)^{-1} \begin{bmatrix} 1 & - & - & - \\ - & - & - & - \\ - & - & - & - \\ - & - & - & - \end{bmatrix} \qquad (14.6.3\text{-}7)$$

Substituting (14.6.3-7) and the second expression from (14.6.3-5) into (15.1.5.4-11) then yields for the earth rate uncertainty variance P_Ω (the 1,1 element of P'):

$$P_\Omega = P_{\Omega_0}\left(1 + \frac{1}{252\, r}\, t^7\, P_{\Omega_0}\right)^{-1} \qquad (14.6.3\text{-}8)$$

Finally, we substitute r from (14.6.1-13) in (14.6.3-8) to obtain:

$$P_{\Omega_{RVib}} = P_{\Omega_0}\left(1 + \frac{1}{252}\, \frac{g^2}{P_{RVib_H}\, T_n}\, t^7\, P_{\Omega_0}\right)^{-1} \qquad (14.6.3\text{-}9)$$

where

$P_{\Omega_{RVib}}$ = Earth rate uncertainty variance produced by random horizontal position vibration during alignment, and in which the Kalman filter is designed assuming zero process noise.

14.6.4 COVARIANCE RESPONSE WITH NO MEASUREMENT NOISE

Let's now consider the situation when the measurement noise $R^\blacklozenge$ is zero for which Section 15.1.5.4.1 provides a general covariance solution procedure, but only for particular constraint conditions. To see whether our (14.6.1-11) - (14.6.1-12) error model satisfies the 15.1.5.4.1 constraints, we partition the (14.6.1-12) matrices to separate states feeding the measurement (through H) from the others. Based on the (14.6.1-12) expressions for the error state vector $\underline{x}'$ and measurement matrix H, partitioned forms for the (14.6.1-12) matrices under zero measurement noise conditions are defined below together with a re-statement of (14.6.1-11):

$$\dot{\underline{x}}' = A\, \underline{x}' + G_P\, \underline{n}'_P \qquad z_n = H\, \underline{x}'_n + G_M\, n_{M_n} \qquad z^\blacklozenge = H\, \underline{x}' + G_M\, n_M^\blacklozenge \qquad (14.6.4\text{-}1)$$

$$H = \begin{bmatrix} 0 & H_{\mathcal{M}} \end{bmatrix} \qquad H_{\mathcal{M}} \equiv g \qquad n_M = \Delta r_{Vib_H} = 0 \qquad n_M^{\blacklozenge} = 0 \qquad R^{\blacklozenge} = 0$$

$$\underline{x}' = \begin{bmatrix} \underline{x}^* \\ x_{\mathcal{M}} \end{bmatrix} \qquad \underline{x}^* \equiv \begin{bmatrix} \delta\omega_{IE_H} \\ \gamma_H \\ \frac{1}{g}\,\delta v_H \end{bmatrix} \qquad x_{\mathcal{M}} \equiv \frac{1}{g}\,\delta\Delta R_H$$

$$A = \begin{bmatrix} A^* & 0 \\ A_{\mathcal{M}*} & A_{\mathcal{M}\mathcal{M}} \end{bmatrix} \qquad A^* \equiv \begin{bmatrix} 0 & 0 & 0 \\ 1 & 0 & 0 \\ 0 & 1 & 0 \end{bmatrix}$$

$$A_{\mathcal{M}*} \equiv \begin{bmatrix} 0 & 0 & 1 \end{bmatrix} \qquad A_{\mathcal{M}\mathcal{M}} = 0$$

$$G_P = \begin{bmatrix} G_{P*} & 0 \\ 0 & G_{P_{\mathcal{M}}} \end{bmatrix} \qquad G_{P*} \equiv \begin{bmatrix} 0 & 0 & 0 \\ -1 & 0 & 0 \\ 0 & -1 & 1 \end{bmatrix} \qquad G_{P_{\mathcal{M}}} \equiv 1 \tag{14.6.4-2}$$

$$\underline{n}'_P = \begin{bmatrix} \underline{n}_{P*} \\ n_{P_{\mathcal{M}}} \end{bmatrix} \qquad \underline{n}_{P*} = \begin{bmatrix} \delta\omega_{Rand_H} \\ \delta\alpha_{Quant_H} \\ \frac{1}{g}\,\delta a_{Rand_H} \end{bmatrix} \qquad n_{P_{\mathcal{M}}} = \frac{1}{g}\,\delta\upsilon_{Quant_H}$$

$$Q_{P_{Dens}} = \begin{bmatrix} Q_{P*_{Dens}} & 0 \\ 0 & Q_{P_{\mathcal{M}}\,Dens} \end{bmatrix}$$

$$Q_{P*_{Dens}} = \begin{bmatrix} q_{\omega Rand} & 0 & 0 \\ 0 & q_{\alpha Quant} & 0 \\ 0 & 0 & \frac{1}{g^2}\,q_{aRand} \end{bmatrix} \qquad Q_{P_{\mathcal{M}}\,Dens} = \frac{1}{g^2}\,q_{\upsilon Quant}$$

where

$\mathcal{M}$, * = Designation for error state $\underline{x}'$ and associated process noise terms directly affecting the measurement ($\mathcal{M}$) and those not directly affecting the measurement (*).

Based on the (14.6.4-2) and (14.6.1-13) definitions we also see that:

$$G_P \, Q_{P_{Dens}} \, G_P^T = \begin{bmatrix} G_{P*} \, Q_{P*_{Dens}} \, G_{P*}^T & 0 \\ 0 & G_{P_{\mathcal{M}}} \, Q_{P_{\mathcal{M}} \, Dens} \, G_{P_{\mathcal{M}}}^T \end{bmatrix}$$

$$G_{P*} \, Q_{P*_{Dens}} \, G_{P*}^T = \begin{bmatrix} 0 & 0 & 0 \\ 0 & q_2 & 0 \\ 0 & 0 & q_3 \end{bmatrix} \qquad \begin{aligned} q_2 &= q_{\omega Rand} \\ q_3 &= q_{\alpha Quant} + \frac{1}{g^2} q_{aRand} \end{aligned} \tag{14.6.4-3}$$

$$G_{P_{\mathcal{M}}} \, Q_{P_{\mathcal{M}} \, Dens} \, G_{P_{\mathcal{M}}}^T = q_4 = \frac{1}{g^2} q_{\upsilon Quant}$$

If we compare (14.6.4-2) with Section 15.1.5.4.1 constraint Equations (15.1.5.4.1-3) and (15.1.5.4.1-4), we see that (14.6.4-2) satisfies the format requirements. Thus, it is valid to apply the Section 15.1.5.4.1 general covariance solution procedure (for zero measurement noise) to our particular zero measurement noise problem.

The zero measurement solution procedure provided at the end of Section 15.1.5.4.1 (the first step of which was completed above), is based on converting the measurement (with zero measurement noise) into a revised measurement defined in (15.1.5.4.1-28) as:

$$\overset{\blacklozenge}{z}_{Rev} = A_{\mathcal{M}*} \, \underline{x}^* + G_{P_{\mathcal{M}}} \, n_{P_{\mathcal{M}}} \tag{14.6.4-4}$$

where

$\overset{\blacklozenge}{z}_{Rev}$ = Revised version of the continuous form $\overset{\blacklozenge}{z}$ measurement (and a scalar in this particular case).

and $A_{\mathcal{M}*}$, $G_{P_{\mathcal{M}}}$, $n_{P_{\mathcal{M}}}$ are as defined in the (14.6.4-2) partitions.

Section 15.1.5.4.1 shows that for a Kalman filter using $\overset{\blacklozenge}{z}$ as the input measurement to estimate $\underline{x}'$, the estimate for the $\underline{x}^*$ partition of $\underline{x}'$ (see Equation (14.6.4-2) for $\underline{x}'$ partition definition) is identical to the $\underline{x}^*$ estimate that would be obtained from a Kalman filter using the revised $\overset{\blacklozenge}{z}_{Rev}$ for the input measurement. Section 15.1.5.4.1 also shows that the Kalman filter estimate for the remaining partition in $\underline{x}'$ (i.e., the $x_{\mathcal{M}}$ partition) equals the true $x_{\mathcal{M}}$ value (i.e., error free). The equivalent statements for the associated $\underline{x}'$ error state uncertainty covariance is given by (15.1.5.4.1-12):

$$P' = \begin{bmatrix} P^* & 0 \\ 0 & 0 \end{bmatrix} \qquad P^* \equiv \mathcal{E}\left(\Delta \underline{x}^* \, \Delta \underline{x}^{*T}\right) \tag{14.6.4-5}$$

where

$\Delta \underline{x}^*$ = Uncertainty in the Kalman filter's estimate of $\underline{x}^*$.

P^* = The estimated $\underline{x}^*$ uncertainty covariance matrix.

$\mathcal{E}(\,)$ = Expected value operator.

The solution procedure outlined at the end of Section 15.1.5.4.1 solves for P* by applying the general covariance solution of Section 15.1.5.4 (Equations (15.1.5.4-7), (15.1.5.4-8) and (15.1.5.4-11)), but based on the (14.6.4-4) measurement. This is achieved by first making the following substitutions in Equations (15.1.5.4-7) and (15.1.5.4-11):

$$\begin{aligned} A(t) &= A^* \qquad & G_P(t)\; Q_{P_{Dens}}(t)\; G_P(t)^T &= G_{P^*}\, Q_{P^*_{Dens}}\, G_{P^*}^T \\ H(t) &= A_{\mathcal{M}^*} \qquad & G_M(t)\; R^{\blacklozenge}(t)\; G_M(t)^T &= G_{P_{\mathcal{M}}}\, Q_{P_{\mathcal{M}\,Dens}}\, G_{P_{\mathcal{M}}}^T \end{aligned} \tag{14.6.4-6}$$

$$P(t) = P^* \qquad P(0) = P^*_0 \tag{14.6.4-7}$$

where

P^*_0 = Initial value for P* in (14.6.4-5).

Then P* is found using (15.1.5.4-7), (15.1.5.4-8) and (15.1.5.4-11) with the result substituted in (14.6.4-5) to determine P′. Let's try this procedure for the (14.6.4-2) - (14.6.4-3) model.

First we apply (14.6.4-6) to (15.1.5.4-7) to find :

$$\begin{bmatrix} \dot{\underline{y}} \\ \dot{\underline{\lambda}} \end{bmatrix} = \begin{bmatrix} -A^{*T} & A_{\mathcal{M}^*}{}^T \left(G_{P_{\mathcal{M}}}\, Q_{P_{\mathcal{M}\,Dens}}\, G_{P_{\mathcal{M}}}^T\right)^{-1} A_{\mathcal{M}^*} \\ \left(G_{P^*}\, Q_{P^*_{Dens}}\, G_{P^*}^T\right) & A^* \end{bmatrix} \begin{bmatrix} \underline{y} \\ \underline{\lambda} \end{bmatrix} \tag{14.6.4-8}$$

Using the appropriate matrix values from (14.6.4-2) - (14.6.4-3), Equation (14.6.4-8) is in component form:

$$\begin{aligned} \dot{y}_1 &= -y_2 \qquad & \dot{\lambda}_1 &= 0 \\ \dot{y}_2 &= -y_3 \qquad & \dot{\lambda}_2 &= q_2\, y_2 + \lambda_1 \\ \dot{y}_3 &= \frac{1}{q_4}\lambda_3 \qquad & \dot{\lambda}_3 &= q_3\, y_3 + \lambda_2 \end{aligned} \tag{14.6.4-9}$$

In principle, Equations (14.6.4-9) can be integrated into the (15.1.5.4-8) state transition solution form and then used with (14.6.4-7) in (15.1.5.4-11) to solve for P*. Substituting P* in (14.6.4-5) then provides P′. In practice (as for Equations (14.6.2-3) that included measurement noise) the analytics are too complex to yield any useful result. So, as in Section 14.6.2, we resort to integrating (14.6.4-9) under individual noise input conditions, the subject of the subsections to follow.

14.6.4.1 RESPONSE TO ACCELEROMETER QUANTIZATION NOISE

Consider the solution to (14.6.4-9) for the case when the accelerometer quantization noise q_4 (see (14.6.4-3) for q_4 definition) is the only active noise source (i.e., the q_2 and q_3 process noise terms are zero). The integration of (14.6.4-9) then directly parallels the Section 14.6.3 integration approach (for active measurement noise and zero process noise). For q_2 and q_3 zero, Equations (14.6.4-9) become:

$$\begin{aligned} \dot{y}_1 &= -y_2 & \dot{\lambda}_1 &= 0 \\ \dot{y}_2 &= -y_3 & \dot{\lambda}_2 &= \lambda_1 \\ \dot{y}_3 &= \frac{1}{q_4}\lambda_3 & \dot{\lambda}_3 &= \lambda_2 \end{aligned} \qquad (14.6.4.1\text{-}1)$$

Equations (14.6.4.1-1) are easily analytically integrated as in Section 14.6.3 (that led to (14.6.3-2)). The result is:

$$\begin{aligned} \lambda_1 &= \lambda_{1_0} \\ \lambda_2 &= \lambda_{2_0} + \lambda_{1_0}\, t \\ \lambda_3 &= \lambda_{3_0} + \lambda_{2_0}\, t + \lambda_{1_0}\frac{1}{2\,!}t^2 \\ y_3 &= y_{3_0} + \frac{1}{q_4}\left(\lambda_{3_0}\, t + \lambda_{2_0}\frac{1}{2\,!}t^2 + \lambda_{1_0}\frac{1}{3\,!}t^3\right) \\ y_2 &= y_{2_0} - y_{3_0}\, t - \frac{1}{q_4}\left(\lambda_{3_0}\frac{1}{2\,!}t^2 + \lambda_{2_0}\frac{1}{3\,!}t^3 + \lambda_{1_0}\frac{1}{4\,!}t^4\right) \\ y_1 &= y_{1_0} - y_{2_0}\, t + y_{3_0}\frac{1}{2\,!}t^2 + \frac{1}{q_4}\left(\lambda_{3_0}\frac{1}{3\,!}t^3 + \lambda_{2_0}\frac{1}{4\,!}t^4 + \lambda_{1_0}\frac{1}{5\,!}t^5\right) \end{aligned} \qquad (14.6.4.1\text{-}2)$$

or in matrix form:

$$\begin{bmatrix} y_1 \\ y_2 \\ y_3 \\ \lambda_1 \\ \lambda_2 \\ \lambda_3 \end{bmatrix} = \begin{bmatrix} 1 & -t & \frac{1}{2!}t^2 & \frac{1}{q_4}\frac{1}{5!}t^5 & \frac{1}{q_4}\frac{1}{4!}t^4 & \frac{1}{q_4}\frac{1}{3!}t^3 \\ 0 & 1 & -t & -\frac{1}{q_4}\frac{1}{4!}t^4 & -\frac{1}{q_4}\frac{1}{3!}t^3 & -\frac{1}{q_4}\frac{1}{2!}t^2 \\ 0 & 0 & 1 & \frac{1}{q_4}\frac{1}{3!}t^3 & \frac{1}{q_4}\frac{1}{2!}t^2 & \frac{1}{q_4}t \\ 0 & 0 & 0 & 1 & 0 & 0 \\ 0 & 0 & 0 & t & 1 & 0 \\ 0 & 0 & 0 & \frac{1}{2!}t^2 & t & 1 \end{bmatrix} \begin{bmatrix} y_{1_0} \\ y_{2_0} \\ y_{3_0} \\ \lambda_{1_0} \\ \lambda_{2_0} \\ \lambda_{3_0} \end{bmatrix} \tag{14.6.4.1-3}$$

Following the procedure outlined in the last paragraph of Section 14.6.4, we next compare (14.6.4.1-3) with the (15.1.5.4-8) general form to see that:

$$\Phi_{yy}(t) = \begin{bmatrix} 1 & -t & \frac{1}{2!}t^2 \\ 0 & 1 & -t \\ 0 & 0 & 1 \end{bmatrix} \qquad \Phi_{y\lambda}(t) = \begin{bmatrix} \frac{1}{q_4}\frac{1}{5!}t^5 & \frac{1}{q_4}\frac{1}{4!}t^4 & \frac{1}{q_4}\frac{1}{3!}t^3 \\ -\frac{1}{q_4}\frac{1}{4!}t^4 & -\frac{1}{q_4}\frac{1}{3!}t^3 & -\frac{1}{q_4}\frac{1}{2!}t^2 \\ \frac{1}{q_4}\frac{1}{3!}t^3 & \frac{1}{q_4}\frac{1}{2!}t^2 & \frac{1}{q_4}t \end{bmatrix}$$

$$\Phi_{\lambda y}(t) = 0 \qquad \Phi_{\lambda\lambda}(t) = \begin{bmatrix} 1 & 0 & 0 \\ t & 1 & 0 \\ \frac{1}{2!}t^2 & t & 1 \end{bmatrix} \tag{14.6.4.1-4}$$

where

$\Phi_{yy}(t)$, $\Phi_{y\lambda}(t)$, $\Phi_{\lambda y}(t)$, $\Phi_{\lambda\lambda}(t)$ = Elements of the state transition matrix associated with the Equation (14.6.4.1-3) state dynamic matrix for propagation of initial conditions on the $\underline{y}$, $\underline{\lambda}$ vectors to the current time.

As in previous sections we simplify the analysis by only considering the case when the earth rate uncertainty variance P_{Ω_0} is the only member of initial P*. Based on the form of $\underline{x}^*$ (14.6.4-2) we then write (as in (14.6.2-13)):

$$P^*_0 = \begin{bmatrix} P_{\Omega_0} & 0 & 0 \\ 0 & 0 & 0 \\ 0 & 0 & 0 \end{bmatrix} \tag{14.6.4.1-5}$$

Continuing the procedure outlined at the end of Section 14.6.4, we use (14.6.4.1-4) and (14.6.4.1-5) with the P(0) conversion expression from (14.6.4-7) in general solution Equations (15.1.5.4-11) to find for particular terms:

$$\Phi_{yy}(t) + \Phi_{y\lambda}(t)\ P^*_0 = \begin{bmatrix} \left(1 + \frac{1}{q_4}\frac{1}{5\,!}\,t^5\,P_{\Omega_0}\right) & -\,t & \frac{1}{2\,!}\,t^2 \\ -\frac{1}{q_4}\frac{1}{4\,!}\,t^4\,P_{\Omega_0} & 1 & -\,t \\ \frac{1}{q_4}\frac{1}{3\,!}\,t^3\,P_{\Omega_0} & 0 & 1 \end{bmatrix}$$

$$\Phi_{\lambda y}(t) + \Phi_{\lambda\lambda}(t)\ P^*_0 = \begin{bmatrix} P_{\Omega_0} & 0 & 0 \\ t\,P_{\Omega_0} & 0 & 0 \\ \frac{1}{2\,!}\,t^2\,P_{\Omega_0} & 0 & 0 \end{bmatrix} \tag{14.6.4.1-6}$$

The determinant of the first expression in (14.6.4.1-6) is:

$$\begin{aligned} Dtr(t) &= 1 + \frac{1}{q_4}\frac{1}{5\,!}\,t^5\,P_{\Omega_0} + \frac{1}{q_4}\frac{1}{3\,!}\,t^5\,P_{\Omega_0} - \frac{1}{q_4}\frac{1}{4\,!}\,t^5\,P_{\Omega_0} - \frac{1}{q_4}\frac{1}{3\,!\,2\,!}\,t^5\,P_{\Omega_0} \\ &= 1 + \frac{1}{q_4}\,t^5\,P_{\Omega_0}\left[\frac{1}{5\,!} + \frac{1}{3\,!} - \frac{1}{4\,!} - \frac{1}{3\,!\,2\,!}\right] \\ &= 1 + \frac{1}{q_4}\,t^5\,P_{\Omega_0}\,\frac{1}{5\,!}\left[1 + 5\times 4 - 5 - \frac{5\times 4}{2\,!}\right] \\ &= 1 + \frac{1}{q_4}\,t^5\,P_{\Omega_0}\left(\frac{6}{5\,!}\right) = 1 + \frac{1}{20\,q_4}\,t^5\,P_{\Omega_0} \end{aligned} \tag{14.6.4.1-7}$$

where

$$\mathrm{Dtr(t)} = \text{Determinant of } \Phi_{yy}(t) + \Phi_{y\lambda}(t)\ P^*_0.$$

with which we find for the inverse of the first expression in (14.6.4.1-6):

$$\left(\Phi_{yy}(t) + \Phi_{y\lambda}(t)\ P^*_0\right)^{-1} = \left(1 + \frac{1}{20\ q_4}\ t^5\ P_{\Omega_0}\right)^{-1} \begin{bmatrix} 1 & - & - \\ - & - & - \\ - & - & - \end{bmatrix} \qquad (14.6.4.1\text{-}8)$$

Substituting (14.6.4.1-8), the second expression from (14.6.4.1-6), and the P(t) conversion expression from (14.6.4-7) into general solution Equation (15.1.5.4-11) then yields for the earth rate uncertainty variance P_Ω (the 1,1 element of P^*):

$$P_\Omega = P_{\Omega_0} \left(1 + \frac{1}{20\ q_4}\ t^5\ P_{\Omega_0}\right)^{-1} \qquad (14.6.4.1\text{-}9)$$

From (14.6.4-5) we see that P_Ω is also the (1,1) element of P'.

Finally, we substitute q_4 from (14.6.4-3) in (14.6.4.1-9) to obtain:

$$P_{\Omega_{\upsilon Quant}} = P_{\Omega_0} \left(1 + \frac{g^2}{q_{\upsilon Quant}}\ \frac{1}{20}\ t^5\ P_{\Omega_0}\right)^{-1} \qquad (14.6.4.1\text{-}10)$$

where

$P_{\Omega_{\upsilon Quant}}$ = Earth rate uncertainty variance produced by accelerometer horizontal quantization noise during alignment, and in which the Kalman filter was designed assuming zero measurement noise and zero for the other process noise sources.

14.6.4.2 RESPONSE TO COMBINED ANGULAR RATE SENSOR QUANTIZATION AND ACCELEROMETER RANDOM OUTPUT NOISE

Equation (14.6.4-9) was developed as a means for finding P^*, and ultimately P', under a zero measurement noise condition (using the procedure outlined at the conclusion of Section 14.6.4). Let's consider the solution to (14.6.4-9) for the case when the combined angular rate sensor quantization and accelerometer random output noise q_3 (see (14.6.4-3)) is the only active noise source (i.e., the q_2 and q_4 process noise terms are zero). From (14.6.4-9) we see that setting q_4 equal to zero introduces a singularity condition that prevents a simple direct integration. However, (14.6.4-9) with zero q_4 can be converted to a singular free form if we adopt the same procedure used in Section 14.6.4 to eliminate the zero measurement noise singularity.

Section 14.6.4 began with the original $\underline{x}'$ error state dynamic equation and the original $z^{\blacklozenge}$ measurement equation, but in the presence of zero original measurement noise $n_M^{\blacklozenge}$. To find the $\underline{x}'$ estimation uncertainty covariance P' under zero $n_M^{\blacklozenge}$ measurement noise, a revised $z_{Rev}^{\blacklozenge}$ measurement was formed that had process noise $np_{\mathcal{M}}$ (with density q_4) as its measurement noise. Use of $z_{Rev}^{\blacklozenge}$ in a revised Kalman filter configuration enabled us to solve for P^*, the uncertainty covariance matrix for the $\underline{x}^*$ partition of $\underline{x}'$. The P^* covariance was then used to find P' under zero $n_M^{\blacklozenge}$ measurement noise. The P' solution was developed analytically in Section 14.6.4.1 for the case when $np_{\mathcal{M}}$ was the only active noise source (i.e., with the remaining process noise sources $\underline{n}_P^*$ and the original measurement noise $n_M^{\blacklozenge}$ set to zero).

In this section we will repeat the Section 14.6.4 procedure, but beginning with the Section 14.6.4 revised $z_{Rev}^{\blacklozenge}$ measurement equation, the $\underline{x}^*$ partition of the $\underline{x}'$ error state dynamic equation, and the condition when $np_{\mathcal{M}}$ (process noise q_4) as well as $n_M^{\blacklozenge}$ measurement noise are zero. We will then form a revised $z_{Rev2}^{\blacklozenge}$ measurement that has coupled process noise $G_{P_{\mathcal{M}/\mathcal{M}}}\, \underline{n}_{P_{\mathcal{M}/\mathcal{M}}}$ (with q_3 density) as its measurement noise. Use of $z_{Rev2}^{\blacklozenge}$ in a revised Kalman filter configuration will enable us to solve for P^{**}, the uncertainty covariance matrix for an $\underline{x}^{**}$ partition of $\underline{x}^*$. The P^{**} covariance will be used to find P^* under conditions of zero $n_M^{\blacklozenge}$ measurement noise and zero $np_{\mathcal{M}}$ (q_4) process noise. Then P^* will be used to find P' under the same conditions. The P' solution will be developed analytically in this section for the case when $G_{P_{\mathcal{M}/\mathcal{M}}}\, \underline{n}_{P_{\mathcal{M}/\mathcal{M}}}$ (density q_3) is the only active noise source. In the next Section 14.6.4.3, we will perform an abbreviated version of the procedure used in this section, but beginning with $z_{Rev2}^{\blacklozenge}$ and the $\underline{x}^{**}$ error state dynamic equation. The result will be the P' response to angular rate sensor output noise (q_2 density) for the case when all other noise sources are zero.

We begin this section with the $\underline{\dot{x}}^*$ partition of $\underline{\dot{x}}$ from (14.6.4-1) as the error state dynamic equation using the (14.6.4-2) partitioning formulas. Revised measurement Equation (14.6.4-4) is used as the starting point measurement equation. Both are repeated below:

$$\dot{\underline{x}}^* = A^* \, \underline{x}^* + G_{P*} \, \underline{n}_{P*}$$

$$z^{\blacklozenge}_{Rev} = H_{Rev} \, \underline{x}^* + G_{M_{Rev}} \, n^{\blacklozenge}_{M_{Rev}} \qquad (14.6.4.2\text{-}1)$$

$$H_{Rev} \equiv A_{\mathcal{M}*} \qquad G_{M_{Rev}} \equiv G_{P_{\mathcal{M}}} \qquad n^{\blacklozenge}_{M_{Rev}} \equiv n_{P_{\mathcal{M}}}$$

where

$H_{Rev}, G_{M_{Rev}}, n^{\blacklozenge}_{M_{Rev}}$ = The revised continuous form measurement matrix, measurement noise coupling matrix and measurement noise.

Using (14.6.4-2) for component values, we reformat the (14.6.4.2-1) participating matrices to separate terms directly affecting the $z^{\blacklozenge}_{Rev}$ measurement from those that do not:

$$H_{Rev} = \begin{bmatrix} 0 & H_{\mathcal{M}/\mathcal{M}} \end{bmatrix} \qquad H_{\mathcal{M}/\mathcal{M}} = 1$$

$$n^{\blacklozenge}_{M_{Rev}} = n_{P_{\mathcal{M}}} = \frac{1}{g} \, \delta\upsilon_{Quant_H} = 0 \qquad R^{\blacklozenge}_{Rev} = Q_{P_{\mathcal{M}} \, Dens} = \frac{1}{g^2} \, q_{\upsilon Quant} = 0$$

$$\underline{x}^* = \begin{bmatrix} \underline{x}^{**} \\ \underline{x}_{\mathcal{M}/\mathcal{M}} \end{bmatrix} \qquad \underline{x}^{**} = \begin{bmatrix} \delta\omega_{IE_H} \\ \gamma_H \end{bmatrix} \qquad x_{\mathcal{M}/\mathcal{M}} = \frac{1}{g} \, \delta v_H$$

$$A^* = \begin{bmatrix} A^{**} & 0 \\ A_{\mathcal{M}/\mathcal{M}**} & A_{\mathcal{M}/\mathcal{M} \, \mathcal{M}/\mathcal{M}} \end{bmatrix} \qquad A^{**} \equiv \begin{bmatrix} 0 & 0 \\ 1 & 0 \end{bmatrix}$$

$$A_{\mathcal{M}/\mathcal{M}**} \equiv \begin{bmatrix} 0 & 1 \end{bmatrix} \qquad A_{\mathcal{M}/\mathcal{M} \, \mathcal{M}/\mathcal{M}} = 0 \qquad (14.6.4.2\text{-}2)$$

$$G_{P*} = \begin{bmatrix} G_{P**} & 0 \\ 0 & G_{P_{\mathcal{M}/\mathcal{M}}} \end{bmatrix} \qquad G_{P**} \equiv \begin{bmatrix} 0 \\ -1 \end{bmatrix} \qquad G_{P_{\mathcal{M}/\mathcal{M}}} \equiv \begin{bmatrix} 1 & 1 \end{bmatrix}$$

$$\underline{n}_{P*} = \begin{bmatrix} n_{P**} \\ \underline{n}_{P_{\mathcal{M}/\mathcal{M}}} \end{bmatrix} \qquad n_{P**} = \delta\omega_{Rand_H} \qquad \underline{n}_{P_{\mathcal{M}/\mathcal{M}}} = \begin{bmatrix} \delta\alpha_{Quant_H} \\ \frac{1}{g} \, \delta a_{Rand_H} \end{bmatrix}$$

$$Q_{P*_{Dens}} = \begin{bmatrix} Q_{P**_{Dens}} & 0 \\ 0 & Q_{P_{\mathcal{M}/\mathcal{M}} \, Dens} \end{bmatrix}$$

$$Q_{P**_{Dens}} = q_{\omega Rand} \qquad Q_{P_{\mathcal{M}/\mathcal{M}} \, Dens} = \begin{bmatrix} q_{\alpha Quant} & 0 \\ 0 & \frac{1}{g^2} \, q_{aRand} \end{bmatrix}$$

where

$**, \mathcal{M}/\mathcal{M}$ = Designation for error state $\underline{x}^*$ and associated process noise terms directly affecting the $z^{\blacklozenge}_{Rev}$ measurement (denoted by $\mathcal{M}/\mathcal{M}$) and those not directly affecting the measurement (denoted by $**$).

Based on the (14.6.4.2-2) definitions we also see that:

$$G_{P*}\, Q_{P*_{Dens}}\, G_{P*}^T = \begin{bmatrix} G_{P**}\, Q_{P**_{Dens}}\, G_{P**}^T & 0 \\ 0 & G_{P_{\mathcal{M}/\mathcal{M}}}\, Q_{P_{\mathcal{M}/\mathcal{M}\,Dens}}\, G_{P_{\mathcal{M}/\mathcal{M}}}^T \end{bmatrix}$$

$$G_{P**}\, Q_{P**_{Dens}}\, G_{P**}^T = \begin{bmatrix} 0 & 0 \\ 0 & q_2 \end{bmatrix} \qquad q_2 = q_{\omega Rand} \tag{14.6.4.2-3}$$

$$G_{P_{\mathcal{M}/\mathcal{M}}}\, Q_{P_{\mathcal{M}/\mathcal{M}\,Dens}}\, G_{P_{\mathcal{M}/\mathcal{M}}}^T = q_3 = q_{\alpha Quant} + \frac{1}{g^2}\, q_{aRand}$$

We now compare (14.6.4.2-2) against Section 15.1.5.4.1 constraint Equations (15.1.5.4.1-3) and (15.1.5.4.1-4), and see that (14.6.4.2-2) satisfies the format requirements. Thus, it is valid to apply the Section 15.1.5.4.1 general covariance solution procedure (for zero measurement noise) to our particular zero revised measurement noise problem. In this case the revised measurement noise $n^{\blacklozenge}_{M_{Rev}}$ is actually process noise $n_{P_{\mathcal{M}}}$, but it appears analytically in the (14.6.4.2-2) revised measurement equation in the form of measurement noise. Thus, for the case of zero revised measurement noise (i.e., zero $n_{P_{\mathcal{M}}}$), we can treat the problem as a zero measurement noise situation for which the Section 15.1.5.4.1 procedure then applies. Remember, however, that the revised measurement in Equation (14.6.4.2-1) was developed in Section 14.6.4 based on zero measurement noise (i.e., the real measurement noise $n^{\blacklozenge}_M$ of Equation (14.6.4-1)). Thus, the problem we are addressing in this section is finding a covariance solution for zero $n^{\blacklozenge}_M$ measurement noise and zero $n_{P_{\mathcal{M}}}$ process noise. Let's continue.

The zero measurement solution to be developed in this case is based on converting the (14.6.4.2-1) measurement $z^{\blacklozenge}_{Rev}$ into a revised measurement of the same form as $z^{\blacklozenge}_{Rev}$ in (15.1.5.4.1-28):

$$z^{\blacklozenge}_{Rev2} = A_{\mathcal{M}/\mathcal{M}**}\, \underline{x}^{**} + G_{P_{\mathcal{M}/\mathcal{M}}}\, \underline{n}_{P_{\mathcal{M}/\mathcal{M}}} \tag{14.6.4.2-4}$$

where

$z^{\blacklozenge}_{Rev2}$ = Revised version of the Equation (14.6.4.2-1) continuous form $z^{\blacklozenge}_{Rev}$ measurement.

and $A_{\mathcal{M}/\mathcal{M}**}$, $G_{P_{\mathcal{M}/\mathcal{M}}}$, $\underline{n}_{P_{\mathcal{M}/\mathcal{M}}}$ are as defined in the (14.6.4.2-2) partitions.

Following the Section 14.6.4 process for the (14.6.4.2-2) model shows as in (14.6.4-5) that:

$$P* = \begin{bmatrix} P** & 0 \\ 0 & 0 \end{bmatrix} \qquad P** \equiv \mathcal{E}\left(\Delta\underline{x}** \, \Delta\underline{x}**^T\right) \tag{14.6.4.2-5}$$

where

$\Delta\underline{x}**$ = Uncertainty in the revised estimate of $\underline{x}**$ (see (14.6.4.2-2) for components) for the Kalman filter that uses $z^{\blacklozenge}_{Rev2}$ as its measurement.

$P*$ = Estimated $\underline{x}*$ uncertainty covariance.

$P**$ = Estimated $\underline{x}**$ uncertainty covariance.

Paralleling Section 14.6.4, we now make the following substitutions in Equations (15.1.5.4-7) and (15.1.5.4-11), but based on the (14.6.4.2-2) $\mathcal{M}/\mathcal{M}$ and ** partitions:

$$\begin{aligned} A(t) &= A** & \qquad G_P(t)\; Q_{P_{Dens}}(t)\; G_P(t)^T &= G_{P**}\; Q_{P**_{Dens}}\; G^T_{P**} \\ H(t) &= A_{\mathcal{M}/\mathcal{M}**} & \qquad G_M(t)\; R^{\blacklozenge}(t)\; G_M(t)^T &= G_{P_{\mathcal{M}/\mathcal{M}}}\; Q_{P_{\mathcal{M}/\mathcal{M}}\,Dens}\; G^T_{P_{\mathcal{M}/\mathcal{M}}} \end{aligned} \tag{14.6.4.2-6}$$

$$P(t) = P** \qquad P(0) = P**_0 \tag{14.6.4.2-7}$$

where

$P**_0$ = Initial value for $P**$ in (14.6.4.2-7).

Then $P**$ is found using (15.1.5.4-7), (15.1.5.4-8) and (15.1.5.4-11) with the result substituted in (14.6.4.2-5) to determine $P*$. For this case of zero $n^{\blacklozenge}_M$ measurement and $n_{P_{\mathcal{M}}}$ process noise, we can then substitute $P*$ into (14.6.4-5) to finally determine P'. The combined (14.6.4.2-5) and (14.6.4-5) operation for P' is:

$$P' = \begin{bmatrix} P** & 0 \\ 0 & 0 \end{bmatrix} \tag{14.6.4.2-8}$$

Applying (14.6.4.2-6) to (15.1.5.4-7) we find :

$$\begin{bmatrix} \dot{\underline{y}} \\ \dot{\underline{\lambda}} \end{bmatrix} = \begin{bmatrix} -A^{**T} & A_{\mathcal{M}/\mathcal{M}^{**}}^{T} \left(G_{P_{\mathcal{M}/\mathcal{M}}} Q_{P_{\mathcal{M}/\mathcal{M}\,Dens}} G_{P_{\mathcal{M}/\mathcal{M}}}^{T} \right)^{-1} A_{\mathcal{M}/\mathcal{M}^{**}} \\ \left(G_{P^{**}} Q_{P^{**}Dens} G_{P^{**}}^{T} \right) & A^{**} \end{bmatrix} \begin{bmatrix} \underline{y} \\ \underline{\lambda} \end{bmatrix}$$

(14.6.4.2-9)

Using the appropriate matrix values from (14.6.4.2-2) - (14.6.4.2-3), Equation (14.6.4.2-9) becomes in component form:

$$\begin{aligned} \dot{y}_1 &= -y_2 & \dot{\lambda}_1 &= 0 \\ \dot{y}_2 &= \frac{1}{q_3}\lambda_2 & \dot{\lambda}_2 &= q_2\, y_2 + \lambda_1 \end{aligned} \qquad (14.6.4.2\text{-}10)$$

Equations (14.6.4.2-10) can now be integrated into the (15.1.5.4-8) state transition solution form and used with (14.6.4.2-7) in (15.1.5.4-11) to solve for P^{**}. Equation (14.6.4.2-8) is then applied to find P'. For the remainder of this section we will find P' for the special case of zero q_2 and the only active noise source being q_3 (i.e., angular rate sensor quantization noise and accelerometer random output noise - See (14.6.4.2-3)). Section 14.6.5 will deal with the solution for P' for non-zero q_2 and q_3 so that results can be compared with the individual non-zero q_3 only and non-zero q_2 only solutions (the latter to be determined in Section 14.6.4.3).

For the zero q_2 case, Equations (14.6.4.2-10) become:

$$\begin{aligned} \dot{y}_1 &= -y_2 & \dot{\lambda}_1 &= 0 \\ \dot{y}_2 &= \frac{1}{q_3}\lambda_2 & \dot{\lambda}_2 &= \lambda_1 \end{aligned} \qquad (14.6.4.2\text{-}11)$$

whose solution is:

$$\begin{aligned} \lambda_1 &= \lambda_{1_0} \\ \lambda_2 &= \lambda_{2_0} + \lambda_{1_0}\, t \\ y_2 &= y_{2_0} + \frac{1}{q_3}\left(\lambda_{2_0}\, t + \lambda_{1_0} \frac{1}{2\,!} t^2 \right) \\ y_1 &= y_{1_0} - y_{2_0}\, t - \frac{1}{q_3}\left(\lambda_{2_0} \frac{1}{2\,!} t^2 + \lambda_{1_0} \frac{1}{3\,!} t^3 \right) \end{aligned} \qquad (14.6.4.2\text{-}12)$$

or in matrix form:

$$\begin{bmatrix} y_1 \\ y_2 \\ \lambda_1 \\ \lambda_2 \end{bmatrix} = \begin{bmatrix} 1 & -t & -\frac{1}{q_3}\frac{1}{3\,!}t^3 & -\frac{1}{q_3}\frac{1}{2\,!}t^2 \\ 0 & 1 & \frac{1}{q_3}\frac{1}{2\,!}t^2 & \frac{1}{q_3}t \\ 0 & 0 & 1 & 0 \\ 0 & 0 & t & 1 \end{bmatrix} \begin{bmatrix} y_{1_0} \\ y_{2_0} \\ \lambda_{1_0} \\ \lambda_{2_0} \end{bmatrix} \tag{14.6.4.2-13}$$

We next compare (14.6.4.2-13) with the (15.1.5.4-8) general form to see that:

$$\Phi_{yy}(t) = \begin{bmatrix} 1 & -t \\ 0 & 1 \end{bmatrix} \qquad \Phi_{y\lambda}(t) = \begin{bmatrix} -\frac{1}{q_3}\frac{1}{3\,!}t^3 & -\frac{1}{q_3}\frac{1}{2\,!}t^2 \\ \frac{1}{q_3}\frac{1}{2\,!}t^2 & \frac{1}{q_3}t \end{bmatrix}$$

$$\Phi_{\lambda y}(t) = 0 \qquad \Phi_{\lambda\lambda}(t) = \begin{bmatrix} 1 & 0 \\ t & 1 \end{bmatrix} \tag{14.6.4.2-14}$$

As in previous sections, we simplify the analysis by considering the case when the earth rate uncertainty variance P_{Ω_0} is the only member of initial P^{**}. Based on the form of $\underline{x}^{**}$ in (14.6.4.2-2) we then write (as in (14.6.4.1-5)):

$$P^{**}_0 = \begin{bmatrix} P_{\Omega_0} & 0 \\ 0 & 0 \end{bmatrix} \tag{14.6.4.2-15}$$

Continuing the procedure outlined following (14.6.4.2-7), we use (14.6.4.2-14) and (14.6.4.2-15) with the P(0) conversion expression from (14.6.4.2-7) in general solution Equations (15.1.5.4-11) to find for particular terms:

$$\Phi_{yy}(t) + \Phi_{y\lambda}(t)\; P^{**}_0 = \begin{bmatrix} \left(1 - \frac{1}{q_3}\frac{1}{3\,!}t^3\, P_{\Omega_0}\right) & -t \\ \frac{1}{q_3}\frac{1}{2\,!}t^2\, P_{\Omega_0} & 1 \end{bmatrix}$$

$$\Phi_{\lambda y}(t) + \Phi_{\lambda\lambda}(t)\; P^{**}_0 = \begin{bmatrix} P_{\Omega_0} & 0 \\ t\, P_{\Omega_0} & 0 \end{bmatrix} \tag{14.6.4.2-16}$$

The determinant of the first expression in (14.6.4.2-16) is:

$$\begin{aligned} Dtr(t) &= 1 - \frac{1}{q_3}\frac{1}{3\,!}\,t^3\,P_{\Omega_0} + \frac{1}{q_3}\frac{1}{2\,!}\,t^3\,P_{\Omega_0} \\ &= 1 + \frac{1}{q_3}\,t^3\,P_{\Omega_0}\left[-\frac{1}{3\,!} + \frac{1}{2\,!}\right] = 1 + \frac{1}{3\,q_3}\,t^3\,P_{\Omega_0} \end{aligned} \qquad (14.6.4.2\text{-}17)$$

where

$$Dtr(t) = \text{Determinant of } \Phi_{yy}(t) + \Phi_{y\lambda}(t)\ P^{**}{}_0.$$

with which we find for the inverse of the first expression in (14.6.4.2-16):

$$\left(\Phi_{yy}(t) + \Phi_{y\lambda}(t)\ P^{**}{}_0\right)^{-1} = \left(1 + \frac{1}{3\,q_3}\,t^3\,P_{\Omega_0}\right)^{-1}\begin{bmatrix} 1 & - \\ - & - \end{bmatrix} \qquad (14.6.4.2\text{-}18)$$

Substituting (14.6.4.2-18), the second expression from (14.6.4.2-16), and the P conversion expression from (14.6.4.2-7) into general solution Equation (15.1.5.4.-11) then yields for the earth rate uncertainty variance P_Ω (the 1,1 element of P^{**}):

$$P_\Omega = P_{\Omega_0}\left(1 + \frac{1}{3\,q_3}\,t^3\,P_{\Omega_0}\right)^{-1} \qquad (14.6.4.2\text{-}19)$$

From (14.6.4.2-8) we see that P_Ω is also the (1,1) element of P'.

Finally, we substitute q_3 from (14.6.4.2-3) in (14.6.4.2-19) to obtain:

$$P_{\Omega\alpha Quant/aRand} = P_{\Omega_0}\left[1 + \frac{1}{\left(q_{\alpha Quant} + \frac{1}{g^2}\,q_{aRand}\right)}\,\frac{1}{3}\,t^3\,P_{\Omega_0}\right]^{-1} \qquad (14.6.4.2\text{-}20)$$

where

$P_{\Omega\alpha\ Quant/aRand}$ = Earth rate uncertainty variance produced by angular rate sensor horizontal quantization noise and accelerometer horizontal random output noise during alignment, and in which the Kalman filter was designed assuming zero measurement noise and zero for the other process noise sources.

14.6.4.3 RESPONSE TO ANGULAR RATE SENSOR RANDOM OUTPUT NOISE

In this section we find the response of the horizontal earth rate uncertainty variance to angular rate sensor random output noise (density q_2) acting alone. By now the reader should be familiar with the general solution approach. For this case we'll take some short cuts by starting from (14.6.4.2-10) (with non-zero q_2 and q_3 as the active noise sources) in comparison with (14.6.4-9) (with non-zero q_2, q_3 and q_4), compared with (14.6.2-3) (with non-zero q_2, q_3, q_4 and r), all repeated below:

$$\begin{aligned} \dot{y}_1 &= -y_2 & \dot{\lambda}_1 &= 0 \\ \dot{y}_2 &= -y_3 & \dot{\lambda}_2 &= q_2\, y_2 + \lambda_1 \\ \dot{y}_3 &= -y_4 & \dot{\lambda}_3 &= q_3\, y_3 + \lambda_2 \\ \dot{y}_4 &= \frac{1}{r}\lambda_4 & \dot{\lambda}_4 &= q_4\, y_4 + \lambda_3 \end{aligned} \qquad (14.6.4.3\text{-}1)$$

$$\begin{aligned} \dot{y}_1 &= -y_2 & \dot{\lambda}_1 &= 0 \\ \dot{y}_2 &= -y_3 & \dot{\lambda}_2 &= q_2\, y_2 + \lambda_1 \\ \dot{y}_3 &= \frac{1}{q_4}\lambda_3 & \dot{\lambda}_3 &= q_3\, y_3 + \lambda_2 \end{aligned} \qquad (14.6.4.3\text{-}2)$$

$$\begin{aligned} \dot{y}_1 &= -y_2 & \dot{\lambda}_1 &= 0 \\ \dot{y}_2 &= \frac{1}{q_3}\lambda_2 & \dot{\lambda}_2 &= q_2\, y_2 + \lambda_1 \end{aligned} \qquad (14.6.4.3\text{-}3)$$

If the development of (14.6.4.2-10) (Equations (14.6.4.3-3)) is reviewed, it will be seen that they originated from (14.6.4-9) (Equations (14.6.4.3-2)) for the singular condition when $q_4 = 0$. Similarly, review of the (14.6.4-9) development (Equations (14.6.4.3-2)) shows that they originated from (14.6.2-3) (Equations (14.6.4.3-1)) for the singular condition when $r = 0$. By direct analogy and inspection of (14.6.4.3-3) compared to (14.6.4.3-2) compared to (14.6.4.3-1), we can immediately write for the singular zero q_3 condition in (14.6.4.3-3):

$$\dot{y}_1 = \frac{1}{q_2}\lambda_1 \qquad \dot{\lambda}_1 = 0 \qquad (14.6.4.3\text{-}4)$$

whose integral solution is:

$$y_1 = y_{1_0} + \frac{1}{q_2}\lambda_{1_0}\, t \qquad \lambda_1 = \lambda_{1_0} \qquad (14.6.4.3\text{-}5)$$

or in matrix form:

$$\begin{bmatrix} y_1 \\ \lambda_1 \end{bmatrix} = \begin{bmatrix} 1 & \frac{1}{q_2} t \\ 0 & 1 \end{bmatrix} \begin{bmatrix} y_{1_0} \\ \lambda_{1_0} \end{bmatrix} \tag{14.6.4.3-6}$$

Comparing (14.6.4.3-6) with the (15.1.5.4-8) general form we get:

$$\Phi_{yy}(t) = 1 \qquad \Phi_{y\lambda}(t) = \frac{1}{q_2} t \qquad \Phi_{\lambda y}(t) = 0 \qquad \Phi_{\lambda\lambda}(t) = 1 \tag{14.6.4.3-7}$$

Extrapolating the next step based on (14.6.4.2-16) (which came from (14.6.4.3-3)), we find from (14.6.4.3-7) in general solution Equations (15.1.5.4-11) for particular terms:

$$\begin{aligned} \Phi_{yy}(t) + \Phi_{y\lambda}(t)\ P^{***}_0 &= 1 + \frac{1}{q_2} t\ P_{\Omega_0} \\ \Phi_{\lambda y}(t) + \Phi_{\lambda\lambda}(t)\ P^{***}_0 &= P_{\Omega_0} \end{aligned} \tag{14.6.4.3-8}$$

where

P^{***}_0 = Initial covariance of the uncertainty in a partition of $\underline{x}^{**}$ in Equations (14.6.4.2-2) containing only the horizontal earth rate estimate earth rate $\delta\omega_{IE_H}$.

From its definition, we see that:

$$P^{***}_0 = P_{\Omega_0} \tag{14.6.4.3-9}$$

Substituting (14.6.4.3-8) and (14.6.4.3-9) into general solution Equation (15.1.5.4-11) then yields for the earth rate uncertainty variance P_Ω:

$$P_\Omega = \left(1 + \frac{1}{q_2} t\ P_{\Omega_0}\right)^{-1} P_{\Omega_0} \tag{14.6.4.3-10}$$

From (14.6.4.2-2) and (14.6.4.2-8) we see that P_Ω is also the (1,1) element of P'. Extrapolating (14.6.4.2-8) to the P^{***} form shows that the remaining elements of P' are zero.

Finally, we substitute q_2 from (14.6.4.2-3) in (14.6.4.3-10) to obtain:

$$P_{\Omega_{\omega Rand}} = \left(1 + \frac{1}{q_{\omega Rand}} t\ P_{\Omega_0}\right)^{-1} P_{\Omega_0} \tag{14.6.4.3-11}$$

where

$P_{\Omega_{\omega Rand}}$ = Earth rate uncertainty variance produced by angular rate sensor horizontal random output noise during alignment, and in which the Kalman filter was designed assuming zero measurement noise and zero for the other process noise sources.

The above development is far from rigorous and is provided in the interests of shortening the process. As an exercise, the reader can verify its validity (as the author has done) by performing the detailed rigorous process of previous sections (e.g., using Section 14.6.4.2 as a template).

14.6.4.4 SUMMARY OF RESPONSES TO MEASUREMENT AND PROCESS NOISE

Equations (14.6.3-9), (14.6.4.1-10), (14.6.4.2-20) and (14.6.4.3-11) are responses of the horizontal earth rate uncertainty to individual random noise error sources present during Fine Alignment, assuming for each that the Kalman filter was designed for only that noise source present (i.e., zero for the other noise sources). These equations are summarized below.

$$\begin{aligned}
P_{\Omega RVib} &= P_{\Omega 0} \left(1 + \frac{1}{252} \frac{g^2}{P_{RVib_H} \, T_n} t^7 \, P_{\Omega 0} \right)^{-1} \\
P_{\Omega \upsilon Quant} &= P_{\Omega 0} \left(1 + \frac{g^2}{q_{\upsilon Quant}} \frac{1}{20} t^5 \, P_{\Omega 0} \right)^{-1} \\
P_{\Omega \alpha Quant/aRand} &= P_{\Omega 0} \left[1 + \frac{1}{\left(q_{\alpha Quant} + \frac{1}{g^2} q_{aRand} \right)} \frac{1}{3} t^3 \, P_{\Omega 0} \right]^{-1} \\
P_{\Omega_{\omega Rand}} &= \left(1 + \frac{1}{q_{\omega Rand}} t \, P_{\Omega 0} \right)^{-1} P_{\Omega 0}
\end{aligned} \qquad (14.6.4.4\text{-}1)$$

Values for the $q_{\alpha Quant}$ and $q_{\upsilon Quant}$ densities can be found using Equations (15.2.1.2-18). Values used for the q_{aRand} and $q_{\omega Rand}$ densities should be based on the characteristics of the particular accelerometers and angular rate sensors being used.

It is also instructive to consider the form of Equations (14.6.4.4-1) under conditions when the initial earth rate uncertainty variance large. Then the equations simplify to:

For Large P_{Ω_0}:

$$P_{\Omega_{RVib}} = \frac{252\ T_n\ P_{RVib_H}}{g^2\ t^7} \qquad P_{\Omega_{\upsilon Quant}} = \frac{20\ q_{\upsilon Quant}}{g^2\ t^5} \tag{14.6.4.4-2}$$

$$P_{\Omega_{\alpha Quant/aRand}} = \frac{3\left(q_{\alpha Quant} + \dfrac{1}{g^2}\ q_{aRand}\right)}{t^3} \qquad P_{\Omega_{\omega Rand}} = \frac{q_{\omega Rand}}{t}$$

14.6.5 COMBINED SOLUTION VERSUS THE SUM OF INDIVIDUAL SOLUTIONS FOR MULTIPLE NOISE SOURCES

In Sections 14.6.3 and 14.6.4 (and its subsections) we found simple closed-form solutions for the estimated horizontal earth rate uncertainty response to individual noise sources (summarized in Section 14.6.4.4), assuming for each that the Kalman filter was designed for only that noise source present (with zero for the other noise sources). In actual practice, the Kalman filter design is based on all noise sources being present (in its covariance/gain calculations). To find the response of the actual filter (designed for all noise sources present) to each noise source, a Fine Alignment suboptimal covariance simulation numerical sensitivity analysis can be performed for each noise source using the actual filter computed Kalman gain matrix (e.g., as in Section 16.2.4). For the Section 14.6.4.4 noise sensitivity results to be useful, they should be reasonably close to the sensitivities obtained with the actual Kalman gains.

An analytical (rather than numerical simulation) assessment can also be made to measure the accuracy to which the Section 14.6.4.4 results match the noise sensitivities of the Fine Alignment process using actual Kalman filter gains. The method is similar to the approach outlined in the previous paragraph. The difference is that the linear sum of the Section 14.6.4.4 results would be compared against the horizontal earth rate uncertainty variance determined analytically with all noise sources present. Unfortunately, as illustrated in Section 14.6.2, finding an analytical solution with all noise sources present is far too complicated compared with the simulation alternative.

As an analytical compromise, the following sections will perform the comparison described in the previous paragraph, but for only two of the noise sources active; horizontal angular rate sensor random noise (q_2) and the q_3 noise group (horizontal angular rate sensor quantization noise and accelerometer random output noise). An analytical solution for the horizontal earth rate estimate uncertainty under simulataneous q_2, q_3 excitation is found fairly easily in Section 14.6.5.1. As an indication of the Section 14.6.2 individual solution accuracies, the 14.6.5.1 result is compared in Section 14.6.5.2 with the sum of the individual q_2 and q_3 results from Section 14.6.4.4. The results compare quite favorably which supports the notion of using the Section 14.6.4.4 sensitivities as a reasonable approximation to the actual sensitivities.

14.6.5.1 RESPONSE TO COMBINED ANGULAR RATE SENSOR RANDOM OUTPUT NOISE, QUANTIZATION NOISE AND ACCELEROMETER RANDOM OUTPUT NOISE

In this section we find the horizontal earth rate uncertainty variance response to simultaneous q_2 and q_3 noise (angular rate sensor output noise, angular rate sensor quantization noise and accelerometer random output noise). The result is determined by returning to Equations (14.6.4.2-10) repeated below:

$$\dot{y}_1 = - y_2 \qquad \dot{\lambda}_1 = 0$$
$$\dot{y}_2 = \frac{1}{q_3} \lambda_2 \qquad \dot{\lambda}_2 = q_2 \, y_2 + \lambda_1 \tag{14.6.5.1-1}$$

We will now find the integral solution to (14.6.5.1-1) and then translate the result to horizontal earth rate uncertainty variance using the Section 14.6.4.2 process following Equations (14.6.4.2-13). The (14.6.5.1-1) integral solution is easily found using Laplace transform techniques. The Laplace transform of (14.6.5.1-1) is:

$$S \, Y_1 - y_{1_0} = - Y_2 \qquad S \, \Lambda_1 - \lambda_{1_0} = 0$$
$$S \, Y_2 - y_{2_0} = \frac{1}{q_3} \Lambda_2 \qquad S \, \Lambda_2 - \lambda_{2_0} = q_2 \, Y_2 + \Lambda_1 \tag{14.6.5.1-2}$$

where

S = Laplace transform parameter.

Y_i, Λ_i = Laplace transforms of y_i, λ_i.

Equations (14.6.5.1-2) are upon rearrangement:

$$\text{a)} \quad Y_1 = \frac{1}{S}\left(- Y_2 + y_{1_0}\right) \qquad \text{c)} \quad \Lambda_1 = \frac{1}{S} \lambda_{1_0}$$
$$\text{b)} \quad Y_2 = \frac{1}{S}\left(\frac{1}{q_3} \Lambda_2 + y_{2_0}\right) \qquad \text{d)} \quad \Lambda_2 = \frac{1}{S}\left(q_2 \, Y_2 + \Lambda_1 + \lambda_{2_0}\right) \tag{14.6.5.1-3}$$

The Λ_1 solution is from (14.6.5.1-3) c) directly. The Y_2 solution is obtained by substituting d) with c) in b), multiplying by S^2, and rearranging:

$$Y_2 = \frac{1}{S}\left(\frac{1}{q_3} \Lambda_2 + y_{2_0}\right) = \frac{1}{S}\left\{\frac{1}{q_3}\left[\frac{1}{S}\left(q_2 \, Y_2 + \frac{1}{S} \lambda_{1_0} + \lambda_{2_0}\right)\right] + y_{2_0}\right\}$$
$$Y_2 \left(S^2 - \frac{q_2}{q_3}\right) = S \, y_{2_0} + \frac{1}{S} \frac{1}{q_3} \lambda_{1_0} + \frac{1}{q_3} \lambda_{2_0} \tag{14.6.5.1-4}$$

or

$$Y_2 = \frac{1}{F}\left(S\, y_{2_0} + \frac{1}{S}\frac{1}{q_3}\lambda_{1_0} + \frac{1}{q_3}\lambda_{2_0}\right) \tag{14.6.5.1-5}$$

with

$$F \equiv S^2 - \alpha^2 \qquad \alpha \equiv \sqrt{\frac{q_2}{q_3}} \tag{14.6.5.1-6}$$

Then Y_1 is found in (14.6.5.1-3) from a) with (14.6.5.1-5) for Y_2; Λ_2 is found from d) with (14.6.5.1-5) for Y_2 and c):

$$Y_1 = \frac{1}{F\,S}\left(F\, y_{1_0} - S\, y_{2_0} - \frac{1}{S}\frac{1}{q_3}\lambda_{1_0} - \frac{1}{q_3}\lambda_{2_0}\right) \tag{14.6.5.1-7}$$

$$\begin{aligned}
\Lambda_2 &= \frac{1}{F\,S}\left[q_2\left(S\, y_{2_0} + \frac{1}{S}\frac{1}{q_3}\lambda_{1_0} + \frac{1}{q_3}\lambda_{2_0}\right) + \frac{F}{S}\lambda_{1_0} + F\,\lambda_{2_0}\right] \\
&= \frac{1}{F\,S}\left[q_2\, S\, y_{2_0} + \left(\frac{F}{S} + \frac{1}{S}\frac{q_2}{q_3}\right)\lambda_{1_0} + \left(F + \frac{q_2}{q_3}\right)\lambda_{2_0}\right] \\
&= \frac{1}{F\,S}\left(q_2\, S\, y_{2_0} + S\,\lambda_{1_0} + S^2\,\lambda_{2_0}\right)
\end{aligned} \tag{14.6.5.1-8}$$

The results in matrix form are:

$$\begin{bmatrix} Y_1 \\ Y_2 \\ \Lambda_1 \\ \Lambda_2 \end{bmatrix} = \frac{1}{F}\begin{bmatrix} \frac{F}{S} & -1 & -\frac{1}{q_3\,S^2} & -\frac{1}{q_3\,S} \\ 0 & S & \frac{1}{S}\frac{1}{q_3} & \frac{1}{q_3} \\ 0 & 0 & \frac{F}{S} & 0 \\ 0 & q_2 & 1 & S \end{bmatrix}\begin{bmatrix} y_{1_0} \\ y_{2_0} \\ \lambda_{1_0} \\ \lambda_{2_0} \end{bmatrix} \tag{14.6.5.1-9}$$

Next, with (14.6.5.1-6) for F, we define the following for particular Laplace formations in (14.6.5.1-9) including their inverse Laplace transforms:

$$\Theta_0 \equiv \frac{1}{S} \qquad \theta_0(t) = 1$$

$$\Theta_1 \equiv \frac{S}{F} = \frac{S}{S^2 - \alpha^2} \qquad \theta_1(t) = \cosh \alpha t$$

$$\Theta_2 \equiv \frac{1}{F} = \frac{1}{S^2 - \alpha^2} \qquad \theta_2(t) = \frac{\sinh \alpha t}{\alpha}$$

$$\Theta_3 \equiv \frac{1}{F\,S} = \frac{1}{\left(S^2 - \alpha^2\right) S} = \frac{1}{\alpha^2}\left(- \frac{1}{S} + \frac{S}{S^2 - \alpha^2}\right) \tag{14.6.5.1-10}$$

$$\theta_3(t) = \frac{1}{\alpha^2}\left(- 1 + \cosh \alpha t\right) = \frac{q_3}{q_2}\left(- 1 + \cosh \alpha t\right)$$

$$\Theta_4 \equiv \frac{1}{F\,S^2} = \frac{1}{\left(S^2 - \alpha^2\right) S^2} \qquad \theta_4(t) = \frac{1}{\alpha^2}\left(\frac{\sinh \alpha t}{\alpha} - t\right) = \frac{q_3}{q_2}\left(\frac{\sinh \alpha t}{\alpha} - t\right)$$

where

$\Theta_i, \theta_i(t)$ = The particular Laplace transform and its inverse Laplace transform.

Using the (14.6.5.1-10) Θ_i definitions, (14.6.5.1-9) is equivalently:

$$\begin{bmatrix} Y_1 \\ Y_2 \\ \Lambda_1 \\ \Lambda_2 \end{bmatrix} = \begin{bmatrix} \Theta_0 & - \Theta_2 & - \Theta_4 \frac{1}{q_3} & - \Theta_3 \frac{1}{q_3} \\ 0 & \Theta_1 & \Theta_3 \frac{1}{q_3} & \Theta_2 \frac{1}{q_3} \\ 0 & 0 & \Theta_0 & 0 \\ 0 & q_2\,\Theta_2 & \Theta_2 & \Theta_1 \end{bmatrix} \begin{bmatrix} y_{1_0} \\ y_{2_0} \\ \lambda_{1_0} \\ \lambda_{2_0} \end{bmatrix} \tag{14.6.5.1-11}$$

whose inverse Laplace transform is:

$$\begin{bmatrix} y_1 \\ y_2 \\ \lambda_1 \\ \lambda_2 \end{bmatrix} = \begin{bmatrix} \theta_0(t) & - \theta_2(t) & - \theta_4(t) \frac{1}{q_3} & - \theta_3(t) \frac{1}{q_3} \\ 0 & \theta_1(t) & \theta_3(t) \frac{1}{q_3} & \theta_2(t) \frac{1}{q_3} \\ 0 & 0 & \theta_0(t) & 0 \\ 0 & q_2\,\theta_2(t) & \theta_2(t) & \theta_1(t) \end{bmatrix} \begin{bmatrix} y_{1_0} \\ y_{2_0} \\ \lambda_{1_0} \\ \lambda_{2_0} \end{bmatrix} \tag{14.6.5.1-12}$$

We next compare (14.6.5.1-12) with the (15.1.5.4-8) general form to see that:

$$\Phi_{yy}(t) = \begin{bmatrix} \theta_0(t) & -\theta_2(t) \\ 0 & \theta_1(t) \end{bmatrix} \qquad \Phi_{y\lambda}(t) = \begin{bmatrix} -\theta_4(t)\dfrac{1}{q_3} & -\theta_3(t)\dfrac{1}{q_3} \\ \theta_3(t)\dfrac{1}{q_3} & \theta_2(t)\dfrac{1}{q_3} \end{bmatrix}$$

$$\Phi_{\lambda y}(t) = \begin{bmatrix} 0 & 0 \\ 0 & q_2\,\theta_2(t) \end{bmatrix} \qquad \Phi_{\lambda\lambda}(t) = \begin{bmatrix} \theta_0(t) & 0 \\ \theta_2(t) & \theta_1(t) \end{bmatrix} \tag{14.6.5.1-13}$$

Using (14.6.5.1-13) and P^{**}_0 from (14.6.4.2-15) finds:

$$\Phi_{yy}(t) + \Phi_{y\lambda}(t)\,P^{**}_0 = \begin{bmatrix} \left(\theta_0(t) - \theta_4(t)\dfrac{1}{q_3}P_{\Omega_0}\right) & -\theta_2(t) \\ \theta_3(t)\dfrac{1}{q_3}P_{\Omega_0} & \theta_1(t) \end{bmatrix}$$

$$\Phi_{\lambda y}(t) + \Phi_{\lambda\lambda}(t)\,P^{**}_0 = \begin{bmatrix} \theta_0(t)\,P_{\Omega_0} & 0 \\ \theta_2(t)\,P_{\Omega_0} & q_2\,\theta_2(t) \end{bmatrix} \tag{14.6.5.1-14}$$

The determinant of the first expression in (14.6.5.1-14) is with the (14.6.5.1-10) inverse Laplace transforms:

$$\begin{aligned} Dtr(t) &= \left(\theta_0(t) - \theta_4(t)\frac{1}{q_3}P_{\Omega_0}\right)\theta_1(t) + \theta_2(t)\,\theta_3(t)\frac{1}{q_3}P_{\Omega_0} \\ &= \left[\left(1 - \left(\frac{\sinh\alpha t}{\alpha} - t\right)\frac{1}{q_2}P_{\Omega_0}\right)\right]\cosh\alpha t + \frac{1}{q_2}\frac{\sinh\alpha t}{\alpha}\left(-1 + \cosh\alpha t\right)P_{\Omega_0} \\ &= \cosh\alpha t - \frac{\sinh\alpha t}{\alpha}\frac{1}{q_2}P_{\Omega_0}\cosh\alpha t + \frac{t}{q_2}P_{\Omega_0}\cosh\alpha t \\ &\quad - \frac{\sinh\alpha t}{q_2\,\alpha}P_{\Omega_0} + \frac{1}{q_2}\frac{\sinh\alpha t}{\alpha}\cosh\alpha t\,P_{\Omega_0} \\ &= \cosh\alpha t + \frac{t}{q_2}P_{\Omega_0}\cosh\alpha t - \frac{\sinh\alpha t}{q_2\,\alpha}P_{\Omega_0} \end{aligned} \tag{14.6.5.1-15}$$

(Continued)

$$
\begin{aligned}
&= \left(1 + \frac{t}{q_2} P_{\Omega_0}\right) \cosh \alpha t - \frac{\sinh \alpha t}{q_2\, \alpha} P_{\Omega_0} \\
&= \cosh \alpha t \left(1 + \frac{t}{q_2} P_{\Omega_0} - \frac{\tanh \alpha t}{q_2\, \alpha} P_{\Omega_0}\right) \\
&= \theta_1(t) \left(1 + \frac{P_{\Omega_0}}{q_2\, \alpha} \left(\alpha t - \tanh \alpha t\right)\right)
\end{aligned}
\qquad \text{(14.6.5.1-15) (Continued)}
$$

where

$$
\mathrm{Dtr}(t) = \text{Determinant of } \Phi_{yy}(t) + \Phi_{y\lambda}(t)\ P^{**}{}_0.
$$

Applying (14.6.5.1-15) obtains for the inverse of the first expression in (14.6.5.1-14):

$$
\left(\Phi_{yy}(t) + \Phi_{y\lambda}(t)\, P^{**}{}_0\right)^{-1} = \left(1 + \frac{P_{\Omega_0}}{q_2\, \alpha} \left(\alpha t - \tanh \alpha t\right)\right)^{-1} \begin{bmatrix} 1 & - \\ - & - \end{bmatrix} \qquad \text{(14.6.5.1-16)}
$$

Substituting (14.6.5.1-16), the second expression from (14.6.5.1-14) with $\theta_0(t)$ from(14.6.5.1-10), and the P conversion expression from (14.6.4.2-7) into general solution Equation (15.1.5.4-11) then yields for the earth rate uncertainty variance P_Ω (the 1,1 element of P^{**}):

$$
P_\Omega = P_{\Omega_0} \left[1 + \frac{P_{\Omega_0}}{q_2\, \alpha} \left(\alpha t - \tanh \alpha t\right)\right]^{-1} \qquad \text{(14.6.5.1-17)}
$$

From (14.6.4.2-8) we see that P_Ω is also the (1,1) element of P'. The α and q_2 terms in (14.6.5.1-17) are from (14.6.5.1-6) and (14.6.4.2-3):

$$
q_2 = q_{\omega Rand} \qquad q_3 = q_{\alpha Quant} + \frac{1}{g^2} q_{aRand} \qquad \alpha \equiv \sqrt{\frac{q_2}{q_3}} \qquad \text{(14.6.5.1-18)}
$$

Equations (14.6.5.1-17) with (14.6.5.1-18) provide a solution for the horizontal earth rate uncertainty variance for simultaneous application of q_2 and q_3 process noise with the other noise sources zero (and the Kalman filter design based on these conditions).

It is also instructive to consider the form of Equation (14.6.5.1-17) under conditions when the initial earth rate uncertainty variance is large. Then the equation simplifies to:

$$
\text{For Large } P_{\Omega_0}: \qquad P_\Omega = \frac{q_2\, \alpha}{\alpha t - \tanh \alpha t} \qquad \text{(14.6.5.1-19)}
$$

14.6.5.2 COMPARISON BETWEEN THE SUM OF INDIVIDUAL SOLUTIONS AND THE COMBINED SOLUTION FOR MULTIPLE NOISE SOURCES

In this section we compare the P_Ω solution from the previous section with the sum of the individual q_2 and q_3 solutions from Equations (14.6.4.4-2), to see whether the sum solution can be used as a reasonable approximation to the previous section Equation (14.6.5.1-19) P_Ω solution.

First we restate the (14.6.5.1-19) solution, renamed to identify it as the response to simultaneous q_2 and q_3:

$$P_{\Omega_{Simult}} = \frac{q_2\,\alpha}{\alpha\,t - \tanh \alpha t} \tag{14.6.5.2-1}$$

where

$P_{\Omega_{Simult}}$ = P_Ω for a Kalman filter designed with simultaneous q_2 and q_3 with no other noise sources.

The individual q_2 and q_3 solutions from (14.6.4.4-2) are with (14.6.5.1-18):

$$P_{\Omega_{\alpha Quant/aRand}} = \frac{3\,q_3}{t^3} = \frac{3\,q_2\,\alpha}{(\alpha t)^3} \qquad P_{\Omega_{\omega Rand}} = \frac{q_{\omega Rand}}{t} = \frac{q_2\,\alpha}{\alpha t} \tag{14.6.5.2-2}$$

The sum of the (14.6.5.2-2) solutions is:

$$\begin{aligned} P_{\Omega_{Sum}} &= P_{\Omega_{\alpha Quant/aRand}} + P_{\Omega_{\omega Rand}} \\ &= \frac{3\,q_2\,\alpha}{(\alpha t)^3} + \frac{q_2\,\alpha}{\alpha t} = q_2\,\alpha\left(\frac{3}{(\alpha t)^3} + \frac{1}{\alpha t}\right) \end{aligned} \tag{14.6.5.2-3}$$

where

$P_{\Omega_{Sum}}$ = The sum of the P_Ω solutions for individual Kalman filters designed with q_2 and with q_3 as the only noise sources.

We now address the question of how accurately (14.6.5.2-3) approximates (14.6.5.2-1) by forming the ratio minus the desired result (i.e., 1):

$$\frac{P_{\Omega_{Sum}}}{P_{\Omega_{Simult}}} - 1 = \left(\frac{3}{(\alpha t)^3} + \frac{1}{\alpha t}\right)\left(\alpha t - \tanh \alpha t\right) - 1 \tag{14.6.5.2-4}$$

Equation (14.6.5.2-4) is tabulated below as a function of αt with α defined in (14.6.5.1-18) as the square root of q_2 / q_3.

$\sqrt{\frac{q_2}{q_3}}\ t$	$\frac{P_{\Omega_{Sum}}}{P_{\Omega_{Simult}}} - 1$
0	0
0.1	- 6.64 E-4
0.3	- 5.78 E-4
1	- 4.64 E-2
3	- 1.09 E-1
10	- 7.30 E-2
∞	0

We see from the table that the maximum for the error function occurs near $q_2 / q_3 = 1$ for which the error is - 4.64% (i.e., the approximate sum solution $P_{\Omega_{Sum}}$ is 4.64% smaller than the true solution $P_{\Omega_{Simult}}$). For large or small values of q_2 / q_3 the error in the $P_{\Omega_{Sum}}$ solution is zero. Thus, we see that use of $P_{\Omega_{Sum}}$ as an approximation to $P_{\Omega_{Simult}}$ has very little error over the full range of q_2 / q_3.

Based on the above results (and general Kalman filter covariance simulation experience), we can extrapolate Equation (14.6.5.2-3) to the general case, that a reasonable approximation for P_Ω with all noise sources present is the sum of all the individual noise responses:

$$P_\Omega \approx P_{\Omega_{RVib}} + P_{\Omega_{\upsilon Quant}} + P_{\Omega_{\alpha Quant/aRand}} + P_{\Omega_{\omega Rand}} \qquad (14.6.5.2\text{-}5)$$

in which the contributing terms are as given by the Equations (14.6.4.4-1) or (14.6.4.4-2) individual noise solutions.

15 Kalman Filtering Techniques

15.0 OVERVIEW

Thus far, we have been primarily concerned with the analytical aspects of "free inertial navigation" in which inertial sensors (angular rate sensors and accelerometers) provide the only input to the navigation system. In this regard the free inertial navigation system is a non-radiating autonomous device, generating its output from internally contained inertial sensors. Two exceptions are worthy of note; the pressure altitude input that is frequently used to control vertical channel error (discussed in Section 4.4.1.2.1), and the quasi-stationary initialization process (discussed in Chapter 6) that relies on the general knowledge that the user vehicle has bounded small amplitude attitude/position motion.

In many applications, the free inertial performance of the INS is not sufficient to achieve all navigation accuracy requirements. To compensate for INS performance deficiencies, other navigation devices have been utilized to attenuate the undesirable INS error characteristics (e.g., unbounded position error growth and position/velocity/attitude Schuler oscillations). Kalman filtering has become a standard method of blending navigation data from the INS and other available navigation sources, to achieve an overall navigation solution that eliminates the undesirable INS error characteristics, while retaining the desirable INS wide-bandwidth/low-noise navigation signal output signature.

In this Chapter we will describe Kalman filtering as a general analytical software process, and as it relates to strapdown inertial navigation applications. Included is a general discussion of the Kalman filter software validation process. Examples are provided illustrating the application of Kalman filtering to INS quasi-stationary initialization (gain determination for the Chapter 6 Fine Alignment process), dynamic moving base INS initialization (i.e., "transfer alignment"), INS updating using a body mounted velocity sensor (e.g., Doppler radar), and INS updating based on GPS (Global Positioning System) range-to-satellite measurements.

Coordinate frames we will use in this chapter include the B, L, N, E and I Frames defined in Section 2.2.

15.1 KALMAN FILTERING IN GENERAL

Kalman filtering is a general analytical process by which inaccessible system parameters can be estimated based on accessible measurements from the system outputs. As applied to inertial navigation, Kalman filters have been utilized to estimate and correct inaccessible INS errors (e.g., attitude error, inertial sensor errors), based on measurements of accessible INS parameters compared to equivalent parameters provided from an alternative navigation data source (e.g., INS computed position compared to calculated position from a GPS receiver aiding device in equivalent formats). The alternative navigation source is typically denoted as the "inertial aiding" device. The process of comparing the accessible INS and aiding device parameters (a subtraction process) generates an error signal (called the "measurement") that contains a composite of errors from the INS and the inertial aid. Processing the measurement through the Kalman filter allows all significant INS and aiding device errors affecting the measurement to be independently estimated. A "control" process is typically incorporated as part of Kalman filter operations to continuously correct the INS errors being estimated.

From an analytical standpoint, let us define an "error state vector" containing as its elements ("error states") the significant INS and aiding device error terms to be accounted for by the Kalman filter. Assuming a linear model for the error states, we can describe their general dynamic characteristics in matrix form by the "error state dynamic equation":

$$\dot{\underline{x}}(t) = A(t)\, \underline{x}(t) + G_P(t)\, \underline{n}_P(t) \tag{15.1-1}$$

where

$\underline{x}(t)$ = Error state vector treated analytically as a column matrix.

$A(t)$ = Error state dynamic matrix.

$\underline{n}_P(t)$ = Vector of independent white "process" noise sources driving $\dot{\underline{x}}(t)$ (treated analytically as a column matrix).

$G_P(t)$ = Process noise dynamic coupling matrix that couples individual $\underline{n}_P(t)$ components into $\dot{\underline{x}}(t)$.

In general, $A(t)$ and $G_P(t)$ are time varying functions of the angular rate, acceleration, attitude, velocity and position parameters calculated within the INS computer. Equation set (12.5.1-1) with Section 12.5.6 for the sensor errors is an example of strapdown inertial navigation system error state propagation equations that fit the (15.1-1) format. For this case, the error states comprising $\underline{x}(t)$ might include $\underline{\psi}^N$, $\delta\underline{V}^N$, $\delta\underline{R}^N$, $\delta\underline{K}_{0Bias}$, $\delta\underline{L}_{0Bias}$. The $\underline{n}_P(t)$ independent process noise vector might include $\delta\underline{\alpha}_{Quant}$, $\delta\underline{\upsilon}_{Quant}$, $\delta\underline{\omega}_{Rand}$, $\delta\underline{a}_{Rand}$. The $A(t)$ and $G_P(t)$ matrices would be the matrix elements in (12.5.1-1) and Section 12.5.6 coupling $\underline{n}_P(t)$ and $\underline{x}(t)$ into $\dot{\underline{x}}(t)$. Additionally, $\underline{x}(t)$ and $\underline{n}_P(t)$ would include the pertinent error effects associated with the aiding device.

The Kalman filter is normally implemented as a repetitive software function at a specified cycle rate in a digital computer. For an aided inertial navigation system, the Kalman filter is typically resident in the inertial navigation system computer. The Kalman filter designed to estimate $\underline{x}$ will operate from the "measurement" input at each filter software cycle time which can be linearly modeled by the linearized "measurement equation":

$$\underline{z}_n = H_n \underline{x}_n + G_{M_n} \underline{n}_{M_n} \tag{15.1-2}$$

where

n = Kalman filter software cycle time index.

$(\ )_n$ = $(\ )$ at the nth Kalman filter cycle time.

$\underline{z}$ = Measurement vector analytically represented as a column matrix.

H = Measurement matrix.

$\underline{n}_M$ = Vector of independent white measurement noise sources (represented analytically as a column matrix). The $\underline{n}_M$ vector represents noise type error effects that may be introduced in the process of taking the measurement.

G_M = Measurement noise dynamic coupling matrix that couples $\underline{n}_M$ into $\underline{z}$.

In general, H and G_M are time varying functions of the navigation parameters calculated in the INS computer.

The measurement equation is a linearized version of a general nonlinear "observation equation" used as the actual input to the Kalman filter:

$$\underline{Z}_{Obs_n} = f(\xi_{INS_n}, \xi_{Aid_n}) \tag{15.1-3}$$

where

ξ_{INS} = INS navigation parameters.

ξ_{Aid} = Aiding device navigation parameters.

$f(\ )$ = Functional operator that compares designated equivalent elements of ξ_{INS} and ξ_{Aid}. The $f(\ )$ operator is designed so that for an error free INS, an error free aiding device, and a perfect (error free) $f(\ )$ software implementation, $f(\ )$ will be zero.

$\underline{Z}_{Obs}$ = Observation vector formed from the comparison between comparable INS and aiding device navigation parameters. For an error free INS, an error free aiding device, and zero error in making the observation, $\underline{Z}_{Obs}$ would equal zero. The linearized form of $\underline{Z}_{Obs}$ is the Equation (15.1-2) measurement vector $\underline{z}$.

We also allow that $\underline{x}$ may be controlled by the Kalman filter to constrain error build-up such that:

$$\underline{x}_n(+_c) = \underline{x}_n(-) + \underline{u}_{c_n} \tag{15.1-4}$$

where

$\underline{u}_{c_n}$ = Control vector derived from the Kalman filter and applied at time t_n to constrain $\underline{x}$.

$(+_c)$ = Designation for parameter value at its designated time stamp (t_n in this case) immediately after ("a posteriori") the application of control resets (c subscript) at the same designated time. Note: The designation $(+_e)$ will also be used in subsequent sections to describe estimation resets in Kalman estimation filter structures.

$(-)$ = Designation for parameter value at its designated time stamp (t_n in this case) prior to ("a priori") the application of any resets (estimation or control) at the same designated time.

The Latin notation "a priori" and "a posteriori" has been adopted in Kalman filter terminology to add an element of "mysterioso".

The process of applying Equation (15.1-4) to the system errors is known as a "control reset" operation. The implementation of (15.1-4) in the INS computer involves applying $\underline{u}_{c_n}$ to the INS (and, in some cases, also to the aiding device) navigation parameters:

$$\begin{aligned} \xi_{INS_n}(+_c) &= \xi_{INS_n}(-) + g_{INS}\left(\xi_{INS_n}(-),\, \underline{u}_{c_n}\right) \\ \xi_{Aid_n}(+_c) &= \xi_{Aid_n}(-) + g_{Aid}\left(\xi_{INS_n}(-),\, \underline{u}_{c_n}\right) \end{aligned} \tag{15.1-5}$$

where

$g_{INS}(\ ),\ g_{Aid}(\ )$ = Non-linear functional operators used to apply $\underline{u}_{c_n}$ to the ξ_{INS}, ξ_{Aid} navigation parameters at time t_n such that the error in these parameters is controlled as specified by Equation (15.1-4).

Equation (15.1-4) represents the linearized version of (15.1-5).

Equations (15.1-4) - (15.1-5) are based on application of the control vector at the Kalman estimation update time t_n. Another possibility is to apply the control reset at an intermediate time between Kalman estimation/update cycle times. In this document we shall primarily consider the (15.1-4) - (15.1-5) forms. Use of the intermediate time control reset is discussed further in Section 15.1.2.4.

15.1.1 DISCRETE FORM OF ERROR STATE PROPAGATION EQUATIONS

For compatibility with the Kalman filter digital computer repetitive software structure, Equation (15.1-1) can be restated in an integrated discrete form for evaluation at the filter cycle times. We first define the integrated solution to (15.1-1) as the sum of two parts:

$$\underline{x}(t) = \underline{x}_{Hmg}(t, t_1) + \underline{x}_{Prt}(t, t_1) \qquad (15.1.1\text{-}1)$$

where

$\underline{x}_{Hmg}(t, t_1)$ = Solution to the homogeneous portion of (15.1-1) from time t_1 to time t (i.e., assuming zero for the forcing process noise $\underline{n}_P(t)$ since time t_1). The $\underline{x}_{Hmg}(t, t_1)$ portion of the $\underline{x}(t)$ solution accounts for initial values on $\underline{x}(t)$ at time t_1 and how they propagate through the (15.1-1) equation in the absence of process noise input.

$\underline{x}_{Prt}(t, t_1)$ = Particular solution to (15.1-1) produced by $\underline{n}_P(t)$ since time t_1 which adds to the $\underline{x}_{Hmg}(t, t_1)$ initial condition propagation portion to form the total $\underline{x}(t)$.

From its definition, the homogeneous portion of (15.1-1) is given by:

$$\dot{\underline{x}}_{Hmg}(t,t_1) = A(t)\, \underline{x}_{Hmg}(t,t_1) \qquad (15.1.1\text{-}2)$$

The $\underline{x}_{Hmg}(t, t_1)$ solution to (15.1.1-2) is structured from the $\underline{x}_{Hmg}(t, t_1)$ definition as the linear form:

$$\underline{x}_{Hmg}(t, t_1) = \Phi(t, t_1)\, \underline{x}(t_1) \qquad (15.1.1\text{-}3)$$

where

$\Phi(t, t_1)$ = Error state transition matrix that propagates $\underline{x}(t)$ from its value at time t_1 to its value at time t in the absence of process noise.

From the definition of $\underline{x}_{Hmg}(t, t_1)$, we see that its value at time $t = t_1$ is $\underline{x}(t_1)$. Thus, Equation (15.1.1-3) evaluated at $t = t_1$ shows that:

$$\Phi(t_1,t_1) = I \qquad (15.1.1\text{-}4)$$

The time derivative of (15.1.1-3) is:

$$\dot{\underline{x}}_{Hmg}(t, t_1) = \dot{\Phi}(t, t_1)\, \underline{x}(t_1) \qquad (15.1.1\text{-}5)$$

Substituting (15.1.1-5) and (15.1.1-3) into (15.1.1-2) provides the differential equation for the error state transition matrix $\Phi(t, t_1)$:

$$\dot{\Phi}(t,t_1) = A(t)\, \Phi(t,t_1) \qquad (15.1.1\text{-}6)$$

The integral solution to (15.1.1-6) from time t_1 with (15.1.1-4) as the initial condition provides $\Phi(t, t_1)$ for evaluation of $\underline{x}_{Hmg}(t, t_1)$ in Equation (15.1.1-3).

A useful characteristic of the error state transition matrix can be derived from Equation (15.1.1-3) for the homogeneous solution at arbitrary times t_2 and t_3 following t_1:

$$\begin{aligned} \underline{x}_{Hmg}(t_2) &= \Phi(t_2,t_1)\, \underline{x}(t_1) \\ \underline{x}_{Hmg}(t_3) &= \Phi(t_3,t_2)\, \underline{x}_{Hmg}(t_2) \end{aligned} \qquad (15.1.1\text{-}7)$$

where

$\underline{x}_{Hmg}(t_i) = \underline{x}_{Hmg}(t, t_1)$ at time t_i following time t_1.

$\Phi(t_i,t_j)$ = Error state transition matrix that propagates $\underline{x}(t)$ from its value at time t_j to its value at time t_i in the absence of process noise (i.e., the homogeneous propagation of $\underline{x}(t)$ from its value at time t_j).

In combination, (15.1.1-7) gives:

$$\underline{x}_{Hmg}(t_3) = \Phi(t_3,t_2)\, \Phi(t_2,t_1)\, \underline{x}(t_1) \qquad (15.1.1\text{-}8)$$

We can also write:

$$\underline{x}_{Hmg}(t_3) = \Phi(t_3,t_1)\, \underline{x}(t_1) \qquad (15.1.1\text{-}9)$$

Equating (15.1.1-8) and (15.1.1-9) then shows that:

$$\Phi(t_3,t_1) = \Phi(t_3,t_2)\, \Phi(t_2,t_1) \qquad (15.1.1\text{-}10)$$

which is a form that will aid in subsequent developments.

The $\underline{x}_{Prt}(t, t_1)$ particular solution to (15.1-1) is obtained by applying $\Phi(t, \tau)$ to the infinitesimal contribution of $\underline{n}_P(t)$ to $\underline{x}(t)$ at time τ (for τ greater than or equal to t_1), and summing (integrating) the contributions from t_1 to t. For $\underline{n}_P(t)$ applied at time τ over an infinitesimal time interval $d\tau$, the contribution to $\underline{x}(t)$ in Equation (15.1-1) at time t is:

$$d\underline{x}_{Prt}(t) = \Phi(t,\tau)\, G_P(\tau)\, \underline{n}_P(\tau)\, d\tau \tag{15.1.1-11}$$

Integrating (15.1.1-11) from t_1 to t then provides the particular solution

$$\underline{x}_{Prt}(t, t_1) = \int_{t_1}^{t} \Phi(t,\tau)\, G_P(\tau)\, \underline{n}_P(\tau)\, d\tau \tag{15.1.1-12}$$

The overall solution to (15.1-1) from time t_1 is the composite of Equations (15.1.1-1), (15.1.1-3), (15.1.1-4), (15.1.1-6) and (15.1.1-12):

$$\begin{aligned} \underline{x}(t) &= \Phi(t, t_1)\, \underline{x}(t_1) + \underline{w}(t, t_1) \\ \Phi(t, t_1) &= I + \int_{t_1}^{t} A(\tau)\, \Phi(\tau, t_1)\, d\tau \\ \underline{w}(t, t_1) &= \int_{t_1}^{t} \Phi(t,\tau)\, G_P(\tau)\, \underline{n}_P(\tau)\, d\tau \end{aligned} \tag{15.1.1-13}$$

where

$\underline{w}(t, t_1)$ = Integrated process noise vector.

If we interpret t_1 as the last Kalman filter update time (cycle n-1 at time t_{n-1}) and t as the current Kalman filter update time (cycle n at time t_n), Equations (15.1.1-13) over a Kalman filter update cycle time interval are then given by:

$$\underline{x}_n = \Phi_n\, \underline{x}_{n-1} + \underline{w}_n \tag{15.1.1-14}$$

with

$$\begin{aligned} \Phi(t, t_{n-1}) &= I + \int_{t_{n-1}}^{t} A(\tau)\, \Phi(\tau, t_{n-1})\, d\tau \\ \Phi_n &\equiv \Phi(t_n, t_{n-1}) \end{aligned} \tag{15.1.1-15}$$

$$\underline{w}_n = \int_{t_{n-1}}^{t_n} \Phi(t_n,\tau)\, G_P(\tau)\, \underline{n}_P(\tau)\, d\tau \tag{15.1.1-16}$$

Equations (15.1.1-14) - (15.1.1-16) represent the equivalent discrete integrated form of (15.1-1) at each filter update cycle time.

15.1.2 KALMAN FILTER CONFIGURATION

A Kalman filter designed to estimate and control $\underline{x}$ can be modeled from Equations (15.1.1-14) and (15.1-2) - (15.1-5) repeated below with more definitive notation:

$$\underline{Z}_{Obs_n} = f\left(\xi_{INS_n}(-), \xi_{Aid_n}(-)\right) \tag{15.1.2-1}$$

$$\underline{x}_n(-) = \Phi_n \, \underline{x}_{n-1}(+_c) + \underline{w}_n \tag{15.1.2-2}$$

$$\underline{z}_n = H_n \, \underline{x}_n(-) + G_{M_n} \underline{n}_{M_n} \tag{15.1.2-3}$$

$$\begin{aligned} \xi_{INS_n}(+_c) &= \xi_{INS_n}(-) + g_{INS}\left(\xi_{INS_n}(-), \underline{u}_{c_n}\right) \\ \xi_{Aid_n}(+_c) &= \xi_{Aid_n}(-) + g_{Aid}\left(\xi_{INS_n}(-), \underline{u}_{c_n}\right) \end{aligned} \tag{15.1.2-4}$$

$$\underline{x}_n(+_c) = \underline{x}_n(-) + \underline{u}_{c_n} \tag{15.1.2-5}$$

Equations (15.1.2-1) - (15.1.2-5) represent the interfaces between the actual navigation parameters (ξ_{INS} and ξ_{Aid}), the Kalman filter observation input ($\underline{Z}_{Obs}$), and the control vector output ($\underline{u}_c$), including the effect on the error state vector $\underline{x}$. The ξ_{INS} and ξ_{Aid} data is available in the navigation computer as is the observation $\underline{Z}_{Obs}$ and the computer generated control vector $\underline{u}_c$. The Kalman filter design problem is: Given $\underline{Z}_{Obs}$ as input and the (15.1.2-2), (15.1.2-3) and (15.1.2-5) linearized analytical model for $\underline{x}$ propagation, measurement and control, estimate $\underline{x}$ and use the $\underline{x}$ estimate to calculate and output the $\underline{x}$ control vector $\underline{u}_c$. A set of Kalman filter and interface equations based on (15.1.2-1) - (15.1.2-5) for performing these operations is represented by the following computation sequence that would be programmed into the navigation computer in the order listed:

$$\underline{Z}_{Obs_n} = f\left(\xi_{INS_n}(-), \xi_{Aid_n}(-)\right) \tag{15.1.2-6}$$

$$\underset{\sim}{\underline{x}}_n(-) = \Phi_n \, \underset{\sim}{\underline{x}}_{n-1}(+_c) \tag{15.1.2-7}$$

$$\underset{\sim}{\underline{z}}_n = H_n \, \underset{\sim}{\underline{x}}_n(-) \tag{15.1.2-8}$$

$$\underset{\sim}{\underline{x}}_n(+_e) = \underset{\sim}{\underline{x}}_n(-) + K_n \left(\underline{Z}_{Obs_n} - \underset{\sim}{\underline{z}}_n\right) \tag{15.1.2-9}$$

$$\underline{u}_{c_n} = \text{function of } \underset{\sim}{\underline{x}}_n(+_e) \tag{15.1.2-10}$$

$$\begin{aligned} \xi_{INS_n}(+_c) &= \xi_{INS_n}(-) + g_{INS}\left(\xi_{INS_n}(-), \underline{u}_{c_n}\right) \\ \xi_{Aid_n}(+_c) &= \xi_{Aid_n}(-) + g_{Aid}\left(\xi_{INS_n}(-), \underline{u}_{c_n}\right) \end{aligned} \tag{15.1.2-11}$$

$$\underset{\sim}{\tilde{x}}_n(+_c) = \underset{\sim}{\tilde{x}}_n(+_e) + \underline{u}_{c_n} \qquad (15.1.2\text{-}12)$$

$$\underline{\tilde{x}}_0 = 0 \qquad \text{Initial Conditions} \qquad (15.1.2\text{-}13)$$

where

$\tilde{\ }$ = Value for parameter estimated (or predicted) by the Kalman filter.

$(+_e)$ = Designation for parameter value at its designated time stamp (t_n in this case) immediately after ("a posteriori") the application of <u>estimation</u> resets (e subscript) at the same designated time.

$(+_c)$ = Designation for parameter value at its designated time stamp (t_n in this case) immediately after ("a posteriori") the application of <u>control</u> resets (c subscript) at the same designated time.

$(-)$ = Designation for parameter value at its designated time stamp (t_n in this case) immediately prior to ("a priori") the application of any resets (estimation or control) at the same designated time.

K_n = Kalman gain matrix. K_n is a time varying function of the statistical inaccuracy in $\underline{\tilde{x}}_n$ (as will be described subsequently).

Equation (15.1.2-7) (known as the estimated error state vector "propagation" operation) is modeled after Equation (15.1.2-2) as the Kalman filter's estimate for what $\underline{\tilde{x}}_n$ will be at the next Kalman cycle based on its estimated value after the last Kalman update. Equation (15.1.2-7) accounts for the error state propagation dynamics but does not account for the $\underline{w}_n$ integrated process noise in (15.1.2-2) other than its average value of zero. In addition, the initial value for $\underline{\tilde{x}}$ is not known, hence, is modeled in Equations (15.1.2-13) as zero. Similarly, Equation (15.1.2-8) is modeled after measurement Equation (15.1.2-3) as the filter's prediction of what the nth measurement will be. Equation (15.1.2-8) accounts for $\underline{\tilde{x}}_n$ and the known form of the measurement matrix, but does not account for the measurement noise $\underline{n}_M$ which it approximates at its average zero value. Equation (15.1.2-9) (known as the "innovations process" or the "estimation update" operation) is the method of compensating for the inaccuracy in the Equations (15.1.2-7) - (15.1.2-8) approximations. The input to Equation (15.1.2-9) is the observation vector $\underline{Z}_{Obs_n}$ whose linearized form is $\underline{z}_n$. The difference between the $\underline{z}_n$ estimate (i.e., $\underline{\tilde{z}}_n$) and $\underline{Z}_{Obs_n}$ in Equation (15.1.2-9) provides the measure (within the measurement noise) of how well the $\underline{\tilde{x}}_n$ estimate is representative of the actual $\underline{x}_n$. The difference vector $\underline{Z}_{Obs_n}$ - $\underline{\tilde{z}}_n$ (called the "measurement residual") is used in feed-back fashion through the Kalman gain matrix K_n, to provide $\underline{\tilde{x}}_n$ updates. After a sufficient number of filter cycles, $\underline{\tilde{x}}_n$ should converge to the correct $\underline{x}_n$ value, within the boundaries of the noise inputs to the system (i.e., $\underline{w}_n$, $\underline{n}_M$ process and measurement noise).

To control $\underline{x}$ to zero, the ideal $\underline{u}_c$ would be set to the negative of $\tilde{\underline{x}}$ at a given cycle time and applied at the same cycle time (i.e., $\underline{u}_{c_n} = - \tilde{\underline{x}}_n$). If this operation were performed immediately following the Kalman estimation update, then $\underline{u}_{c_n} + \tilde{\underline{x}}_n(+_e) = 0$ and we see from Equations (15.1.2-12) and (15.1.2-7) that the propagated $\tilde{\underline{x}}$ (i.e., $\tilde{\underline{x}}_n(-)$) would also be controlled to zero. Thus, an ideal control would be effected in which the estimated error state vector is maintained at zero both following and before the Kalman measurement/estimation reset operations. Assuming an accurate estimation process (i.e., K_n selection) the actual error state vector $\underline{x}$ will also be controlled in this manner within the limits of the process and measurement noise affecting the estimation process. Unfortunately, except for a few limited exceptions, applying an ideal control in this manner is not possible because of real time constraints present in actual applications.

The fundamental real time constraint in a large Kalman filter (i.e., with many error state components) is the time interval required to execute the operations implied by Equations (15.1.2-6) - (15.1.2-12), the most time consuming of which are the calculations of Φ_n and K_n. The cycle time interval for a large Kalman filter is generally sized based on providing sufficient time to calculate Φ_n and K_n. Thus, by the time that execution of Equations (15.1.2-6) - (15.1.2-12) are completed, a full Kalman cycle has expired, and $\tilde{\underline{x}}_n(+)$ in these equations would not be calculated until time t_{n+1}. Therefore, the ideal control of setting $\underline{u}_{c_n} = - \tilde{\underline{x}}_n(+_e)$ and applying it at t_n cannot be executed because $\tilde{\underline{x}}_n$ is not yet known at t_n, and will not be known until t_{n+1}. A possible solution to this dilemma might be to pre compute Φ_n and K_n one cycle earlier so they will be ready for application when the measurement is made. Given that Φ_n and K_n have been pre computed, the time to execute (15.1.2-6) - (15.1.2-12) is generally fast enough to be considered instantaneous so that $\tilde{\underline{x}}_n(+_e)$ will, in effect, be known at t_n. In this case we can then set and apply $\underline{u}_{c_n} = - \tilde{\underline{x}}_n(+_e)$ at t_n, thereby achieving our ideal control reset operation. To pre compute Φ_n and K_n, their computation must be completed by t_n. Unfortunately, except for a few exceptions (e.g., the quasi-stationary ground alignment problem discussed in Section 15.2.1), Φ_n and K_n are functions of dynamic navigation parameters measured over the t_{n-1} to t_n time interval, hence, their computation cannot be initiated until t_n.

The method of "delayed control resets" is a general technique that has been successfully applied to deal with the computation delay in a real time Kalman filter. With this approach, the Kalman filter operations are similar to Equations (15.1.2-6) - (15.1.2-13), except that the control function is applied one cycle later, thereby allowing one full cycle time to execute the real time Kalman filter computations. The equivalent to the Equations (15.1.2-1) - (15.1.2-5) design model for a delayed control reset Kalman filter would be as follows:

$$\xi_{INS_n}(+_c) = \xi_{INS_n}(-) + g_{INS}\left(\xi_{INS_n}(-), \underline{u}_{c_n}\right)$$
$$\xi_{Aid_n}(+_c) = \xi_{Aid_n}(-) + g_{Aid}\left(\xi_{INS_n}(-), \underline{u}_{c_n}\right) \qquad (15.1.2\text{-}14)$$

$$\underline{Z}_{Obs_n} = f\left(\xi_{INS_n}(+_c), \xi_{Aid_n}(+_c)\right) \qquad (15.1.2\text{-}15)$$

$$\underline{x}_n(-) = \Phi_n\, \underline{x}_{n-1}(+_c) + \underline{w}_n \qquad (15.1.2\text{-}16)$$

$$\underline{x}_n(+_c) = \underline{x}_n(-) + \underline{u}_{c_n} \qquad (15.1.2\text{-}17)$$

$$\underline{z}_n = H_n\, \underline{x}_n(+_c) + G_{M_n}\, \underline{n}_{M_n} \qquad (15.1.2\text{-}18)$$

The delayed control reset Kalman filter and interface equations based on (15.1.2-14) - (15.1.2-18) would then be represented by the following computation sequence programmed into the navigation computer in the order listed:

$$\xi_{INS_n}(+_c) = \xi_{INS_n}(-) + g_{INS}\left(\xi_{INS_n}(-), \underline{u}_{c_n}\right)$$
$$\xi_{Aid_n}(+_c) = \xi_{Aid_n}(-) + g_{Aid}\left(\xi_{INS_n}(-), \underline{u}_{c_n}\right) \qquad (15.1.2\text{-}19)$$

$$\underline{Z}_{Obs_n} = f\left(\xi_{INS_n}(+_c), \xi_{Aid_n}(+_c)\right) \qquad (15.1.2\text{-}20)$$

$$\tilde{\underline{x}}_n(-) = \Phi_n\, \tilde{\underline{x}}_{n-1}(+_e) \qquad (15.1.2\text{-}21)$$

$$\tilde{\underline{x}}_n(+_c) = \tilde{\underline{x}}_n(-) + \underline{u}_{c_n} \qquad (15.1.2\text{-}22)$$

$$\tilde{\underline{z}}_n = H_n\, \tilde{\underline{x}}_n(+_c) \qquad (15.1.2\text{-}23)$$

$$\tilde{\underline{x}}_n(+_e) = \tilde{\underline{x}}_n(+_c) + K_n\left(\underline{Z}_{Obs_n} - \tilde{\underline{z}}_n\right) \qquad (15.1.2\text{-}24)$$

$$\underline{u}_{c_{n+1}} = \text{function of } \tilde{\underline{x}}_n(+_e) \qquad (15.1.2\text{-}25)$$

$$\tilde{\underline{x}}_0 = 0 \qquad \text{Initial Conditions} \qquad (15.1.2\text{-}26)$$

If Equations (15.1.2-19) - (15.1.2-26) are compared with Equations (15.1.2-6) - (15.1.2-13), it should be apparent that they differ in one fundamental way: in Equations (15.1.2-6) - (15.1.2-13) the $\underline{u}_c$ control vector is applied (in (15.1.2-11) - (15.1.2-12)) at cycle time n based on the $\tilde{\underline{x}}$ estimate at cycle time n (See Equation (15.1.2-10)); for delayed control reset Equations (15.1.2-19) - (15.1.2-26), the $\underline{u}_c$ control vector is also based on the $\tilde{\underline{x}}$ estimate at cycle time n (See Equation (15.1.2-25)), but is then applied at the next $n+1^{th}$ cycle time (in Equations (15.1.2-19) and (15.1.2-22)).

For Equations (15.1.2-19) - (15.1.2-26) we will also set the control vector equal to the negative of the error states being controlled, but to be applied during the next n+1th cycle; $\underline{u}_{c_{n+1}} = - \tilde{\underline{x}}_n(+_e)$ or equivalently, $\underline{u}_{c_n} = - \tilde{\underline{x}}_{n-1}(+_e)$. With this control law we see from (15.1.2-21) and (15.1.2-22) that:

$$\tilde{\underline{x}}_n(+_c) = \tilde{\underline{x}}_n(-) + \underline{u}_{c_n} = \Phi_n \, \tilde{\underline{x}}_{n-1}(+_e) - \tilde{\underline{x}}_{n-1}(+_e) = \left(\Phi_n - I\right) \tilde{\underline{x}}_{n-1}(+_e) \qquad (15.1.2\text{-}27)$$

Thus, the delayed control reset leaves a residual error in controlling the error state vector to zero. The residual control error is generally small because $\Phi_n \approx I$ and because the Kalman filter estimation loop maintains $\tilde{\underline{x}}_{n-1}(+_e)$ at a small value. Note, that the residual control error is not an error in estimating $\tilde{\underline{x}}$, only in controlling $\tilde{\underline{x}}$ to the ideal zero value. Equation (15.1.2-27) is the filter's estimate of what the error state vector will be in response to delayed resets. The actual error state vector will behave similarly within the boundaries of input process and measurement noise. Thus, (15.1.2-27) is not an error in the sense of an unknown uncertainty (i.e., it is known explicitly in the Kalman filter). In principle, the estimated error state vector $\tilde{\underline{x}}_n$ can be used to correct the navigation data output (with a one Kalman cycle delay) if such a set of corrected output data was beneficial. The approach would be as follows after the $\tilde{\underline{x}}_n(+_c)$ vector in Equation (15.1.2-22) has been calculated:

$$\begin{aligned} \underline{u}_{cOut_n} &= - \tilde{\underline{x}}_n(+_c) \\ \xi_{INS/Out_n}(+_c) &= \xi_{INS_n}(+_c) + g_{INS}\left(\xi_{INS_n}(+_c), \underline{u}_{cOut_n}\right) \\ \xi_{Aid/Out_n}(+_c) &= \xi_{Aid_n}(+_c) + g_{Aid}\left(\xi_{INS_n}(+_c), \underline{u}_{cOut_n}\right) \end{aligned} \qquad (15.1.2\text{-}28)$$

where

$\underline{u}_{cOut_n}$ = Equivalent to control vector $\underline{u}_c$ for output data.

$\xi_{INS/Out_n}(+_c)$, $\xi_{Aid/Out_n}(+_c)$ = ξ_{INS}, ξ_{Aid} outputs based on zeroing the estimated error state vector $\tilde{\underline{x}}$ using the control function. The "time stamp" for these outputs would correspond to the n cycle time stamp after applying the control vector in Equations (15.1.2-19). The outputs would typically be available one filter cycle later than cycle n.

A variation to Equations (15.1.2-19) - (15.1.2-26) applies the state transition matrix Φ_n to the $\underline{u}_{c_n}$ control vector prior to its application in Equations (15.1.2-19) and (15.1.2-22). Analytically, this can be represented by an adjunct expression preceding (15.1.2-19) of the form $\underline{u}_{c\Phi_n} = \Phi_n \, \underline{u}_{c_n}$, with $\underline{u}_{c\Phi_n}$ then used for $\underline{u}_{c_n}$ in (15.1.2-19) and (15.1.2-22). If Φ_n could be calculated instantaneously (an impossibility in practice), $\underline{u}_{c\Phi_n}$ would be applied at the cycle n time, resulting in zero for $\underline{\tilde{x}}_n(+_c)$ (i.e., an ideal control reset). The reader can verify this by applying the same procedure leading to Equation (15.1.2-27). Due to the time delay in calculating Φ_n, however, the $\underline{u}_{c\Phi_n}$ control cannot be applied at cycle time n, but at a later time (call it n′) after Φ_n has been calculated. To properly account for the Φ_n computational delay, n′ would be included in the Kalman filter computational structure such that the error state vector $\underline{\tilde{x}}$ is first propagated to Kalman cycle n, the $\underline{Z}_{Obs_n}$ observation is made for the $\underline{\tilde{x}}$ estimate update, then $\underline{\tilde{x}}$ is propagated to n′ when $\underline{u}_{c\Phi_n}$ is applied. The net result is that $\underline{\tilde{x}}$ at the n′ control reset time will have a residual of the form given in Equation (15.1.2-27), but with Φ_n replaced by the error state transition matrix that propagates $\underline{\tilde{x}}$ from n to n′.

Finally, it should be noted that Equations (15.1.2-6) - (15.1.2-13) (or (15.1.2-19) - (15.1.2-26)) represent "Kalman filter" estimation/control operations, but are also representative of the larger class of error model feed-back filters. What distinguishes the Kalman filter from other filter configurations is the form of the gain matrix K_n. For a Kalman filter, the K_n gain matrix is a time varying function of the statistical characteristics of the estimated compared to the actual error state vector. Specifically, the Kalman filter gain matrix is designed to minimize the statistical variance of each element in $\underline{\tilde{x}}$ from the true $\underline{x}$ error state vector element values (the so-called "minimum variance" approach).

The following subsections develop the equations for computing the Kalman gain matrix and discuss the process of making the measurement for Kalman filter input, applying the Kalman filter control vector to the inertial navigation parameters, and accounting for timing and synchronization of the Kalman filter estimation cycle time with the observation equation data comparison.

15.1.2.1 KALMAN GAIN CALCULATION

This section deals with the design of the K_n gain matrix for the Equation (15.1.2-9) (or (15.1.2-24)) estimation process using the Kalman filter "minimum variance" approach. The Kalman filter design problem is to select the K_n matrix for Equation (15.1.2-9) (or (15.1.2-24)) that will minimize the variance of the error in $\underline{\tilde{x}}$ after each update. The error in $\underline{\tilde{x}}$ is denoted as the estimated error state vector "uncertainty". To formulate the problem analytically, we first define:

$$\Delta\underline{x} \equiv \underline{\tilde{x}} - \underline{x} \qquad (15.1.2.1\text{-}1)$$

where

$\Delta\underline{x}$ = Estimated error state vector uncertainty defined as the difference between the estimated and actual error state vectors.

At time t_n, using the notation of the previous section, the estimation error before and after update operations is for the ideal control reset filter:

$$\begin{aligned}
\Delta\underline{x}_n(-) &= \tilde{\underline{x}}_n(-) - \underline{x}_n(-) \\
\Delta\underline{x}_n(+_e) &= \tilde{\underline{x}}_n(+_e) - \underline{x}_n(-) \\
\Delta\underline{x}_n(+_c) &= \tilde{\underline{x}}_n(+_c) - \underline{x}_n(+_c)
\end{aligned} \qquad (15.1.2.1\text{-}2)$$

Note that the $\Delta\underline{x}_n(+_e)$ expression in (15.1.2.1-2) uses $\underline{x}_n(-)$ for the actual error state vector at time t_n because for the ideal filter, there is no difference in the true value immediately before or immediately after the estimation update, and the estimation update occurs before control application (see Equations (15.1.2-7) - (15.1.2-9) and (15.1.2-12)).

For the delayed control filter, the equivalent to (15.1.2.1-2) is:

$$\begin{aligned}
\Delta\underline{x}_n(-) &= \tilde{\underline{x}}_n(-) - \underline{x}_n(-) \\
\Delta\underline{x}_n(+_c) &= \tilde{\underline{x}}_n(+_c) - \underline{x}_n(+_c) \\
\Delta\underline{x}_n(+_e) &= \tilde{\underline{x}}_n(+_e) - \underline{x}_n(+_c)
\end{aligned} \qquad (15.1.2.1\text{-}3)$$

Note that the $\Delta\underline{x}_n(+_e)$ expression in (15.1.2.1-3) uses $\underline{x}_n(+_c)$ for the actual error state vector at time t_n because for the delayed control filter, there is no difference in the true value immediately before or immediately after the estimation update, and the estimation update occurs following control application (See Equations (15.1.2-21) - (15.1.2-24)).

We now also define the covariance matrix associated with $\Delta\underline{x}$ as:

$$P \equiv \mathcal{E}\left(\Delta\underline{x}\ \Delta\underline{x}^T\right) \qquad (15.1.2.1\text{-}4)$$

where

P = Estimated error state uncertainty covariance matrix.

$\mathcal{E}(\ )$ = The expected value operator (i.e., average statistical value).

It is easily verified by expanding Equation (15.1.2.1-4) that:

$$P = \begin{bmatrix} \mathcal{E}\left(\Delta x_1^2\right) & \mathcal{E}\left(\Delta x_1\, \Delta x_2\right) & \mathcal{E}\left(\Delta x_1\, \Delta x_3\right) & \cdots \\ \mathcal{E}\left(\Delta x_2\, \Delta x_1\right) & \mathcal{E}\left(\Delta x_2^2\right) & \mathcal{E}\left(\Delta x_2\, \Delta x_3\right) & \cdots \\ \mathcal{E}\left(\Delta x_3\, \Delta x_1\right) & \mathcal{E}\left(\Delta x_3\, \Delta x_2\right) & \mathcal{E}\left(\Delta x_3^2\right) & \cdots \\ \vdots & \vdots & \vdots & \end{bmatrix} \qquad (15.1.2.1\text{-}5)$$

where

$\Delta x_1, \Delta x_2, \Delta x_3 \cdots$ = Elements 1, 2, 3, $\cdots$ of $\Delta \underline{x}$.

Equation (15.1.2.1-5) shows that the diagonal elements of P equal the variances of the elements of $\Delta \underline{x}$ (i.e., the mean squared values) and the off-diagonal terms equal the covariances. It should also be apparent that P is a symmetrical matrix, hence, it is equal to its transpose:

$$P^T = P \qquad (15.1.2.1\text{-}6)$$

The covariance matrix concept has been introduced as the measure of uncertainty in $\Delta \underline{x}$ (i.e., the statistics of its error characteristics). The basis for selecting a Kalman gain matrix K_n in Equations (15.1.2-9) (or (15.1.2-24)) will be to minimize P_n after the update. We now return to Equations (15.1.2-8) - (15.1.2-9) (or (15.1.2-22) - (15.1.2-24)) to derive an expression for $P_n(+_e)$ (P at filter cycle n, after the estimation update) in terms of $P_n(-)$ (P at filter cycle n, before applying updates), the statistics of the measurement noise, and the general gain matrix K_n.

We begin with the Equations (15.1.2-6) - (15.1.2-13) idealized Kalman filter configuration (without delayed control resets) and calculate $\Delta \underline{x}_n(+_e)$, the error state uncertainty after estimation updating, as a function of $\Delta \underline{x}_n(-)$. We will then show that the result so obtained is identical for the delayed reset filter. Proceeding, we subtract $\underline{x}_n(-)$ from both sides of the Equation (15.1.2-9) update expression, substitute the linearized $\underline{z}_n$ form of $\underline{Z}_{Obs_n}$ from (15.1.2-3), and apply (15.1.2-8) for $\underset{\sim}{\underline{z}}_n$:

$$\begin{aligned} \underset{\sim}{\underline{x}}_n(+_e) - \underline{x}_n(-) &= \underset{\sim}{\underline{x}}_n(-) + K_n \left(\underline{z}_n - \underset{\sim}{\underline{z}}_n\right) - \underline{x}_n(-) \\ &= \underset{\sim}{\underline{x}}_n(-) - \underline{x}_n(-) + K_n\, H_n \left(\underline{x}_n(-) - \underset{\sim}{\underline{x}}_n(-)\right) + K_n\, G_{M_n}\, \underline{n}_{M_n} \qquad (15.1.2.1\text{-}7) \\ &= \left(I - K_n\, H_n\right)\left(\underset{\sim}{\underline{x}}_n(-) - \underline{x}_n(-)\right) + K_n\, G_{M_n}\, \underline{n}_{M_n} \end{aligned}$$

or, with the first expression in (15.1.2.1-2):

$$\Delta \underline{x}_n(+_e) = \left(I - K_n\, H_n\right) \Delta \underline{x}_n(-) + K_n\, G_{M_n}\, \underline{n}_{M_n} \qquad (15.1.2.1\text{-}8)$$

The equivalent relationship for the delayed control reset Kalman filter (Equations (15.1.2-19) - (15.1.2-26)) is developed by proceeding as above for the idealized filter, but first subtracting $\underline{x}_n(+_c)$ from Equation (15.1.2-24). The result is as in (15.1.2.1-8), but with $\Delta\underline{x}_n(-)$ replaced by $\Delta\underline{x}_n(+_c)$ defined in (15.1.2.1-3):

$$\Delta\underline{x}_n(+_e) = \left(I - K_n H_n\right) \Delta\underline{x}_n(+_c) + K_n G_{M_n} \underline{n}_{M_n} \qquad (15.1.2.1\text{-}9)$$

We now develop an expression for the estimated error state uncertainty in the delayed reset filter after the control update $\Delta\underline{x}_n(+_c)$ in terms of $\Delta\underline{x}_n(-)$, the error state uncertainty prior to application of the update. From Equations (15.1.2.1-3), (15.1.2-22) and (15.1.2-17) we write:

$$\begin{aligned} \Delta\underline{x}_n(+_c) &= \tilde{\underline{x}}_n(+_c) - \underline{x}_n(+_c) \\ &= \tilde{\underline{x}}_n(-) + \underline{u}_{c_n} - \left(\underline{x}_n(-) + \underline{u}_{c_n}\right) = \tilde{\underline{x}}_n(-) - \underline{x}_n(-) \end{aligned} \qquad (15.1.2.1\text{-}10)$$

or, with the first expression in (15.1.2.1-3):

$$\Delta\underline{x}_n(+_c) = \Delta\underline{x}_n(-) \qquad (15.1.2.1\text{-}11)$$

Thus, for the delayed control reset Kalman filter, application of the control vector has no effect on the estimated error state uncertainty. We will find this to be a general rule for all Kalman filter configurations. Using (15.1.2.1-11) in (15.1.2.1-9), we obtain the identical (15.1.2.1-8) result for the delayed reset filter. Thus, (15.1.2.1-8) is equally valid for the delayed or idealized control reset Kalman filter configurations.

Equation (15.1.2.1-8) shows how the uncertainty in $\tilde{\underline{x}}$ is impacted by estimation reset Equations (15.1.2-8) - (15.1.2-9) (or control/estimation Equations (15.1.2-22) - (15.1.2-24)). We now utilize the definition for the $\Delta\underline{x}$ covariance matrix to develop the statistical equivalent to (15.1.2.1-8). Substituting (15.1.2.1-8) into (15.1.2.1-4) at filter cycle n after estimation updating yields:

$$\begin{aligned} P_n(+_e) &= \mathcal{E}\left\{\left[\left(I - K_n H_n\right) \Delta\underline{x}_n(-) + K_n G_{M_n} \underline{n}_{M_n}\right]\left[\left(I - K_n H_n\right) \Delta\underline{x}_n(-) + K_n G_{M_n} \underline{n}_{M_n}\right]^T\right\} \\ &= \mathcal{E}\left[\left(I - K_n H_n\right) \Delta\underline{x}_n(-)\, \Delta\underline{x}_n^T(-) \left(I - K_n H_n\right)^T\right] + \mathcal{E}\left(K_n G_{M_n} \underline{n}_{M_n} \underline{n}_{M_n}^T G_{M_n}^T K_n^T\right) \\ &+ \mathcal{E}\left[\left(I - K_n H_n\right) \Delta\underline{x}_n(-)\, \underline{n}_{M_n}^T G_{M_n}^T K_n^T\right] + \mathcal{E}\left[K_n G_{M_n} \underline{n}_{M_n} \Delta\underline{x}_n^T(-) \left(I - K_n H_n\right)^T\right] \qquad (15.1.2.1\text{-}12) \\ &= \left(I - K_n H_n\right) \mathcal{E}\left(\Delta\underline{x}_n(-)\, \Delta\underline{x}_n^T(-)\right) \left(I - K_n H_n\right)^T + K_n G_{M_n} \mathcal{E}\left(\underline{n}_{M_n} \underline{n}_{M_n}^T\right) G_{M_n}^T K_n^T \\ &\quad + \left(I - K_n H_n\right) \mathcal{E}\left(\Delta\underline{x}_n(-)\, \underline{n}_{M_n}^T\right) G_{M_n}^T K_n^T + K_n G_{M_n} \mathcal{E}\left(\underline{n}_{M_n} \Delta\underline{x}_n^T(-)\right) \left(I - K_n H_n\right)^T \end{aligned}$$

Using the (15.1.2.1-4) definition, the $\mathcal{E}\left(\Delta\underline{x}_n(-)\ \Delta\underline{x}_n^T(-)\right)$ term in (15.1.2.1-12) is identified as $P_n(-)$, the $\Delta\underline{x}$ covariance matrix at cycle n prior to updating. The $\mathcal{E}\left(\underline{n}_{M_n}\,\underline{n}_{M_n}^T\right)$ expression in (15.1.2.1-12) is defined as:

$$R_n \equiv \mathcal{E}\left(\underline{n}_{M_n}\,\underline{n}_{M_n}^T\right) \qquad (15.1.2.1\text{-}13)$$

where

R_n = Independent measurement noise covariance matrix. Because $\underline{n}_M$ has been defined to contain independent elements, the expected value of the off-diagonal products in R_n are zero (Use (15.1.2.1-5) as a guide for the expanded form of R_n). Hence, R_n is a diagonal matrix with each element equal to the variance of the corresponding $\underline{n}_M$ element.

In order to evaluate the $\mathcal{E}\left(\Delta\underline{x}_n(-)\ \underline{n}_{M_n}^T\right)$ and $\mathcal{E}\left(\underline{n}_{M_n}\,\Delta\underline{x}_n^T(-)\right)$ terms in (15.1.2.1-12) we have to specify the cycle to cycle correlation characteristics of the measurement noise $\underline{n}_M$. We assume that $\underline{n}_M$ is a "white" sequence (in n) (i.e., $\underline{n}_M$ at cycle n is uncorrelated with $\underline{n}_M$ at previous cycle times). Since $\underline{n}_{M_n}$ is uncorrelated from past values of $\underline{n}_M$, past measurements ($\underline{z}$ in Equation (15.1.2-3) or (15.1.2-18) containing $\underline{n}_M$) are also uncorrelated with $\underline{n}_{M_n}$. Since past measurements and controls were used to generate $\underline{x}_n$ and $\tilde{\underline{x}}_n(-)$ (in Equations (15.1.2-2), (15.1.2-5), (15.1.2-7) and (15.1.2-12) (or (15.1.2-16), (15.1.2-17), (15.1.2-21) and (15.1.2-22)), we can conclude with (15.1.2.1-2) (or (15.1.2.1-3)) that $\Delta\underline{x}_n(-)$ is also uncorrelated with $\underline{n}_{M_n}$. Thus, we can write for the (15.1.2.1-12) terms in question:

$$\mathcal{E}\left(\Delta\underline{x}_n(-)\ \underline{n}_{M_n}^T\right) = 0 \qquad \mathcal{E}\left(\underline{n}_{M_n}\,\Delta\underline{x}_n^T(-)\right) = 0 \qquad (15.1.2.1\text{-}14)$$

Using the previous results, the Equation (15.1.2.1-12) error state vector covariance estimation update equation becomes the simplified form:

$$P_n(+_e) = \left(I - K_n\,H_n\right)P_n(-)\left(I - K_n\,H_n\right)^T + K_n\,G_{M_n}\,R_n\,G_{M_n}^T\,K_n^T \qquad (15.1.2.1\text{-}15)$$

Equation (15.1.2.1-15) relates the statistical uncertainty in $\tilde{\underline{x}}$ after the estimation update with the statistical uncertainty in $\tilde{\underline{x}}$ before applying updates, as a result of applying update Equations (15.1.2-6) and (15.1.2-8) - (15.1.2-9) (or (15.1.2-19) - (15.1.2-20) and (15.1.2-22) - (15.1.2-24)) using the observation $\underline{Z}_{Obs}$ (or, the linearized equivalent measurement $\underline{z}$) containing measurement noise. The $\tilde{\underline{x}}$ uncertainty is represented by the covariance matrix P_n and the noise characteristics of the measurement are contained in the R_n covariance matrix. Equation (15.1.2.1-15) is the statistical equivalent of Equation (15.1.2.1-8). It is important to recognize that Equations (15.1.2.1-8) and (15.1.2.1-15) are completely general at this point, being valid for any gain matrix K_n used in estimation update Equation (15.1.2-9) (or (15.1.2-24)). General Equation (15.1.2.1-15) is known as the "Joseph's form" for error state vector uncertainty covariance matrix updating.

The Kalman gain is defined analytically as the "optimal" K_n that minimizes $P_n(+_e)$ in Equation (15.1.2.1-15) (i.e., minimizes the statistical uncertainty in $\underline{\tilde{x}}$ after the update, or equivalently, minimizes the variance of the uncertainty in $\underline{\tilde{x}}$ after the update). To determine the optimal K_n that minimizes $P_n(+_e)$, we first expand (15.1.2.1-15) as follows using (15.1.2.1-6):

$$\begin{aligned} P_n(+_e) &= P_n(-) - K_n\, H_n\, P_n(-) - P_n(-)\, (K_n\, H_n)^T \\ &\quad + K_n\, H_n\, P_n(-)\, (K_n\, H_n)^T + K_n\, G_{M_n}\, R_n\, G_{M_n}^T\, K_n^T \\ &= P_n(-) + K_n \left(H_n\, P_n(-)\, H_n^T + G_{M_n}\, R_n\, G_{M_n}^T\right) K_n^T \\ &\quad - K_n\, H_n\, P_n^T(-) - P_n(-)\, H_n^T\, K_n^T \\ &= P_n(-) + K_n \left(H_n\, P_n(-)\, H_n^T + G_{M_n}\, R_n\, G_{M_n}^T\right) K_n^T \\ &\quad - K_n \left(P_n(-)\, H_n^T\right)^T - \left(P_n(-)\, H_n^T\right) K_n^T \end{aligned} \qquad (15.1.2.1\text{-}16)$$

To simplify the algebra, the coefficients in (15.1.2.1-16) are defined as:

$$A_n \equiv H_n\, P_n(-)\, H_n^T + G_{M_n}\, R_n\, G_{M_n}^T \qquad\qquad B_n \equiv P_n(-)\, H_n^T \qquad (15.1.2.1\text{-}17)$$

so that (15.1.2.1-16) simplifies to:

$$P_n(+_e) = P_n(-) + K_n\, A_n\, K_n^T - K_n\, B_n^T - B_n\, K_n^T \qquad (15.1.2.1\text{-}18)$$

We now make an observation on the form of (15.1.2.1-18) as contrasted with a term of the form:

$$(K_n - D_n)\, C_n\, (K_n - D_n)^T = K_n\, C_n\, K_n^T - K_n\, C_n\, D_n^T - D_n\, C_n\, K_n^T + D_n\, C_n\, D_n^T \qquad (15.1.2.1\text{-}19)$$

or, for C_n symmetrical such that $C_n = C_n^T$:

$$(K_n - D_n)\, C_n\, (K_n - D_n)^T = K_n\, C_n\, K_n^T - K_n\, (D_n\, C_n)^T - (D_n\, C_n)\, K_n^T + D_n\, C_n\, D_n^T \qquad (15.1.2.1\text{-}20)$$

If (15.1.2.1-18) is compared with (15.1.2.1-20) it should be clear that the two are identical in form, except for the $D_n C_n D_n^T$ and $P_n(-)$ terms. That is, for C_n and D_n defined as follows, the two expressions are equivalent if (15.1.2.1-18) is corrected for $P_n(-)$ and $D_n C_n D_n^T$.

$$C_n = A_n \qquad D_n C_n = B_n \tag{15.1.2.1-21}$$

or

$$D_n = B_n C_n^{-1} = B_n A_n^{-1} \tag{15.1.2.1-22}$$

We must now verify that $C_n = A_n$ is symmetrical since Equation (15.1.2.1-20) was based on this assumption. A look at (15.1.2.1-17) reveals that this is indeed the case. A_n is composed of a symmetrical matrix (covariance matrix R_n) modified by G_{M_n} and $G_{M_n}^T$ plus a symmetrical matrix $P_n(-)$ modified by H_n and H_n^T. It is easily verified that $G_{M_n} R_n G_{M_n}^T$ and $H_n P_n(-) H_n^T$ are each symmetrical by proving that each equals its transpose:

$$\begin{aligned}\left(G_{M_n} R_n G_{M_n}^T\right)^T &= G_{M_n} R_n^T G_{M_n}^T = G_{M_n} R_n G_{M_n}^T \\ \left(H_n P_n(-) H_n^T\right)^T &= H_n P_n^T(-) H_n^T = H_n P_n(-) H_n^T\end{aligned} \tag{15.1.2.1-23}$$

Hence, since both elements of A_n are symmetrical, A_n is symmetrical. We now use (15.1.2.1-21) to rewrite (15.1.2.1-18) as:

$$P_n(+_e) = P_n(-) + K_n C_n K_n^T - K_n (D_n C_n)^T - D_n C_n K_n^T \tag{15.1.2.1-24}$$

which, with the (15.1.2.1-20) identity is:

$$P_n(+_e) = P_n(-) + (K_n - D_n) C_n (K_n - D_n)^T - D_n C_n D_n^T \tag{15.1.2.1-25}$$

Substituting (15.1.2.1-22) for D_n and (15.1.2.1-21) for $D_n C_n$, Equation (15.1.2.1-25) becomes:

$$P_n(+_e) = P_n(-) - B_n \left(B_n A_n^{-1}\right)^T + \left(K_n - B_n A_n^{-1}\right) A_n \left(K_n - B_n A_n^{-1}\right)^T \tag{15.1.2.1-26}$$

Equation (15.1.2.1-26) is in a form that can now be used by inspection to define the optimum K_n that minimizes $P_n(+_e)$. Before this is done, however, the properties of the last term in (15.1.2.1-26) must be clearly understood. We will soon show that this term always has positive terms along the diagonal. Hence, because it is added to $P_n(-)$ in (15.1.2.1-26) to form $P_n(+_e)$, this term increases the magnitude of the diagonal elements in P_n. Since the diagonal elements in P_n represent the variances of the $\underset{-}{\tilde{x}}$ element uncertainties, we wish the diagonal elements in $P_n(+_e)$ to be minimized through the updating process. Because K_n only appears in the last term of Equation (15.1.2.1-26), and because the last term only increases $P_n(+_e)$, we can conclude that the optimum value for K_n that minimizes $P_n(+_e)$ is that value that sets the last term in (15.1.2.1-26) to zero. From (15.1.2.1-26), this value is seen by inspection to be:

$$K_n = B_n A_n^{-1} \tag{15.1.2.1-27}$$

or with (15.1.2.1-17):

$$K_n = P_n(-) H_n^T \left(H_n P_n(-) H_n^T + G_{M_n} R_n G_{M_n}^T \right)^{-1} \tag{15.1.2.1-28}$$

Equation (15.1.2.1-28) is the optimal Kalman gain that will generate a minimum variance estimate for $\underset{-}{\tilde{x}}$ for each update application of Equation (15.1.2-9) (or (15.1.2-24)).

We now go back a step and prove that the last term in (15.1.2.1-26) does indeed always have positive diagonal elements as stipulated in our logic for selecting K_n. If we define the $K_n - B_n A_n^{-1}$ term as L_n for simplicity, the last term in (15.1.2.1-26) is, with (15.1.2.1-17):

$$\begin{aligned} \left(K_n - B_n A_n^{-1}\right) A_n \left(K_n - B_n A_n^{-1}\right)^T &= L_n A_n L_n^T \\ &= L_n \left(H_n P_n(-) H_n^T + G_{M_n} R_n G_{M_n}^T \right) L_n^T \\ &= L_n H_n P_n(-) H_n^T L_n^T + L_n G_{M_n} R_n G_{M_n}^T L_n^T \\ &= (L_n H_n) P_n(-) (L_n H_n)^T + \left(L_n G_{M_n}\right) R_n \left(L_n G_{M_n}\right)^T \end{aligned} \tag{15.1.2.1-29}$$

Each of the two terms in the above expression consists of a covariance matrix ($P_n(-)$ or R_n) pre and post multiplied by a matrix and its transpose. Let's look at the $\left(L_n G_{M_n}\right) R_n \left(L_n G_{M_n}\right)^T$ term as an example and reintroduce the (15.1.2.1-13) definition for R_n:

$$\begin{aligned} \left(L_n G_{M_n}\right) R_n \left(L_n G_{M_n}\right)^T &= \left(L_n G_{M_n}\right) \mathcal{E}\left(\underline{n}_{M_n} \underline{n}_{M_n}^T \right) \left(L_n G_{M_n}\right)^T \\ &= \mathcal{E}\left[\left(L_n G_{M_n} \underline{n}_{M_n}\right) \left(L_n G_{M_n} \underline{n}_{M_n}\right)^T \right] \end{aligned} \tag{15.1.2.1-30}$$

The $L_n G_{M_n} \underline{n}_{M_n}$ term in the above expression is also a vector (say $\underline{Y}_n$) so that

$$\left(L_n G_{M_n}\right) R_n \left(L_n G_{M_n}\right)^T = \mathcal{E}\left[\left(L_n G_{M_n} \underline{n}_{M_n}\right) \left(L_n G_{M_n} \underline{n}_{M_n}\right)^T \right] = \mathcal{E}\left(\underline{Y}_n \underline{Y}_n^T \right) \tag{15.1.2.1-31}$$

If the right side of (15.1.2.1-31) is expanded in component form (as we did for P_n previously) it will be obvious that the diagonal elements are the variances (or mean squared values) of the $\underline{Y}_n$ elements. Hence, the diagonal elements are positive. A similar argument also applies for the $\left(L_n\, H_n\right) P_n(-) \left(L_n\, H_n\right)^T$ term in (15.1.2.1-29), hence, its diagonal elements are also positive. It is concluded that the sum of these terms (the last term in (15.1.2.1-26)) must, therefore, also have positive diagonal elements, thereby, validating the assumption used previously in selecting the optimal Kalman K_n matrix value.

Kalman gain Equation (15.1.2.1-28) is a function of the independent measurement noise covariance matrix R, the measurement noise dynamic coupling matrix G_M, the measurement matrix H and the error state vector uncertainty covariance matrix before updates P(-). As we see from Equation (15.1.2.1-15), application of Kalman filter updates affects the covariance matrix P. Additionally, the P covariance matrix is affected by approximations in estimated error state vector propagation Equation (15.1.2-7) or (15.1.2-21) applied between estimation updates. The net result is that in order to provide values for P in Kalman gain Equation (15.1.2.1-28), a separate calculation must be carried out in parallel to compute P due to the repeated application of the propagation/update process in Equations (15.1.2-6) - (15.1.2-12) (or (15.1.2-19) - (15.1.2-25)). The next section defines the equations used to compute P for the Kalman filter gain calculation.

15.1.2.1.1 Covariance Matrix Calculation

Calculation of the error state vector uncertainty covariance matrix P for Kalman gain Equation (15.1.2.1-28) involves two steps that are repeated for each estimation cycle; covariance update and covariance propagation. The covariance matrix update calculation adjusts P (from the P(-) value to the $P(+_e)$ value) for the error state vector uncertainty reduction afforded by the application of estimation update Equations (15.1.2-6) and (15.1.2-8) - (15.1.2-9) for the idealized control reset filter (or control/estimation update Equations (15.1.2-19) - (15.1.2-20) and (15.1.2-22) - (15.1.2-24) for the delayed control filter). The covariance matrix propagation calculation adjusts P for uncertainties introduced by applying error state vector propagation Equation (15.1.2-21) for the delayed control reset Kalman filter (or propagation/control Equations (15.1.2-7) and (15.1.2-11) - (15.1.2-12) for the idealized filter) with their approximation of neglected process noise between update cycles. As such, covariance propagation computes $P_n(-)$ from $P_{n-1}(+_e)$.

The covariance matrix update operation is represented by the Equation (15.1.2.1-15) Joseph's form repeated below:

$$P_n(+_e) = \left(I - K_n\, H_n\right) P_n(-) \left(I - K_n\, H_n\right)^T + K_n\, G_{M_n}\, R_n\, G_{M_n}^T\, K_n^T \qquad (15.1.2.1.1\text{-}1)$$

It is also to be noted that if the gain for the filter is calculated according to Equation (15.1.2.1-28), Equation (15.1.2.1.1-1) for $P_n(+_e)$ can be simplified. The derivation of the simplified form is easily achieved from Equation (15.1.2.1-26) using (15.1.2.1-27) for the optimal gain condition and (15.1.2.1-17) for B_n:

$$\begin{aligned} P_n(+_e) &= P_n(-) - B_n \left(B_n\, A_n^{-1} \right)^T = P_n(-) - P_n(-)\, H_n^T\, K_n^T \\ &= P_n(-) - \left(K_n\, H_n\, P_n^T(-) \right)^T \end{aligned} \qquad (15.1.2.1.1\text{-}2)$$

Introducing the symmetry characteristic of P (i.e., P equals P transpose) into (15.1.2.1.1-2) yields:

$$P_n(+_e) = P_n(-) - \left(K_n\, H_n\, P_n(-) \right)^T \qquad (15.1.2.1.1\text{-}3)$$

We then take the transpose of (15.1.2.1.1-3) with P transpose equal to P to obtain the desired result:

$$P_n(+_e) = P_n(-) - K_n\, H_n\, P_n(-) = \left(I - K_n\, H_n \right) P_n(-) \qquad (15.1.2.1.1\text{-}4)$$

Equation (15.1.2.1.1-4) is equivalent to Equation (15.1.2.1.1-1) for cases when K_n satisfies Equation (15.1.2.1-28). In applying (15.1.2.1.1-4), it is important to recognize that it is based on an exact computation of (15.1.2.1-28) and its application as defined by estimation Equation (15.1.2-9) or (15.1.2-24). For the more general case when K_n is not exactly computed according to (15.1.2.1-28) or when the exactly computed gains are not exactly applied in (15.1.2-9) or (15.1.2-24), Equation (15.1.2.1.1-1) should be used. An example of the previous situation is a case when there is doubt regarding the dynamic model for a particular error state, but when it is still to be accounted for statistically in the error state vector covariance and gain matrix calculations. For such a situation, the attempt to estimate this error state in Equation (15.1.2-9) or (15.1.2-24) might not be performed, which is equivalent to setting the elements of K_n for the error state to zero. The error state is still accounted for statistically in the covariance matrix P, and P is used to calculate K_n with (15.1.2.1-28). However, before applying the calculated K_n, the gain elements for the error state in question are set to zero, thereby applying a gain matrix in (15.1.2-9) or (15.1.2-24) that does not satisfy (15.1.2.1-28). For this case, Equation (15.1.2.1.1-1) must be used to represent the covariance updating process with K_n equal to the value after zeroing the appropriate gains. This method of dealing with error states with questionable dynamic characteristics is known as the "considered variable" approach.

Equation (15.1.2.1.1-1) or (15.1.2.1.1-4) defines the covariance $P_n(+_e)$ following an estimation update as a function of the covariance $P_n(-)$ before applying the update. What we now seek is the covariance propagation expression defining $P_n(-)$ in terms of the covariance $P_{n-1}(+_e)$ after the previous filter update. For the delayed control reset Kalman filter (Equations

(15.1.2-19) - (15.1.2-26)), this is easily obtained from propagation Equation (15.1.2-16) and the equivalent estimation filter form (15.1.2-21). Taking the difference between these expressions and applying the first and third expressions in (15.1.2.1-3) yields for the error state uncertainty propagation:

$$\Delta \underline{x}_n(-) = \Phi_n \, \Delta \underline{x}_{n-1}(+_e) - \underline{w}_n \qquad (15.1.2.1.1\text{-}5)$$

We now show that the uncertainty propagation equation for the idealized control reset Equation (15.1.2-6) - (15.1.2-13) Kalman filter is identical to uncertainty propagation Equation (15.1.2.1.1-5) for the delayed control reset filter. To determine the effect of error state propagation on error state uncertainty for the idealized control reset filter, we take the difference between Equation (15.1.2-2) and the equivalent idealized estimation filter form (15.1.2-7), and apply the first and third expressions in (15.1.2.1-2):

$$\Delta \underline{x}_n(-) = \Phi_n \, \Delta \underline{x}_{n-1}(+_c) - \underline{w}_n \qquad (15.1.2.1.1\text{-}6)$$

The estimated error state uncertainty following control reset for the idealized control reset filter is from (15.1.2-5), (15.1.2-12) and the third expression in (15.1.2.1-2):

$$\begin{aligned} \Delta \underline{x}_n(+_c) &= \tilde{\underline{x}}_n(+_c) - \underline{x}_n(+_c) \\ &= \tilde{\underline{x}}_n(+_e) + \underline{u}_{c_n} - \left(\underline{x}_n(-) + \underline{u}_{c_n}\right) = \tilde{\underline{x}}_n(+_e) - \underline{x}_n(-) \end{aligned} \qquad (15.1.2.1.1\text{-}7)$$

or, with the middle expression in (15.1.2.1-2):

$$\Delta \underline{x}_n(+_c) = \Delta \underline{x}_n(+_e) \qquad (15.1.2.1.1\text{-}8)$$

Hence, the error state uncertainty is unaffected by application of the controls as has been previously stipulated. Using (15.1.2.1.1-8) in (15.1.2.1.1-6), we obtain the identical (15.1.2.1.1-5) result for the idealized control reset Kalman filter. Thus, (15.1.2.1.1-5) is equally valid for the delayed or idealized control reset Kalman filter configurations.

The equivalent covariance matrix expression associated with the (15.1.2.1.1-5) error uncertainty propagation equation is from (15.1.2.1-4):

$$\begin{aligned} P_n(-) &= \mathcal{E}\left(\Delta \underline{x}_n(-) \, \Delta \underline{x}_n^T(-)\right) = \mathcal{E}\left[\left(\Phi_n \, \Delta \underline{x}_{n-1}(+_e) - \underline{w}_n\right)\left(\Phi_n \, \Delta \underline{x}_{n-1}(+_e) - \underline{w}_n\right)^T\right] \\ &= \Phi_n \, \mathcal{E}\left(\Delta \underline{x}_{n-1}(+_e) \, \Delta \underline{x}_{n-1}^T(+_e)\right) \Phi_n^T + \mathcal{E}\left(\underline{w}_n \, \underline{w}_n^T\right) \\ &\quad - \Phi_n \, \mathcal{E}\left(\Delta \underline{x}_{n-1}(+_e) \, \underline{w}_n^T\right) - \mathcal{E}\left(\underline{w}_n \, \Delta \underline{x}_{n-1}^T(+_e)\right) \Phi_n^T \end{aligned} \qquad (15.1.2.1.1\text{-}9)$$

The first expected value term in the above expression should be recognized as the covariance of $\underline{x}_n$ after the last filter estimation update (i.e., $P_{n-1}(+_e)$). The second term is the covariance matrix associated with the driving noise. We define:

$$Q_n \equiv \mathcal{E}\left(\underline{w}_n \, \underline{w}_n^T\right) \tag{15.1.2.1.1-10}$$

where

Q_n = Integrated process noise matrix.

In order to evaluate the $\mathcal{E}\left(\Delta\underline{x}_{n-1}(+_e)\, \underline{w}_n^T\right)$ and $\mathcal{E}\left(\underline{w}_n \, \Delta\underline{x}_{n-1}^T(+_e)\right)$ terms in (15.1.2.1.1-9) we have to specify the correlation between $\Delta\underline{x}_{n-1}(+_e)$ and $\underline{w}_n$. This is readily accomplished from Equation (15.1.1-16) which shows that for the assumed independent white characteristic of the process noise vector $\underline{n}_P$, the integrated white noise vector $\underline{w}_n$ is only a function of independent noise inputs following t_{n-1}. The $\Delta\underline{x}_{n-1}(+_e)$ vector, on the other hand, is created at t_{n-1} and is produced by events prior to t_{n-1}, before the creation of $\underline{w}_n$. Thus, $\underline{w}_n$ and $\Delta\underline{x}_{n-1}(+_e)$ are independent of each other and we can write:

$$\mathcal{E}\left(\Delta\underline{x}_{n-1}(+_e)\, \underline{w}_n^T\right) = 0 \qquad \mathcal{E}\left(\underline{w}_n \, \Delta\underline{x}_{n-1}^T(+_e)\right) = 0 \tag{15.1.2.1.1-11}$$

Substituting (15.1.2.1.1-10) - (15.1.2.1.1-11) into (15.1.2.1.1-9) then yields the covariance propagation relationship between $P_n(-)$ and $P_{n-1}(+_e)$:

$$P_n(-) = \Phi_n \, P_{n-1}(+_e)\, \Phi_n^T + Q_n \tag{15.1.2.1.1-12}$$

Equations (15.1.2.1.1-12) and (15.1.2.1.1-1) or (15.1.2.1.1-4) describe the reset and propagation of the covariance matrix P_n over an estimation filter processing cycle. These equations together with Equation (15.1.2.1-28) enable the optimal gain matrix K_n to be calculated for each filter cycle for application in estimation Equation (15.1.2-9) or (15.1.2-24). Typical algorithms for calculating Φ_n and Q_n in Equation (15.1.2.1.1-12) are described in Sections 15.1.2.1.1.1 and 15.1.2.1.1.2.

It is to be noted from Equations (15.1.2.1.1-12) and (15.1.2.1.1-1) or (15.1.2.1.1-4), that the optimal gain determination requires an updating of the covariance matrix P based on its value for the previous interval. An integration process is implied by this operation that must be initialized at the start of the estimation filter computation process. The initial value of P (i.e., P_0) is determined by our best estimate (on a root-mean-square basis) of the variances (and covariances) associated with the uncertainties in the estimated error state vector $\underline{\tilde{x}}$ at the start of the estimation process. One of the advantages (and shortcomings) of the minimum variance approach is that it is based on knowledge of the initial uncertainty in $\underline{\tilde{x}}$ (as manifested in P_0). In addition, knowledge of the statistics of the driving and measurement noise (as manifested in Q_n and R_n) is required. If these statistical parameters are known (and they usually are), the Equation (15.1.2.1-28) gain formula yields excellent filter performance. On the other hand, if Q_n, R_n and P_0 are unknown (or have large uncertainties), performance deficiencies can be introduced.

We conclude this section with the development of the continuous differential equation form of covariance propagation Equation (15.1.2.1.1-12) that will be useful in a later section for developing an algorithm for computing the integrated process noise matrix Q_n. The continuous form of the covariance propagation equation is derived by first generalizing Equations (15.1.1-15), (15.1.1-16), (15.1.2.1.1-10) and (15.1.2.1.1-12) to represent propagation between two arbitrary time points (t followed by t_1), both within the t_{n-1} to t_n time interval:

$$P(t_1) = \Phi(t_1,t)\, P(t)\, \Phi(t_1,t)^T + Q(t_1,t) \qquad (15.1.2.1.1\text{-}13)$$

$$Q(t_1,t) = \mathcal{E}\left(\underline{w}(t_1,t)\, \underline{w}(t_1,t)^T\right) \qquad (15.1.2.1.1\text{-}14)$$

$$\underline{w}(t_1,t) = \int_t^{t_1} \Phi(t_1,\tau)\, G_P(\tau)\, \underline{n}_P(\tau)\, d\tau \qquad (15.1.2.1.1\text{-}15)$$

$$\Phi(t_1,t) = I + \int_t^{t_1} A(\tau)\, \Phi(\tau,t)\, d\tau \qquad (15.1.2.1.1\text{-}16)$$

where

$P(t_1), P(t)$ = P at times t_1 and t.

For t_1 very close to t, (15.1.2.1.1-15) and (15.1.2.1.1-16) can be approximated to first order by:

$$\underline{w}(t_1,t) \approx G_P(t) \int_t^{t_1} \underline{n}_P(\tau)\, d\tau \qquad (15.1.2.1.1\text{-}17)$$

$$\Phi(t_1,t) \approx I + A(t)\, \Delta t \qquad (15.1.2.1.1\text{-}18)$$

with

$$\Delta t \equiv t_1 - t \qquad (15.1.2.1.1\text{-}19)$$

Substituting (15.1.2.1.1-18) and (15.1.2.1.1-19) into (15.1.2.1.1-13) for small Δt gives, after rearrangement, to first order:

$$\frac{P(t+\Delta t) - P(t)}{\Delta t} \approx A(t)\,P(t) + P(t)\,A(t)^T + \frac{1}{\Delta t}\,Q(t+\Delta t,t) \qquad (15.1.2.1.1\text{-}20)$$

The $Q(t+\Delta t,t)$ term in (15.1.2.1.1-20) is evaluated by substituting (15.1.2.1.1-17) and (15.1.2.1.1-19) into (15.1.2.1.1-14):

$$Q(t+\Delta t,t) \approx G_P(t)\,\mathcal{E}\left(\int_t^{t+\Delta t} \underline{n}_P(\tau)\,d\tau \int_t^{t+\Delta t} \underline{n}_P(\tau)^T\,d\tau\right) G_P(t)^T \qquad (15.1.2.1.1\text{-}21)$$

The expected value term in (15.1.2.1.1-21) can be rearranged into the following equivalent form:

$$\begin{aligned}
\mathcal{E}\left(\int_t^{t+\Delta t} \underline{n}_P(\tau)\,d\tau \int_t^{t+\Delta t} \underline{n}_P(\tau)^T\,d\tau\right) &= \mathcal{E}\left(\int_t^{t+\Delta t} \underline{n}_P(\tau_\alpha)\,d\tau_\alpha \int_t^{t+\Delta t} \underline{n}_P(\tau_\beta)^T\,d\tau_\beta\right) \\
&= \mathcal{E}\left[\int_t^{t+\Delta t}\left(\int_t^{t+\Delta t} \underline{n}_P(\tau_\alpha)\,\underline{n}_P(\tau_\beta)^T\,d\tau_\alpha\right) d\tau_\beta\right] \qquad (15.1.2.1.1\text{-}22) \\
&= \int_t^{t+\Delta t}\left(\int_t^{t+\Delta t} \mathcal{E}\left[\underline{n}_P(\tau_\alpha)\,\underline{n}_P(\tau_\beta)^T\right] d\tau_\alpha\right) d\tau_\beta
\end{aligned}$$

where

τ_α, τ_β = Running time parameters that range from t to t $+\Delta t$.

To develop the equation for the P time derivative, we now let Δt go to zero in the limit so that (15.1.2.1.1-20) - (15.1.2.1.1-22) become:

$$\left[\frac{P(t+\Delta t) - P(t)}{\Delta t}\right]_{\lim \Delta t \to 0} = \dot{P}(t) = A(t)\,P(t) + P(t)\,A(t)^T + G_P(t)\,Q_0(t)\,G_P(t)^T \qquad (15.1.2.1.1\text{-}23)$$

with

$$Q_0(t) \equiv \left\{\frac{1}{\Delta t}\int_t^{t+\Delta t}\left(\int_t^{t+\Delta t} \mathcal{E}\left[\underline{n}_P(\tau_\alpha)\,\underline{n}_P(\tau_\beta)^T\right] d\tau_\alpha\right) d\tau_\beta\right\}_{\lim \Delta t \to 0} \qquad (15.1.2.1.1\text{-}24)$$

Equation (15.1.2.1.1-24) can be simplified by invoking the definition for the process noise as a vector of independent components, each being white, hence, uncorrelated with themselves at two different time points. Stated mathematically, the component independence constraint is:

$$\mathcal{E}(n_{P_i}\, n_{P_j}) = 0 \qquad \text{for } i \neq j \tag{15.1.2.1.1-25}$$

where

n_{P_i}, n_{P_j} = Elements i and j of $\underline{n}_P$.

The white noise constraint on each element of $\underline{n}_P$ can be defined by:

$$\mathcal{E}\left(n_{P_i}(\tau_\alpha)\, n_{P_i}(\tau_\beta)\right) = q_{PDens_i}(\tau_\beta)\, \delta(\tau_\alpha - \tau_\beta) \tag{15.1.2.1.1-26}$$

with

$$\delta(\tau_\alpha - \tau_\beta) = 0 \qquad \text{For} \quad \tau_\alpha \neq \tau_\beta$$
$$\int_{\tau_\alpha < \tau_\beta}^{\tau_\alpha > \tau_\beta} \delta(\tau_\alpha - \tau_\beta)\, d\tau_\alpha = 1 \tag{15.1.2.1.1-27}$$

where

$q_{PDens_i}(\tau_\beta)$ = White noise density associated with n_{P_i} at time τ_β.

$\delta(\tau_\alpha - \tau_\beta)$ = Dirac delta function.

Substituting (15.1.2.1.1-26) - (15.1.2.1.1-27) with (15.1.2.1.1-25) into the (15.1.2.1.1-24) inner integral then gives:

$$\int_t^{t+\Delta t} \mathcal{E}\left[\underline{n}_P(\tau_\alpha)\, \underline{n}_P(\tau_\beta)^T\right] d\tau_\alpha = Q_{PDens}(\tau_\beta) \tag{15.1.2.1.1-28}$$

where

$Q_{PDens}(\tau_\beta)$ = Process noise density matrix at time τ_β. By virtue of Equation (15.1.2.1.1-25), $Q_{PDens}(\tau_\beta)$ is diagonal with each element equal to $q_{PDens_i}(\tau_\beta)$.

We now apply (15.1.2.1.1-28) in (15.1.2.1.1-24) yielding:

$$Q_0(t) = \left.\left(\frac{1}{\Delta t}\int_t^{t+\Delta t} Q_{PDens}(\tau_\beta)\, d\tau_\beta\right)\right|_{\lim\ \Delta t \rightarrow 0} = Q_{PDens}(t) \tag{15.1.2.1.1-29}$$

Substituting (15.1.2.1.1-29) into (15.1.2.1.1-23) obtains the sought after differential equation for covariance propagation between Kalman filter update cycles:

$$\dot{P}(t) = A(t)\,P(t) + P(t)\,A(t)^T + G_P(t)\,Q_{P_{Dens}}(t)\,G_P(t)^T \qquad (15.1.2.1.1\text{-}30)$$

The integral of Equation (15.1.2.1.1-30) between time t_{n-1} and t_n (with $P_{n-1}(+_e)$ as the initial condition on P(t) at $t = t_{n-1}$) yields the Equation (15.1.2.1.1-12) discrete equivalent form if we equate $P_n(-)$ to $P(t_n)$.

Finally, it is sometimes useful to recognize the equivalency between the noise density for a stationary white noise random process and the power spectral density for the noise process as defined in Chapter 10 by Equation (10.2.2-21). Recall from Section 10.2.2 (Equations (10.2.2-9) - (10.2.2-10)) that a stationary white random process n(t) would have an auto-correlation function $\phi(t, \tau)$ for the time interval τ that is independent of time and given by:

$$\phi(t,\tau) \equiv \mathcal{E}\big(n(t)\,n(t+\tau)\big) \qquad \phi(t,\tau) = \phi(\tau) \qquad (15.1.2.1.1\text{-}31)$$

Equation (10.2.2-17) shows that $\phi(\tau)$ is a symmetrical function of τ (i.e., $\phi(\tau) = \phi(-\tau)$). From Equation (10.2.2-21) (based on $\phi(\tau)$ symmetry) we have defined the power spectral density for the stationary white noise process as:

$$G(\omega) \equiv \frac{2}{\pi}\int_0^{\infty} \phi(\tau)\cos\omega\tau\, d\tau = \frac{1}{\pi}\int_{-\infty}^{\infty} \phi(\tau)\cos\omega\tau\, d\tau \qquad (15.1.2.1.1\text{-}32)$$

where

$G(\omega)$ = Power spectral density for the white noise process n(t).

For the stationary n(t) we can also write from Equation (15.1.2.1.1-26):

$$\mathcal{E}\big(n(t)\,n(t+\tau)\big) = q_{PDen}\,\delta(\tau)$$

$$\delta(\tau) = 0 \quad \text{for } \tau \neq 0 \qquad \int_{\tau<0}^{\tau>0} \delta(\tau)\,d\tau = 1 \qquad (15.1.2.1.1\text{-}33)$$

where

q_{PDen} = Process noise density for the stationary white noise process.

Substituting (15.1.2.1.1-31) with (15.1.2.1.1-33) in (15.1.2.1.1-32) shows that:

$$G(\omega) = \frac{1}{\pi}\,q_{PDen} \qquad (15.1.2.1.1\text{-}34)$$

Thus, based on the (15.1.2.1.1-32) definition, the power spectral density for a stationary white noise process is constant and equal to the process noise density divided by π.

15.1.2.1.1.1 Error State Transition Matrix Computation Algorithm

In order to calculate the error state transition matrix Φ_n for Equations (15.1.2-7), (15.1.2-21) and (15.1.2.1.1-12), Equation (15.1.1-15) must be integrated from t_{n-1} to t_n. Because the A matrix in (15.1.1-15) may contain significant high frequency terms (particularly in strapdown applications in which $\underline{a}_{SF}$, $\underline{\omega}_{IB}$ and C_B^N are typically part of the error state dynamic matrix A - See Equations (12.5.1-1)), a high repetition rate for the associated integration algorithm is typically required. Direct integration of Equation (15.1.1-15) at high repetition rate is generally not practical from a throughput standpoint for current state-of-the-art navigation computers. Consequently, alternative means are typically utilized for calculating Φ_n. For example, consider implementing the Equation (15.1.1-15) Φ_n integration algorithm by repetitive application of Equation (15.1.1-10):

$$\Phi(t_m, t_{n-1}) = \Phi(t_m, t_{m-1})\,\Phi(t_{m-1}, t_{n-1}) \qquad \Phi(t_{n-1}, t_{n-1}) = I$$
$$\Phi_n \equiv \Phi(t_n, t_{n-1}) = \Phi(t_m, t_{n-1}) \quad \text{At } t_m = t_n \qquad (15.1.2.1.1.1\text{-}1)$$

where

m = Repetition cycle index for a computation loop faster than the Kalman filter cycle rate (e.g., 1 second for the m loop compared to 5 seconds for the Kalman loop).

If the m cycle period is selected short enough, the $\Phi(t_m, t_{m-1})$ term in (15.1.2.1.1.1-1) can be approximated by the integral of Equation (15.1.1-6) with (15.1.1-4) to first order:

$$\Phi(t_m, t_{m-1}) \approx I + \Delta\Phi_m \qquad \Delta\Phi_m \equiv \int_{t_{m-1}}^{t_m} A(t)\,dt \qquad (15.1.2.1.1.1\text{-}2)$$

Equation (15.1.2.1.1.1-2) can be extended in accuracy for A(t) being constant. From Equation (15.1.1-6) and (15.1.1-4) we write:

$$\dot{\Phi}(t, t_{m-1}) = A\,\Phi(t, t_{m-1}) \qquad \Phi(t_{m-1}, t_{m-1}) = I \qquad (15.1.2.1.1.1\text{-}3)$$

in which A is now assumed constant. We then write a trial solution to (15.1.2.1.1.1-3) that satisfies the identity initial condition constraint:

$$\Phi(t, t_{m-1}) = I + B_1\,(t - t_{m-1}) + B_2\,(t - t_{m-1})^2 + B_3\,(t - t_{m-1})^3 + \cdots \qquad (15.1.2.1.1.1\text{-}4)$$

The derivative of (15.1.2.1.1.1-4) is:

$$\dot{\Phi}(t, t_{m-1}) = B_1 + 2\, B_2\, (t - t_{m-1}) + 3\, B_3\, (t - t_{m-1})^2 + 4\, B_4\, (t - t_{m-1})^3 + \cdots \qquad (15.1.2.1.1.1\text{-}5)$$

Substituting (15.1.2.1.1.1-4) and (15.1.2.1.1.1-5) into (15.1.2.1.1.1-3) gives:

$$\begin{aligned} & B_1 + 2\, B_2\, (t - t_{m-1}) + 3\, B_3\, (t - t_{m-1})^2 + 4\, B_4\, (t - t_{m-1})^3 + \cdots \\ & \qquad = A \left(I + B_1\, (t - t_{m-1}) + B_2\, (t - t_{m-1})^2 + B_3\, (t - t_{m-1})^3 + \cdots \right) \end{aligned} \qquad (15.1.2.1.1.1\text{-}6)$$

Equating coefficients of equal powers of $(t - t_{m-1})$ yields:

$$\begin{aligned} & B_1 = A \qquad B_2 = \frac{1}{2}\, A\, B_1 = \frac{1}{2}\, A^2 = \frac{1}{2\,!}\, A^2 \\ & B_3 = \frac{1}{3}\, A\, B_2 = \frac{1}{3\,!}\, A^3 \qquad \text{Etc.} \end{aligned} \qquad (15.1.2.1.1.1\text{-}7)$$

With (15.1.2.1.1.1-7), $\Phi(t, t_{m-1})$ in Equation (15.1.2.1.1.1-4) becomes:

$$\Phi(t, t_{m-1}) = I + A\, (t - t_{m-1}) + \frac{1}{2\,!}\, A^2\, (t - t_{m-1})^2 + \frac{1}{3\,!}\, A^3\, (t - t_{m-1})^3 + \cdots \qquad (15.1.2.1.1.1\text{-}8)$$

or at time $t = t_m$:

$$\Phi(t_m, t_{m-1}) = I + A\, T_m + \frac{1}{2\,!} (A\, T_m)^2 + \frac{1}{3\,!} (A\, T_m)^3 + \cdots \qquad (15.1.2.1.1.1\text{-}9)$$

where

T_m = Time interval from t_{m-1} to t_m.

But from the definition in (15.1.2.1.1.1-2), we also know that for the constant A case:

$$A\, T_m = \Delta\Phi_m \qquad (15.1.2.1.1.1\text{-}10)$$

With (15.1.2.1.1.1-10), Equation (15.1.2.1.1.1-9) becomes:

$$\begin{aligned} & \Phi(t_m, t_{m-1}) = e^{\Delta\Phi_m} \\ & e^{\Delta\Phi_m} \equiv I + \Delta\Phi_m + \frac{1}{2\,!}\, \Delta\Phi_m^2 + \frac{1}{3\,!}\, \Delta\Phi_m^3 + \cdots \end{aligned} \qquad (15.1.2.1.1.1\text{-}11)$$

Finally, we use the more general (15.1.2.1.1.1-2) form for calculating $\Delta\Phi_m$:

$$\Delta\Phi_m \equiv \int_{t_{m-1}}^{t_m} A(t)\, dt \qquad (15.1.2.1.1.1\text{-}12)$$

Equations (15.1.2.1.1.1-11) and (15.1.2.1.1.1-12) describe an algorithm for calculating $\Phi(t_m,t_{m-1})$ for Equation(15.1.2.1.1.1-1) that is accurate to first order for general A(t) and is exact for constant A(t). The T_m time interval would be selected to enable truncation of (15.1.2.1.1.1-11) at a reasonable point for computer throughput considerations. The digital integration of A(t) to obtain $\Delta\Phi_m$ in (15.1.2.1.1.1-12) would typically be performed individually on each element of A(t) using a suitable algorithm at the iteration rate required to handle the expected dynamics of the particular element (i.e., faster than the m cycle rate). Typically, many of the elements of A(t) are zero, hence, the associated integrals are also zero and need not be computed. Many of the elements of A(t) will have very little change over the T_m integration interval and their integral can, therefore, be accurately approximated by the average of the value at the start and end of the interval multiplied by T_m. The overall result is that the computer throughput required to integrate A(t) in obtaining $\Delta\Phi_m$ can generally be reduced to a level that is acceptable on the basis of overall computation time resources.

15.1.2.1.1.2 Integrated Process Noise Matrix Algorithm

An algorithm for calculating the Q_n integrated process noise matrix in Equation (15.1.2.1.1-12) is developed in Section 15.1.2.1.1.3. Section 15.1.2.1.1.3 treats the general problem of propagating the covariance matrix and estimated error state vector between Kalman estimation updates as a repetitive algorithm integration process (at an m cycle rate between Kalman estimation n cycles). This contrasts with the single update forms of Equation (15.1.2.1.1-12) for covariance propagation and (15.1.2-7) or (15.1.2-21) for estimated error state vector propagation. Section 15.1.2.1.1.3 then develops the Q_n algorithm for (15.1.2.1.1-12) as a limiting case of the general m cycle covariance propagation algorithm in which only one m cycle is used for covariance propagation between n cycles. The Q_n algorithm so derived is imbedded in Equation (15.1.2.1.1.3-37), a revised form of Equation (15.1.2.1.1-12) for Kalman filter implementation.

15.1.2.1.1.3 Covariance And Estimated Error State Vector Propagation By Iterative Algorithm Processing Between Kalman Update Cycles

For high frequency dynamic environments and/or situations in which there is a lengthy time period between Kalman update cycles, to preserve integration accuracy it may be necessary to propagate the covariance matrix P and estimated error state vector using a repetitive discrete processing routine within the basic Kalman cycle. This section provides an example of such a repetitive algorithm based on sequential propagation of an m cycle computation loop imbedded within the basic n cycle Kalman estimation/update loop. In the process of developing the covariance propagation algorithm, we will also develop the estimated error state vector propagation algorithm as well as an algorithm for calculating the integrated process noise matrix Q_n in Equation (15.1.2.1.1-12) (for covariance propagation at the n cycle rate, i.e., when m = n).

The m cycle covariance propagation algorithm is based on the integrated form of the continuous covariance differential propagation Equation (15.1.2.1.1-30) from the m-1 to m cycle times. The algorithm derivation begins with Equation (15.1.2.1.1-30) repeated below:

$$\dot{P}(t) = A(t)\,P(t) + P(t)\,A(t)^T + G_P(t)\,Q_{P_{Dens}}(t)\,G_P(t)^T \tag{15.1.2.1.1.3-1}$$

Equation (15.1.2.1.1.3-1) can be integrated over an m cycle using the method introduced in Reference 6 - Section 4.6. We first define transformation variables $\underline{\lambda}(t)$ and $\underline{y}(t)$ such that:

$$\underline{\lambda}(t) = P(t)\,\underline{y}(t) \tag{15.1.2.1.1.3-2}$$

and

$$\dot{\underline{y}}(t) = -\,A(t)^T\,\underline{y}(t) \tag{15.1.2.1.1.3-3}$$

Differentiating (15.1.2.1.1.3-2) and substituting (15.1.2.1.1.3-1) and (15.1.2.1.1.3-3) yields:

$$\begin{aligned}\dot{\underline{\lambda}}(t) &= \dot{P}(t)\,\underline{y}(t) + P(t)\,\dot{\underline{y}}(t)\\ &= \left(A(t)\,P(t) + P(t)\,A(t)^T + G_P(t)\,Q_{P_{Dens}}(t)\,G_P(t)^T\right)\underline{y}(t)\\ &\qquad - P(t)\,A(t)^T\,\underline{y}(t)\\ &= A(t)\,P(t)\,\underline{y}(t) + G_P(t)\,Q_{P_{Dens}}(t)\,G_P(t)^T\,\underline{y}(t)\end{aligned} \tag{15.1.2.1.1.3-4}$$

Substituting (15.1.2.1.1.3-2) finds:

$$\dot{\underline{\lambda}}(t) = A(t)\,\underline{\lambda}(t) + G_P(t)\,Q_{P_{Dens}}(t)\,G_P(t)^T\,\underline{y}(t) \tag{15.1.2.1.1.3-5}$$

Equations (15.1.2.1.1.3-3) and (15.1.2.1.1.3-5) can be expressed in the equivalent form:

$$\begin{bmatrix}\dot{\underline{y}}(t)\\ \dot{\underline{\lambda}}(t)\end{bmatrix} = \begin{bmatrix}-A(t)^T & 0\\ G_P(t)\,Q_{P_{Dens}}(t)\,G_P(t)^T & A(t)\end{bmatrix}\begin{bmatrix}\underline{y}(t)\\ \underline{\lambda}(t)\end{bmatrix} \tag{15.1.2.1.1.3-6}$$

Equation (15.1.2.1.1.3-6) is a homogeneous linear vector differential equation whose general solution from time t_{m-1} to t_m (as in Section 15.1.1) has the form:

$$\begin{bmatrix}\underline{y}_m\\ \underline{\lambda}_m\end{bmatrix} = \begin{bmatrix}\Phi_{yy_m} & 0\\ \Phi_{\lambda y_m} & \Phi_{\lambda\lambda_m}\end{bmatrix}\begin{bmatrix}\underline{y}_{m-1}\\ \underline{\lambda}_{m-1}\end{bmatrix} \tag{15.1.2.1.1.3-7}$$

where

$\Phi_{yy_m}, \Phi_{\lambda y_m}, \Phi_{\lambda\lambda_m}$ = State transition matrix elements associated with the Equation (15.1.2.1.1.3-6) state dynamic matrix (in square brackets) that propagates $\underline{y}$, $\underline{\lambda}$ from their values at t_{m-1} to their values at t_m.

Note that the upper right element in the state transition matrix has been set to zero because from Equation (15.1.2.1.1.3-6), $\underline{\lambda}$ has no coupling into $\underline{y}$.

Expanding (15.1.2.1.1.3-7) in component form obtains:

$$\underline{y}_m = \Phi_{yy_m}\, \underline{y}_{m-1} \qquad \underline{\lambda}_m = \Phi_{\lambda y_m}\, \underline{y}_{m-1} + \Phi_{\lambda\lambda_m}\, \underline{\lambda}_{m-1} \tag{15.1.2.1.1.3-8}$$

Applying (15.1.2.1.1.3-2) at cycles m-1 and m:

$$\underline{\lambda}_{m-1} = P_{m-1}\, \underline{y}_{m-1} \qquad \underline{\lambda}_m = P_m\, \underline{y}_m \tag{15.1.2.1.1.3-9}$$

Combining (15.1.2.1.1.3-8) and (15.1.2.1.1.3-9):

$$\underline{\lambda}_m = P_m\, \underline{y}_m = P_m\, \Phi_{yy_m}\, \underline{y}_{m-1} = \Phi_{\lambda y_m}\, \underline{y}_{m-1} + \Phi_{\lambda\lambda_m}\, P_{m-1}\, \underline{y}_{m-1} \tag{15.1.2.1.1.3-10}$$

Since (15.1.2.1.1.3-10) is valid for any $\underline{y}_{m-1}$, it follows that:

$$P_m\, \Phi_{yy_m} = \Phi_{\lambda y_m} + \Phi_{\lambda\lambda_m}\, P_{m-1}$$

or

$$P_m = \Phi_{\lambda\lambda_m}\, P_{m-1}\, \Phi_{yy_m}^{-1} + \Phi_{\lambda y_m}\, \Phi_{yy_m}^{-1} \tag{15.1.2.1.1.3-11}$$

Using classical state vector theory (as in Section 15.1.1, Equations (15.1.1-2) - (15.1.1-4) and (15.1.1-6)), Equation (15.1.2.1.1.3-6) can be used to develop the means for evaluating the Φ terms in (15.1.2.1.1.3-11):

$$\dot{\Phi}(t) = M(t)\;\; \Phi(t) \qquad \Phi(t_{m-1}) = I$$

$$M(t) \equiv \begin{bmatrix} -\,A(t)^{\,T} & 0 \\ G_P(t)\;\, Q_{P_{Dens}}(t)\;\, G_P(t)^{\,T} & A(t) \end{bmatrix} \tag{15.1.2.1.1.3-12}$$

$$\Phi(t) = \begin{bmatrix} \Phi_{yy}(t) & \Phi_{y\lambda}(t) \\ \Phi_{\lambda y}(t) & \Phi_{\lambda\lambda}(t) \end{bmatrix}$$

where

$\Phi_{yy}(t),\ \Phi_{y\lambda}(t),\ \Phi_{\lambda y}(t),\ \Phi_{\lambda\lambda}(t)$ = State transition matrix elements that propagate $\underline{y}$, $\underline{\lambda}$ from their values at t_{m-1} to their values at t.

The upper right component of (15.1.2.1.1.3-12) is:

$$\dot{\Phi}_{y\lambda}(t) = -A(t)^T \Phi_{y\lambda}(t) \qquad \Phi_{y\lambda}(t_{m-1}) = 0$$

for which we write the trivial solution:

$$\Phi_{y\lambda}(t) = 0 \qquad (15.1.2.1.1.3\text{-}13)$$

as stipulated earlier in (15.1.2.1.1.3-7). With (15.1.2.1.1.3-13), the remaining components of (15.1.2.1.1.3-12) become:

$$\dot{\Phi}_{yy}(t) = -A(t)^T \Phi_{yy}(t) \qquad \Phi_{yy}(t_{m-1}) = I \qquad (15.1.2.1.1.3\text{-}14)$$

$$\dot{\Phi}_{\lambda\lambda}(t) = A(t)\,\Phi_{\lambda\lambda}(t) \qquad \Phi_{\lambda\lambda}(t_{m-1}) = I$$

$$\dot{\Phi}_{\lambda y}(t) = G_P(t)\,Q_{P_{Dens}}(t)\,G_P(t)^T \Phi_{yy}(t) + A(t)\,\Phi_{\lambda y}(t) \qquad (15.1.2.1.1.3\text{-}15)$$
$$\Phi_{\lambda y}(t_{m-1}) = 0$$

A useful relationship between $\Phi_{yy}(t)$ and $\Phi_{\lambda\lambda}(t)$ is revealed from the transpose of $\dot{\Phi}_{yy}(t)$ in (15.1.2.1.1.3-14):

$$\dot{\Phi}_{yy}(t)^T = -\Phi_{yy}(t)^T A(t) \qquad \Phi_{yy}(t_{m-1})^T = I \qquad (15.1.2.1.1.3\text{-}16)$$

From (15.1.2.1.1.3-16), the $\dot{\Phi}_{\lambda\lambda}(t)$ expression in (15.1.2.1.1.3-15), and the initial conditions in (15.1.2.1.1.3-14) - (15.1.2.1.1.3-15), we see that:

$$\frac{d}{dt}\left(\Phi_{yy}(t)^T \Phi_{\lambda\lambda}(t)\right) = \dot{\Phi}_{yy}(t)^T \Phi_{\lambda\lambda}(t) + \Phi_{yy}(t)^T \dot{\Phi}_{\lambda\lambda}(t) = 0$$
$$\Phi_{yy}(t_{m-1})^T \Phi_{\lambda\lambda}(t_{m-1}) = I \qquad (15.1.2.1.1.3\text{-}17)$$

The general solution to (15.1.2.1.1.3-17) is the interesting property:

$$\Phi_{yy}(t)^T \Phi_{\lambda\lambda}(t) = I$$

or, upon taking the transpose and multiplying on the right by $\Phi_{yy}(t)$ inverse:

$$\Phi_{yy}(t)^{-1} = \Phi_{\lambda\lambda}(t)^{T} \tag{15.1.2.1.1.3-18}$$

Use of (15.1.2.1.1.3-18) allows (15.1.2.1.1.3-11) to be simplified. The Φ parameters in (15.1.2.1.1.3-11) equate to their values at $t = t_m$, thus (15.1.2.1.1.3-11) becomes the simpler form:

$$P_m = \Phi_{\lambda\lambda_m} P_{m-1} \Phi_{\lambda\lambda_m}^{T} + \Phi_{\lambda y_m} \Phi_{\lambda\lambda_m}^{T} \tag{15.1.2.1.1.3-19}$$

The $\Phi_{\lambda\lambda_m}$, $\Phi_{\lambda y_m}$ terms in (15.1.2.1.1.3-19) are the integral of Equations (15.1.2.1.1.3-15) evaluated at time t_m. A first order solution to the (15.1.2.1.1.3-15) integral is obtained by setting the right sides to the initial conditions and integrating from t_{m-1} to t_m:

$$\begin{aligned} \Phi_{\lambda\lambda_m} &\approx I + \int_{t_{m-1}}^{t_m} A(t)\, dt = I + \Delta\Phi_{\lambda\lambda_m} \\ \Phi_{\lambda y_m} &\approx \int_{t_{m-1}}^{t_m} G_P(t)\, Q_{P_{Dens}}(t)\, G_P(t)^T\, dt = \Delta\Phi_{\lambda y_m} \end{aligned} \tag{15.1.2.1.1.3-20}$$

with

$$\Delta\Phi_{\lambda\lambda_m} \equiv \int_{t_{m-1}}^{t_m} A(t)\, dt \qquad \Delta\Phi_{\lambda y_m} \equiv \int_{t_{m-1}}^{t_m} G_P(t)\, Q_{P_{Dens}}(t)\, G_P(t)^T\, dt \tag{15.1.2.1.1.3-21}$$

If $A(t)$ and $G_P(t)\ Q_{P_{Dens}}(t)\ G_P(t)^T$ are approximated as constant, the classical complete solution to (15.1.2.1.1.3-15) is obtained (as in Section 15.1.2.1.1.1) as elements 2,1 and 2,2 of the general solution to (15.1.2.1.1.3-12):

$$\begin{bmatrix} \Phi_{yy_m} & \Phi_{y\lambda_m} \\ \Phi_{\lambda y_m} & \Phi_{\lambda\lambda_m} \end{bmatrix} = e^{M} \tag{15.1.2.1.1.3-22}$$

with from (15.1.2.1.1.3-12):

$$M = \begin{bmatrix} -A^T & 0 \\ G_P\, Q_{P_{Dens}}\, G_P^T & A \end{bmatrix} T_m \tag{15.1.2.1.1.3-23}$$

where

T_m = Time interval from t_{m-1} to t_m.

The e functional in (15.1.2.1.1.3-22) is by definition (as in (15.1.2.1.1.1-11)):

$$e^{M} = I + M + \frac{1}{2\,!} M^2 + \frac{1}{3\,!} M^3 + \cdots \qquad (15.1.2.1.1.3\text{-}24)$$

For the constant A(t), $G_P(t)$, $Q_{P_{Dens}}(t)$ case, Equations (15.1.2.1.1.3-21) can be introduced to convert (15.1.1.1.2-23) to the equivalent form:

$$M = \begin{bmatrix} -\Delta\Phi_{\lambda\lambda_m}^{T} & 0 \\ \Delta\Phi_{\lambda y_m} & \Delta\Phi_{\lambda\lambda_m} \end{bmatrix} \qquad (15.1.2.1.1.3\text{-}25)$$

It is easily verified by substituting (15.1.2.1.1.3-25) into (15.1.2.1.1.3-24) that Equation (15.1.2.1.1.3-22) for $\Phi_{\lambda\lambda_m}$, $\Phi_{\lambda y_m}$ is, to first order, equal to the Equation (15.1.2.1.1.3-20) result. Thus, the (15.1.2.1.1.3-25) form is actually more general, being accurate to first order for general time varying A(t) and $G_P(t)\ Q_{P_{Dens}}(t)\ G_P(t)^T$ matrices, and being exact for the case when M can be approximated as constant.

A recursive relationship can be written for (15.1.2.1.1.3-24) - (15.1.2.1.1.3-25) based on generation of the Taylor series terms:

$$
\begin{aligned}
M^{j} &= M^{j-1} \begin{bmatrix} -\Delta\Phi_{\lambda\lambda_m}^{T} & 0 \\ \Delta\Phi_{\lambda y_m} & \Delta\Phi_{\lambda\lambda_m} \end{bmatrix} \\
M^{0} &= I \\
e_j^{M} &= e_{j-1}^{M} + \frac{1}{j\,!} M^{j} \\
e_0^{M} &= I
\end{aligned}
\qquad (15.1.2.1.1.3\text{-}26)
$$

where

j = Number of terms carried in the e^M expansion.

M^j = M to the power of j.

e_j^M = e^M truncated at j terms in (15.1.2.1.1.3-24).

We also define:

$$\begin{bmatrix} --- & --- \\ M_{\lambda y}^{j} & M_{\lambda\lambda}^{j} \end{bmatrix} \equiv M^{j} \qquad (15.1.2.1.1.3\text{-}27)$$

With (15.1.2.1.1.3-21), (15.1.2.1.1.3-22), (15.1.2.1.1.3-26) and (15.1.2.1.1.3-27), the overall Equation (15.1.2.1.1.3-19) discrete covariance matrix propagation algorithm then becomes the iterative process:

Initialize computational parameters:

$$M_{\lambda\lambda}^{0} = I \qquad M_{\lambda y}^{0} = 0 \qquad \Phi_{\lambda\lambda}^{0} = I \qquad \Phi_{\lambda y}^{0} = 0$$

Input:

$$\Delta\Phi_{\lambda\lambda_m} = \int_{t_{m-1}}^{t_m} A(t)\, dt \qquad \Delta\Phi_{\lambda y_m} = \int_{t_{m-1}}^{t_m} G_P(t)\, Q_{P_{Dens}}(t)\, G_P(t)^T\, dt$$

DO for j =1 to the specified $\Phi_{\lambda\lambda}$ expansion order:

IF j is less than the specified $\Phi_{\lambda y}$ expansion order, THEN:

$$M_{\lambda y}^{j} = - M_{\lambda y}^{j-1}\, \Delta\Phi_{\lambda\lambda_m}^{T} + M_{\lambda\lambda}^{j-1}\, \Delta\Phi_{\lambda y_m}$$

$$\Phi_{\lambda y}^{j} = \Phi_{\lambda y}^{j-1} + \frac{1}{j\,!} M_{\lambda y}^{j} \qquad (15.1.2.1.1.3\text{-}28)$$

ENDIF

IF j is equal to the specified $\Phi_{\lambda y}$ expansion order: $B = \Phi_{\lambda y}\, \Phi_{\lambda\lambda}^{T}$

$$M_{\lambda\lambda}^{j} = M_{\lambda\lambda}^{j-1}\, \Delta\Phi_{\lambda\lambda_m}$$

$$\Phi_{\lambda\lambda}^{j} = \Phi_{\lambda\lambda}^{j-1} + \frac{1}{j\,!} M_{\lambda\lambda}^{j}$$

ENDDO

$$\Phi_m = \Phi_{\lambda\lambda} \qquad Q_m = \frac{1}{2}\left(B + B^T\right)$$

$$P_m = \Phi_m\, P_{m-1}\, \Phi_m^T + Q_m$$

where

$\Phi_{\lambda y}^{j}, \Phi_{\lambda\lambda}^{j}$ = $\Phi_{\lambda y}, \Phi_{\lambda\lambda}$ in (15.1.2.1.1.3-22) but truncated after the M^j term in the (15.1.2.1.1.3-24) expansion series.

Φ_m = Error state transition matrix that propagates the error state vector from t_{m-1} to t_m. If the (15.1.2.1.1.3-28) algorithm for Φ_m is compared with Equations (15.1.2.1.1.1-11) - (15.1.2.1.1.1-12), it should be apparent that Φ_m and $\Delta\Phi_{\lambda\lambda_m}$ of this section are identical to $\Phi(t_m,t_{m-1})$ and $\Delta\Phi_m$, the t_{m-1} to t_m error state transition matrix of Section 15.1.2.1.1.1 with its associated integrated error state dynamic matrix.

Q_m = Integrated effect of process noise on P from t_{m-1} to t_m.

B = Intermediate parameter used to assure that Q_m is symmetric.

Note in Equations (15.1.2.1.1.3-28) that the B term has the same form as the term on the right in (15.1.2.1.1.3-19). From the form of Equation (15.1.2.1.1-12) compared to (15.1.2.1.1.3-19) we can identify this term as Q_m, the integrated effect of process noise on the covariance P over an m cycle. However, because Equations (15.1.2.1.1.3-28) are truncated, the B so computed will not exactly equal the true Q_m. Of particular concern are errors in B that make it asymmetric as contrasted with the true Q_m that is always symmetric. To compensate for B asymmetry, Q_m is calculated in Equations (15.1.2.1.1.3-28) as $\frac{1}{2}\left(B + B^T\right)$ which is always symmetric (i.e., it equals its transpose), and exactly equals B when B is symmetric. Also note that the (15.1.2.1.1.3-28) algorithm for B is based on the assumption that the selected $\Phi_{\lambda y}$ expansion order will be less than or equal to the selected $\Phi_{\lambda\lambda}$ expansion order. Generally speaking, for comparable accuracy in the P_m equation, the expansion orders for Φ_m and Q_m should be equal. From (15.1.2.1.1.3-28) we see that this condition corresponds with the expansion orders for $\Phi_{\lambda y}$ and $\Phi_{\lambda\lambda}$ being the same. For some applications, however, it is desirable to have a higher expansion order for Φ_m than for Q_m to assure an accurate estimated error state vector propagation (to be discussed subsequently).

In Equations (15.1.2.1.1.3-28), the covariance matrix for the covariance propagation algorithm is initialized at the computed P value following the last Kalman estimation reset (i.e., $P_{n-1}(+_e)$). The covariance matrix used for the next Kalman cycle gain calculation (i.e., $P_n(-)$) is then set to the output of the covariance propagation algorithm at the next Kalman cycle time t_n. Thus, the initial condition and output for the (15.1.2.1.1.3-28) algorithm is:

$$\begin{aligned} P_m &= P_{n-1}(+_e) \qquad \text{At } t = t_{n-1} \\ P_n(-) &= P_m \qquad \text{At } t = t_n \end{aligned} \qquad (15.1.2.1.1.3\text{-}29)$$

The companion to Equations (15.1.2.1.1.3-28) - (15.1.2.1.1.3-29) for Kalman filter estimated error state vector m cycle propagation, is easily written by inspection of (15.1.2-7) for the idealized control reset case using Φ_m from (15.1.2.1.1.3-28):

$$\begin{aligned} \underline{\tilde{x}}_m &= \Phi_m \, \underline{\tilde{x}}_{m-1} \\ \underline{\tilde{x}}_m &= \underline{\tilde{x}}_{n-1}(+_c) \qquad \text{At } t = t_{n-1} \\ \underline{\tilde{x}}_n(-) &= \underline{\tilde{x}}_m \qquad \text{At } t = t_n \end{aligned} \qquad (15.1.2.1.1.3\text{-}30)$$

For the delayed reset case, the equivalent to (15.1.2.1.1.3-30) is from (15.1.2-21):

$$\underline{\tilde{x}}_m = \Phi_m \, \underline{\tilde{x}}_{m-1}$$

$$\underline{\tilde{x}}_m = \underline{\tilde{x}}_{n-1}(+_e) \qquad \text{At } t = t_{n-1} \qquad (15.1.2.1.1.3\text{-}31)$$

$$\underline{\tilde{x}}_n(-) = \underline{\tilde{x}}_m \qquad \text{At } t = t_n$$

Equations (15.1.2.1.1.3-28) - (15.1.2.1.1.3-31) provide algorithms for propagating the covariance matrix and estimated error state vector between Kalman estimation cycles. The number of terms carried in the (15.1.2.1.1.3-28) expansion series depends on the width of the m cycle time interval, the frequency content of the A(t) and $G_P(t)\ Q_{P_{Dens}}(t)\ G_P(t)^T$ terms in (15.1.2.1.1.3-28) and the number of integrators linking each error state to the measurement. Simulation studies are generally required to verify that the series expansion order is sufficient for proper characterization of the Φ_m and Q_m matrices.

The integration of A(t) and $G_P(t)\ Q_{P_{Dens}}(t)\ G_P(t)^T$ in (15.1.2.1.1.3-28) to obtain $\Delta\Phi_{\lambda\lambda_m}$ and $\Delta\Phi_{\lambda y_m}$ would be performed digitally, typically on the individual elements of A(t) and $G_P(t)\ Q_{P_{Dens}}(t)\ G_P(t)^T$, using a suitable algorithm at the iteration rate required to handle the expected dynamics of the particular element. Normally, many of the elements of A(t) and $G_P(t)\ Q_{P_{Dens}}(t)\ G_P(t)^T$ are zero, hence, the associated integral is also zero and need not be computed. Many of the elements of A(t) and $G_P(t)\ Q_{P_{Dens}}(t)\ G_P(t)^T$ have very little change over the integration interval (i.e., the m cycle) and their integral can, therefore, be accurately approximated by the average of the value at the start and end of the m cycle interval multiplied by the m cycle time. Generally, the $Q_{P_{Dens}}(t)$ matrix is constant and by definition, diagonal. Many of the rows of $G_P(t)$ have only one element (i.e., no noise element cross-coupling). For strapdown applications, some of the elements of $G_P(t)$ represent direction cosine matrices (e.g., $C_B^N(t)$) that couple orthogonal inertial sensor triad output white random noise into attitude (for angular rate sensors) or velocity (for accelerometers - See Equations (12.5.1-1)). For this situation, if the associated three elements of $Q_{P_{Dens}}(t)$ coupled by $C_B^N(t)$ are equal (i.e., if each of the three orthogonal inertial sensors are characterized as having equal output random noise densities), the associated product elements of $G_P(t)\ Q_{P_{Dens}}(t)\ G_P(t)^T$ will be:

$$C_B^N(t)\ Q_{P_{Dens/Assoc}}(t)\left(C_B^N(t)\right)^T = C_B^N(t)\ V_{P_{Dens/Assoc}}(t)\ I\left(C_B^N(t)\right)^T$$

$$= V_{P_{Dens/Assoc}}(t)\ C_B^N(t)\ I\left(C_B^N(t)\right)^T \qquad (15.1.2.1.1.3\text{-}32)$$

$$= V_{P_{Dens/Assoc}}(t)\ I = Q_{P_{Dens/Assoc}}(t)$$

where

$QP_{Dens/Assoc}(t)$ = 3 by 3 diagonal matrix containing the diagonal elements of $QP_{Dens}(t)$ associated with the $C_B^N(t)$ coupling from $GP(t)$.

$VP_{Dens/Assoc}(t)$ = Noise densities of each element of $QP_{Dens/Assoc}(t)$ (all assumed equal).

I = Identity matrix.

Thus, in the above situation, the significantly time varying $C_B^N(t)$ term does not appear at all in the coupled noise density matrix $GP(t)\ QP_{Dens}(t)\ GP(t)^T$. The overall result is that the computer throughput required for the $\Delta\Phi_{\lambda\lambda_m}$ and $\Delta\Phi_{\lambda y_m}$ integration process can generally be reduced to a level that has negligible penalty on overall computation time resources.

We conclude this section with an evaluation of the covariance propagation algorithm based on a two term expansion series for $\Phi_{\lambda\lambda}$ and $\Phi_{\lambda y}$. Carrying out the Equation (15.1.2.1.1.3-28) operations for a two term expansion shows that:

$$\Phi_{\lambda\lambda} = I + \Delta\Phi_{\lambda\lambda_m} + \frac{1}{2}\,\Delta\Phi_{\lambda\lambda_m}^2$$

$$\begin{aligned} B &\approx \Delta\Phi_{\lambda y_m} + \frac{1}{2}\left(\Delta\Phi_{\lambda y_m}\,\Delta\Phi_{\lambda\lambda_m}^T + \Delta\Phi_{\lambda\lambda_m}\,\Delta\Phi_{\lambda y_m}\right) \qquad (15.1.2.1.1.3\text{-}33)\\ &\approx \frac{1}{2}\,\Delta\Phi_{\lambda y_m} + \frac{1}{2}\left(I + \Delta\Phi_{\lambda\lambda_m}\right)\Delta\Phi_{\lambda y_m}\left(I + \Delta\Phi_{\lambda\lambda_m}\right)^T\\ &\approx \frac{1}{2}\,\Delta\Phi_{\lambda y_m} + \frac{1}{2}\,\Phi_{\lambda\lambda_m}\,\Delta\Phi_{\lambda y_m}\,\Phi_{\lambda\lambda_m}^T \end{aligned}$$

In the Equations (15.1.2.1.1.3-33) B calculation, the approximation has been made that $\Delta\Phi$ products are on the order of the error in the second order expansion utilized, hence, can be deleted (or added) with negligible accuracy alteration.

Recognizing from (15.1.2.1.1.3-21) that $\Delta\Phi_{\lambda y_m} = \Delta\Phi_{\lambda y_m}^T$, we see from (15.1.2.1.1.3-33) and (15.1.2.1.1.3-28) that:

$$\Phi_m = \Phi_{\lambda\lambda} \qquad Q_m \approx \frac{1}{2}\,\Delta\Phi_{\lambda y_m} + \frac{1}{2}\,\Phi_m\,\Delta\Phi_{\lambda y_m}\,\Phi_m^T \qquad (15.1.2.1.1.3\text{-}34)$$

Substituting (15.1.2.1.1.3-34) into the (15.1.2.1.1.3-28) covariance propagation equation then obtains (with (15.1.2.1.1.3-21) for $\Delta\Phi_{\lambda y_m}$):

$$P_m = \Phi_m \left(P_{m-1} + \frac{1}{2} Q_{1_m} \right) \Phi_m^T + \frac{1}{2} Q_{1_m}$$

$$Q_{1_m} = \int_{t_{m-1}}^{t_m} G_P(t)\; Q_{P_{Dens}}(t)\; G_P(t)^T \, dt \qquad (15.1.2.1.1.3\text{-}35)$$

where

Q_{1_m} = Value for Q_m that would have been obtained using a first order expansion for $\Phi_{\lambda y}$ in (15.1.2.1.1.3-28).

From Equation (15.1.2.1.1.3-22) using (15.1.2.1.1.3-21) for $\Delta\Phi_{\lambda\lambda_m}$ and $\Phi_{\lambda\lambda} = \Phi_m$ (as in (15.1.2.1.1.3-28)), we can also write the more general form (as in (15.1.2.1.1.1-11) - (15.1.2.1.1.1-12)):

$$\Phi_m = e^{\Delta\Phi_m} \qquad \Delta\Phi_m \equiv \int_{t_{m-1}}^{t_m} A(t)\, dt \qquad (15.1.2.1.1.3\text{-}36)$$

Equations (15.1.2.1.1.3-35) - (15.1.2.1.1.3-36) can be viewed as a general covariance propagation algorithm for the selected expansion for Φ_m and a second order expansion for Q_m. We finally note that if only one m cycle is used, then m is the Kalman estimation cycle n, and (15.1.2.1.1.3-36) with (15.1.2.1.1.3-35) becomes the equivalent of (15.1.2.1.1-12):

$$P_n(-) = \Phi_n \left(P_{n-1}(+_e) + \frac{1}{2} Q_{1_n} \right) \Phi_n^T + \frac{1}{2} Q_{1_n}$$

$$Q_{1_n} = \int_{t_{n-1}}^{t_n} G_P(t)\, Q_{P_{Dens}}(t)\, G_P(t)^T \, dt \qquad (15.1.2.1.1.3\text{-}37)$$

$$\Phi_n = e^{\Delta\Phi_n} \qquad \Delta\Phi_n \equiv \int_{t_{n-1}}^{t_n} A(t)\, dt$$

Equations (15.1.2.1.1.3-37) represent an approximation algorithm that can be considered for implementing covariance propagation Equation (15.1.2.1.1-12) in the INS computer. In many applications, the $G_P(t)\; Q_{P_{Dens}}(t)\; G_P(t)^T$ matrix is approximated as constant (see discussion in second paragraph following Equations (15.1.2.1.1.3-31)) in which case, Q_{1_n} in (15.1.2.1.1.3-37) simplifies to:

$$Q_{1_n} = G_P(t)\, Q_{P_{Dens}}(t)\, G_P(t)^T \, T_n \approx \text{Constant} \qquad (15.1.2.1.1.3\text{-}38)$$

where

T_n = Kalman filter estimation cycle time interval.

The Equation (15.1.2.1.1.3-38) approximation can also be applied to Equations (15.1.2.1.1.3-35) with T_n replaced by T_m.

15.1.2.1.1.4 Covariance Matrix Numerical Conditioning Control

Some of the intrinsic characteristics of the covariance matrix (P) are that it is symmetric (element in row i, column j equals element in row j, column i), each diagonal element is positive, and the matrix is positive definite. The positive definite characteristic of P is mathematically defined as the requirement that the operator $\underline{V}^T \, P \, \underline{V}$ be greater than zero for all real $\underline{V}$ in which $\underline{V}$ is an arbitrary vector (Note: If the operator is less than zero for any $\underline{V}$, then P is said to be negative definite). The positive definite requirement embodies the positive diagonal P element constraint in addition to the requirement that $P_{ii} \, P_{jj}$ be not less than $\left| P_{ij} \, P_{ji} \right|$ (Note: P_{ij} is the element of P in row i, column j). As part of Kalman filter software background computations involved in the calculation of P, control algorithms are frequently utilized to preserve the positive definite characteristic of P in the presence of numerical round-off which could drive P to be negative definite. Negative definite values for P can produce instabilities in the Kalman filter which utilizes P as a basic building block in the Kalman estimation process.

The symmetric P characteristic is easily satisfied by periodically setting the off-diagonal elements of P equal to $\frac{1}{2}\left(P + P^T\right)$. A very simple algorithm that can be used to assure that the diagonal elements of P remain positive and greater than designer selected reasonable minimum values, is the following applied periodically:

$$\text{IF } \left(P_{ii} \le P_{ii_{Min}}\right) \text{ THEN } \qquad P_{ii} = P_{ii_{Min}} \tag{15.1.2.1.1.4-1}$$

where

P_{ii} = Diagonal element of P in row i, column i.

$P_{ii_{Min}}$ = Minimum value of P_{ii} based on designer judgment of the uncertainties in the Kalman filter analytical models used for estimation.

A more sophisticated approach for maintaining P positive definite is to augment Equation (15.1.2.1.1.4-1) with a P adjustment based on the more general characteristic that:

$$\left| P_{ij} \, P_{ji} \right| \le P_{ii} \, P_{jj} \tag{15.1.2.1.1.4-2}$$

where

P_{ij} = Element of P in row i, column j.

A control algorithm to realize (15.1.2.1.1.4-2) (following (15.1.2.1.1.4-1)) is:

$$\begin{aligned} &\text{IF } \left\{ \left| P_{ij}(-)\ P_{ji}(-) \right| > P_{ii}\, P_{jj} \right\} \text{ THEN} \\ &\qquad P_{ij}(+) = P_{ji}(+) = \sqrt{P_{ii}\, P_{jj}}\ \text{Sign}\left(P_{ij}(-) + P_{ji}(-) \right) \end{aligned} \tag{15.1.2.1.1.4-3}$$

where

(-), (+) = Designation for the element before and after the correction.

Sign ( ) = 1 for ( ) $\geq$ 0 and -1 for ( ) $<$ 0.

If the diagonal elements of P become greater than is reasonable based on engineering judgment, their growth rate can be terminated by setting the appropriate rows/columns of the integrated coupled process noise matrix Q_n to zero:

$$\text{IF } \left(P_{ii} \geq P_{ii_{Max}} \right) \text{ THEN} \qquad Q_{ij} = Q_{ji} = 0 \ \text{ FOR ALL j} \tag{15.1.2.1.1.4-4}$$

where

$P_{ii_{Max}}$ = Maximum reasonable value for P_{ii}.

Q_{ij} = Element in row i, column j of Q_n.

A more sophisticated approach for controlling the maximum value of the P diagonal utilizes a compressed form of the Joseph's form covariance reset Equation (15.1.2.1.1-1):

$$P(+) = J\, P(-)\ J^T \tag{15.1.2.1.1.4-5}$$

where

J = Diagonal matrix with elements selected to control the diagonal elements of P to be less than $P_{ii_{Max}}$.

The algorithm for J is derived from the diagonal components of (15.1.2.1.1.4-5):

$$P_{ii}(+) = J_{ii}^2\, P_{ii}(-) \tag{15.1.2.1.1.4-6}$$

where

J_{ii} = Element in row i, column i of diagonal matrix J.

If $P_{ii}(-)$ is greater or equal to $P_{ii_{Max}}$, we set J_{ii} in (15.1.2.1.1.4-6) so that $P_{ii}(+)$ equals $P_{ii_{Max}}$, thus:

IF $\left(P_{ii}(-) \geq P_{ii_{Max}}\right)$ THEN

$$J_{ii} = \sqrt{\frac{P_{ii_{Max}}}{P_{ii}(-)}} \qquad (15.1.2.1.1.4\text{-}7)$$

ELSE

$J_{ii} = 1$

ENDIF

The J matrix so calculated is then used in (15.1.2.1.1.4-5) to correct P so that the diagonal elements remain below $P_{ii_{Max}}$. Note, that the complete (15.1.2.1.1.4-5) form is required to assure that P will remain positive definite after the correction. The simpler, but dangerous alternative of reducing only the diagonal elements P, can cause P to go negative definite. The simple Equation (15.1.2.1.1.4-1) algorithm used to control P_{ii} to be greater than $P_{ii_{Min}}$ is allowable because increasing the P diagonal makes P more positive definite.

15.1.2.2 THE OBSERVATION AND THE MEASUREMENT EQUATION

Measurement Equation (15.1-2) shows the measurement to be a linear function of the error state vector. In practice, Equation (15.1-2) represents a linearized form of an "observation equation" formed from the comparison of two identical functions of the navigation parameters obtained from two navigation devices (as described in generalized format by Equations (15.1-3)). For example, if the navigation devices are an INS and a GPS receiver, the navigation function might be the range ρ from the GPS receiver antenna to a particular GPS satellite (a function of the INS position location). An observation would be the difference between ρ computed from INS position data and ρ calculated from a GPS satellite-to-receiver signal transmission time measurement. The observation vector in this case would be the vector of such observations to several satellites. The observation vector taken in this manner would be a function of the INS position, the INS computed position error, GPS receiver error (e.g., clock frequency and phase) and other external error effects such as satellite position uncertainties and atmospheric effects on satellite-to-receiver signal transmission time. The linearized form of the observation equation is the Equation (15.1-2) measurement equation. The linearization process used to derive the measurement equation must include any conversion required to express the measurement as a function of the selected error state vector parameters. This step is necessary because the error in the navigation parameters used in forming the observation equation (e.g., position errors in C_N^E and altitude h - See Section 4.4) are not necessarily directly equal to the error state vector parameters (e.g., position error $\delta \underline{R}^N$ - See Section 12.2.3). Hence, when the linearization process is performed (by taking the differential of the observation equation) the immediate result is generally a function of error parameters that need conversion to their equivalent error state vector forms.

To provide a specific example illustrating the development of the observation and measurement equation, consider a horizontal position measurement formed from the INS navigation-to-earth Frame direction cosine matrix C_N^E and latitude/longitude provided from another separate navigation device. Let us consider the observation to be formed as the difference between $\underline{u}_{ZN}^E$ computed from INS data and the equivalent vector calculated from position data provided by the other navigation device where:

E = Earth fixed coordinate frame as defined in Section 2.2.

$\underline{u}_{ZN}^E$ = E Frame projection of a unit vector along the INS N Frame Z axis which, from Section 2.2, lies along the upward geodetic vertical at the INS position location.

Since $\underline{u}_{ZN}^E$ computed with data from the INS and the other navigation device will be very close to one another, the difference will lie primarily in a plane perpendicular to $\underline{u}_{ZN}^E$ (i.e., in a horizontal plane). To simplify the observation to have only two significant (horizontal) components, we include a transformation into the INS locally horizontal N Frame in forming the observation equation:

$$\underline{Z}_{POS} = \left[\left(C_{N_{INS}}^E \right)^T \left(\underline{u}_{ZN_{INS}}^E - \underline{u}_{ZN_{OTH}}^E \right) \right]_H \qquad (15.1.2.2\text{-}1)$$

where

INS = Subscript designation for parameter value calculated using INS inertially calculated computer data.

$\underline{Z}_{POS}$ = Observation for the position measurement.

$\underline{u}_{ZN_{INS}}^E, \underline{u}_{ZN_{OTH}}^E$ = $\underline{u}_{ZN}^E$ formed from the INS and other navigation device output data.

$C_{N_{INS}}^E$ = C_N^E calculated in the INS computer (i.e., containing error).

N = Locally level navigation coordinates as defined in Section 2.2.

H = Horizontal components of the designated vector.

The $\underline{u}_{ZN_{INS}}^E$ contribution to $\underline{Z}_{POS}$ in (15.1.2.2-1) is zero because it lies by definition along the INS defined vertical:

$$\left[\left(C^E_{N_{INS}}\right)^T \underline{u}^E_{ZN_{INS}}\right]_H = \left(\underline{u}^N_{ZN_{INS}}\right)_H = 0 \qquad (15.1.2.2\text{-}2)$$

The $\underline{u}^E_{ZN_{OTH}}$ term in (15.1.2.2-1) is calculated from the other navigation device output data as a vector along the upward geodetic vertical at the other navigation device position location, plus a correction to account for lever arm separation from the INS. An equation for the correction terms can be found using $\delta\underline{u}^E_{ZN}$ in Section 12.2.3, the change in $\underline{u}^E_{ZN}$ produced by horizontal position error $\delta\underline{R}^E_H$. Combining (12.2.3-11) for $\delta\underline{u}^E_{ZN}$ with $\left(\underline{\varepsilon}^N \times \underline{u}^N_{ZN}\right)$ from the rearranged form of (12.2.3-16) and substituting (12.2.4-6) shows that $\delta\underline{u}^E_{ZN} = \frac{1}{R}\,\delta\underline{R}^E_H$. A similar expression can be used for correcting $\underline{u}^E_{ZN_{OTH}}$ in (15.1.2.2-1) using the horizontal lever arm component in place of $\delta\underline{R}^E_H$. Thus:

$$\underline{u}^E_{ZN_{OTH}} = \underline{u}^E_{Up_{OTH}} + \frac{1}{R}\,\underline{l}^E_H \qquad (15.1.2.2\text{-}3)$$

where

$\underline{u}^E_{Up_{OTH}}$ = E Frame components of a unit vector along the other navigation device position geodetic vertical as calculated with position data from the other navigation device.

$\underline{l}$ = Lever arm from the INS to the other navigation device.

The $\underline{l}^E$ term in (15.1.2.2-3) can be described in terms of projections along vehicle reference axes as in Section 8.3:

$$\underline{l}^E = C^E_B\, C^B_M\, C^M_{VRF}\, \underline{l}^{VRF} \qquad (15.1.2.2\text{-}4)$$

where

B = INS sensor (or "body") reference axes as defined in Section 2.2.

M = INS mount coordinate frame aligned to the INS mount.

VRF = Vehicle reference axis coordinate frame having a specified fixed alignment to the vehicle carrying the INS and aiding device.

With (15.1.2.2-2) - (15.1.2.2-4) and application of generalized Equation (3.1.1-47) (with H being equivalent to $\perp$, representing vector components perpendicular to a unit vector along the local vertical), Equation (15.1.2.2-1) becomes:

$$\underline{Z}_{POS} = -\left[\left(C^{E}_{N_{INS}}\right)^{T}\left(\underline{u}^{E}_{Up_{OTH}} + \frac{1}{R}\,\underline{l}^{E}_{H}\right)\right]_{H} \qquad (15.1.2.2\text{-}5)$$

or

$$\underline{Z}_{POS} = -\left(C^{E\ \ T}_{N_{INS}}\right)_{H}\underline{u}^{E}_{Up_{OTH}} - \frac{1}{R}\,\underline{l}^{N}_{H} \qquad (15.1.2.2\text{-}6)$$

with (15.1.2.2-4) in the N Frame:

$$\underline{l}^{N} = C^{N}_{B}\, C^{B}_{M}\, C^{M}_{VRF}\, \underline{l}^{VRF} \qquad (15.1.2.2\text{-}7)$$

where

$\left(C^{E\ \ T}_{N_{INS}}\right)_{H}$ = Horizontal component of $\left(C^{E}_{N_{INS}}\right)^{T}$ which, from the Section 2.2 definition for the N Frame having Z axis vertical, is $\left(C^{E}_{N_{INS}}\right)^{T}$ with the third row set to zero.

Equations (15.1.2.2-6) - (15.1.2.2-7) constitute the "observation equation" for the position measurement. The $\underline{Z}_{POS}$ observation so calculated would be used in Equation (15.1-3) as the position measurement portion of the $\underline{Z}_{Obs}$ observation vector input to the Kalman filter. The "measurement equation" is now derived in the Equation (15.1-2) linearized format by taking the differential of $\underline{Z}_{POS}$ Equation (15.1.2.2-1):

$$\underline{z}_{POS} \equiv \delta\underline{Z}_{POS} = \left[\delta\left(C^{E}_{N_{INS}}\right)^{T}\left(\underline{u}^{E}_{ZN_{INS}} - \underline{u}^{E}_{ZN_{OTH}}\right) + \left(C^{E}_{N_{INS}}\right)^{T}\left(\delta\underline{u}^{E}_{ZN_{INS}} - \delta\underline{u}^{E}_{ZN_{OTH}}\right)\right]_{H}$$

(15.1.2.2-8)

Because the INS and other navigation device have comparable accuracy position data, $\underline{u}^{E}_{ZN_{INS}}$ will approximately equal $\underline{u}^{E}_{ZN_{OTH}}$. Consequently, the product of $\left(\underline{u}^{E}_{ZN_{INS}} - \underline{u}^{E}_{ZN_{OTH}}\right)$ with $\delta\left(C^{E}_{N_{INS}}\right)^{T}$ in (15.1.2.2-8) is second order, hence, negligible and we write:

$$\underline{z}_{POS} \approx \left[\left(C^{E}_{N}\right)^{T}\left(\delta\underline{u}^{E}_{ZN_{INS}} - \delta\underline{u}^{E}_{ZN_{OTH}}\right)\right]_{H} \qquad (15.1.2.2\text{-}9)$$

In Equation (15.1.2.2-9) we have, for simplicity, deleted the INS subscript on C^{E}_{N}. For the remainder of this section, we will also drop the INS subscript on all direction cosine matrices with the general understanding that they are based on INS computed navigation parameters.

The $\delta \underline{u}^{E}_{ZN_{INS}}$ term in (15.1.2.2-9) is expanded using (12.2.3-11):

$$\delta \underline{u}^{E}_{ZN_{INS}} = C^{E}_{N} \left(\underline{\varepsilon}^{N}_{INS} \times \underline{u}^{N}_{ZN_{INS}} \right) \tag{15.1.2.2-10}$$

If $\delta \underline{R}^{N}_{H}$ is being used as the position error state rather than $\underline{\varepsilon}^{N}$, equivalency Equation (12.2.3-24) is incorporated by which:

$$\underline{\varepsilon}^{N} \times \underline{u}^{N}_{ZN} = \frac{1}{R} \, \delta \underline{R}^{N}_{H} \tag{15.1.2.2-11}$$

thereby making (15.1.2.2-10):

$$\delta \underline{u}^{E}_{ZN_{INS}} = \frac{1}{R} \, C^{E}_{N} \, \delta \underline{R}^{N}_{INS_{H}} \tag{15.1.2.2-12}$$

The $\delta \underline{u}^{E}_{ZN_{OTH}}$ term in (15.1.2.2-9) is expanded from (15.1.2.2-3) taking care to use the Section 3.5.4 formalism for vector differentials:

$$\delta \underline{u}^{E}_{ZN_{OTH}} = \delta \underline{u}^{E}_{Up_{OTH}} + \frac{1}{R} \, \delta \underline{l}^{E}_{H_{E}} \tag{15.1.2.2-13}$$

where

$\delta \underline{l}^{E}_{H_{E}}$ = The differential of $\underline{l}_{H}$ taken in the E Frame (subscript) and projected on the E Frame (superscript) (as in generalized Equation (3.5.4-1)).

An expression for $\delta \underline{u}^{E}_{Up_{OTH}}$ can be determined similar to (15.1.2.2-12) by which:

$$\delta \underline{u}^{E}_{UP_{OTH}} = \frac{1}{R} \, \delta \underline{R}^{E}_{OTH_{H}} \tag{15.1.2.2-14}$$

where

$\delta \underline{R}^{E}_{OTH_{H}}$ = Horizontal position error associated with the other navigation device in reporting its own position, as projected on E Frame axes.

The $\delta \underline{l}^{E}_{H_{E}}$ term in (15.1.2.2-13) is evaluated from the differential of E Frame version of (15.1.2.2-7), treating C^{M}_{VRF} as an error free reference:

$$\delta \underline{l}^{E}_{H_{E}} = \left[\delta \left(C^{E}_{B} \, C^{B}_{M} \right) C^{M}_{VRF} \, \underline{l}^{VRF} + C^{E}_{B} \, C^{B}_{M} \, C^{M}_{VRF} \, \delta \underline{l}^{VRF}_{VRF} \right]_{H} \tag{15.1.2.2-15}$$

where

$\delta \underline{l}_{VRF}^{VRF}$ = Error in $\underline{l}$ evaluated in the VRF Frame (subscript) and projected on VRF Frame axes (superscript).

The $\delta\left(C_B^E\, C_M^B\right)$ term in (15.1.2.2-15) is found from generalized Equations (3.5.2-47) to be:

$$\delta\left(C_B^E\, C_M^B\right) = -\left[\left(\underline{\alpha}_{M\,to\,B}^{E} + \underline{\alpha}_{B\,to\,E}^{E}\right)\times\right] C_B^E\, C_M^B \tag{15.1.2.2-16}$$

with:

$$\left(\underline{\alpha}_{M\,to\,B}^{B}\times\right) = -\,\delta C_M^B \left(C_M^B\right)^T \qquad \underline{\alpha}_{M\,to\,B}^{E} = C_B^E\, \underline{\alpha}_{M\,to\,B}^{B} \tag{15.1.2.2-17}$$

$$\left(\underline{\alpha}_{B\,to\,E}^{E}\times\right) = -\,\delta C_B^E \left(C_B^E\right)^T \tag{15.1.2.2-18}$$

From (12.2.1-3):

$$\left(\underline{\psi}^{E}\times\right) = -\,\delta C_B^E \left(C_B^E\right)^T \tag{15.1.2.2-19}$$

Thus, comparing (15.1.2.2-18) and (15.1.2.2-19), we see that:

$$\underline{\alpha}_{B\,to\,E}^{E} = \underline{\psi}^{E} \tag{15.1.2.2-20}$$

From the transpose of (8.3-2) we have to first order:

$$C_M^B = I - \left(\underline{J}\times\right) \tag{15.1.2.2-21}$$

where

$\underline{J}$ = INS to mount misalignment calibration vector.

Substituting (15.1.2.2-21) and its differential into (15.1.2.2-17) shows that to first order:

$$\underline{\alpha}_{M\,to\,B}^{B} \approx \delta\underline{J} \tag{15.1.2.2-22}$$

and

$$\underline{\alpha}_{M\,to\,B}^{E} = C_B^E\, \delta\underline{J} \tag{15.1.2.2-23}$$

Substituting (15.1.2.2-20) and (15.1.2.2-23) into (15.1.2.2-16) then yields:

$$\delta\left(C_B^E\, C_M^B\right) = -\left[\left(\underline{\psi}^{E} + C_B^E\, \delta\underline{J}\right)\times\right] C_B^E\, C_M^B \tag{15.1.2.2-24}$$

The $\delta \underline{l}_{VRF}^{VRF}$ term in (15.1.2.2-15) can be expanded into a term representing a potential error state portion plus a random portion attributed to vibration induced bending effects:

$$\delta \underline{l}_{VRF}^{VRF} = \delta \underline{l}_{Stat}^{VRF} + \delta \underline{l}_{Vib}^{VRF} \qquad (15.1.2.2\text{-}25)$$

where

$\delta \underline{l}_{Stat}^{VRF}$ = Portion of $\delta \underline{l}_{VRF}^{VRF}$ that is slowly changing (or constant), hence, potentially modelable as an error state vector component.

$\delta \underline{l}_{Vib}^{VRF}$ = Random portion of $\delta \underline{l}_{VRF}^{VRF}$ induced by vibration.

Substituting (15.1.2.2-24) and (15.1.2.2-25) in (15.1.2.2-15) then yields:

$$\delta \underline{l}_{H_E}^{E} = \left\{ - \left[\left(\underline{\psi}^E + C_B^E \, \delta \underline{J} \right) \times \right] C_B^E \, C_M^B \, C_{VRF}^M \, \underline{l}^{VRF} + C_B^E \, C_M^B \, C_{VRF}^M \left(\delta \underline{l}_{Stat}^{VRF} + \delta \underline{l}_{Vib}^{VRF} \right) \right\}_H \qquad (15.1.2.2\text{-}26)$$

or upon compression and rearrangement:

$$\delta \underline{l}_{H_E}^{E} = \left[\left(\underline{l}^E \times \right) \left(\underline{\psi}^E + C_B^E \, \delta \underline{J} \right) + C_{VRF}^E \left(\delta \underline{l}_{Stat}^{VRF} + \delta \underline{l}_{Vib}^{VRF} \right) \right]_H \qquad (15.1.2.2\text{-}27)$$

We now substitute Equations (15.1.2.2-12) - (15.1.2.2-14) and (15.1.2.2-27) into (15.1.2.2-9) to find:

$$\begin{aligned} \underline{z}POS \approx \Bigg[\left(C_N^E \right)^T \Bigg\{ & \frac{1}{R} \, C_N^E \, \delta \underline{R}_{INS_H}^N - \frac{1}{R} \, \delta \underline{R}_{OTH_H}^E \\ & - \frac{1}{R} \left[\left(\underline{l}^E \times \right) \left(\underline{\psi}^E + C_B^E \, \delta \underline{J} \right) + C_{VRF}^E \left(\delta \underline{l}_{Stat}^{VRF} + \delta \underline{l}_{Vib}^{VRF} \right) \right]_H \Bigg\} \Bigg]_H \end{aligned} \qquad (15.1.2.2\text{-}28)$$

The following application of generalized Equation (3.1.1-47) (with H equivalent to $\perp$) will prove useful for the $\delta \underline{l}$ terms in (15.1.2.2-28), recognizing that the X, Y axes of the N Frame are horizontal:

$$\begin{aligned} \left(C_N^E \right)^T \left[C_{VRF}^E \, \delta \underline{l}^{VRF} \right]_H &= C_E^N \, C_{VRF}^E \, \delta \underline{l}_H^{VRF} = C_{VRF}^N \, \delta \underline{l}_H^{VRF} \\ &= \left(C_{VRF}^N \, \delta \underline{l}^{VRF} \right)_H = \left(C_{VRF}^N \right)_H \delta \underline{l}^{VRF} \end{aligned} \qquad (15.1.2.2\text{-}29)$$

Applying (15.1.2.2-29) to (15.1.2.2-28) then yields the final form for the measurement equation:

$$\underline{z}_{POS} = \frac{1}{R}\,\delta\underline{R}^{N}_{INS_H} - \frac{1}{R}\left(C^{E\,T}_{N}\right)_H \delta\underline{R}^{E}_{OTH_H} - \frac{1}{R}\left(\underline{l}^{N}\times\right)_H\left(\underline{\psi}^{N} + C^{N}_{B}\,\delta\underline{J}\right)$$
$$- \frac{1}{R}\left(C^{N}_{VRF}\right)_H \delta\underline{l}^{VRF}_{Stat} - \frac{1}{R}\left(C^{N}_{VRF}\right)_H \delta\underline{l}^{VRF}_{Vib} \tag{15.1.2.2-30}$$

with from (15.1.2.2-7) and (15.1.2.2-21), and neglecting $\underline{J}$ compared to the zero order terms:

$$\underline{l}^{N} = C^{N}_{VRF}\,\underline{l}^{VRF} \qquad C^{N}_{VRF} \approx C^{N}_{B}\,C^{M}_{VRF} \tag{15.1.2.2-31}$$

where

$\left(C^{E\,T}_{N}\right)_H, \left(C^{N}_{VRF}\right)_H = \left(C^{E}_{N}\right)^{T}, C^{N}_{VRF}$ with the third row set to zero.

The $\delta\underline{R}^{N}_{INS_H}$, $\delta\underline{R}^{E}_{OTH_H}$, $\underline{\psi}^{N}$, $\delta\underline{J}$, $\delta\underline{l}^{VRF}_{Stat}$ terms in (15.1.2.2-30) represent components of the error state vector $\underline{x}$, and $\delta\underline{l}^{VRF}_{Vib}$ represents the measurement noise $\underline{n}_M$ associated with the $\underline{z}_{POS}$ measurement. If we compare (15.1.2.2-30) with the generalized form (15.1-2) we see that they are equivalent; the coefficients of $\delta\underline{R}^{N}_{INS_H}$, $\delta\underline{R}^{E}_{OTH_H}$, $\underline{\psi}^{N}$, $\delta\underline{J}$, $\delta\underline{l}^{VRF}_{Stat}$ represent the elements of the measurement matrix H associated with $\underline{z}_{POS}$ and the coefficient of $\delta\underline{l}^{VRF}_{Vib}$ represents the measurement noise dynamic coupling matrix G_M for $\underline{z}_{POS}$. Knowledge of the error characteristics of the other navigation device would enable $\delta\underline{R}^{E}_{OTH_H}$ to be expressed in terms of error states that better define $\delta\underline{R}^{E}_{OTH_H}$. On the other hand, if the error model for $\delta\underline{R}^{E}_{OTH_H}$ is not well understood, the $\left(C^{E\,T}_{N}\right)_H \delta\underline{R}^{E}_{OTH_H}$ term in (15.1.2.2-30) might be represented as simply $\delta\underline{R}^{N}_{OTH_H}$ and treated as a part of the measurement noise. In the latter case, the estimation filter cycle rate would have to be set slow enough to assure that the $\delta\underline{R}^{N}_{OTH_H}$ error will be random from measurement cycle to measurement cycle (i.e., consistent with the assumptions underlying Equation (15.1-2) for which the measurement noise vector is white (i.e., uncorrelated from cycle to cycle)).

In practice, the $\underline{\psi}^{N}$, $\delta\underline{J}$ terms multiplying $\underline{l}^{N}$ in measurement Equation (15.1.2.2-30) typically have negligible impact on the measurement, hence, they and supporting Equations (15.1.2.2-31) can be ignored. They were included in the development of (15.1.2.2-30) to illustrate the overall measurement equation development process. With this simplification, the (15.1.2.2-30) equation for position measurements would then become:

$$\underline{z}_{POS} \approx \frac{1}{R}\,\delta\underline{R}^{N}_{INS_H} - \frac{1}{R}\left(C^{E^T}_{N}\right)_H \delta\underline{R}^{E}_{OTH_H} - \frac{1}{R}\left(C^{N}_{VRF}\right)_H \delta\underline{l}^{VRF}_{Stat} - \frac{1}{R}\left(C^{N}_{VRF}\right)_H \delta\underline{l}^{VRF}_{Vib}$$

(15.1.2.2-32)

15.1.2.3 CONTROL VECTOR SELECTION AND APPLICATION

The $\underline{u}_c$ control vector in Equations (15.1.2-5) and (15.1.2-11) - (15.1.2-12) (or (15.1.2-17), (15.1.2-19) and (15.1.2-22)), is constructed to null the error state vector components that are computer controllable; i.e., errors in computer calculated parameters that are permitted to be adjusted for improved accuracy. A fundamental benefit derived from such controls is that the associated error state components remain small, hence, second order error effects (i.e., products of error state vector components) are reduced. This can be an important consideration when applying a Kalman filter because its inherent structure is based on neglecting second order errors. Minimizing the time period between estimation/control updates provides additional benefit in this regard.

In some applications, safety and/or computer accessibility considerations do not allow the basic inertial navigation parameters to be adjusted for fear of inadvertently corrupting the navigation data with incorrect data. In such applications, the estimated navigation error states are used to correct the inertial navigation output data (See Section 15.1.2.3.1). It is to be noted, however, that this approach leaves residual second order errors in the corrected navigation outputs because the basic Kalman filter estimated error states used for correction are only accurate to first order. The most accurate correction method is to directly update the basic navigation parameters using the $\underline{u}_c$ control vector, which drives the errors toward null, thereby rendering second order error effects negligible.

Construction of $\underline{u}_c$ to null selected error states consists of setting the appropriate components of $\underline{u}_c$ to the negative of the last estimated value for the error states being controlled. Based on the discussion in Section 15.1.2, the idealized Kalman reset which uses Equations (15.1.2-6) - (15.1.2-13), would then have:

$$\underline{u}_{c_n} = - \tilde{\underline{x}}_{c_n}(+_e) \qquad (15.1.2.3\text{-}1)$$

where

$(+_e)$ = Designation for the parameter value immediately following the <u>estimation</u> update.

$\tilde{\underline{x}}_{c_n}(+_e)$ = The estimated error state vector $\tilde{\underline{x}}_n(+_e)$, but with the uncontrolled components set to zero.

From Equation (15.1.2-12) and (15.1.2-7), use of the (15.1.2.3-1) control law sets the estimated components of $\underline{\tilde{x}}$ to zero for the error states being controlled. For the case when real time constraints do not permit use of the idealized control reset, the delayed reset technique can be applied using Kalman filter Equations (15.1.2-19) - (15.1.2-26). Then based on the discussion in Section 15.1.2, we would set:

$$\underline{u}_{c_n} = - \underline{\tilde{x}}_{c_{n-1}}(+_e) \qquad (15.1.2.3\text{-}2)$$

Application of the (15.1.2.3-1) or (15.1.2.3-2) control laws to the system navigation parameters consists of two steps; conversion of the control vector components to a form that is directly representative of the navigation parameters in the system computer, then application of the converted control vector to the navigation parameters (as in (15.1.2-11) or (15.1.2-19)). Equation (15.1.2-12) with (15.1.2.3-1), or (15.1.2-22) with (15.1.2.3-2), would then be processed so that the estimated error state vector reflects application of the control vector to the navigation parameters. Application of the control vector to the navigation parameters has been described previously by the general non-linear Equation (15.1-5). As in the previous section, we now provide an illustrative example of the above control reset procedure.

Let us consider the $\delta\underline{R}^N_{INS_H}$, $\delta\underline{R}^E_{OTH_H}$, $\underline{\psi}^N$, $\delta\underline{J}$, $\delta\underline{l}^{VRF}_{Stat}$ error terms in Equation (15.1.2.2-30) as components of the $\underline{x}$ error state vector, and that we choose to control $\delta\underline{R}^N_{INS_H}$, $\underline{\psi}^N$, and $\delta\underline{J}$ in the INS navigation parameters to zero. Let us also assume that observation Equation (15.1.2.2-6) with (15.1.2.2-7) is implemented in the INS computer so that $\underline{l}^{VRF}$ is an accessible navigation parameter. Therefore, $\delta\underline{l}^{VRF}_{Stat}$ is controllable by adjusting $\underline{l}^{VRF}$. Let us further assume that an input interface to the other navigation device is not available, hence, control of $\delta\underline{R}^E_{OTH_H}$ is not possible. For this situation, the Equation (15.1.2.3-1) or (15.1.2.3-2) control vector $\underline{u}_c$ would be formed from the negative of the estimated $\delta\underline{R}^N_{INS_H}$, $\underline{\psi}^N$, $\delta\underline{J}$, $\delta\underline{l}^{VRF}_{Stat}$ error state vector components. The $\underline{u}_c$ so formed would be utilized directly in Equation (15.1.2-12) or (15.1.2-22) following application of the control vector to the navigation parameters (in (15.1.2-11) or (15.1.2-19)) as outlined below.

Adopting the Equation (15.1.2.3-1) and (15.1.2.3-2) nomenclature, let us define the non-zero elements of $\underline{u}_c$ where:

$\delta\underline{R}^N_{INS/H_c}$, $\underline{\psi}^N_c$, $\delta\underline{J}_c$, $\delta\underline{l}^{VRF}_{Stat_c}$ = Elements of $\underline{u}_c$ used to control error state vector components $\delta\underline{R}^N_{INS_H}$, $\underline{\psi}^N$, $\delta\underline{J}$, $\delta\underline{l}^{VRF}_{Stat}$ as prescribed by Equation (15.1.2.3-1) or (15.1.2.3-2).

Application of the control vector to the $\underline{J}$ and $\underline{l}^{VRF}$ parameters in Equations (15.1.2-11) or (15.1.2-19) is trivial:

$$\underline{J}(+_c) = \underline{J}(-) + \delta\underline{J}_c \qquad \underline{l}^{VRF}(+_c) = \underline{l}^{VRF}(-) + \delta\underline{l}^{VRF}_{Stat_c} \tag{15.1.2.3-3}$$

where

$(-)$ = Designation for parameter value immediately preceding the control reset.

$(+_c)$ = Designation for parameter value immediately following the control reset.

Application of the $\delta\underline{R}^N_{INS/H_c}$, $\underline{\psi}^N_c$ controls is generally more involved because the position/attitude navigation parameters may not be in a form that permits direct control application. For example, let us consider the case when the basic INS horizontal position parameter is C^E_N and the basic attitude parameter is C^L_B (by integrating Equations (12.1-1) and (12.1-10)). We can also write:

$$C^N_B = C^N_L \, C^L_B \tag{15.1.2.3-4}$$

or upon inversion:

$$C^L_B = \left(C^N_L\right)^T C^N_B \tag{15.1.2.3-5}$$

The errors in C^N_B and C^E_N are characterized from Equation (12.2.1-9) and (12.2.1-11) by:

$$\delta C^E_N = C^E_N \left(\underline{\varepsilon}^N \times\right) \tag{15.1.2.3-6}$$

$$\delta C^N_B = -\left(\underline{\gamma}^N \times\right) C^N_B \tag{15.1.2.3-7}$$

where

$\underline{\gamma}^N$ = Rotation angle error vector associated with the $\hat{C}^N_B$ matrix considering the N Frame to be misaligned, as projected on Frame N axes.

$\underline{\varepsilon}^N$ = Rotation angle error vector associated with the $\hat{C}^E_N$ matrix considering the N Frame to be misaligned, as projected on Frame N axes.

Because C^N_L is constant, the differential of (15.1.2.3-5) with (15.1.2.3-7), (15.1.2.3-4) and generalized Equation (3.1.1-39) gives for the error in C^L_B:

$$\delta C_B^L = -\left(C_L^N\right)^T \left(\underline{\gamma}^N \times\right) C_L^N \, C_B^L = -\left(\underline{\gamma}^L \times\right) C_B^L \tag{15.1.2.3-8}$$

$$\underline{\gamma}^L = \left(C_L^N\right)^T \underline{\gamma}^N \tag{15.1.2.3-9}$$

Equations (15.1.2.3-6) and (15.1.2.3-8) - (15.1.2.3-9) show how the errors in C_N^E and C_B^L can be calculated in terms of the $\underline{\varepsilon}^N$ and $\underline{\gamma}^L$ error parameters. The $\underline{\varepsilon}^N$, $\underline{\gamma}^L$ terms can also be interpreted as control correction parameters in which δC_N^E, δC_B^L represent control corrections to C_N^E, C_B^L corresponding to $\underline{\varepsilon}^N$, $\underline{\gamma}^L$ control parameters. The associated control application equations would then be:

$$\begin{aligned} C_N^E(+_c) &= C_N^E(-) + \delta C_N^E = C_N^E(-)\left[I + \left(\underline{\varepsilon}_c^N \times\right)\right] \\ C_B^L(+_c) &= C_B^L(-) + \delta C_B^L = \left[I - \left(\underline{\gamma}_c^L \times\right)\right] C_B^L(-) \end{aligned} \tag{15.1.2.3-10}$$

with, from (15.1.2.3-9);

$$\underline{\gamma}_c^L = \left(C_L^N\right)^T \underline{\gamma}_c^N \tag{15.1.2.3-11}$$

where

$\underline{\varepsilon}_c^N$, $\underline{\gamma}_c^N$ = Control vector equivalents for adjusting C_N^E, C_B^L.

For the example problem with which we are dealing, the position/attitude control corrections are represented by $\delta \underline{R}_{INS/H_c}^N$, $\underline{\psi}_c^N$ rather than the $\underline{\varepsilon}_c^N$, $\underline{\gamma}_c^N$ form. Equivalency Equations (12.2.1-17) and (12.2.3-25) provide the means for converting $\delta \underline{R}_{INS/H_c}^N$, $\underline{\psi}_c^N$ to the $\underline{\varepsilon}_c^N$, $\underline{\gamma}_c^N$ application form:

$$\underline{\varepsilon}_c^N = \frac{1}{R}\left(\underline{u}_{ZN_{INS}}^N \times \delta \underline{R}_{INS/H_c}^N\right) \qquad \underline{\gamma}_c^N = \underline{\psi}_c^N + \underline{\varepsilon}_c^N \tag{15.1.2.3-12}$$

Note in (15.1.2.3-12) that the approximation has been made of neglecting the vertical ε_{ZN} term in (12.2.3-25).

Equations (15.1.2.3-10) - (15.1.2.3-12) would be the means by which the position/attitude control vector components $\delta \underline{R}_{INS/H_c}^N$, $\underline{\psi}_c^N$ would be applied to the INS position/attitude navigation parameters C_N^E, C_B^L. Note that Equations (15.1.2.3-10) represent first order accuracy forms which could produce second order orthogonality/normalization errors in C_N^E, C_B^L. In order to eliminate the orthogonality/normalization error effects, the following revised form of (15.1.2.3-10) can be used based on Equations (7.1.1.2-1), (7.1.1.2-3), (7.3.1-6) and (7.3.1-8):

$$C_B^L(+_c) = \left[I - \frac{\sin \gamma_c}{\gamma_c} \left(\underline{\gamma}_c^L \times \right) + \frac{(1 - \cos \gamma_c)}{\gamma_c^2} \left(\underline{\gamma}_c^L \times \right)^2 \right] C_B^L(-)$$

$$C_N^E(+_c) = C_N^E(-) \left[I + \frac{\sin \varepsilon_c}{\varepsilon_c} \left(\underline{\varepsilon}_c^N \times \right) + \frac{(1 - \cos \varepsilon_c)}{\varepsilon_c^2} \left(\underline{\varepsilon}_c^N \times \right)^2 \right] \tag{15.1.2.3-13}$$

For a general Kalman filter application, all of the Equations (12.5.1-1) navigation error terms might be included as error states. Then the associated error state vector would be of the form:

$$\underline{x} = \left[\left(\underline{\psi}^N\right)^T, \left(\delta \underline{V}^N\right)^T, \left(\delta \underline{R}^N\right)^T, \cdots\cdots \right]^T \tag{15.1.2.3-14}$$

The associated $\underline{u}_c$ control vector calculated from error state vector estimates would then be:

$$\underline{u}_c = \left[\left(\underline{\psi}_c^N\right)^T, \left(\delta \underline{V}_c^N\right)^T, \left(\delta \underline{R}_c^N\right)^T, \cdots\cdots \right]^T \tag{15.1.2.3-15}$$

Assuming that the INS navigation parameters are integrated in N Frame coordinates (e.g., Equations (12.1-1) - (12.1-12)), the (15.1.2.3-15) control vector components must first be converted to an equivalent compatible form for application to the INS "foreground" navigation parameters. The term "foreground" is frequently applied to represent the basic navigation integration operations in an INS that are executed to calculate the fundamental INS navigation parameters (i.e., the ξ_{INS} term in Equation (15.1-3)). We will continue to use the term "foreground" in this section based on this meaning. Using equivalency Equation (12.2.3-25), we approximate the foreground C_N^E position matrix control reset angle as in (15.1.2.3-12):

$$\underline{\varepsilon}_c^N = \frac{1}{R} \left(\underline{u}_{ZN}^N \times \delta \underline{R}_c^N \right) \tag{15.1.2.3-16}$$

Note in (15.1.2.3-16) that the approximation has been made of neglecting the vertical ε_{ZN} term in (12.2.3-25). This is because ε_{ZN} is not definable from the selected (15.1.2.3-14) error states, hence, we represent it as zero in the control reset formulation. As we shall see subsequently, this simplification has absolutely no detrimental impact whatsoever in achieving the desired goal of resetting the selected (15.1.2.3-14) error states in accordance with (15.1.2.3-15) when applying the foreground navigation parameter corrections.

Using (15.1.2.3-16) and equivalency Equations (12.2.1-17), (12.2.2-5) and (12.2.3-26), the foreground C_B^N attitude matrix, $\underline{v}^N$ velocity and h altitude foreground control signals are obtained from the (15.1.2.3-15) components of $\underline{u}_c$:

$$\underline{\gamma}_c^N = \underline{\psi}_c^N + \underline{\varepsilon}_c^N \tag{15.1.2.3-17}$$

$$\delta\underline{v}_c^N = \delta\underline{V}_c^N - \underline{\varepsilon}_c^N \times \underline{v}^N \tag{15.1.2.3-18}$$

$$\delta h_c = \underline{u}_{ZN}^N \cdot \delta\underline{R}_c^N \tag{15.1.2.3-19}$$

Applying (15.1.2.3-16) - (15.1.2.3-19) to the foreground parameters using (15.1.2.3-11) and (15.1.2.3-13) for the matrix controls then obtains:

$$\begin{aligned}
\underline{\gamma}_c^L &= \left(C_L^N\right)^T \underline{\gamma}_c^N \\
C_B^L(+_c) &= \left[I - \frac{\sin \gamma_c}{\gamma_c}\left(\underline{\gamma}_c^L \times\right) + \frac{(1 - \cos \gamma_c)}{\gamma_c^2}\left(\underline{\gamma}_c^L \times\right)^2 \right] C_B^L(-) \\
\underline{v}^N(+_c) &= \underline{v}^N(-) + \delta\underline{v}_c^N \\
C_N^E(+_c) &= C_N^E(-) \left[I + \frac{\sin \varepsilon_c}{\varepsilon_c}\left(\underline{\varepsilon}_c^N \times\right) + \frac{(1 - \cos \varepsilon_c)}{\varepsilon_c^2}\left(\underline{\varepsilon}_c^N \times\right)^2 \right] \\
h(+_c) &= h(-) + \delta h_c
\end{aligned} \tag{15.1.2.3-20}$$

We now demonstrate, as previously stipulated, that application of control Equations (15.1.2.3-16) - (15.1.2.3-20) provides an accurate reset of the navigation error states in (15.1.2.3-14). For example, consider the $\underline{\psi}^N$ attitude error state vector component. We know from equivalency Equation (12.2.1-17) that:

$$\underline{\psi}^N = \underline{\gamma}^N - \underline{\varepsilon}^N \tag{15.1.2.3-21}$$

from which:

$$\underline{\psi}^N(+_c) = \underline{\gamma}^N(+_c) - \underline{\varepsilon}^N(+_c) \tag{15.1.2.3-22}$$

$$\underline{\psi}^N(-) = \underline{\gamma}^N(-) - \underline{\varepsilon}^N(-) \tag{15.1.2.3-23}$$

We can also write the differential form of the (15.1.2.3-20) foreground controls as:

$$\underline{\gamma}^N(+_c) = \underline{\gamma}^N(-) + \underline{\gamma}_c^N \qquad (15.1.2.3\text{-}24)$$

$$\underline{\varepsilon}^N(+_c) = \underline{\varepsilon}^N(-) + \underline{\varepsilon}_c^N \qquad (15.1.2.3\text{-}25)$$

Substituting (15.1.2.3-24) - (15.1.2.3-25) into (15.1.2.3-22), applying (15.1.2.3-17) for $\underline{\gamma}_c^N$, and then substituting $\underline{\psi}^N(-)$ from (15.1.2.3-23) shows that:

$$\begin{aligned}\underline{\psi}^N(+_c) &= \left(\underline{\gamma}^N(-) + \underline{\gamma}_c^N\right) - \left(\underline{\varepsilon}^N(-) + \underline{\varepsilon}_c^N\right) \\ &= \left(\underline{\gamma}^N(-) + \underline{\psi}_c^N + \underline{\varepsilon}_c^N\right) - \left(\underline{\varepsilon}^N(-) + \underline{\varepsilon}_c^N\right) \\ &= \underline{\gamma}^N(-) - \underline{\varepsilon}^N(-) + \underline{\psi}_c^N = \underline{\psi}^N(-) + \underline{\psi}_c^N\end{aligned} \qquad (15.1.2.3\text{-}26)$$

Thus, execution of the stipulated $\underline{\gamma}_c^N$, $\underline{\varepsilon}_c^N$ controls to the foreground achieves the goal of completely resetting $\underline{\psi}^N$ as specified.

Performing the same analysis for the velocity and position controls we first find from (12.2.2-5) and (12.2.3-24):

$$\begin{aligned}\delta\underline{V}^N &= \delta\underline{v}^N + \underline{\varepsilon}^N \times \underline{v}^N \\ \delta\underline{R}^N &= R\left(\underline{\varepsilon}^N \times \underline{u}_{ZN}^N\right) + \delta h \, \underline{u}_{ZN}^N\end{aligned} \qquad (15.1.2.3\text{-}27)$$

whence:

$$\begin{aligned}\delta\underline{V}^N(+_c) &= \delta\underline{v}^N(+_c) + \underline{\varepsilon}^N(+_c) \times \underline{v}^N \\ \delta\underline{V}^N(-) &= \delta\underline{v}^N(-) + \underline{\varepsilon}^N(-) \times \underline{v}^N \\ \delta\underline{R}^N(+_c) &= R\left(\underline{\varepsilon}^N(+_c) \times \underline{u}_{ZN}^N\right) + \delta h(+_c) \, \underline{u}_{ZN}^N \\ \delta\underline{R}^N(-) &= R\left(\underline{\varepsilon}^N(-) \times \underline{u}_{ZN}^N\right) + \delta h(-) \, \underline{u}_{ZN}^N\end{aligned} \qquad (15.1.2.3\text{-}28)$$

The differential form of the (15.1.2.3-20) foreground velocity and altitude controls is:

$$\begin{aligned}\delta\underline{v}^N(+_c) &= \delta\underline{v}^N(-) + \delta\underline{v}_c^N \\ \delta h(+_c) &= \delta h(-) + \delta h_c\end{aligned} \qquad (15.1.2.3\text{-}29)$$

Combining (15.1.2.3-16), (15.1.2.3-28) - (15.1.2.3-29), (15.1.2.3-25) and (15.1.2.3-18) - (15.1.2.3-19) then shows (upon using the (3.1.1-16) vector triple cross product identity) that:

$$
\begin{aligned}
\delta \underline{V}^N(+_c) &= \left(\delta \underline{v}^N(-) + \delta \underline{v}_c^N\right) + \left(\underline{\varepsilon}^N(-) + \underline{\varepsilon}_c^N\right) \times \underline{v}^N \\
&= \left(\delta \underline{v}^N(-) + \delta \underline{V}_c^N - \underline{\varepsilon}_c^N \times \underline{v}^N\right) + \left(\underline{\varepsilon}^N(-) + \underline{\varepsilon}_c^N\right) \times \underline{v}^N \\
&= \delta \underline{v}^N(-) + \underline{\varepsilon}^N(-) \times \underline{v}^N + \delta \underline{V}_c^N = \delta \underline{V}^N(-) + \delta \underline{V}_c^N
\end{aligned}
$$

(15.1.2.3-30)

$$
\begin{aligned}
\delta \underline{R}^N(+_c) &= R \left[\left(\underline{\varepsilon}^N(-) + \underline{\varepsilon}_c^N\right) \times \underline{u}_{ZN}^N\right] + \left(\delta h(-) + \delta h_c\right) \underline{u}_{ZN}^N \\
&= R \left(\underline{\varepsilon}^N(-) \times \underline{u}_{ZN}^N\right) + \delta h(-)\, \underline{u}_{ZN}^N + R \left[\frac{1}{R}\left(\underline{u}_{ZN}^N \times \delta \underline{R}_c^N\right) \times \underline{u}_{ZN}^N\right] + \left(\underline{u}_{ZN}^N \cdot \delta \underline{R}_c^N\right) \underline{u}_{ZN}^N \\
&= R \left(\underline{\varepsilon}^N(-) \times \underline{u}_{ZN}^N\right) + \delta h(-)\, \underline{u}_{ZN}^N + \delta \underline{R}_c^N - \left(\underline{u}_{ZN}^N \cdot \delta \underline{R}_c^N\right) \underline{u}_{ZN}^N + \left(\underline{u}_{ZN}^N \cdot \delta \underline{R}_c^N\right) \underline{u}_{ZN}^N \\
&= \delta \underline{R}^N(-) + \delta \underline{R}_c^N
\end{aligned}
$$

Thus, execution of the stipulated $\underline{\varepsilon}_c^N$, $\delta \underline{v}_c^N$, δh_c controls to the foreground achieves the goal of completely resetting $\delta \underline{V}^N$, $\delta \underline{R}^N$ as specified.

15.1.2.3.1 Navigation Error Reduction By External Correction

For situations when resetting navigation parameters is not permitted, navigation error corrections can be achieved by setting the control vector to zero in (15.1.2-10) - (15.1.2-12) or (15.1.2-19), (15.1.2-22), (15.1.2-25), and correcting the navigation outputs. For the example in the previous section, Equations (15.1.2.3-10) or (15.1.2.3-13) would be implemented as an output function, vis. (for (15.1.2.3-13)):

$$
\begin{aligned}
C_{B_{Out}}^L &= \left[I - \frac{\sin \gamma_c}{\gamma_c} \left(\underline{\gamma}_c^L \times\right) + \frac{(1 - \cos \gamma_c)}{\gamma_c^2} \left(\underline{\gamma}_c^L \times\right)^2 \right] C_B^L \\
C_{N_{Out}}^E &= C_N^E \left[I + \frac{\sin \varepsilon_c}{\varepsilon_c} \left(\underline{\varepsilon}_c^N \times\right) + \frac{(1 - \cos \varepsilon_c)}{\varepsilon_c^2} \left(\underline{\varepsilon}_c^N \times\right)^2 \right]
\end{aligned}
\qquad (15.1.2.3.1\text{-}1)
$$

where

$C_{N_{Out}}^E$, $C_{B_{Out}}^L$ = Corrected forms of C_N^E, C_B^L used for calculating position and attitude output parameters.

The basic C_N^E, C_B^L parameters in the INS computer would not be adjusted by this process, hence, would continue to contain error as represented in the Kalman filter estimated error state vector.

15.1.2.4 TIMING AND SYNCHRONIZATION OF THE OBSERVATION/ MEASUREMENT AND KALMAN FILTER CYCLE

For the Kalman filter configurations in the ideal control reset format (Equations (15.1.2-6) - (15.1.2-13)) and in the delayed reset format (Equations (15.1.2-19) - (15.1.2-26)), we have implicitly assumed that the observation/measurement is taken and the controls applied at the same time instant t_n (i.e., the start of the Kalman filter n^{th} computation cycle). In many applications, the control cannot be applied precisely at the observation/measurement time due to time delay from when the control message is issued by the Kalman filter until it is received and applied (particularly for controls applied to external aiding devices). For measurements taken from an external aiding device, the data from the aiding device, although sampled at a specified time point, may have a "time stamp" (i.e., time for which the data corresponds) that is earlier than the sample time. Additionally, Section 15.1.2 (following Equation (15.1.2-28)) discusses a variation to the delayed control reset method (for improved control reset accuracy) in which control application is delayed until the state transition matrix is calculated and the control vector has been propagated to the last (most recent) Kalman cycle time. For simplicity, the latter variation will not be included in this section's discussion. The following discussion will be based on the assumption that the Kalman filter resides within the INS computer, hence, has immediate access to INS navigation data parameters.

To avoid the possibility of Kalman filter performance degradation (and potential instability), it is important that the Kalman filter model accurately reflects the actual time that the controls are applied, that the Kalman observation $\underline{Z}_{Obs_n}$ (Equation (15.1-3)) be formed from a comparison between INS and aiding device parameters (ξ_{INS} and ξ_{Aid}) with corresponding time stamps, and that the Kalman measurement model is based on the observation time stamp. Since the INS parameters at the Kalman data sample time generally have a time stamp equal to the sample time (assuming regular high frequency INS parameter updating), the time stamp for the ξ_{INS} and ξ_{Aid} sampled data will generally not correspond, with the ξ_{INS} time stamp being at a later time. In order to create time stamp correspondence, two alternatives are possible; extrapolating the ξ_{Aid} data to the ξ_{INS} time stamp, or interpolating the ξ_{INS} data to the ξ_{Aid} time stamp. The extrapolation method may seem appealing because it results in data correspondence at the computer specified sample time, which can then be synchronized to occur at the Kalman cycle time. Then Kalman filter Equations (15.1.2-19) - (15.1.2-26) could be applied to the observation formed from the sampled ξ_{INS} and extrapolated ξ_{Aid} data.

Unfortunately, extrapolation of ξ_{Aid} data is typically a noisy process, hence, would generally introduce unacceptable measurement noise into $\underline{Z}_{Obs_n}$. Therefore, the interpolation method is preferred in which a past saved ξ_{INS} data time history is interpolated to the ξ_{Aid} time stamp once the time stamp is known at the ξ_{Aid} sample time. This typically involves saving an INS data time history in a past time window that is as long (in time) as the maximum anticipated time difference (latency) from the ξ_{Aid} data time stamp to the ξ_{Aid} data sample time. Because INS data is typically relatively noise free, and because interpolation is generally a smoothing process, forming $\underline{Z}_{Obs}$ from ξ_{Aid} and interpolated ξ_{INS} data results in an observation vector with relatively low measurement noise. However, $\underline{Z}_{Obs}$ now has a time stamp corresponding to the sampled ξ_{Aid} time stamp, a time in the past, and a time which can vary from measurement to measurement.

To accommodate the interpolation form of the observation and to allow for latency from the time that control commands are issued to the time they are received and applied, the Kalman filter structures described in Section 15.1.2 must be revised. For example, consider the reformulated Kalman filter configuration represented by the following calculations in the sequence shown:

At t_{n+s}, Sample INS Saved Data History And Aiding Device Data For The Kalman Filter (15.1.2.4-1)

$$\underline{Z}_{Obs_n} = f\left(\xi_{INS_n}, \xi_{Aid_n}\right) \quad \text{Based On Interpolated Saved Data} \tag{15.1.2.4-2}$$

$$\underline{\tilde{x}}_{n-1+c}(-) = \Phi_{n-1+c,\,n-1}\, \underline{\tilde{x}}_{n-1}(+_e) \tag{15.1.2.4-3}$$

$$\underline{\tilde{x}}_{n-1+c}(+_c) = \underline{\tilde{x}}_{n-1+c}(-) + \underline{u}_{c_{n-1+c}} \tag{15.1.2.4-4}$$

$$\underline{\tilde{x}}_n(-) = \Phi_{n,\,n-1+c}\, \underline{\tilde{x}}_{n-1+c}(+_c) \tag{15.1.2.4-5}$$

$$\underline{\tilde{z}}_n = H_n\, \underline{\tilde{x}}_n(-) \tag{15.1.2.4-6}$$

$$\underline{\tilde{x}}_n(+_e) = \underline{\tilde{x}}_n(-) + K_n\left(\underline{Z}_{Obs_n} - \underline{\tilde{z}}_n\right) \tag{15.1.2.4-7}$$

$$\underline{u}_{c_{n+c}} = \text{function of } \underline{\tilde{x}}_n(+_e) \tag{15.1.2.4-8}$$

At t_{n+u}, Transmit Control Vector And Its Specified Application Time (t_{n+c}) To Aiding Device (15.1.2.4-9)

$$\xi_{INS_{n+c}}(+c) = \xi_{INS_{n+c}}(-) + g_{INS}\left(\xi_{INS_{n+c}}(-),\, \underline{u}_{c_{n+c}}\right)$$
$$\xi_{Aid_{n+c}}(+c) = \xi_{Aid_{n+c}}(-) + g_{Aid}\left(\xi_{INS_{n+c}}(-),\, \underline{u}_{c_{n+c}}\right) \qquad (15.1.2.4\text{-}10)$$

$$\tilde{\underline{x}}_0 = 0 \qquad (15.1.2.4\text{-}11)$$

where

n = Kalman filter cycle time index which is now defined to correspond in time with the ξ_{Aid} time stamp.

$n{+}s$ = Subscript indicating the time point following t_n (the ξ_{Aid} time stamp) that data for the Kalman filter is sampled.

$n{+}u$ = Subscript indicating the time point following t_{n+s} (the Kalman filter data sample time) that the n cycle Kalman filter computations are completed including the last step of calculating the control vector $\underline{u}_c$.

$n{+}c$ = Subscript indicating the time point following t_{n+u} that the control vector is applied.

$n{-}1{+}c$ = Subscript indicating the time point following t_{n-1+s} (the previous ξ_{Aid} sample time) that the control vector was applied.

$\Phi_{n-1+c,\,n-1}$ = State transition matrix that propagates the error state vector from time t_{n-1} to the following control time t_{n-1+c}.

$\Phi_{n,\,n-1+c}$ = State transition matrix that propagates the error state vector from the t_{n-1+c} control time to t_n.

Equations (15.1.2.4-1) - (15.1.2.4-11) represent a revised form of delayed reset equations (15.1.2-19) - (15.1.2-26) that explicitly account for observation/measurement interpolation and associated delayed control resets. In particular, the data sampling, control transmission and control application times (t_{n+s}, t_{n+u}, and t_{n+c}) in Equations (15.1.2.4-1) - (15.1.2.4-11) are now defined to occur at three distinct time points (no longer at the same time instant). Note, also, that sampling of the Kalman filter input data in Equation (15.1.2.4-1) and applying the controls in Equations (15.1.2.4-10) are the only operations in Equations (15.1.2.4-1) - (15.1.2.4-11) that have to be performed "instantaneously" in real time. The remainder of the computations, including calculation of the observation vector and the Kalman gain for Equation (15.1.2.4-7) as prescribed in Section 15.1.2.1, can be performed leisurely in the order indicated for completion by the next sample time. With such an organization, the sample time can be defined to occur after completion of Kalman processing and control operations.

The control vector $\underline{u}_c$ for (15.1.2.4-10) and (15.1.2.4-4) is computed following Kalman update operations based on the value for the estimated error state vector at the last measurement time stamp (i.e., the current Kalman n cycle start time). Note in Equation (15.1.2.4-9) that the future control application time (t_{n+c}) is transmitted with the control vector $\underline{u}_c$. This allows the Kalman filter to specify the control time, including allowances to accommodate potential time delay from when the control message is sent to when it is received and applied. The net result is that $\underline{u}_c$ will be applied precisely at the Kalman specified t_{n+c} in accordance with the Equation (15.1.2.4-4) model.

Equation (15.1.2.4-2) for the observation is based on interpolating the INS data to the past aiding device time stamp (as described previously) which, by definition, is also the Kalman filter cycle time n. Implementation of Equations (15.1.2.4-1) - (15.1.2.4-10) in the order indicated is only possible through the saved past INS data interpolation process used to form the observation (and state transition and process noise matrices as discussed above). To assure accurate interpolation for any aiding device time stamp, it is important that INS data save operations be initiated following the previous specified control vector application time t_{n-1+c} (to avoid the presence of control reset transients in the data set used for interpolation), and the saved INS data time window must be longer than the maximum potential latency in the aiding device time stamp. Accurate interpolation to any past aiding device time stamp is thereby assured.

Equations (15.1.2.4-3) - (15.1.2.4-6) reflect the effect of the control reset operation at t_{n+c} on error state vector propagation from n-1 to n-1+c to n, and forming the observation vector at cycle time n. Calculation of the $\Phi_{n-1+c,n-1}$ and $\Phi_{n,n-1+c}$ matrices for Equations (15.1.2.4-3) and (15.1.2.4-5) can be performed (as in Section 15.1.2.1.1.1) by m cycle propagation from n-1 to n-1+c and from n-1+c to n. The method would be as follows. We first define the running error state dynamic matrix integral from the start of Kalman filter operations:

$$\mathrm{IA}(t) \equiv \int_0^t A(t)\, dt \qquad (15.1.2.4\text{-}12)$$

and then note from (15.1.2.1.1.1-12) that:

$$\Delta\Phi_m \equiv \int_{t_{m-1}}^{t_m} A(t)\, dt = \mathrm{IA}(t_m) - \mathrm{IA}(t_{m-1}) \qquad (15.1.2.4\text{-}13)$$

Equation (15.1.2.4-13) shows how $\Delta\Phi_m$ can be calculated as the difference between the m-1 and m values of the running integral IA(t). The $\Delta\Phi_m$ term is used in Equations (15.1.2.1.1.1-11) and (15.1.2.1.1.1-1) to calculate the Φ_n state transition matrix by propagation across m cycles. The same method can be used to calculate $\Phi_{n-1+c,n-1}$ and $\Phi_{n,n-1+c}$, except that the m cycle propagations would be from the previous Kalman cycle time t_{n-1} (i.e., the aiding device previous output data time stamp) to the previous control time (t_{n-1+c}), and from t_{n-1+c} to the current Kalman cycle time t_n. Using the (15.1.2.4-13) structure to compute $\Delta\Phi_m$, the $IA(t_m)$ terms used would be calculated at specified fixed time interval m cycle points between Kalman cycles, and at the t_n Kalman cycle times (the t_{n-1+c} time point would be specified to occur on an m cycle). The $IA(t_n)$ Kalman cycle value would be calculated from saved IA(t) data that was sampled at the n+s sample time, then interpolated to t_n, the aiding device output data time stamp (i.e., the same process used for interpolating the INS data in forming the observation vector). Once $\Phi_{n-1+c,n-1}$ and $\Phi_{n,n-1+c}$ are evaluated by this method, the Φ_n matrix for error state uncertainty covariance propagation (e.g., Equation (15.1.2.1.1.3-37)) can be calculated from (15.1.1-10) as:

$$\Phi_n = \Phi_{n,n-1+c}\,\Phi_{n-1+c,n-1} \tag{15.1.2.4-14}$$

If the more elaborate m cycle P covariance and $\underline{x}$ error state vector propagation approach is required as discussed in Section 15.1.2.1.1.3 (e.g., Equations (15.1.2.1.1.3-28) - (15.1.2.1.1.3-31)), the previous technique can also be applied to the Equation (15.1.2.1.1.3-21) terms, vis.:

$$\Delta\Phi_{\lambda\lambda_m} \equiv IA(t_m) - IA(t_{m-1}) \qquad IA(t) \equiv \int_0^t A(t)\, dt$$
$$\Delta\Phi_{\lambda y_m} \equiv IQ(t_m) - IQ(t_{m-1}) \qquad IQ(t) \equiv \int_0^t G_P(t)\, Q_{P_{Dens}}(t)\, G_P(t)^T\, dt \tag{15.1.2.4-15}$$

With Equations (15.1.2.4-15), the $IA(t_m)$ and $IQ(t_m)$ would be calculated at specified m cycle time points between Kalman cycles, and at the t_n Kalman cycle times. The $IA(t_n)$, $IQ(t_n)$ values would be calculated from saved IA(t), IQ(t) data that was sampled at the n+s sample time, then interpolated to the aiding device time stamp.

15.1.3 SUBOPTIMAL KALMAN FILTERS

The general Kalman filter configurations discussed in Section 15.1.2 were based on the implied assumption that analytical models for the error states accurately characterize the effects present on the input observation vector. In a practical Kalman filter design, the analytical models for error state propagation, estimation, control and measurement are approximations to the true error state dynamics (known as the "real world" model). The principal reason that approximations are used is to reduce the dimensionality of the error state vector to minimize computer throughput associated with Kalman filter matrix computational operations. Thus, we find, for example, that the error states included in the Kalman filter are only those that have significant impact on Kalman filter performance objectives. Nevertheless, the computed gain matrix is still based on the "optimal" form of Equation (15.1.2.1-28). Because of approximations in the error state vector dynamic and measurement models (including the linearization process used in the error model development that neglects second order error effects), the performance of the actual Kalman filter will be less accurate than if the exact "real world" model was utilized. Thus, the performance will be sub-optimum, hence, the actual filter based on the approximate model is known as a "sub-optimal" Kalman filter.

The first step in the design of a Kalman filter is to write the error state dynamic and measurement equations for the "real world" model. Note, that even the "real world" model is not the actual true world model due to the limited knowledge of the design analyst regarding all pertinent error effects and their analytical formulation. Once the "real world" model is defined, the associated "optimal" Kalman filter configuration is evaluated by simulation of the real world with the Kalman gains based on the real world model. The "optimal" performance results so obtained then serve as a reference from which to judge "sub-optimal" filter performance based on the approximate error-state/measurement model. The Kalman filter design process, thereby, becomes a "cut-and-try" operation in which the approximate model is iteratively adjusted, then tested, until the achieved sub-optimal performance is reasonably close to the ideal optimal performance. The simulation program used in the design iteration process is based on covariance error analysis techniques (as described in Section 16.2.7), which generally provides sensitivity outputs that help the designer identify the dominant terms affecting Kalman filter performance. Use of the sensitivity data provides a basis for modifications to the approximate error-state/measurement model for improved Kalman filter performance.

Some of the techniques used to simplify the error state dynamic/measurement model for the suboptimal Kalman filter include error state reduction, approximating neglected error states by process noise, approximating the integrated coupled process noise matrix as a constant, and truncation of the state transition and integrated coupled process noise matrix expansion series. Added process noise is also used at times to approximately account for second order error effects in the Kalman filter error model. In the "considered variable" approach (Section 15.1.2.1.1 following Equation (15.1.2.1.1-4)), suboptimal performance is the result of intentionally not estimating certain error states (by zeroing their rows in the Kalman gain matrix), while still accounting for their presence in the Kalman filter covariance matrix (used in cases when there is sufficient uncertainty in the error state dynamics that estimation implementation may be too inaccurate). An important rule to bear in mind during the design process is that error states to be included are those that significantly affect the measurement (i.e., error states that are "observable"). Not including such error states can result in mis-estimating error states that have been included in the filter's error state model (due to "filter misinterpretation" of measurement data input signatures). On the other hand, if two error states have identical dynamic signatures on the measurement, including both as separate error states can lead to mis-estimating each (also classified as an "observability" problem). In this case, a combined error state is sometimes incorporated in which one error state is used to represent the combined effect of both error states. In some cases, it may be desired to estimate certain error states that may not have good individual observability under normal operating conditions.

Intentional maneuvers can be introduced in such cases to enhance observability so that these error effects can be more quickly estimated (e.g., the transfer alignment problem discussed in Section 15.2.2).

In the final analysis, previous experience and a good understanding of error state propagation characteristics are important factors for the Kalman filter designer to assure a successful and expeditious Kalman filter design process.

15.1.4 KALMAN FILTER SOFTWARE VALIDATION

Although a Kalman filter is generally a complex software package, its validation process can be fairly straight-forward because of its fundamental underlying structure. The Kalman filter elements are well defined analytically and can be validated individually based on their intrinsic properties. Once the elements are validated, the proper operation of the filter is assured through its theoretical structure.

As an example, consider the following operations that can be performed using specialized test simulators for validating the Kalman filter algorithms defined by Equations (15.1.2-6) - (15.1.2-13):

- The state transition matrix Φ_n, measurement $\underline{z}_n$, and observation $\underline{Z}_{Obs_n}$ algorithms can be validated by operating Equations (15.1.2-6) - (15.1.2-13) "open loop" (i.e., setting the Kalman gain K_n and control vector $\underline{u}_c$ to zero) using simulators for ξ_{INS_n} and ξ_{Aid_n}. The ξ_{INS_n} simulator would consist of the strapdown inertial navigation algorithms upon which Φ_n is based. The ξ_{Aid_n} simulator would be built onto a previously validated trajectory generator (e.g., Chapter 17); the trajectory generator would also provide the strapdown inertial sensor inputs to ξ_{INS_n}. The Kalman filter error state vector $\tilde{\underline{x}}_n$ components would be initialized to some arbitrary non-zero value; the same error values would be inserted into the ξ_{INS_n}, ξ_{Aid_n} parameters. Under these conditions, the Kalman filter measurement $\underline{z}_n$ calculated with (15.1.2-8) should track the observation vector $\underline{Z}_{Obs_n}$ computed with (15.1.2-6), resulting in a zero value for the measurement residual $\underline{Z}_{Obs_n}$ - $\tilde{\underline{z}}_n$ (within the fundamental linearization error in $\underline{z}_n$). A zero measurement residual validates the Φ_n, $\underline{z}_n$ and $\underline{Z}_{Obs_n}$ algorithms and associated timing structure in the simulation implementation.
- The covariance propagation algorithm (with process noise set to zero) can be validated as part of the previous process by initially setting the covariance matrix equal to the

arbitrarily defined $\underline{\tilde{x}}_n$ error state vector times its transpose. The covariance matrix would then be propagated without resets using the Equation (15.1.2.1.1-12) algorithm or a Section 15.1.2.1.1.3 equivalent. The propagated covariance matrix should then equal the propagated error state vector times its transpose.

- The algorithms for calculating the Kalman gain matrix K_n and resetting the covariance matrix can be validated by comparing the covariance reset algorithm output with the output from an equivalent alternative algorithm based on the analytical form of K_n (e.g., the Equation (15.1.2.1.1-1) Joseph's form compared with the Equation (15.1.2.1.1-4) optimal form). The results should be identical.

- The basic estimation capability of the Kalman filter can be validated by disabling the control vector (setting $\underline{u}_c$ to zero) and allowing the Kalman filter to estimate $\underline{\tilde{x}}_n$ in the presence of selected values for the error state components initially imbedded in ξ_{INS_n} and ξ_{Aid_n}. For this test, the process and measurement noise matrices in the Kalman filter covariance propagation/reset routines would be set to zero to heighten sensitivity (and better account for the error condition being simulated).

- Kalman filter estimation capability in the presence of process and measurement noise can be validated by repeating the previous test, but with random noise (from a software noise generator at the Kalman filter specified white noise source amplitudes) applied appropriately to the ξ_{INS_n}, ξ_{Aid_n} models (for process noise) and to the $\underline{Z}_{Obs_n}$ routine (for measurement noise). The Kalman filter process and measurement noise matrices would also be active for this test. In parallel, a "truth model" error state vector history would be generated using the same noise and initial conditions applied to a simulated version of error state dynamic Equation (15.1-1). The uncertainty in the Kalman filter estimated error state vector is evaluated by comparing the filter error state vector estimate with the "truth model" error state vector. Repeated runs with different random noise generator initial "seeds" provides an ensemble history of the error state uncertainty. The ensemble average of the uncertainty times its transpose (at common time points) should match the corresponding filter covariance matrix history.

- The control vector $\underline{u}_c$ interface in control reset Equations (15.1.2-11) and (15.1.2-12) can be validated by assigning an arbitrary value to $\underline{u}_c$ and applying it to the previous equations. If the control reset equations and the measurement/observation algorithms are consistent, the measurement residual $\underline{Z}_{Obs_n} - \underline{\tilde{z}}_n$ should be unaffected by the control reset application.

15.1.5 SUPPLEMENTARY TOPICS

In this section we will develop some additional fundamental material that can be generally useful in deriving analytic solutions to Kalman filter estimation problems. The principal results from this section form the basis for the analytical solutions developed in Chapter 14, Section 14.6 for the quasi-stationary fine alignment Kalman filter.

The additional fundamental background material to be presented derives a continuous form Kalman filter error covariance propagation/reset equation based on an alternate form of the Kalman gain equation (derived using the so-called "matrix inversion lemma"). A general solution to the continuous Kalman filter equation is then developed based on the solution approach outlined in Reference 6, Section 4.6.

15.1.5.1 THE MATRIX INVERSION LEMMA

The matrix inversion states that:

$$\left(A^{-1} + B^T C^{-1} B\right)^{-1} = A - A B^T \left(C + B A B^T\right)^{-1} B A \qquad (15.1.5.1\text{-}1)$$

or equivalently:

$$\left(A^{-1} + B^T C^{-1} B\right)\left[A - A B^T \left(C + B A B^T\right)^{-1} B A\right] = I \qquad (15.1.5.1\text{-}2)$$

where

A, C = Square non-singular matrices (i.e., non-zero determinant).

B = An arbitrary matrix of appropriate dimensions.

To prove the validity of Equation (15.1.5.1-2) we first expand the left side as:

$$\begin{aligned}\left(A^{-1} + B^T C^{-1} B\right)&\left[A - A B^T \left(C + B A B^T\right)^{-1} B A\right] \\ &= I - B^T \left(C + B A B^T\right)^{-1} B A + B^T C^{-1} B A \\ &\quad - B^T C^{-1} B A B^T \left(C + B A B^T\right)^{-1} B A\end{aligned} \qquad (15.1.5.1\text{-}3)$$

where

I = Identity matrix.

The farthest right term in (15.1.5.1-3) can be further expanded as follows:

$$\begin{aligned}B^T C^{-1} B A B^T &\left(C + B A B^T\right)^{-1} B A \\ &= B^T C^{-1} \left(- C + C + B A B^T\right)\left(C + B A B^T\right)^{-1} B A \\ &= - B^T \left(C + B A B^T\right)^{-1} B A + B^T C^{-1} B A\end{aligned} \qquad (15.1.5.1\text{-}4)$$

Substituting (15.1.5.1-4) into (15.1.5.1-3) shows that (15.1.5.1-3) equals I which proves (15.1.5.1-2) (and (15.1.5.1-1)).

15.1.5.2 ALTERNATE KALMAN GAIN EXPRESSION

The basic equation for the Kalman gain is given by (15.1.2.1-28) repeated below:

$$K_n = P_n(-)\, H_n^T \left(H_n\, P_n(-)\, H_n^T + G_{M_n}\, R_n\, G_{M_n}^T\right)^{-1} \qquad (15.1.5.2\text{-}1)$$

where

- n = Subscript indicating value for parameter at Kalman filter computer cycle n.
- K = Kalman gain matrix.
- $P(-)$ = Error state covariance matrix immediately prior to processing the measurement.
- H = Measurement matrix.
- G_M = Measurement noise dynamic coupling matrix.
- R = Covariance matrix of independent measurement noise sources.

An alternative form for K_n can be developed using the (15.1.5.1-1) matrix inversion lemma with A, B and C identified as P, H and $G_M\, R\, G_M^T$ respectively:

$$\left[P^{-1} + H^T \left(G_M\, R\, G_M^T\right)^{-1} H\right]^{-1} = P - P\, H^T \left(G_M\, R\, G_M^T + H\, P\, H^T\right)^{-1} H\, P \qquad (15.1.5.2\text{-}2)$$

Multiplying (15.1.5.2-2) by $H^T \left(G_M\, R\, G_M^T\right)^{-1}$ finds with expansion:

$$\left[P^{-1} + H^T \left(G_M\, R\, G_M^T\right)^{-1} H\right]^{-1} H^T \left(G_M\, R\, G_M^T\right)^{-1}$$

$$= \left[P - P\, H^T \left(G_M\, R\, G_M^T + H\, P\, H^T\right)^{-1} H\, P\right] H^T \left(G_M\, R\, G_M^T\right)^{-1}$$

$$= P\, H^T \left(G_M\, R\, G_M^T\right)^{-1} - P\, H^T \left(G_M\, R\, G_M^T + H\, P\, H^T\right)^{-1} H\, P\, H^T \left(G_M\, R\, G_M^T\right)^{-1} \qquad (15.1.5.2\text{-}3)$$

$$= P\, H^T \left(G_M\, R\, G_M^T\right)^{-1}$$
$$\quad - P\, H^T \left(H\, P\, H^T + G_M\, R\, G_M^T\right)^{-1} \left(H\, P\, H^T + G_M\, R\, G_M^T - G_M\, R\, G_M^T\right) \left(G_M\, R\, G_M^T\right)^{-1}$$

$$= P\, H^T \left(H\, P\, H^T + G_M\, R\, G_M^T\right)^{-1}$$

Augmenting the (15.1.5.2-3) result with n subscripts and (-) designation for P then, upon comparison with (15.1.5.2-1, obtains an alternative equivalent expression for the Kalman gain:

$$K_n = \left[P_n(-)^{-1} + H_n^T \left(G_{M_n} R_n G_{M_n}^T \right)^{-1} H_n \right]^{-1} H_n^T \left(G_{M_n} R_n G_{M_n}^T \right)^{-1} \qquad (15.1.5.2\text{-}4)$$

15.1.5.3 THE CONTINUOUS FORM KALMAN FILTER

The following sections derive the covariance and error state vector propagation/reset differential time rate equations for a continuous form Kalman filter that matches the performance characteristics of the conventional discrete Kalman filter discussed previously. The method will be to first present the equations for a hypothetical discrete Kalman filter having an update rate (m cycle) that is faster than the discrete filter (at n cycle rate) we are attempting to match. The continuous filter is then determined by letting the m cycle filter update time period go to zero in the limit. In the process, equivalencies will be developed for the noise parameters in the continuous filter compared to the conventional discrete n cycle Kalman filter to achieve equivalent performance.

15.1.5.3.1 Covariance Propagation/Reset Rate Equation

Consider a Kalman filter operating at a faster cycle time than the n cycle rate. The error state covariance propagation/reset equations for the faster filter are from (15.1.2.1.1-4) and (15.1.2.1.1-12):

$$\begin{aligned} P_m(-) &= \Phi_m P_{m-1}(+) \Phi_m^T + Q_m \\ P_m(+) &= P_m(-) - K_{Fst_m} H_m P_m(-) \end{aligned} \qquad (15.1.5.3.1\text{-}1)$$

where

m = Subscript indicating values for the parameters at the faster filter m cycle rate.

$P(-)$, $P(+)$ = Error state covariance matrix immediately before (-) and immediately after (+) processing the measurement.

Φ = Error state transition matrix.

Q = Integrated process noise covariance matrix.

K_{Fst} = Kalman gain matrix for the faster Kalman filter.

Let us define the following:

$$\begin{aligned} \Delta P_{\Phi Q_m} &\equiv \Phi_m P_{m-1}(+) \Phi_m^T + Q_m - P_{m-1}(+) \\ \Delta P_{K_m} &\equiv - K_{Fst_m} H_m P_m(-) \end{aligned} \qquad (15.1.5.3.1\text{-}2)$$

where

$\Delta P_{\Phi Q_m}, \Delta P_{K_m}$ = Change in P over cycle m caused by P propagation ($\Delta P_{\Phi Q_m}$) and Kalman reset (ΔP_{K_m}).

From the $\Delta P_{\Phi Q_m}, \Delta P_{K_m}$ definitions, we see from (15.1.5.3.1-1) that:

$$P_m(-) = P_{m-1}(+) + \Delta P_{\Phi Q_m} \qquad P_m(+) = P_m(-) + \Delta P_{K_m} \tag{15.1.5.3.1-3}$$

or in combination:

$$P_m(+) = P_{m-1}(+) + \Delta P_{\Phi Q_m} + \Delta P_{K_m} \tag{15.1.5.3.1-4}$$

The total change in P over an m cycle is given by:

$$P_m(+) - P_{m-1}(+) = \Delta P_{\Phi Q_m} + \Delta P_{K_m} \tag{15.1.5.3.1-5}$$

For the Kalman gain matrix in the (15.1.5.3.1-2) ΔP_{K_m} expression, we use the alternative (15.1.5.2-4) form:

$$K_{Fst_m} = \left[P_m(-)^{-1} + H_m^T \left(G_{M_m} R_{Fst_m} G_{M_m}^T \right)^{-1} H_m \right]^{-1} H_m^T \left(G_{M_m} R_{Fst_m} G_{M_m}^T \right)^{-1} \tag{15.1.5.3.1-6}$$

where

Fst = Designation for parameter value associated with the m cycle faster Kalman filter.

R_{Fst_m} = Covariance matrix of independent measurement noise sources associated with the m cycle faster Kalman filter measurement. We anticipate the m cycle measurement noise R_{Fst} to be different from, though related to, the n cycle measurement noise R.

We substitute (15.1.5.3.1-6) into the (15.1.5.3.1-2) ΔP_{K_m} expression and expand:

$$\begin{aligned} \Delta P_{K_m} &= - \left[P_m(-)^{-1} + H_m^T \left(G_{M_m} R_{Fst_m} G_{M_m}^T \right)^{-1} H_m \right]^{-1} H_m^T \left(G_{M_m} R_{Fst_m} G_{M_m}^T \right)^{-1} H_m P_m(-) \\ &= - \left[P_m(-)^{-1} + H_m^T \left(G_{M_m} R_{Fst_m} G_{M_m}^T \right)^{-1} H_m \right]^{-1} \left[- P_m(-)^{-1} \right. \\ &\qquad \left. + P_m(-)^{-1} + H_m^T \left(G_{M_m} R_{Fst_m} G_{M_m}^T \right)^{-1} H_m \right] P_m(-) \end{aligned} \tag{15.1.5.3.1-7}$$

(Continued)

$$= - P_m(-) + \left[P_m(-)^{-1} + H_m^T \left(G_{M_m} R_{Fst_m} G_{M_m}^T \right)^{-1} H_m \right]^{-1}$$

$$= - P_m(-) + \left\{ P_m(-)^{-1} \left[I + P_m(-) H_m^T \left(G_{M_m} R_{Fst_m} G_{M_m}^T \right)^{-1} H_m \right] \right\}^{-1} \quad \text{(15.1.5.3.1-7) (Continued)}$$

$$= \left\{ \left[I + P_m(-) H_m^T \left(G_{M_m} R_{Fst_m} G_{M_m}^T \right)^{-1} H_m \right]^{-1} - I \right\} P_m(-)$$

Post-multiplying the (15.1.5.3.1-7) result by $P_m(-)^{-1}$ and rearranging yields:

$$I + \Delta P_{K_m} P_m(-)^{-1} = \left[I + P_m(-) H_m^T \left(G_{M_m} R_{Fst_m} G_{M_m}^T \right)^{-1} H_m \right]^{-1} \quad \text{(15.1.5.3.1-8)}$$

or taking the inverse:

$$\left(I + \Delta P_{K_m} P_m(-)^{-1} \right)^{-1} = I + P_m(-) H_m^T \left(G_{M_m} R_{Fst_m} G_{M_m}^T \right)^{-1} H_m \quad \text{(15.1.5.3.1-9)}$$

We now make the approximation that for the short m cycle time, ΔP_{K_m} will be small compared to $P_m(-)$, hence, $\Delta P_{K_m} P_m(-)^{-1}$ will be small compared to the identity matrix I. Then by truncated binomial series expansion we can approximate:

$$\left(I + \Delta P_{K_m} P_m(-)^{-1} \right)^{-1} \approx I - \Delta P_{K_m} P_m(-)^{-1} \quad \text{(15.1.5.3.1-10)}$$

with which (15.1.5.3.1-9) becomes:

$$\Delta P_{K_m} P_m(-)^{-1} \approx - P_m(-) H_m^T \left(G_{M_m} R_{Fst_m} G_{M_m}^T \right)^{-1} H_m \quad \text{(15.1.5.3.1-11)}$$

or

$$\Delta P_{K_m} \approx - P_m(-) H_m^T \left(G_{M_m} R_{Fst_m} G_{M_m}^T \right)^{-1} H_m P_m(-) \quad \text{(15.1.5.3.1-12)}$$

Equation (15.1.5.3.1-12) is equivalently:

$$\Delta P_{K_m} = - P_m(-) H_m^T \left(G_{M_m} R_{Fst_m} T_m G_{M_m}^T \right)^{-1} H_m P_m(-) T_m \quad \text{(15.1.5.3.1-13)}$$

We define:

$$R^{\blacklozenge}{}_m \equiv R_{Fst_m} T_m \quad \text{(15.1.5.3.1-14)}$$

with which (15.1.5.3.1-13) becomes:

$$\Delta P_{K_m} = - P_m(-)\, H_m^T \left(G_{M_m}\, R^{\blacklozenge}{}_m\, G_{M_m}^T \right)^{-1} H_m\, P_m(-)\, T_m \qquad (15.1.5.3.1\text{-}15)$$

where

T_m = Time period for the m Kalman filter propagation/update cycle.

$R^{\blacklozenge}$ = High speed Kalman filter equivalent measurement noise matrix.

Equation (15.1.5.3.1-15) can be used to establish an approximate equivalency relationship between $R^{\blacklozenge}{}_m$ and R_n, the discrete n cycle Kalman filter measurement noise matrix. The equivalency is obtained by summing (15.1.5.3.1-15) over an n cycle and setting the result to the equivalent ΔP_K obtained from the n cycle filter based on the same approximations (using (15.1.5.3.1-12) with m = n and $R_{Fst} = R$):

$$\Delta P_{K_n} = \sum_{m=1}^{r} \Delta P_{K_m} = - \sum_{m=1}^{r} P_m(-)\, H_m^T \left(G_{M_m}\, R^{\blacklozenge}{}_m\, G_{M_m}^T \right)^{-1} H_m\, P_m(-)\, T_m$$

$$= - P_n(-)\, H_n^T \left(G_{M_n}\, R_n\, G_{M_n}^T \right)^{-1} H_n\, P_n(-) \qquad (15.1.5.3.1\text{-}16)$$

where

r = Number of m cycles in an n cycle.

Over the t_{n-1} to t_n time interval, we then approximate:

$$P_m(-) \approx P_n(-) \qquad (15.1.5.3.1\text{-}17)$$

and set:

From t_{n-1} To t_n:

$$G_{M_m} = G_{M_n} \qquad H_m = H_n \qquad R^{\blacklozenge}{}_m = \text{Constant} \equiv R^{\blacklozenge}{}_n \qquad (15.1.5.3.1\text{-}18)$$

Then, recognizing that $r\, T_m = T_n$, where

T_n = Time period for the n Kalman filter propagation/update cycle.

Equation (15.1.5.3.1-16) becomes:

$$P_n(-)\, H_n^T \left(G_{M_n}\, R^{\blacklozenge}{}_n\, G_{M_n}^T \right)^{-1} H_n\, P_n(-)\, r\, T_m$$

$$= P_n(-)\, H_n^T \left(G_{M_n}\, \frac{R^{\blacklozenge}{}_n}{T_n}\, G_{M_n}^T \right)^{-1} H_n\, P_n(-) \qquad (15.1.5.3.1\text{-}19)$$

$$= P_n(-)\, H_n^T \left(G_{M_n}\, R_n\, G_{M_n}^T \right)^{-1} H_n\, P_n(-)$$

Equation (15.1.5.3.1-19) yields the approximate equivalency relationship between $R^{\blacklozenge}$ and R from which we see that:

$$R^{\blacklozenge}{}_n = R_n \, T_n \qquad (15.1.5.3.1\text{-}20)$$

Returning to our original task of developing the continuous Kalman filter covariance propagation/reset equation, we divide (15.1.5.3.1-15) by the m cycle time (which will eventually become a time derivative as we let the m cycle time go to zero in the limit):

$$\frac{\Delta P_{K_m}}{T_m} = - P_m(-) \, H_m^T \left(G_{M_m} \, R^{\blacklozenge}{}_m \, G_{M_m}^T \right)^{-1} H_m \, P_m(-) \qquad (15.1.5.3.1\text{-}21)$$

Returning to (15.1.5.3.1-5) and dividing by T_m finds:

$$\frac{P_m(+) - P_{m-1}(+)}{T_m} = \frac{\Delta P_{\Phi Q_m}}{T_m} + \frac{\Delta P_{K_m}}{T_m} \qquad (15.1.5.3.1\text{-}22)$$

The equation for the continuous Kalman filter is obtained from (15.1.5.3.1-22) by substituting $\Delta P_{\Phi Q_m}$ from (15.1.5.3.1-2), Equation (15.1.5.3.1-21) for $\frac{\Delta P_{K_m}}{T_m}$, letting T_m go to zero in the limit, and applying the limit procedure used in Section 15.1.2.1.1 for the $\frac{\Delta P_{\Phi Q_m}}{T_m}$ term. The final result so obtained is:

$$\begin{aligned} \dot{P}(t) = \; & A(t) \, P(t) + P(t) \, A(t)^T + G_P(t) \, Q_{P_{Dens}}(t) \, G_P(t)^T \\ & - P(t) \, H(t)^T \left(G_M(t) \, R^{\blacklozenge}(t) \, G_M(t)^T \right)^{-1} H(t) \, P(t) \end{aligned} \qquad (15.1.5.3.1\text{-}23)$$

where

$A(t)$ = Error state dynamic matrix.

$Q_{P_{Dens}}(t)$ = Process noise density matrix.

$G_P(t)$ = Process noise dynamic coupling matrix.

$R^{\blacklozenge}(t)$ = Continuous form Kalman filter uncorrelated measurement noise matrix.

Equation (15.1.5.3.1-23) is the continuous form Kalman filter covariance propagation/reset equation. The analytical form of (15.1.5.3.1-23) is of a general mathematical type known as the matrix Ricatti equation which has classical solution approaches (such as outlined in Section

15.1.5.4). We further note based on Equations (15.1.5.3.1-18) and (15.1.5.3.1-20) that $R^{\blacklozenge}(t)$, $H(t)$ and $G_M(t)$ in (15.1.5.3.1-23) can be approximated as continuous functions with:

$$R^{\blacklozenge}(t_{n-1/2}) = R_n T_n \qquad H(t_{n-1/2}) = H_n \qquad G_M(t_{n-1/2}) = G_{M_n} \tag{15.1.5.3.1-24}$$

Finally, as an interpretation of $R^{\blacklozenge}(t)$ and its relation to R_n, we can also define $R^{\blacklozenge}(t)$ as the density of a white noise process by considering the state dynamic equation:

$$\dot{\underline{v}}_M(t) \equiv \frac{1}{T_n}\, \underline{n}_M^{\blacklozenge}(t) \qquad \underline{v}_M(t_{n-1}) = 0 \tag{15.1.5.3.1-25}$$

where

$\underline{n}_M^{\blacklozenge}(t)$ = White noise measurement vector whose density is $R^{\blacklozenge}(t)$.

$\underline{v}_M(t)$ = Integral of $\underline{n}_M^{\blacklozenge}(t)$ scaled by $1 / T_n$.

The integral of (15.1.5.3.1-25) over the t_{n-1} to t_n time interval is:

$$\underline{v}_M(t_n) = \frac{1}{T_n} \int_{t_{n-1}}^{t_n} \underline{n}_M^{\blacklozenge}(t)\, dt \tag{15.1.5.3.1-26}$$

Thus, $\underline{v}_M(t)$ evaluated at t_n is the average over the t_{n-1} to t_n time interval of the white noise vector $\underline{n}_M^{\blacklozenge}(t)$. We also know from the integral of generalized Equation (15.1.2.1.1-30) (by setting $A(t) = 0$, $G_P(t) = 1 / T_n$, $Q_{P_{Dens}}(t) = R^{\blacklozenge}(t)$ and $\underline{v}_M(t_{n-1}) = 0$ to correspond with (15.1.5.3.1-25)) that the covariance of $\underline{v}_M(t_n)$ is:

$$\mathcal{E}\left(\underline{v}_M(t_n)\, \underline{v}_M(t_n)^T\right) = \frac{1}{T_n^2} \int_{t_{n-1}}^{t_n} R^{\blacklozenge}(t)\, dt = \frac{1}{T_n^2} R^{\blacklozenge}{}_n T_n = \frac{R^{\blacklozenge}{}_n}{T_n} \tag{15.1.5.3.1-27}$$

which with (15.1.5.3.1-20) shows that:

$$\mathcal{E}\left(\underline{v}_M(t_n)\, \underline{v}_M(t_n)^T\right) = R_n \tag{15.1.5.3.1-28}$$

But from Equation (15.1.2.1-13):

$$R_n = \mathcal{E}\left(\underline{n}_{M_n}\, \underline{n}_{M_n}^T\right) \tag{15.1.5.3.1-29}$$

where

$\underline{n}_{M_n}$ = Discrete Kalman filter measurement noise at cycle n.

Hence, from (15.1.5.3.1-28) - (15.1.5.3.1-29) and (15.1.5.3.1-26):

$$\underline{n}_{M_n} = \frac{1}{T_n} \int_{t_{n-1}}^{t_n} \underline{n}_M^{\blacklozenge}(t)\, dt \qquad (15.1.5.3.1\text{-}30)$$

We see from (15.1.5.3.1-29) - (15.1.5.3.1-30), that (15.1.5.3.1-20) is equivalent to representing the n cycle Kalman filter measurement noise $\underline{n}_{M_n}$ as the average value of the integral over the Kalman filter n cycle update interval, of a white process noise vector $\underline{n}_M^{\blacklozenge}(t)$ having density $R^{\blacklozenge}{}_n$ given by (15.1.5.3.1-20), or as the continuous $R^{\blacklozenge}(t)$ form with the $R^{\blacklozenge}(t_{n-1/2})$ points defined by (15.1.5.3.1-24).

15.1.5.3.2 Error State Vector Propagation/Reset Rate Equation

The continuous form Kalman filter error state vector propagation/reset rate equations have three elements; error state vector propagation/reset, estimated error state vector propagation/reset and estimated error state uncertainty vector propagation/reset. Let's deal with element one first. For our fast m cycle Kalman filter using the idealized control reset Kalman filter configuration as a base, we can write from (15.1.2-2) and (15.1.2-5) in combination:

$$\underline{x}_m(+_c) = \Phi_m\, \underline{x}_{m-1}(+_c) + \underline{w}_m + \underline{u}_{cFst_m} \qquad (15.1.5.3.2\text{-}1)$$

where

$(+_c)$ = Designation for value immediately after a control reset.

$\underline{u}_{cFst}$ = Control vector for the fast m cycle Kalman filter.

The change in $\underline{x}$ over an update cycle divided by the m cycle time is from (15.1.5.3.2-1):

$$\frac{\underline{x}_m(+_c) - \underline{x}_{m-1}(+_c)}{T_m} = \frac{1}{T_m}\left[\left(\Phi_m - I\right) \underline{x}_{m-1}(+_c) + \underline{w}_m\right] + \frac{1}{T_m}\, \underline{u}_{cFst_m} \qquad (15.1.5.3.2\text{-}2)$$

Applying (15.1.2.1.1-17) - (15.1.2.1.1-18) and letting Tm go to zero in the limit then finds for the continuous form error state vector propagation/reset equation:

$$\dot{\underline{x}}(t) = A(t)\, \underline{x}(t) + G_P(t)\, \underline{n}_P(t) + \underline{u}_c^{\blacklozenge}(t) \qquad (15.1.5.3.2\text{-}3)$$

with

$$\underline{u}_c^{\blacklozenge}(t) \equiv \left(\frac{1}{T_m}\,\underline{u}_{cFst_m}\right)_{T_m \to 0} \qquad (15.1.5.3.2\text{-}4)$$

where

$\underline{u}_c^{\blacklozenge}(t)$ = Continuous form Kalman filter control vector.

The continuous form estimated error state vector propagation/reset equation is developed from the m cycle version of (15.1.2-7) - (15.1.2-9), (15.1.2-12) and (15.1.2-3) using the linearized $\underline{z}$ in place of $\underline{Z}_{Obs}$:

$$\begin{aligned}
\tilde{\underline{x}}_m(-) &= \Phi_m\,\tilde{\underline{x}}_{m-1}(+_c) \\
\tilde{\underline{z}}_{Fst_m} &= H_m\,\tilde{\underline{x}}_m(-) \\
\tilde{\underline{x}}_m(+_e) &= \tilde{\underline{x}}_m(-) + K_{Fst_m}\left(\underline{z}_{Fst_m} - \tilde{\underline{z}}_{Fst_m}\right) \\
\tilde{\underline{x}}_m(+_c) &= \tilde{\underline{x}}_m(+_e) + \underline{u}_{cFst_m}
\end{aligned} \qquad (15.1.5.3.2\text{-}5)$$

with

$$\underline{z}_{Fst_m} = H_m\,\underline{x}_m(-) + G_{M_m}\,\underline{n}_{MFst_m} \qquad (15.1.5.3.2\text{-}6)$$

where

$(-), (+_e)$ = Designation for values immediately before (-) and immediately after $(+_e)$ an estimation (e) update.

$\underline{z}_{Fst}, \tilde{\underline{z}}_{Fst}$ = Linearized measurement and Kalman filter expected value of the measurement for the fast m cycle filter.

$\underline{n}_{MFst_m}$ = Measurement noise for the fast m cycle Kalman filter, which from Section 15.1.5.3.1, has covariance matrix R_{Fst_m}.

Equations (15.1.5.3.2-5) combined are:

$$\tilde{\underline{x}}_m(+_c) = \Phi_m\,\tilde{\underline{x}}_{m-1}(+_c) + K_{Fst_m}\left(\underline{z}_{Fst_m} - H_m\,\tilde{\underline{x}}_m(-)\right) + \underline{u}_{cFst_m} \qquad (15.1.5.3.2\text{-}7)$$

The change in $\tilde{\underline{x}}$ over an update cycle divided by the m cycle time is from (15.1.5.3.2-7):

$$\frac{\tilde{\underline{x}}_m(+_c) - \tilde{\underline{x}}_{m-1}(+_c)}{T_m} = \frac{1}{T_m}\left(\Phi_m - I\right)\tilde{\underline{x}}_{m-1}(+_c) + \frac{1}{T_m}\,K_{Fst_m}\left(\underline{z}_{Fst_m} - H_m\,\tilde{\underline{x}}_m(-)\right) + \frac{1}{T_m}\,\underline{u}_{cFst_m}$$

(15.1.5.3.2-8)

The $\frac{1}{T_m} K_{Fst_m}$ term in (15.1.5.3.2-8) is handled by comparing (15.1.5.3.1-15) with the ΔP_{K_m} expression in (15.1.5.3.1-2) which shows that:

$$\frac{1}{T_m} K_{Fst_m} = K^{\blacklozenge}{}_m \tag{15.1.5.3.2-9}$$

in which

$$K^{\blacklozenge}{}_m \equiv P_m(-)\, H_m^T \left(G_{M_m} R^{\blacklozenge}{}_m\, G_{M_m}^T \right)^{-1} \tag{15.1.5.3.2-10}$$

where

$K^{\blacklozenge}_m$ = High speed Kalman filter equivalent Kalman gain matrix at cycle m.

The $\underline{z}_{Fst_m}$ term in (15.1.5.3.2-8) is provided by (15.1.5.3.2-6), but with the $\underline{n}_{MFst_m}$ term replaced by an equivalency relation based on the $\underline{n}^{\blacklozenge}_M(t)$ white noise type measurement vector concept introduced at the end of Section 15.1.5.3.1 (which defined $\underline{n}^{\blacklozenge}_M(t)$ as a white noise vector whose density is $R^{\blacklozenge}(t)$). We know from Equation (15.1.5.3.1-14) that $\underline{n}_{MFst_m}$ must have a covariance matrix R_{Fst_m} equal to $\frac{1}{T_m} R^{\blacklozenge}{}_m$. Based on the form of (15.1.5.3.1-26), let's try the following and see if it fits the previous covariance requirement:

$$\underline{n}_{MFst_m} = \frac{1}{T_m} \int_{t_{m-1}}^{t_m} \underline{n}^{\blacklozenge}_M(t)\, dt \tag{15.1.5.3.2-11}$$

Using (15.1.5.3.2-11) and the approach leading to (15.1.5.3.1-27), we see that:

$$R_{Fst_m} \equiv \mathcal{E}\left(\underline{n}_{MFst_m}\, \underline{n}_{MFst_m}^T \right) = \frac{1}{T_m^2} R^{\blacklozenge}{}_m\, T_m = \frac{1}{T_m} R^{\blacklozenge}{}_m \tag{15.1.5.3.2-12}$$

which matches the (15.1.5.3.1-14) requirement. Thus, (15.1.5.3.2-11) is a valid representation of $\underline{n}_{MFst_m}$.

We now substitute (15.1.5.3.2-9) in (15.1.5.3.2-8), (15.1.5.3.2-11) in (15.1.5.3.2-6), let T_m go to zero in the limit, and define $\underline{z}_{Fst_m}$ in the limit as the continuous form Kalman filter measurement. The result with (15.1.5.3.2-10) is the continuous form estimated error state vector propagation/reset equation:

$$\dot{\tilde{\underline{x}}}(t) = \left(A(t) - K^{\blacklozenge}(t)\, H(t)\right) \tilde{\underline{x}}(t) + K^{\blacklozenge}(t)\, \underline{z}^{\blacklozenge}(t) + \underline{u}_c^{\blacklozenge}(t) \qquad (15.1.5.3.2\text{-}13)$$

in which

$$\underline{z}^{\blacklozenge}(t) = H(t)\, \underline{x}(t) + G_M(t)\, \underline{n}_M^{\blacklozenge}(t)$$
$$K^{\blacklozenge}(t) = P(t)\, H(t)^T \left(G_M(t)\, R^{\blacklozenge}(t)\, G_M(t)^T\right)^{-1} \qquad (15.1.5.3.2\text{-}14)$$

where

$\underline{z}^{\blacklozenge}(t)$ = Continuous form Kalman filter measurement vector.

with $H(t)$, $G_M(t)$, $R^{\blacklozenge}(t)$ provided by (15.1.5.3.1-24) and $\underline{u}_c^{\blacklozenge}(t)$ as defined in (15.1.5.3.2-4).

Using the (15.1.2.1-1) error state uncertainty definition

$$\Delta\underline{x}(t) \equiv \tilde{\underline{x}}(t) - \underline{x}(t) \qquad (15.1.5.3.2\text{-}15)$$

we subtract (15.1.5.3.2-3) from (15.1.5.3.2-13) to obtain the continuous form error state uncertainty vector propagation/reset equation:

$$\Delta\dot{\underline{x}}(t) = \left(A(t) - K^{\blacklozenge}(t)\, H(t)\right) \Delta\underline{x}(t) - G_P(t)\, \underline{n}_P(t) + K^{\blacklozenge}(t)\, G_M(t)\, \underline{n}_M^{\blacklozenge}(t) \qquad (15.1.5.3.2\text{-}16)$$

Note, as usual, that the previous error state uncertainty vector propagation/reset rate equation is independent of the control vector $\underline{u}_c^{\blacklozenge}(t)$, as is error state uncertainty covariance rate Equation (15.1.5.3.1-23) which represents the statistical equivalent of (15.1.5.3.2-16).

We conclude this section with a discussion of the continuous form error state vector and estimated error state vector rate equations under two control vector conditions: no control which corresponds to a pure estimation problem, and complete idealized closed-loop control which controls the estimated error state vector to zero. Under the former conditions, $\underline{u}_c^{\blacklozenge}(t) = 0$, and Equations (15.1.5.3.2-3) and (15.1.5.3.2-13) become:

<u>For A Pure Estimator</u>:

$$\dot{\underline{x}}(t) = A(t)\, \underline{x}(t) + G_P(t)\, \underline{n}_P(t) \qquad (15.1.5.3.2\text{-}17)$$
$$\dot{\tilde{\underline{x}}}(t) = \left(A(t) - K^{\blacklozenge}(t)\, H(t)\right) \tilde{\underline{x}}(t) + K^{\blacklozenge}(t)\, \underline{z}^{\blacklozenge}(t)$$

with (15.1.5.3.2-14) for $K^{\blacklozenge}(t)$ and $\underline{z}^{\blacklozenge}(t)$.

For the complete idealized closed-loop control case, Equation (15.1.2.3-1) applies which for the high speed filter m cycle rate is:

$$\underline{u}_{cFst_m} = -\, \tilde{\underline{x}}_m(+_e) \qquad (15.1.5.3.2\text{-}18)$$

Then from (15.1.5.3.2-5):

$$\tilde{\underline{x}}_m(+_c) = \tilde{\underline{x}}_m(+_e) + \underline{u}_{cFst_m} = 0 \quad \rightarrow \quad \tilde{\underline{x}}_m(-) = \Phi_m\, \tilde{\underline{x}}_{m-1}(+_c) = 0 \qquad (15.1.5.3.2\text{-}19)$$

or in the limit:

$$\tilde{\underline{x}}(t) = 0 \qquad (15.1.5.3.2\text{-}20)$$

Substituting (15.1.5.3.2-20) in Equation (15.1.5.3.2-13) shows that:

$$\underline{u}_c^{\blacklozenge}(t) = -\, K^{\blacklozenge}(t)\, \underline{z}^{\blacklozenge}(t) \qquad (15.1.5.3.2\text{-}21)$$

Finally, we substitute (15.1.5.3.2-21) in (15.1.5.3.2-3) to obtain:

<u>For Complete Closed-Loop Control:</u>

$$\dot{\underline{x}}(t) = A(t)\, \underline{x}(t) + G_P(t)\, \underline{n}_P(t) - K^{\blacklozenge}(t)\, \underline{z}^{\blacklozenge}(t) \qquad (15.1.5.3.2\text{-}22)$$

with (15.1.5.3.2-14) for $K^{\blacklozenge}(t)$ and $\underline{z}^{\blacklozenge}(t)$.

An alternative version of (15.1.5.3.2-22) substitutes $\underline{z}^{\blacklozenge}(t)$ from (15.1.5.3.2-14) yielding:

$$\dot{\underline{x}}(t) = \left(A(t) - K^{\blacklozenge}(t)\, H(t)\right) \underline{x}(t) + G_P(t)\, \underline{n}_P(t) - K^{\blacklozenge}(t)\, G_M(t)\, \underline{n}_M^{\blacklozenge}(t) \qquad 15.1.5.3.2\text{-}23)$$

Note from (15.1.5.3.2-15) and (15.1.5.3.2-20) that the effect of the closed-loop control is to set $\underline{x}(t) = -\, \Delta\underline{x}(t)$. If we now substitute $\underline{x}(t) = -\, \Delta\underline{x}(t)$ and its derivative in (15.1.5.3.2-3), we find that the result is Equation (15.1.5.3.2-16).

15.1.5.4 GENERAL SOLUTION TO THE CONTINUOUS KALMAN FILTER COVARIANCE EQUATION

Equation (15.1.5.3.1-23) for the continuous form Kalman filter is a non-linear differential equation for the error state covariance matrix P(t). The analytical form of the continuous Kalman filter is of the mathematical type known as the "matrix Ricatti equation". Its general solution can be found by applying the method of Reference 6 - Section 4.6 as outlined below (as in Section 15.1.2.1.1.3, but expanded to include the W(t) term):

We first define:

$$\underline{\lambda}(t) \equiv P(t)\, \underline{y}(t) \tag{15.1.5.4-1}$$

$$\dot{\underline{y}}(t) \equiv -A(t)^T\, \underline{y}(t) + H(t)^T \left(G_M(t)\, R^{\blacklozenge}(t)\, G_M(t)^T\right)^{-1} H(t)\, P(t)\, \underline{y}(t) \tag{15.1.5.4-2}$$

where

$\underline{\lambda}(t)$, $\underline{y}(t)$ = Column matrix vectors of the same dimension as P(t).

Substituting (15.1.5.4-1) in (15.1.5.4-2) obtains:

$$\dot{\underline{y}}(t) = -A(t)^T\, \underline{y}(t) + H(t)^T \left(G_M(t)\, R^{\blacklozenge}(t)\, G_M(t)^T\right)^{-1} H(t)\, \underline{\lambda}(t) \tag{15.1.5.4-3}$$

Taking the derivative of (15.1.5.4-1) finds:

$$\dot{\underline{\lambda}}(t) = \dot{P}(t)\, \underline{y}(t) + P(t)\, \dot{\underline{y}}(t) \tag{15.1.5.4-4}$$

Substituting (15.1.5.4-2) and (15.1.5.3.1-23) in (15.1.5.4-4) yields:

$$\begin{aligned}\dot{\underline{\lambda}}(t) = &\Big[A(t)\, P(t) + P(t)\, A(t)^T + G_P(t)\, Q_{P_{Dens}}(t)\, G_P(t)^T \\ &\quad - P(t)\, H(t)^T \left(G_M(t)\, R^{\blacklozenge}(t)\, G_M(t)^T\right)^{-1} H(t)\, P(t)\Big]\, \underline{y}(t) \\ &+ P(t)\Big[-A(t)^T\, \underline{y}(t) + H(t)^T \left(G_M(t)\, R^{\blacklozenge}(t)\, G_M(t)^T\right)^{-1} H(t)\, P(t)\, \underline{y}(t)\Big] \\ = &\; A(t)\, P(t)\, \underline{y}(t) + G_P(t)\, Q_{P_{Dens}}(t)\, G_P(t)^T\, \underline{y}(t)\end{aligned} \tag{15.1.5.4-5}$$

or with (15.1.5.4-1):

$$\dot{\underline{\lambda}}(t) = G_P(t)\, Q_{P_{Dens}}(t)\, G_P(t)^T\, \underline{y}(t) + A(t)\, \underline{\lambda}(t) \tag{15.1.5.4-6}$$

Equations (15.1.5.4-3) and (15.1.5.4-6) in matrix form are:

$$\begin{bmatrix} \dot{\underline{y}}(t) \\ \dot{\underline{\lambda}}(t) \end{bmatrix} = \begin{bmatrix} -A(t)^T & H(t)^T \left(G_M(t)\; R^{\blacklozenge}(t)\; G_M(t)^T\right)^{-1} H(t) \\ G_P(t)\; Q_{P_{Dens}}(t)\; G_P(t)^T & A(t) \end{bmatrix} \begin{bmatrix} \underline{y}(t) \\ \underline{\lambda}(t) \end{bmatrix} \tag{15.1.5.4-7}$$

Equation (15.1.5.4-7) is a linear homogeneous differential equation for the $\underline{y}(t)$, $\underline{\lambda}(t)$ vector parameters whose solution is readily obtained by classical Section 15.1.1 techniques:

$$\begin{bmatrix} \underline{y}(t) \\ \underline{\lambda}(t) \end{bmatrix} = \begin{bmatrix} \Phi_{yy}(t) & \Phi_{y\lambda}(t) \\ \Phi_{\lambda y}(t) & \Phi_{\lambda\lambda}(t) \end{bmatrix} \begin{bmatrix} \underline{y}(0) \\ \underline{\lambda}(0) \end{bmatrix} \tag{15.1.5.4-8}$$

where

$\underline{y}(0)$, $\underline{\lambda}(0)$ = Initial values for $\underline{y}(t)$, $\underline{\lambda}(t)$ at time t = 0.

$\Phi_{yy}(t)$, $\Phi_{y\lambda}(t)$, $\Phi_{\lambda y}(t)$, $\Phi_{\lambda\lambda}(t)$ = Elements of the state transition matrix associated with the Equation (15.1.5.4-7) state dynamic matrix for propagation of initial conditions on the $\underline{y}(t)$, $\underline{\lambda}(t)$ vectors to the current time.

The error state covariance matrix P(t) is then determined by application of (15.1.5.4-1) to the individual $\underline{y}(t)$ and $\underline{\lambda}(t)$ rows of (15.1.5.4-8):

$$\begin{aligned} \underline{\lambda}(t) &= \Phi_{\lambda y}(t)\, \underline{y}(0) + \Phi_{\lambda\lambda}(t)\, \underline{\lambda}(0) = \Phi_{\lambda y}(t)\, \underline{y}(0) + \Phi_{\lambda\lambda}(t)\, P(0)\, \underline{y}(0) \\ &= \left(\Phi_{\lambda y}(t) + \Phi_{\lambda\lambda}(t)\, P(0)\right) \underline{y}(0) \\ &= P(t)\, \underline{y}(t) = P(t) \left(\Phi_{yy}(t)\, \underline{y}(0) + \Phi_{y\lambda}(t)\, P(0)\, \underline{y}(0)\right) \\ &= P(t) \left(\Phi_{yy}(t) + \Phi_{y\lambda}(t)\, P(0)\right) \underline{y}(0) \end{aligned} \tag{15.1.5.4-9}$$

where

P(0) = P(t) at time t = 0.

Since (15.1.5.4-9) is valid for any $\underline{y}(0)$, it follows that:

$$P(t) \left(\Phi_{yy}(t) + \Phi_{y\lambda}(t)\, P(0)\right) = \Phi_{\lambda y}(t) + \Phi_{\lambda\lambda}(t)\, P(0) \tag{15.1.5.4-10}$$

Rearrangement then solves for P(t):

$$P(t) = \left(\Phi_{\lambda y}(t) + \Phi_{\lambda\lambda}(t)\, P(0)\right) \left(\Phi_{yy}(t) + \Phi_{y\lambda}(t)\, P(0)\right)^{-1} \tag{15.1.5.4-11}$$

Equation (15.1.5.4-11) with (15.1.5.4-7) - (15.1.5.4-8) for state transition matrix component definition is the general solution to Equation (15.1.5.3.1-23) for the continuous form Kalman filter. These generally apply for situations in which the coupled-measurement-noise matrix $G_M(t)\, R^{\blacklozenge}(t)\, G_M(t)^T$ is non-singular (see (15.1.5.4-7)). For cases when $G_M(t)\, R^{\blacklozenge}(t)\, G_M(t)^T$ is singular (i.e., zero determinant) these equations have no general solution, however, under particular conditions, can be converted to an equivalent non-singular form that can be solved. The following subsection deals with the singular coupled-measurement-noise case in which $R^{\blacklozenge}(t)$ is zero under particular constraint conditions.

15.1.5.4.1 General Covariance Response With Zero Measurement Noise Under Particular Constraint Conditions

In this section we will seek the general solution to continuous Kalman filter covariance rate Equation (15.1.5.3.1-23) for a singular coupled-measurement-noise condition in which the following particular constraint conditions exist: 1) The measurement noise $R^{\blacklozenge}(t)$ is zero, 2) The error states directly affecting the measurement have no coupling into the other error states, 3) The process noise feeding the non-measurement input error states has no coupling to the measurement-input error states, and 4) The process noise feeding the measurement-input error states has no coupling to the non-measurement-input error states. Analytically, the conditions we impose follow, including a re-statement of the associated generalized continuous form error state dynamic and measurement equations from (15.1.5.3.2-17) and (15.1.5.3.2-14):

$$\dot{\underline{x}}(t) = A(t)\,\underline{x}(t) + G_P(t)\,\underline{n}_P(t) \qquad (15.1.5.4.1\text{-}1)$$

$$\underline{z}^{\blacklozenge}(t) = H(t)\,\underline{x}(t) + G_M(t)\,\underline{n}_M^{\blacklozenge}(t) \qquad (15.1.5.4.1\text{-}2)$$

$$H(t) = \begin{bmatrix} 0 & H_{\mathcal{M}}(t) \end{bmatrix} \qquad \underline{n}_M^{\blacklozenge}(t) = 0 \qquad R^{\blacklozenge}(t) = 0$$

$$\underline{x}(t) \equiv \begin{bmatrix} \underline{x}^*(t) \\ \underline{x}_{\mathcal{M}}(t) \end{bmatrix} \qquad A(t) = \begin{bmatrix} A^*(t) & 0 \\ A_{\mathcal{M}*}(t) & A_{\mathcal{M}\mathcal{M}}(t) \end{bmatrix}$$

$$G_P(t) = \begin{bmatrix} G_{P*}(t) & 0 \\ 0 & G_{P_{\mathcal{M}}}(t) \end{bmatrix} \qquad \underline{n}_P(t) = \begin{bmatrix} \underline{n}_{P*}(t) \\ \underline{n}_{P_{\mathcal{M}}}(t) \end{bmatrix} \qquad (15.1.5.4.1\text{-}3)$$

$$Q_{P_{Dens}}(t) = \begin{bmatrix} Q_{P*_{Dens}}(t) & 0 \\ 0 & Q_{P_{\mathcal{M}\,Dens}}(t) \end{bmatrix}$$

where

$\underline{x}_{\mathcal{M}}(t)$ = Error states directly affecting the measurement.

$\underline{x}^*(t)$ = Error states not directly affecting the measurement.

$\underline{n}_{P*}(t)$ = Portion of $\underline{n}_P(t)$ that through $G_{P*}(t)$, feeds $\underline{x}^*(t)$ but not $\underline{x}_{\mathcal{M}}(t)$.

$\underline{n}_{P_{\mathcal{M}}}(t)$ = Portion of $\underline{n}_P(t)$ that through $G_{P_{\mathcal{M}}}(t)$, feeds $\underline{x}_{\mathcal{M}}(t)$ but not $\underline{x}^*(t)$.

$\underline{n}_M^{\blacklozenge}(t)$ = The continuous form measurement noise vector defined as the white noise vector whose density matrix is $R^{\blacklozenge}(t)$, the continuous Kalman filter measurement noise matrix.

$A^*(t)$, $A_{\mathcal{M}*}(t)$, $A_{\mathcal{M}\mathcal{M}}(t)$, $G_{P*}(t)$, $G_{P_{\mathcal{M}}}(t)$, $Q_{P*_{Dens}}(t)$, $Q_{P_{\mathcal{M}\,Dens}}(t)$, $H_{\mathcal{M}}(t)$ =
Partitions of $A(t)$, $G_P(t)$, $Q_{P_{Dens}}(t)$ and $H(t)$ as defined in (15.1.5.4.1-3) based on the indicated $\underline{x}(t)$ error state vector and $\underline{n}_P(t)$ process noise partitions.

Note that the error state vector $\underline{x}(t)$ is defined in (15.1.5.4.1-3) as having $\underline{x}^*(t)$ and $\underline{x}_{\mathcal{M}}(t)$ as separate columns at the top and bottom of $\underline{x}(t)$. Since the ordering of the error states in $\underline{x}(t)$ is arbitrary, the form assumed for $\underline{x}(t)$ in (15.1.5.4.1-3) suffers no loss in generality.

In addition to (15.1.5.4.1-3), we will constrain the measurement vector to have the same or greater than the number of elements in $\underline{x}_{\mathcal{M}}(t)$ (call it $n/x\mathcal{M}$) with the measurement matrix containing at least one partition of $n/x\mathcal{M}$ rows that is non-singular. Thus, we also require as a minimum for at least one of the $n/x\mathcal{M}$ row partitions:

$$\mathrm{Dtr}\left(H_{\mathcal{M}_{n/x\mathcal{M}}}(t)\right) \neq 0 \quad \text{For At Least one of the } H_{\mathcal{M}_{n/x\mathcal{M}}}(t)\text{'s} \qquad (15.1.5.4.1\text{-}4)$$

where

$n/x\mathcal{M}$ = The number of elements in $\underline{x}_{\mathcal{M}}(t)$.

$(\ )_{n/x\mathcal{M}}$ = Square matrix formed from $n/x\mathcal{M}$ rows of ( ). It is assumed that the number of rows in ( ) is greater or equal to $n/x\mathcal{M}$. Thus, $(\ )_{n/x\mathcal{M}}$ can be any of the set of $n/x\mathcal{M}$ rows of ( ).

Dtr ( ) = Determinant of the square matrix ( ).

From the general form of continuous covariance rate Equation (15.1.5.3.1-23) (and its transformed linear form (15.1.5.4-1) and (15.1.5.4-7)), we see that for zero $R^{\blacklozenge}(t)$, the equations are singular. By imposing the (15.1.5.4.1-3) and (15.1.5.4.1-4) constraints, we will show in this section how these equations can be converted to an equivalent singular free form in which the $H(t)^T \left(G_M(t)\ R^{\blacklozenge}(t)\ G_M(t)^T\right)^{-1} H(t)$ term is replaced by a similar term, but containing process noise. The form of the singular free equations is identical to the original Equations (15.1.5.4-1) and (15.1.5.4-7), hence, the form of (15.1.5.4-8) and (15.1.5.4-11), the original equation solution for $P(t)$ will also apply for the revised singular free equations.

We begin our analysis by defining a partitioned error state uncertainty vector based on $\underline{x}(t)$ in (15.1.5.4.1-3) as:

$$\Delta\underline{x}(t) \equiv \begin{bmatrix} \Delta\underline{x}^*(t) \\ \Delta\underline{x}_{\mathcal{M}}(t) \end{bmatrix} \qquad (15.1.5.4.1\text{-}5)$$

where

$$\Delta\underline{x}^*(t),\ \Delta\underline{x}_{\mathcal{M}}(t) \ = \text{Uncertainty in the Kalman filter's estimate of } \underline{x}^*(t),\ \underline{x}_{\mathcal{M}}(t)\ .$$

For our fast m cycle Kalman filter of Sections 15.1.5.3.1 and 15.1.5.3.2, the general error uncertainty Kalman update equation is from (15.1.2.1-8) with (15.1.2.1-28):

$$\Delta\underline{x}_m(+) \ = \ \left(I - K_{Fst_m}\, H_m\right)\Delta\underline{x}_m(-) + K_{Fst_m}\, G_{M_m}\, \underline{n}_{M_m}$$
$$K_{Fst_m} \ = \ P_m(-)\, H_m^T \left(H_m\, P_m(-)\, H_m^T + G_{M_m}\, R_{Fst_m}\, G_{M_m}^T\right)^{-1} \qquad (15.1.5.4.1\text{-}6)$$

For zero noise, $\underline{n}_{M_m}$ and R_{Fst_m} are zero, and Equations (15.1.5.4.1-6) combined become:

$$\Delta\underline{x}_m(+) \ = \ \left(I - K_{Fst_m}\, H_m\right)\Delta\underline{x}_m(-) \ = \ \left[I - P_m(-)\, H_m^T \left(H_m\, P_m(-)\, H_m^T\right)^{-1} H_m\right]\Delta\underline{x}_m(-) \qquad (15.1.5.4.1\text{-}7)$$

Now multiply (15.1.5.4.1-7) by H_m:

$$H_m\ \Delta\underline{x}_m(+) \ = \ \left[H_m - H_m\, P_m(-)\ H_m^T \left(H_m\, P_m(-)\ H_m^T\right)^{-1} H_m\right]\Delta\underline{x}_m(-)$$
$$= \ \left(H_m - H_m\right)\Delta\underline{x}_m(-) \ = \ 0 \qquad (15.1.5.4.1\text{-}8)$$

or with (15.1.5.4.1-3) for H(t):

$$H_{\mathcal{M}_m}\, \Delta\underline{x}_{\mathcal{M}_m}(+) \ = \ 0 \qquad (15.1.5.4.1\text{-}9)$$

or in the limit for the equivalent continuous (infinitely fast m cycle) Kalman filter:

$$H_{\mathcal{M}}(t)\ \Delta\underline{x}_{\mathcal{M}}(t) \ = \ 0 \qquad (15.1.5.4.1\text{-}10)$$

Based on the (15.1.5.4.1-4) constraint, Equation (15.1.5.4.1-10) shows that:

$$\Delta\underline{x}_{\mathcal{M}}(t) \ = \ 0 \qquad (15.1.5.4.1\text{-}11)$$

Thus, for the hypothesized zero measurement noise conditions, the uncertainty in the error states directly feeding the measurement is zero. Equation (15.1.5.4.1-11) makes sense intuitively when one recognizes from (15.1.5.4.1-2), that for this particular problem, the $\underline{z}^{\blacklozenge}(t)$ measurement vector is equal to measurement noise plus the measurement matrix $H_{\mathcal{M}}(t)$ multiplied by $\underline{x}_{\mathcal{M}}(t)$. For zero coupled-measurement-noise and our (15.1.5.4.1-4) constraint on $H_{\mathcal{M}}(t)$, the $\underline{x}_{\mathcal{M}}(t)$ error state vector is completely observable, hence, can be estimated exactly without error.

General Equation (15.1.2.1-4) defines the covariance matrix P(t) as the expected value of $\Delta\underline{x}(t)$ multiplied by its transpose. Using (15.1.5.4.1-5) with (15.1.5.4.1-11) for $\Delta\underline{x}_{\mathcal{M}}(t)$, we see then that the partitioned form of P(t) is:

$$
\begin{aligned}
P(t) &= \mathcal{E}\left(\begin{bmatrix} \Delta\underline{x}^*(t) \\ \Delta\underline{x}_{\mathcal{M}}(t) \end{bmatrix} \begin{bmatrix} \Delta\underline{x}^*(t)^T & \Delta\underline{x}_{\mathcal{M}}(t)^T \end{bmatrix} \right) \\
&= \begin{bmatrix} \mathcal{E}\left(\Delta\underline{x}^*(t)\ \Delta\underline{x}^*(t)^T\right) & \mathcal{E}\left(\Delta\underline{x}^*(t)\ \Delta\underline{x}_{\mathcal{M}}(t)^T\right) \\ \mathcal{E}\left(\Delta\underline{x}_{\mathcal{M}}(t)\ \Delta\underline{x}^*(t)^T\right) & \mathcal{E}\left(\Delta\underline{x}_{\mathcal{M}}(t)\ \Delta\underline{x}_{\mathcal{M}}(t)^T\right) \end{bmatrix} = \begin{bmatrix} P^*(t) & 0 \\ 0 & 0 \end{bmatrix}
\end{aligned} \qquad (15.1.5.4.1\text{-}12)
$$

where

$P^*(t)$ = Estimated error state $\underline{x}^*(t)$ uncertainty covariance matrix.

Let's apply (15.1.5.4.1-12) in (15.1.5.4-1). We first introduce partitioning for $\underline{\lambda}(t)$ and $\underline{y}(t)$:

$$
\underline{\lambda}(t) \equiv \begin{bmatrix} \underline{\lambda}^*(t) \\ \underline{\lambda}_{\mathcal{M}}(t) \end{bmatrix} \qquad \underline{y}(t) \equiv \begin{bmatrix} \underline{y}^*(t) \\ \underline{y}_{\mathcal{M}}(t) \end{bmatrix} \qquad (15.1.5.4.1\text{-}13)
$$

where

$\underline{\lambda}^*(t)$, $\underline{y}^*(t)$ = Top elements in $\underline{y}(t)$ and $\underline{\lambda}(t)$ of length equal to the length of $\underline{x}^*(t)$.

$\underline{\lambda}_{\mathcal{M}}(t)$, $\underline{y}_{\mathcal{M}}(t)$ = Bottom elements in $\underline{y}(t)$ and $\underline{\lambda}(t)$ of length equal to the length of $\underline{x}_{\mathcal{M}}(t)$.

Substituting (15.1.5.4.1-13) with (15.1.5.4.1-12) for P(t) in (15.1.5.4-1) shows that the equivalent to (15.1.5.4-1) for zero measurement noise is:

$$
\underline{\lambda}^*(t) = P^*(t)\ \underline{y}^*(t) \qquad (15.1.5.4.1\text{-}14)
$$

$$
\underline{\lambda}_{\mathcal{M}}(t) = 0 \qquad (15.1.5.4.1\text{-}15)
$$

We now address Equation (15.1.5.4-6) by first substituting the (15.1.5.4.1-3) partitioned forms of $G_P(t)$ and $Q_{P_{Dens}}(t)$ to find:

$$
G_P(t)\ Q_{P_{Dens}}(t)\ G_P(t)^T = \begin{bmatrix} G_{P^*}(t)\ Q_{P^*_{Dens}}(t)\ G_{P^*}(t)^T & 0 \\ 0 & G_{P_{\mathcal{M}}}(t)\ Q_{P_{\mathcal{M}\,Dens}}(t)\ G_{P_{\mathcal{M}}}(t)^T \end{bmatrix} \qquad (15.1.5.4.1\text{-}16)
$$

Substituting (15.1.5.4.1-16) with A(t) from (15.1.5.4.1-3), $\underline{y}(t)$ and $\underline{\lambda}(t)$ from (15.1.5.4.1-13), and $\underline{\lambda}_{\mathcal{M}}(t)$ from (15.1.5.4.1-15) in (15.1.5.4-6) yields:

$$\dot{\underline{\lambda}}^*(t) = \left(G_{P*}(t)\ Q_{P*_{Dens}}(t)\ G_{P*}(t)^T\right) \underline{y}^*(t) + A^*(t)\ \ \underline{\lambda}^*(t) \qquad (15.1.5.4.1\text{-}17)$$

$$\dot{\underline{\lambda}}_{\mathcal{M}}(t) = 0$$
$$= \left(G_{P_{\mathcal{M}}}(t)\ Q_{P_{\mathcal{M}\,Dens}}(t)\ G_{P_{\mathcal{M}}}(t)^T\right) \underline{y}_{\mathcal{M}}(t) + A_{\mathcal{M}*}(t)\ \underline{\lambda}^*(t) + A_{\mathcal{M}\mathcal{M}}(t)\ \underline{\lambda}_{\mathcal{M}}(t) \qquad (15.1.5.4.1\text{-}18)$$
$$= \left(G_{P_{\mathcal{M}}}(t)\ Q_{P_{\mathcal{M}\,Dens}}(t)\ G_{P_{\mathcal{M}}}(t)^T\right) \underline{y}_{\mathcal{M}}(t) + A_{\mathcal{M}*}(t)\ \underline{\lambda}^*(t)$$

Equation (15.1.5.4.1-18) rearranged is:

$$\underline{y}_{\mathcal{M}}(t) = -\left(G_{P_{\mathcal{M}}}(t)\ Q_{P_{\mathcal{M}\,Dens}}(t)\ G_{P_{\mathcal{M}}}(t)^T\right)^{-1} A_{\mathcal{M}*}(t)\ \underline{\lambda}^*(t) \qquad (15.1.5.4.1\text{-}19)$$

Lastly, we address Equation (15.1.5.4-3) by first noting from the definitions for A(t) and H(t) in (15.1.5.4.1-3) that:

$$A(t)^T \equiv \begin{bmatrix} A^*(t)^T & A_{\mathcal{M}*}(t)^T \\ 0 & A_{\mathcal{M}\mathcal{M}}(t)^T \end{bmatrix}$$

$$H(t)^T\left(G_M(t)\ R^{\blacklozenge}(t)\ G_M(t)^T\right)^{-1} H(t) = \begin{bmatrix} 0 & 0 \\ 0 & H_{\mathcal{M}}(t)^T\left(G_M(t)\ R^{\blacklozenge}(t)\ G_M(t)^T\right)^{-1} H_{\mathcal{M}}(t) \end{bmatrix} \qquad (15.1.5.4.1\text{-}20)$$

Substituting (15.1.5.4.1-20) in (15.1.5.4-3) gives for $\dot{\underline{y}}^*(t)$:

$$\dot{\underline{y}}^*(t) = -A^*(t)^T\ \underline{y}^*(t) - A_{\mathcal{M}*}(t)^T\ \underline{y}_{\mathcal{M}}(t) \qquad (15.1.5.4.1\text{-}21)$$

or with (15.1.5.4.1-19) for $\underline{y}_{\mathcal{M}}(t)$:

$$\dot{\underline{y}}^*(t) = -\ A^*(t)^T\ \underline{y}^*(t) + A_{\mathcal{M}*}(t)^T\left(G_{P_{\mathcal{M}}}(t)\ Q_{P_{\mathcal{M}\,Dens}}(t)\ G_{P_{\mathcal{M}}}(t)^T\right)^{-1} A_{\mathcal{M}*}(t)\ \underline{\lambda}^*(t) \qquad (15.1.5.4.1\text{-}22)$$

Equations (15.1.5.4.1-14), (15.1.5.4.1-17) and (15.1.5.4.1-22) comprise our final result and are summarized below in matrix form:

$$\underline{\lambda}^*(t) \equiv P^*(t)\, \underline{y}^*(t) \qquad (15.1.5.4.1\text{-}23)$$

$$\begin{bmatrix} \dot{\underline{y}}^*(t) \\ \dot{\underline{\lambda}}^*(t) \end{bmatrix} = \begin{bmatrix} -A^*(t)^T & A_{\mathcal{M}^*}(t)^T \left(G_{P_{\mathcal{M}}}(t)\, Q_{P_{\mathcal{M}\,Dens}}(t)\, G_{P_{\mathcal{M}}}(t)^T\right)^{-1} A_{\mathcal{M}^*}(t) \\ \left(G_{P^*}(t)\, Q_{P^*_{Dens}}(t)\, G_{P^*}(t)^T\right) & A^*(t) \end{bmatrix} \begin{bmatrix} \underline{y}^*(t) \\ \underline{\lambda}^*(t) \end{bmatrix}$$

(15.1.5.4.1-24)

If we compare Equations (15.1.5.4.1-23) and (15.1.5.4.1-24) with the non-partitioned Equations (15.1.5.4-1) and (15.1.5.4-7), we see that they are of identical form. In our revised Equations (15.1.5.4.1-23) and (15.1.5.4.1-24), the $\underline{y}^*(t)$, $\underline{\lambda}^*(t)$, $P^*(t)$, $A^*(t)$ and $G_{P^*}(t)\; Q_{P^*_{Dens}}(t)\; G_{P^*}(t)^T$ partitions are used in place of the non-partitioned $\underline{y}(t)$, $\underline{\lambda}(t)$, $P(t)$, $A(t)$ and $G_P(t)\; Q_{P_{Dens}}(t)\; G_P(t)^T$. Of primary significance, however, is that $A_{\mathcal{M}^*}(t)$ is used in place of the measurement matrix $H(t)$, and $G_{P_{\mathcal{M}}}(t)\; Q_{P_{\mathcal{M}\,Dens}}(t)\; G_{P_{\mathcal{M}}}(t)^T$ is used in place of the coupled-measurement-noise matrix $G_M(t)\; R^{\blacklozenge}(t)\; G_M(t)^T$. Thus, we have eliminated the singularity condition caused by $R^{\blacklozenge}(t)$ in (15.1.5.4-7). We can now write the continuous form covariance equation corresponding to (15.1.5.4.1-23) and (15.1.5.4.1-24). The result is obtained by inspection of the non-partitioned covariance rate Equation (15.1.5.3.1-23) and the relationship between its contributing terms and the matrix elements of (15.1.5.4-7). Then from (15.1.5.4.1-24) compared with (15.1.5.4-7) and (15.1.5.3.1-23) we directly write:

$$\begin{aligned} \dot{P}^*(t) = {} & A^*(t)\, P^*(t) + P^*(t)\, A^*(t)^T + G_{P^*}(t)\, Q_{P^*_{Dens}}(t)\, G_{P^*}(t)^T \\ & - P^*(t)\, H_{Rev}(t)^T \left(G_{M_{Rev}}(t)\, R^{\blacklozenge}_{Rev}(t)\, G_{M_{Rev}}(t)^T\right)^{-1} H_{Rev}(t)\, P^*(t) \end{aligned} \qquad (15.1.5.4.1\text{-}25)$$

$$H_{Rev}(t) = A_{\mathcal{M}^*}(t) \qquad G_{M_{Rev}}(t) = G_{P_{\mathcal{M}}}(t) \qquad R^{\blacklozenge}_{Rev}(t) = Q_{P_{\mathcal{M}\,Dens}}(t)$$

where

$H_{Rev}(t)$, $G_{M_{Rev}}(t)$, $R^{\blacklozenge}_{Rev}(t)$ = Revised versions of the continuous form $H(t)$, $G_M(t)$, $R^{\blacklozenge}(t)$ parameters.

Equation (15.1.5.4.1-24) can be derived from (15.1.5.4.1-25) following the identical procedure used to derive (15.1.5.4-7) from (15.1.5.3.1-23).

Similarly, we can write the corresponding continuous form error state uncertainty equation by comparing (15.1.5.4.1-25) with (15.1.5.3.1-23) and its relationship to (15.1.5.3.2-16) and (15.1.5.3.2-14).

$$\Delta\dot{\underline{x}}^*(t) = \left(A^*(t) - K^\blacklozenge_{Rev}(t)\, H_{Rev}(t)\right) \Delta\underline{x}^*(t) - G_{P*}(t)\, \underline{n}_{P*}(t) + K^\blacklozenge_{Rev}(t)\, G_{M_{Rev}}(t)\, \underline{n}^\blacklozenge_{M_{Rev}}(t)$$

$$K^\blacklozenge_{Rev}(t) = P^*(t)\, H_{Rev}(t)^T \left(G_{M_{Rev}}(t)\, R^\blacklozenge_{Rev}(t)\, G_{M_{Rev}}(t)^T\right)^{-1} \qquad (15.1.5.4.1\text{-}26)$$

$$\underline{n}^\blacklozenge_{M_{Rev}}(t) = \underline{n}_{P_{\mathcal{M}}}(t)$$

where

$K^\blacklozenge_{Rev}(t)$ = Revised version of the continuous form $K^\blacklozenge(t)$ Kalman gain matrix.

$\underline{n}^\blacklozenge_{M_{Rev}}(t)$ = Revised version of the continuous form $\underline{n}^\blacklozenge_M(t)$ measurement noise vector corresponding to the $R^\blacklozenge_{Rev}(t)$ continuous form measurement noise matrix definition in Equations (15.1.5.4.1-25).

Finally, by comparison of (15.1.5.4.1-26) with (15.1.5.3.2-16) and its relationship to (15.1.5.3.2-17) and (15.1.5.3.2-14), we can also write the equations for a pure Kalman error state estimator configuration having the (15.1.5.4.1-26) error state uncertainty propagation/reset model:

$$\dot{\underline{x}}^*(t) = A^*(t)\ \underline{x}^*(t) + G_{P*}(t)\ \underline{n}_{P*}(t) \qquad (15.1.5.4.1\text{-}27)$$

$$\underline{z}^\blacklozenge_{Rev}(t) = H_{Rev}(t)\ \underline{x}^*(t) + G_{M_{Rev}}(t)\ \underline{n}^\blacklozenge_{M_{Rev}}(t)$$

$$H_{Rev}(t) = A_{\mathcal{M}*}(t) \qquad G_{M_{Rev}}(t) = G_{P_{\mathcal{M}}}(t) \qquad \underline{n}^\blacklozenge_{M_{Rev}}(t) = \underline{n}_{P_{\mathcal{M}}}(t) \qquad (15.1.5.4.1\text{-}28)$$

$$\dot{\tilde{\underline{x}}}^*(t) = \left(A^*(t) - K^\blacklozenge_{Rev}(t)\, H_{Rev}(t)\right) \tilde{\underline{x}}^*(t) + K^\blacklozenge_{Rev}(t)\, \underline{z}^\blacklozenge_{Rev}(t)$$

where

$\underline{z}^\blacklozenge_{Rev}(t)$ = Revised version of the continuous form $\underline{z}^\blacklozenge(t)$ measurement vector.

The equivalent discrete n cycle measurement equation version of (15.1.5.4.1-28) is with (15.1.5.3.1-30):

$$\underline{z}_{Rev_n} = H_{Rev_n}\, \underline{x}^*_n + G_{M/Rev_n}\, \underline{n}_{M/Rev_n} \qquad \underline{n}_{M/Rev_n} = \frac{1}{T_n}\int_{t_{n-1}}^{t_n} \underline{n}^\blacklozenge_{M_{Rev}}(t)\, dt \qquad (15.1.5.4.1\text{-}29)$$

where

$\underline{z}_{Rev_n}$ = Revised n cycle Kalman filter measurement.

$\underline{n}_{M/Rev_n}$ = Revised n cycle Kalman filter measurement noise.

Equation (15.1.5.4.1-29) is the measurement equation for a revised Kalman filter whose covariance is P^*_n, the upper left partition of the P_n covariance associated with the $\underline{x}^*(t)$ error state partition of $\underline{x}(t)$ at Kalman filter cycle n. Equation (15.1.5.4.1-28) is the corresponding continuous form measurement equation and (15.1.5.4.1-27) is the associated error state dynamic equation. Note from $\dot{\underline{x}}(t)$ in (15.1.5.4.1-1) using the (15.1.5.4.1-3) partition definition, that Equation (15.1.5.4.1-27) for $\dot{\underline{x}}^*(t)$ is merely the upper ($\underline{x}^*(t)$) partition of $\dot{\underline{x}}(t)$. Thus, we can conclude that for zero measurement noise (and compatibility with the (15.1.5.4.1-3) - (15.1.5.4.1-4) constraints), the $P^*(t)$ covariance partition in $P(t)$ can be determined using a revised Kalman filter configuration based on the original error state dynamic equation, but whose measurement equation is (15.1.5.4.1-29) (or (15.1.5.4.1-28) for the continuous form), instead of the original $\underline{z}^{\blacklozenge}(t)$ in (15.1.5.4.1-2). Additionally, from (15.1.5.4.1-12), the remaining elements of P(t) for the constrained zero measurement case will be zero. The identical conclusions apply for Kalman filter estimates; the $\underline{x}(t)$ estimate for the constrained zero measurement noise case has its $\underline{x}^*(t)$ partition equal to the estimate obtained from a Kalman filter operating from the revised $\underline{z}^{\blacklozenge}_{Rev}(t)$ measurement, with the remaining partition in $\underline{x}(t)$ (i.e., the $\underline{x}_{\mathcal{M}}(t)$ partition) having its estimate equal to the true $\underline{x}_{\mathcal{M}}(t)$ value (i.e., error free).

Based on the above results and Equation (15.1.5.4.1-25) compared to (15.1.5.3.1-23), we conclude that the following simple procedure can be used for determining P(t) for the case of zero measurement noise, based on the general P(t) covariance solution in Section 15.1.5.4:

1) Partition the (15.1.5.4.1-1) - (15.1.5.4.1-2) error state dynamic and measurement equations and verify that the partitioned forms fit the (15.1.5.4.1-3) - (15.1.5.4.1-4) constraint conditions.

2) Assuming that the constraint conditions are satisfied, apply the following conversion formulas for the matrices in (15.1.5.3.1-23):

$$A(t) = A^*(t) \qquad G_P(t) = G_{P^*}(t) \qquad Q_{P_{Dens}}(t) = Q_{P^*_{Dens}}(t)$$

$$H(t) = A_{\mathcal{M}^*}(t) \qquad G_M(t) = G_{P_{\mathcal{M}}}(t) \qquad R^{\blacklozenge}(t) = Q_{P_{\mathcal{M}}\,Dens}(t)$$

3) Using the matrix definitions from 2), solve (15.1.5.3.1-23) for P(t) by applying the Section 15.1.5.4 general covariance solution Equations (15.1.5.4-7), (15.1.5.4-8) and (15.1.5.4-11) with P(0) in (15.1.5.4-11) set equal to P*(0), the initial value for P*(t).

4) Identify the P(t) solution from (15.1.5.4-11) in 3) as P*(t).

5) Find the desired P(t) solution by setting the P*(t) partition to the 4) result and equating the remaining elements of P(t) to zero.

15.2 EXAMPLES OF KALMAN FILTERING APPLIED TO STRAPDOWN INERTIAL NAVIGATION

The following subsections provide examples illustrating the application of Kalman filtering to INS quasi-stationary initialization (gain determination for the Chapter 6 Fine Alignment process), dynamic moving base INS initialization (i.e., "transfer alignment"), INS updating using a body mounted velocity sensor (e.g., Doppler radar), and INS updating based on GPS (Global Positioning System) range-to-satellite measurements.

15.2.1 KALMAN FILTERING APPLIED TO THE QUASI-STATIONARY FINE ALIGNMENT PROBLEM

The Fine Alignment process discussed in Section 6.1.2 is typically structured using feed-back gains derived from Kalman filter theory. The Kalman gain calculations are derived from the N Frame form of Fine Alignment process propagation Equations (6.1.2-1) (with zero gains - i.e., no Kalman filter estimation/control) and with $\underline{\omega}_{IL}^{L}$, $\underline{\omega}_{IE}^{N}$ from (6.1.2-2):

$$\begin{aligned}
\dot{C}_B^N &= C_B^N \left(\underline{\omega}_{IB}^B \times\right) - \left(\underline{\omega}_{IE}^N \times\right) C_B^N \\
\underline{\omega}_{IE}^N &= \underline{\omega}_{IE_H}^N + \underline{u}_{ZN}^N \, \omega_e \sin l \\
\dot{\underline{v}}_H^N &= \left(C_B^N \, \underline{a}_{SF}^B\right)_H \\
\Delta\dot{\underline{R}}_H^N &= \underline{v}_H^N
\end{aligned} \qquad (15.2.1\text{-}1)$$

The Fine Alignment process consists of integrating "foreground" Equations (6.1.2-1) (or their N Frame equivalent as defined by (15.2.1-1)), periodically sampling the $\Delta\underline{R}_H^N$ integrated horizontal velocity term (the "position divergence"), and using $\Delta\underline{R}_H^N$ in feed-back fashion through appropriate gains to level the C_B^N body direction cosine matrix and to estimate the $\underline{\omega}_{IE_H}^N$ horizontal earth rate components. The estimated horizontal earth components define the C_B^N matrix heading relative to true north. At Fine Alignment completion, the estimated horizontal earth components are either used for heading initialization in the C_N^E position direction cosine matrix, or to rotate the C_B^N matrix to the proper true north heading (as described in Sections 6.2.1 and 6.2.2).

Note that the integral of (15.2.1-1) includes an integration of velocity $\underline{v}_H^N$ into position $\Delta\underline{R}_H^N$. When structuring the software used to perform the velocity-into-position integration, it is important that the repetition rate for the integration algorithm be fast enough that position algorithm folding effect errors are not introduced (i.e., folding of linear vibration excitation of the strapdown sensor assembly during the alignment process). Chapter 10, Sections 10.1.3.2.4, 10.4.2, 10.6.2 and the position error response in Equations (10.3-20) and (10.6.1-25) describe various methods for evaluating the position integration algorithm folding effect error as a function of algorithm repetition rate and input vibration level. Sections 7.4 and 7.4.1 of Chapter 7 provides further discussion on the impact of folding on horizontal earth rate estimation accuracy. In this section, we will assume that the algorithm repetition rate is sufficiently fast that folding effects are negligible in the Kalman filter error model for the alignment process.

To cast the Fine Alignment process in a Kalman filter framework, we first define the observation equation (as in Section 15.1):

$$\underline{Z}_{Obs} = \Delta\underline{R}_H^N - \Delta\underline{R}_{Ref_H}^N \qquad \Delta\underline{R}_{Ref_H}^N = 0 \tag{15.2.1-2}$$

where

$\Delta\underline{R}_{Ref_H}^N$ = Reference position divergence approximated as zero (i.e., zero net horizontal motion except for small bounded quasi-stationary random vibration type movement caused, for example in an aircraft INS, by wind gusts, fuel-stores loading, crew/passenger movement).

To construct the Fine Alignment Kalman filter, we now write the error form of foreground Equations (15.2.1-1). Using Equations (14.2-18) (the error form of Equations (14.1-1)) as a template (but without the Kalman feedback), the result is:

$$\begin{aligned}
\dot{\gamma}_{ZN} &= -\underline{u}_{ZN}^N \cdot \left(C_B^N\, \delta\underline{\omega}_{IB}^B\right) - \underline{u}_{ZN}^N \cdot \left(\underline{\omega}_{IE_H}^N \times \underline{\gamma}_H^N\right) + \underline{u}_{ZN}^N \cdot \left[C_B^N \left(\underline{\omega}_{Vib}^B \times \delta\underline{\alpha}_{Quant}\right)\right] \\
\delta\dot{\underline{\omega}}_{IE_H}^N &= 0 \\
\dot{\underline{\gamma}}_H^N &= -\left(C_B^N\, \delta\underline{\omega}_{IB}^B\right)_H - \gamma_{ZN}\, \underline{\omega}_{IE_H}^N \times \underline{u}_{ZN}^N - \left(\underline{u}_{ZN}^N \times \underline{\gamma}_H^N\right) \omega_e \sin l + \delta\underline{\omega}_{IE_H}^N \\
&\quad + \left(C_B^N\right)_H \left(\underline{\omega}_{Vib}^B \times \delta\underline{\alpha}_{Quant}\right)
\end{aligned} \tag{15.2.1-3}$$

(Continued)

$$\delta\underline{\omega}_{IB}^{B} = \delta K_{Scal/Mis}\left(C_B^N\right)^T \underline{\omega}_{IE}^{N} + \delta\underline{K}_{Bias} + \delta\underline{\omega}_{Rand}$$

$$\delta\dot{\underline{v}}_{H}^{N} = \left(C_B^N\, \delta\underline{a}_{SF}{}^{B}\right)_H + g\left(\underline{u}_{ZN}^{N} \times \underline{\gamma}_{H}^{N}\right) - g\left(\underline{u}_{ZN}^{N}\times\right) C_B^N\, \delta\underline{\alpha}_{Quant}$$

$$- \left(C_B^N\right)_H \left(\underline{a}_{Vib}^{B} \times \delta\underline{\alpha}_{Quant} + \underline{\omega}_{Vib}^{B} \times \delta\underline{\upsilon}_{Quant}\right) \qquad (15.2.1\text{-}3)\ \text{(Continued)}$$

$$\delta\underline{a}_{SF}^{B} = g\, \delta L_{Scal/Mis}\left(C_B^N\right)^T \underline{u}_{ZN}^{N} + \delta\underline{L}_{Bias} + \delta\underline{a}_{Rand}$$

$$\delta\Delta\dot{\underline{R}}_{H}^{N} = \delta\underline{v}_{H}^{N} + \left(C_B^N\, \delta\underline{\upsilon}_{Quant}\right)_H$$

where

$\delta(\ )$ = Error in the bracketed quantity.

For Kalman filter gain determination, Equations (15.2.1-3) can be simplified by recognizing the horizontal-tilt/earth-rate and vibration/quantization-noise products to be negligibly small. We further recognize (from experience and knowledge of error effects on Kalman filter performance) that for this quasi-stationary application (also assuming a quasi-stationary attitude), the Kalman filter will not be able to distinguish constant inertial sensor errors from horizontal tilt and earth rate component uncertainties. Thus, we only include the random component of the (15.2.1-3) sensor errors in the Kalman filter error model. Since the N Frame is an arbitrary locally level frame used for navigation integration, we can arbitrarily define it to have its initial heading correspond with the C_B^N heading at Fine Alignment initiation. This selection defines γ_{ZN} to zero at the start of Fine Alignment. We also assume that Fine Alignment will be completed in a reasonably short time period such that the heading error γ_{ZN} will not have sufficient time to build from its starting zero value due to sensor error. Hence, the γ_{ZN} coupling term in the horizontal tilt rate equation can also be ignored. With these approximations, Equations (15.2.1-3) simplify to the following form for Kalman filter design:

$$\delta\dot{\underline{\omega}}_{IE_H}^{N} = 0$$

$$\dot{\underline{\gamma}}_{H}^{N} = -\left(C_B^N\, \delta\underline{\omega}_{Rand}\right)_H + \delta\underline{\omega}_{IE_H}^{N} \qquad (15.2.1\text{-}4)$$

$$\delta\dot{\underline{v}}_{H}^{N} = \left(C_B^N\, \delta\underline{a}_{Rand}\right)_H + g\left(\underline{u}_{ZN}^{N} \times \underline{\gamma}_{H}^{N}\right) - g\left(\underline{u}_{ZN}^{N}\times\right) C_B^N\, \delta\underline{\alpha}_{Quant}$$

$$\delta\Delta\dot{\underline{R}}_{H}^{N} = \delta\underline{v}_{H}^{N} + \left(C_B^N\, \delta\underline{\upsilon}_{Quant}\right)_H$$

We also must not forget that (15.2.1-3) (the basis for (15.2.1-4)) does not allow for error effects produced by computational algorithm error (e.g., from inadequate update rates) which can add additional noise-like terms. If these effects are significant compared to the (15.2.1-3) noise terms, they should be accounted for in (15.2.1-4).

Equations (15.2.1-4) can be reformatted for Kalman filter error state dynamic equation equivalency by noting that for any arbitrary vector $\underline{V}$:

$$\left(C_B^N \underline{V}\right)_H = I_H \, C_B^N \, \underline{V} \qquad (15.2.1\text{-}5)$$

where

I_H = Identity matrix with the 3,3 element (vertical component) set to zero.

By inference, parameters with the H notation in (15.2.1-4) - (15.2.1-5) represent column or square matrices with row 3 set to zero. In formulating the Kalman filter error state vector we want to eliminate the null rows. Therefore, for greater specificity, we will apply (15.2.1-5) to (15.2.1-4) to obtain the following more explicit form of the quasi-stationary Fine Alignment Kalman filter error state dynamic equations that identifies the dimensions of each matrix element and which deletes the null rows in (15.2.1-4):

$$\begin{aligned}
\delta\dot{\underline{\omega}}_{IE\,2x1}^{N} &= 0 \\
\dot{\underline{\gamma}}_{2x1}^{N} &= \delta\underline{\omega}_{IE\,2x1}^{N} - I_{2x3}\, C_B^N \, \delta\underline{\omega}_{Rand} \\
\delta\dot{\underline{v}}_{2x1}^{N} &= g\left(\underline{u}_{ZN}^{N}\times\right)_{2x2} \underline{\gamma}_{2x1}^{N} - g\left(\underline{u}_{ZN}^{N}\times\right)_{2x3} C_B^N \, \delta\underline{\alpha}_{Quant} + I_{2x3}\, C_B^N \, \delta\underline{a}_{Rand} \\
\delta\Delta\dot{\underline{R}}_{2x1}^{N} &= \delta\underline{v}_{2x1}^{N} + I_{2x3}\, C_B^N \, \delta\underline{\upsilon}_{Quant}
\end{aligned} \qquad (15.2.1\text{-}6)$$

where

2x1 = Subscript designation on vector ( ) indicating that $(\;)_{2x1}$ is a two element column matrix (one column) equal to the top two elements of the associated three element column matrix ( ).

2x3 = Subscript designation on matrix ( ) indicating that $(\;)_{2x3}$ has two rows and three columns equal to the top two rows of the associated ( ) three-by-three square matrix.

2x2 = Subscript designation on matrix ( ) indicating that $(\;)_{2x2}$ has two rows and two columns equal to the upper left two-by-two matrix partition in the associated ( ) three-by-three square matrix.

I_{2x3} = First two rows of the three-by-three identity matrix.

Vector and matrix elements in (15.2.1-6) without specific row/column designation are normal notation three element vectors or three-by-three matrices.

The measurement equation for the Kalman filter is the differential of (15.2.1-2):

$$\underline{z}^N = \delta\underline{Z}_{Obs} = \delta\Delta\underline{R}_H^N - \delta\Delta\underline{R}_{Ref_H}^N \tag{15.2.1-7}$$

where

$\delta\Delta\underline{R}_{Ref_H}^N$ = Error in $\Delta\underline{R}_{Ref_H}^N$ caused by its approximation in (15.2.1-2) as zero.

The $\delta\Delta\underline{R}_{Ref_H}^N$ error can be defined as the difference between its true value and the value utilized in the strapdown INS computer software:

$$\delta\Delta\underline{R}_{Ref_H}^N \equiv \Delta\underline{R}_{Ref_H}^N - \Delta\underline{R}_{True_H}^N \tag{15.2.1-8}$$

where

$\Delta\underline{R}_{True_H}^N$ = Actual horizontal position motion during initial alignment.

We attribute $\Delta\underline{R}_{True_H}^N$ to quasi-stationary random vibration type motion, hence:

$$\Delta\underline{R}_{True_H}^N = \Delta\underline{R}_{Vib_H}^N \tag{15.2.1-9}$$

where

$\Delta\underline{R}_{Vib_H}^N$ = Horizontal random position disturbance during alignment from a nominally stationary position.

Then with (15.2.1-9) and (15.2.1-2) for $\Delta\underline{R}_{Ref_H}^N$, Equation (15.2.1-8) is:

$$\delta\Delta\underline{R}_{Ref_H}^N = -\Delta\underline{R}_{Vib_H}^N \tag{15.2.1-10}$$

and (15.2.1-7) becomes:

$$\underline{z}^N = \delta\Delta\underline{R}_H^N + \Delta\underline{R}_{Vib_H}^N \tag{15.2.1-11}$$

With Kalman filter row/column notation reformatting as in (15.2.1-6), Equation (15.2.1-11) is equivalently:

$$\underline{z}_n^N = \delta\Delta\underline{R}^N_{2x1_n} + \Delta\underline{R}^N_{Vib/2x1_n} \quad (15.2.1\text{-}12)$$

Error state dynamic Equations (15.2.1-6) and measurement Equation (15.2.1-12) can now be used to define the quasi-stationary Fine Alignment Kalman filter for leveling and control of foreground Equations (15.2.1-1), and for estimation of horizontal earth rate for initial heading determination. From (15.2.1-6) compared with general Equation (15.1-1), we first define the error state vector, the error state dynamic matrix, the process noise vector, and the process noise dynamic coupling matrix for the Kalman filter as:

$$\underline{x} = \left[\left(\delta\underline{\omega}^N_{IE_{2x1}}\right)^T, \left(\underline{\gamma}^N_{2x1}\right)^T, \left(\delta\underline{v}^N_{2x1}\right)^T, \left(\delta\Delta\underline{R}^N_{2x1}\right)^T\right]^T \quad (15.2.1\text{-}13)$$

$$\underline{n}_P = \left[\left(\delta\underline{\omega}_{Rand}\right)^T, \left(\delta\underline{\alpha}_{Quant}\right)^T, \left(\delta\underline{a}_{Rand}\right)^T, \left(\delta\underline{\upsilon}_{Quant}\right)^T\right]^T \quad (15.2.1\text{-}14)$$

$$A = \begin{bmatrix} 0 & 0 & 0 & 0 \\ I_{2x2} & 0 & 0 & 0 \\ 0 & g\left(\underline{u}^N_{ZN}\times\right)_{2x2} & 0 & 0 \\ 0 & 0 & I_{2x2} & 0 \end{bmatrix} \quad (15.2.1\text{-}15)$$

$$G_P = \begin{bmatrix} 0 & 0 & 0 & 0 \\ -\,I_{2x3}\,C^N_B & 0 & 0 & 0 \\ 0 & -\,g\left(\underline{u}^N_{ZN}\times\right)_{2x3} C^N_B & I_{2x3}\,C^N_B & 0 \\ 0 & 0 & 0 & I_{2x3}\,C^N_B \end{bmatrix} \quad (15.2.1\text{-}16)$$

where

I_{2x2} = Two by two identity matrix.

We assume that the three components for each of the random and quantization noise vectors in (15.2.1-14) have equal densities. Then, from its definition in Equations (15.1.2.1.1-25) - (15.1.2.1.1-28), we can write for the $\underline{n}_P$ process noise density matrix $Q_{P_{Dens}}$:

$$Q_{P_{Dens}} = \begin{bmatrix} q_{\omega Rand}\, I & 0 & 0 & 0 \\ 0 & q_{\alpha Quant}\, I & 0 & 0 \\ 0 & 0 & q_{aRand}\, I & 0 \\ 0 & 0 & 0 & q_{\upsilon Quant}\, I \end{bmatrix} \qquad (15.2.1\text{-}17)$$

where

$q_{\omega Rand}, q_{\alpha Quant}, q_{aRand}, q_{\upsilon Quant}$ = Noise density for each component of $\delta\underline{\omega}_{Rand}$, $\delta\underline{\alpha}_{Quant}$, $\delta\underline{a}_{Rand}$, $\delta\underline{\upsilon}_{Quant}$.

I = Three-by-three identity matrix.

Equation (15.2.1-15) shows that for the Fine Alignment Kalman filter, the error state dynamic matrix is constant. This particularly simple form of the error state dynamic matrix allows the state transition matrix Φ_n to be calculated exactly in closed-form by analytical integration of the elements of general Equation (15.1.1-15). The result is:

$$\Phi_n = \begin{bmatrix} I_{2x2} & 0 & 0 & 0 \\ T_n\, I_{2x2} & I_{2x2} & 0 & 0 \\ \frac{1}{2}\, g\, T_n^2 \left(\underline{u}_{ZN}^N \times\right)_{2x2} & g\, T_n \left(\underline{u}_{ZN}^N \times\right)_{2x2} & I_{2x2} & 0 \\ \frac{1}{6}\, g\, T_n^3 \left(\underline{u}_{ZN}^N \times\right)_{2x2} & \frac{1}{2}\, g\, T_n^2 \left(\underline{u}_{ZN}^N \times\right)_{2x2} & T_n\, I_{2x2} & I_{2x2} \end{bmatrix} \qquad (15.2.1\text{-}18)$$

where

T_n = Kalman filter estimation time cycle interval.

Thus, we see that for this problem, the error state transition matrix is constant and independent of attitude, position location, or time in alignment. This allows Φ_n to be calculated once as part of the Fine Alignment initialization process, and then applied during Fine Alignment operations as required.

The coupled process noise density matrix ($G_P\, Q_{P_{Dens}}\, G_P^T$) for covariance propagation (see Equation (15.1.2.1.1-30)) is obtained by combining Equations (15.2.1-16) and (15.2.1-17) while recognizing that:

$$C_B^N \left(C_B^N\right)^T = I \qquad I_{2x3}\, I_{2x3}^T = I_{2x2} \qquad (15.2.1\text{-}19)$$

In the process we also note that the transpose of $\left(\underline{u}^{N}_{ZN}\times\right)$ is its negative so that (13.1-8) with our previous definitions is:

$$\left(\underline{u}^{N}_{ZN}\times\right)\left(\underline{u}^{N}_{ZN}\times\right)^{T} = I_H \qquad \left(\underline{u}^{N}_{ZN}\times\right) = \begin{bmatrix} \left(\underline{u}^{N}_{ZN}\times\right)_{2x3} \\ 0 \end{bmatrix} \qquad I_H = \begin{bmatrix} I_{2x2} & 0 \\ 0 & 0 \end{bmatrix} \tag{15.2.1-20}$$

Thus,

$$\left(\underline{u}^{N}_{ZN}\times\right)\left(\underline{u}^{N}_{ZN}\times\right)^{T} = \begin{bmatrix} \left(\underline{u}^{N}_{ZN}\times\right)_{2x3}\left(\underline{u}^{N}_{ZN}\times\right)^{T}_{2x3} & 0 \\ 0 & 0 \end{bmatrix} = \begin{bmatrix} I_{2x2} & 0 \\ 0 & 0 \end{bmatrix} \tag{15.2.1-21}$$

Hence,

$$\left(\underline{u}^{N}_{ZN}\times\right)_{2x3}\left(\underline{u}^{N}_{ZN}\times\right)^{T}_{2x3} = I_{2x2} \tag{15.2.1-22}$$

Combining Equations (15.2.1-16) - (15.2.1-17) while applying (15.2.1-19) and (15.2.1-22) then yields for $G_P\, Q_{P_{Dens}}\, G_P^T$:

$$G_P\, Q_{P_{Dens}}\, G_P^T = \begin{bmatrix} 0 & 0 & 0 & 0 \\ 0 & q_{\omega Rand}\, I_{2x2} & 0 & 0 \\ 0 & 0 & \left(g^2\, q_{\alpha Quant} + q_{aRand}\right) I_{2x2} & 0 \\ 0 & 0 & 0 & q_{\upsilon Quant}\, I_{2x2} \end{bmatrix} \tag{15.2.1-23}$$

Thus, the coupled process noise density matrix $G_P\, Q_{P_{Dens}}\, G_P^T$ as well as the error state dynamic matrix A (from Equation (15.2.1-15)) are constant and independent of attitude, position location, or time in alignment. This allows the integrated process noise matrix Q_n in (15.1.2.1.1-12) (computed from $G_P\, Q_{P_{Dens}}\, G_P^T$ and A) to be pre-calculated once as part of the Fine Alignment initialization process; then used as required during Fine Alignment operations. A second order approximation for Q_n can be obtained from Equations (15.1.2.1.1.3-37) compared to (15.1.2.1.1-12):

$$Q_{1_n} = G_P\, Q_{P_{Dens}}\, G_P^T\, T_n \qquad Q_n \approx \frac{1}{2} Q_{1_n} + \frac{1}{2} \Phi_n\, Q_{1_n}\, \Phi_n^T \tag{15.2.1-24}$$

However, because Q_n need only be calculated once during Fine Alignment initialization, computer throughput is not an issue, and Q_n can also be evaluated to higher order using the more elaborate Equations (15.1.2.1.1.3-28) algorithm for m = n with $\Delta\Phi_{\lambda y_n}$ and $\Delta\Phi_{\lambda\lambda_n}$ from (15.1.2.1.1.3-21):

$$\Delta\Phi_{\lambda\lambda_n} = A\, T_n \qquad \Delta\Phi_{\lambda y_n} = G_P\, Q_{P_{Dens}}\, G_P^T\, T_n \tag{15.2.1-25}$$

The measurement matrix, measurement noise vector, and measurement noise dynamic coupling matrix can be identified from Equations (15.2.1-12) and (15.2.1-13) compared with general measurement Equation (15.1-2):

$$H_n = \begin{bmatrix} 0 & 0 & 0 & I_{2x2} \end{bmatrix} \qquad \underline{n}_{M_n} = \Delta R^N_{Vib_{2x1}} \qquad G_{M_n} = I_{2x2} \tag{15.2.1-26}$$

Assuming that the elements of $\underline{n}_M$ are uncorrelated with equal variance, we use the (15.1.2.1-13) definition to write for the measurement noise matrix:

$$R_n \equiv \mathcal{E}\left(\underline{n}_{M_n}\, \underline{n}^T_{M_n}\right) = P_{RVib_H}\, I_{2x2} \tag{15.2.1-27}$$

where

P_{RVib_H} = Horizontal quasi-stationary position random motion variance.

Thus, H_n, G_{M_n} and R_n are also constant and independent of initial attitude, position location, and time in Fine Alignment.

Using the pre-calculated value for Φ_n, Q_n as described previously, and H_n, G_{M_n} and R_n from (15.2.1-26) - (15.2.1-27) (all constant and independent of time in Fine Alignment), Equations (15.1.2.1.1-12), (15.1.2.1-28) and (15.1.2.1.1-4) then directly apply for covariance propagation, Kalman gain determination, and covariance reset:

$$\begin{aligned} P_n(-) &= \Phi_n\, P_{n-1}(+_e)\, \Phi^T_n + Q_n \\ K_n &= P_n(-)\, H^T_n \left(H_n\, P_n(-)\, H^T_n + G_{M_n}\, R_n\, G^T_{M_n}\right)^{-1} \\ P_n(+_e) &= \left(I - K_n\, H_n\right) P_n(-) \end{aligned} \tag{15.2.1-28}$$

Equations (15.2.1-28) processed repetitively allow the gain matrix K_n to be calculated for Fine Alignment estimation and control reset operations. Note, however, that because all the coefficient matrices in (15.2.1-28) are constant and independent of time in Fine Alignment, Equations (15.2.1-28) can be pre-computed during the previous n cycle so that the gain matrix K_n is available for application at time t_n. This allows the ideal control reset Kalman filter configuration of Equations (15.1.2-6) - (15.1.2-13) to be realized for which we write for the ideal control vector:

$$\underline{u}_{c_n} = -\underline{\tilde{x}}_n(+_e) \qquad (15.2.1\text{-}29)$$

Substituting (15.2.1-29) in (15.1.2-12) and (15.1.2-7) - (15.1.2-9) (with (15.1.2-13)) finds:

$$\begin{aligned}
\underline{\tilde{x}}_{n-1}(+_c) &= \underline{\tilde{x}}_{n-1}(+_e) + \underline{u}_{c_{n-1}} = 0 \\
\underline{\tilde{x}}_n(-) &= \Phi_n\, \underline{\tilde{x}}_{n-1}(+_c) = 0 \\
\underline{\tilde{z}}_n &= H_n\, \underline{\tilde{x}}_n(-) = 0 \\
\underline{\tilde{x}}_n(+_e) &= \underline{\tilde{x}}_n(-) + K_n\left(\underline{Z}_{Obs_n} - \underline{\tilde{z}}_n\right) = K_n\, \underline{Z}_{Obs_n}
\end{aligned} \qquad (15.2.1\text{-}30)$$

The observation in matrix dimension form is from (15.2.1-2):

$$\underline{Z}_{Obs_n} = \Delta\underline{R}^N_{2x1_n} \qquad (15.2.1\text{-}31)$$

With (15.2.1-29) - (15.2.1-31) we see then that:

$$\underline{u}_{c_n} = -K_n\, \underline{Z}_{Obs_n} = -K_n\, \Delta\underline{R}^N_{2x1_n} \qquad (15.2.1\text{-}32)$$

Equation (15.2.1-32) using K_n from Equations (15.2.1-28) represents the total Kalman filter computational requirement, with the resulting control vector $\underline{u}_{c_n}$ then only applied to the foreground Equation (15.2.1-1) navigation parameters. The method for applying $\underline{u}_{c_n}$ to the foreground is developed by first defining the individual components of $\underline{u}_c$ as:

$$\underline{u}_c \equiv \left[\left(\delta\underline{\omega}^N_{IE/2x1_c}\right)^T, \left(\underline{\gamma}^N_{2x1_c}\right)^T, \left(\delta\underline{v}^N_{2x1_c}\right)^T, \left(\delta\Delta\underline{R}^N_{2x1_c}\right)^T\right]^T \qquad (15.2.1\text{-}33)$$

where

c = Subscript designation for the $\underline{u}_c$ control vector component to be applied to the indicated (15.2.1-13) error state vector component.

With the (15.2.1-33) definition and using Section 15.1.2.3 (particularly Equation (15.1.2.3-13)) as a guide, application of $\underline{u}_{c_n}$ to the Equation (15.2.1-1) parameters is then given in matrix dimension form by:

$$\underline{\omega}^{N}_{IE/2x1_n}(+c) = \underline{\omega}^{N}_{IE/2x1_n}(-) + \delta\underline{\omega}^{N}_{IE/2x1/c_n}$$

$$\underline{\gamma}^{N}_{Hc_n} = \begin{bmatrix} \underline{\gamma}^{N}_{2x1/c_n} \\ 0 \end{bmatrix}$$

$$C^{N}_{B_n}(+c) = \left[I - \frac{\sin \gamma_{Hc_n}}{\gamma_{Hc_n}} \left(\underline{\gamma}^{N}_{Hc_n}\times\right) + \frac{1 - \cos \gamma_{Hc_n}}{\gamma^{2}_{Hc_n}} \left(\underline{\gamma}^{N}_{Hc_n}\times\right)^{2} \right] C^{N}_{B_n}(-) \qquad (15.2.1\text{-}34)$$

$$\underline{v}^{N}_{2x1_n}(+c) = \underline{v}^{N}_{2x1_n}(-) + \delta\underline{v}^{N}_{2x1/c_n}$$

$$\Delta\underline{R}^{N}_{2x1_n}(+c) = \Delta\underline{R}^{N}_{2x1_n}(-) + \delta\Delta\underline{R}^{N}_{2x1/c_n}$$

Equations (15.2.1-34) constitute the operation represented symbolically by Equation (15.1.2-11).

The overall Fine Alignment process consists of integrating foreground Equations (15.2.1-1) while applying control Equations (15.2.1-32) - (15.2.1-34) at the Kalman cycle time using the Kalman gain matrix calculated from covariance propagation/reset Equations (15.2.1-28). The matrix coefficient terms in (15.2.1-28) are constant as defined by Equations (15.2.1-18) and (15.2.1-24) - (15.2.1-27). By the (15.2.1-13) definition for the error state vector $\underline{x}$, eight error states are identified, two horizontal components for each vector element. Thus, by implication, Equations (15.2.1-28) would involve an 8 by 8 covariance matrix and associated matrix products. The interesting aspect about this particular problem is that by rearrangement of the $\underline{x}$ elements so that they are organized by horizontal axis, Equations (15.2.1-28) partition into two independent sets with identical dynamics and solutions, each consisting of 4 simultaneous equations. Recognition of this concept allows Equations (15.2.1-28) to be reduced to a series of 4 by 4 matrix operations, with the resulting Kalman gain matrix then applied separately through control Equations (15.2.1-32) - (15.2.1-34) to each horizontal axis of the (15.2.1-1) foreground navigation parameters. The following subsections develop the partitioned form of the quasi-stationary fine alignment Kalman filter including initialization of the foreground and Kalman filter parameters.

15.2.1.1 PARTITIONED VERSION OF THE QUASI-STATIONARY FINE ALIGNMENT KALMAN FILTER

If we expand the general error state dynamic Equation (15.1-1) into its individual scalar components using (15.2.1-13) - (15.2.1-16) for $\underline{x}$, $\underline{n}_P$, A and G_P, it will be seen after reordering, rearrangement and regrouping, that they are equivalent to the following alternative partitioned forms:

$$\dot{\underline{x}}_a = A_a \underline{x}_a + G_{P_a} \underline{n}_{P_a} \qquad \dot{\underline{x}}_b = A_b \underline{x}_b + G_{P_b} \underline{n}_{P_b} \tag{15.2.1.1-1}$$

in which

$$\underline{x}_a \equiv \left(- \delta\omega_{IE_{NY}}, - \gamma_{NY}, \delta v_{NX}, \delta\Delta R_{NX}\right)^T$$

$$\underline{x}_b \equiv \left(\delta\omega_{IE_{NX}}, \gamma_{NX}, \delta v_{NY}, \delta\Delta R_{NY}\right)^T$$

$$A_a = A_b = \begin{bmatrix} 0 & 0 & 0 & 0 \\ 1 & 0 & 0 & 0 \\ 0 & g & 0 & 0 \\ 0 & 0 & 1 & 0 \end{bmatrix} \tag{15.2.1.1-2}$$

and

$$\underline{n}_{P_a} \equiv \left(\delta\omega_{Rand_{NY}}, \delta\alpha_{Quant_{NY}}, \delta a_{Rand_{NX}}, \delta\upsilon_{Quant_{NX}}\right)^T$$

$$\underline{n}_{P_b} \equiv \left(\delta\omega_{Rand_{NX}}, \delta\alpha_{Quant_{NX}}, \delta a_{Rand_{NY}}, \delta\upsilon_{Quant_{NY}}\right)^T$$

$$G_{P_a} = \begin{bmatrix} 0 & 0 & 0 & 0 \\ 1 & 0 & 0 & 0 \\ 0 & g & 1 & 0 \\ 0 & 0 & 0 & 1 \end{bmatrix} \qquad G_{P_b} = \begin{bmatrix} 0 & 0 & 0 & 0 \\ -1 & 0 & 0 & 0 \\ 0 & -g & 1 & 0 \\ 0 & 0 & 0 & 1 \end{bmatrix} \tag{15.2.1.1-3}$$

where

$\underline{x}_a$, $\underline{x}_b$, A_a, A_b, $\underline{n}_{P_a}$, $\underline{n}_{P_b}$, G_{P_a}, G_{P_b} = Rearranged four component partitions of $\underline{x}$, A, $\underline{n}_P$, G_P defined in Equations (15.2.1-13) - (15.2.1-16).

NX, NY = Subscripts denoting N Frame horizontal components (X, Y).

$\delta\omega_{IE_{Ni}}$, γ_{Ni}, δv_{Ni}, $\delta\Delta R_{Ni}$ = Components of the original Equation (15.2.1-13) $\underline{x}$ error state vector along N Frame axis i.

$\delta\omega_{Rand_{Ni}}$, $\delta\alpha_{Quant_{Ni}}$, $\delta a_{Rand_{Ni}}$, $\delta\upsilon_{Quant_{Ni}}$ = N Frame i axis components of the original Equation (15.2.1-14) B Frame process noise vector ($\underline{n}_P$), as projected on the N Frame.

The process noise density matrices associated with $G_{P_a} \underline{n}_{P_a}$ and $G_{P_b} \underline{n}_{P_b}$ in Equations (15.2.1.1-1) can be found by first writing $\delta\omega_{IE_{Ni}}$, γ_{Ni}, δv_{Ni}, $\delta\Delta R_{Ni}$ from the previous definitions in vector form, and restating the (15.2.1-14) definition for $\underline{n}_P$:

$$\underline{n}_P^N \equiv \begin{bmatrix} \delta\underline{\omega}_{Rand}^N \\ \delta\underline{\alpha}_{Quant}^N \\ \delta\underline{a}_{Rand}^N \\ \delta\underline{\upsilon}_{Quant}^N \end{bmatrix} = \begin{bmatrix} C_B^N\, \delta\underline{\omega}_{Rand} \\ C_B^N\, \delta\underline{\alpha}_{Quant} \\ C_B^N\, \delta\underline{a}_{Rand} \\ C_B^N\, \delta\underline{\upsilon}_{Quant} \end{bmatrix} \qquad \underline{n}_P \equiv \begin{bmatrix} \delta\underline{\omega}_{Rand} \\ \delta\underline{\alpha}_{Quant} \\ \delta\underline{a}_{Rand} \\ \delta\underline{\upsilon}_{Quant} \end{bmatrix} \tag{15.2.1.1-4}$$

in which it is understood by their original definitions in Sections 12.4, 8.1.1.1 and 8.1.1.2 that $\delta\underline{\omega}_{Rand}$, $\delta\underline{\alpha}_{Quant}$, $\delta\underline{a}_{Rand}$, $\delta\underline{\upsilon}_{Quant}$ (without superscripts) are implicitly defined (to simplify notation) as being in B Frame coordinates, and where:

$\underline{n}_P^N$ = Process noise vector $\underline{n}_P$, but with components projected on N Frame axes.

Recognizing the components of $\underline{n}_P$ as defined to be independent, that $C_B^N \left(C_B^N\right)^T = I$, and assuming for each of the $\underline{n}_P$ component vectors that their elements have equal densities, we see from (15.2.1.1-4):

$$\mathcal{E}\left[\underline{n}_P^N \left(\underline{n}_P^N\right)^T\right] = \mathcal{E}\left[\underline{n}_P \left(\underline{n}_P\right)^T\right] = \begin{bmatrix} \mathcal{E}_1 & 0 & 0 & 0 \\ 0 & \mathcal{E}_2 & 0 & 0 \\ 0 & 0 & \mathcal{E}_3 & 0 \\ 0 & 0 & 0 & \mathcal{E}_4 \end{bmatrix} \tag{15.2.1.1-5}$$

$$\mathcal{E}_1 = \mathcal{E}\left[\delta\underline{\omega}_{Rand}\left(\delta\underline{\omega}_{Rand}\right)^T\right] \qquad \mathcal{E}_2 = \mathcal{E}\left[\delta\underline{\alpha}_{Quant}\left(\delta\underline{\alpha}_{Quant}\right)^T\right]$$

$$\mathcal{E}_3 = \mathcal{E}\left[\delta\underline{a}_{Rand}\left(\delta\underline{a}_{Rand}\right)^T\right] \qquad \mathcal{E}_4 = \mathcal{E}\left[\delta\underline{\upsilon}_{Quant}\left(\delta\underline{\upsilon}_{Quant}\right)^T\right]$$

where

$\mathcal{E}(\,)$ = Expected value operator.

Equation (15.2.1.1-5) with Equations (15.1.2.1.1-25) - (15.1.2.1.1-28) show that the process noise densities for $\underline{n}_P^N$ and $\underline{n}_P$ are equal. Thus, with (15.2.1-17), we can write for the $\underline{n}_P^N$ process noise density matrix:

$$Q_{P_{Dens}}^N = \begin{bmatrix} q_{\omega Rand}\, I & 0 & 0 & 0 \\ 0 & q_{\alpha Quant}\, I & 0 & 0 \\ 0 & 0 & q_{aRand}\, I & 0 \\ 0 & 0 & 0 & q_{\upsilon Quant}\, I \end{bmatrix} \tag{15.2.1.1-6}$$

where

$$Q^{N}_{P_{Dens}} = \text{White noise density matrix associated with } \underline{n}^{N}_{P}.$$

As defined in (15.2.1.1-3), the $\underline{n}_{P_a}$, $\underline{n}_{P_b}$ vectors are partitions of $\underline{n}^{N}_{P}$. By matching the individual components of $\underline{n}_{P_a}$, $\underline{n}_{P_b}$ with $\underline{n}^{N}_{P}$ we can identify the associated white noise densities using (15.2.1.1-6). The obvious result is:

$$Q_{P_{Dens/a}} = Q_{P_{Dens/b}} = \begin{bmatrix} q_{\omega Rand} & 0 & 0 & 0 \\ 0 & q_{\alpha Quant} & 0 & 0 \\ 0 & 0 & q_{aRand} & 0 \\ 0 & 0 & 0 & q_{\upsilon Quant} \end{bmatrix} \qquad (15.2.1.1\text{-}7)$$

where

$$Q_{P_{Dens/a}}, Q_{P_{Dens/b}} = \text{White noise densities associated with } \underline{n}_{P_a}, \underline{n}_{P_b}.$$

Using (15.2.1.1-7) and G_{P_a}, G_{P_b} from (15.2.1.1-3), the coupled process noise density matrices associated with $G_{P_a}\,\underline{n}_{P_a}$ and $G_{P_b}\,\underline{n}_{P_b}$ in Equations (15.2.1.1-1) for $\underline{x}_a$, $\underline{x}_b$ covariance propagation (see Equation (15.1.2.1.1-30)) are then:

$$G_{P_a}\, Q_{P_{Dens/a}}\, G^{T}_{P_a} = G_{P_b}\, Q_{P_{Dens/b}}\, G^{T}_{P_b}$$

$$= \begin{bmatrix} 0 & 0 & 0 & 0 \\ 0 & q_{\omega Rand} & 0 & 0 \\ 0 & 0 & g^2\, q_{\alpha Quant} + q_{aRand} & 0 \\ 0 & 0 & 0 & q_{\upsilon Quant} \end{bmatrix} \qquad (15.2.1.1\text{-}8)$$

where

$$\left(G_{P_a}\, Q_{P_{Dens/a}}\, G^{T}_{P_a}\right), \left(G_{P_b}\, Q_{P_{Dens/b}}\, G^{T}_{P_b}\right) = \text{Coupled process noise density matrices associated with } G_{P_a}\,\underline{n}_{P_a} \text{ and } G_{P_b}\,\underline{n}_{P_b} \text{ in Equations (15.2.1.1-1).}$$

We also write the components of the Kalman filter measurement equation in the generalized (15.1-2) format based on the (15.2.1.1-1) $\underline{x}_a$, $\underline{x}_b$ definitions, and using (15.2.1-26) - (15.2.1-27) for H, $\underline{n}_M$, G_M and R:

$$z_{a_n} = H_a \, \underline{x}_{a_n} + \Delta r_{Vib_{NX/n}} \qquad z_{b_n} = H_b \, \underline{x}_{b_n} + \Delta r_{Vib_{NY/n}}$$

$$H_a = H_b = \begin{bmatrix} 0 & 0 & 0 & 1 \end{bmatrix} \qquad (15.2.1.1\text{-}9)$$

$$R_{a_n} \equiv \mathcal{E}\left(\Delta r^2_{Vib_{NX/n}}\right) = R_{b_n} \equiv \mathcal{E}\left(\Delta r^2_{Vib_{NY/n}}\right) = P_{RVib_H}$$

where

z_a, z_b = N Frame X, Y (horizontal) components of the measurement vector $\underline{z}$.

$\Delta r_{Vib_{NX}}, \Delta r_{Vib_{NY}}$ = N Frame X, Y components of the horizontal position vibration disturbance vector $\Delta \underline{R}^N_{Vib_H}$.

P_{RVib_H} = Variance for each of the horizontal position disturbance vibration components.

From (15.2.1.1-1) - (15.2.1.1-3) and (15.2.1.1-8) - (15.2.1.1-9), we see that the $\underline{x}_a$, z_a and $\underline{x}_b$, z_b equation sets are completely uncoupled with identical error state dynamic matrices (A_a and A_b), identical coupled process noise densities ($G_{P_a} \, Q_{P_{Dens/a}} \, G^T_{P_a}$ and $G_{P_b} \, Q_{P_{Dens/b}} \, G^T_{P_b}$), identical measurement matrices (H_a, H_b), and identical measurement noise variances (R_a, R_b). Therefore, Kalman filter operations associated with $\underline{x}_a$, z_a will also be uncoupled from $\underline{x}_b$, z_b Kalman filter operations, resulting in two parallel four-state Kalman filters for separately estimating and controlling the four components in $\underline{x}_a$ and in $\underline{x}_b$, each filter having identical covariance propagation/reset dynamics, process noise and measurement noise. Finally, if we stipulate that the initial uncertainties in $\underline{x}_a$ and $\underline{x}_b$ will be identical, the covariance propagation/reset equations in each filter will be initialized identically, thereby generating identical Kalman gain profiles. The net result is that we need only implement a single four-state filter covariance propagation/reset operation, using the resulting Kalman gains to then separately estimate $\underline{x}_a$ and $\underline{x}_b$. The single four-state Kalman filter will be based on error state dynamic and measurement equations of the general (15.2.1.1-1) - (15.2.1.1-3) and (15.2.1.1-8) - (15.2.1.1-9) form:

$$\dot{\underline{x}}^* = A^* \, \underline{x}^* + G_{P^*} \, \underline{n}_{P^*} \qquad z^*_n = H^* \, \underline{x}^*_n + G_{M^*} \, n_{M^*_n} \qquad (15.2.1.1\text{-}10)$$

with

$$\underline{x}^* = \left(\delta\omega_{IE_H}, \gamma_H, \delta v_H, \delta\Delta R_H\right)^T \qquad A^* = \begin{bmatrix} 0 & 0 & 0 & 0 \\ 1 & 0 & 0 & 0 \\ 0 & g & 0 & 0 \\ 0 & 0 & 1 & 0 \end{bmatrix}$$

$$\underline{n}_{P^*} = \left(\delta\omega_{Rand_H}, \delta\alpha_{Quant_H}, \delta a_{Rand_H}, \delta\upsilon_{Quant_H}\right)^T \qquad G_{P^*} = \begin{bmatrix} 0 & 0 & 0 & 0 \\ -1 & 0 & 0 & 0 \\ 0 & -g & 1 & 0 \\ 0 & 0 & 0 & 1 \end{bmatrix}$$

$$Q_{P^*Dens} = \begin{bmatrix} q_{\omega Rand} & 0 & 0 & 0 \\ 0 & q_{\alpha Quant} & 0 & 0 \\ 0 & 0 & q_{aRand} & 0 \\ 0 & 0 & 0 & q_{\upsilon Quant} \end{bmatrix} \tag{15.2.1.1-11}$$

$$G_{P^*}\, Q_{P^*Dens}\, G_{P^*}^T = \begin{bmatrix} 0 & 0 & 0 & 0 \\ 0 & q_{\omega Rand} & 0 & 0 \\ 0 & 0 & g^2\, q_{\alpha Quant} + q_{aRand} & 0 \\ 0 & 0 & 0 & q_{\upsilon Quant} \end{bmatrix}$$

$$H^* = \begin{bmatrix} 0 & 0 & 0 & 1 \end{bmatrix} \qquad G_M{}^* = 1$$

$$R^* \equiv \mathcal{E}\left(n_{M^*_n}^2\right) = P_{RVib_H} \qquad n_M{}^* = \Delta r_{Vib_H}$$

where

- $*$ = Designation for the equivalent to the a, b versions of the same parameters in Equations (15.2.1.1-1) - (15.2.1.1-3) and (15.2.1.1-8) - (15.2.1.1-9).
- Q_{P^*Dens} = White noise density associated with $\underline{n}_{P^*}$.
- H = Designation for horizontal component in general (either X or Y).

The associated covariance propagation/reset and Kalman gain equations then are as in (15.2.1-28) with (15.2.1-18), but for the four-state filter:

$$P^*_n(-) = \Phi^*\, P^*_{n-1}(+_e)\, \Phi^{*T} + Q^*$$
$$K^*_n = P^*_n(-)\, H^{*T}\left(H^*\, P^*_n(-)\, H^{*T} + P_{RVib_H}\right)^{-1} \tag{15.2.1.1-12}$$
$$P^*_n(+_e) = \left(I - K^*_n\, H^*\right) P^*_n(-)$$

with

$$\Phi^* = \begin{bmatrix} 1 & 0 & 0 & 0 \\ T_n & 1 & 0 & 0 \\ \frac{1}{2} g\, T_n^2 & g\, T_n & 1 & 0 \\ \frac{1}{6} g\, T_n^3 & \frac{1}{2} g\, T_n^2 & T_n & 1 \end{bmatrix} \tag{15.2.1.1-13}$$

where

P^* = Error state covariance matrix for the partitioned four-state Kalman filter.

Q^* = Integrated process noise matrix for the four-state Kalman filter.

K^* = Four-state filter Kalman gain matrix.

Φ^* = Four-state Kalman filter error state transition matrix.

To second order accuracy, the Q^* matrix in (15.2.1.1-12) can be calculated (as in (15.2.1-24)) from:

$$Q_{1^*} = \left(G_{P^*}\, Q_{P^*_{Dens}}\, G_{P^*}^T \right) T_n \qquad Q^* \approx \frac{1}{2} Q_{1^*} + \frac{1}{2} \Phi^*\, Q_{1^*}\, \Phi^{*T} \tag{15.2.1.1-14}$$

or Q^* can be calculated to higher order accuracy using the more elaborate Equations (15.1.2.1.1.3-28) algorithm with m = n, $Q^* = Q_n$, and from (15.1.2.1.1.3-21):

$$\Delta\Phi_{\lambda\lambda_n} = A^*\, T_n \qquad \Delta\Phi_{\lambda y_n} = \left(G_{P^*}\, Q_{P^*_{Dens}}\, G_{P^*}^T \right) T_n \tag{15.2.1.1-15}$$

The K^* Kalman gain matrix from the (15.2.1.1-12) four-state partitioned covariance propagation/reset operations is used to set the control vectors for the $\underline{x}_a$ and $\underline{x}_b$ error state vectors. The $\underline{x}_a$ and $\underline{x}_b$ control vectors can then be recombined into $\underline{u}_c$, the unpartitioned $\underline{x}$ control vector form, for control of $\underline{x}$. Based on the forms of $\underline{x}_a$ and $\underline{x}_b$ in (15.2.1.1-2) and the $\underline{u}_c$ definition in (15.2.1-33), we define:

$$\underline{u}_{c_a} = \left(-\, \delta\omega_{IE/c_{NY}},\ -\, \gamma_{c_{NY}},\ \delta v_{c_{NX}},\ \delta\Delta R_{NX} \right)^T$$
$$\underline{u}_{c_b} = \left(\delta\omega_{IE/c_{NX}},\ \gamma_{c_{NX}},\ \delta v_{c_{NY}},\ \delta\Delta R_{c_{NY}} \right)^T \tag{15.2.1.1-16}$$

where

$\underline{u}_{c_a}, \underline{u}_{c_b}$ = Control vectors for the $\underline{x}_a$ and $\underline{x}_b$ error state vectors.

We also write for the components of the (15.2.1-31) observation equation, the non-linear version of the linearized (15.2.1.1-9) measurement equation:

$$Z_{Obs/a_n} = \Delta R_{NX_n} \qquad Z_{Obs/b_n} = \Delta R_{NY_n} \tag{15.2.1.1-17}$$

where

$Z_{Obs/a}, Z_{Obs/b}$ = N Frame X, Y (horizontal) components of the observation vector $\underline{Z}_{Obs}$.

$\Delta R_{NX}, \Delta R_{NY}$ = N Frame X, Y (horizontal) components of $\Delta \underline{R}_H^N$.

As in (15.2.1-29), the $\underline{u}_{c_a}$, $\underline{u}_{c_b}$ control vectors are set to control the estimated values of $\underline{x}_a$ and $\underline{x}_b$ to zero. As in (15.2.1-32), this corresponds to:

$$\underline{u}_{c/a_n} = - K^*_n \, \Delta R_{NX_n} \qquad \underline{u}_{c/b_n} = - K^*_n \, \Delta R_{NY_n} \tag{15.2.1.1-18}$$

Now define:

$$K^* \equiv \left(K_{\delta\omega IE}, K_\gamma, K_{\delta v}, K_{\delta\Delta R}\right)^T \tag{15.2.1.1-19}$$

where

$K_{\delta\omega IE}, K_\gamma, K_{\delta v}, K_{\delta\Delta R}$ = Components of K^*.

Then substitute (15.2.1.1-19) in (15.2.1.1-18), equate the result to (15.2.1.1-16), condense into matrix dimension form, and incorporate the definition of $\left(\underline{u}_{ZN}^N \times\right)_{2x2}$ from (15.2.1-6). The result is the elements of the (15.2.1-33) $\underline{u}_c$ control vector in terms of $\Delta \underline{R}_{2x1}^N$ observation and K^* gain components:

$$\delta\underline{\omega}_{IE/2x1/c_n}^N = K_{\delta\omega IE_n} \left(\underline{u}_{ZN}^N \times\right)_{2x2} \Delta \underline{R}_{2x1_n}^N \qquad \underline{\gamma}_{2x1/c_n}^N = K_{\gamma_n} \left(\underline{u}_{ZN}^N \times\right)_{2x2} \Delta \underline{R}_{2x1_n}^N$$
$$\delta\underline{v}_{2x1/c_n}^N = - K_{\delta v_n} \Delta \underline{R}_{2x1_n}^N \qquad \delta\Delta\underline{R}_{2x1/c_n}^N = - K_{\delta\Delta R_n} \Delta \underline{R}_{2x1_n}^N \tag{15.2.1.1-20}$$

Foreground control Equations (15.2.1-34) remain valid using (15.2.1.1-20) for the control inputs.

Summarized below, are the previous Fine Alignment processing routines described by Equations (15.2.1-1), (15.2.1.1-11) - (15.2.1.1-15), (15.2.1.1-19) - (15.2.1.1-20) and (15.2.1-34):

FOREGROUND:

$$\dot{C}_B^N = C_B^N \left(\underline{\omega}_{IB}^B \times\right) - \left(\underline{\omega}_{IE}^N \times\right) C_B^N$$

$$\underline{\omega}_{IE}^N = \underline{\omega}_{IE_H}^N + \underline{u}_{ZN}^N \, \omega_e \sin l$$

$$\dot{\underline{v}}_H^N = \left(C_B^N \, \underline{a}_{SF}^B\right)_H$$

$$\Delta\dot{\underline{R}}_H^N = \underline{v}_H^N$$

GAIN CALCULATION:

$$P^*_n(-) = \Phi^* \, P^*_{n-1}(+_e) \, \Phi^{*T} + Q^*$$

$$K^*_n \equiv \left(K_{\delta\omega IE}, K_\gamma, K_{\delta v}, K_{\delta\Delta R}\right)_n^T = P^*_n(-) \, H^{*T} \left(H^* \, P^*_n(-) \, H^{*T} + P_{RVib_H}\right)^{-1}$$

$$P^*_n(+_e) = \left(I - K^*_n \, H^*\right) P^*_n(-)$$

with the following constant matrices set-up once at the start of Fine Alignment:

$$\Phi^* = \begin{bmatrix} 1 & 0 & 0 & 0 \\ T_n & 1 & 0 & 0 \\ \frac{1}{2} g \, T_n^2 & g \, T_n & 1 & 0 \\ \frac{1}{6} g \, T_n^3 & \frac{1}{2} g \, T_n^2 & T_n & 1 \end{bmatrix} \qquad (15.2.1.1\text{-}21)$$

$$H^* \equiv \begin{bmatrix} 0 & 0 & 0 & 1 \end{bmatrix}$$

$$G_{P^*} \, Q_{P^*_{Dens}} \, G_{P^*}^T = \begin{bmatrix} 0 & 0 & 0 & 0 \\ 0 & q_{\omega Rand} & 0 & 0 \\ 0 & 0 & g^2 \, q_{\alpha Quant} + q_{aRand} & 0 \\ 0 & 0 & 0 & q_{\upsilon Quant} \end{bmatrix}$$

(Continued)

Q* To Second Order Accuracy:

$$Q_{1*} = \left(G_{P*} \, Q_{P*_{Dens}} \, G_{P*}^{T} \right) T_n \qquad Q^* \approx \frac{1}{2} Q_{1*} + \frac{1}{2} \Phi^* \, Q_{1*} \, \Phi^{*T}$$

Q* To Higher Order Accuracy:

Equations (15.1.2.1.1.3-28) with $m = n$, $Q^* = Q_n$, and from (15.1.2.1.1.3-21):

$$\Delta\Phi_{\lambda\lambda_n} = A^* \, T_n \qquad \Delta\Phi_{\lambda y_n} = \left(G_{P*} \, Q_{P*_{Dens}} \, G_{P*}^{T} \right) T_n$$

FOREGROUND CONTROL:

(15.2.1.1-21)
(Continued)

$$\underline{\omega}^{N}_{IE/2x1_n}(+c) = \underline{\omega}^{N}_{IE/2x1_n}(-) + K_{\delta\omega IE_n} \left(\underline{u}^{N}_{ZN} \times \right)_{2x2} \Delta\underline{R}^{N}_{2x1_n}$$

$$\underline{\gamma}^{N}_{Hc_n} = \begin{bmatrix} K_{\gamma_n} \left(\underline{u}^{N}_{ZN} \times \right)_{2x2} \Delta\underline{R}^{N}_{2x1_n} \\ 0 \end{bmatrix}$$

$$C^{N}_{B_n}(+c) = \left[I - \frac{\sin \gamma_{Hc_n}}{\gamma_{Hc_n}} \left(\underline{\gamma}^{N}_{Hc_n} \times \right) + \frac{1 - \cos \gamma_{Hc_n}}{\gamma_{Hc_n}^{2}} \left(\underline{\gamma}^{N}_{Hc_n} \times \right)^{2} \right] C^{N}_{B_n}(-)$$

$$\underline{v}^{N}_{2x1_n}(+c) = \underline{v}^{N}_{2x1_n}(-) - K_{\delta v_n} \, \Delta\underline{R}^{N}_{2x1_n}$$

$$\Delta\underline{R}^{N}_{2x1_n}(+c) = \Delta\underline{R}^{N}_{2x1_n}(-) - K_{\delta\Delta R_n} \, \Delta\underline{R}^{N}_{2x1_n}$$

Note that the Foreground Control equations in (15.2.1.1-21) are equivalent to the continuous form Fine Alignment process Equations (6.1.2-2) discussed in Section 6.1.2. Equivalency is achieved by equating the K_1 - K_4 gains in (6.1.2-2) to the $K^{\blacklozenge}(t)$ continuous form of the discrete $K_{\delta\omega IE}$, K_{γ}, $K_{\delta v}$, $K_{\delta\Delta R}$ gains using Section 15.1.5.3.2, Equation (15.1.5.3.2-14) for $K^{\blacklozenge}(t)$.

15.2.1.2 FOREGROUND AND KALMAN FILTER PARAMETER INITIALIZATION

We complete this section with a discussion of foreground and Kalman filter parameter initialization for the Section 15.2.1.1 partitioned quasi-stationary Fine Alignment Kalman filter configuration. Kalman filter parameters requiring initialization are the 4 by 4 P* covariance matrix, the vibration disturbance variance P_{RVib_H}, the elements of the coupled process noise density matrix $\left(G_{P*} \, Q_{P*_{Dens}} \, G_{P*}^{T} \right)$ and the Kalman control reset time interval T_n.

Initialization of the foreground navigation parameters in Equations (15.2.1.1-21) is performed at completion of Coarse Leveling (Section 6.1.1) as follows:

Foreground Initialization:

$$\underline{\omega}_{IE_H}^N = 0 \qquad \underline{v}_H^N = 0 \qquad \Delta\underline{R}_H^N = 0 \tag{15.2.1.2-1}$$

$$C_B^N = \text{Value at completion of Coarse Leveling}$$

The P* covariance matrix is initialized to reflect the error state uncertainties of Equations (15.2.1.2-1) along the horizontal N Frame axes. We assume that the initial horizontal earth rate and tilt errors in (15.2.1.2-1) are uncorrelated from each other and from the other initial position/velocity errors, which then defines the form of the initial P* matrix. Using the (15.2.1.1-2) and (15.2.1.1-11) ordering for the error state vector components, the initial P* matrix thereby has the following form:

$$P^*_0 = \begin{bmatrix} P_{\delta\omega IE/H_0} & 0 & 0 & 0 \\ 0 & P_{\gamma/H_0} & 0 & 0 \\ 0 & 0 & P_{\delta v/H_0} & P_{\delta v\delta\Delta R/H_0} \\ 0 & 0 & P_{\delta\Delta R\delta v/H_0} & P_{\delta\Delta R/H_0} \end{bmatrix} \tag{15.2.1.2-2}$$

where

P^*_0 = Initial value for P*.

$P_{\delta\omega IE/H_0}$, P_{γ/H_0}, $P_{\delta v/H_0}$, $P_{\delta v\delta\Delta R/H_0}$, $P_{\delta\Delta R\delta v/H_0}$, $P_{\delta\Delta R/H_0}$ = Initial values of the Equation (15.2.1.1-2) $\underline{x}_a$ or $\underline{x}_b$ error state vector uncertainty covariance elements.

From the $\underline{\omega}_{IE}^N$ earth rate vector expression in (6.1.2-2), since the vertical component of $\underline{\omega}_{IE}^N$ is $\omega_e \sin l$ and the $\underline{\omega}_{IE}^N$ magnitude is ω_e, the magnitude of the horizontal $\underline{\omega}_{IE}^N$ component, is $\omega_e \cos l$. Because either horizontal N Frame axis (i) can be at any arbitrary heading relative to true north ψ_{T_i}, the component of true earth rate along one of the horizontal axes ω_{IE_i} is in general:

$$\omega_{IE_i} = \omega_e \cos l \cos \psi_{T_i} \tag{15.2.1.2-3}$$

The initial uncertainty in the i axis earth rate $\delta\omega_{IE_i}$ is the difference between the initial foreground value in (15.2.1.2-1) and the true (15.2.1.2-3) value:

$$\delta\omega_{IE_i} = -\,\omega_e \cos l \cos \psi_{T_i} \tag{15.2.1.2-4}$$

The $P_{\delta\omega IE/H_0}$ variance in (15.2.1.2-2) is the expected value of (15.2.1.2-4) squared. If we assume that the initial heading ψ_{T_i} is equally likely to be at any value over all 360 degrees of possible headings, the expected value of $\cos^2\psi_{T_i}$ is 0.5, hence:

$$P_{\delta\omega IE/H_0} = \frac{1}{2}\,\omega_e^2 \cos^2 l \qquad (15.2.1.2\text{-}5)$$

The initial tilt variance P_{γ/H_0} in (15.2.1.2-2) is based on the expected residual tilt error following Coarse Leveling (Section 6.1.1). A conservative estimate (good general practice in setting Kalman filter matrix numerical values) is to set P_{γ/H_0} to correspond with 1 degree, or $1 \times 1 / (57.3^2) = 0.000305$ radians squared.

The $P_{\delta v/H_0}$, $P_{\delta v \delta\Delta R/H_0}$, $P_{\delta\Delta R \delta v/H_0}$, $P_{\delta\Delta R/H_0}$ initial horizontal velocity/position covariance elements are evaluated based on an assumed dynamic model for the quasi-stationary vibration type motion along each horizontal i axis, e.g.:

$$\dot{v}_{Vib_i} = -k\,R_{Vib_i} - c\,v_{Vib_i} + n_{Vib_i} \qquad \dot{R}_{Vib_i} = v_{Vib_i} \qquad (15.2.1.2\text{-}6)$$

where

R_{Vib_i}, v_{Vib_i} = Quasi-stationary position, velocity motion along horizontal axis i.

n_{Vib_i} = White noise quasi-stationary random acceleration forcing function input along horizontal axis i (e.g., wind gusts, stores/fuel loading, crew/passenger motion).

k, c = Spring, damping coefficients associated with the structure that prevents the vehicle carrying the INS from translating (e.g., for an aircraft, the landing gear connecting the fuselage to the wheels and runway, when the parking brake is on or wheels chocked).

Equations (15.2.1.2-6) when combined have the classic form:

$$\ddot{R}_{Vib_i} + 2\,\zeta\,\omega_n\,\dot{R}_{Vib_i} + \omega_n^2\,R_{Vib_i} = n_{Vib_i}$$
$$k = \omega_n^2 \qquad c = 2\,\zeta\,\omega_n \qquad (15.2.1.2\text{-}7)$$

where

ω_n, ζ = Undamped natural frequency and damping ratio of the (15.2.1.2-7) dynamic response characteristic.

Using the (15.2.1.2-7) definitions for k and c, Equations (15.2.1.2-6) are equivalently:

$$\dot{v}_{Vib_i} = -\omega_n^2\,R_{Vib_i} - 2\,\zeta\,\omega_n\,v_{Vib_i} + n_{Vib_i} \qquad \dot{R}_{Vib_i} = v_{Vib_i} \qquad (15.2.1.2\text{-}8)$$

The uncertainties δv_i, $\delta\Delta R_i$ in the initial foreground horizontal velocity/position are defined as the difference between the (15.2.1.2-1) horizontal velocity/position and the true horizontal velocity/position of Equations (15.2.1.2-8), hence:

$$\delta v_i = - v_{Vib_i} \qquad \delta\Delta R_i = - R_{Vib_i} \qquad (15.2.1.2\text{-}9)$$

Thus, $P_{\delta v/H_0}$, $P_{\delta v \delta\Delta R/H_0}$, $P_{\delta\Delta R \delta v/H_0}$, $P_{\delta\Delta R/H_0}$ (the initial expected value of the δv_i, $\delta\Delta R_i$ uncertainty covariance elements), equal the initial covariance of v_{Vib_i}, R_{Vib_i}. The initial covariance of v_{Vib_i}, R_{Vib_i} can be defined as the steady state solution to the covariance form of Equations (15.2.1.2-8) which, using (15.1-1) and (15.1.2.1.1-30) as templates, is:

$$\dot{P}_{Vib} = A_{Vib}\, P_{Vib} + P_{Vib}\, A_{Vib}^T + Q_{Vib/Dens}$$

$$A_{Vib} = \begin{bmatrix} -2\,\zeta\,\omega_n & -\omega_n^2 \\ 1 & 0 \end{bmatrix} \qquad Q_{Vib/Dens} = \begin{bmatrix} q_{Vib/Dens} & 0 \\ 0 & 0 \end{bmatrix} \qquad (15.2.1.2\text{-}10)$$

$$P_{Vib} \equiv \begin{bmatrix} P_{vv/Vib} & P_{vR/Vib} \\ P_{Rv/Vib} & P_{RR/Vib} \end{bmatrix}$$

where

P_{Vib} = Covariance matrix associated with the state vector $\left(v_{Vib_i}, R_{Vib_i}\right)^T$.

$q_{Vib/Dens}$ = Density of n_{Vib_i} (assumed equal for each axis i).

In the steady state (assuming convergence of (15.2.1.2-8) to a stable random solution), $\dot{P}_{Vib}$ in (15.2.1.2-10) is zero and we find from $\dot{P}_{vR/Vib}$, $\dot{P}_{Rv/Vib}$ and $\dot{P}_{RR/Vib}$ that:

$$P_{vv/Vib} - \omega_n^2\, P_{RR/Vib} - 2\,\zeta\,\omega_n\, P_{vR/Vib} = 0 \qquad (15.2.1.2\text{-}11)$$

$$P_{vv/Vib} - \omega_n^2\, P_{RR/Vib} - 2\,\zeta\,\omega_n\, P_{Rv/Vib} = 0 \qquad (15.2.1.2\text{-}12)$$

$$P_{vR/Vib} + P_{Rv/Vib} = 0 \qquad (15.2.1.2\text{-}13)$$

Summing (15.2.1.2-11) - (15.2.1.2-12) and substituting (15.2.1.2-13) shows that:

$$P_{vv/Vib} = \omega_n^2\, P_{RR/Vib} \qquad (15.2.1.2\text{-}14)$$

Differencing (15.2.1.2-11) - (15.2.1.2-12) and substituting (15.2.1.2-13) shows that:

$$P_{vR/Vib} = P_{Rv/Vib} = 0 \qquad (15.2.1.2\text{-}15)$$

We now identify $P_{RR/Vib}$ as identically the same parameter as in the Equation (15.2.1-27) measurement noise variance P_{RVib_H}, hence, with (15.2.1.2-14) we can write:

$$P_{RR/Vib} = P_{RVib_H} \qquad P_{vv/Vib} = \omega_n^2 \, P_{RVib_H} \tag{15.2.1.2-16}$$

Based on the sentence following Equation (15.2.1.2-9), we see from (15.2.1.2-15) and (15.2.1.2-16) that the initial velocity/position error covariance elements in (15.2.1.2-2) are given by:

$$P_{\delta v/H_0} = \omega_n^2 \, P_{RVib_H} \qquad P_{\delta \Delta R/H_0} = P_{RVib_H}$$
$$P_{\delta v \delta \Delta R/H_0} = P_{\delta \Delta R \delta v/H_0} = 0 \tag{15.2.1.2-17}$$

Equations (15.2.1.2-17) allow the initial horizontal velocity uncertainty variance $P_{\delta v/H_0}$ to be expressed in terms of the quasi-stationary horizontal position disturbance motion variance in P_{RVib_H} and the undamped natural frequency associated with the random translational motion process. The Equations (15.2.1.2-17) form for $P_{\delta v/H_0}$ is advantageous because the value of P_{RVib_H} has to be frequently chosen based on a very rudimentary understanding of the quasi-stationary process. The random position amplitude squared (P_{RVib_H}) can generally be estimated reasonably well by intuitive judgment. The undamped natural frequency parameter ω_n can be estimated reasonably well by intuition. We also allow a safety factor of two squared in setting P_{RVib_H}. As an example, for an aircraft/landing-gear model, we might estimate the position disturbance root-mean-square amplitude at 0.1 inch and the undamped natural frequency at 6.28 radians per second (i.e., 1 Hz). Then, providing the safety factor of 2 squared; $P_{RVib_H} = (0.1/12)^2 \times 2^2 = 2.78\text{E-}4 \ \text{ft}^2$ and $P_{\delta v/H_0} = 6.28^2 \times 2.78\text{E-}4 = 0.0109 \ (\text{fps})^2$.

The $q_{\omega Rand}$, q_{aRand} elements of the coupled process noise density matrix $\left(G_{P*} \, Q_{P*_{Dens}} \, G_{P*}^T\right)$ are equated to the values that characterize the particular angular rate sensors and accelerometers being used. The $q_{\alpha Quant}$, $q_{\upsilon Quant}$ terms in $\left(G_{P*} \, Q_{P*_{Dens}} \, G_{P*}^T\right)$ are calculated from a model for the quantization error. If the pulse size is ε for say, an accelerometer, then without quantization compensation (See Section 8.1.3), the computed velocity obtained by summing accelerometer pulses will, at a given point in time, be in error by a random number x_ε that can lie anywhere in the interval 0 to ε with equal likelihood. If quantization compensation is included in the system software, ε can be interpreted as the fraction of a pulse error that remains after compensation is applied.

We can characterize the x_ε error as being uniformly distributed from 0 to ε with a mean value of 0.5 ε. The probability density p_x of x_ε is constant (i.e., a uniform distribution). The integral of p_x over the x_ε range of 0 to ε is one (i.e., the probability is one that x_ε lies in this range). Therefore, $p_x = 1 / \varepsilon$. The uncertainty in a random parameter is typically measured by the variance about its mean defined as the expected value of the square of the difference between the parameter and its mean value. The variance of x_ε about its 0.5 ε mean (i.e., the "quantization error") can be calculated as the integral of p_x times $(x_\varepsilon - 0.5\ \varepsilon)^2$ which the reader can verify to be $\frac{1}{12}\varepsilon^2$.

The velocity including the quantization error is integrated by a sample/summing process to calculate position. Each velocity sample will contain the x_ε random quantization error, each being independent from the previous sample. Then, the position integral will contain an integrated quantization error equal to the sum of the x_ε's on the sampled velocities multiplied by the integration update time interval Δt. Each x_ε thereby increases the position error by $x_\varepsilon\ \Delta t$, corresponding to an increase in the position error variance by the variance in x_ε multiplied by the square of the integration update time interval (i.e., $\frac{1}{12}\varepsilon^2\ \Delta t^2$). The position variance is thereby increased at a rate equal to $\frac{1}{12}\varepsilon^2\ \Delta t^2$ divided by Δt, or at $\frac{1}{12}\varepsilon^2\ \Delta t = \frac{1}{12}\frac{\varepsilon^2}{f_{Intg}}$ in which f_{Intg} is the position algorithm integration rate (i.e., the reciprocal of Δt). The $\frac{1}{12}\frac{\varepsilon^2}{f_{Intg}}$ quantity represents $q_{\upsilon Quant}$ in $\left(G^*_P\ Q_{P^*Dens}\ G^{*T}_P\right)$. From this reasoning we then write for $q_{\alpha Quant}$ and $q_{\upsilon Quant}$:

$$q_{\alpha Quant} = \frac{1}{12}\frac{\varepsilon_\alpha^2}{f_{Att/Vel}} \qquad q_{\upsilon Quant} = \frac{1}{12}\frac{\varepsilon_\upsilon^2}{f_{Vel/Pos}} \qquad (15.2.1.2\text{-}18)$$

where

$\varepsilon_\alpha, \varepsilon_\upsilon$ = Angular rate sensor and accelerometer output pulse sizes (or equivalent remaining after applying quantization compensation if used) (measured in radians and feet per second).

$f_{Att/Vel}$ = Frequency that attitude is used in the foreground calculations for acceleration-transformation/velocity-integration update (Hz).

$f_{Vel/Pos}$ = Frequency that velocity is used in the foreground calculations for position integration update (Hz).

We finally discuss the selection of the T_n Kalman filter update time interval. For the quasi-stationary Fine Alignment problem, the Kalman filter update time is selected long enough to validate the fundamental assumption implied in the measurement noise model; namely, that the measurement noise will be uncorrelated with itself from measurement to measurement. The measurement noise is the quasi-stationary horizontal position motion (per Equation (15.2.1-26)), hence, its dynamic model along each horizontal axis is as defined by Equations (15.2.1.2-8). The T_n used should be long enough that the transient response of (15.2.1.2-8) to the R_{Vib_i} position at the previous measurement time, has decayed to near zero at the current measurement time. The characteristic roots of Equation (15.2.1.2-8) contain a transient time constant of $\frac{1}{\zeta\, \omega_n}$. Hence, to assure a valid Kalman filter structure, T_n should be selected to be two or three times $\frac{1}{\zeta\, \omega_n}$. Estimation of the damping ratio ζ for T_n determination is usually based on intuitive judgment (e.g., 0.5 for the aircraft/landing-gear structure). If the undamped natural frequency in the previous example is 1 Hz (or 6.28 radians per second for ω_n), the corresponding time constant would be $\frac{1}{0.5 \times 6.28} = 0.318$ seconds. Allowing three time constants for safety would set $T_n = 1$ second.

15.2.2 KALMAN FILTERING APPLIED TO DYNAMIC MOVING BASE INS INITIAL ALIGNMENT

In some circumstances, INS initial alignment is required under dynamic motion in which the INS to be initialized is located on a dynamically moving vehicle (i.e., "moving base" INS alignment). For such a scenario, a reference aiding device is required in the observation calculation for Kalman filter input (unlike the quasi-stationary alignment process discussed in Section 15.2.1 for which the aiding device was replaced by knowledge that the INS was quasi-stationary). For the moving base alignment, the aiding device is typically another INS located in the same vehicle. A classic example of moving base alignment is the initialization of an INS in an air-launched missile prior to missile launch, using the launching aircraft INS as the reference aiding device. In effect, the missile INS alignment initialization operation transfers the aircraft INS attitude/heading into the missile INS, thus the terminology "transfer alignment" to describe the moving base alignment process when using a reference INS for the aiding device.

The basic principle used in defining the observation equation for the moving base transfer alignment Kalman filter is that INS attitude/heading errors affect the calculation of navigation frame acceleration, thereby developing velocity errors in the inertial navigation computation. A comparison between velocity (or integrated velocity) data between the INS and reference aiding device, therefore, provides a measurement of the effect of attitude/heading error. Use of such a measurement is sometimes referred to as "velocity matching" or "integrated velocity matching", depending on the form of the observation utilized.

In the following subsections we will describe two forms of the integrated velocity matching moving base alignment filter; a version based on an E Frame observation and a version based on an N Frame observation. The final subsection discusses the alternative of velocity matching compared to the integrated velocity matching methods described.

15.2.2.1 MOVING BASE ALIGNMENT USING AN E FRAME OBSERVATION

In this section we describe an integrated velocity matching scheme in which the observation is formed in the earth fixed E Frame. The rationale for choosing the E Frame is that it is well defined without singularities for all INS position locations, and is known in both the INS and aiding reference, hence, is a convenient frame for vector data communication between navigation devices.

An observation equation for moving base alignment can be developed in the E Frame by considering the relationship between the position location of the INS that is being aligned (initialized), and the position location of the reference aiding device:

$$\underline{R}^E_{INS} = \underline{R}^E_{REF} + \underline{l}^E \qquad (15.2.2.1\text{-}1)$$

where

$\underline{R}^E_{INS}$ = E Frame position vector from earth's center to the INS being initialized.

$\underline{R}^E_{REF}$ = E Frame position vector from earth's center to the reference aiding device.

$\underline{l}^E$ = E Frame components of the lever arm from the reference aiding device to the INS being initialized.

The derivative of (15.2.2.1-1) is:

$$\dot{\underline{R}}^E_{INS} = \dot{\underline{R}}^E_{REF} + \dot{\underline{l}}^E \qquad (15.2.2.1\text{-}2)$$

Using the Equation (4.3-1) definition for velocity relative to the earth, we see that (15.2.2.1-2) is equivalently:

$$\underline{v}^E_{INS} = \underline{v}^E_{REF} + \dot{\underline{l}}^E \qquad (15.2.2.1\text{-}3)$$

From (4.3-1), we can integrate Equation (15.2.2.1-2) from the start of the moving base alignment computational process to obtain the equivalent position change form at time t:

$$\Delta \underline{R}^E_{INS}(t) = \Delta \underline{R}^E_{REF}(t) + \underline{l}^E(t) - \underline{l}^E_0 \qquad (15.2.2.1\text{-}4)$$

with

$$\Delta \underline{R}^E_{INS}(t) = \int_0^t \underline{v}^E_{INS}\, dt \qquad \Delta \underline{R}^E_{REF}(t) = \int_0^t \underline{v}^E_{REF}\, dt \qquad (15.2.2.1\text{-}5)$$

where

$\Delta \underline{R}^E_{INS}(t)$, $\Delta \underline{R}^E_{REF}(t)$ = Position change relative to the earth of the INS and the reference aiding device at time t into alignment operations.

$\underline{v}^E_{INS}$, $\underline{v}^E_{REF}$ = Velocity relative to the earth in the E Frame of the INS and reference aiding device.

0 = Subscript designating initial value at the start of the moving base alignment process.

Equations (15.2.2.1-4) - (15.2.2.1-5) are the basis for the moving base alignment observation equation which we now state as:

$$\underline{Z}_{Obs_n} = \widehat{\underline{M}}^E_n - \left(\hat{\underline{l}}^E_n - \hat{\underline{l}}^E_0\right) \qquad (15.2.2.1\text{-}6)$$

$$\widehat{\underline{M}}^E(t) = \Delta \widehat{\underline{R}}^E_{INS}(t) - \Delta \widehat{\underline{R}}^E_{REF}(t) = \int_0^t \left(\widehat{C}^E_N\, \widehat{\underline{v}}^N - \widehat{\underline{v}}^E_{REF}\right) dt \qquad (15.2.2.1\text{-}7)$$

where

$\underline{Z}_{Obs}$ = Observation vector.

$\widehat{\ }$ = Designates parameters calculated in the INS or aiding device computer, hence, contain errors.

$\widehat{\underline{v}}^N$ = $\widehat{\underline{v}}^N_{INS}$ with the INS label dropped for simplicity. For the remainder of this section, the INS subscript will be dropped for all INS computed parameters.

The horizontal components of the (15.2.2.1-6) observation equation are generally sufficient for a successful transfer alignment. These can be obtained by transforming (15.2.2.1-6) to the N Frame and deleting the N Frame Z axis component. However, the accompanying analytics to derive the corresponding measurement equation (the differential of the observation equation) then becomes more involved. In the interests of analytical expediency, this section will be based on the full (15.2.2.1-6) vector form for the observation equation. Derivation of the equivalent horizontal component form will be left as an exercise for the interested reader.

Equation (15.2.2.1-6) can be clarified by defining the lever arm components in the strapdown sensor axis ("body") B Frame in which they are stationary, thus:

$$\underline{\hat{l}}_n^E = \hat{C}_{B_n}^E \, \underline{\hat{l}}^B \qquad \underline{\hat{l}}_0^E = \hat{C}_{B_0}^E \, \underline{\hat{l}}^B \tag{15.2.2.1-8}$$

With (15.2.2.1-8), Equation (15.2.2.1-6) becomes:

$$\underline{Z}_{Obs_n} = \underline{\hat{M}}_n^E - \left(\hat{C}_{B_n}^E - \hat{C}_{B_0}^E\right) \underline{\hat{l}}^B \tag{15.2.2.1-9}$$

Assuming a set of foreground navigation equations in which the L Frame is used for B Frame referencing and the N Frame for velocity/position referencing, the direction cosine matrices in (15.2.2.1-9) would be calculated from:

$$\begin{aligned} \hat{C}_{B_n}^E &= \hat{C}_{N_n}^E \, C_L^N \, \hat{C}_{B_n}^L \\ \hat{C}_{B_0}^E &= \text{Provided as input to the INS as attitude initialization data.} \end{aligned} \tag{15.2.2.1-10}$$

Equations (15.2.2.1-7) and (15.2.2.1-9) - (15.2.2.1-10) constitute the observation equations for the moving base alignment. The corresponding measurement equation is obtained as the differential of observation Equation (15.2.2.1-9):

$$\underline{z}_n = \delta \underline{M}_n^E - \left(\delta C_{B_n}^E - \delta C_{B_0}^E\right) \underline{\hat{l}}^B - \left(\hat{C}_{B_n}^E - \hat{C}_{B_0}^E\right) \delta \underline{l}_n^B \tag{15.2.2.1-11}$$

The direction cosine error terms in (15.2.2.1-11) can be related to attitude error vector equivalents using Equation (12.2.1-3). We also assume that the attitude uncertainty during the moving base alignment process approximately equals the initial attitude uncertainty (assuming that Kalman filter attitude estimate updates will also be used to update the initial attitude error uncertainty). Thus:

$$\begin{aligned} \delta C_{B_n}^E &= -\left(\underline{\psi}_n^E \times\right) \hat{C}_{B_n}^E \\ \delta C_{B_0}^E &= -\left(\underline{\psi}_0^E \times\right) \hat{C}_{B_0}^E \approx -\left(\underline{\psi}_n^E \times\right) \hat{C}_{B_0}^E \end{aligned} \tag{15.2.2.1-12}$$

and from (12.2.1-7):

$$\underline{\psi}_n^E = \hat{C}_{N_n}^E \, \underline{\psi}_n^N \tag{15.2.2.1-13}$$

The $\delta \underline{l}^B$ lever arm uncertainty in (15.2.2.1-11) is the difference between $\underline{\hat{l}}^B$ and the true lever arm value $\underline{l}^B$. The true lever arm can be modeled as a constant plus a dynamic flexing component:

$$\underline{l}^B = \underline{l}^B_{Cnst} + \underline{l}^B_{Flex} \qquad \dot{\underline{l}}^B_{Cnst} = 0 \tag{15.2.2.1-14}$$

where

$\underline{l}^B_{Cnst}, \underline{l}^B_{Flex}$ = Constant and flexing portions of portion $\underline{l}^B$.

The computer's estimate for the lever arm ($\hat{\underline{l}}^B$) is modeled as a constant:

$$\hat{\underline{l}}^B = \hat{\underline{l}}^B_{Cnst} \qquad \dot{\hat{\underline{l}}}^B_{Cnst} = 0 \tag{15.2.2.1-15}$$

The $\delta\underline{l}^B$ lever arm uncertainty in (15.2.2.1-11) is (15.2.2.1-15) minus (15.2.2.1-14):

$$\delta\underline{l}^B = \hat{\underline{l}}^B - \underline{l}^B = \delta\underline{l}^B_{Cnst} - \underline{l}^B_{Flex} \qquad \delta\dot{\underline{l}}^B_{Cnst} = 0 \tag{15.2.2.1-16}$$

With (15.2.2.1-12), (15.2.2.1-13) and (15.2.2.1-16) in (15.2.2.1-11), we obtain the equivalent measurement equation form:

$$z_n = \delta\underline{M}^E_n - \left\{\left[\left(\hat{C}^E_{B_n} - \hat{C}^E_{B_0}\right)\hat{\underline{l}}^B\right]\times\right\}\hat{C}^E_{N_n}\,\underline{\psi}^N_n - \left(\hat{C}^E_{B_n} - \hat{C}^E_{B_0}\right)\delta\underline{l}^B_{Cnst} + \left(\hat{C}^E_{B_n} - \hat{C}^E_{B_0}\right)\underline{l}^B_{Flex_n} \tag{15.2.2.1-17}$$

Using (15.2.2.1-17) for the measurement equation, we see from the (15.1-2) standard form that $\underline{l}^B_{Flex}$ is the measurement noise for the moving base alignment Kalman filter defined thus far. This is a very convenient result because an estimate for flexure position vibration amplitude (for the Kalman filter measurement noise matrix) is generally fairly easy to assess by engineering judgment for a given application. In fact, knowing that $\underline{l}^B_{Flex}$ would be the measurement noise was part of the rationale in choosing integrated velocity matching to form the observation.

The $\delta\underline{M}^E_n$ term in (15.2.2.1-17) can be related to velocity error equivalents (defined in the E Frame) from the differential of (15.2.2.1-7) evaluated at t_n, using the (12.2.2-1) and (12.2.2-3) velocity error definitions:

$$\delta\underline{M}^E(t) = \int_0^t \delta\dot{\underline{M}}^E\,dt \tag{15.2.2.1-18}$$

$$\begin{aligned}\delta\dot{\underline{M}}^E &= \delta\left(\hat{C}^E_N\,\hat{\underline{v}}^N - \hat{\underline{v}}^E_{REF}\right) = \delta\hat{\underline{v}}^E - \delta\hat{\underline{v}}^E_{REF} \\ &= \delta\underline{V}^E - \delta\underline{V}^E_{REF} = \hat{C}^E_N\,\delta\underline{V}^N - \delta\underline{V}^E_{REF}\end{aligned} \tag{15.2.2.1-19}$$

Note in (15.2.2.1-18), that the initial value of $\delta \underline{M}^E(t)$ is zero. This is justified from the formal definition of $\delta \underline{M}^E$ as the difference between $\widehat{\underline{M}}^E$ in (15.2.2.1-7) and its true value; i.e., $\delta \underline{M}^E \equiv \widehat{\underline{M}}^E - \left(\Delta \underline{R}^E_{INS} - \Delta \underline{R}^E_{Ref}\right)$. We see from (15.2.2.1-7), that $\widehat{\underline{M}}^E$ is zero at time $t = 0$ as is $\Delta \underline{R}^E_{INS} - \Delta \underline{R}^E_{Ref}$ based on (15.2.2.1-4). Thus, $\delta \underline{M}^E$ is also zero at $t = 0$ as stipulated in (15.2.2.1-18).

For the dynamic moving base alignment problem, accurate synchronization between the INS and reference aiding device is critical in forming the Equations (15.2.2.1-7) and (15.2.2.1-9) observation input to the Kalman filter. Assuming that the (15.2.2.1-7) integration is performed in the INS computer, synchronization inaccuracy between the INS and aiding device can be attributed to timing uncertainty in the $\hat{\underline{v}}^E_{REF}$ data input to the INS. Inaccuracy in the $\hat{\underline{v}}^E_{REF}$ input due to timing uncertainty (known as "senescence" error) is part of the $\delta \underline{V}^E_{Ref}$ error in (15.2.2.1-19) which can be expanded to read:

$$\delta \underline{V}^E_{Ref} = \delta \underline{V}^E_{Ref/Sen} + \delta \underline{V}^E_{Ref/Other} \tag{15.2.2.1-20}$$

where

$\delta \underline{V}^E_{Ref/Sen}$ = Reference velocity error due to $\hat{\underline{v}}^E_{REF}$ time stamp uncertainty data senescence.

$\delta \underline{V}^E_{Ref/Other}$ = Remaining errors in $\delta \underline{V}^E_{Ref}$ other than $\delta \underline{V}^E_{Ref/Sen}$.

The $\delta \underline{V}^E_{Ref/Sen}$ error is created by velocity change (acceleration) in the presence of data senescence for which we can write:

$$\delta \underline{V}^E_{Ref/Sen} = \dot{\hat{\underline{v}}}^E_{REF} \, \delta\tau_{Sen} \tag{15.2.2.1-21}$$

where

$\delta\tau_{Sen}$ = Data senescence time stamp uncertainty.

The $\dot{\hat{\underline{v}}}^E_{REF}$ term in (15.2.2.1-21) can be calculated from INS data using the derivative of (15.2.2.1-3), rearranged:

$$\dot{\hat{\underline{v}}}^E_{REF} = \dot{\hat{\underline{v}}}^E - \ddot{\hat{\underline{l}}}^E \tag{15.2.2.1-22}$$

The $\dot{\hat{\underline{v}}}^E$ term in (15.2.2.1-22) is the derivative of:

$$\hat{\underline{v}}^E = \hat{C}^E_N \, \hat{\underline{v}}^N \tag{15.2.2.1-23}$$

Taking the derivative of (15.2.2.1-23) and applying generalized Equation (3.3.2-6) yields:

$$\dot{\hat{\underline{v}}}^E = \hat{C}^E_N \, \dot{\hat{\underline{v}}}^N + \dot{\hat{C}}^E_N \, \hat{\underline{v}}^N = \hat{C}^E_N \left(\dot{\hat{\underline{v}}}^N + \hat{\underline{\omega}}^N_{EN} \times \hat{\underline{v}}^N \right) \tag{15.2.2.1-24}$$

The $\ddot{\hat{\underline{l}}}^E$ term in (15.2.2.1-22) is evaluated from the double derivative of the transformed B Frame lever arm components, while recognizing $\hat{\underline{l}}^B$ to be approximately constant:

$$\ddot{\hat{\underline{l}}}^E = \ddot{\hat{C}}^E_B \, \hat{\underline{l}}^B \tag{15.2.2.1-25}$$

The $\ddot{\hat{C}}^E_B$ term in (15.2.2.1-25) is determined by successive application of generalized Equation (3.3.2-6) and its derivative, and approximating the angular rate of the B Frame relative to the earth as the B Frame rate relative to inertial space:

$$\begin{aligned} \ddot{\hat{C}}^E_B &= \frac{d}{dt} \dot{\hat{C}}^E_B = \frac{d}{dt}\left[\hat{C}^E_B \left(\hat{\underline{\omega}}^B_{EB} \times \right) \right] = \dot{\hat{C}}^E_B \left(\hat{\underline{\omega}}^B_{EB} \times \right) + \hat{C}^E_B \left(\dot{\hat{\underline{\omega}}}^B_{EB} \times \right) \\ &= \hat{C}^E_B \left(\hat{\underline{\omega}}^B_{EB} \times \right)^2 + \hat{C}^E_B \left(\dot{\hat{\underline{\omega}}}^B_{EB} \times \right) = \hat{C}^E_B \left[\left(\hat{\underline{\omega}}^B_{EB} \times \right)^2 + \left(\dot{\hat{\underline{\omega}}}^B_{EB} \times \right) \right] \\ &\approx \hat{C}^E_B \left[\left(\tilde{\underline{\omega}}^B_{IB} \times \right)^2 + \left(\dot{\tilde{\underline{\omega}}}^B_{IB} \times \right) \right] \end{aligned} \tag{15.2.2.1-26}$$

where

$\sim$ = Designation for sensor inputs containing errors (in this case, the angular rate sensors).

With (15.2.2.1-24) - (15.2.2.1-26) and (15.2.2.1-21) - (15.2.2.1-22), Equation (15.2.2.1-20) for the $\delta \underline{V}^E_{Ref}$ term in (15.2.2.1-19) becomes:

$$\delta \underline{V}^E_{Ref} = \left\{ \hat{C}^E_N \left(\dot{\hat{\underline{v}}}^N + \hat{\underline{\omega}}^N_{EN} \times \hat{\underline{v}}^N \right) - \hat{C}^E_B \left[\left(\tilde{\underline{\omega}}^B_{IB} \times \right)^2 + \left(\dot{\tilde{\underline{\omega}}}^B_{IB} \times \right) \right] \hat{\underline{l}}^B \right\} \delta\tau_{Sen} + \delta \underline{V}^E_{Ref/Other} \tag{15.2.2.1-27}$$

The $\delta\tau_{Sen}$ senescence error in (15.2.2.1-27) might be modeled as the sum of a random walk changing constant $\delta\tau_{Sen/RndCnst}$, a first order Markov process $\delta\tau_{Sen/Mark}$ (see Section 12.5.6 for definition), plus a completely random jitter component $\delta\tau_{Sen/Jit}$:

$$\delta\tau_{Sen} = \delta\tau_{Sen/RndCnst} + \delta\tau_{Sen/Mark} + \delta\tau_{Sen/Jit} \qquad (15.2.2.1\text{-}28)$$

$$\begin{aligned} \dot{\delta\tau}_{Sen/RndCnst} &= n_{Sen/RndCnst} \\ \dot{\delta\tau}_{Sen/Mrk} &= - C_{Sen/Mrk}\, \delta\tau_{Sen/Mrk} + n_{Sen/Mrk} \end{aligned} \qquad (15.2.2.1\text{-}29)$$

Simplified versions of (15.2.2.1-28) - (15.2.2.1-29) are also possible. For example, $\delta\tau_{Sen/RndCnst}$ and $\delta\tau_{Sen/Mark}$ can be merged into a combined single error state that uses the $\dot{\delta\tau}_{Sen/RndCnst}$ model initially (when $\delta\tau_{Sen/RndCnst}$ is large), and then switches to the $\dot{\delta\tau}_{Sen/Mrk}$ model when the $\delta\tau_{Sen/RndCnst}$ variance is reduced by the Kalman filter estimation process to the $\delta\tau_{Sen/Mark}$ initial uncertainty.

We now substitute (15.2.2.1-27) with (15.2.2.1-28) into (15.2.2.1-19) to obtain for $\delta\dot{\underline{M}}$:

$$\begin{aligned} \delta\dot{\underline{M}}^E &= \hat{C}_N^E\, \delta\underline{V}^N \\ &- \left\{\hat{C}_N^E \left(\dot{\hat{\underline{v}}}^N + \underline{\omega}_{EN}^N \times \hat{\underline{v}}^N\right) - \hat{C}_B^E \left[\left(\tilde{\underline{\omega}}_{IB}^B \times\right)^2 + \left(\dot{\tilde{\underline{\omega}}}_{IB}^B \times\right)\right] \hat{\underline{l}}^B\right\} \left(\delta\tau_{Sen/RndCnst} + \delta\tau_{Sen/Mark}\right) \\ &- \left\{\hat{C}_N^E \left(\dot{\hat{\underline{v}}}^N + \underline{\omega}_{EN}^N \times \hat{\underline{v}}^N\right) - \hat{C}_B^E \left[\left(\tilde{\underline{\omega}}_{IB}^B \times\right)^2 + \left(\dot{\tilde{\underline{\omega}}}_{IB}^B \times\right)\right] \hat{\underline{l}}^B\right\} \delta\tau_{Sen/Jit} - \delta\underline{V}_{Ref/Other}^E \end{aligned} \qquad (15.2.2.1\text{-}30)$$

Let us summarize the results we have obtained thus far before continuing the discussion. The observation equation for the moving base alignment Kalman filter is given by (15.2.2.1-9) - (15.2.2.1-10) with input from Equations (15.2.2.1-7). The measurement equation is given by (15.2.2.1-17) with input from (15.2.2.1-16), (15.2.2.1-18), (15.2.2.1-29) and (15.2.2.1-30). Both equation sets are repeated below for easy reference:

<u>Observation Equation</u>:

$$\begin{aligned} \underline{Z}_{Obs_n} &= \hat{\underline{M}}_n^E - \left(\hat{C}_{B_n}^E - \hat{C}_{B_0}^E\right) \hat{\underline{l}}^B \\ \hat{C}_{B_n}^E &= \hat{C}_{N_n}^E\, C_L^N\, \hat{C}_{B_n}^L \\ \hat{C}_{B_0}^E &= \text{Provided as input to the INS as attitude initialization data.} \\ \hat{\underline{M}}^E(t) &= \int_0^t \left(\hat{C}_N^E\, \hat{\underline{v}}^N - \hat{\underline{v}}_{REF}^E\right) dt \end{aligned} \qquad (15.2.2.1\text{-}31)$$

(Continued)

Measurement Equation:

$$\underline{z}_n = \delta\underline{M}_n^E - \left\{\left[\left(\hat{C}_{B_n}^E - \hat{C}_{B_0}^E\right)\underline{\hat{l}}^B\right]\times\right\}\hat{C}_{N_n}^E\,\underline{\psi}_n^N - \left(\hat{C}_{B_n}^E - \hat{C}_{B_0}^E\right)\delta\underline{l}_{Cnst}^B + \left(\hat{C}_{B_n}^E - \hat{C}_{B_0}^E\right)\underline{l}_{Flex_n}^B$$

$$\delta\dot{\underline{l}}_{Cnst}^B = 0$$

$$\delta\underline{M}^E(t) = \int_0^t \delta\dot{\underline{M}}^E\,dt \qquad (15.2.2.1\text{-}31)\ \text{(Continued)}$$

$$\delta\dot{\underline{M}}^E = \hat{C}_N^E\,\delta\underline{V}^N$$

$$- \left\{\hat{C}_N^E\left(\dot{\hat{\underline{v}}}^N + \underline{\omega}_{EN}^N \times \hat{\underline{v}}^N\right) - \hat{C}_B^E\left[\left(\underline{\tilde{\omega}}_{IB}^B\times\right)^2 + \left(\dot{\underline{\tilde{\omega}}}_{IB}^B\times\right)\right]\underline{\hat{l}}^B\right\}\left(\delta\tau_{Sen/RndCnst} + \delta\tau_{Sen/Mark}\right)$$

$$- \left\{\hat{C}_N^E\left(\dot{\hat{\underline{v}}}^N + \underline{\omega}_{EN}^N \times \hat{\underline{v}}^N\right) - \hat{C}_B^E\left[\left(\underline{\tilde{\omega}}_{IB}^B\times\right)^2 + \left(\dot{\underline{\tilde{\omega}}}_{IB}^B\times\right)\right]\underline{\hat{l}}^B\right\}\delta\tau_{Sen/Jit} - \delta\underline{V}_{Ref/Other}^E$$

$$\delta\dot{\tau}_{Sen/RndCnst} = n_{Sen/RndCnst}$$

$$\delta\dot{\tau}_{Sen/Mrk} = -\,C_{Sen/Mrk}\,\delta\tau_{Sen/Mrk} + n_{Sen/Mrk}$$

The measurement equation in (15.2.2.1-31) identifies the error states required from the Kalman filter error models to include $\delta\underline{M}^E$, $\delta\underline{V}^N$, $\underline{\psi}^N$, $\delta\underline{l}_{Cnst}^B$, $\delta\tau_{Sen/RndCnst}$, $\delta\tau_{Sen/Mark}$ and $\delta\underline{V}_{Ref/Other}^E$. The error state vector for the moving base alignment filter must include these error effects to properly account for potential errors in the observation vector $\underline{Z}_{Obs}$ input to the Kalman filter.

Error state dynamic equations for $\delta\underline{V}^N$ and $\underline{\psi}^N$ can be provided from the INS error model such as described by Equations (12.5.1-1) (or a simplified equivalent). Classical models for the inertial sensor errors in Equations (12.5.1-1) are discussed in Section 12.5.6. The $\delta\underline{M}^E$, $\delta\underline{l}_{Cnst}^B$, $\delta\tau_{Sen/RndCnst}$, $\delta\tau_{Sen/Mark}$ error state dynamic models are the $\delta\dot{\underline{M}}^E$, $\delta\dot{\underline{l}}_{Cnst}^B$, $\delta\dot{\tau}_{Sen/RndCnst}$, $\delta\dot{\tau}_{Sen/Mrk}$ expressions in (15.2.2.1-31). The error model for $\delta\underline{V}_{Ref/Other}^E$ typically assumes a perfect reference, except for random output errors approximated as white noise (which then would be considered as process noise in the (15.2.2.1-31) $\delta\dot{\underline{M}}^E$ expression). If the reference aiding device is another inertial navigation system, $\delta\underline{V}_{Ref/Other}^E$ can be equated to the accelerometer quantization error in the reference INS velocity data, augmented in magnitude to also account for other random effects such as angular rate sensor error. If the reference aiding device is an inertial navigation system with an error model similar to the INS being aligned, a more sophisticated approach can be considered; $\delta \underline{V}^{E}_{Ref/Other}$ can be approximated by the reference device accelerometer quantization error, with the remaining contributing process noise terms to $\delta \underline{V}^{E}_{Ref/Other}$ then absorbed into the process noise contributing to the $\delta \underline{V}^{N}$ error state. This latter approach is possible due to the form of the $\delta \dot{\underline{M}}^{E}$ expression in (15.2.2.1-31) which contains as input, the direct difference between $\delta \underline{V}^{E}_{Ref/Other}$ and transformed $\delta \underline{V}^{N}$.

Control resets for the moving base alignment filter can be designed for the measurement $\delta \underline{M}^{E}$, the INS navigation and inertial sensor errors, and for the lever arm $\delta \underline{l}^{N}_{Cnst}$. The techniques described in Section 15.1.2.3 can be utilized to implement the foreground control reset process. For example, if all of the Equations (12.5.1-1) navigation error terms are included as error states, the error state vector would be of the form:

$$\underline{x} = \left[\left(\delta \underline{M}^{E}\right)^{T}, \left(\underline{\psi}^{N}\right)^{T}, \left(\delta \underline{V}^{N}\right)^{T}, \left(\delta \underline{R}^{N}\right)^{T}, \underline{SensErr}^{T}, \left(\delta \underline{l}^{B}_{Cnst}\right)^{T}, \delta\tau_{Sen/RndCnst}, \delta\tau_{Sen/Mark} \right]^{T} \tag{15.2.2.1-32}$$

where

$\underline{SensErr}$ = Vector of inertial sensor error states.

The associated $\underline{u}_{c}$ control vector calculated from error state vector estimates would then be:

$$\underline{u}_{c} = \left[\left(\delta \underline{M}^{E}_{c}\right)^{T}, \left(\underline{\psi}^{N}_{c}\right)^{T}, \left(\delta \underline{V}^{N}_{c}\right)^{T}, \left(\delta \underline{R}^{N}_{c}\right)^{T}, \underline{SensErr}^{T}_{c}, \left(\delta \underline{l}^{B}_{Cnst_{c}}\right)^{T}, 0, 0 \right]^{T} \tag{15.2.2.1-33}$$

The control vector would be set to the negative of the estimated error states being controlled to drive the controlled error states to zero (as discussed in Section 15.1.2). If the method of delayed resets is being used as applied in Section 15.1.2.4 with Equation (15.1.2.4-8), we would then set:

$$\underline{u}_{c_{n+c}} = - \tilde{\underline{x}}_{Cntrld_{n}}(+_{e}) \tag{15.2.2.1-34}$$

where

$\tilde{\underline{x}}_{Cntrld}$ = The estimated error state vector $\tilde{\underline{x}}$, but substituting zero for the components not being controlled.

Assuming that the INS navigation parameters are integrated by the foreground in N Frame coordinates (e.g., Equations (12.1-1) - (12.1-12)), Equations (15.1.2.3-16) - (15.1.2.3-20) would be used for control reset of the foreground navigation parameters. Additionally, the $\hat{C}^{E}_{B_0}$

matrix, provided as input initialization data, would be updated for the $\underline{\psi}_c^N$ control resets based on Equations (15.2.2.1-12) - (15.2.2.1-13) using (15.1.2.3-20) for the form of matrix update operation:

$$\underline{\psi}_c^E = \hat{C}_{N_n}^E \underline{\psi}_c^N \qquad \hat{C}_{B_0}^E(+) = \left[I - \frac{\sin \psi_c}{\psi_c} \left(\underline{\psi}_c^E \times \right) + \frac{(1 - \cos \psi_c)}{\psi_c^2} \left(\underline{\psi}_c^E \times \right)^2 \right] \hat{C}_{B_0}^E(-) \quad (15.2.2.1\text{-}35)$$

The measurement, lever arm and inertial sensor error controls would be applied as:

$$\begin{aligned} \delta \underline{M}^E(+) &= \delta \underline{M}^E(-) + \delta \underline{M}_c^E \\ \hat{\underline{l}}^B(+) &= \hat{\underline{l}}^B(-) + \delta \underline{l}_{Cnst_c}^B \\ \underline{\widehat{Coef}}(+) &= \underline{\widehat{Coef}}(-) + \underline{SensErr}_c \end{aligned} \qquad (15.2.2.1\text{-}36)$$

where

$\underline{\widehat{Coef}}$ = Inertial sensor compensation coefficients whose errors are represented by $\underline{SensErr}$.

Based on the foregoing analytical discussion, the moving base alignment Kalman filter is then easily constructed from the defined error states in (15.2.2.1-32), their associated error state dynamic equations (from (12.5.1-1) and (15.2.2.1-31) with format Equation (15.1-1)), the (15.2.2.1-31) observation and measurement equations (with format Equation (15.1-2)) and the previously described control reset implementation. Equations (15.1.2.4-1) - (15.1.2.4-11) can be used for the overall Kalman filter configuration with the error state transition and Kalman gain matrices calculated as described in Sections 15.1.2.1, 15.1.2.1.1 and 15.1.2.1.1.1 - 15.1.2.1.1.3.

Foreground initialization for the previous moving base alignment process initializes $\hat{\underline{M}}^E(t)$ to zero, $\hat{\underline{l}}^B$, $\underline{\widehat{Coef}}$ at their last calibrated values, and the attitude, velocity, position navigation parameters ($\hat{C}_B^L$, $\hat{\underline{v}}^N$, $\hat{C}_N^E$, $\hat{h}$) based on inputs from the aiding device or other systems in the vehicle executing the overall alignment process. The initial N Frame definition for this process would be selected for consistency with the available navigation input data format (e.g., North/East/Vertical). The initial L Frame orientation is set once the N Frame is selected (from the L Frame definition as provided numerically by Equation (4.1.1-2)). The initial $\hat{C}_N^E$, $\hat{h}$ position data can generally be set to the equivalent initial position data provided. For enhanced accuracy, the initial position can be adjusted for the known lever arm between the INS and the device providing the initial position input. In the case of the velocity $\hat{\underline{v}}^N$ initialization using aiding device input velocity, the $\hat{\underline{l}}^B$ lever arm effect should generally be taken into account. For example, using Equation (15.2.2.1-3) transformed to the N Frame with generalized Equation (3.4-4), and approximating $\hat{\underline{l}}^B$ as constant

$$\begin{aligned}\hat{\underline{v}}^N_{INS} &= \hat{C}^N_E\, \hat{\underline{v}}^E_{REF} + \hat{C}^N_E\, \dot{\hat{\underline{l}}}^E = \hat{C}^N_E\, \hat{\underline{v}}^E_{REF} + \hat{C}^N_E\, \hat{C}^E_B \left(\hat{\underline{\omega}}^B_{EB} \times \hat{\underline{l}}^B\right) \\ &\approx \hat{C}^N_E\, \hat{\underline{v}}^E_{REF} + \hat{C}^N_B \left(\hat{\underline{\omega}}^B_{IB} \times \hat{\underline{l}}^B\right)\end{aligned} \quad (15.2.2.1\text{-}37)$$

The error state vector would be initialized at zero (i.e., as in Equation (15.1.2.4-11)) and the initial value for the covariance matrix would be based on the error in the foreground initialization. Based on the (15.2.2.1-31) definition for $\delta\underline{M}^E(t)$ as a closed integral from time $t = 0$, the initial value for the $\delta\underline{M}^E(t)$ uncertainty would be equated to zero. Initial uncertainties selected for SensErr would be based on knowledge of the inertial sensor calibration accuracy and stability since calibration. The initial uncertainty in $\delta\underline{l}^B_{Cnst}$ would be based on an estimate for the $\hat{\underline{l}}^B$ determination accuracy. The uncertainty in the initial attitude error $\underline{\psi}^N$ would be based on an estimate for the alignment uncertainty between the INS and the device providing the attitude initialization data, including vehicle flexing under acceleration (e.g., 1 degree as a conservative estimate). The initial uncertainty in the velocity error $\delta\underline{V}^N$ can be based on an estimate for lever arm dynamic flexing between the INS and referencing aiding device using a simplified dynamic flexure model (e.g., similar to Equations (15.2.1.2-8) and (15.2.1.2-10) - (15.2.1.2-16) with R_{Vib} and v_{Vib} interpreted as dynamic position/velocity flexure), plus the effect of $\hat{\underline{l}}^B$ uncertainty in (15.2.2.1-37) which can be approximated as $\hat{C}^N_B \left(\hat{\underline{\omega}}^B_{IB} \times \delta\underline{l}^B_{Cnst}\right)$. The initial position uncertainty can be approximated as zero. Even though it is recognized that the uncertainty between some error states may be correlated (e.g., initial lever arm and velocity error uncertainties), it is conservative to approximate them as being initially uncorrelated (i.e., setting the initial covariance matrix off-diagonal elements to zero).

The process noise source density matrix $Q_{P_{Dens}}$ for the moving base alignment Kalman filter would be set based on engineering knowledge of the process noise terms in (12.5.1-1) and (15.2.2.1-31). The (12.5.1-1) inertial sensor $Q_{P_{Dens}}$ terms would be set as described in Section 15.2.1.

Initialization of the Equation (15.2.2.1-31) $\delta\tau_{Sen/Mrk}$ first order Markov process noise density would be based on the steady state $\delta\tau_{Sen/Mrk}$ uncertainty variance. As in Section 15.2.1.2 (Equations (15.2.1.2-10) - (15.2.1.2-16)), we write from (15.1-1), (15.1.2.1.1-30) and (15.2.2.1-31) for $\delta\tau_{Sen/Mrk}$:

$$\dot{P}_{Sen/Mrk} = -2\,C_{Sen/Mrk}\,P_{Sen/Mrk} + Q_{Sen/Mrk} = 0$$
$$\Rightarrow \quad Q_{Sen/Mrk} = 2\,C_{Sen/Mrk}\,P_{Sen/Mrk/Std} \qquad (15.2.2.1\text{-}38)$$

where

$P_{Sen/Mrk}$, $P_{Sen/Mrk/Std}$, $Q_{Sen/Mrk}$ = Instantaneous variance, steady state variance, and process noise density for the uncertainty in $\delta\tau_{Sen/Mrk}$.

The $\delta\tau_{Sen/RndCnst}$ process noise density ($Q_{Sen/RndCnst}$) can be calculated from (15.2.2.1-31) for $\delta\tau_{Sen/RndCnst}$ using (15.1.2.1.1-30), as the estimated potential change in the $\delta\tau_{Sen/RndCnst}$ variance ($\Delta P_{Sen/RndCnst}$) over the moving base alignment period T_{Align}:

$$\dot{P}_{Sen/RndCnst} = Q_{Sen/RndCnst} \quad \Rightarrow \quad Q_{Sen/RndCnst} = \Delta P_{Sen/RndCnst} / T_{Align} \qquad (15.2.2.1\text{-}39)$$

The measurement noise variance for the (15.2.2.1-31) $\underline{z}_n$ measurement equation would be set to the estimated variance of the lever arm flexure $\underline{l}^B_{Flex}$.

In an actual implementation of the above described Kalman filter, approximations would probably be incorporated such as neglecting the angular-rate/quantization-noise product terms in Equations (12.5.1-1), and excluding the $\delta\underline{R}^N$ error state as having negligible impact on the alignment process. It is common to neglect senescence and lever arm error states in moving base alignment implementations, although not necessarily advisable.

A disadvantage in the (15.2.2.1-31) observation equation is that to minimize error build-up in the $\widehat{\underline{M}}(t)$ digital integration of $\hat{\underline{v}}^E_{REF}$, a high data rate would typically be required for $\hat{\underline{v}}^E_{REF}$ data transfer to the INS under dynamic acceleration/vibration conditions. To minimize the rate at which reference device data must be transferred, the (15.2.2.1-31) observation equation can be restructured as follows:

$$\underline{z}_{Obs_n} = \Delta\widehat{\underline{R}}^E_n - \Delta\widehat{\underline{R}}^E_{REF_n} - \left(\widehat{C}^E_{B_n} - \widehat{C}^E_{B_0}\right)\hat{\underline{l}}^B$$

$$\Delta\widehat{\underline{R}}^E(t) = \int_0^t \widehat{C}^E_N\,\hat{\underline{v}}^N\,dt \qquad \Delta\widehat{\underline{R}}^E_{REF}(t) = S\hat{\underline{v}}^E_{REF}(t_{REF}=t) - S\hat{\underline{v}}^E_{REF}(t_{REF}=0) \qquad (15.2.2.1\text{-}40)$$

$$S\hat{\underline{v}}^E_{REF}(t_{REF}) = \int^{t_{REF}} \hat{\underline{v}}^E_{REF}\,dt$$

where

t = Time since alignment initiation as determined by the INS being aligned.

t_{REF} = Equivalent time parameter for the reference device that is nominally equal to the t time parameter, but may differ due to senescence uncertainty error.

$S\hat{\underline{v}}_{REF}^{E}(t_{REF})$ = Running integral of $\hat{\underline{v}}_{REF}^{E}$ computed in the reference device since before t = 0.

$\Delta\hat{\underline{R}}_{REF_n}^{E}$ = Integral of $\hat{\underline{v}}_{REF}^{E}$ since the start of alignment.

Note, from the (15.2.2.1-1) - (15.2.2.1-5), that the $\Delta\hat{\underline{R}}_{n}^{E}$ and $\Delta\hat{\underline{R}}_{REF_n}^{E}$ terms in (15.2.2.1-40) are actual position vector $\underline{R}^E$ changes, hence, can also be computed from the INS and reference aiding device position data, providing they have sufficient resolution and accuracy.

Equations (15.2.2.1-40) allow the $\hat{\underline{v}}_{REF}^{E}$ integration to be performed in the reference aiding device computer and delivered to the INS in the form of $S\hat{\underline{v}}_{REF}^{E}(t_{REF})$. The $\Delta\hat{\underline{R}}_{REF_n}^{E}$ input to the $\underline{Z}_{Obs_n}$ is set to $S\hat{\underline{v}}_{REF}^{E}(t_{REF})$ minus its value at the start of alignment. Note that for this type of observation, reference data senescence uncertainty has a different effect on the observation/measurement than in the previous example. For the (15.2.2.1-40) case, data senescence introduces an error in $\Delta\hat{\underline{R}}_{REF_n}^{E}$ during position change (i.e., velocity) equal to $\hat{\underline{v}}_{REF}^{E}\,\delta\tau_{Sen}$ which directly impacts the observation. The impact on the $\underline{z}_n$ measurement equation in (15.2.2.1-31) is to eliminate the senescence terms in the $\delta\dot{\underline{M}}^{E}$ expression and replace them with a $\hat{\underline{v}}_{REF}^{E}\,\delta\tau_{Sen}$ term in $\underline{z}_n$. The revised measurement equation thereby becomes:

$$
\begin{aligned}
\underline{z}_n = \; & \delta\underline{M}_n^E - \hat{\underline{v}}_{REF_n}^{E}\left(\delta\tau_{Sen/RndCnst} + \delta\tau_{Sen/Mark_n}\right) - \left\{\left[\left(\hat{C}_{B_n}^{E} - \hat{C}_{B_0}^{E}\right)\hat{\underline{l}}^{B}\right]\times\right\}\hat{C}_{N_n}^{E}\,\underline{\psi}_n^N \\
& - \left(\hat{C}_{B_n}^{E} - \hat{C}_{B_0}^{E}\right)\delta\underline{l}_{Cnst}^{B} - \hat{\underline{v}}_{REF_n}^{E}\,\delta\tau_{Sen/Jit_n} + \left(\hat{C}_{B_n}^{E} - \hat{C}_{B_0}^{E}\right)\underline{l}_{Flex_n}^{B}
\end{aligned}
\qquad (15.2.2.1\text{-}41)
$$

$$
\delta\underline{M}^E(t) = \delta\underline{M}_0^E + \int_0^t \delta\dot{\underline{M}}^E\,dt \qquad \delta\dot{\underline{M}}^E = \hat{C}_N^E\,\delta\underline{V}^N - \delta\underline{V}_{Ref/Other}^{E}
$$

where

$\delta\underline{M}_0^E$ = Initial value for $\delta\underline{M}^E(t)$.

Based on the analytical definition for $\underline{Z}_{Obs_n}$ in (15.2.2.1-40), we see that the initial value is identically zero. Hence, $\underline{z}_n$ in (15.2.2.1-41) (the differential of $\underline{Z}_{Obs_n}$) should also be initially zero. The $\underline{z}_n$ constraint in (15.2.2.1-41) with $\delta\underline{M}^E(t) = \delta\underline{M}_0^E$ then sets the following requirement on $\delta\underline{M}_0^E$:

$$\delta\underline{M}_0^E = \hat{\underline{v}}^E_{REF_0}\left(\delta\tau_{Sen/RndCnst} + \delta\tau_{Sen/Mark_0}\right) + \hat{\underline{v}}^E_{REF_0}\,\delta\tau_{Sen/Jit_0} \qquad (15.2.2.1\text{-}42)$$

where

0 = Subscript reference to value for the parameter at time t = 0.

Note in Equation (15.2.2.1-41), that the $\delta\tau_{Sen/Jit}$ term now becomes part of the measurement noise, while in (15.2.2.1-31), it was part of the process noise. The error models for $\delta\tau_{Sen/RndCnst}$, $\delta\tau_{Sen/Mark}$ in (15.2.2.1-41) would be as in (15.2.2.1-31). A fundamental difference in the (15.2.2.1-41) measurement compared to the $\underline{z}_n$ measurement equation in (15.2.2.1-31) relates to the initial value for $\delta\underline{M}^E(t)$ uncertainty. For the (15.2.2.1-31) measurement, the initial value for $\delta\underline{M}^E(t)$ uncertainty is zero, which satisfies the $\underline{z}_n = 0$ condition at $t = 0$ in the (15.2.2.1-31) $\underline{z}_n$ equation. In contrast, for the (15.2.2.1-41) measurement, Equation (15.2.2.1-42) is required for $\delta\underline{M}_0^E$ which accounts for initial senescence error in $S\hat{\underline{v}}^E_{REF}(t_{REF}=0)$ when forming $\Delta\hat{\underline{R}}^E_{REF_n}$ in (15.2.2.1-40). The initial uncertainty in $\delta\underline{M}_0^E$ then goes toward initialization of the covariance matrix, including initial correlation of $\delta\underline{M}_0^E$ with the initial uncertainties in $\delta\tau_{Sen/RndCnst}$ and $\delta\tau_{Sen/Mark_0}$.

The previous approaches to the moving base alignment problem generate a set of error models that are well defined analytically (with only minor approximations), and which are consistent with the error models usually employed in strapdown inertial navigation Kalman filter configurations.

15.2.2.2 MOVING BASE ALIGNMENT USING AN N FRAME OBSERVATION

In this section we will describe an integrated velocity matching scheme in which the observation is formed in the locally level navigation N Frame. The N Frame observation approach has been more commonly used in practice because the velocity data provided from typical INS reference devices is usually formatted in a locally level coordinate frame.

The equation for the N Frame observation is derived beginning with Equation (15.2.2.1-3) of the previous section transformed to the N Frame:

$$\underline{v}^N_{INS} = \underline{v}^N_{REF} + C^N_E\,\underline{l}^E \qquad (15.2.2.2\text{-}1)$$

Applying generalized Equation (3.4-6), the lever arm term is:

$$C_E^N \, \dot{\underline{l}}^E = \dot{\underline{l}}^N + \underline{\omega}_{EN}^N \times \underline{l}^N \approx \dot{\underline{l}}^N \tag{15.2.2.2-2}$$

With (15.2.2.2-2), the integral of (15.2.2.2-1) becomes:

$$\int_0^t \underline{v}_{INS}^N \, dt = \int_0^t \underline{v}_{REF}^N \, dt + \underline{l}^N(t) - \underline{l}_0^N \tag{15.2.2.2-3}$$

Following the same process leading to (15.2.2.1-9), we then structure the N Frame observation equation from (15.2.2.2-3) as:

$$\begin{gathered}
\underline{Z}_{Obs_n} = \widehat{\underline{M}}_n^N - \left(\hat{\underline{l}}_n^N - \hat{\underline{l}}_0^N\right) = \widehat{\underline{M}}_n^N - \left(\widehat{C}_{B_n}^N - \widehat{C}_{B_0}^N\right) \hat{\underline{l}}^B \\
\widehat{\underline{M}}^N(t) \equiv \int_0^t \left(\widehat{\underline{v}}^N - \widehat{\underline{v}}_{REF}^N\right) dt \\
\widehat{C}_{B_n}^N = C_L^N \, \widehat{C}_{B_n}^L \qquad \widehat{C}_{B_0}^N = C_L^N \, \widehat{C}_{B_0}^L
\end{gathered} \tag{15.2.2.2-4}$$

The horizontal (i.e., X, Y) components of the (15.2.2.2-4) observation equation are generally sufficient for a successful transfer alignment. For analytical expediency, this section will be based on the full (15.2.2.2-4) vector form for the observation equation, with the derivation of the equivalent horizontal component form left as an exercise for the interested reader.

The associated measurement equation is derived as the differential of (15.2.2.2-4) using (12.2.1-9) for attitude error definition:

$$\begin{gathered}
\underline{z}_n = \delta \underline{M}_n^N - \left(\delta C_{B_n}^N - \delta C_{B_0}^N\right) \hat{\underline{l}}^B - \left(\widehat{C}_{B_n}^N - \widehat{C}_{B_0}^N\right) \delta \underline{l}^B \\
\delta C_{B_n}^N = - \left(\underline{\gamma}_n^N \times\right) \widehat{C}_{B_n}^N \qquad \delta C_{B_0}^N = - \left(\underline{\gamma}_0^N \times\right) \widehat{C}_{B_0}^N \approx - \left(\underline{\gamma}_n^N \times\right) \widehat{C}_{B_0}^N
\end{gathered} \tag{15.2.2.2-5}$$

Combining and incorporating the (15.2.2.1-16) lever arm error model then obtains for the N Frame measurement equation:

$$\underline{z}_n = \delta \underline{M}_n^N - \left\{ \left[\left(\widehat{C}_{B_n}^N - \widehat{C}_{B_0}^N\right) \hat{\underline{l}}^B \right] \times \right\} \underline{\gamma}_n^N - \left(\widehat{C}_{B_n}^N - \widehat{C}_{B_0}^N\right) \delta \underline{l}_{Cnst}^B + \left(\widehat{C}_{B_n}^N - \widehat{C}_{B_0}^N\right) \underline{l}_{Flex_n}^B \tag{15.2.2.2-6}$$

The $\delta \underline{M}_n^N$ term in (15.2.2.2-6) is $\delta \underline{M}^N(t)$ at t_n which, from the differential of (15.2.2.2-4) using (12.2.2-2) for velocity error definition, is given by:

$$\delta\underline{M}^N(t) = \int_0^t \delta\dot{\underline{M}}^N \, dt \qquad (15.2.2.2\text{-}7)$$

$$\delta\dot{\underline{M}}^N \equiv \delta\underline{v}^N - \delta\underline{v}^N_{REF} \qquad (15.2.2.2\text{-}8)$$

As in Equations (15.2.2.1-20) - (15.2.2.1-21), the $\delta\underline{v}^N_{REF}$ term in (15.2.2.2-8) can be modeled to account for data senescence error:

$$\delta\underline{v}^N_{Ref} = \delta\underline{v}^N_{Ref/Sen} + \delta\underline{v}^N_{Ref/Other} \qquad \delta\underline{v}^N_{Ref/Sen} = \dot{\hat{\underline{v}}}^N_{REF}\,\delta\tau_{Sen} \qquad (15.2.2.2\text{-}9)$$

The $\dot{\hat{v}}^N_{REF}$ in (15.2.2.2-9) is derived from the derivative of (15.2.2.2-1) with (15.2.2.2-2) using the (15.2.2.1-25) - (15.2.2.1-26) approach for the lever arm term:

$$\dot{\hat{\underline{v}}}^N_{REF} \approx \dot{\hat{\underline{v}}}^N - \ddot{\hat{\underline{l}}}^N \approx \dot{\hat{\underline{v}}}^N - \hat{C}^N_B\left[\left(\tilde{\underline{\omega}}^B_{IB}\times\right)^2 + \left(\dot{\tilde{\underline{\omega}}}^B_{IB}\times\right)\right]\hat{\underline{l}}^B \qquad (15.2.2.2\text{-}10)$$

Using (15.2.2.2-10) and the (15.2.2.1-28) model for $\delta\tau_{Sen}$ in (15.2.2.2-9), and substitution of the result in (15.2.2.2-8) then obtains for $\delta\dot{\underline{M}}^N$:

$$\begin{aligned}\delta\dot{\underline{M}}^N = {} & \delta\underline{v}^N - \left\{\dot{\hat{\underline{v}}}^N - \hat{C}^N_B\left[\left(\tilde{\underline{\omega}}^B_{IB}\times\right)^2 + \left(\dot{\tilde{\underline{\omega}}}^B_{IB}\times\right)\right]\hat{\underline{l}}^B\right\}\left(\delta\tau_{Sen/RndCnst} + \delta\tau_{Sen/Mark}\right) \\ & - \left\{\dot{\hat{\underline{v}}}^N - \hat{C}^N_B\left[\left(\tilde{\underline{\omega}}^B_{IB}\times\right)^2 + \left(\dot{\tilde{\underline{\omega}}}^B_{IB}\times\right)\right]\hat{\underline{l}}^B\right\}\delta\tau_{Sen/Jit} - \delta\underline{v}^N_{Ref/Other}\end{aligned} \qquad (15.2.2.2\text{-}11)$$

The models for $\delta\tau_{Sen/RndCnst}$ and $\delta\tau_{Sen/Mark}$ are provided by (15.2.2.1-29).

Moving base observation/measurement Equations (15.2.2.2-4), (15.2.2.2-6) - (15.2.2.2-7) and (15.2.2.2-11) are the N Frame equivalents to E Frame Equations (15.2.2.1-31). The fundamental difference between these equations is the form of the navigation error states contained in the measurement; for the (15.2.2.1-31) E Frame version, the navigation error states appearing in $\underline{z}_n$ are $\underline{\psi}^N$ and $\delta\underline{V}^N$; for the N Frame version, the navigation error states appearing in $\underline{z}_n$ are $\underline{\gamma}^N$ and $\delta\underline{v}^N$. Thus, for the N Frame version, the more appropriate navigation error model would be Equations (12.5.2-1) containing $\underline{\gamma}^N$, $\delta\underline{v}^N$ as contrasted with navigation error Equations (12.5.1-1) containing $\underline{\psi}^N$, $\delta\underline{V}^N$ for the E Frame observation. In practice, the moving base alignment Kalman filter would probably continue to be used for general inertial aiding operations following alignment completion. If the general Kalman filter applications are based on the $\underline{\psi}^N$, $\delta\underline{V}^N$ Equations (12.5.1-1) error model, a disparity exists if the N Frame measurement is used for alignment based on the $\underline{\gamma}^N$, $\delta\underline{v}^N$ Equation (12.5.2-1) model. To reconcile this disparity, the approximation can be made of using $\underline{\psi}^N$, $\delta\underline{V}^N$ for $\underline{\gamma}^N$, $\delta\underline{v}^N$ in N Frame measurement Equations (15.2.2.2-6) and (15.2.2.2-11). From equivalency Equations (12.2.1-17) and (12.2.2-5) we see that this is equivalent to approximating $\underline{\varepsilon}^N$ as zero which, from equivalency Equation (12.2.3-24), approximates the horizontal component of $\delta\underline{R}^N$ as zero during moving base alignment.

We also note that for N Frame observation Equation (15.2.2.2-4), the reference velocity input data is in the N Frame which then implicitly sets the N Frame for the INS to correspond with the N Frame of the reference aiding device. The C_B^L, $\underline{v}^N$, C_N^E foreground navigation parameters in the INS must then be initialized to reflect this N Frame selection.

As in Section 15.2.2.1 Equations (15.2.2.1-40) - (15.2.2.1-42), a revised version of the N Frame observation is also possible for which the reference velocity data is pre-integrated in the reference device computer. Starting from (15.2.2.2-4) and (15.2.2.2-6), and following the same steps used for development of the Section 15.2.2.1 E Frame version, the results obtained for the N Frame equivalent are as follows:

$$\underline{Z}_{Obs_n} = \Delta\hat{\underline{Sv}}^N_n - \Delta\hat{\underline{Sv}}^N_{REF_n} - \left(\hat{C}^N_{B_n} - \hat{C}^N_{B_0}\right)\hat{\underline{l}}^B$$

$$\Delta\hat{\underline{Sv}}^N(t) \equiv \int_0^t \hat{\underline{v}}^N \, dt \qquad \Delta\hat{\underline{Sv}}^N_{REF}(t) = \hat{\underline{Sv}}^N_{REF}(t_{REF}=t) - \hat{\underline{Sv}}^N_{REF}(t_{REF}=0)$$

$$\hat{\underline{Sv}}^N_{REF}(t_{REF}) \equiv \int_0^{t_{REF}} \hat{\underline{v}}^N_{REF} \, dt$$

$$\underline{z}_n = \delta\underline{M}^N_n - \hat{\underline{v}}^N_{REF_n}\left(\delta\tau_{Sen/RndCnst} + \delta\tau_{Sen/Mark_n}\right) - \left\{\left[\left(\hat{C}^N_{B_n} - \hat{C}^N_{B_0}\right)\hat{\underline{l}}^B\right]\times\right\}\underline{\gamma}^N_n \qquad (15.2.2.2\text{-}12)$$
$$- \left(\hat{C}^N_{B_n} - \hat{C}^N_{B_0}\right)\delta\underline{l}^B_{Cnst} - \hat{\underline{v}}^N_{REF}\,\delta\tau_{Sen/Jit_n} + \left(\hat{C}^N_{B_n} - \hat{C}^N_{B_0}\right)\underline{l}^B_{Flex_n}$$

$$\delta\underline{M}^N(t) = \delta\underline{M}^N_0 + \int_0^t \delta\dot{\underline{M}}^N \, dt \qquad \delta\dot{\underline{M}}^N = \delta\underline{v}^N - \delta\underline{v}^N_{Ref/Other}$$

$$\delta\underline{M}^N_0 = \hat{\underline{v}}^N_{REF_0}\left(\delta\tau_{Sen/RndCnst} + \delta\tau_{Sen/Mark_0}\right) + \hat{\underline{v}}^N_{REF_0}\,\delta\tau_{Sen/Jit_0}$$

Control reset operations for the N Frame observation moving base Kalman filter directly parallel those described in Section 15.2.2.1 for the E Frame observation configuration (i.e., the discussion surrounding Equations (15.2.2.1-33) - (15.2.2.1-36)). Control resets would be designed for the measurement $\delta \underline{M}^N$, the INS navigation and inertial sensor errors, and for the lever arm $\delta \underline{l}^N_{Cnst}$. The techniques described in Section 15.1.2.3 can be utilized to implement the foreground control reset process. For example, if all of the Equations (12.5.2-1) navigation error terms are included as error states, the error state vector would be of the form:

$$\underline{x} = \left[\left(\delta \underline{M}^N\right)^T, \left(\underline{\gamma}^N\right)^T, \left(\delta \underline{v}^N\right)^T, \left(\underline{\varepsilon}^N\right)^T, \delta h, \underline{SensErr}^T, \left(\delta \underline{l}^B_{Cnst}\right)^T, \delta\tau_{Sen/RndCnst}, \delta\tau_{Sen/Mark} \right]^T \tag{15.2.2.2-13}$$

The associated $\underline{u}_c$ control vector calculated from error state vector estimates would then be:

$$\underline{u}_c = \left[\left(\delta \underline{M}^N_c\right)^T, \left(\underline{\gamma}^N_c\right)^T, \left(\delta \underline{v}^N_c\right)^T, \left(\underline{\varepsilon}^N_c\right)^T, \delta h_c, \underline{SensErr}^T_c, \left(\delta \underline{l}^B_{Cnst_c}\right)^T, 0, 0 \right]^T \tag{15.2.2.2-14}$$

The control vector would be set to the negative of the estimated error states being controlled to drive the controlled error states to zero (as discussed in Section 15.1.2). If the method of delayed resets is being used as applied in Section 15.1.2.4, Equation (15.2.2.1-34) of the previous section would apply for setting the control vector.

Assuming that the INS navigation parameters are integrated by the foreground in N Frame coordinates (e.g., Equations (12.1-1) - (12.1-12)), Equations (15.1.2.3-20) would be used for control reset of the foreground navigation parameters. Additionally, the $\hat{C}^N_{B_0}$ matrix which was provided as input initialization data, would be updated for the $\underline{\gamma}^N_c$ control resets based on Equations (15.2.2.2-5) using (15.1.2.3-20) for the form of matrix update operation:

$$\hat{C}^N_{B_0}(+) = \left[I - \frac{\sin \gamma_c}{\gamma_c} \left(\underline{\gamma}^N_c \times\right) + \frac{(1 - \cos \gamma_c)}{\gamma_c^2} \left(\underline{\gamma}^N_c \times\right)^2 \right] \hat{C}^N_{B_0}(-) \tag{15.2.2.2-15}$$

The measurement, lever arm and inertial sensor error controls would be applied as in Equations (15.2.2.1-36) of the previous section.

The discussion in Section 15.2.2.1 regarding foreground and Kalman filter parameter initialization also applies for the N Frame observation version.

15.2.2.3 VELOCITY VERSUS INTEGRATED VELOCITY MATCHING

Sections 15.2.2.1 and 15.2.2.2 describe moving base alignment techniques based on using integrated velocity matching for the observation/measurement. We could also have formulated the observation as a velocity matching process. For example, for an N Frame velocity matching observation, we can have from Equation (15.2.2.2-1) with $\hat{\dot{\underline{l}}}^B = 0$ (as in (15.2.2.1-15)) and application of generalized Equation (3.4-4):

$$\begin{aligned}
\underline{Z}_{Vel/Obs_n} &= \hat{\underline{v}}^N - \hat{\underline{v}}_{REF}^N - \hat{C}_E^N \hat{\dot{\underline{l}}}^E = \hat{\underline{v}}^N - \hat{\underline{v}}_{REF}^N - \hat{C}_E^N \hat{C}_B^E \left(\hat{\dot{\underline{l}}}^B + \hat{\underline{\omega}}_{EB}^B \times \hat{\underline{l}}^B \right) \\
&\approx \hat{\underline{v}}^N - \hat{\underline{v}}_{REF}^N - \hat{C}_B^N \left(\hat{\dot{\underline{l}}}^B + \tilde{\underline{\omega}}_{IB}^B \times \hat{\underline{l}}^B \right) \qquad (15.2.2.3\text{-}1) \\
&= \hat{\underline{v}}^N - \hat{\underline{v}}_{REF}^N - \hat{C}_B^N \left(\tilde{\underline{\omega}}_{IB}^B \times \hat{\underline{l}}^B \right)
\end{aligned}$$

where

$\underline{Z}_{Vel/Obs_n}$ = Velocity matching observation.

The corresponding measurement equation is the differential of the (15.2.2.3-1) expression following the ≈ sign using (15.2.2.1-16) for $\delta \underline{l}^B$, whence:

$$\underline{z}_{Vel_n} = \delta \underline{v}_n^N + \cdots + \hat{C}_B^N \dot{\underline{l}}_{Flex_n}^B \qquad (15.2.2.3\text{-}2)$$

where

$\underline{z}_{Vel_n}$ = Velocity matching measurement.

For analysis purposes, we now substitute the integral of $\delta \dot{\underline{v}}^N$ from (12.5.2-1) for $\delta \underline{v}_n^N$ while making the approximation that $\underline{\gamma}^N$ is essentially constant over the alignment period. The result is:

$$\underline{z}_{Vel_n} = \Delta \underline{v}_{SF_n}^N \times \underline{\gamma}^N + \cdots + \hat{C}_B^N \dot{\underline{l}}_{Flex_n}^B \qquad (15.2.2.3\text{-}3)$$

with

$$\Delta \underline{v}_{SF_n}^N \equiv \int_0^{t_n} \underline{a}_{SF}^N \, dt \qquad (15.2.2.3\text{-}4)$$

where

$\Delta \underline{v}_{SF_n}^N$ = Equivalent to velocity change produced by specific force acceleration since the start of alignment.

Equation (15.2.2.3-3) illustrates the relationship between the principal error state to be estimated during alignment (i.e., the $\hat{C}_B^L$ attitude error $\underline{\gamma}$) and the principal measurement noise term $\dot{\underline{l}}^{B}_{Flex_n}$. The equation shows that the observability of $\underline{\gamma}$ for Kalman estimation is dependent on the $\Delta\underline{v}^{N}_{SF_n}$ specific force velocity change. Under non-maneuvering conditions, $\Delta\underline{v}^{N}_{SF_n}$ is a large vertical linear ramping function caused by the balancing of gravity (downward) by 1 g specific force acceleration (upward). In Equation (15.2.2.3-3), the vertical $\Delta\underline{v}^{N}_{SF_n}$ cross-product with $\underline{\gamma}$ amplifies the horizontal components of $\underline{\gamma}$ (i.e., the $\hat{C}^{L}_{B}$ attitude "tilt" errors) in the measurement. To amplify the vertical component of $\underline{\gamma}$ (i.e., the $\hat{C}^{L}_{B}$ heading error), a horizontal component in $\Delta\underline{v}^{N}_{SF_n}$ is required, which is typically generated by an intentional maneuver. The magnitude of the required maneuver depends on the accuracy to which the heading error is to be estimated compared with the magnitude of the (15.2.2.3-3) flexure rate measurement noise.

As an example, let us estimate the required $\Delta\underline{v}^{N}_{SF_n}$ based on the magnitude of $\dot{\underline{l}}^{B}_{Flex}$ using a simple second order vibration model for flexure such as described in Section 15.2.1.2. Using the square root of Equation (15.2.1.2-14) as a guide for the relationship between vibration amplitude and rate, the magnitude of the flexure rate $\dot{\underline{l}}^{B}_{Flex}$ (call it $\dot{l}_{Flex}$) is related to the $\underline{l}^{B}_{Flex}$ flexure amplitude (call it l_{Flex}) by:

$$\dot{l}_{Flex} = \omega_n \, l_{Flex} \tag{15.2.2.3-5}$$

where

ω_n = Undamped natural frequency of the lever arm structure connecting the INS being aligned to the INS reference device.

For a flexure amplitude l_{Flex} of 0.1 ft and a flexure frequency of 2 Hz (e.g., an air launched missile mounted on an aircraft pylon in flight), we see from (15.2.2.3-5) that $\dot{l}_{Flex} = 2\,\pi \times 2 \times 0.1 = 1.3$ fps. To determine the required $\Delta\underline{v}^{N}_{SF_n}$ maneuver, we first stipulate that the accuracy to which the $\Delta\underline{v}^{N}_{SF_n} \times \underline{\gamma}^{N}$ product in (15.2.2.3-3) can be discriminated from $\dot{\underline{l}}^{B}_{Flex}$ with a single measurement is on the order of $\dot{l}_{Flex}$.(i.e., a "signal-to-noise" ratio of 1 in the measurement). We then set $\underline{\gamma}^{N}$ to the allowable estimation error and solve for $\Delta\underline{v}^{N}_{SF_n}$. If we wish to estimate the heading error to an accuracy of 1 milli-radian with a single measurement, $\Delta\underline{v}^{N}_{SF_n}$ would thereby require a horizontal component on the order of 1.3 fps / 0.001 rad = 1,300 fps. Mission constraints generally do not allow such an extreme maneuver, however, the equivalent result can be achieved with a smaller maneuver by processing successive measurements in the Kalman filter (in effect, filtering the flexure noise).

To retain observability, the horizontal component of $\Delta \underline{v}_{SF_n}^N$ must be maintained at an acceptable magnitude during the resulting alignment process. For compatibility with mission requirements, the typical result for an aircraft application is the need for a repetitive oscillating high g horizontal velocity maneuver across the average flight path direction during the alignment period (i.e., a sequence of the so-called "S Turn" maneuver consisting of repetitive segments of roll right - turn right - roll left - turn left).

Now, let's look at the equivalent treatment for the N Frame integrated velocity measurement $\underline{z}_n$. From Equations (15.2.2.2-6) - (15.2.2.2-8) we see that:

$$\underline{z}_n = \int_0^{t_n} \delta \underline{v}^N \, dt + \cdots + \left(\hat{C}_{B_n}^N - \hat{C}_{B_0}^N \right) \underline{l}_{Flex_n}^B \qquad (15.2.2.3\text{-}6)$$

and with $\delta \dot{\underline{v}}^N$ from (12.5.2-1):

$$\underline{z}_n = \Delta \underline{R}_{SF_n}^N \times \underline{\gamma}^N + \cdots + \left(\hat{C}_{B_n}^N - \hat{C}_{B_0}^N \right) \underline{l}_{Flex_n}^B \qquad (15.2.2.3\text{-}7)$$

with

$$\Delta \underline{R}_{SF_n}^N \equiv \int_0^{t_n} \int_0^{t} \underline{a}_{SF}^N \, d\tau \, dt \qquad (15.2.2.3\text{-}8)$$

where

$\Delta \underline{R}_{SF_n}^N$ = Equivalent to position change produced by specific force acceleration since the start of alignment.

Equation (15.2.2.3-7) shows that the accuracy to which $\underline{\gamma}$ can be estimated is dependent on the magnitude of the position change $\Delta \underline{R}_{SF_n}^N$ in the presence of the measurement flexure noise $\underline{l}_{Flex}^B$. With the same procedure used for the velocity matching maneuver requirement we see from (15.2.2.3-7) that the single measurement maneuver requirement for 1 milli-rad heading estimation accuracy (with the same 0.1 ft flexure vibration) is 0.1 ft / 0.001 rad = 100 ft. In other words, for integrated velocity matching, a horizontal position change maneuver of only 100 feet allows the heading error to be estimated to the same accuracy for which velocity matching required a 1,300 fps horizontal velocity maneuver. Moreover, for successive filter estimation cycles, the position maneuver executed at the start of alignment remains and does not have to be repeated. For the velocity matching technique, the lateral velocity maneuver has to be a continuing cyclic pattern to assure an average velocity in the desired flight direction.

The previous discussion illustrates the obvious advantage of integrated velocity matching over velocity matching for the moving base alignment Kalman filter application. The main advantage for the velocity matching technique is that the integrated velocity error state is not required, thereby reducing the overall Kalman filter error state vector and related matrix dimensions.

The above simplified discussion should be viewed as only qualitative in nature. The actual maneuver requirements for a given application are also dependent on the magnitude of the other effects appearing in the measurement (i.e., the $\cdots$ terms in Equations (15.2.2.3-3) and (15.2.2.3-7)). Covariance simulation numerical analysis techniques are generally used to accurately determine overall requirements (See Chapter 16).

15.2.3 INS KALMAN FILTER AIDING USING A BODY MOUNTED VELOCITY SENSOR

This section discusses a Kalman filter measurement using reference velocity data provided in body B Frame axes (e.g., similar to the form of data measured by a Doppler radar), to illustrate the form of the observation and measurement equation. For this type of reference data, we might form the observation from the following modified form of (15.2.2.2-1):

$$\underline{v}^N_{INS} = C^N_B \, \underline{v}^B_{REF} + C^N_E \, \dot{\underline{l}}^E \qquad (15.2.3\text{-}1)$$

The observation equation derived from (15.2.3-1) would be:

$$\underline{Z}_{Obs} = \hat{\underline{v}}^N - \hat{C}^N_B \, \hat{\underline{v}}^B_{REF} - \hat{C}^N_E \, \hat{\dot{\underline{l}}}^E \qquad (15.2.3\text{-}2)$$

or following the approach that led to (15.2.2.3-1):

$$\begin{aligned} \underline{Z}_{Obs} &\approx \hat{\underline{v}}^N - \hat{C}^N_B \, \hat{\underline{v}}^B_{REF} - \hat{C}^N_B \left(\hat{\dot{\underline{l}}}^B + \tilde{\underline{\omega}}^B_{IB} \times \hat{\underline{l}}^B \right) \\ &= \hat{\underline{v}}^N - \hat{C}^N_B \, \hat{\underline{v}}^B_{REF} - \hat{C}^N_B \left(\tilde{\underline{\omega}}^B_{IB} \times \hat{\underline{l}}^B \right) \end{aligned} \qquad (15.2.3\text{-}3)$$

The associated measurement equation is obtained from the differential of the (15.2.3-3) expression following the $\approx$ sign. Using the (15.2.2.1-16) lever arm error model for $\delta \underline{l}^B$, applying (12.2.1-9) for the $\hat{C}^N_B$ error (i.e., $\delta C^N_B = - \left(\underline{\gamma}^N \times \right) C^N_B \approx - \left(\underline{\gamma}^N \times \right) \hat{C}^N_B$), and dropping $\delta C^N_B \left(\hat{\dot{\underline{l}}}^B + \tilde{\underline{\omega}}^B_{IB} \times \hat{\underline{l}}^B \right)$ and $\hat{C}^N_B \left(\delta \underline{\omega}^B_{IB} \times \hat{\underline{l}}^B \right)$ as negligibly small, we obtain for the measurement:

$$\begin{aligned}
\underline{z} &= \delta\underline{v}^N - \delta C_B^N \hat{\underline{v}}_{REF}^B - \hat{C}_B^N \, \delta\underline{v}_{REF}^B - \hat{C}_B^N \left(\delta\dot{\underline{l}}^B + \tilde{\underline{\omega}}_{IB}^B \times \delta\underline{l}^B \right) \\
&= \delta\underline{v}^N + \left(\underline{\gamma}^N \times \right) \hat{C}_B^N \hat{\underline{v}}_{REF}^B - \hat{C}_B^N \, \delta\underline{v}_{REF}^B - \hat{C}_B^N \left[- \dot{\underline{l}}_{Flex}^B + \tilde{\underline{\omega}}_{IB}^B \times \left(\delta\underline{l}_{Cnst}^B - \underline{l}_{Flex}^B \right) \right] \qquad (15.2.3\text{-}4) \\
&= \delta\underline{v}^N - \hat{\underline{v}}_{REF}^N \times \underline{\gamma}^N - \hat{C}_B^N \, \delta\underline{v}_{REF}^B - \hat{C}_B^N \left(\tilde{\underline{\omega}}_{IB}^B \times \delta\underline{l}_{Cnst}^B \right) + \hat{C}_B^N \left(\dot{\underline{l}}_{Flex}^B + \tilde{\underline{\omega}}_{IB}^B \times \underline{l}_{Flex}^B \right)
\end{aligned}$$

The $\hat{\underline{v}}_{REF}^N$ term in (15.2.3-4) is calculated from (15.2.3-1) using the approximations applied in (15.2.3-3):

$$\hat{\underline{v}}_{REF}^N = \hat{\underline{v}}^N - \hat{C}_B^N \left(\tilde{\underline{\omega}}_{IB}^B \times \hat{\underline{l}}^B \right) \qquad (15.2.3\text{-}5)$$

Substituting (15.2.3-5) into (15.2.3-4) then yields:

$$\begin{aligned}
\underline{z} = \delta\underline{v}^N - \left[\hat{\underline{v}}^N - \hat{C}_B^N \left(\tilde{\underline{\omega}}_{IB}^B \times \hat{\underline{l}}^B \right) \right] \times \underline{\gamma}^N - \hat{C}_B^N \, \delta\underline{v}_{REF}^B \\
- \hat{C}_B^N \left(\tilde{\underline{\omega}}_{IB}^B \times \delta\underline{l}_{Cnst}^B \right) + \hat{C}_B^N \left(\dot{\underline{l}}_{Flex}^B + \hat{\underline{\omega}}_{IB}^B \times \underline{l}_{Flex}^B \right)
\end{aligned} \qquad (15.2.3\text{-}6)$$

Equation (15.2.3-6) shows the presence of a $\hat{\underline{v}}^N \times \underline{\gamma}^N$ term in the measurement which is characteristic of B Frame reference velocity measurements. Note also that if we include $\delta\underline{l}_{Cnst}^B$ in our error state vector, the $\hat{\underline{l}}^B$ term in (15.2.3-6) can be chosen to equal zero (i.e., the rigid lever arm effect thereby being totally accounted for in $\delta\underline{l}_{Cnst}^B$). Under these conditions, we can apply equivalency Equations (12.2.1-17) and (12.2.2-5) in Equation (15.2.3-6) to obtain the equivalent measurement equation in terms of $\underline{\psi}^N$, $\delta\underline{V}^N$ error parameters:

$$\begin{aligned}
\underline{z} &= \delta\underline{V}^N - \hat{\underline{v}}^N \times \left(\underline{\gamma}^N - \underline{\varepsilon}^N \right) - \hat{C}_B^N \, \delta\underline{v}_{REF}^B - \hat{C}_B^N \left(\tilde{\underline{\omega}}_{IB}^B \times \delta\underline{l}_{Cnst}^B \right) + \hat{C}_B^N \left(\dot{\underline{l}}_{Flex}^B + \tilde{\underline{\omega}}_{IB}^B \times \underline{l}_{Flex}^B \right) \\
&= \delta\underline{V}^N - \hat{\underline{v}}^N \times \underline{\psi}^N - \hat{C}_B^N \, \delta\underline{v}_{REF}^B - \hat{C}_B^N \left(\tilde{\underline{\omega}}_{IB}^B \times \delta\underline{l}_{Cnst}^B \right) \qquad (15.2.3\text{-}7) \\
&\quad + \hat{C}_B^N \left(\dot{\underline{l}}_{Flex}^B + \tilde{\underline{\omega}}_{IB}^B \times \underline{l}_{Flex}^B \right)
\end{aligned}$$

15.2.4 KALMAN FILTERING APPLIED TO GPS - INS POSITION AIDING

Use of the outputs from a GPS (Global Positioning System) receiver provides the ability to obtain accurate measurements of position relative to the earth derived from range measurements to orbiting GPS satellites. Range is obtained from the measured time interval for a satellite signal to travel from the satellite to the user GPS receiver antenna. Satellite orbit ephemeris data provided on the signal transmission allows receiver determination of satellite position relative to the earth. Range measurements to the satellites coupled with the satellite-to-earth data allows determination of vehicle position relative to the earth. For an unambiguous fix, range data from three separate satellites is required. Data from a fourth satellite is required for receiver clock calibration against the synchronized satellite transmission clock system.

The observation for the GPS measurement is the difference between the range to each satellite obtained from the GPS receiver data, compared with comparable data derived from INS inertially derived range, the latter corrected for lever arm displacement between the GPS receiver antenna and INS:

$$z_{Obs/i} = \hat{\rho}_i - \tilde{\rho}_{GPSi} \qquad (15.2.4\text{-}1)$$

where

i = Satellite number for the measurement.

$z_{Obs/i}$ = Observation vector component for the ith GPS satellite range observation.

$\hat{\rho}_i$ = Range from the GPS receiver antenna to the ith satellite as determined using INS data and satellite ephemeris data.

$\tilde{\rho}_{GPSi}$ = Range from the user GPS receiver antenna to the ith satellite as determined from the satellite-to-GPS receiver antenna signal transmission time measurement.

The GPS derived range data in Equations (15.2.4-1) is calculated from the measured time interval for a GPS satellite transmission signal to reach the user receiver.

$$\tilde{\rho}_{GPSi} = c\left(\tilde{t}_{RECi} - \tilde{t}_{GPSi}\right) \qquad (15.2.4\text{-}2)$$

where

t_{GPSi} = The time of signal transmission from the ith GPS satellite.

t_{RECi} = The GPS receiver clock time when the ith satellite transmission signal is received. Nominally, the GPS receiver clock time is synchronous with the GPS satellite clock time.

c = Speed of light.

The GPS measurement equation for the Kalman filter is obtained from the differential form of Equations (15.2.4-1) - (15.2.4-2):

$$z_i = \delta\rho_i - c\left(\delta t_{RECi} - \delta t_{GPSi}\right) \qquad (15.2.4\text{-}3)$$

where

z_i = i^{th} GPS satellite range measurement.

δ = Error in the indicated quantities.

The $\delta\rho_i$ term in Equation (15.2.4-3) can be expanded in terms of INS and GPS satellite ephemeris errors by first defining the range vector from the GPS receiver antenna to the satellite in earth fixed E Frame coordinates as calculated by the INS:

$$\underline{\hat{\rho}}_i^E = \underline{\hat{R}}_{Si}^E - \underline{\hat{R}}^E - \underline{\hat{l}}^E \qquad (15.2.4\text{-}4)$$

where

$\underline{\hat{\rho}}_i^E$ = Range vector from the GPS antenna to satellite i.

$\underline{\hat{R}}_{Si}^E$ = Vector from the center of the earth to satellite i based on transmitted satellite i ephemeris data.

$\underline{\hat{R}}^E$ = Vector from the center of the earth to the INS based on inertially computed data (e.g., using Section 4.4.2.2 - Equations (4.4.2.2-2) and (4.4.2.2-4) - (4.4.2.2-5) for conversion of altitude and latitude/longitude position to $\underline{R}^E$).

$\underline{\hat{l}}^E$ = Vector from the INS to the GPS antenna as calculated with INS data.

The computed range to satellite i ($\hat{\rho}_i$) is the magnitude of $\underline{\hat{\rho}}_i^E$. Its square, then, is given by:

$$\hat{\rho}_i^2 = \left(\underline{\hat{\rho}}_i^E\right)^T \underline{\hat{\rho}}_i^E \qquad (15.2.4\text{-}5)$$

An analytical expression for the range error ($\delta\rho_i$) is obtained from the differential of (15.2.4-5):

$$2\,\hat{\rho}_i\,\delta\rho_i = \left(\delta\underline{\rho}_i^E\right)^T \underline{\hat{\rho}}_i^E + \left(\underline{\hat{\rho}}_i^E\right)^T \delta\underline{\rho}_i^E = 2\left(\underline{\hat{\rho}}_i^E\right)^T \delta\underline{\rho}_i^E \qquad (15.2.4\text{-}6)$$

or

$$\delta\rho_i = \frac{\left(\underline{\hat{\rho}}_i^E\right)^T \delta\underline{\rho}_i^E}{\hat{\rho}_i} \qquad (15.2.4\text{-}7)$$

or, with the differential of (15.2.4-4) for $\delta\underline{\rho}_i^E$:

$$\delta\rho_i = \frac{\left(\underline{\hat{\rho}}_i^E\right)^T \left(\delta\underline{R}_{Si}^E - \delta\underline{R}^E - \delta\underline{l}^E\right)}{\hat{\rho}_i} \tag{15.2.4-8}$$

Since $\delta\rho_i$ is calculated from a vector dot product, its value is identical (invariant) when evaluated in any coordinate frame (See generalized Equation (3.1.1-29)). Using the locally level navigation N Frame to evaluate $\delta\rho_i$, we thereby obtain:

$$\delta\rho_i = \frac{\left(\hat{C}_E^N \, \underline{\hat{\rho}}_i^E\right)^T \left(\hat{C}_E^N \, \delta\underline{R}_{Si}^E - \delta\underline{R}^N - \delta\underline{l}^N\right)}{\hat{\rho}_i} \tag{15.2.4-9}$$

The $\delta\underline{l}^N$ term in the previous expression is expanded using (12.2.1-9):

$$\delta\underline{l}^N = \delta\left(\hat{C}_B^N \, \underline{\hat{l}}^B\right) = \delta C_B^N \, \underline{\hat{l}}^B + \hat{C}_B^N \, \delta\underline{l}^B = \left(\hat{C}_B^N \, \underline{\hat{l}}^B\right) \times \underline{\gamma}^N + \hat{C}_B^N \, \delta\underline{l}^B \tag{15.2.4-10}$$

With (15.2.4-10) in (15.2.4-9), the expression for $\delta\rho_i$ in Equation (15.2.4-3) is determined:

$$\delta\rho_i = \frac{\left(\hat{C}_E^N \, \underline{\hat{\rho}}_i^E\right)^T \left(\hat{C}_E^N \, \delta\underline{R}_{Si}^E - \delta\underline{R}^N - \left(\hat{C}_B^N \, \underline{\hat{l}}^B\right) \times \underline{\gamma}^N - \hat{C}_B^N \, \delta\underline{l}^B\right)}{\hat{\rho}_i} \tag{15.2.4-11}$$

The range terms in (15.2.4-11) can be calculated with Equations (15.2.4-4) - (15.2.4-5).

Summarizing, Equations (15.2.4-1) and (15.2.4-3) with (15.2.4-2), (15.2.4-4), (15.2.4-5), and (15.2.4-11) comprise a set of observation/measurement equations for GPS - INS aiding for each (i) of the GPS satellite measurements. An error model for the INS navigation error terms ($\delta\underline{R}^N$, $\delta\underline{V}^N$, $\underline{\gamma}^N$) in these equations is provided by Equations (12.5.3-1). Equations (15.2.2.1-16) can be used to model the lever arm error $\delta\underline{l}^B$. We conclude this section with a discussion of error models for the δt_{RECi}, δt_{GPSi}, $\delta\underline{R}_{Si}^E$ terms in (15.2.4-3) and (15.2.4-11) associated with the GPS measurement.

For the Kalman filter GPS aiding of an INS, it is typically assumed that the satellite position locations and transmission data are perfect, thus, $\delta\underline{R}_{Si}^E$ and δt_{GPSi} would be equated to zero. For suboptimal covariance simulation performance analysis (See Section 16.1.1.1) $\delta\underline{R}_{Si}^E$ and δt_{GPSi} might be included in the real world error state vector based on classical GPS satellite system error models (e.g., Reference 27 - Chapter 11).

The δt_{RECi} error in (15.2.4-3) can be defined from a model of the GPS receiver clock time generation process, assuming a common clock for all (i) satellite measurements. The GPS receiver clock time is generated by counting pulses from a GPS receiver high frequency source. Hence, the receiver clock time is, in effect, given by:

$$\hat{t}_{REC_n} = \hat{t}_{REC_{n-1}} + T_{Osc_0} = \hat{t}_{REC_{n-1}} + \frac{1}{f_{Osc_0}} \qquad (15.2.4\text{-}12)$$

where

n = Receiver clock frequency source pulse count cycle.

T_{Osc_0} = Nominal time interval between receiver clock frequency source pulses.

f_{Osc_0} = Nominal receiver frequency source pulse rate.

Rearranging (15.2.4-12) and dividing by the actual time between clock pulses obtains:

$$\frac{\hat{t}_{REC_n} - \hat{t}_{REC_{n-1}}}{\Delta t} = \frac{1}{f_{Osc_0}\,\Delta t} \qquad (15.2.4\text{-}13)$$

The $\frac{1}{\Delta t}$ term on the right in (15.2.4-13) is the actual frequency of the receiver clock input oscillator. With this substitution, Equation (15.2.4-13) in the limit as $\Delta t \rightarrow 0$, is given by the equivalent differential equation:

$$\dot{\hat{t}}_{REC} = \frac{\tilde{f}_{Osc}}{f_{Osc_0}} \qquad (15.2.4\text{-}14)$$

where

$\tilde{f}_{Osc}$ = Actual GPS receiver clock input oscillator pulse frequency.

A δt_{RECi} error state model for (15.2.4-3) can be derived as the differential of (15.2.4-14) plus a random term (uncorrelated from satellite-to-satellite and cycle-to-cycle measurement) due to frequency source jitter and pulse quantization:

$$\dot{\delta t}_f = \frac{\delta f_{Osc}}{f_{Osc_0}} \qquad \delta t_{RECi} = \delta t_f + n_t \qquad (15.2.4\text{-}15)$$

where

δt_f = GPS receiver clock time error exclusive of random jitter/pulse quantization.

n_t = Random uncorrelated GPS receiver clock time jitter and quantization error.

δf_{Osc} = GPS receiver clock input oscillator frequency error.

The receiver clock oscillator frequency error can be modeled as a randomly varying constant plus a first order Markov process:

$$\delta f_{Osc} = \delta f_{Osc/RndCnst} + \delta f_{Osc/Mark}$$

$$\delta \dot{f}_{Osc/RndCnst} = n_{Osc/RndCnst} \qquad (15.2.4\text{-}16)$$

$$\delta \dot{f}_{Osc/Mark} = - C_{Osc/Mark}\, \delta f_{Osc/Mark} + n_{Osc/Mark}$$

where

$\delta f_{Osc/RndCnst}$, $\delta f_{Osc/Mark}$ = Receiver clock frequency random constant and first order Markov process errors.

$C_{Osc/Mark}$, $n_{Osc/Mark}$ = Correlation frequency and white source noise input for the GPS receiver oscillator frequency error first order Markov process.

$n_{Osc/RndCnst}$ = White source noise input for the GPS receiver oscillator frequency random constant error model.

16 Covariance Simulation Programs

16.0 OVERVIEW

Covariance simulation programs are commonly used to provide numerical time histories depicting the accuracy of a given system configuration in terms of the covariance of its associated linearized error state vector. For a Kalman filter aided system, the covariance simulation is also utilized as a basic design tool during the synthesis and test of the suboptimal Kalman filter configuration used in the actual system. The suboptimal Kalman filter configuration is typically based on a simplified error state dynamic/measurement model (compared to the "real world" error state dynamics/measurements) with numerical values for its defining matrix elements that may differ from real world values. The covariance simulation is used to evaluate the performance of the suboptimal filter operating in a real world environment, and to provide the design engineer with useful sensitivities for identifying sources of undesirable performance characteristics during the design process. As part of the suboptimal Kalman filter design process, the covariance simulation is also typically structured to provide an optimal solution based on the hypothetical "optimal real world" Kalman gain matrix based on the "real world" model, for use as a yardstick to measure suboptimal filter performance.

In the following sections we will define the analytical basis for suboptimal covariance analysis, the structure of typical covariance simulation programs, and the use of covariance simulation programs in the design and performance analysis of suboptimal Kalman filters.

16.1 COVARIANCE SIMULATION ANALYTICAL DEFINITION

The covariance simulation contains error state dynamic and measurement models for the real world and for the simplified world of the suboptimal Kalman filter that operates from real world measurements. This section develops real world and suboptimal Kalman filter analytical models for covariance simulation based on the analytical forms presented in 15.1.2. This section is divided into two parts; Section 16.1.1 (and its subsections) based on idealized control resets, and Section 16.1.2 (and its subsections) based on delayed control resets.

16.1.1 FORMULATION BASED ON IDEALIZED CONTROL RESETS

Using (15.1.2-2), (15.1.2-3) and (15.1.2-5) as a template for the idealized control reset case, we first define a general real world model as the following computations executed in the indicated sequence:

$$\underline{x}_n(-) = \Phi_{xx_n}\,\underline{x}_{n-1}(+_c) + \Phi_{xy_n}\,\underline{y}_{n-1}(+_c) + \underline{w}_{x_n} \qquad (16.1.1\text{-}1)$$

$$\underline{y}_n(-) = \Phi_{yx_n}\,\underline{x}_{n-1}(+_c) + \Phi_{yy_n}\,\underline{y}_{n-1}(+_c) + \underline{w}_{y_n} \qquad (16.1.1\text{-}2)$$

$$\underline{z}_n = H_{x_n}\,\underline{x}_n(-) + H_{y_n}\,\underline{y}_n(-) + G_{M_n}\,\underline{n}_{M_n} \qquad (16.1.1\text{-}3)$$

$$\underline{x}_n(+_c) = \underline{x}_n(-) + \underline{u}_{c_n} \qquad (16.1.1\text{-}4)$$

$$\underline{y}_n(+_c) = \underline{y}_n(-) \qquad (16.1.1\text{-}5)$$

where

$(+_c)$ = Designation for parameter value at its designated time stamp (t_n in this case) immediately after ("a posteriori") the application of control resets (c subscript) at the same designated time.

$(-)$ = Designation for parameter value at its designated time stamp (t_n in this case) prior to the application of any resets (estimation or control) at the same designated time.

$\underline{x}$ = Real world error state vector components that are used as the basis for the suboptimal Kalman filter error state vector model.

$\underline{y}$ = Real world error state vector components that are not accounted for in the suboptimal Kalman filter error state model. The total real world error state vector is the composite of $\underline{x}$ and $\underline{y}$.

$\Phi_{xx}, \Phi_{xy}, \Phi_{yx}, \Phi_{yy}$ = Elements of the real world error state transition matrix Φ defined analytically by (15.1.1-15). The suboptimal Kalman filter is based on the Φ_{xx} component.

$\underline{w}_x, \underline{w}_y$ = Elements of the real world integrated process noise vector defined analytically by (15.1.1-16). The suboptimal Kalman filter is based on the $\underline{w}_x$ component.

H_x, H_y = Elements of the real world measurement matrix H. The suboptimal Kalman filter is based on the H_x component.

$\underline{u}_c$ = Control vector calculated within the suboptimal Kalman filter based on its estimate of $\underline{x}$.

The equivalent version of (16.1.1-1) - (16.1.1-5) contained within the suboptimal Kalman filter is from (15.1.2-6) - (15.1.2-13):

$$\underline{\tilde{x}}_n(-) = \Phi^*_{xx_n} \underline{\tilde{x}}_{n-1}(+_c) \tag{16.1.1-6}$$

$$\underline{\tilde{z}}_n = H^*_{x_n} \underline{\tilde{x}}_n(-) \tag{16.1.1-7}$$

$$\underline{\tilde{x}}_n(+_e) = \underline{\tilde{x}}_n(-) - K_n \left(\underline{\tilde{z}}_n - \underline{z}_n\right) \tag{16.1.1-8}$$

$$\underline{u}_{c_n} = \text{function of } \underline{\tilde{x}}_n(+_e) \tag{16.1.1-9}$$

$$\underline{\tilde{x}}_n(+_c) = \underline{\tilde{x}}_n(+_e) + \underline{u}_{c_n} \tag{16.1.1-10}$$

$$\underline{\tilde{x}}_0 = 0 \tag{16.1.1-11}$$

$$\underline{\tilde{y}}_n = 0 \tag{16.1.1-12}$$

where

$*$ = Superscript designator for value of the parameter used in the suboptimal Kalman filter which may differ from the correct real world value.

$(+_e)$ = Designation for parameter value at its designated time stamp (t_n in this case) immediately after ("a posteriori") the application of estimation resets (e subscript) within the suboptimal Kalman filter at the same designated time.

Note in Equation (16.1.1-8) that we have used the real world measurement $\underline{z}_n$ (i.e., the linearized form of the observation vector $\underline{Z}_{Obs_n}$) to approximate $\underline{Z}_{Obs_n}$ in (15.1.2-9). Also note that we have added Equation (16.1.1-12) as a formal statement that the filter is only designed to account for the $\underline{x}$ error states.

As in (15.1.2.1-1) we now define the suboptimal Kalman filter estimated error state vector uncertainty as:

$$\Delta\underline{x} \equiv \underline{\tilde{x}} - \underline{x} \tag{16.1.1-13}$$

where

$\Delta\underline{x}$ = Uncertainty in $\underline{\tilde{x}}$, the Kalman filter estimate for $\underline{x}$.

For the idealized control reset filter, we have for $\Delta\underline{x}$ from (15.1.2.1-2):

$$\begin{aligned} \Delta\underline{x}_n(-) &= \underline{\tilde{x}}_n(-) - \underline{x}_n(-) \\ \Delta\underline{x}_n(+_e) &= \underline{\tilde{x}}_n(+_e) - \underline{x}_n(-) \\ \Delta\underline{x}_n(+_c) &= \underline{\tilde{x}}_n(+_c) - \underline{x}_n(+_c) \end{aligned} \tag{16.1.1-14}$$

The value for $\Delta\underline{x}$ immediately following an estimation update is found by combining suboptimal filter Equations (16.1.1-7) - (16.1.1-8) with the real world $\underline{z}_n$ measurement formula (16.1.1-3) using the first and second equation in (16.1.1-14) for the $\Delta\underline{x}$'s:

$$\begin{aligned}
\Delta\underline{x}_n(+_e) &= \tilde{\underline{x}}_n(+_e) - \underline{x}_n(-) = \tilde{\underline{x}}_n(-) - K_n\left(\tilde{\underline{z}}_n - \underline{z}_n\right) - \underline{x}_n(-) \\
&= \tilde{\underline{x}}_n(-) - \underline{x}_n(-) - K_n\left(H^*_{x_n}\tilde{\underline{x}}_n(-) - H_{x_n}\underline{x}_n(-) - H_{y_n}\underline{y}_n(-) - G_{M_n}\underline{n}_{M_n}\right) \\
&= \Delta\underline{x}_n(-) - K_n\left(H^*_{x_n}\tilde{\underline{x}}_n(-) - H^*_{x_n}\underline{x}_n(-) + H^*_{x_n}\underline{x}_n(-) - H_{x_n}\underline{x}_n(-)\right. \\
&\qquad \left. - H_{y_n}\underline{y}_n(-) - G_{M_n}\underline{n}_{M_n}\right) \\
&= \Delta\underline{x}_n(-) - K_n\left[H^*_{x_n}\Delta\underline{x}_n(-) + \left(H^*_{x_n} - H_{x_n}\right)\underline{x}_n(-) - H_{y_n}\underline{y}_n(-) - G_{M_n}\underline{n}_{M_n}\right]
\end{aligned} \tag{16.1.1-15}$$

or

$$\begin{aligned}
\Delta\underline{x}_n(+_e) = {} & \left(I_x - K_n H^*_{x_n}\right)\Delta\underline{x}_n(-) - K_n\left(H^*_{x_n} - H_{x_n}\right)\underline{x}_n(-) \\
& + K_n H_{y_n}\underline{y}_n(-) + K_n G_{M_n}\underline{n}_{M_n}
\end{aligned} \tag{16.1.1-16}$$

where

I_x = Identity matrix with same dimension as $\underline{x}$.

The $\underline{x}$ error state uncertainty immediately following a control update is developed by combining suboptimal filter Equation (16.1.1-10) with real world Equation (16.1.1-4) using the second and third equation in (16.1.1-14) for the $\Delta\underline{x}$'s:

$$\begin{aligned}
\Delta\underline{x}_n(+_c) &= \tilde{\underline{x}}_n(+_c) - \underline{x}_n(+_c) \\
&= \tilde{\underline{x}}_n(+_e) + \underline{u}_{c_n} - \left(\underline{x}_n(-) + \underline{u}_{c_n}\right) = \tilde{\underline{x}}_n(+_e) - \underline{x}_n(-) = \Delta\underline{x}_n(+_e)
\end{aligned} \tag{16.1.1-17}$$

The $\underline{x}$ error state uncertainty immediately preceding the estimation update is derived by combining suboptimal filter Equation (16.1.1-6) with real world Equation (16.1.1-1) using the first and third equation in (16.1.1-14) for the $\Delta\underline{x}$'s:

$$\begin{aligned}
\Delta\underline{x}_n(-) &= \tilde{\underline{x}}_n(-) - \underline{x}_n(-) \\
&= \Phi^*_{xx_n}\tilde{\underline{x}}_{n-1}(+_c) - \Phi_{xx_n}\underline{x}_{n-1}(+_c) - \Phi_{xy_n}\underline{y}_{n-1}(+_c) - \underline{w}_{x_n} \\
&= \Phi^*_{xx_n}\tilde{\underline{x}}_{n-1}(+_c) - \Phi^*_{xx_n}\underline{x}_{n-1}(+_c) + \Phi^*_{xx_n}\underline{x}_{n-1}(+_c) \\
&\qquad - \Phi_{xx_n}\underline{x}_{n-1}(+_c) - \Phi_{xy_n}\underline{y}_{n-1}(+_c) - \underline{w}_{x_n} \\
&= \Phi^*_{xx_n}\Delta\underline{x}_{n-1}(+_c) + \left(\Phi^*_{xx_n} - \Phi_{xx_n}\right)\underline{x}_{n-1}(+_c) - \Phi_{xy_n}\underline{y}_{n-1}(+_c) - \underline{w}_{x_n}
\end{aligned} \tag{16.1.1-18}$$

Substituting (16.1.1-17), (16.1.1-4) and (16.1.1-5) into (16.1.1-18) gives:

$$\begin{aligned}\Delta\underline{x}_n(-) &= \overset{*}{\Phi}_{xx_n}\,\Delta\underline{x}_{n-1}(+_e) + \left(\overset{*}{\Phi}_{xx_n} - \Phi_{xx_n}\right)\left(\underline{x}_{n-1}(-) + \underline{u}_{c_{n-1}}\right) \\ &\quad - \Phi_{xy_n}\,\underline{y}_{n-1}(-) - \underline{w}_{x_n}\end{aligned} \tag{16.1.1-19}$$

It is convenient at this point to hypothesize a form for the control vector $\underline{u}_c$. For inertial navigation estimation applications, $\underline{u}_c$ in (16.1.1-9) is typically a linear function of the suboptimal filter estimated error state vector $\underline{\tilde{x}}_n(+_e)$ with phasing to control $\underline{\tilde{x}}_n(+_e)$ to zero. Hence, with the second equation in (16.1.1-14):

$$\underline{u}_{c_n} = -L_{x_n}\,\underline{\tilde{x}}_n(+_e) = -L_{x_n}\left(\Delta\underline{x}_n(+_e) + \underline{x}_n(-)\right) \tag{16.1.1-20}$$

where

L_x = The "control matrix" for $\underline{x}$ error state vector control.

We then substitute (16.1.1-20) into (16.1.1-19) to obtain:

$$\begin{aligned}\Delta\underline{x}_n(-) &= \left[\overset{*}{\Phi}_{xx_n} - \left(\overset{*}{\Phi}_{xx_n} - \Phi_{xx_n}\right)L_{x_{n-1}}\right]\Delta\underline{x}_{n-1}(+_e) \\ &\quad + \left(\overset{*}{\Phi}_{xx_n} - \Phi_{xx_n}\right)\left(I_x - L_{x_{n-1}}\right)\underline{x}_{n-1}(-) - \Phi_{xy_n}\,\underline{y}_{n-1}(-) - \underline{w}_{x_n}\end{aligned} \tag{16.1.1-21}$$

The (16.1.1-1) - (16.1.1-2) real world error state propagation formula can also be expanded by first substituting (16.1.1-4) and (16.1.1-5):

$$\begin{aligned}\underline{x}_n(-) &= \Phi_{xx_n}\left(\underline{x}_{n-1}(-) + \underline{u}_{c_{n-1}}\right) + \Phi_{xy_n}\,\underline{y}_{n-1}(-) + \underline{w}_{x_n} \\ \underline{y}_n(-) &= \Phi_{yx_n}\left(\underline{x}_{n-1}(-) + \underline{u}_{c_{n-1}}\right) + \Phi_{yy_n}\,\underline{y}_{n-1}(-) + \underline{w}_{y_n}\end{aligned} \tag{16.1.1-22}$$

and then applying (16.1.1-20):

$$\begin{aligned}\underline{x}_n(-) &= -\Phi_{xx_n}L_{x_{n-1}}\,\Delta\underline{x}_{n-1}(+_e) + \Phi_{xx_n}\left(I_x - L_{x_{n-1}}\right)\underline{x}_{n-1}(-) + \Phi_{xy_n}\,\underline{y}_{n-1}(-) + \underline{w}_{x_n} \\ \underline{y}_n(-) &= -\Phi_{yx_n}L_{x_{n-1}}\,\Delta\underline{x}_{n-1}(+_e) + \Phi_{yx_n}\left(I_x - L_{x_{n-1}}\right)\underline{x}_{n-1}(-) + \Phi_{yy_n}\,\underline{y}_{n-1}(-) + \underline{w}_{y_n}\end{aligned} \tag{16.1.1-23}$$

The pertinent results of the above development are provided by Equations (16.1.1-21), (16.1.1-23) and (16.1.1-16) summarized (and slightly rearranged) below:

$$\Delta\underline{x}_n(-) = \left[\overset{*}{\Phi}_{xx_n} - \left(\overset{*}{\Phi}_{xx_n} - \Phi_{xx_n}\right) L_{x_{n-1}}\right] \Delta\underline{x}_{n-1}(+_e)$$
$$- \Phi_{xy_n}\, \underline{y}_{n-1}(-) + \left(\overset{*}{\Phi}_{xx_n} - \Phi_{xx_n}\right)\left(I_x - L_{x_{n-1}}\right) \underline{x}_{n-1}(-) - \underline{w}_{x_n}$$
$$\underline{y}_n(-) = - \Phi_{yx_n} L_{x_{n-1}} \Delta\underline{x}_{n-1}(+_e) + \Phi_{yy_n}\, \underline{y}_{n-1}(-) + \Phi_{yx_n}\left(I_x - L_{x_{n-1}}\right) \underline{x}_{n-1}(-) + \underline{w}_{y_n} \qquad (16.1.1\text{-}24)$$
$$\underline{x}_n(-) = - \Phi_{xx_n} L_{x_{n-1}} \Delta\underline{x}_{n-1}(+_e) + \Phi_{xy_n}\, \underline{y}_{n-1}(-) + \Phi_{xx_n}\left(I_x - L_{x_{n-1}}\right) \underline{x}_{n-1}(-) + \underline{w}_{x_n}$$

$$\Delta\underline{x}_n(+_e) = \left(I_x - K_n \overset{*}{H}_{x_n}\right) \Delta\underline{x}_n(-) + K_n H_{y_n}\, \underline{y}_n(-)$$
$$- K_n \left(\overset{*}{H}_{x_n} - H_{x_n}\right) \underline{x}_n(-) + K_n G_{M_n} \underline{n}_{M_n} \qquad (16.1.1\text{-}25)$$

It will prove beneficial for subsequent optimal Kalman filter formulations to also introduce the concept of a $\underline{y}$ estimate and associated uncertainty as:

$$\Delta\underline{y} \equiv \tilde{\underline{y}} - \underline{y} \qquad (16.1.1\text{-}26)$$

where

$\tilde{\underline{y}}$ = Kalman filter estimate for $\underline{y}$.

$\Delta\underline{y}$ = Uncertainty in $\tilde{\underline{y}}$.

For the suboptimal Kalman filter, $\underline{y}$ is not accounted for, hence, from (16.1.1-12) and (16.1.1-26):

For The Suboptimal Kalman Filter:

$$\tilde{\underline{y}} = 0 \quad \Rightarrow \quad \underline{y} = - \Delta\underline{y} \qquad (16.1.1\text{-}27)$$

With (16.1.1-27), Equations (16.1.1-24) and (16.1.1-25) can now be converted to a more familiar form if we define an augmented error state vector with associated error state transition matrix, integrated process noise vector, measurement model, and estimation gain matrix as follows:

$$\underline{x}' \equiv \begin{bmatrix} \Delta\underline{x} \\ \Delta\underline{y} \\ -\underline{x} \end{bmatrix} \qquad \underline{w}' \equiv \begin{bmatrix} \underline{w}_x \\ \underline{w}_y \\ \underline{w}_x \end{bmatrix}$$

$$\Phi_n' \equiv \begin{bmatrix} \left[\Phi^*_{xx_n} - \left(\Phi^*_{xx_n} - \Phi_{xx_n}\right) L_{x_{n-1}}\right] & \Phi_{xy_n} & -\left(\Phi^*_{xx_n} - \Phi_{xx_n}\right)\left(I_x - L_{x_{n-1}}\right) \\ \Phi_{yx_n} L_{x_{n-1}} & \Phi_{yy_n} & \Phi_{yx_n}\left(I_x - L_{x_{n-1}}\right) \\ \Phi_{xx_n} L_{x_{n-1}} & \Phi_{xy_n} & \Phi_{xx_n}\left(I_x - L_{x_{n-1}}\right) \end{bmatrix} \qquad (16.1.1\text{-}28)$$

$$H_n' \equiv \begin{bmatrix} H^*_{x_n} & H_{y_n} & -\left(H^*_{x_n} - H_{x_n}\right) \end{bmatrix}$$

$$K_n' \equiv \begin{bmatrix} K_n \\ 0 \\ 0 \end{bmatrix} \qquad I' \equiv \begin{bmatrix} I_x & 0 & 0 \\ 0 & I_y & 0 \\ 0 & 0 & I_x \end{bmatrix}$$

where

$\underline{x}'$ = Augmented error state vector having $\Delta\underline{x}$, $\Delta\underline{y}$, and $-\underline{x}$ for elements as shown in (16.1.1-28).

$\underline{w}'$, Φ', H', K', I' = Integrated process noise vector, error state transition matrix, measurement matrix, estimation gain matrix and identity matrix associated with $\underline{x}'$ in Equations (16.1.1-28).

I_y = Identity matrix with same dimension as $\underline{y}$.

Using (16.1.1-27) - (16.1.1-28), Equations (16.1.1-24) - (16.1.1-25) reduce to the following familiar forms:

$$\underline{x}_n'(-) = \Phi_n' \, \underline{x}_{n-1}'(+_e) - \underline{w}_n' \qquad (16.1.1\text{-}29)$$

$$\underline{x}_n'(+_e) = \left(I' - K_n' \, H_n'\right) \underline{x}_n'(-) + K_n' \, G_{M_n} \, \underline{n}_{M_n} \qquad (16.1.1\text{-}30)$$

where

$\underline{x}_n'(+_e) = \underline{x}'$ as defined in (16.1.1-28) with $\Delta\underline{x} = \Delta\underline{x}(+_e)$, $\Delta\underline{y} = \Delta\underline{y}(+_e)$ and $\underline{x} = \underline{x}(-)$.

Because Equations (16.1.1-29) - (16.1.1-30) are identical in form to Equations (15.1.2.1.1-5) and (15.1.2.1-8), the covariance equivalents of (16.1.1-29) - (16.1.1-30) should also be identical in form to Equations (15.1.2.1.1-12) and (15.1.2.1-15) (the covariance equivalents of (15.1.2.1.1-5) and (15.1.2.1-8)). Thus, with (15.1.2.1-4):

$$P_n^{'} \equiv \mathcal{E}\left(\underline{x}_n^{'}\, \underline{x}_n^{'\,T}\right) \tag{16.1.1-31}$$

$$P_n^{'}(-) = \Phi_n^{'}\, P_{n-1}^{'}(+_e)\, \Phi_n^{'\,T} + Q_n^{'} \tag{16.1.1-32}$$

$$P_n^{'}(+_e) = \left(I' - K_n^{'} H_n^{'}\right) P_n^{'}(-) \left(I' - K_n^{'} H_n^{'}\right)^T + K_n^{'}\, G_{M_n}\, R_n\, G_{M_n}^T\, K_n^{'\,T} \tag{16.1.1-33}$$

with from (15.1.2.1-13), (15.1.2.1.1-10) and (16.1.1-28):

$$R_n = \mathcal{E}\left(\underline{n}_{M_n}\, \underline{n}_{M_n}^T\right)$$

$$Q_n^{'} = \mathcal{E}\left(\underline{w}_n^{'}\, \underline{w}_n^{'\,T}\right) = \mathcal{E}\left(\begin{bmatrix} \underline{w}_{x_n} \\ \underline{w}_{y_n} \\ \underline{w}_{x_n} \end{bmatrix} \begin{bmatrix} \underline{w}_{x_n}^T & \underline{w}_{y_n}^T & \underline{w}_{x_n}^T \end{bmatrix}\right) \tag{16.1.1-34}$$

$$= \begin{bmatrix} \mathcal{E}\left(\underline{w}_x\, \underline{w}_x^T\right) & \mathcal{E}\left(\underline{w}_x\, \underline{w}_y^T\right) & \mathcal{E}\left(\underline{w}_x\, \underline{w}_x^T\right) \\ \mathcal{E}\left(\underline{w}_y\, \underline{w}_x^T\right) & \mathcal{E}\left(\underline{w}_y\, \underline{w}_y^T\right) & \mathcal{E}\left(\underline{w}_y\, \underline{w}_x^T\right) \\ \mathcal{E}\left(\underline{w}_x\, \underline{w}_x^T\right) & \mathcal{E}\left(\underline{w}_x\, \underline{w}_y^T\right) & \mathcal{E}\left(\underline{w}_x\, \underline{w}_x^T\right) \end{bmatrix}_n \equiv \begin{bmatrix} Q_{xx} & Q_{xy} & Q_{xx} \\ Q_{yx} & Q_{yy} & Q_{yx} \\ Q_{xx} & Q_{xy} & Q_{xx} \end{bmatrix}_n$$

where

Q' = Equation (16.1.1-28) $\underline{w}'$ integrated process noise vector covariance.

The augmented error state transition matrix $\Phi_n^{'}$ in (16.1.1-32) and the integrated process noise matrix $Q_n^{'}$ in (16.1.1-34) can be calculated using the methods defined in Sections 15.1.2.1.1.1 and 15.1.2.1.1.2. Alternative versions of the (16.1.1-32) covariance propagation equation can be defined using the methods of 15.1.2.1.1.3. The suboptimal Kalman filter gain matrix K_n in (16.1.1-28) is calculated using the techniques of Section 15.1.2.1 based on the suboptimal filter covariance propagation/reset versions of Equations (15.1.2.1-15) and (15.1.2.1.1-12) with (15.1.2.1-28):

$$P_n^{*}(-) = \Phi_{xx_n}^{*}\, P_{n-1}^{*}(+_e)\, \Phi_{xx_n}^{*\,T} + Q_{xx_n}^{*} \tag{16.1.1-35}$$

$$K_n = I_{Est}\, P_n^{*}(-)\, H_{x_n}^{*\,T} \left(H_{x_n}^{*}\, P_n^{*}(-)\, H_{x_n}^{*\,T} + G_{M_n}^{*}\, R_n^{*}\, G_{M_n}^{*\,T}\right)^{-1} \tag{16.1.1-36}$$

$$P^*_n(+_e) = \left(I_X - K_n H^*_{x_n}\right) P^*_n(-) \left(I_X - K_n H^*_{x_n}\right)^T + K_n G^*_{M_n} R^*_n G^{*\,T}_{M_n} K_n^T \qquad (16.1.1\text{-}37)$$

where

P^* = Covariance matrix calculated in the actual Kalman filter for evaluating the suboptimal Kalman gain matrix.

I_{Est} = "Estimation matrix" defined as a diagonal matrix with unity on the diagonal of each row for which the corresponding error state vector component is being estimated (or zero otherwise). If all error state vector components are being estimated, $I_{Est} = I_X$. The I_{Est} matrix provides the mechanism for implementing the "considered variable" approach in the suboptimal estimator (See Section 15.1.2.1.1 following Equation (15.1.2.1.1-4) for further detail).

It is to be noted from Sections 15.1.2.1 and 15.1.2.1.1, that Equations (16.1.1-35) - (16.1.1-37) for suboptimal covariance analysis represent the suboptimal gain calculation for the idealized control reset filter configuration discussed in this section, as well as for the delayed control reset Kalman filter to be discussed subsequently in Section 16.1.2.

Equations (16.1.1-32) - (16.1.1-33) with (16.1.1-28) define the covariance effects in the real world of the suboptimal Kalman filter (represented by covariance Equations (16.1.1-35) - (16.1.1-37)) operating from real world inputs and providing control feedback to the real world through control matrix L_X based on idealized control resets. A covariance analysis program implements these equations in a digital computer simulation to generate numerical solutions as a function of time that define suboptimal Kalman filter performance in the real world. The following subsections discuss additional details associated with these equations.

16.1.1.1 SUBOPTIMAL KALMAN FILTER PERFORMANCE EVALUATION

The inaccuracy in the suboptimal Kalman filter real world performance is defined as the uncertainty in the Kalman filter's estimate of the $\underline{x}$ error state vector (i.e., by the error state uncertainty vector $\Delta\underline{x}$). The covariance of $\Delta\underline{x}$ is generally used for suboptimal Kalman filter performance evaluation. From (16.1.1-31) with (16.1.1-28) for $\underline{x}'$, we see that the covariance of $\Delta\underline{x}$ equals $P_{\Delta x\, \Delta x}$, the upper left partition of P':

$$P' = \mathcal{E}\left(\underline{x}'\,\underline{x}'^{T}\right) = \mathcal{E}\left(\begin{bmatrix} \Delta\underline{x} \\ \Delta\underline{y} \\ -\underline{x} \end{bmatrix} \begin{bmatrix} \Delta\underline{x}^{T} & \Delta\underline{y}^{T} & -\underline{x}^{T} \end{bmatrix}\right)$$

$$= \begin{bmatrix} \mathcal{E}\left(\Delta\underline{x}\,\Delta\underline{x}^{T}\right) & \mathcal{E}\left(\Delta\underline{x}\,\Delta\underline{y}^{T}\right) & -\mathcal{E}\left(\Delta\underline{x}\,\underline{x}^{T}\right) \\ \mathcal{E}\left(\Delta\underline{y}\,\Delta\underline{x}^{T}\right) & \mathcal{E}\left(\Delta\underline{y}\,\Delta\underline{y}^{T}\right) & -\mathcal{E}\left(\Delta\underline{y}\,\underline{x}^{T}\right) \\ -\mathcal{E}\left(\underline{x}\,\Delta\underline{x}^{T}\right) & -\mathcal{E}\left(\underline{x}\,\Delta\underline{y}^{T}\right) & \mathcal{E}\left(\underline{x}\,\underline{x}^{T}\right) \end{bmatrix} \equiv \begin{bmatrix} P_{\Delta x\,\Delta x} & P_{\Delta x\,\Delta y} & -P_{\Delta x\,x} \\ P_{\Delta y \Delta x} & P_{\Delta y \Delta y} & -P_{\Delta y\,x} \\ -P_{x\,\Delta x} & -P_{x\,\Delta y} & P_{xx} \end{bmatrix} \qquad (16.1.1.1\text{-}1)$$

where

P' = Covariance of $\underline{x}'$.

16.1.1.2 COVARIANCE MATRIX INITIALIZATION

In the covariance simulation, the Equation (16.1.1.1-1) P' covariance matrix is initialized based on Equations (16.1.1-11) - (16.1.1-13), (16.1.1-26) and (16.1.1-27):

$$\Delta\underline{x}_0 = \tilde{\underline{x}}_0 - \underline{x}_0 = -\underline{x}_0$$
$$\Delta\underline{y}_0 = -\underline{y}_0 \qquad (16.1.1.2\text{-}1)$$

where

0 = Subscript denoting initial condition at time t = 0.

Then, with (16.1.1.2-1), we see from (16.1.1.1-1) that:

$$P_0' = \begin{bmatrix} P_{xx\,0} & P_{xy0} & P_{xx\,0} \\ P_{yx0} & P_{yy0} & P_{yx0} \\ P_{xx\,0} & P_{xy0} & P_{xx\,0} \end{bmatrix} \qquad (16.1.1.2\text{-}2)$$

The Suboptimal Kalman filter P* covariance matrix is initialized based on the particular Kalman filter initialization implementation (e.g., Sections 15.2.1.2 and 15.2.2.1).

16.1.1.3 OPTIMAL KALMAN FILTER PERFORMANCE EVALUATION

The $P_{\Delta x\,\Delta x}$ performance in (16.1.1.1-1) will be suboptimal because of variations in the suboptimal Kalman filter error models compared to the real world. The degree of suboptimality in $P_{\Delta x\,\Delta x}$ is measured by the difference between $P_{\Delta x\,\Delta x}$ and a reference covariance matrix based on an optimal Kalman filter formulation. The optimal Kalman filter is defined as one whose error model matches the real world error model including all real world error states, and whose Kalman gain is based on the full real world model. As such, the optimal Kalman filter provides the theoretically best performance possible with the real world model. To define the optimal Kalman filter, we first introduce the following real world error model definitions based on the parameters in Equations (16.1.1-1) - (16.1.1-5) and the Section 16.1.1 notation:

$$\underline{\chi} \equiv \begin{bmatrix} \underline{x} \\ \underline{y} \end{bmatrix} \qquad \Phi \equiv \begin{bmatrix} \Phi_{xx} & \Phi_{xy} \\ \Phi_{yx} & \Phi_{yy} \end{bmatrix} \qquad \underline{w} \equiv \begin{bmatrix} \underline{w}_x \\ \underline{w}_y \end{bmatrix}$$

$$H \equiv \begin{bmatrix} H_x & H_y \end{bmatrix} \qquad I = \begin{bmatrix} I_x & 0 \\ 0 & I_y \end{bmatrix} \tag{16.1.1.3-1}$$

where

$\underline{\chi}$ = Error state vector containing all real world error states and used to formulate the optimal Kalman filter configuration.

Using the (16.1.1.3-1) definitions, we first rewrite Equations (16.1.1-1) - (16.1.1-5) in the more compact notation:

$$\underline{\chi}_n(-) = \Phi_n \, \underline{\chi}_{n-1}(+_c) + \underline{w}_n \tag{16.1.1.3-2}$$

$$\underline{z}_n = H_n \, \underline{\chi}_n(-) + G_{M_n} \, \underline{n}_{M_n} \tag{16.1.1.3-3}$$

$$\underline{\chi}_n(+_c) = \underline{\chi}_n(-) + \underline{u}_{cOpt_n} \tag{16.1.1.3-4}$$

where

$\underline{u}_{cOpt}$ = Control vector calculated in the optimal filter to control $\underline{\chi}$.

The optimal Kalman filter is based on a full exact implementation of the (16.1.1.3-2) - (16.1.1.3-4) model using Equations (15.1.2-7) - (15.1.2-10) and (15.1.2-12) - (15.1.2-13) (for ideal control resets) with the $\underline{Z}_{Obs}$ observation replaced by the linearized equivalent measurement $\underline{z}$ (as in Section 16.1.1):

$$\tilde{\underline{\chi}}_n(-) = \Phi_n \, \tilde{\underline{\chi}}_{n-1}(+_c) \tag{16.1.1.3-5}$$

$$\tilde{\underline{z}}_{Opt_n} = H_n \, \tilde{\underline{\chi}}_n(-) \tag{16.1.1.3-6}$$

$$\tilde{\underline{\chi}}_n(+_e) = \tilde{\underline{\chi}}_n(-) - K_{Opt_n}\left(\tilde{\underline{z}}_{Opt_n} - \underline{z}_n\right) \tag{16.1.1.3-7}$$

$$\underline{u}_{cOpt_n} = \text{function of } \tilde{\underline{\chi}}_n(+_e) \qquad (16.1.1.3\text{-}8)$$

$$\tilde{\underline{\chi}}_n(+_c) = \tilde{\underline{\chi}}_n(+_e) + \underline{u}_{cOpt_n} \qquad (16.1.1.3\text{-}9)$$

$$\tilde{\underline{\chi}}_0 = 0 \qquad (16.1.1.3\text{-}10)$$

where

$\tilde{\underline{\chi}}$ = Optimal Kalman filter estimate for $\underline{\chi}$.

K_{Opt} = Optimal Kalman gain matrix based on the $\underline{\chi}$ error model.

Following the same steps that led to Equations (16.1.1-24) - (16.1.1-25), the previous real world and Kalman filter models translate into the following optimal Kalman filter estimated error state uncertainty propagation/reset equations:

$$\Delta\underline{\chi} \equiv \tilde{\underline{\chi}} - \underline{\chi} \qquad (16.1.1.3\text{-}11)$$

$$\Delta\underline{\chi}_n(-) = \Phi_n \, \Delta\underline{\chi}_{n-1}(+_e) - \underline{w}_n \qquad (16.1.1.3\text{-}12)$$

$$\Delta\underline{\chi}_n(+_e) = \left(I - K_{Opt_n} H_n\right) \Delta\underline{\chi}_n(-) + K_{Opt_n} G_{M_n} \underline{n}_{M_n} \qquad (16.1.1.3\text{-}13)$$

where

$\Delta\underline{\chi}$ = Uncertainty in the optimal Kalman filter's estimate for $\underline{\chi}$.

The covariance equivalent to (16.1.1.3-11) - (16.1.1.3-13) is then obtained as was Equations (15.1.2.1-15) and (15.1.2.1.1-12) from Equations (15.1.2.1-8) and (15.1.2.1.1-5):

$$P_{Opt_n} \equiv E\left(\Delta\underline{\chi}_n \, \Delta\underline{\chi}_n^T\right) \qquad (16.1.1.3\text{-}14)$$

$$P_{Opt_n}(-) = \Phi_n \, P_{Opt_{n-1}}(+_e) \, \Phi_n^T + Q_{RW_n} \qquad (16.1.1.3\text{-}15)$$

$$P_{Opt_n}(+_e) = \left(I - K_{Opt_n} H_n\right) P_{Opt_n}(-) \left(I - K_{Opt_n} H_n\right)^T + K_{Opt_n} G_{M_n} R_n G_{M_n}^T K_{Opt_n}^T \qquad (16.1.1.3\text{-}16)$$

where

P_{Opt} = Optimal Kalman filter error state uncertainty covariance matrix.

Q_{RW} = Real world integrated process noise covariance matrix.

with from (15.1.2.1-13), (15.1.2.1.1-10) and (16.1.1.3-1):

$$R_n = \mathcal{E}\left(\underline{n}_{M_n} \underline{n}_{M_n}^T\right)$$

$$Q_{RW_n} \equiv \mathcal{E}\left(\underline{w}_n \underline{w}_n^T\right) = \mathcal{E}\left(\begin{bmatrix} \underline{w}_{x_n} \\ \underline{w}_{y_n} \end{bmatrix} \begin{bmatrix} \underline{w}_{x_n}^T & \underline{w}_{y_n}^T \end{bmatrix}\right) \tag{16.1.1.3-17}$$

$$= \begin{bmatrix} \mathcal{E}\left(\underline{w}_x \underline{w}_x^T\right) & \mathcal{E}\left(\underline{w}_x \underline{w}_y^T\right) \\ \mathcal{E}\left(\underline{w}_y \underline{w}_x^T\right) & \mathcal{E}\left(\underline{w}_y \underline{w}_y^T\right) \end{bmatrix}_n \equiv \begin{bmatrix} Q_{xx} & Q_{xy} \\ Q_{yx} & Q_{yy} \end{bmatrix}_n$$

The K_{Opt} Kalman gain for the previous development is obtained directly from the (16.1.1.3-15) $P_{Opt}(-)$ solution using the (15.1.2.1-28) format:

$$K_{Opt_n} = I_{OptEst}\ P_{Opt_n}(-)\ H_n^T \left(H_n\ P_{Opt_n}(-)\ H_n^T + G_{M_n}\ R_n\ G_{M_n}^T\right)^{-1} \tag{16.1.1.3-18}$$

where

I_{OptEst} = "Optimal estimation matrix" defined as a diagonal matrix with unity on the diagonal of each row for which the corresponding error state vector component is being optimally estimated (or zero otherwise). For all error state components being optimally estimated, $I_{OptEst} = I$.

The I_{OptEst} matrix provides the mechanism for implementing the "considered variable" approach in the optimal estimator (See Section 15.1.2.1.1 for further detail) for situations when it is known that the associated error effect cannot actually be estimated in practice. An example would be a vibration sensitive sensor error whose error coefficient may be constant but whose state dynamic coupling element is random vibration acceleration squared . The random vibration appears as a constant average mean squared value in the (16.1.1.3-15) error state transition matrix. In a real application, the vibration squared value would not be available to an actual Kalman filter as it truly represents the instantaneous random vibration being experienced in real time. We account for the effect in Equations (16.1.1.3-14) - (16.1.1.3-16) by representing the vibration as the average mean squared value, but not allowing the "optimal" Kalman filter to estimate the sensor error coefficient.

Initialization of P_{Opt} for Equations (16.1.1.3-14) - (16.1.1.3-16) is based on (16.1.1.3-10) and (16.1.1.3-11):

$$\Delta\underline{\chi}_0 = \tilde{\underline{\chi}}_0 - \underline{\chi}_0 = -\underline{\chi}_0 \tag{16.1.1.3-19}$$

Then, with (16.1.1.3-19), we find from (16.1.1.3-1) and (16.1.1.3-14) that:

$$P_{Opt_0} = \mathcal{E}\left(\Delta\underline{\chi}_0\, \Delta\underline{\chi}_0^T\right) = \mathcal{E}\left(\underline{\chi}_0\, \underline{\chi}_0^T\right) = \mathcal{E}\left(\begin{bmatrix} \underline{x}_0 \\ \underline{y}_0 \end{bmatrix} \begin{bmatrix} \underline{x}_0^T & \underline{y}_0^T \end{bmatrix}\right)$$

$$= \begin{bmatrix} \mathcal{E}\left(\underline{x}_0\, \underline{x}_0^T\right) & \mathcal{E}\left(\underline{x}_0\, \underline{y}_0^T\right) \\ \mathcal{E}\left(\underline{y}_0\, \underline{x}_0^T\right) & \mathcal{E}\left(\underline{y}_0\, \underline{y}_0^T\right) \end{bmatrix} \equiv \begin{bmatrix} P_{xx_0} & P_{xy_0} \\ P_{yx_0} & P_{yy_0} \end{bmatrix} \tag{16.1.1.3-20}$$

We also see from (16.1.1.3-1) and (16.1.1.3-11) that in general:

$$P_{Opt} = \mathcal{E}\left(\Delta\underline{\chi}\, \Delta\underline{\chi}^T\right) = \mathcal{E}\left(\begin{bmatrix} \Delta\underline{x}_{Opt} \\ \Delta\underline{y}_{Opt} \end{bmatrix} \begin{bmatrix} \Delta\underline{x}_{Opt}^T & \Delta\underline{y}_{Opt}^T \end{bmatrix}\right)$$

$$= \begin{bmatrix} \mathcal{E}\left(\Delta\underline{x}_{Opt}\, \Delta\underline{x}_{Opt}^T\right) & \mathcal{E}\left(\Delta\underline{x}_{Opt}\, \Delta\underline{y}_{Opt}^T\right) \\ \mathcal{E}\left(\Delta\underline{y}_{Opt}\, \Delta\underline{x}_{Opt}^T\right) & \mathcal{E}\left(\Delta\underline{y}_{Opt}\, \Delta\underline{y}_{Opt}^T\right) \end{bmatrix} \equiv \begin{bmatrix} P_{Opt\,\Delta x\,\Delta x} & P_{Opt\,\Delta x\,\Delta y} \\ P_{Opt\,\Delta y\,\Delta x} & P_{Opt\,\Delta y\,\Delta y} \end{bmatrix} \tag{16.1.1.3-21}$$

where

$\Delta\underline{x}_{Opt}, \Delta\underline{y}_{Opt}$ = Uncertainties in $\underline{\tilde{x}}$ and $\underline{\tilde{y}}$ when using the optimal Kalman filter.

Equation (16.1.1.3-21) shows the format of the optimal Kalman filter covariance matrix for comparison against the equivalent data format in (16.1.1.1-1) for suboptimal Kalman filter performance evaluation. The upper left 2 by 2 suboptimal covariance array in (16.1.1.1-1) should be compared against the (16.1.1.3-21) optimal covariance equivalent to evaluate the degree of suboptimality in the Section 16.1.1 suboptimal Kalman filter.

Note in Equation (16.1.1.1-1) that the $P_{\Delta x\,\Delta x}, P_{\Delta x\,\Delta y}, P_{\Delta y\Delta x}, P_{\Delta y\Delta y}$ upper two left covariance partitions for the suboptimal Kalman filter in the real world (based on Equations (16.1.1-28), (16.1.1-32) - (16.1.1-37) and (16.1.1.2-2)), can be converted to Equation (16.1.1.3-21) for optimal Kalman filter real world performance evaluation (i.e., based on Equations (16.1.1.3-1), (16.1.1.3-14) - (16.1.1.3-18) and (16.1.1.3-20)), by setting $L_x = I_x$, setting $H_x^* = H_x$, and replacing the suboptimal Kalman gain matrix K' with the optimal Kalman gain matrix equivalent $\begin{bmatrix} K_{Opt} \\ 0 \end{bmatrix}$. Application of the previous observation allows common software code to be used in the covariance simulation program for calculating the suboptimal P' and optimal P_{Opt} covariance performance matrices.

16.1.1.4 SIMPLIFIED COVARIANCE ANALYSIS EQUATIONS

A fundamental problem associated with the implementation of Equations (16.1.1-32) - (16.1.1-33) in a covariance analysis digital computer simulation program is the throughput for execution of the matrix operations. In this regard, the dimension of the P' covariance matrix is the governing factor, which, as can be seen from (16.1.1.1-1), is 2 a + b where:

a = Dimension of $\underline{x}$.

b = Dimension of $\underline{y}$.

If we only require the $\Delta\underline{x}$, $\Delta\underline{y}$ covariance for output, the dimension of P' can be reduced to a + b by excluding the $\underline{x}$ terms, but only for applications with zero coupling of $\underline{x}$ into $\Delta\underline{x}$ and $\Delta\underline{y}$ (See Equation (16.1.1.1-1)). From Equations (16.1.1-28) - (16.1.1-30) we see that there will be zero coupling of $\underline{x}$ into $\Delta\underline{x}$, $\Delta\underline{y}$ under the following Φ' and H' matrix conditions:

$$\left(\overset{*}{\Phi}_{XX} - \Phi_{XX}\right)\left(I_X - L_X\right) = 0 \qquad \Phi_{YX}\left(I_X - L_X\right) = 0 \qquad \overset{*}{H}_X - H_X = 0 \qquad (16.1.1.4\text{-}1)$$

Equations (16.1.1.4-1) are satisfied by:

<u>Conditions For Zero x Coupling Into $\Delta\underline{x}$, $\Delta\underline{y}$:</u>

$$\overset{*}{H}_X = H_X \qquad (16.1.1.4\text{-}2)$$

$$\text{IF } \left(L_X(i,i) \neq 1\right) \text{ THEN } \overset{*}{\Phi}_{XX}(i,j) = \Phi_{XX}(i,j) \text{ And } \Phi_{YX}(i,j) = 0 \text{ For All } j$$

where

i, j = Designation of element in row i, column j of the associated matrix.

The (16.1.1.4-2) conditions are satisfied for a broad range of Kalman filter performance analysis problems requiring covariance simulation analysis for suboptimal performance evaluation. Under Equations (16.1.1.4-2), Equations (16.1.1-28) and (16.1.1-31) - (16.1.1-34) simplify to:

$$\underline{x}'' \equiv \begin{bmatrix} \Delta\underline{x} \\ \Delta\underline{y} \end{bmatrix} \qquad \underline{w} \equiv \begin{bmatrix} \underline{w}_X \\ \underline{w}_Y \end{bmatrix}$$

$$\Phi \equiv \begin{bmatrix} \Phi_{XX} & \Phi_{XY} \\ \Phi_{YX} & \Phi_{YY} \end{bmatrix} \qquad H = \begin{bmatrix} H_X & H_Y \end{bmatrix} \qquad (16.1.1.4\text{-}3)$$

$$K'' \equiv \begin{bmatrix} K \\ 0 \end{bmatrix} \qquad I \equiv \begin{bmatrix} I_X & 0 \\ 0 & I_Y \end{bmatrix}$$

$$P'' \equiv \mathcal{E}\left(\underline{x}''\ \underline{x}''^T\right)$$

$$P''_n(-) = \Phi_n P''_{n-1}(+_e)\ \Phi_n^{\ T} + Q_n$$

$$P''_n(+_e) = \left(I - K''_n H_n\right) P''_n(-) \left(I - K''_n H_n\right)^T + K''_n G_{M_n} R_n G^T_{M_n} K''^T_n \qquad (16.1.1.4\text{-}4)$$

$$R_n = \mathcal{E}\left(\underline{n}_{M_n} \underline{n}^T_{M_n}\right) \qquad Q_n = \mathcal{E}\left(\underline{w}_n\ \underline{w}^T_n\right) = \begin{bmatrix} Q_{xx} & Q_{xy} \\ Q_{yx} & Q_{yy} \end{bmatrix}_n$$

where

" = Equivalent to augmented ' parameter indicator, but simplified by elimination of $\underline{x}$ components.

Suboptimal Kalman filter covariance Equations (16.1.1-35) - (16.1.1-37) would still be as shown for calculation of the Suboptimal Kalman gain matrix K in (16.1.1.4-3), but with substitution of the (16.1.1.4-2) $\overset{*}{H}_x$ condition:

$$\overset{*}{P}_n(-) = \overset{*}{\Phi}_{xx_n} \overset{*}{P}_{n-1}(+_e)\ \overset{*}{\Phi}{}^T_{xx_n} + \overset{*}{Q}_{xx_n}$$

$$K_n = I_{Est}\ \overset{*}{P}_n(-)\ H_{x_n}^{\ T}\left(H_{x_n} \overset{*}{P}_n(-)\ H_{x_n}^{\ T} + \overset{*}{G}_{M_n} \overset{*}{R}_n \overset{*}{G}{}^T_{M_n}\right)^{-1} \qquad (16.1.1.4\text{-}5)$$

$$\overset{*}{P}_n(+_e) = \left(I_x - K_n H_{x_n}\right) \overset{*}{P}_n(-) \left(I_x - K_n H_{x_n}\right)^T + K_n \overset{*}{G}_{M_n} \overset{*}{R}_n \overset{*}{G}{}^T_{M_n} K_n^T$$

Suboptimal Kalman filter performance from (16.1.1.4-4) would be obtained from $P_{\Delta x \Delta x}$ in the resulting reduced form of (16.1.1.1-1):

$$P'' = \begin{bmatrix} P_{\Delta x \Delta x} & P_{\Delta x \Delta y} \\ P_{\Delta y \Delta x} & P_{\Delta y \Delta y} \end{bmatrix} \qquad (16.1.1.4\text{-}6)$$

Initialization of the P" matrix would be based on the resulting reduced form of (16.1.1.2-2):

$$P''_0 = \begin{bmatrix} P_{xx_0} & P_{xy_0} \\ P_{yx_0} & P_{yy_0} \end{bmatrix} \qquad (16.1.1.4\text{-}7)$$

Note that the simplified version Equations (16.1.1.4-3) - (16.1.1.4-4) and (16.1.1.4-6) - (16.1.1.4-7) for suboptimal Kalman filter performance in the real world, is identical to Equations (16.1.1.3-1), (16.1.1.3-14) - (16.1.1.3-17) and (16.1.1.3-20) for the optimal Kalman filter real world performance, if we substitute the optimal Kalman gain matrix K_{Opt} (from Equation (16.1.1.3-18)) for the suboptimal equivalent K". Thus, the same software code can be applied in a covariance simulation program for either optimal or suboptimal covariance performance evaluation by using Equation (16.1.1.3-18) for the optimal gain matrix, or K" as in (16.1.1.4-3) with K from Equations (16.1.1.4-5) for the suboptimal gain matrix calculation.

16.1.2 FORMULATION BASED ON DELAYED CONTROL RESETS

In Section 16.1.1 and its subsections we derived equations to be used for suboptimal covariance analysis of Kalman filters based on the Equations (15.1.2-1) - (15.1.2-13) idealized control reset formulation. In this section we perform a similar analysis for the delayed control reset configuration defined by Equations (15.1.2-14) - (15.1.2-26). We begin by using (15.1.2-16) - (15.1.2-18) as a template for the delayed control reset case, to first define the following general real world model (in the computation sequence indicated) based on delaying the control application until after error state vector propagation:

$$\underline{x}_n(-) = \Phi_{xx_n}\,\underline{x}_{n-1}(+_c) + \Phi_{xy_n}\,\underline{y}_{n-1}(+_c) + \underline{w}_{x_n} \qquad (16.1.2\text{-}1)$$

$$\underline{y}_n(-) = \Phi_{yx_n}\,\underline{x}_{n-1}(+_c) + \Phi_{yy_n}\,\underline{y}_{n-1}(+_c) + \underline{w}_{y_n} \qquad (16.1.2\text{-}2)$$

$$\underline{x}_n(+_c) = \underline{x}_n(-) + \underline{u}_{c_n} \qquad (16.1.2\text{-}3)$$

$$\underline{y}_n(+_c) = \underline{y}_n(-) \qquad (16.1.2\text{-}4)$$

$$\underline{z}_n = H_{x_n}\,\underline{x}_n(+_c) + H_{y_n}\,\underline{y}_n(+_c) + G_{M_n}\,\underline{n}_{M_n} \qquad (16.1.2\text{-}5)$$

The equivalent version of (16.1.2-1) - (16.1.2-5) contained within the delayed control reset Kalman filter configuration is from Equations (15.1.2-19) - (15.1.2-26) and the Section 16.1.1 notation:

$$\underline{\tilde{x}}_n(-) = \overset{*}{\Phi}_{xx_n}\,\underline{\tilde{x}}_{n-1}(+_e) \qquad (16.1.2\text{-}6)$$

$$\underline{\tilde{x}}_n(+_c) = \underline{\tilde{x}}_n(-) + \underline{u}_{c_n} \qquad (16.1.2\text{-}7)$$

$$\underline{\tilde{z}}_n = \overset{*}{H}_{x_n}\,\underline{\tilde{x}}_n(+_c) \qquad (16.1.2\text{-}8)$$

$$\underline{\tilde{x}}_n(+_e) = \underline{\tilde{x}}_n(+_c) - K_n\left(\underline{\tilde{z}}_n - \underline{z}_n\right) \qquad (16.1.2\text{-}9)$$

$$\underline{u}_{c_{n+1}} = \text{function of } \underline{\tilde{x}}_n(+_e) \qquad (16.1.2\text{-}10)$$

$$\underline{\tilde{x}}_0 = 0 \qquad (16.1.2\text{-}11)$$

$$\underline{\tilde{y}}_n = 0 \qquad (16.1.2\text{-}12)$$

In Equation (16.1.2-9) we have used the real world measurement $\underline{z}_n$ (i.e., the linearized form of the observation vector $\underline{Z}_{Obs_n}$) to approximate $\underline{Z}_{Obs_n}$ in (15.1.2-24), and have added Equation (16.1.2-12) as a formal statement that the filter is only designed to account for the $\underline{x}$ error states.

As in Section 16.1.1 - Equation (16.1.1-14), but for the delayed control reset filter, we then write from (15.1.2.1-3) for $\Delta\underline{x}$ as defined in (16.1.1-13):

$$\begin{aligned} \Delta\underline{x}_n(-) &= \tilde{\underline{x}}_n(-) - \underline{x}_n(-) \\ \Delta\underline{x}_n(+_c) &= \tilde{\underline{x}}_n(+_c) - \underline{x}_n(+_c) \\ \Delta\underline{x}_n(+_e) &= \tilde{\underline{x}}_n(+_e) - \underline{x}_n(+_c) \end{aligned} \tag{16.1.2-13}$$

Following the development steps that led to (16.1.1-16), the estimated error state uncertainty $\Delta\underline{x}$ following the estimation update is obtained by combining suboptimal filter Equations (16.1.2-8) - (16.1.2-9) with the real world measurement $\underline{z}_n$ from (16.1.2-5) using the second and third equation in (16.1.2-13) for the $\Delta\underline{x}$'s:

$$\begin{aligned} \Delta\underline{x}_n(+_e) = &\left(I_x - K_n \overset{*}{H}_{x_n}\right) \Delta\underline{x}_n(+_c) - K_n \left(\overset{*}{H}_{x_n} - H_{x_n}\right) \underline{x}_n(+_c) \\ &+ K_n H_{y_n} \underline{y}_n(+_c) + K_n G_{M_n} \underline{n}_{M_n} \end{aligned} \tag{16.1.2-14}$$

Equation (16.1.2-14) relates the uncertainty after the estimation update to the uncertainty after the control update, at the current Kalman n cycle. From the $\Delta\underline{x}$ uncertainty definition in the second and third equation of (16.1.2-13), the uncertainty after the control update can be related to the previous cycle uncertainty following estimation update using (16.1.2-7), (16.1.2-3), (16.1.2-6) and (16.1.2-1):

$$\begin{aligned} \Delta\underline{x}_n(+_c) &= \tilde{\underline{x}}_n(+_c) - \underline{x}_n(+_c) = \tilde{\underline{x}}_n(-) + \underline{u}_{c_n} - \left(\underline{x}_n(-) + \underline{u}_{c_n}\right) = \tilde{\underline{x}}_n(-) - \underline{x}_n(-) \\ &= \overset{*}{\Phi}_{xx_n} \tilde{\underline{x}}_{n-1}(+_e) - \Phi_{xx_n} \underline{x}_{n-1}(+_c) - \Phi_{xy_n} \underline{y}_{n-1}(+_c) - \underline{w}_{x_n} \\ &= \overset{*}{\Phi}_{xx_n} \tilde{\underline{x}}_{n-1}(+_e) - \overset{*}{\Phi}_{xx_n} \underline{x}_{n-1}(+_c) + \overset{*}{\Phi}_{xx_n} \underline{x}_{n-1}(+_c) - \Phi_{xx_n} \underline{x}_{n-1}(+_c) \\ &\quad - \Phi_{xy_n} \underline{y}_{n-1}(+_c) - \underline{w}_{x_n} \\ &= \overset{*}{\Phi}_{xx_n} \Delta\underline{x}_{n-1}(+_e) + \left(\overset{*}{\Phi}_{xx_n} - \Phi_{xx_n}\right) \underline{x}_{n-1}(+_c) - \Phi_{xy_n} \underline{y}_{n-1}(+_c) - \underline{w}_{x_n} \end{aligned} \tag{16.1.2-15}$$

The equivalent to Equation (16.1.2-15) for the real world error state vector $\underline{x}$ component can be determined by first introducing a hypothesized form of the control vector $\underline{u}_c$. As in Section 16.1.1, we model the control vector from (16.1.2-10) as a linear function of $\tilde{\underline{x}}_n(+_e)$ with phasing to control $\underline{\tilde{x}}_n(+_e)$ to zero, and to be applied at the next Kalman cycle time. Hence, with the third equation in (16.1.2-13):

$$\underline{u}_{c_{n+1}} = - L_{x_{n+1}} \underline{\tilde{x}}_n(+_e) = - L_{x_{n+1}} \left(\Delta\underline{x}_n(+_e) + \underline{x}_n(+_c)\right) \qquad (16.1.2\text{-}16)$$

Substituting (16.1.2-16) and (16.1.2-1) into (16.1.2-3), we now derive an expression for the current cycle real world $\underline{x}$ error state vector following control update, in terms of the error state uncertainty and real world error state values after the previous cycle estimation/control update:

$$\begin{aligned} \underline{x}_n(+_c) &= \underline{x}_n(-) + \underline{u}_{c_n} = \underline{x}_n(-) - L_{x_n}\left(\Delta\underline{x}_{n-1}(+_e) + \underline{x}_{n-1}(+_c)\right) \\ &= \Phi_{xx_n}\,\underline{x}_{n-1}(+_c) + \Phi_{xy_n}\,\underline{y}_{n-1}(+_c) + \underline{w}_{x_n} - L_{x_n}\left(\Delta\underline{x}_{n-1}(+_e) + \underline{x}_{n-1}(+_c)\right) \\ &= - L_{x_n}\,\Delta\underline{x}_{n-1}(+_e) + \left(\Phi_{xx_n} - L_{x_n}\right)\underline{x}_{n-1}(+_c) + \Phi_{xy_n}\,\underline{y}_{n-1}(+_c) + \underline{w}_{x_n} \end{aligned} \qquad (16.1.2\text{-}17)$$

In a similar manner, using (16.1.2-4) with (16.1.2-2), the expression for the $(+_c)$ value of $\underline{y}$ in terms of previous $(+_c)$ error state components at the current Kalman cycle is:

$$\underline{y}_n(+_c) = \underline{y}_n(-) = \Phi_{yx_n}\,\underline{x}_{n-1}(+_c) + \Phi_{yy_n}\,\underline{y}_{n-1}(+_c) + \underline{w}_{y_n} \qquad (16.1.2\text{-}18)$$

The results of the previous development are provided by Equations (16.1.2-15), (16.1.2-17) - (16.1.2-18) and (16.1.2-14) summarized (and slightly rearranged) below:

$$\begin{aligned} \Delta\underline{x}_n(+_c) &= \overset{*}{\Phi}_{xx_n}\,\Delta\underline{x}_{n-1}(+_e) - \Phi_{xy_n}\,\underline{y}_{n-1}(+_c) + \left(\overset{*}{\Phi}_{xx_n} - \Phi_{xx_n}\right)\underline{x}_{n-1}(+_c) - \underline{w}_{x_n} \\ \underline{y}_n(+_c) &= \Phi_{yy_n}\,\underline{y}_{n-1}(+_c) + \Phi_{yx_n}\,\underline{x}_{n-1}(+_c) + \underline{w}_{y_n} \\ \underline{x}_n(+_c) &= - L_{x_n}\,\Delta\underline{x}_{n-1}(+_e) + \Phi_{xy_n}\,\underline{y}_{n-1}(+_c) + \left(\Phi_{xx_n} - L_{x_n}\right)\underline{x}_{n-1}(+_c) + \underline{w}_{x_n} \end{aligned} \qquad (16.1.2\text{-}19)$$

$$\begin{aligned} \Delta\underline{x}_n(+_e) = {} & \left(I_x - K_n\,\overset{*}{H}_{x_n}\right)\Delta\underline{x}_n(+_c) + K_n\,H_{y_n}\,\underline{y}_n(+_c) \\ & - K_n\left(\overset{*}{H}_{x_n} - H_{x_n}\right)\underline{x}_n(+_c) + K_n\,G_{M_n}\,\underline{n}_{M_n} \end{aligned} \qquad (16.1.2\text{-}20)$$

As in Section 16.1.1 we introduce the concept of a $\underline{y}$ estimate and associated uncertainty as:

$$\Delta\underline{y} \equiv \underline{\tilde{y}} - \underline{y} \qquad (16.1.2\text{-}21)$$

For the suboptimal Kalman filter, $\underline{y}$ is not accounted for, hence, from (16.1.2-12) and (16.1.2-21):

For The Suboptimal Kalman Filter:

$$\underline{\tilde{y}} = 0 \qquad \Rightarrow \qquad \underline{y} = - \Delta\underline{y} \qquad (16.1.2\text{-}22)$$

With (16.1.2-22), Equations (16.1.2-19) and (16.1.2-20) can now be converted to a more familiar form (as in Section 16.1.1) using an augmented error state vector with associated error state transition matrix, integrated process noise vector, measurement model, and estimation gain matrix as follows:

$$\underline{x}' \equiv \begin{bmatrix} \Delta\underline{x} \\ \Delta\underline{y} \\ -\underline{x} \end{bmatrix} \qquad \underline{w}' \equiv \begin{bmatrix} \underline{w}_x \\ \underline{w}_y \\ \underline{w}_x \end{bmatrix}$$

$$\Phi' \equiv \begin{bmatrix} \Phi^*_{xx} & \Phi_{xy} & -\left(\Phi^*_{xx} - \Phi_{xx}\right) \\ 0 & \Phi_{yy} & \Phi_{yx} \\ L_x & \Phi_{xy} & \left(\Phi_{xx} - L_x\right) \end{bmatrix} \qquad (16.1.2\text{-}23)$$

$$H' \equiv \begin{bmatrix} H^*_x & H_y & -\left(H^*_x - H_x\right) \end{bmatrix}$$

$$K' \equiv \begin{bmatrix} K \\ 0 \\ 0 \end{bmatrix} \qquad I' \equiv \begin{bmatrix} I_x & 0 & 0 \\ 0 & I_y & 0 \\ 0 & 0 & I_x \end{bmatrix}$$

Using (16.1.2-22) - (16.1.2-23) with definitions for the parameters provided in Section 16.1.1, Equations (16.1.2-19) - (16.1.2-20) reduce to the following familiar forms:

$$\underline{x}'_n(+_c) = \Phi'_n \, \underline{x}'_{n-1}(+_e) - \underline{w}'_n \qquad (16.1.2\text{-}24)$$

$$\underline{x}'_n(+_e) = \left(I' - K'_n H'_n\right) \underline{x}'_n(+_c) + K'_n \, G_{M_n} \, \underline{n}_{M_n} \qquad (16.1.2\text{-}25)$$

where

$$\underline{x}'_n(+_e) = \underline{x}' \text{ as defined in (16.1.2-23) with } \Delta\underline{x} = \Delta\underline{x}(+_e),\ \Delta\underline{y} = \Delta\underline{y}(+_e) \text{ and } \underline{x} = \underline{x}(+_c).$$

As in Sections 16.1.1 and 16.1.1.1, the covariance equivalent to (16.1.2-24) - (16.1.2-25) then follows as:

$$P' \equiv \mathcal{E}\left(\underline{x}' \underline{x}'^T\right) = \begin{bmatrix} \mathcal{E}\left(\Delta\underline{x}\, \Delta\underline{x}^T\right) & \mathcal{E}\left(\Delta\underline{x}\, \Delta\underline{y}^T\right) & -\mathcal{E}\left(\Delta\underline{x}\, \underline{x}^T\right) \\ \mathcal{E}\left(\Delta\underline{y}\, \Delta\underline{x}^T\right) & \mathcal{E}\left(\Delta\underline{y}\, \Delta\underline{y}^T\right) & -\mathcal{E}\left(\Delta\underline{y}\, \underline{x}^T\right) \\ -\mathcal{E}\left(\underline{x}\, \Delta\underline{x}^T\right) & -\mathcal{E}\left(\underline{x}\, \Delta\underline{y}^T\right) & \mathcal{E}\left(\underline{x}\, \underline{x}^T\right) \end{bmatrix}$$

$$\equiv \begin{bmatrix} P_{\Delta x\, \Delta x} & P_{\Delta x\, \Delta y} & -P_{\Delta x\, x} \\ P_{\Delta y \Delta x} & P_{\Delta y \Delta y} & -P_{\Delta y\, x} \\ -P_{x\, \Delta x} & -P_{x\, \Delta y} & P_{xx} \end{bmatrix} \tag{16.1.2-26}$$

$$P_n'(+_c) = \Phi_n' P_{n-1}'(+_e)\, \Phi_n'^T + Q_n' \tag{16.1.2-27}$$

$$P_n'(+_e) = \left(I' - K_n' H_n'\right) P_n'(+_c) \left(I' - K_n' H_n'\right)^T + K_n' G_{M_n} R_n G_{M_n}^T K_n'^T \tag{16.1.2-28}$$

For the delayed control reset filter configuration, the R_n, Q_n' matrices in (16.1.2-27) - (16.1.2-28) and the suboptimal Kalman gain matrix K_n in Equations (16.1.2-23) are calculated as in Section 16.1.1 (for idealized control resets) using Equations (16.1.1-34) - (16.1.1-37). Initialization of the P' covariance matrix for delayed control resets and the suboptimal performance evaluation from P' in Equations (16.1.2-26), is identical to Equation (16.1.1.2-2) and the discussion in Section 16.1.1.1 for the idealized control reset filter.

The following subsections discuss optimal Kalman filter performance evaluation for delayed control resets, and potential simplifications to delayed control reset suboptimal covariance analysis Equations (16.1.2-27) - (16.1.2-28).

16.1.2.1 OPTIMAL KALMAN FILTER PERFORMANCE EVALUATION

The delayed control reset optimal Kalman filter is based on the compact version of real world model Equations (16.1.2-1) - (16.1.2-5) using the Equation (16.1.1.3-1) notation.

$$\underline{\chi}_n(-) = \Phi_n\, \underline{\chi}_{n-1}(+_c) + \underline{w}_n \tag{16.1.2.1-1}$$

$$\underline{\chi}_n(+_c) = \underline{\chi}_n(-) + \underline{u}_{cOpt_n} \tag{16.1.2.1-2}$$

$$\underline{z}_n = H_n\, \underline{\chi}_n(+_c) + G_{M_n}\, \underline{n}_{M_n} \tag{16.1.2.1-3}$$

where

$\underline{\chi}$ = Error state vector containing all real world error states and used to formulate the optimal Kalman filter configuration.

$\underline{u}_{cOpt}$ = Control vector calculated in the optimal filter to control $\underline{\chi}$.

The optimal delayed control reset Kalman filter is based on a full exact implementation of the (16.1.2.1-1) - (16.1.2.1-3) model using Equations (15.1.2-21) - (15.1.2-26) with the $\underline{Z}_{Obs}$ observation replaced by the linearized equivalent measurement $\underline{z}$ (as in Equation (16.1.2-9)):

$$\tilde{\underline{\chi}}_n(-) = \Phi_n \, \tilde{\underline{\chi}}_{n-1}(+_e) \tag{16.1.2.1-4}$$

$$\tilde{\underline{\chi}}_n(+_c) = \tilde{\underline{\chi}}_n(-) + \underline{u}_{cOpt_n} \tag{16.1.2.1-5}$$

$$\tilde{\underline{z}}_{Opt_n} = H_n \, \tilde{\underline{\chi}}_n(+_c) \tag{16.1.2.1-6}$$

$$\tilde{\underline{\chi}}_n(+_e) = \tilde{\underline{\chi}}_n(+_c) - K_{Opt_n} \left(\tilde{\underline{z}}_{Opt_n} - \underline{z}_n \right) \tag{16.1.2.1-7}$$

$$\underline{u}_{cOpt_{n+1}} = \text{function of } \tilde{\underline{\chi}}_n(+_e) \tag{16.1.2.1-8}$$

$$\tilde{\underline{\chi}}_0 = 0 \tag{16.1.2.1-9}$$

where, as in Section 16.1.1.3:

$\tilde{\underline{\chi}}$ = Optimal Kalman filter estimate for $\underline{\chi}$.

K_{Opt} = Optimal Kalman gain matrix based on the $\underline{\chi}$ error model.

Following the same steps that led to Equations (16.1.2-19) - (16.1.2-20), the previous real world and Kalman filter models translate into the following optimal delayed control Kalman filter estimated error state uncertainty propagation/reset equations:

$$\Delta\underline{\chi} \equiv \tilde{\underline{\chi}} - \underline{\chi} \tag{16.1.2.1-10}$$

$$\Delta\underline{\chi}_n(+_c) = \Phi_n \, \Delta\underline{\chi}_{n-1}(+_e) - \underline{w}_n \tag{16.1.2.1-11}$$

$$\Delta\underline{\chi}_n(+_e) = \left(I - K_{Opt_n} H_n \right) \Delta\underline{\chi}_n(+_c) + K_{Opt_n} G_{M_n} \underline{n}_{M_n} \tag{16.1.2.1-12}$$

where

$\Delta\underline{\chi}$ = Uncertainty in the optimal Kalman filter's estimate for $\underline{\chi}$.

The covariance equivalent to (16.1.2.1-11) - (16.1.2.1-12) is as in Section 16.1.1.3:

$$P_{Opt_n} \equiv \mathcal{E}\left(\Delta\underline{\chi}_n \, \Delta\underline{\chi}_n^T \right) \tag{16.1.2.1-13}$$

$$P_{Opt_n}(+_c) = \Phi_n P_{Opt_{n-1}}(+_e) \Phi_n^T + Q_{RW_n} \qquad (16.1.2.1\text{-}14)$$

$$P_{Opt_n}(+_e) = \left(I - K_{Opt_n} H_n\right) P_{Opt_n}(+_c) \left(I - K_{Opt_n} H_n\right)^T + K_{Opt_n} G_{M_n} R_n G_{M_n}^T K_{Opt_n}^T \qquad (16.1.2.1\text{-}15)$$

where

P_{Opt} = Optimal Kalman filter error state uncertainty covariance matrix.

The K_{Opt} Kalman gain for the previous development is obtained (as in Equation (16.1.1.3-18)) from the (16.1.2.1-14) $P_{Opt}(+_c)$ solution using the (15.1.2.1-28) format while recognizing from (15.1.2.1-11) that the (-) and ($+_c$) values of P are identical for delayed control reset filters at the same time points:

$$K_{Opt_n} = I_{OptEst}\, P_{Opt_n}(+_c)\, H_n^T \left(H_n P_{Opt_n}(+_c)\, H_n^T + G_{M_n} R_n G_{M_n}^T\right)^{-1} \qquad (16.1.2.1\text{-}16)$$

Values for the R_n, Q_{RW_n}, matrices and P_{Opt} initialization for Equations (16.1.2.1-14) - (16.1.2.1-15) are as defined for the idealized optimal control reset Kalman in Equations (16.1.1.3-17) and (16.1.1.3-20).

As in Equation (16.1.1.3-21), the partitioned form of P_{Opt} for the delayed control reset Kalman filter is given by:

$$P_{Opt} = \begin{bmatrix} \mathcal{E}\left(\Delta \underline{x}_{Opt}\, \Delta \underline{x}_{Opt}^T\right) & \mathcal{E}\left(\Delta \underline{x}_{Opt}\, \Delta \underline{y}_{Opt}^T\right) \\ \mathcal{E}\left(\Delta \underline{y}_{Opt}\, \Delta \underline{x}_{Opt}^T\right) & \mathcal{E}\left(\Delta \underline{y}_{Opt}\, \Delta \underline{y}_{Opt}^T\right) \end{bmatrix} \equiv \begin{bmatrix} P_{Opt\,\Delta x\,\Delta x} & P_{Opt\,\Delta x\,\Delta y} \\ P_{Opt\,\Delta y\,\Delta x} & P_{Opt\,\Delta y\,\Delta y} \end{bmatrix} \qquad (16.1.2.1\text{-}17)$$

Equation (16.1.2.1-17) shows the format of the optimal delayed control reset Kalman filter covariance matrix for comparison against the equivalent data format in (16.1.2-26) for the suboptimal delayed control reset Kalman filter. The upper left 2 by 2 suboptimal covariance array in (16.1.2-26) should be compared against the (16.1.2.1-17) optimal covariance equivalent to evaluate the degree of suboptimality in the Section 16.1.2 delayed control reset suboptimal Kalman filter.

Note in Equation (16.1.2-26) that the $P_{\Delta x\,\Delta x}$, $P_{\Delta x\,\Delta y}$, $P_{\Delta y \Delta x}$, $P_{\Delta y \Delta y}$ upper two left covariance partitions for the suboptimal delayed control reset Kalman filter in the real world (based on Equations (16.1.2-23), (16.1.2-27) - (16.1.2-28) and (16.1.1.2-2)), can be converted to Equation (16.1.2.1-17) for optimal delayed control reset Kalman filter real world performance evaluation (i.e., based on Equations (16.1.1.3-1) and (16.1.2.1-13) - (16.1.2.1-16), (16.1.1.3-17) and (16.1.1.3-20)), by setting $\Phi_{xx}^* = \Phi_{xx}$, $H_x^* = H_x$, interchanging the first and third partitions in the second partition row of Φ', and replacing the suboptimal Kalman gain matrix K' with the optimal Kalman gain matrix equivalent $\begin{bmatrix} K_{Opt} \\ 0 \end{bmatrix}$. Application of the previous observation allows common software code to be used in the covariance simulation program for calculating the suboptimal P' and optimal P_{Opt} covariance performance matrices.

16.1.2.2 SIMPLIFIED COVARIANCE ANALYSIS EQUATIONS

As in Section 16.1.1.4 for the idealized control reset Kalman filter, Equations (16.1.2-27) - (16.1.2-28) for the delayed control reset Kalman filter can be reduced in dimension by excluding the $\underline{x}$ terms for applications with zero coupling of $\underline{x}$ into $\Delta\underline{x}$, $\Delta\underline{y}$. From Equations (16.1.2-23) - (16.1.2-25) we see that there will be zero $\underline{x}$ coupling into $\Delta\underline{x}$, $\Delta\underline{y}$ under the following Φ' and H' matrix conditions:

<u>Conditions For Zero x Coupling Into $\Delta\underline{x}$, $\Delta\underline{y}$:</u>

$$\Phi^{*}_{xx} = \Phi_{xx} \qquad \Phi_{yx} = 0 \qquad H^{*}_{x} = H_x \tag{16.1.2.2-1}$$

The (16.1.2.2-1) conditions are satisfied for a broad range of delayed control reset Kalman filter performance analysis problems requiring covariance simulation analysis for suboptimal performance evaluation. Under Equations (16.1.2.2-1), Equations (16.1.2-23), (16.1.2-26) - (16.1.2-28) and (16.1.1-34) simplify to:

$$\underline{x}'' \equiv \begin{bmatrix} \Delta\underline{x} \\ \Delta\underline{y} \end{bmatrix} \qquad \underline{w} \equiv \begin{bmatrix} \underline{w}_x \\ \underline{w}_y \end{bmatrix}$$

$$\Phi \equiv \begin{bmatrix} \Phi_{xx} & \Phi_{xy} \\ 0 & \Phi_{yy} \end{bmatrix} \qquad H = \begin{bmatrix} H_x & H_y \end{bmatrix} \tag{16.1.2.2-2}$$

$$K'' \equiv \begin{bmatrix} K \\ 0 \end{bmatrix} \qquad I \equiv \begin{bmatrix} I_x & 0 \\ 0 & I_y \end{bmatrix}$$

$$P'' \equiv \mathcal{E}\left(\underline{x}'' \, \underline{x}''^{T}\right)$$

$$P''_n(+_c) = \Phi_n \, P''_{n-1}(+_e) \, \Phi_n^{T} + Q_n$$

$$P''_n(+_e) = \left(I - K''_n H_n\right) P''_n(+_c) \left(I - K''_n H_n\right)^{T} + K''_n \, G_{M_n} \, R_n \, G^{T}_{M_n} \, K''^{T}_n \qquad (16.1.2.2\text{-}3)$$

$$R_n = \mathcal{E}\left(\underline{n}_{M_n} \, \underline{n}^{T}_{M_n}\right) \qquad Q_n = \mathcal{E}\left(\underline{w}_n \, \underline{w}^{T}_n\right) = \begin{bmatrix} Q_{xx} & Q_{xy} \\ Q_{yx} & Q_{yy} \end{bmatrix}_n$$

Suboptimal Kalman filter covariance Equations (16.1.1-35) - (16.1.1-37) are applicable for calculation of the Suboptimal Kalman gain matrix K in (16.1.2.2-3). Under the (16.1.2.2-1) conditions, these equations become:

$$\text{Same As (16.1.1.4-5) But With } \Phi^{*}_{xx} = \Phi_{xx} \qquad (16.1.2.2\text{-}4)$$

Suboptimal Kalman filter performance from (16.1.2.2-3) would be obtained from $P_{\Delta x \, \Delta x}$ as shown in (16.1.1.4-6). Initialization of the P" matrix would be as presented in Equation (16.1.1.4-7).

Note that the simplified version Equations (16.1.2.2-2) - (16.1.2.2-4) with (16.1.1.4-6) - (16.1.1.4-7) for suboptimal delayed control reset Kalman filter performance in the real world is identical to Equations (16.1.1.3-1), (16.1.2.1-13) - (16.1.2.1-15), (16.1.1.3-17) and (16.1.1.3-20) (with $\Phi_{yx} = 0$ as in (16.1.2.2-1)) for the optimal delayed control reset Kalman filter real world performance, if we substitute the optimal Kalman gain matrix K_{Opt} (from Equation (16.1.2.1-16)) for the suboptimal equivalent K". Thus, the same software code can be applied in a covariance simulation program for either optimal or suboptimal covariance performance evaluation by using either Equation (16.1.2.1-16) for the optimal gain matrix, or K" as in (16.1.2.2-2) with K from Equations (16.1.2.2-4) for the suboptimal gain matrix calculation.

16.2 SUBOPTIMAL COVARIANCE ANALYSIS SIMULATION PROGRAM CONFIGURATION

In this section we discuss the basic structure of the suboptimal covariance analysis simulation program built around the analytical framework of Section 16.1.1. Topics covered include a summary of the typical Section 16.1.1 equations implemented; methods for calculating the key terms in these equations; performance evaluation output routines; generating performance sensitivities to initial covariance values, process/measurement noise terms and step transients; error budget analysis; general program structure; and use of the program in suboptimal Kalman filter design and/or performance evaluation.

16.2.1 BASIC SUBOPTIMAL COVARIANCE SIMULATION ANALYSIS PROGRAM EQUATIONS

A suboptimal covariance simulation implements the equations developed in Section 16.1.1 to enable performance evaluation of Kalman aided systems and to assist in the Kalman filter design process. In particular, for the idealized control reset Kalman filter configuration, the equations implemented would typically consist of a version of (16.1.1-28), (16.1.1-32) - (16.1.1-37), (16.1.1.2-2), (16.1.1.3-1), (16.1.1.3-15) - (16.1.1.3-18) and (16.1.1.3-20) repeated below:

Suboptimal Kalman Filter Covariance Performance Calculations

$$P_n'(-) = \Phi_n' P_{n-1}'(+_e) \Phi_n'^T + Q_n' \tag{16.2.1-1}$$

$$P_n'(+_e) = \left(I' - K_n' H_n'\right) P_n'(-) \left(I' - K_n' H_n'\right)^T + K_n' G_{M_n} R_n G_{M_n}^T K_n'^T \tag{16.2.1-2}$$

$$P_0' = \begin{bmatrix} P_{xx_0} & P_{xy_0} & P_{xx_0} \\ P_{yx_0} & P_{yy_0} & P_{yx_0} \\ P_{xx_0} & P_{xy_0} & P_{xx_0} \end{bmatrix} \tag{16.2.1-3}$$

$$\Phi_n' \equiv \begin{bmatrix} \left[\Phi_{xx_n}^* - \left(\Phi_{xx_n}^* - \Phi_{xx_n}\right) L_{x_{n-1}}\right] & \Phi_{xy_n} & -\left(\Phi_{xx_n}^* - \Phi_{xx_n}\right)\left(I_x - L_{x_{n-1}}\right) \\ \Phi_{yx_n} L_{x_{n-1}} & \Phi_{yy_n} & \Phi_{yx_n}\left(I_x - L_{x_{n-1}}\right) \\ \Phi_{xx_n} L_{x_{n-1}} & \Phi_{xy_n} & \Phi_{xx_n}\left(I_x - L_{x_{n-1}}\right) \end{bmatrix}$$

$$Q_n' \equiv \begin{bmatrix} Q_{xx} & Q_{xy} & Q_{xx} \\ Q_{yx} & Q_{yy} & Q_{yx} \\ Q_{xx} & Q_{xy} & Q_{xx} \end{bmatrix}_n \tag{16.2.1-4}$$

$$H_n' \equiv \begin{bmatrix} H_{x_n}^* & H_{y_n} & -\left(H_{x_n}^* - H_{x_n}\right) \end{bmatrix}$$

$$K_n' \equiv \begin{bmatrix} K_n \\ 0 \\ 0 \end{bmatrix} \qquad I' \equiv \begin{bmatrix} I_x & 0 & 0 \\ 0 & I_y & 0 \\ 0 & 0 & I_x \end{bmatrix}$$

Suboptimal Kalman Filter Gain Calculations

$$P^*_n(-) = \Phi^*_{XX_n} P^*_{n-1}(+_e) \Phi^{*T}_{XX_n} + Q^*_{XX_n} \tag{16.2.1-5}$$

$$K_n = I_{Est} P^*_n(-) H^{*T}_{X_n} \left(H^*_{X_n} P^*_n(-) H^{*T}_{X_n} + G^*_{M_n} R^*_n G^{*T}_{M_n} \right)^{-1} \tag{16.2.1-6}$$

$$P^*_n(+_e) = \left(I_X - K_n H^*_{X_n} \right) P^*_n(-) \left(I_X - K_n H^*_{X_n} \right)^T + K_n G^*_{M_n} R^*_n G^{*T}_{M_n} K_n^T \tag{16.2.1-7}$$

Optimal Kalman Filter Covariance Performance Calculations

$$P_{Opt_n}(-) = \Phi_n P_{Opt_{n-1}}(+_e) \Phi_n^T + Q_{RW_n} \tag{16.2.1-8}$$

$$K_{Opt_n} = I_{OptEst} P_{Opt_n}(-) H_n^T \left(H_n P_{Opt_n}(-) H_n^T + G_{M_n} R_n G_{M_n}^T \right)^{-1} \tag{16.2.1-9}$$

$$P_{Opt_n}(+_e) = \left(I - K_{Opt_n} H_n \right) P_{Opt_n}(-) \left(I - K_{Opt_n} H_n \right)^T + K_{Opt_n} G_{M_n} R_n G_{M_n}^T K_{Opt_n}^T \tag{16.2.1-10}$$

$$P_{Opt_0} \equiv \begin{bmatrix} P_{XX_0} & P_{XY_0} \\ P_{YX_0} & P_{YY_0} \end{bmatrix} \tag{16.2.1-11}$$

$$\Phi_n \equiv \begin{bmatrix} \Phi_{XX} & \Phi_{XY} \\ \Phi_{YX} & \Phi_{YY} \end{bmatrix}_n \qquad Q_{RW_n} \equiv \begin{bmatrix} Q_{XX} & Q_{XY} \\ Q_{YX} & Q_{YY} \end{bmatrix}_n$$

$$I \equiv \begin{bmatrix} I_X & 0 \\ 0 & I_Y \end{bmatrix} \qquad H_n \equiv \begin{bmatrix} H_X & H_Y \end{bmatrix}_n \tag{16.2.1-12}$$

Suboptimal and optimal Kalman filter performance would be evaluated for output from the covariance simulation program from P' and P_{Opt} in the previous equations as described in Sections 16.1.1.1 and 16.1.1.3.

The equivalent to Equations (16.2.1-1) - (16.2.1-12) for the delayed control reset Kalman filter configuration is provided in Sections 16.1.2 and 16.1.2.1.

Simplified versions of the previous equations for the idealized control reset and delayed control reset Kalman filters are provided in Sections 16.1.1.4 and 16.1.2.2 subject to specified error model configuration conditions defined in these sections.

16.2.2 EXTENDED COVARIANCE PROPAGATION CYCLE

The covariance propagation equations in Section 16.2.1 are based on one propagation cycle per Kalman update cycle. In many applications, Kalman filter estimation updates are infrequent (or nonexistent in the case of a free-inertial navigation system). In order to reduce the power series expansion order in calculating the error state transition matrix (Φ) and propagated process noise matrix (Q) (i.e., as described in Sections 15.1.2.1.1.1 and 15.1.2.1.1.2), the covariance simulation program propagation algorithm in general applications is typically based on successive intermediate propagations between Kalman updates (as described in Section 15.1.2.1.1.3). This also allows the covariance propagation cycle time to be tailored to the particular application so that longer propagation times can be used during benign trajectory segments in which the error state dynamic matrix elements have slow rates of change. The expansion order for the error state transition and integrated process noise matrices can thereby still be maintained at a reasonable level. The net result is a faster overall covariance simulation program execution time for a given trajectory profile.

When using intermediate covariance updates between Kalman updates, the control reset matrix (L_X) in Equations (16.2.1-4) must only be applied for the propagation cycles following the corresponding control times. This would typically be designed to occur for the covariance propagation cycle following the Kalman estimation update cycle. For the intermediate covariance propagation cycles, L_X would be set to zero.

16.2.3 SPECIFYING ERROR MODELS

Implementing the equations described in Section 16.2.1 in a covariance simulation program generally requires that several key terms be provided, either by input, or for the most part, by computation within the simulation program:

- Real world error state transition and integrated process noise matrix elements Φ_{XX}, Φ_{XY}, Φ_{YX}, Φ_{YY}, Q_{XX}, Q_{XY}, Q_{YX}, Q_{YY}.

- Real world measurement and measurement coupling/noise matrix elements H_X, H_Y, G_M, R.

- Real world initial error state covariance matrix values P_{XX_0}, P_{XY_0}, P_{YX_0}, P_{YY_0}.

- Suboptimal Kalman filter error state transition and integrated process noise matrices $\overset{*}{\Phi}_{XX}$, $\overset{*}{Q}_{XX}$.

- Suboptimal Kalman filter measurement matrix and measurement coupling/noise matrices H_x^*, G_M^*, R^*.

- Suboptimal Kalman filter initial error state covariance matrix P_0^*.

- Kalman filter error state control matrix L_x.

The real world and Kalman filter initial error state covariance matrices P_{xx_0}, P_{xy_0}, P_{yx_0}, P_{yy_0}, P_0^* would be provided by the user to the covariance simulation program as part of input/initialization operations. The same is generally true for the Kalman filter error state control matrix L_x, except for the unusual situation when it is time varying, in which case it must be programmed into the simulation as a function of time. For a generalized simulation program designed to handle a broad class of system error model configurations, it is generally difficult to define a simple classical input interface for setting values for the remaining terms. The typical result is that the covariance simulation program must be structured to allow users to directly, but easily, program their error model configuration into subroutines assigned for error model definition. Using this approach, the simulation can be designed for user system error model definition in its most easily represented form, with the simulation program then calculating parameters for the Section 16.2.1 equations from this model form. The basic model representing error state behavior is defined by Equations (15.1-1) - (15.1-2) which for the real world error state vector are:

$$\dot{\underline{x}}_{RW} = A\,\underline{x}_{RW} + G_P\,\underline{n}_P \tag{16.2.3-1}$$

$$\underline{z}_n = H_{RW_n}\,\underline{x}_{RW_n} + G_{M_n}\,\underline{n}_{M_n} \tag{16.2.3-2}$$

with, as in Section 16.1.1.3:

$$\underline{x}_{RW} \equiv \begin{bmatrix} \underline{x} \\ \underline{y} \end{bmatrix} \qquad H_{RW} \equiv \begin{bmatrix} H_x & H_y \end{bmatrix} \tag{16.2.3-3}$$

where

$\underline{x}_{RW}$ = Real world complete error state vector.

H_{RW} = Real world complete measurement matrix.

It is also be useful to define the partitions of A and G_P as:

$$A \equiv \begin{bmatrix} A_{xx} & A_{xy} \\ A_{yx} & A_{yy} \end{bmatrix} \qquad G_P \equiv \begin{bmatrix} G_{P_x} \\ G_{P_y} \end{bmatrix} \tag{16.2.3-4}$$

The $\underline{n}_P$ vector of independent white process noise sources in (16.2.3-1) would be characterized by the diagonal elements of its density matrix Q_{PDens}. The $\underline{n}_M$ vector of independent white sequence measurement noise sources in (16.2.3-2) would be characterized by the diagonal elements of its covariance matrix R. The A, G_P, Q_{PDens}, H_{RW}, G_M and R matrices for (16.2.3-1) - (16.2.3-2) would be programmed into the simulation by the user (generally as a function of time). Note, as discussed following Equation (15.1.2.1.1.3-31), that because G_P is only used in the form $G_P \, Q_{PDens} \, G_P^T$ to calculate Q, many elements of G_P cancel when forming $G_P \, Q_{PDens} \, G_P^T$. This allows the user programming of G_P to be simplified by setting these aforementioned G_P elements to the corresponding identity matrix. Note also, that A, G_P, H_{RW}, G_M are generally functions of angular rate, acceleration, attitude, velocity and position navigation data parameters that vary as a function of time. Thus, to program A, G_P, H_{RW}, G_M into the simulation, a trajectory generator must also be interfaced to the covariance simulation to provide the required navigation data inputs. A trajectory generator is a computer program that simulates the kinematic angular-rate/acceleration history along a user specified trajectory and calculates the corresponding attitude, velocity, position history for output. Section 17.0 describes the structure of trajectory generators.

With A, G_P, Q_{PDens} programmed into the covariance simulation by the user, standard subroutines built into the simulation can then be used to calculate the error state transition matrix Φ (with its defined elements Φ_{xx}, Φ_{xy}, Φ_{yx}, Φ_{yy}) and the integrated coupled process noise matrix Q (with its defined elements Q_{xx}, Q_{xy}, Q_{yx}, Q_{yy}) for the Section 16.2.1 equations (for example, as described in Sections 15.1.2.1.1.1 and 15.1.2.1.1.2). Note that the option to replace Equation (16.2.1-1) and (16.2.1-8) with the iterative processing algorithm between Kalman estimation cycles (as described in Section 15.1.2.1.1.3), is also based on using A, G_P, Q_{PDens} for input. The programmed H_{RW}, G_M, R would be used directly in the Section 16.2.1 equations.

The suboptimal Kalman filter error model for the covariance simulation would be defined similarly using the basic Equations (15.1-1) - (15.1-2) format with the Section 16.1.1 and 16.2.3 nomenclature:

$$\dot{\underline{x}}^* = \overset{*}{A}_{xx} \, \underline{x}^* + \overset{*}{G}_{P_x} \, \overset{*}{\underline{n}}_P \tag{16.2.3-5}$$

$$\overset{*}{\underline{z}}_n = \overset{*}{H}_{x_n} \, \overset{*}{\underline{x}}_n + \overset{*}{G}_{M_n} \, \overset{*}{\underline{n}}_{M_n} \tag{16.2.3-6}$$

where

$\underline{x}^*$ = Error state vector definition for the suboptimal Kalman filter (i.e., neglecting $\underline{y}$).

$\underline{z}^*$ = Measurement model used for the suboptimal Kalman filter.

$*$ = Designation for values used in the suboptimal Kalman filter for the equivalent Equation (16.2.3-1) - (16.2.3-4) parameters.

The $\overset{*}{\underline{n}}_P$ vector of independent white process noise sources in (16.2.3-5) would be characterized by the diagonal elements of its density matrix $\overset{*}{Q}_{PDens}$. The $\overset{*}{\underline{n}}_M$ vector of independent white sequence measurement noise sources in (16.2.3-6) would be characterized by the diagonal elements of its covariance matrix R*. The $\overset{*}{A}_{XX}$, $\overset{*}{G}_{P_X}$, $\overset{*}{Q}_{P_{Dens}}$, $\overset{*}{H}_X$, $\overset{*}{G}_M$ and R* matrices for (16.2.3-5) - (16.2.3-6) would be programmed into the simulation by the user (generally as a function of time).

With $\overset{*}{A}_{XX}$, $\overset{*}{G}_{P_X}$, $\overset{*}{Q}_{P_{Dens}}$ programmed into the covariance simulation by the user, standard subroutines built into the simulation can then be used to calculate the suboptimal filter error state transition matrix $\overset{*}{\Phi}_{XX}$ and the integrated coupled process noise matrix $\overset{*}{Q}_{XX}$ for the Section 16.2.1 equations (for example, as described in Sections 15.1.2.1.1.1 and 15.1.2.1.1.2). Note that the option to replace Equation (16.2.1-5) with the iterative processing algorithm between Kalman estimation cycles (as described in Section 15.1.2.1.1.3), is also based on using $\overset{*}{A}_{XX}$, $\overset{*}{G}_{P_X}$, $\overset{*}{Q}_{P_{Dens}}$ for input. Alternatively, the actual algorithms planned for calculating $\overset{*}{\Phi}_{XX}$, $\overset{*}{Q}_{XX}$ and propagating P* in the suboptimal Kalman filter can be programmed directly into the covariance simulation (in lieu of programming $\overset{*}{A}_{XX}$, $\overset{*}{G}_{P_X}$, $\overset{*}{Q}_{P_{Dens}}$ and using Sections 15.1.2.1.1.1 - 15.1.2.1.1.2 or Section 15.1.2.1.1.3). The programmed $\overset{*}{H}_X$, $\overset{*}{G}_M$, R* would be used directly in the Section 16.2.1 equations.

The following subsections discuss modeling of particular elements of the error state dynamic equation including process noise, acceleration squared error effects and gravity error.

16.2.3.1 PROCESS NOISE ERROR MODELS

For the most part, the definition for the process noise density matrix elements (for Q_{P_X}, Q_{P_y}, $\overset{*}{Q}_{P_X}$) are easily definable from the error state dynamic equations containing the associated $\underline{n}_P$ and $\overset{*}{\underline{n}}_P$ process noise components. In general, the process noise density components are defined directly in terms of the variance of their integrals divided by the integration time (e.g., the noise density for angular rate sensor output white noise is typically measured in terms of its integral in degrees squared per hour of integration time). For a first order Markov process having the general form:

$$\dot{x} = - C\, x + G_P\, n \tag{16.2.3.1-1}$$

the process noise density associated with the white noise n is sometimes defined in terms of the steady state variance for the error state x. The steady state variance for x is analytically obtained from (16.2.3.1-1) using Equations (15.1.2.1.1-30) and (15.1-1) as a template, with $\dot{P}$ set to zero:

$$\dot{P}_{Stdy\ State} = 0 = - 2\, C\, P_{Stdy\ State} + G_P^2\, Q_{PDens} \tag{16.2.3.1-2}$$

or

$$Q_{PDens} = \frac{2\, C\, P_{Stdy\ State}}{G_P^2} \tag{16.2.3.1-3}$$

Process noise densities for the inertial sensor quantization noise terms (e.g., in Equations (12.5.1-1)) are specified using the Section 15.2.1.2 technique leading to Equation (15.2.1.2-18). To properly account for each of these terms, the Equation (12.6-2) and (12.6-3) vibration terms should be included which results in the following revised (12.5.1-1) form (showing only the quantization terms):

$$\begin{aligned}
\dot{\underline{\psi}}^N &= \cdots + C_B^N \left(\underline{\omega}_{IB}^B \times \delta\underline{\alpha}_{Quant}\right) + C_B^N \left(\underline{\omega}_{Vib}^B \times \delta\underline{\alpha}_{Quant}\right) \\
\delta\dot{\underline{V}}^N &= \cdots - \left(\underline{a}_{SF}^N \times\right) C_B^N\, \delta\underline{\alpha}_{Quant} - \left[\left(C_B^N\, \underline{\omega}_{IB}^B\right) \times\right] C_B^N\, \delta\underline{\upsilon}_{Quant} \\
&\quad - C_B^N \left(\underline{a}_{Vib}^B \times \delta\underline{\alpha}_{Quant} + \underline{\omega}_{Vib}^B \times \delta\underline{\upsilon}_{Quant}\right) \\
\delta\dot{\underline{R}}^N &= \cdots + C_B^N\, \delta\underline{\upsilon}_{Quant}
\end{aligned} \tag{16.2.3.1-4}$$

If we assume that the attitude update, acceleration-transformation/velocity-update and position update frequencies are identical, the leading $\delta\underline{\alpha}_{Quant}$ noise terms in the (16.2.3.1-4) $\dot{\underline{\psi}}^N$, $\delta\dot{\underline{V}}^N$ expressions and the leading $\delta\underline{\upsilon}_{Quant}$ noise terms in the $\delta\dot{\underline{V}}^N$, $\delta\dot{\underline{R}}^N$ expression have component noise densities as defined by Equation (15.2.1.2-18):

$$q_{\psi V\,\alpha Quant} = \frac{1}{12}\,\frac{\varepsilon_\alpha^2}{f_{Att/Vel}} \qquad q_{VR\,\upsilon Quant} = \frac{1}{12}\,\frac{\varepsilon_\upsilon^2}{f_{Vel/Pos}} \tag{16.2.3.1-5}$$

where

$q_{\psi V\,\alpha Quant}$ = $\delta\underline{\alpha}_{Quant}$ process noise density elements for the $\left(\underline{a}_{SF}^N \times\right) C_B^N\, \delta\underline{\alpha}_{Quant}$ term in the $\delta\dot{\underline{V}}^N$ expression and for the $C_B^N \left(\underline{\omega}_{IB}^B \times \delta\underline{\alpha}_{Quant}\right)$ term in the $\dot{\underline{\psi}}^N$ expression.

$q_{VR\upsilon Quant}$ = $\delta\underline{\upsilon}_{Quant}$ process noise density elements for the $C_B^N \, \delta\underline{\upsilon}_{Quant}$ term in the $\delta\dot{\underline{R}}^N$ expression and for the $\left[\left(C_B^N \, \underline{\omega}_{IB}^B\right)\times\right] C_B^N \, \delta\underline{\upsilon}_{Quant}$ term in the $\delta\dot{\underline{V}}^N$ expression.

If the attitude update, acceleration-transformation/velocity-update and position update frequencies are different, we must also address the more fundamental question of how to deal with the same noise source applied at different rates to different error state dynamic equations. For example, consider the case when the attitude rate is updated at twice the acceleration-transformation/velocity-update rate, and how $\delta\underline{\alpha}_{Quant}$ process noise should then be modeled in the Equation (16.2.3.1-4) $\dot{\underline{\psi}}^N$, $\delta\dot{\underline{V}}^N$ expressions (exclusive of the vibration terms in these expressions - To be discussed subsequently). For this situation, we consider the noise source in question to be composed of the sum of two independent noise sources; the $\delta\underline{\alpha}_{Quant}$ portion affecting $\dot{\underline{\psi}}^N$ due to the attitude update cycle between velocity updates, and the remaining $\delta\underline{\alpha}_{Quant}$ portion affecting both $\dot{\underline{\psi}}^N$ and $\delta\dot{\underline{V}}^N$ due to the common time point attitude update and acceleration-transformation/velocity-update cycle. Then the $\dot{\underline{\psi}}^N$, $\delta\dot{\underline{V}}^N$ expressions in (16.2.3.1-4) become:

$$
\begin{aligned}
\dot{\underline{\psi}}^N &= \cdots + C_B^N \left[\underline{\omega}_{IB}^B \times \left(\delta\underline{\alpha}_{\psi Quant} + \delta\underline{\alpha}_{\psi VQuant}\right)\right] + \cdots \\
\delta\dot{\underline{V}}^N &= \cdots - \left(\underline{a}_{SF}^N \times\right) C_B^N \, \delta\underline{\alpha}_{\psi VQuant} + \cdots
\end{aligned}
\qquad (16.2.3.1\text{-}6)
$$

where

$\delta\underline{\alpha}_{\psi Quant}$ = Portion of $\delta\underline{\alpha}_{Quant}$ affecting only $\dot{\underline{\psi}}^N$ due to the attitude update cycle between velocity updates.

$\delta\underline{\alpha}_{\psi VQuant}$ = Portion of $\delta\underline{\alpha}_{Quant}$ affecting both $\dot{\underline{\psi}}^N$ and $\delta\dot{\underline{V}}^N$ due to the common cycle attitude update and acceleration-transformation/velocity-update.

The associated elements of error state dynamic Equation (16.2.3.1-4) have the form:

$$
\begin{bmatrix} \dot{\underline{\psi}}^N \\ \delta\dot{\underline{V}}^N \end{bmatrix} = \cdots + G_{P_{\psi V}} \begin{bmatrix} \delta\underline{\alpha}_{\psi Quant} \\ \delta\underline{\alpha}_{\psi VQuant} \end{bmatrix} + \cdots \qquad (16.2.3.1\text{-}7)
$$

$$
G_{P_{\psi V}} = \begin{bmatrix} C_B^N \left(\underline{\omega}_{IB}^B \times\right) & C_B^N \left(\underline{\omega}_{IB}^B \times\right) \\ 0 & -\left(\underline{a}_{SF}^N \times\right) C_B^N \end{bmatrix} \qquad (16.2.3.1\text{-}8)
$$

where

$Gp_{\psi V}$ = Portion of Gp that couples the Equation (16.2.3.1-4) $\delta\underline{\alpha}_{\psi Quant}$ process noise term into $\underline{\dot{\psi}}^N$ and $\delta\underline{\dot{V}}^N$ (but not including the vibration - $\delta\underline{\alpha}_{Quant}$ product terms).

Because the $\delta\underline{\alpha}_{\psi Quant}$ and $\delta\underline{\alpha}_{\psi VQuant}$ process noise vectors have been defined to be independent members of the same noise process $\delta\underline{\alpha}_{Quant}$, and because there are equal amounts of $\delta\underline{\alpha}_{\psi Quant}$ and $\delta\underline{\alpha}_{\psi VQuant}$ (because the attitude update rate is assumed to be twice the acceleration-transformation/velocity-update rate), the process noise density matrix associated with $\begin{bmatrix} \delta\underline{\alpha}_{\psi Quant} \\ \delta\underline{\alpha}_{\psi VQuant} \end{bmatrix}$ is easily defined (as for Equation (16.2.3.1-5)) to be:

$$Q_{\psi V\alpha Quant} \equiv \begin{bmatrix} q_{\psi V\alpha Quant}\, I & 0 \\ 0 & q_{\psi V\alpha Quant}\, I \end{bmatrix} \tag{16.2.3.1-9}$$

where

$Q_{\psi V\alpha Quant}$ = Process noise density matrix for $\begin{bmatrix} \delta\underline{\alpha}_{\psi Quant} \\ \delta\underline{\alpha}_{\psi VQuant} \end{bmatrix}$.

with $q_{\psi V\alpha Quant}$ as calculated in Equation (16.2.3.1-5).

A similar analysis applies for the $\delta\underline{\upsilon}_{Quant}$ noise in the (16.2.3.1-4) $\delta\underline{\dot{V}}^N, \delta\underline{\dot{R}}^N$ expressions (exclusive of vibration effects - To be treated subsequently). For example, consider the case when the velocity update frequency is higher the position update frequency. The equivalent treatment for $\delta\underline{\upsilon}_{Quant}$ in the $\delta\underline{\dot{V}}^N$, $\delta\underline{\dot{R}}^N$ expressions begins from (16.2.3.1-4) with:

$$\begin{bmatrix} \delta\underline{\dot{V}}^N \\ \delta\underline{\dot{R}}^N \end{bmatrix} = \cdots + Gp_{VR} \begin{bmatrix} \delta\underline{\upsilon}_{VQuant} \\ \delta\underline{\upsilon}_{VRQuant} \end{bmatrix} + \cdots \tag{16.2.3.1-10}$$

$$Gp_{VR} = \begin{bmatrix} -\left[\left(C_B^N\, \underline{\omega}_{IB}^B\right)\times\right] C_B^N & -\left[\left(C_B^N\, \underline{\omega}_{IB}^B\right)\times\right] C_B^N \\ 0 & C_B^N \end{bmatrix} \tag{16.2.3.1-11}$$

where

$\delta\underline{\upsilon}_{VQuant}$ = Portion of $\delta\underline{\upsilon}_{Quant}$ affecting only $\delta\dot{\underline{V}}^N$ due to the acceleration-transformation/velocity-update cycle between position updates.

$\delta\underline{\upsilon}_{VRQuant}$ = Portion of $\delta\underline{\upsilon}_{Quant}$ affecting both $\delta\dot{\underline{V}}^N$ and $\delta\dot{\underline{R}}^N$ due to the common cycle acceleration-transformation/velocity-update/position-update.

$G_{P_{VR}}$ = Portion of G_P that couples the Equation (16.2.3.1-4) $\delta\underline{\upsilon}_{Quant}$ process noise term into $\delta\dot{\underline{V}}^N$ and $\delta\dot{\underline{R}}^N$ (but not including the vibration - $\delta\underline{\upsilon}_{Quant}$ product terms to be discussed subsequently).

For this case we write the more general form for the process noise density matrix associated with $\begin{bmatrix} \delta\underline{\upsilon}_{VQuant} \\ \delta\underline{\upsilon}_{VRQuant} \end{bmatrix}$ as:

$$Q_{VR\,\upsilon Quant} \equiv \begin{bmatrix} (r-1)\; q_{VR\,\upsilon Quant}\, I & 0 \\ 0 & q_{VR\,\upsilon Quant}\, I \end{bmatrix} \tag{16.2.3.1-12}$$

in which $q_{VR\,\upsilon Quant}$ is calculated as in Equation (16.2.3.1-5) and where:

r = Number of acceleration-transformation/velocity-update cycles per position update cycle.

The r-1 term in (16.2.3.1-12) arises because there are r-1 contributions of $\delta\underline{\upsilon}_{VQuant}$ to $\delta\dot{\underline{V}}^N$ between position update cycles.

Let us now address the process noise modeling for the vibration terms in (16.2.3.1-4). For example, consider the $\underline{\omega}^B_{Vib} \times \delta\underline{\alpha}_{Quant}$ term in the $\dot{\underline{\psi}}^N$ expression. We assume that $\underline{\omega}^B_{Vib}$ is a random vector. As such, in the (16.2.3.1-4) $\dot{\underline{\psi}}^N$ expression, the composite noise vector $\underline{\omega}^B_{Vib} \times \delta\underline{\alpha}_{Quant}$ can be considered uncorrelated with the other quantization term $\underline{\omega}^B_{IB} \times \delta\underline{\alpha}_{Quant}$ treated previously. Hence, $\underline{\omega}^B_{Vib} \times \delta\underline{\alpha}_{Quant}$ can be treated as an independent process noise vector. The X axis component of $\underline{\omega}^B_{Vib} \times \delta\underline{\alpha}_{Quant}$ is given by:

$$\left(\underline{\omega}^B_{Vib} \times \delta\underline{\alpha}_{Quant}\right)_X = \omega_{Vib_Y}\, \delta\alpha_{Quant_Z} - \omega_{Vib_Z}\, \delta\alpha_{Quant_Y} \tag{16.2.3.1-13}$$

where

$\omega_{Vib_i}, \delta\alpha_{Quant_i}$ = Components of $\underline{\omega}^B_{Vib}, \delta\underline{\alpha}_{Quant}$ along B Frame axis i.

If we assume that $\underline{\delta\alpha}_{Quant}$ and $\underline{\omega}^{B}_{Vib}$ are independent, that the $\delta\alpha_{Quant_i}$ components are uncorrelated, and approximate the ω_{Vib_i} components as being uncorrelated, we can write for the variance of the X component of $\underline{\omega}^{B}_{Vib} \times \underline{\delta\alpha}_{Quant}$:

$$
\begin{aligned}
\mathcal{E}\left\{\left[\left(\underline{\omega}^{B}_{Vib} \times \underline{\delta\alpha}_{Quant}\right)_X\right]^2\right\} &= \mathcal{E}\left[\left(\omega_{Vib_Y}\, \delta\alpha_{Quant_Z} - \omega_{Vib_Z}\, \delta\alpha_{Quant_Y}\right)^2\right] \\
&= \mathcal{E}\left(\omega^2_{Vib_Y}\, \delta\alpha^2_{Quant_Z} + \omega^2_{Vib_Z}\, \delta\alpha^2_{Quant_Y}\right) \qquad (16.2.3.1\text{-}14) \\
&= \mathcal{E}\left(\omega^2_{Vib_Y}\right) \mathcal{E}\left(\delta\alpha^2_{Quant_Z}\right) + \mathcal{E}\left(\omega^2_{Vib_Z}\right) \mathcal{E}\left(\delta\alpha^2_{Quant_Y}\right)
\end{aligned}
$$

or, from the discussion leading to (15.2.1.2-18), and assuming equal quantization noise variances per axis:

$$
\mathcal{E}\left\{\left[\left(\underline{\omega}^{B}_{Vib} \times \underline{\delta\alpha}_{Quant}\right)_X\right]^2\right\} = \frac{1}{12}\left(\sigma^2_{\omega_Y Vib} + \sigma^2_{\omega_Z Vib}\right) \varepsilon^2_{\alpha} \qquad (16.2.3.1\text{-}15)
$$

where

$\sigma_{\omega_i Vib}$ = Root-mean-square angular rate vibration magnitude around axis i.

We can also define for $\sigma_{\omega_Y Vib}$ and $\sigma_{\omega_Z Vib}$:

$$
\sigma_{\omega_Y Vib} = \mu_{\omega_Y Vib}\, \sigma_{\omega Vib} \qquad \sigma_{\omega_Z Vib} = \mu_{\omega_Z Vib}\, \sigma_{\omega Vib} \qquad (16.2.3.1\text{-}16)
$$

where

$\sigma_{\omega Vib}$ = Root-mean-square angular vibration rate vector magnitude, the parameter that is typically used to specify angular vibration.

$\mu_{\omega_Y Vib}, \mu_{\omega_Z Vib}$ = Ratios of $\sigma_{\omega_Y Vib}$ and $\sigma_{\omega_Z Vib}$ to $\sigma_{\omega Vib}$.

Following the same process, we would find that the covariances between the components of $\underline{\omega}^{B}_{Vib} \times \underline{\delta\alpha}_{Quant}$ (e.g., $\mathcal{E}\left[\left(\underline{\omega}^{B}_{Vib} \times \underline{\delta\alpha}_{Quant}\right)_X \left(\underline{\omega}^{B}_{Vib} \times \underline{\delta\alpha}_{Quant}\right)_Y\right]$) equal zero. Hence, generalizing from (16.2.3.1-15) - (16.2.3.1-16) to arbitrary axis i, the components of the process noise density matrix associated with $\underline{\omega}^{B}_{Vib} \times \underline{\delta\alpha}_{Quant}$ would be as in Equation (15.2.1.2-18):

$$
\begin{aligned}
q_{\psi\alpha VibQuant_{ii}} &= \frac{1}{12}\left(\mu^2_{\omega_{i+1} Vib} + \mu^2_{\omega_{i+2} Vib}\right) \frac{\varepsilon^2_{\alpha}}{f_{Att/Att}}\, \sigma^2_{\omega Vib} \\
q_{\psi\alpha VibQuant_{ij}} &= 0 \quad \text{For } i \neq j
\end{aligned}
\qquad (16.2.3.1\text{-}17)
$$

with in general:

$$\sigma_{\omega_i Vib} = \mu_{\omega_i Vib}\, \sigma_{\omega Vib} \qquad (16.2.3.1\text{-}18)$$

where

$q_{\psi\alpha VibQuant_{ij}}$ = Element in row i column j of the $\underline{\omega}^B_{Vib} \times \delta\underline{\alpha}_{Quant}$ process noise density matrix.

$\mu_{\omega_i Vib}$ = Ratio of $\sigma_{\omega_i Vib}$ to $\sigma_{\omega Vib}$.

$f_{Att/Att}$ = Frequency that attitude is used for attitude updating (Hz).

and in which i+1 or i+2 equal to 4 is interpreted as 1, and i+1 or i+2 equal to 5 is interpreted as 2. We also note from its definition that:

$$\mu^2_{\omega_X Vib} + \mu^2_{\omega_Y Vib} + \mu^2_{\omega_Z Vib} = 1 \qquad (16.2.3.1\text{-}19)$$

The component of GP that couples $\underline{\omega}^B_{Vib} \times \delta\underline{\alpha}_{Quant}$ into $\dot{\underline{\psi}}^N$ is C^N_B as seen in (16.2.3.1-4).

The remaining vibration/quantization process noise densities and coupling matrices for Equations (16.2.3.1-4) are obtained similarly. The result is:

$$\begin{aligned}
q_{V\alpha VibQuant_{ii}} &= \frac{1}{12}\left(\mu^2_{a_{i+1} Vib} + \mu^2_{a_{i+2} Vib}\right)\frac{\varepsilon^2_\alpha}{f_{Att/Vel}}\,\sigma^2_{aVib} \\
q_{V\alpha VibQuant_{ij}} &= 0 \quad \text{For } i \neq j \\
q_{V\upsilon VibQuant_{ii}} &= \frac{1}{12}\left(\mu^2_{\omega_{i+1} Vib} + \mu^2_{\omega_{i+2} Vib}\right)\frac{\varepsilon^2_\upsilon}{f_{Att/Vel}}\,\sigma^2_{\omega Vib} \\
q_{V\upsilon VibQuant_{ij}} &= 0 \quad \text{For } i \neq j
\end{aligned} \qquad (16.2.3.1\text{-}20)$$

with

$$\sigma_{a_i Vib} = \mu_{a_i Vib}\, \sigma_{aVib} \qquad \mu^2_{a_X Vib} + \mu^2_{a_Y Vib} + \mu^2_{a_Z Vib} = 1 \qquad (16.2.3.1\text{-}21)$$

where

$q_{V\alpha VibQuant_{ij}}$ = Element in row i column j of the $\underline{a}^B_{Vib} \times \delta\underline{\alpha}_{Quant}$ process noise density matrix associated with the $\underline{a}^B_{Vib} \times \delta\underline{\alpha}_{Quant}$ term in the (16.2.3.1-4) $\delta\dot{\underline{V}}^N$ expression. The associated GP coupling term is $- C^N_B$.

$q_{V\upsilon VibQuant_{ij}}$ = Element in row i column j of the $\underline{\omega}_{Vib}^{B} \times \delta\underline{\upsilon}_{Quant}$ process noise density matrix associated with the $\underline{\omega}_{Vib}^{B} \times \delta\underline{\upsilon}_{Quant}$ term in the (16.2.3.1-4) $\delta\dot{\underline{V}}^{N}$ expression. The associated GP coupling term is $- C_B^N$.

$\sigma_{a_i Vib}$ = Root-mean-square specific force acceleration vibration along axis i.

σ_{aVib} = Acceleration vibration vector magnitude, the parameter that is typically used to specify linear vibration.

$\mu_{a_i Vib}$ = Ratio of $\sigma_{a_i Vib}$ to σ_{aVib}.

$f_{Att/Vel}$ = Frequency that attitude is used in the foreground calculations for acceleration-transformation/velocity-integration update (Hz).

We also assume that $\underline{\omega}_{Vib}^{B} \times \delta\underline{\alpha}_{Quant}$, $\underline{a}_{Vib}^{B} \times \delta\underline{\alpha}_{Quant}$, and $\underline{\omega}_{Vib}^{B} \times \delta\underline{\upsilon}_{Quant}$ can be treated as independent noise sources. This implies that the components of $\underline{\omega}_{Vib}^{B}$ and $\underline{a}_{Vib}^{B}$ are uncorrelated. This is, at best, an approximation since in practice, both $\underline{\omega}_{Vib}^{B}$ and $\underline{a}_{Vib}^{B}$ are created from the same input vibration sources. However, in the interests of simplifying the analysis, we will accept the minor deficiencies of this modeling inaccuracy.

The $\sigma_{\omega Vib}$ and σ_{aVib} terms in the previous development represent angular and linear vibrations of the strapdown inertial sensor assembly. These vibrations are produced from user vehicle vibrations transmitted through the INS mount and into the sensor assembly (generally across elastomeric isolators used to isolate the sensor assembly from external thermal/vibration/shock environments). The vehicle vibrations are generally specified in terms of a linear specific force acceleration power spectrum (versus frequency), which is transmitted into the sensor assembly as linear acceleration vibration along the input acceleration axis, and as angular vibration across the input acceleration axis (due to sensor assembly mount imbalance).

Numerical values for $\sigma_{\omega Vib}$ and σ_{aVib} can be evaluated using the methods of Chapter 10 Section 10.6 as a function of input acceleration vibration and sensor assembly mount dynamic response characteristics. For example, from Equations (10.6.1-25) we see that:

$$\mathcal{E}\left(\overline{a_{SF}(t)^2}\right) = \int_0^\infty B_A^2(\omega)\, G_{aVib}(\omega)\, d\omega$$

$$\mathcal{E}\left(\overline{\theta(t)^2}\right) = \int_0^\infty B_\vartheta^2(\omega)\, G_{aVib}(\omega)\, d\omega \qquad (16.2.3.1\text{-}22)$$

$$\mathcal{E}\left(\overline{a_{Vib}(t)^2}\right) = \int_0^\infty G_{aVib}(\omega)\, d\omega$$

where

$G_{aVib}(\omega)$ = Random acceleration vibration power spectral density input to the sensor assembly mount.

$B_\vartheta(\omega)$, $B_A(\omega)$ = Sensor-assembly/mount-isolator dynamic response characteristics calculated as $B_\vartheta(\Omega)$, $B_A(\Omega)$ in Equation (10.6.1-8) with $\Omega = \omega$ (and for worst case analysis, ε's set to have the same polarity).

$\mathcal{E}\left(\overline{a_{SF}(t)^2}\right)$, $\mathcal{E}\left(\overline{\theta(t)^2}\right)$ = Mean squared values of the sensor assembly linear acceleration and angle vibrations induced by the vibration input to the sensor assembly mount.

$\mathcal{E}\left(\overline{a_{Vib}(t)^2}\right)$ = Mean squared value of the random vibration acceleration input to the sensor assembly mount.

Recognizing that the sensor assembly angular rate response to vibration is the derivative of its angle vibration response, the methods of Chapter 10 easily show that the angular rate vibration response equivalent to the (16.2.3.1-22) $\mathcal{E}\left(\overline{\theta(t)^2}\right)$ expression is:

$$\mathcal{E}\left(\overline{\omega(t)^2}\right) = \int_0^\infty \omega^2\, B_\vartheta^2(\omega)\, G_{aVib}(\omega)\, d\omega \qquad (16.2.3.1\text{-}23)$$

where

$\mathcal{E}\left(\overline{\omega(t)^2}\right)$ = Mean squared value of the vibration induced sensor assembly angular rate oscillation.

Then, using the definitions for $\sigma_{\omega Vib}$ and σ_{aVib}, we see that:

$$\sigma_{\omega Vib} = \sqrt{\mathcal{E}\left(\overline{\omega(t)^2}\right)} \qquad \sigma_{aVib} = \sqrt{\mathcal{E}\left(\overline{a_{SF}(t)^2}\right)} \qquad (16.2.3.1\text{-}24)$$

We can also define:

$$a_{RMSVibIn} \equiv \sqrt{\mathcal{E}\left(a_{Vib}(t)^2\right)} \qquad (16.2.3.1\text{-}25)$$

where

$a_{RMSVibIn}$ = Root-mean-square (RMS) input vibration acceleration.

From Equations (16.2.3.1-22) - (16.2.3.1-25), we see that $\sigma_{\omega Vib}$ and σ_{aVib} vary linearly with the input mean squared vibration acceleration $a_{RMSVibIn}$. As such, normalized values of $\sigma_{\omega Vib}$, σ_{aVib} can be obtained by dividing the (16.2.3.1-24) values by $a_{RMSVibIn}$:

$$G_{\omega/a} \equiv \frac{\sigma_{\omega Vib}}{a_{RMSVibIn}} \qquad G_{a/a} \equiv \frac{\sigma_{aVib}}{a_{RMSVibIn}} \qquad (16.2.3.1\text{-}26)$$

where

$G_{\omega/a}$, $G_{a/a}$ = Transmissibility of input specific force acceleration RMS vibration into strapdown sensor assembly RMS angular rate and linear specific force acceleration vibration.

Once calculated, the $G_{\omega/a}$, $G_{a/a}$ vibration transmission coefficients can be used in the covariance simulation to calculate the current sensor vibration effects:

$$\sigma_{\omega Vib} = G_{\omega/a}\, a_{RMSVibIn} \qquad \sigma_{aVib} = G_{a/a}\, a_{RMSVibIn} \qquad (16.2.3.1\text{-}27)$$

with $a_{RMSVibIn}$ programmed as a function of time to simulate the expected vibration history for the particular trajectory profile used in the simulation.

16.2.3.2 ACCELERATION SQUARED ERROR EFFECTS MODELING

Some error effects in strapdown inertial navigation systems are excited by the product of acceleration components along the same or different B Frame axes. Examples are angular rate sensor and accelerometer "g^2 error" effects, and residual (uncompensated) coning/sculling (or pseudo coning/sculling) error.

Residual coning/sculling (or pseudo coning/sculling) error effects can be represented by:

$$\begin{aligned} \delta Con_i &= K_{PsCon_i}\, Con_i + \delta ConAlg_i \\ \delta Scul_i &= L_{PsScul_i}\, Scul_i + \delta SculAlg_i \end{aligned} \qquad (16.2.3.2\text{-}1)$$

where

δCon_i, $\delta Scul_i$ = Total residual coning, sculling error along B Frame axis i.

Con_i, $Scul_i$ = Actual average coning, sculling magnitude around (along) axis i (a function of input vibration and sensor assembly mount characteristics).

K_{PsCon_i}, L_{PsScul_i} = Pseudo-coning, pseudo-sculling error coefficients for B Frame axis i defined as a fraction of average per axis true coning, sculling magnitude.

$\delta ConAlg_i$, $\delta SculAlg_i$ = Coning, sculling algorithm error.

To facilitate numerical evaluation, Equations (16.2.3.2-1) are first expanded as:

$$
\begin{aligned}
\delta Con_i &= a_{RMSVibIn}^2 \left(\frac{a_{RMSVibIn_i}}{a_{RMSVibIn}} \right)^2 \left(K_{PsCon_i} \frac{Con_i}{a_{RMSVibIn_i}^2} + \frac{\delta ConAlg_i}{a_{RMSVibIn_i}^2} \right) \\
&= a_{RMSVibIn}^2 \, \eta_{a_i Vib}^2 \left(K_{PsCon_i} \frac{Con_i}{a_{RMSVibIn_i}^2} + \frac{\delta ConAlg_i}{a_{RMSVibIn_i}^2} \right)
\end{aligned}
\qquad (16.2.3.2\text{-}2)
$$

$$
\begin{aligned}
\delta Scul_i = a_{RMSVibIn}^2 \Bigg[L_{PsScul_i} \Bigg(& \frac{Scul_{ij}}{a_{RMSVibIn_j}^2} \frac{a_{RMSVibIn_j}^2}{a_{RMSVibIn}^2} \\
& + \frac{Scul_{ik}}{a_{RMSVibIn_k}^2} \frac{a_{RMSVibIn_k}^2}{a_{RMSVibIn}^2} \Bigg) + \frac{\delta SculAlg_{ij}}{a_{RMSVibIn_j}^2} \frac{a_{RMSVibIn_j}^2}{a_{RMSVibIn}^2} + \frac{\delta SculAlg_{ik}}{a_{RMSVibIn_k}^2} \frac{a_{RMSVibIn_k}^2}{a_{RMSVibIn}^2} \Bigg] \\
= a_{RMSVibIn}^2 \Bigg[L_{PsScul_i} \Bigg(& \frac{Scul_{ij}}{a_{RMSVibIn_j}^2} \eta_{a_j Vib}^2 + \frac{Scul_{ik}}{a_{RMSVibIn_k}^2} \eta_{a_k Vib}^2 \Bigg) \\
& + \frac{\delta SculAlg_{ij}}{a_{RMSVibIn_j}^2} \eta_{a_j Vib}^2 + \frac{\delta SculAlg_{ik}}{a_{RMSVibIn_k}^2} \eta_{a_k Vib}^2 \Bigg]
\end{aligned}
\qquad (16.2.3.2\text{-}3)
$$

where

$a_{RMSVibIn}$ = Root-mean-square value for the input acceleration vibration <u>vector</u> magnitude.

$a_{RMSVibIn\, i,j,k}$ = Root-mean-square value for the input acceleration vibration vector <u>component</u> along sensor assembly axis i, j, or k.

$\eta_{a_{i,\,j,\,k}Vib}$ = Ratio of $a_{RMSVibIn\;i,\,j,\,k}$ over $a_{RMSVibIn}$.

j, k = Mutually perpendicular sensor assembly axes perpendicular to axis i.

$Scul_{ij}$, $Scul_{ik}$, $\delta SculAlg_{ij}$, $\delta SculAlg_{ik}$ = Portions of $Scul_i$ and $\delta SculAlg_i$ generated by linear vibration along axes j and k.

Equations (16.2.3.2-2) and (16.2.3.2-3) are based on the assumptions that acceleration vibration components along sensor assembly axes are uncorrelated from each other, and that i axis acceleration vibration produces j and k axis angular vibration (due to mounting imbalances), but negligible i axis angular vibration. The j, k axis angular vibrations then generate i axis coning δCon_i. Sculling along sensor axis i ($\delta Scul_i$) is produced by j axis angular vibration created by k axis linear vibration, and by k axis angular vibration created by j axis linear vibration.

As shown in Chapter 10, coning, sculling and associated algorithm error terms vary linearly with the input mean squared vibration acceleration (e.g., see Equations (10.6.1-23)). Based on this observation, we can define the coning/sculling ratio terms in the (16.2.3.2-2) - (16.2.3.2-3) results as normalized values. If we assume equal dynamic and imbalance response characteristics for each sensor assembly axis, we can make the further assumption that the normalized values are the same for each axis. In this manner (16.2.3.2-2) - (16.2.3.2-3) simplify to the following:

$$\begin{aligned} \delta Con_i &= a_{RMSVibIn}^2\, \eta_{a_j Vib}^2 \left(K_{PsCon_i}\, Con_{Norm} + \delta ConAlg_{Norm} \right) \\ \delta Scul_i &= a_{RMSVibIn}^2 \left(\eta_{a_j Vib}^2 + \eta_{a_k Vib}^2 \right) \left(L_{PsScul_i}\, Scul_{Norm} + \delta SculAlg_{Norm} \right) \end{aligned} \qquad (16.2.3.2\text{-}4)$$

or equivalently:

$$\delta Con_i = a_{RMSVibIn}^2\, K_{ResCon_i} \qquad \delta Scul_i = a_{RMSVibIn}^2\, L_{ResScul_i} \qquad (16.2.3.2\text{-}5)$$

with

$$\begin{aligned} K_{ResCon_i} &= \eta_{a_j Vib}^2 \left(K_{PsCon_i}\, Con_{Norm} + \delta ConAlg_{Norm} \right) \\ L_{ResScul_i} &= \left(\eta_{a_j Vib}^2 + \eta_{a_k Vib}^2 \right) \left(L_{PsScul_i}\, Scul_{Norm} + \delta SculAlg_{Norm} \right) \end{aligned} \qquad (16.2.3.2\text{-}6)$$

where

K_{ResCon_i}, $L_{ResScul_i}$ = Composite residual coning, sculling error coefficients.

$Con_{Norm}, Scul_{Norm}, \delta ConAlg_{Norm}, \delta SculAlg_{Norm}$ = Normalized values for Con_i, $Scul_{ij}$, $Scul_{ik}$, $\delta ConAlg_i$, $\delta SculAlg_{ij}$ and $\delta SculAlg_{ik}$ in Equations (16.2.3.2-2) - (16.2.3.2-3).

Numerical values for $Con_{Norm}, Scul_{Norm}, \delta ConAlg_{Norm}, \delta SculAlg_{Norm}$ in (16.2.3.2-6) can be evaluated using the methods of Chapter 10 by dividing the equivalent Chapter 10 calculated values (e.g., based on Equations (10.6.1-25)) by the associated input mean squared vibration acceleration:

$$Con_{Norm} \equiv \frac{\mathcal{E}\left(\dot{\Phi}_{Con_Z}\right)}{a_{RMSVibIn}^2} \qquad Scul_{Norm} \equiv \frac{\mathcal{E}\left(\dot{v}_{SF/Scul_Z}\right)}{a_{RMSVibIn}^2}$$

$$\delta ConAlg_{Norm} \equiv \frac{\mathcal{E}\left(\delta\dot{\Phi}_{Algo\text{-}m_Z}\right)}{a_{RMSVibIn}^2} \qquad \delta SculAlg_{Norm} \equiv \frac{\mathcal{E}\left(\delta\dot{v}_{SF/Algo\text{-}m_Z}\right)}{a_{RMSVibIn}^2} \qquad (16.2.3.2\text{-}7)$$

$$a_{RMSVibIn}^2 = \mathcal{E}\left(\overline{a_{Vib}(t)^2}\right)$$

where

$\mathcal{E}\left(\dot{\Phi}_{Con_Z}\right)$, $\mathcal{E}\left(\delta\dot{\Phi}_{Algo\text{-}m_Z}\right)$, $\mathcal{E}\left(\dot{v}_{SF/Scul_Z}\right)$, $\mathcal{E}\left(\delta\dot{v}_{SF/Algo\text{-}m_Z}\right)$, $\mathcal{E}\left(\overline{a_{Vib}(t)^2}\right)$ = Coning, coning algorithm error, sculling, sculling algorithm error and average acceleration vibration input squared calculated using Equations (10.6.1-25) of Chapter 10 with the vibration input set to the input vibration vector magnitude power spectrum.

It is useful to evaluate the potential error in Equations (16.2.3.2-5) - (16.2.3.2-6) if their underlying assumption of uncorrelated linear vibration components is invalid. For example, as a worst case, consider a situation when the linear vibration vector is along a fixed direction in the sensor frame, not necessarily along i, j or k. The magnitude of the coning error coefficient in this case (call it K_{ResCon_0}) can be calculated from K_{ResCon_i} in (16.2.3.2-6) by setting $\eta_{a_i Vib}$ equal to 1. If the vibration was along axis i, this would be the correct solution (with zero for the K_{ResCon_j}, K_{ResCon_k} components). However, if the vibration is along another general axis direction, say one which projects equal linear vibrations along i, j and k (i.e., $\eta_{a_i Vib} = \eta_{a_j Vib} = \eta_{a_k Vib} = \frac{1}{\sqrt{3}}$), the (16.2.3.2-6) solution (for equal K_{PsCon} pseudo-coning coefficients per axis) would be $K_{ResCon_i} = K_{ResCon_j} = K_{ResCon_k} = \frac{1}{3} K_{ResCon_0}$. The corresponding vector magnitude is the square root of the sum of the squares or $\frac{1}{\sqrt{3}} K_{ResCon_0}$, hence, somewhat smaller than the actual magnitude K_{ResCon_0}.

For the sculling case, consider the same worst case scenario in which the linear vibration vector is along a fixed direction in the sensor frame, say along axis k. The magnitude of the i axis sculling error coefficient in this case (call it $L_{ResScul_0}$) can be calculated from $L_{ResScul_i}$ in (16.2.3.2-6) by setting $\eta_{a_k Vib}$ equal to 1 and $\eta_{a_j Vib}$ to zero. For equal L_{PsScul} pseudo-sculling coefficients per axis, the j axis composite residual sculling error coefficient ($L_{ResScul_j}$) would also equal $L_{ResScul_0}$. The k axis coefficient ($L_{ResScul_k}$) would be zero. If the vibration was actually along axis k, this would be the correct solution for the i, j and k axis $L_{ResScul}$ components. However, if the vibration is along another general axis direction, say one which projects equal linear vibrations along i, j and k (i.e., $\eta_{a_i Vib} = \eta_{a_j Vib} = \eta_{a_k Vib} = \frac{1}{\sqrt{3}}$), the (16.2.3.2-6) solution (for equal L_{PsScul} pseudo-sculling coefficients per axis) would be $L_{ResScul_i} = L_{ResScul_j} = L_{ResRes_k} = \frac{2}{3} L_{ResScul_0}$. The corresponding vector magnitude is the square root of the sum of the squares or $\frac{2}{\sqrt{3}} L_{ResScul_0}$. The actual $L_{ResScul}$ coefficients in this case would equal $L_{ResScul_0}$ along each of the two axes perpendicular to the vibration vector with zero along the vibration axis. The associated actual $L_{ResScul}$ vector magnitude would then be $\sqrt{2}\, L_{ResScul_0}$, somewhat larger than the $\frac{2}{\sqrt{3}} L_{ResScul_0}$ magnitude resulting from the Equation (16.2.3.2-6) solution. The ratio between the two is $\sqrt{2} / \left(2 / \sqrt{3}\right) = \sqrt{3/2}$.

From the previous discussion we see that the assumption of uncorrelated linear acceleration vibration input components produces a smaller estimate for coning/sculling effects than for correlated linear vibration components. On the other hand, the Chapter 10 solutions for the $\mathcal{E}\left(\dot{\Phi}_{Con_z}\right)$, $\mathcal{E}\left(\delta\dot{\Phi}_{Algo\text{-}m_z}\right)$, $\mathcal{E}\left(\dot{v}_{SF/Scul_z}\right)$, $\mathcal{E}\left(\delta\dot{v}_{SF/Algo\text{-}m_z}\right)$, $\mathcal{E}\left(\overline{a_{Vib}(t)^2}\right)$ terms in (16.2.3.2-7) (used to calculate the (16.2.3.2-6) normalized coefficients) were based on conservative worst case phasing analysis. Thus, the combination of both assumptions in (16.2.3.2-6) results in a more realistic approximation.

Using the previous formulation, δCon_i, $\delta Scul_i$ would be treated as additions to $\delta\underline{\omega}_{IB}^{B}$, $\delta\underline{a}_{SF}^{B}$ in the strapdown inertial navigation error state dynamic equations (e.g., Equations (12.5.1-1)) with the K_{ResCon_i}, $L_{ResScul_i}$ coefficients treated as unknown constant error states and $a_{RMSVibIn}^2$ being part of the error state dynamic matrix. In practice, the variances for each i axis component of K_{ResCon_i}, $L_{ResScul_i}$ would be set to the same value. In a covariance simulation program, $a_{RMSVibIn}$ would typically be programmed as a function of time to simulate the expected vibration history for the particular trajectory profile used in the simulation. Note, that Equations (16.2.3.2-5) would generally not be included in a real Kalman filter error model because the $a^2_{RMSVibIn}$ coefficient is generally not available. In the real world model of a covariance simulation using (16.2.3.2-5), the "optimal" Kalman filter (see Section 16.1.1.3 or 16.1.2.1) would then be configured to not estimate K_{ResCon_i}, $L_{ResScul_i}$ by setting the appropriate elements in I_{OptEst} to zero (in Equation (16.1.1.3-18) or (16.1.2.1-16)). If $a^2_{RMSVibIn}$ was measured in the actual INS, then K_{ResCon_i}, $L_{ResScul_i}$ could in principle also be included in the suboptimal Kalman filter. However, due to the uncertainty in the basic model for K_{ResCon_i}, $L_{ResScul_i}$ (as represented by (16.2.3.2-6) using the simplified models of Chapter 10), if K_{ResCon_i}, $L_{ResScul_i}$ is to be included in the suboptimal filter, the "considered variable" approach would probably be taken (by which K_{ResCon_i}, $L_{ResScul_i}$ are not estimated) by setting the appropriate Kalman gain elements to zero (i.e., in the I_{Est} matrix of Equation (16.1.1-36)).

Let us now consider an inertial sensor error characteristic of the form:

$$\delta_i = K_{jk_i} \, a_{SFTot_j} \, a_{SFTot_k} \qquad (16.2.3.2\text{-}8)$$

where

δ_i = Inertial sensor error along B frame axis i.

a_{SFTot_j}, a_{SFTot_k} = Total specific force acceleration along B Frame axes j and k. Axes i, j, k may be the same axis or different axes, depending on the particular sensor error.

K_{jk_i} = General specific force acceleration product sensor error coefficient (for general j, k axis acceleration products).

In general, the total specific force acceleration along an arbitrary B Frame axis i can be defined as:

$$a_{SFTot_i} = a_{SF_i} + a_{Vib_i} \qquad (16.2.3.2\text{-}9)$$

where

a_{SF_i} = Specific force acceleration along B Frame axis i caused by all effects except vibration.

a_{Vib_i} = Specific force acceleration vibration along B Frame axis i.

With (16.2.3.2-9), we see that the acceleration product term in (16.2.3.2-8) is:

$$\begin{aligned} a_{SFTot_j} \, a_{SFTot_k} &= \left(a_{SF_j} + a_{Vib_j}\right)\left(a_{SF_k} + a_{Vib_k}\right) \\ &= a_{SF_j} \, a_{SF_k} + a_{SF_j} \, a_{Vib_k} + a_{Vib_j} \, a_{SF_k} + a_{Vib_j} \, a_{Vib_k} \end{aligned} \qquad (16.2.3.2\text{-}10)$$

The average value of (16.2.3.2-10) in (16.2.3.2-8) produces a systematic error in δ_i that leads to attitude/velocity error build-up. Considering the vibration terms to be random, the average value of (16.2.3.2-10) is given by:

$$\left(a_{SFTot_j}\, a_{SFTot_k}\right)_{Avg} = a_{SF_j}\, a_{SF_k} + \left(a_{Vib_j}\, a_{Vib_k}\right)_{Avg} \tag{16.2.3.2-11}$$

where

Avg = Subscript designation for the average value of the associated term in brackets.

As a worst case, we consider the i and j axis vibration terms to be perfectly correlated (i.e., in a fixed proportion and in phase with one another, which would be the case for a vibration vector input to the INS along a fixed direction in the B Frame). Then the vibration term in (16.2.3.2-11) is given by:

$$\left(a_{Vib_j}\, a_{Vib_k}\right)_{Avg} = \sigma_{a_jVib}\, \sigma_{a_kVib} \tag{16.2.3.2-12}$$

where

$\sigma_{a_jVib}, \sigma_{a_kVib}$ = Root-mean-square specific force acceleration vibration along axes j and k.

We can also write as in (16.2.3.1-21):

$$\sigma_{a_jVib} = \mu_{a_jVib}\, \sigma_{aVib} \qquad \sigma_{a_kVib} = \mu_{a_kVib}\, \sigma_{aVib}$$
$$\mu_{a_XVib}^2 + \mu_{a_YVib}^2 + \mu_{a_ZVib}^2 = 1 \tag{16.2.3.2-13}$$

where

σ_{aVib} = Root-mean-square specific force acceleration vector magnitude.

μ_{a_iVib} = Ratio of σ_{a_iVib} to σ_{aVib}.

With (16.2.3.2-11) - (16.2.3.2-13), the average value of Equation (16.2.3.2-8) becomes:

$$\delta_{i_{Avg}} = \left(a_{SF_j}\, a_{SF_k} + \mu_{a_jVib}\, \mu_{a_kVib}\, \sigma_{aVib}^2\right) K_{jk_i} \tag{16.2.3.2-14}$$

For simulation purposes, σ_{aVib} can be evaluated using Equation (16.2.3.1-27) in the previous section with $a_{RMSVibIn}$ provided as a programmed function of time for the trajectory profile being simulated.

Depending on whether it was an angular rate sensor or accelerometer error, $\delta_{i_{Avg}}$ would be treated as an addition to $\delta\underline{\omega}_{IB}^{B}$ or $\delta\underline{a}_{SF}^{B}$ in the strapdown inertial navigation error state dynamic equations (e.g., Equations (12.5.1-1)) with the K_{jk_i} coefficient treated as an unknown constant error state and $\left(a_{SF_j}\, a_{SF_k} + \mu_{a_j Vib}\, \mu_{a_k Vib}\, \sigma_{aVib}^{2}\right)$ being part of the error state dynamic matrix. In general practice, δ_i type error effects are not usually modeled in the actual suboptimal Kalman filter, hence, Equation (16.2.3.2-14) would normally represent a real world effect that would only be used in the real world covariance simulation error state model. If δ_i type errors are to be included in the actual (suboptimal) Kalman filter error state vector, means must be provided for determining the associated vibration effects for the error state dynamic matrix. The basic choices are to use (16.2.3.2-14) with a predetermined estimate for σ_{aVib}, or to calculate the average vibration product $\left(a_{Vib_j}\, a_{Vib_k}\right)_{Avg}$ in (16.2.3.2-11) using actual acceleration measurements (high-passed to calculate the vibration components). The former technique adds an uncertainty to the error state dynamic matrix (i.e., the difference between Φ_{XX}^{*} and Φ_{XX} in Equations (16.1.1-28) for suboptimal filter performance assessment); the author has no knowledge of the latter technique ever being used. The comments following Equation (16.2.3.2-7) regarding use of the "considered variable" approach may also apply for δ_i type errors, depending on the degree of uncertainty in the δ_i error model.

16.2.3.3 GRAVITY ERROR MODELING

In the development of the strapdown inertial navigation error state dynamic equations in Chapter 12, we have included a gravity error term $\delta\underline{g}_{Mdl}$ representing the difference between the gravity model used in the inertial navigation software and true gravity. For covariance simulation analysis (and at times, but rarely, in the actual suboptimal Kalman filter), $\delta\underline{g}_{Mdl}$ is included as part of the error state dynamic equation. The error model for $\delta\underline{g}_{Mdl}$ is typically designed to statistically account for gravity anomalies (in magnitude and angular deflection) caused by local earth mass density/distribution aberrations. A simple and effective gravity error model considers $\delta\underline{g}_{Mdl}$ to vary as a first order Markov process in linear distance traveled:

$$\frac{d\left(\delta\underline{g}_{Mdl}\right)}{ds} = -\frac{1}{l_g}\,\delta\underline{g}_{Mdl} + \underline{n}_{\delta g} \qquad (16.2.3.3\text{-}1)$$

where

ds = Infinitesimal distance traveled relative to the earth.

$d\left(\delta\underline{g}_{Mdl}\right)$ = Infinitesimal change in $\delta\underline{g}_{Mdl}$ over ds.

l_g = Gravity model error correlation distance.

$\underline{n}_{\delta g}$ = White process noise input vector.

The analytical solution to (16.2.3.3-1) can be shown to equal $\delta \underline{g}_{Mdl\,0}\, e^{-\frac{\Delta s}{l_g}}$ + (Integrated Effect of $\underline{n}_{\delta g}$ Over Δs), in which $\delta \underline{g}_{Mdl\,0}$ equals $\delta \underline{g}_{Mdl}$ at some initial location and Δs is the linear distance traveled relative to the earth from that location. From this analytical form we see that l_g can also be defined as the value of Δs that attenuates the effect of $\delta \underline{g}_{Mdl\,0}$ on $\delta \underline{g}_{Mdl}$ by e^{-1}.

The equivalent distance derivative form of generalized time derivative error state dynamic Equation (15.1-1) and its associated covariance rate Equation (15.1.2.1.1-30), is:

$$\frac{d}{ds}\underline{x} = A'\,\underline{x} + G'_P\,\underline{n}'_P$$
$$\frac{dP}{ds} = A'\,P + P\,A'^T + G'_P\,Q'_{PDens}\,G'^T_P \qquad (16.2.3.3\text{-}2)$$

where

' = Designation for parameters based on the distance derivative formulation.

Using (16.2.3.3-2), we can write the equivalent covariance form of (16.2.3.3-1) as:

$$\frac{d\,P_{\delta g}}{ds} = -\frac{2}{l_g}\,P_{\delta g} + Q_{\delta gDens} \qquad (16.2.3.3\text{-}3)$$

where

$P_{\delta g}$ = Covariance of $\delta \underline{g}_{Mdl}$.

$Q_{\delta gDens}$ = Process noise density matrix associated with $\underline{n}_{\delta g}$ in (16.2.3.3-1). Note that since (16.2.3.3-1) is a distance differential equation, the units of $Q_{\delta gDens}$ are g^2 per ft.

But we can also write:

$$\frac{d\,P_{\delta g}}{ds} = \frac{d\,P_{\delta g}}{dt}\,\frac{dt}{ds} = \dot{P}_{\delta g}\,\frac{1}{v} \qquad (16.2.3.3\text{-}4)$$

where

v = Magnitude of the velocity relative to the earth.

Substituting (16.2.3.3-4) into (16.2.3.3-3) then yields after rearrangement:

$$\dot{P}_{\delta g} = -\frac{2\,v}{l_g} P_{\delta g} + v\, Q_{\delta g Dens} \qquad (16.2.3.3\text{-}5)$$

Equation (16.2.3.3-5) can now be used for covariance simulation purposes based on time integration for the gravity modeling error. The initial value for $P_{\delta g}$ would be based on known statistics of the gravity anomalies at the simulation trajectory profile starting point. The $Q_{\delta g Dens}$ process noise density matrix would be set to match the steady state value of (16.2.3.3-3) (i.e., when $\frac{d\,P_{\delta g}}{ds} = 0$) for which:

$$Q_{\delta g Dens} = \frac{2}{l_g} P_{\delta g_{StdSt}} \qquad (16.2.3.3\text{-}6)$$

where

$P_{\delta g_{StdSt}}$ = Steady state value for $P_{\delta g}$.

Note from Equation (16.2.3.3-5) compared with generalized covariance time derivative rate Equation (15.1.2.1.1-30) (and its (15.1-1) error state dynamic equation equivalent), that the time derivative form error state dynamic equation for the gravity model uncertainty can be written as:

$$\dot{\underline{\delta g}}_{Mdl} = -\frac{v}{l_g} \underline{\delta g}_{Mdl} + \sqrt{v}\, \underline{n}_{\delta g} \qquad (16.2.3.3\text{-}7)$$

We finally note that to be more general, we can represent l_g and $P_{\delta g_{StdSt}}$ as time varying functions along the particular trajectory we are simulating. For more generality, Equations (16.2.3.3-5) - (16.2.3.3-6) can also be written independently for each component of $\underline{\delta g}_{Mdl}$ along each axis of the navigation coordinate frame in which $\underline{\delta g}_{Mdl}$ is used for input to the velocity error rate equation. With this approach, individual values of l_g can be assigned for each navigation axis. We also have the option of representing ds in the previous derivation as horizontal distance or as distance along a particular navigation frame axis, for which the v parameter in (16.2.3-3-5) would be then be set to horizontal velocity magnitude or the magnitude of the velocity component along the particular navigation frame axis. The horizontal velocity magnitude approach is commonly used for applications involving long distance trajectories near the earth's surface.

16.2.4 SENSITIVITIES AND ERROR BUDGETS

To effectively utilize a covariance simulation program in the design and evaluation of Kalman filter aided (or unaided) inertial navigation systems, it is beneficial to include the capability of evaluating the sensitivity of the error state vector uncertainty components to factors affecting their values. These include initial error state values, process noise inputs, measurement noise input during the estimation update process, and error state uncertainty transients that may appear at arbitrary time points. It is also helpful to translate the sensitivities into their individual numerical impact on the error state uncertainties based on numerical values for the error state uncertainty input sources. This is sometimes denoted as an error budget.

The sensitivity of the generalized error state vector $\underline{x}'$ in Equations (16.1.1-28) to initial values of $\underline{x}'$ is readily determined by application of Equations (16.1.1-29) - (16.1.1-30) with the process/measurement noise terms set to zero. Consider the response of these equations to a unity initial condition on one of the $\underline{x}'$ components (with zero for the other components):

$$\underline{S}_{j_n}(-) = \Phi'_n \, \underline{S}_{j_{n-1}}(+e) \qquad \underline{S}_{j_n}(+e) = \left(I' - K'_n \, H'_n\right) \underline{S}_{j_n}(-)$$
$$S_{ij_0} = 1 \quad \text{For } i = j \qquad S_{ij_0} = 0 \quad \text{For } i \neq j \tag{16.2.4-1}$$

where

$\underline{S}_{j_n}$ = Response of $\underline{x}'$ at propagation cycle n to unity initial value on the jth element of $\underline{x}'$ with initial values of zero for the other $\underline{x}'$ elements.

S_{ij_n} = Element i of $\underline{S}_{j_n}$.

As defined above, $\underline{S}_{j_n}$ is the error state vector initial condition sensitivity vector relating the sensitivity of $\underline{x}'$ to initial values of $\underline{x}'$ component i. We can also define an initial condition sensitivity matrix as the collection of $\underline{S}_{j_n}$'s for all j's:

$$S_{IC} \equiv \left[\; \underline{S}_1 \quad \underline{S}_2 \quad \cdot \; \cdot \quad \cdot \; \cdot \quad \underline{S}_{m-1} \quad \underline{S}_m \; \right] \tag{16.2.4-2}$$

where

S_{IC} = Error state vector initial condition sensitivity matrix whose element in row i, column j is S_{ij}.

m = Dimension of $\underline{x}'$.

Combining (16.2.4-1) - (16.2.4-2) obtains the dynamic equations for S_{IC}:

$$S_{IC_n}(-) = \Phi'_n \, S_{IC_{n-1}}(+e) \qquad S_{IC_n}(+e) = \left(I' - K'_n \, H'_n\right) S_{IC_n}(-)$$
$$S_{IC_0} = I' \tag{16.2.4-3}$$

Equations (16.2.4-3) can be computed in parallel with the (16.2.1-1) - (16.2.1-2) to evaluate the initial condition sensitivity matrix corresponding to the covariance matrix P'.

Equation (16.2.4-3) can also be utilized to investigate the sensitivity of $\underline{x}'$ to step transients in the error state components applied at a selected n cycle time:

$$S_{Tr_n}(-) = \Phi_n' \, S_{Tr_{n-1}}(+_e) \qquad S_{Tr_n}(+_e) = \left(I' - K_n' \, H_n'\right) S_{Tr_n}(-)$$
$$S_{Tr_0} = 0 \qquad S_{Tr_{nTrans}} = I \tag{16.2.4-4}$$

where

nTrans = Cycle number when the transient step in $\underline{x}'$ occurs.

S_{Tr} = Error state vector transient input sensitivity matrix whose element in row i, column j is the response of $\underline{x}'$ element i to a unit step change in $\underline{x}'$ element j at cycle time nTrans.

The sensitivity of the error states to individual process and measurement noise terms can be obtained as the covariance response to the associated covariance process noise density/measurement noise matrix elements. As such, parallel versions of Equations (16.2.1-1) - (16.2.1-2) are executed with zero for the initial P' value, and unity for the particular process noise density/measurement noise elements whose sensitivity on P' is to be determined. For the process noise density sensitivity runs, the Q' value in (16.2.1-1) would be calculated using the methods of Sections 15.1.2.1.1.1 and 15.1.2.1.1.2 with unity for the particular process noise density terms whose sensitivity is to be determined (and zero for the remaining noise density elements). In order to keep the number of parallel versions of (16.2.1-1) - (16.2.1-2) to a minimum, common process noise densities with equal values can be run as a group in a single run (e.g., the sensitivity of P' to the combined effect of X, Y, Z angular rate sensor random output noise can be assessed in a single run by setting each of their associated process noise densities to unity. The same is true for common measurement noise forms along different measurement axes.).

Equations (16.2.4-1) - (16.2.4-4) and the previous paragraph are based on the idealized control reset Kalman filter formulation. A similar set of sensitivities and procedures can be derived for the delayed control reset Kalman filter configuration from Equations (16.1.2-24) - (16.1.2-25) for initial-condition/transient sensitivities and Equations (16.1.2-27) - (16.1.2-28) for process/measurement noise sensitivities.

Once the sensitivities are evaluated for each error source as described above, an error budget matrix can be prepared depicting a breakdown of all error contributions to each element of $\underline{x}'$. In general, the basic error budget does not include the transient effects described above, as these are considered to be abnormal effects requiring individual error budget analysis. The analytics for the basic error budget are developed by first writing the deterministic relationship for the ith element in $\underline{x}'$ as a sum of its contributing elements:

$$x'_i = S_{IC_i}\, \underline{x}'_0 + x'_{Proc_i} + x'_{Meas_i} \tag{16.2.4-5}$$

where

x'_i = Element i of $\underline{x}'$.

$\underline{x}'_0$ = Initial value of $\underline{x}'$.

S_{IC_i} = Row i of S_{IC}.

x'_{Proc_i} = Portion of x'_i produced by process noise.

x'_{Meas_i} = Portion of x'_i produced by measurement noise during the estimation reset process.

The variance of x'_i is the expected value of its square which, assuming that initial conditions, transients, process noise and measurement noise terms are uncorrelated, is from (16.2.4-5):

$$\sigma^2_{x'_i} \equiv \mathcal{E}\left(x'^2_i\right) = \mathcal{E}\left[\left(S_{IC_i}\,\underline{x}'_0\right)\left(S_{IC_i}\,\underline{x}'_0\right)^T\right] + \mathcal{E}\left(x'^2_{Proc_i}\right) + \mathcal{E}\left(x'^2_{Meas_i}\right) \tag{16.2.4-6}$$

where

$\sigma_{x'_i}$ = Root-mean-square value for x'_i.

The initial condition term in (16.2.4-6) can be expanded as follows:

$$\begin{aligned}\mathcal{E}\left[\left(S_{IC_i}\,\underline{x}'_0\right)\left(S_{IC_i}\,\underline{x}'_0\right)^T\right] &= \sum_j \mathcal{E}\left(S^2_{IC_{ij}}\, x'^2_{j0}\right) + 2\sum_j \sum_{k>j} \mathcal{E}\left(S_{IC_{ij}}\, S_{IC_{ik}}\, x'_{j0}\, x'_{k0}\right) \\ &= \sum_j S^2_{IC_{ij}}\, \mathcal{E}\left(x'^2_{j0}\right) + 2\sum_j \sum_{k>j} S_{IC_{ij}}\, S_{IC_{ik}}\, \mathcal{E}\left(x'_{j0}\, x'_{k0}\right)\end{aligned} \tag{16.2.4-7}$$

where

$S_{IC_{ij}}$ = Element in row i, column j of S_{IC}.

Equation (16.2.4-7) is equivalently:

$$\mathcal{E}\left[\left(S_{IC_i}\,\underline{x}'_0\right)\left(S_{IC_i}\,\underline{x}'_0\right)^T\right] = \sum_j \left(S_{IC_{ij}}\, \sigma_{x'_{j0}}\right)^2 + 2\sum_j \sum_{k>j} S_{IC_{ij}}\, S_{IC_{ik}}\, P'_{jk_0} \tag{16.2.4-8}$$

where

$\sigma_{x'_{j0}}$ = Root-mean-square value for x'_{j0}, the initial value of x'_j.

P'_{jk_0} = Element in row j, column k of the initial P' covariance matrix.

The process and measurement noise terms in (16.2.4-6) can be defined as the sum of the contributions from the individual common noise groupings for which process/measurement noise covariance sensitivities were determined (see previous discussion on process/measurement noise sensitivities):

$$\mathcal{E}\left(x'^{\,2}_{Proc_i}\right) = \sum_l P'_{Proc/ii_l}\, q_{PDens_l} = \sum_l \left(\sqrt{P'_{Proc/ii_l}}\,\sigma_{PDens_l}\right)^2$$
$$\mathcal{E}\left(x'^{\,2}_{Meas_i}\right) = \sum_m P'_{Meas/ii_m}\, r_{Meas_m} = \sum_m \left(\sqrt{P'_{Meas/ii_m}}\,\sigma_{Meas_m}\right)^2 \qquad (16.2.4\text{-}9)$$

where

P'_{Proc/ii_l} = Element in row i, column i of the P' covariance response to unity process noise density for all process noise elements in process noise sensitivity run group *l*.

q_{PDens_l} = Process noise density for each independent process noise element in process noise sensitivity group *l*.

P'_{Meas/ii_m} = Element in row i, column i of the P' covariance response to unity measurement noise density for all measurement noise elements in measurement noise sensitivity run group m.

r_{Meas_m} = Measurement noise variance for each independent measurement noise element in measurement noise sensitivity group m.

σ_{PDens_l}, σ_{Meas_m} = Square root of q_{PDens_l}, r_{Meas_m}.

We now substitute (16.2.4-8) - (16.2.4-9) into (16.2.4-6) to obtain:

$$\sigma^2_{x'_i} = \sum_j \left(S_{IC_{ij}}\,\sigma_{x'j_0}\right)^2 + \sum_l \left(\sqrt{P'_{Proc/ii_l}}\,\sigma_{PDens_l}\right)^2 + \sum_m \left(\sqrt{P'_{Meas/ii_m}}\,\sigma_{Meas_m}\right)^2 + \sigma^2_{Misc_i} \qquad (16.2.4\text{-}10)$$

with

$$\sigma^2_{Misc_i} = 2\sum_j \sum_{k>j} S_{IC_{ij}}\, S_{IC_{ik}}\, P'_{jk_0} + \begin{array}{c}\text{Small Process Noise, Measurement}\\ \text{Noise Terms Not Included}\\ \text{In The } l,\ \text{m Summations}\end{array} \qquad (16.2.4\text{-}11)$$

where

σ_{Misc_i} = Composite of miscellaneous contributions to $\sigma_{x'_i}$, each of which are considered too small for individual identification.

In general, the P'_{jk_0} cross-correlation terms in (16.2.4-11) are small or zero, hence, have fairly minor impact on $\sigma^2_{x'_i}$. Note, that the Equation (16.2.4-11) definition for σ_{Misc_i} allows for the fact that, to reduce simulation running time, a complete set of process/measurement noise sensitivities (i.e., a full set of *l*, m runs) might not have been obtained, hence, the effect of those not run has been included in σ_{Misc_i}.

From its definition, $\sigma^2_{x'_i}$ in (16.2.4-10) should equal P'_{ii} where:

P'_{ii} = Element in row i, column i of P'.

Setting $\sigma^2_{x'_i}$ Equation (16.2.4-10) to P'_{ii} with rearrangement provides an expression for calculating $\sigma^2_{Misc_i}$:

$$\sigma^2_{Misc_i} = P'_{ii} - \sum_j \left(S_{IC_{ij}}\,\sigma_{x'_{j0}}\right)^2 - \sum_l \left(\sqrt{P'_{Proc/ii_l}}\,\sigma_{PDens_l}\right)^2 - \sum_m \left(\sqrt{P'_{Meas/ii_m}}\,\sigma_{Meas_m}\right)^2 \qquad (16.2.4\text{-}12)$$

Equation (16.2.4-10) with (16.2.4-12) is now in a convenient form for generating an error budget matrix such as the matrix mapping of these equations in Table 16.2.4-1. Such an error budget table would be prepared for each x'_i element of interest.

The error source sensitivity column entries and P'_{ii} (for the miscellaneous calculation) in Table 16.2.4-1 would be the numerical results from the covariance and sensitivity simulation runs. The error source value column entries would be the numerical values used in the P' covariance run. Note, that if the miscellaneous term is specified as an allowance (an assigned numerical value), P'_{ii} would not be required. As a result, the entire error source value column can then be treated as an independent numerical input for which the table can be used to calculate the resulting $\sigma_{x'_i}$ value. For such a case, the sensitivity column would be determined once from a set of sensitivity simulation runs.

Table 16.2.4-1 Basic Error Budget For Error Parameter x'_i

ERROR SOURCE	ERROR SOURCE VALUE	ERROR SOURCE SENSITIVITY	ERROR SOURCE VALUE TIMES SENSITIVITY	(ERROR SOURCE VALUE TIMES SENSITIVITY) - SQUARED
Initial Uncertainties (For All j)	$\sigma_{x'j0}$	$S_{IC_{ij}}$	$S_{IC_{ij}} \sigma_{x'j0}$	$\left(S_{IC_{ij}} \sigma_{x'j0}\right)^2$
Process Noise (For All l)	σ_{PDens_l}	$\sqrt{P'_{Proc/ii_l}}$	$\sqrt{P'_{Proc/ii_l}}\ \sigma_{PDens_l}$	$\left(\sqrt{P'_{Proc/ii_l}}\ \sigma_{PDens_l}\right)^2$
Measurement Noise (For All m)	σ_{Meas_m}	$\sqrt{P'_{Meas/ii_m}}$	$\sqrt{P'_{Meas/ii_m}}\ \sigma_{Meas_m}$	$\left(\sqrt{P'_{Meas/ii_m}}\ \sigma_{Meas_m}\right)^2$
Miscellaneous	---	---	---	P'_{ii} Minus Sum Of All Above
SUM OF SQUARES $\left(\sigma^2_{x'_i}\right)$	---	---	---	Sum Of All Above
ROOT-SUM-SQUARE $\left(\sigma_{x'_i}\right)$	---	---	---	Square Root Of Above Element

An error budget can also be prepared for the portion of x'_i produced by transients at a selected nTrans cycle time. Starting with the equivalent to (16.2.4-5) we can write:

$$x'_{Tr_i} = S_{Tr_i}\, \partial \underline{x}'_{Tr_{nTrans}} \tag{16.2.4-13}$$

where

x'_{Tr_i} = Portion of x'_i produced from a step transient in $\underline{x}'$ occurring at the selected nTrans cycle time.

$\partial \underline{x}'_{Tr_{nTrans}}$ = Step transient in $\underline{x}'$ at the nTrans cycle time.

S_{Tr_i} = Row i of S_{Tr}.

As in Equation (16.2.4-6), we then define the variance in x'_{Tr_i} from (16.2.4-13) as:

$$\sigma^2_{x'\,Tr_i} \equiv \mathcal{E}\left(x'^{\,2}_{Tr_i}\right) = \mathcal{E}\left[\left(S_{Tr_i}\, \partial \underline{x}'_{Tr_{nTrans}}\right)\left(S_{Tr_i}\, \partial \underline{x}'_{Tr_{nTrans}}\right)^T\right] \tag{16.2.4-14}$$

where

$$\sigma_{x'\,Tr_i} = \text{Root-mean-square value for } x'_{Tr_i}.$$

Expansion of (16.2.4-14) yields:

$$\sigma^2_{x'\,Tr_i} = \sum_j \left(S_{Tr_{ij}}\, \sigma_{x'_{j\,nTrans}}\right)^2 + 2 \sum_j \sum_{k>j} S_{Tr_{ij}}\, S_{Tr_{ik}}\, \partial P'_{jk_{nTrans}} \qquad (16.2.4\text{-}15)$$

where

$S_{Tr_{ij}}$ = Element in row i, column j of S_{Tr}.

$\sigma_{x'_{j\,nTrans}}$ = Root-mean-square value of the jth element in $\partial \underline{x}'_{Tr_{nTrans}}$

$\partial P'_{jk_{nTrans}}$ = Element in row j, column k of the covariance of $\partial \underline{x}'_{Tr_{nTrans}}$.

For simplicity, we restrict the transient analyses to conditions when $\partial P'_{jk_{nTrans}}$ is zero. Then (16.2.4-15) reduces to:

$$\sigma^2_{x'\,Tr_i} = \sum_j \left(S_{Tr_{ij}}\, \sigma_{x'_{j\,nTrans}}\right)^2 \qquad (16.2.4\text{-}16)$$

Equation (16.2.4-16) is in a convenient form for error budget analysis using the tabular technique of Table 16.2.4-1.

16.2.5 PERFORMANCE EVALUATION OUTPUT ROUTINES

As discussed in Section 16.1.1.1, the performance of the suboptimal filter in the real world is characterized by the upper left partition of the P' matrix defined as $P_{\Delta x\, \Delta x}$. The square root of the diagonal elements in $P_{\Delta x\, \Delta x}$ provides the root-mean-square (RMS) value for the uncertainty in each element of the suboptimal filter estimated error state vector $\underline{\tilde{x}}$. The RMS values for the remaining real world error states not accounted for in the suboptimal filter (i.e., the $\underline{y}$ portion of the real world error state vector) are provided by the square root of the diagonal elements of the $P_{\Delta y \Delta y}$ partition of P' (See Equation (16.1.1.1-1)). The RMS values so obtained are in terms of the error parameters in the error state vector.

A common requirement is that suboptimal Kalman filter attitude, velocity or position performance be expressed in terms of error parameters that differ from those used to represent attitude/velocity/position error in the error state vector. A conversion routine is then required to transform the relevant $P_{\Delta x\, \Delta x}$ elements into an equivalent output covariance in terms of the desired output error definitions.

As an example, consider the case when the Kalman filter error state vector uses $\underline{\psi}^N$ to represent attitude (see Equations (12.2.1-4) and (12.2.1-7) for definition) but the desired output is attitude in terms of roll, pitch, true heading error ($\delta\phi$, $\delta\theta$, $\delta\psi_T$) (as defined in Section 12.2.1). The required covariance conversion formula is derived from Equations (12.2.1-43) - (12.2.1-45) which, for the corresponding error state uncertainties, when combined and cast in matrix form is given by:

$$\begin{bmatrix} \Delta\delta\phi \\ \Delta\delta\theta \\ \Delta\delta\psi_T \end{bmatrix} = \begin{bmatrix} -\sec\theta\sin\psi_P & -\sec\theta\cos\psi_P & 0 & -\frac{1}{R}\sec\theta\cos\psi_P & \frac{1}{R}\sec\theta\sin\psi_P \\ -\cos\psi_P & \sin\psi_P & 0 & \frac{1}{R}\sin\psi_P & \frac{1}{R}\cos\psi_P \\ -\tan\theta\sin\psi_P & -\tan\theta\cos\psi_P & 1 & \frac{1}{R}\begin{pmatrix} -\tan\theta\cos\psi_P \\ +\tan l\cos\alpha \end{pmatrix} & \frac{1}{R}\begin{pmatrix} \tan\theta\sin\psi_P \\ -\tan l\sin\alpha \end{pmatrix} \end{bmatrix} \begin{bmatrix} \Delta\psi_{XN} \\ \Delta\psi_{YN} \\ \Delta\psi_{ZN} \\ \Delta\delta R_{XN} \\ \Delta\delta R_{YN} \end{bmatrix} \tag{16.2.5-1}$$

where

$\Delta(\ ) =$ Uncertainty in ( ).

Equation (16.2.5-1) has the form:

$$\underline{a}_{Out} = B\,\underline{x}' \tag{16.2.5-2}$$

where

$\underline{a}_{Out} =$ Vector of desired output uncertainty parameter forms.

$B =$ Conversion matrix.

Equation (16.2.5-2) is a general form representing the calculation of output parameters in general as functions of $\underline{x}'$ vector elements.

We now determine the covariance of $\underline{a}_{Out}$ in terms of the covariance of $\underline{x}'$. Using the (15.1.2.1-4) definition we find:

$$P_{aOut} = \mathcal{E}\left(\underline{a}_{Out}\,\underline{a}_{Out}^T\right) = \mathcal{E}\left[B\,\underline{x}'\,(B\,\underline{x}')^T\right] = \mathcal{E}\left[B\,\underline{x}'\,(\underline{x}')^T\,B^T\right] = B\,\mathcal{E}\left(\underline{x}'\,(\underline{x}')^T\right)B^T \tag{16.2.5-3}$$

or

$$P_{aOut} = B\,P'\,B^T \tag{16.2.5-4}$$

where

$P_{aOut} =$ Covariance of $\underline{a}_{Out}$.

Once P_{aOut} is calculated from (16.2.5-4), the root-mean-square (RMS) values for the desired output parameter can be obtained as the square root of the corresponding P_{aOut} diagonal elements.

As another example that fits into the previous format, consider the case when the Kalman filter error state vector uses $\delta\underline{V}^N$ to represent velocity error (see Equations (12.2.2-1) and (12.2.2-3) for definition) but the desired output is velocity error uncertainty in locally level east/north/up geographic coordinates as represented by the $\delta\underline{\mathcal{V}}^{Geo}$ velocity error parameter (see Equation (12.2.2-17) for definition). An equation in the (16.2.5-2) format for $\delta\underline{\mathcal{V}}^{Geo}$ uncertainty in terms of $\delta\underline{V}^N$ (and $\delta\underline{R}_H^N$) uncertainty can be developed by combining Equations (12.2.2-19) - (12.2.2-20) and (12.2.2-29) - (12.2.2-30), with the error state components treated as uncertainties.

As a final example, consider the case when the Kalman filter error state vector uses $\underline{\psi}^N$ and $\delta\underline{V}^N$ to represent attitude and velocity error (see Equations (12.2.1-4), (12.2.1-7), (12.2.2-1) and (12.2.2-3) for definitions) but the desired output is attitude/velocity error uncertainty in terms of $\underline{\gamma}^N$ and $\delta\underline{v}^N$ attitude/velocity error parameters (see Equations (12.2.1-10) and (12.2.2-2) for definition). The associated (16.2.5-2) format formula is derived from the converse of Equations (12.2.1-17) and (12.2.2-5) relating $\underline{\psi}^N$, $\delta\underline{V}^N$ and $\underline{\gamma}^N$, $\delta\underline{v}^N$:

$$\underline{\gamma}^N = \underline{\psi}^N + \underline{\varepsilon}^N \qquad \delta\underline{v}^N = \delta\underline{V}^N - \underline{\varepsilon}^N \times \underline{v}^N \tag{16.2.5-5}$$

Conversion Equations (16.2.5-5) require the $\underline{\varepsilon}^N$ position parameter for input (see Equation (12.2.3-4) for definition). Let's assume that position error parameters are included in the suboptimal filter error state vector in the form of $\delta\underline{R}^N$ as defined in Equations (12.2.3-1) - (12.2.3-2). From Equation (12.2.3-25) we see that:

$$\underline{\varepsilon}^N = \frac{1}{R}\left(\underline{u}_{ZN}^N \times \delta\underline{R}^N\right) + \varepsilon_{ZN}\, \underline{u}_{ZN}^N \tag{16.2.5-6}$$

Thus, to calculate $\underline{\varepsilon}^N$ in terms of $\delta\underline{R}^N$, the ε_{ZN} vertical component of $\underline{\varepsilon}^N$ is required. Let us assume that ε_{ZN} is not included in the suboptimal Kalman filter error state vector. Then to evaluate $\underline{\varepsilon}^N$ in (16.2.5-6) we have two choices; assume ε_{ZN} is small enough to be neglected in (16.2.5-6), or add it to the $\underline{y}$ portion of the real world error state vector. Let us assume we adopt the latter approach in which case we will incorporate the error state dynamic equation for ε_{ZN} (as a function of $\delta\underline{R}^N$ from Equations (12.3.5-29)) into the $\underline{y}$ error state dynamic equation:

$$\varepsilon_{ZN} = -\frac{1}{R}\left(\underline{\omega}_{EN_H}^{N} \cdot \delta\underline{R}_H^{N}\right) + \delta\rho_{ZN}$$

$$\delta\rho_{ZN} = 0 \qquad \text{For Wander Azimuth Implementation} \tag{16.2.5-7}$$

$$\delta\rho_{ZN} = -\frac{1}{R}\,\underline{\omega}_{IE_H}^{N} \cdot \delta\underline{R}_H^{N} \qquad \text{For Free Azimuth Implementation}$$

Let us now combine Equations (16.2.5-5) - (16.2.5-6) to obtain:

$$\begin{aligned}\underline{\gamma}^N &= \underline{\psi}^N + \frac{1}{R}\left(\underline{u}_{ZN}^N \times\right)\delta\underline{R}^N + \underline{u}_{ZN}^N\,\varepsilon_{ZN}\\ \delta\underline{v}^N &= \delta\underline{V}^N + \frac{1}{R}\left(\underline{v}^N\times\right)\left(\underline{u}_{ZN}^N\times\right)\delta\underline{R}^N + \left(\underline{v}^N \times \underline{u}_{ZN}^N\right)\varepsilon_{ZN}\end{aligned} \tag{16.2.5-8}$$

or in matrix form:

$$\begin{bmatrix}\underline{\gamma}^N\\ \delta\underline{v}^N\end{bmatrix} = \begin{bmatrix} I & 0 & \frac{1}{R}\left(\underline{u}_{ZN}^N\times\right) & -\underline{u}_{ZN}^N\\ 0 & I & \frac{1}{R}\left(\underline{v}^N\times\right)\left(\underline{u}_{ZN}^N\times\right) & -\left(\underline{v}^N \times \underline{u}_{ZN}^N\right)\end{bmatrix}\begin{bmatrix}\underline{\psi}^N\\ \delta\underline{V}^N\\ \delta\underline{R}^N\\ -\varepsilon_{ZN}\end{bmatrix} \tag{16.2.5-9}$$

Note in (16.2.5-9) that we have represented ε_{ZN} by its negative in the column vector on the right. This anticipates compatibility with the covariance simulation analysis error state vector $\underline{x}'$ as defined in (16.1.1-28) with (16.1.1-27). Using the (16.1.1-13) and (16.1.1-26) definitions, the uncertainty in Equation (16.2.5-9) is then written as:

$$\begin{bmatrix}\Delta\underline{\gamma}^N\\ \Delta\left(\delta\underline{v}^N\right)\end{bmatrix} = \begin{bmatrix} I & 0 & \frac{1}{R}\left(\underline{u}_{ZN}^N\times\right) & -\underline{u}_{ZN}^N\\ 0 & I & \frac{1}{R}\left(\underline{v}^N\times\right)\left(\underline{u}_{ZN}^N\times\right) & -\left(\underline{v}^N \times \underline{u}_{ZN}^N\right)\end{bmatrix}\begin{bmatrix}\Delta\underline{\psi}^N\\ \Delta\left(\delta\underline{V}^N\right)\\ \Delta\left(\delta\underline{R}^N\right)\\ \Delta\left(-\varepsilon_{ZN}\right)\end{bmatrix} \tag{16.2.5-10}$$

Equation (16.2.5-10) is in the (16.2.5-2) format for application of covariance conversion Equation (16.2.5-4).

Sensitivity expressions can also be developed for output parameters from conversion Equation (16.2.5-2). For the initial condition, process noise and measurement effects contributing to the basic error budget, the equivalent output parameter sensitivity expressions are derived beginning with the vector form of (16.2.4-5):

$$\underline{x}' = S_{IC}\, \underline{x}'_0 + \underline{x}'_{Proc} + \underline{x}'_{Meas} \qquad (16.2.5\text{-}11)$$

Substituting (16.2.5-11) into (16.2.5-2) finds for the output parameter vector $\underline{a}_{Out}$:

$$\underline{a}_{Out} = B\, \underline{x}' = B\, S_{IC}\, \underline{x}'_0 + B\, \underline{x}'_{Proc} + B\, \underline{x}'_{Meas} \qquad (16.2.5\text{-}12)$$

From (16.2.5-12) we then define:

$$S_{aOut/IC} \equiv B\, S_{IC} \qquad (16.2.5\text{-}13)$$

$$\underline{a}_{Out/Proc} \equiv B\, \underline{x}'_{Proc} \qquad \underline{a}_{Out/Meas} \equiv B\, \underline{x}'_{Meas} \qquad (16.2.5\text{-}14)$$

where

$S_{aOut/IC}$ = Output parameter vector sensitivity matrices to $\underline{x}'$ initial conditions.

$\underline{a}_{Out/Proc}$, $\underline{a}_{Out/Meas}$ = Portions of $\underline{a}_{Out}$ produced by process and measurement noise.

with which (16.2.5-12) becomes:

$$\underline{a}_{Out} = S_{aOut/IC}\, \underline{x}'_0 + \underline{a}_{Out/Proc} + \underline{a}_{Out/Meas} \qquad (16.2.5\text{-}15)$$

The covariance form of (16.2.5-14) is:

$$P_{aOut/Proc} = B\, P'_{Proc}\, B^T \qquad P_{aOut/Meas} = B\, P'_{Meas}\, B^T \qquad (16.2.5\text{-}16)$$

where

$P_{aOut/Proc}$, $P_{aOut/Meas}$ = Covariance of $\underline{a}_{Out/Proc}$, $\underline{a}_{Out/Meas}$.

P'_{Proc}, P'_{Meas} = Covariance of $\underline{x}'_{Proc}$, $\underline{x}'_{Meas}$.

As in (16.2.4-9), the P'_{Proc}, P'_{Meas} covariance matrices can be constructed as a sum of the noise sensitivity run results multiplied by the noise values for the noise sensitivity groups:

$$P'_{Proc} = \sum_{l} P'_{Proc_l}\, q_{PDens_l} \qquad P'_{Meas} = \sum_{m} P'_{Meas_m}\, r_{Meas_m} \qquad (16.2.5\text{-}17)$$

From (16.2.5-17) and (16.2.5-16) we define:

$$P_{aOut/Proc_l} \equiv B\, P'_{Proc_l}\, B^T \qquad P_{aOut/Meas_m} \equiv B\, P'_{Meas_m}\, B^T \qquad (16.2.5\text{-}18)$$

where

$P_{aOut/Proc_l}$, $P_{aOut/Meas_m}$ = Sensitivity of $P_{aOut/Proc}$, $P_{aOut/Meas}$ to q_{PDens_l}, q_{PDens_m}.

Then with (16.2.5-16) - (16.2.5-18) we see that:

$$P_{aOut/Proc} = \sum_l P_{aOut/Proc_l}\, q_{PDens_l} \qquad P_{aOut/Meas} = \sum_m P_{aOut/Meas_m}\, r_{Meas_m} \tag{16.2.5-19}$$

Using (16.2.5-19), we are now in a position to write the equation for the variance of the ith component of $\underline{a}_{Out}$ in (16.2.5-15). By direct analogy, the result is obtained from inspection of Equations (16.2.4-10) and (16.2.4-12) derived from (16.2.4-5):

$$\begin{aligned} \sigma^2_{aOut_i} = & \sum_j \left(S_{aOut/IC_{ij}}\, \sigma_{x'_{j0}}\right)^2 + \sum_l \left(\sqrt{P_{aOut/Proc/ii_l}}\,\sigma_{PDens_l}\right)^2 \\ & + \sum_m \left(\sqrt{P_{aOut/Meas/ii_m}}\,\sigma_{Meas_m}\right)^2 + \sigma^2_{aOut/Misc_i} \end{aligned} \tag{16.2.5-20}$$

$$\begin{aligned} \sigma^2_{aOut/Misc_i} = & P_{aOut/ii} - \sum_j \left(S_{aOut/IC_{ij}}\, \sigma_{x'_{j0}}\right)^2 - \sum_l \left(\sqrt{P_{aOut/Proc/ii_l}}\,\sigma_{PDens_l}\right)^2 \\ & - \sum_m \left(\sqrt{P_{aOut/Meas/ii_m}}\,\sigma_{Meas_m}\right)^2 \end{aligned} \tag{16.2.5-21}$$

where

σ_{aOut_i} = Root-mean-square value for a_{Out_i}, the ith component of $\underline{a}_{Out}$.

$S_{aOut/IC_{ij}}$ = Element in row i, column j of $S_{aOut/IC}$ calculated with Equation (16.2.5-13).

$P_{aOut/Proc/ii_l}$, $P_{aOut/Meas/ii_m}$, $P_{aOut/ii}$ = Elements in row i, column i of $P_{aOut/Proc_l}$, $P_{aOut/Meas_m}$, P_{aOut} calculated with Equations (16.2.5-18) and (16.2.5-4).

A basic error budget table equivalent to Equations (16.2.5-20) - (16.2.5-21) can also be generated by analogy to Table 16.2.4-1 by replacing the x'_i error sensitivities in the right three columns of Table 16.2.4-1 with the a_{Out_i} sensitivities, and P'_{ii} in the miscellaneous calculation with $P_{aOut/ii}$.

An output parameter error budget can also be prepared for the portion of a_{Out_i} produced by transients at the nTrans cycle time. We begin with the vector form of (16.2.4-13):

$$\underline{x}'_{Tr} = S_{Tr}\, \partial \underline{x}'_{Tr_{nTrans}} \tag{16.2.5-22}$$

Substituting (16.2.5-22) into (16.2.5-2) finds for the portion of $\underline{a}_{Out}$ produced by transients:

$$\underline{a}_{Out_{Tr}} = B\ \underline{x}'_{Tr} = B\ S_{Tr}\ \partial\underline{x}'_{Tr_{nTrans}} \tag{16.2.5-23}$$

where

$\underline{a}_{Out_{Tr}}$ = Portion of $\underline{a}_{Out}$ produced by $\partial\underline{x}'_{Tr_{nTrans}}$ transients at the nTrans cycle time.

From (16.2.5-23) we then define:

$$S_{aOut/Tr} \equiv B\ S_{Tr} \tag{16.2.5-24}$$

with which (16.2.5-23) becomes:

$$\underline{a}_{Out_{Tr}} = S_{aOut/Tr}\ \partial\underline{x}'_{Tr_{nTrans}} \tag{16.2.5-25}$$

By direct analogy, the variance of the ith component of $\underline{a}_{Out_{Tr}}$ in (16.2.5-25) is obtained through inspection of Equation (16.2.4-16) derived from (16.2.4-13):

$$\sigma^2_{aOut/Tr_i} = \sum_j \left(S_{aOut/Tr_{ij}}\ \sigma_{x'_{j\,nTrans}} \right)^2 \tag{16.2.5-26}$$

where

σ_{aOut/Tr_i} = Root-mean-square value for $a_{Out/Tr\,i}$, the ith component of $\underline{a}_{Out_{Tr}}$.

$S_{aOut/Tr_{ij}}$ = Element in row i, column j of $S_{aOut/Tr}$ calculated with Equation (16.2.5-24).

16.2.6 GENERAL COVARIANCE SIMULATION PROGRAM STRUCTURE

This section summarizes the material covered in Sections 16.2.1 - 16.2.5 and provides an overall description of the principal functions and capabilities that can be incorporated in a covariance simulation program for generalized usage.

16.2.6.1 BASIC COVARIANCE PROPAGATION/RESET EQUATIONS

Equations (16.2.1-1) - (16.2.1-4) define the general operations implemented in a covariance simulation program for initializing, propagating and resetting the augmented covariance matrix P' (for real world and suboptimal Kalman filter error state uncertainties), based on idealized Kalman filter control resets. For a delayed control reset Kalman filter, the equivalent equations are (16.1.2-23), (16.1.2-27) - (16.1.2-28) and (16.2.1-3). The covariance simulation program also executes Equations (16.2.1-5) - (16.2.1-7) defining the equivalent operations performed in the suboptimal Kalman filter to evaluate the P^* actual (suboptimal) covariance matrix including generation of the suboptimal Kalman gain matrix K (used in the augmented P' calculations). For the idealized control reset Kalman filter, Equations (16.2.1-8) - (16.2.1-12) would be processed in the covariance simulation for evaluating the theoretical optimal ("best achievable") Kalman filter performance in the form of the P_{Opt} covariance matrix. For the delayed control reset Kalman filter, the P_{Opt} covariance matrix would be calculated with Equations (16.1.2.1-14) - (16.1.2.1-16) and (16.2.1-11).

In a typical covariance simulation program, the actual (suboptimal) Kalman filter equations would be executed first over the input trajectory profile, with the computed suboptimal gains K saved as a function of trajectory time when computed. The augmented covariance P' propagation/reset equations would then be run over the same trajectory profile using the suboptimal saved gains K for input. For reference performance comparison, the P_{Opt} covariance propagation/reset equations can then be run for the same trajectory profile to calculate the ideal best achievable performance for P' comparison. A more sophisticated simulation program would automatically configure itself for a P^*, P', or P_{Opt} run based on operator input command.

We also note in passing, that the covariance simulation can also be used to evaluate unaided inertial navigation system performance by operation of the program without Kalman resets. The unaided inertial mode can be configured from a user specified input condition that calls the covariance propagation equation in the optimal Kalman filter performance routines.

As part of the covariance propagation/reset operations described above, positive definite tests can be incorporated (for output status information) to measure the deleterious effect of numerical finite word length round-off on P', P^* and P_{Opt}. These tests will also measure the effect on covariance positive definiteness of integrated process noise matrix algorithm power series expansion truncation error. Simplified covariance positive definite tests can be structured from Equation (15.1.2.1.1.4-2) as:

$$\varepsilon_{ij} = P_{ii}\, P_{jj} - \left| P_{ij}\, P_{ji} \right| \qquad (16.2.6.1\text{-}1)$$

where

ε_{ij} = Test parameter for covariance elements P_{ii}, P_{jj}, P_{ij}, P_{ji}. The ε_{ij} parameter should remain positive if P is positive definite. A typical simulation program might output the ε_{ij}'s that become negative.

Controls can also be incorporated in the simulation program to assure that the covariance matrices satisfy basic inherent characteristics of the processing algorithms (e.g., to remain symmetrical, the controls for which are described in Section 15.1.2.1.1.4).

16.2.6.2 SIMPLIFIED COVARIANCE SIMULATION PROGRAM CONFIGURATIONS

To reduce throughput in a covariance simulation program, the P' covariance computations are often simplified based on the simplifying assumptions of Section 16.1.1.4 (for idealized control resets) or Section 16.1.2.2 (for delayed control resets). In either case, the simplifying assumptions eliminate the presence of the control matrix L_X in the covariance propagation/update equations. A more sophisticated covariance simulation program might be structured to automatically reconfigure itself from the general case to the simplified version based on user input instructions.

16.2.6.3 ERROR STATE CONFIGURATION

The assignment of error states to their selected location in the error state vector is part of the initialization process in a covariance simulation program. A useful feature is the ability to assign (or reassign) error state locations arbitrarily without the need to eliminate error state locations not being used. The basic covariance propagation/update software structure would then be configured to eliminate the null arithmetic associated with unused (null) error state locations. The advantage of this approach is the ability to easily define the suboptimal (reduced) error state vector from the complete real world error state model without reordering vector locations, and the ability to easily convert a previous covariance error state model configuration to a new one by adding or deleting error states in arbitrary locations, without the need for reordering. The initialization process with such an approach would identify the numerical assignment for the active states in the error state vector for a particular simulation run configuration.

16.2.6.4 ESTIMATION/CONTROL CONFIGURATION

The general ability to select which error states are to be estimated and which are to be controlled is a useful feature to be built into a covariance simulation program. This can be implemented by operator selection of the required states to be estimated/controlled which would then be translated by the simulation software into the equivalent I_{Est} (or I_{OptEst} for an optimal performance run) and L_X matrix configurations (see Equations (16.2.1-4), (16.2.1-6) and (16.2.1-9) for idealized control resets or (16.1.2-23), (16.2.1-6) and (16.1.2.1-16) for delayed control resets). For simulations incorporating several types of Kalman measurements, a different estimation and control matrix might be required for each measurement type.

When the general P' updating equations are implemented, the control matrix L_X is applied during the covariance propagation cycle following the Kalman estimation reset. The application of L_X is manifested in the Φ' matrix (see Equation (16.2.1-4) for idealized control resets or

(16.1.2-23) for delayed control resets). For simulation configurations in which several P' propagation cycles are utilized between Kalman updates, L_X should be applied during the P' propagation cycle immediately following the Kalman estimation update. Otherwise L_X should be treated as equal to zero.

16.2.6.5 COVARIANCE PROPAGATION TIMING STRUCTURE

For most covariance simulation programs, the covariance matrix is propagated through several cycles of different time duration between Kalman update cycles. For example, during dynamic maneuver segments of the trajectory profile, the time interval for the covariance propagation cycle might be reduced to improve the accuracy of the state transition and propagated process noise truncated power series expansion algorithms. Conversely, for benign trajectory profile segments (e.g., straight and level aircraft flight), the propagation cycle might be increased to reduce simulation run time. If the trajectory profile is created by a trajectory generator with discrete profile segments (see Chapter 17), for improved accuracy, it might be required that the covariance propagation update include the trajectory generator segment transition time points. A basic covariance propagation time point at fixed time intervals is usually implemented corresponding to the user desired data output times. The covariance propagation cycle time points must, of course, also include the Kalman update times which may be arbitrarily spaced depending on the measurement type, and for situations when several different measurements are being simulated. A generalized covariance simulation program will be typically configured to automatically select the covariance propagation intervals based on the previous (and possibly other) considerations.

16.2.6.6 KALMAN ESTIMATION RESET TIMING STRUCTURE

A covariance simulation program must provide the ability for operator selection of the update times (and/or frequency) for Kalman estimation resets. For an application involving several types of measurements applied at different times, it is convenient to embody all measurements into a single measurement equation (i.e., Equation (16.2.3-2) for the real world and (16.2.3-6) for the suboptimal Kalman filter world) in which all measurements are considered to occur simultaneously whenever a Kalman estimation update is made. With this approach, simulation of a particular measurement at its assigned measurement time is achieved by retaining the rows of the measurement matrix (H_{RW} and H_x^*) associated with that measurement and zeroing the remaining rows. When using the integrated measurement approach, it is important that the full measurement noise matrices be retained for all measurements (i.e., G_M, R, G_M^*, R^*); otherwise singularities may be created in the Kalman gain calculation (Equations (16.2.1-6) or (16.2.1-9)) from the matrix inverse operation.

16.2.6.7 ERROR MODEL SPECIFICATION

As discussed in Section 16.2.3, specification of the error model in a given covariance simulation program is typically handled by direct user programming of the continuous form equation matrices A, G_P, Q_{PDens}, H_{RW}, G_M, R, A^*_{XX}, $G^*_{P_X}$, $Q^*_{P_{Dens}}$, H^*_X, G^*_M and R*. Sections 16.2.3.1 - 16.2.3.3 provide examples of analytical error models that can be used to characterize different process noise effects, acceleration squared effects and gravity uncertainty. For ease in programming, it is convenient if the covariance simulation program architecture assigns separate subroutines for programming the A, G_P, Q_{PDens}, H_{RW}, G_M, R, A^*_{XX}, $G^*_{P_X}$, $Q^*_{P_{Dens}}$, H^*_X, G^*_M and R* matrices. Calculation of the error state transition matrix and integrated process noise matrix for covariance propagation (Φ, Φ^*_{XX}, Q, Q^*_{XX}) would then be performed within the covariance simulation from A, G_P, Q_{PDens}, A^*_{XX}, $G^*_{P_X}$, $Q^*_{P_{Dens}}$ using the power series expansion algorithms of Sections 15.1.2.1.1.1 - 15.1.2.1.1.3.

The covariance simulation program architecture should allow operator selection of the expansion order for the error state transition matrix and integrated process noise matrix (Φ, Φ^*_{XX}, Q, Q^*_{XX}) power series expansion algorithms. The number of terms carried in the expansions may vary from application to application and includes a tradeoff between added computation time for increasing expansion order versus increased simulation covariance propagation time interval for more expansion terms. In the case of the integrated process noise matrix expansions, it is useful to employ a positive definite test on the result (e.g., Equation (16.2.6.1-1) applied to elements of Q or Q*) as an operator alert that sufficient expansion terms may have not been carried. Most integrated process noise expansion algorithms employ a final routine that forces the result to have the proper symmetric structure (e.g., Equations (15.1.2.1.1.3-28)). In the case of the error state transition matrix, a general ground rule to follow is that the computation expansion order be equal to or greater than the maximum number of integrators between the error state vector elements and the measurement.

16.2.6.8 TRAJECTORY GENERATOR INTERFACE

The structure of the covariance program must contain an interface with a trajectory generator computer program supplying angular rate, acceleration, attitude, velocity and position data for calculating the elements of the A, G_P, H_{RW}, G_M, A^*_{XX}, $G^*_{P_X}$, H^*_X, G^*_M matrices. The simulated time data rates from the trajectory generator must be sufficient to enable accurate calculation of Φ, Φ^*_{XX}, Q, Q^*_{XX} from A, G_P, Q_{PDens}, A^*_{XX}, $G^*_{P_X}$, $Q^*_{P_{Dens}}$ using the covariance simulation computation algorithms. Chapter 17 describes the general structure of a typical trajectory generator program.

16.2.6.9 SENSITIVITY AND ERROR BUDGET OUTPUTS

To aid in the design/evaluation of the suboptimal Kalman filter, the covariance simulation program is typically structured to provide sensitivity and error budget output information. The sensitivity outputs define the sensitivity of each element of the error state vector to unity values for initial error state component uncertainties, transients on the error state components, process noise groupings and measurement noise groupings. The basic error budget provides a breakout of the individual contributions to each error state vector component of the effects of initial error state component uncertainties, process noise groupings and measurement noise groupings for their assigned numerical values in the particular simulation run. A separate budget can also be calculated providing a breakout of the individual contributions to each error state vector component of transients on the error states. Section 16.2.4 describes how the sensitivities and error budget parameters are calculated.

The sensitivities to initial error state uncertainties and transients (S_{IC} and S_{Tr}) are evaluated with Equations (16.2.4-3) and (16.2.4-4). These equations can be processed in parallel with the basic P' covariance propagation/update operations for little computation time penalty because their inputs (Φ', K', H') are already evaluated as part of the P' calculations. The sensitivities to the process noise and measurement noise groupings are obtained (as explained in Section 16.2.4) by running a parallel P' propagation/reset calculation starting with zero P' initial condition for each process noise and measurement noise group for which the sensitivity is to be obtained. For several noise sensitivity groupings, the added parallel noise sensitivity computations can add considerably to the processing time, however, it is less than proportional to the basic P' computation time because Φ', K', H' do not have to be recomputed for the sensitivity calculations.

The error budget for each error state is evaluated from the sensitivities and error source values using the approach described in Section 16.2.4 (e.g., Table 16.2.4-1 for preparation of the basic error budget). Usually the error budget is evaluated only for the time points that the output is desired.

16.2.6.10 OUTPUT ROUTINES

Section 16.1.1.1 defines how the P' augmented covariance matrix is interpreted to evaluate suboptimal Kalman filter performance. Section 16.2.5 describes the analytical conversion process for evaluating error parameters other than those directly represented in the P' covariance. The conversion process applies to error parameters that are linearly related to the P' error state vector elements. Section 16.2.5 describes the conversion process to obtain the equivalent output parameter covariance elements, sensitivities and error budget breakout.

16.2.7 COVARIANCE SIMULATION PROGRAM USE IN SUBOPTIMAL KALMAN FILTER DESIGN AND/OR PERFORMANCE EVALUATION

The initial design/verification process for a suboptimal Kalman filter requires the use of a covariance simulation program as the basic design tool. The covariance simulation allows the design engineer to numerically determine suboptimal Kalman filter performance from the appropriate elements of the augmented P' covariance matrix. It also provides the designer with a best achievable reference Kalman filter performance profile in the form of the "optimal" filter covariance P_{Opt}. Suboptimal filter performance is evaluated by comparing P' with P_{Opt} for a selected group of "representative" trajectory profiles. Suboptimal Kalman filter design is the iterative numerical process by which the suboptimal filter is adjusted (in the P^* propagation/reset equations resident in the covariance simulation) until P' performance approaches P_{Opt} goals. Adjustments to the suboptimal filter configuration are based on the simulation sensitivity/error budget outputs and the designer's knowledge of error propagation effects and their impact on Kalman aided system performance. Adjustments to the Kalman filter during the design iteration process typically consist of additions or deletions of error states, and increasing process and/or measurement noise to artificially account for deleted error states. The object is to obtain reasonable filter performance with as few error states as possible to minimize the final impact on target computer throughput (in which the suboptimal Kalman filter covariance propagation/reset equations will be implemented in real time).

The basic error budget is typically the principal tool for identifying the cause of observed suboptimal Kalman filter anomalous behavior. A typical analysis of the basic error budget entails identification of the elements having the largest impact on deficient suboptimal filter performance. The primary factors producing anomalies are usually deleted error states. However, the obvious solution of including the principal deleted states having the greatest impact does not always resolve the problem. There are times when deleting an error state (e.g., because of its small impact on the measurement) causes the filter to erroneously attribute its effect on the measurement to a large uncertainty in one of the included error states. For this situation, a small increase in process or measurement noise (to have the same magnitude effect on the measurement as the deleted error state) can sometimes prevent the filter from attributing the measurement variation to the included states. Based on engineering judgment, the likely change to the suboptimal filter is then implemented and run in the covariance simulation to evaluate the expected performance improvement achieved. The process continues until reasonable performance for an acceptable number of included error states is achieved (hence, acceptable impact on target computer throughput). In the final analysis, design of the suboptimal Kalman filter is a cut-and-try process whose final result is significantly influenced by the experience and analytical capability of the design engineer.

The transient effects error budget is typically used to assess the effects of anomalous behavior on system performance after the Kalman filter design has been completed. In one interesting application in which the author participated, a Kalman filter was used as an external monitor of strapdown INS performance during stationary fine alignment (using the same inputs provided to the Fine Alignment Kalman filter resident in the actual INS). Monitoring of the estimated error states provided a measure of system/sensor performance that could be compared against previous signatures from the same system. Deviations from past performance identified the presence of anomalies. During the design of the monitor software, transient sensitivities and error budgets were used to verify that transient shifts in the INS sensor error characteristics during fine alignment would register measurable signatures on the monitor error state estimates.

17 Trajectory Generators

17.0 OVERVIEW

A trajectory generator is a software simulator used to create a numerical data history of navigation parameters corresponding to a user specified trajectory profile. The trajectory generator provides input data to various simulation programs used to analyze performance and validate software operations associated with inertial navigation systems. Classical applications for trajectory generators are to provide input data for covariance analysis programs (See Chapter 16) and for data input to Kalman filter software validation programs (See Section 15.1.4). Basic outputs from trajectory generators used for strapdown inertial navigation applications are the components of specific force acceleration and inertial angular rate (i.e., the input to the strapdown angular rate sensors and accelerometers) in addition to their associated attitude, velocity and position navigation data. In general, the trajectory generator must be designed to provide for two basic functions; trajectory shaping and trajectory regeneration.

The trajectory shaping function allows the user to create trajectories corresponding to selected application characteristics. In general, this translates into the capability for creating specific-force/angular-rate profiles that when integrated, generate an attitude/velocity/position history representative of the user's specified application trajectory profile. The trajectory shaping function is generally implemented in segments. Linking the segments in a continuous time sequence then creates the desired trajectory profile. Due to the diverse nature of typical trajectory segment specifications, the segment shaping function invariably consists of an interactive process by which the user selects an acceleration/angular-rate profile for the segment, monitors its effect on the trajectory, and adjusts the acceleration/angular-rate components until the generated trajectory satisfies requirements. That segment is then recorded and the next segment is shaped. The process continues until the complete trajectory is formed. To facilitate the segment shaping process, the trajectory generator software can be designed to provide shaping aids that allow the user to create particular characteristics over a segment (e.g., a specified average specific force acceleration component over the segment, a specified attitude at completion of the segment, a specified position location or velocity condition at completion of the segment). Once a particular segment has been declared acceptable, it is advantageous if it can be characterized by a minimum number of parameters for storage.

The trajectory regeneration function consists of retrieving the saved trajectory segment characteristic parameters in the time sequence in which they were created, and using the retrieved parameters to recreate the trajectory profile. This generally consists of converting the segment parameters to the corresponding specific-force/angular-rate profile for each segment, then integrating the specific-force/angular-rate to generate the corresponding attitude/velocity/position data. A smoothing function may also be incorporated to assure that the angular-rate/acceleration data is free of discontinuities at the segment junctions (e.g., a sharp step change in the angular-rate/acceleration from one segment to the next). The integration process is typically based on the strapdown inertial navigation routines normally implemented in the strapdown INS computer (i.e., as described in differential equation format in Chapter 4 and by the equivalent digital processing algorithms in Chapter 7). The result is a time history of the basic navigation parameters (e.g., $\underline{a}_{SF}^{B}$, $\underline{\omega}_{IB}^{B}$, C_{B}^{L}, $\underline{v}^{N}$, C_{N}^{E}, h) generated at an output rate selected by the software program requiring trajectory data input (Note that the basic form of the acceleration/angular-rate output data would typically be in the form of the integral of $\underline{a}_{SF}^{B}$, $\underline{\omega}_{IB}^{B}$ over the selected output time interval). Using the conversion routines of Chapter 3, other classic navigation parameters can also be generated for output (e.g., attitude quaternion, position vector, north/east geographic coordinate referenced components) from the basic navigation parameters.

As a specific detailed example, the remainder of this chapter describes the design of a particular trajectory generator for applications in which the vehicle longitudinal axis is generally in the direction of the velocity vector, hence, the velocity vector profile approximates the attitude history of the longitudinal axis. Applications that fit the previous constraint include aircraft, missiles, underwater vehicles and oil-well-survey probes. The trajectory generator shaping function is designed to create the trajectory segments based on the vehicle velocity/longitudinal-axis constraint. Once the profile segments are shaped, capabilities are added for smoothing the interface between trajectory segments, generating simulated strapdown inertial sensor input specific force acceleration and angular rate, changing the orientation of the inertial sensor axes from the velocity direction (to simulate, for example, angle of attack/sideslip effects or mounting the sensor assembly on a rotating platform), simulating a sensor assembly mount at a specified lever arm location in the vehicle, adding wind gust induced aerodynamic force accelerations, and adding high frequency angular/linear motion effects. These added capabilities are incorporated as part of the trajectory regeneration function.

Coordinate frames used in this chapter are the B, N, L, Geo, E, and I Frames defined in Section 2.2 plus specialized coordinate frames defined in the particular sections in which they are introduced.

17.1 TRAJECTORY SHAPING FUNCTION

The trajectory shaping function is divided into three basic operations:

- Segment parameter selection
- Quick-look projection
- End-of-segment data generation
- Real world error state transition and integrated process noise matrix elements

For the trajectory generator being described, each trajectory segment is characterized by five basic parameters; the time duration of the segment, the change in signed velocity magnitude across the trajectory segment ("signed" to be clarified subsequently), and three components of a rotation vector defining the angular rotation of the velocity vector across the trajectory segment. The rotation vector is specified in a coordinate frame that rotates with the velocity vector. It is further assumed in the definition of the segment parameters that the angular rate components and the rate of change of the velocity magnitude are constant across the segment. Selection of the segment parameters can be performed directly, or through the assistance of computational aids used to create desired trajectory segment characteristics.

Once a trial set of segment parameters are selected, the Quick-Look Projection option quickly calculates and displays the trajectory segment characteristics produced by application of the selected parameters (e.g., attitude/velocity/position at segment end, average specific force acceleration across the segment, angle of attack/sideslip effects during forthcoming trajectory regeneration). In order to minimize computation time, the Quick-Look calculations are based on closed-form equations that rapidly provide the desired segment characteristics for user assessment. Although the closed-form equations are exact for the end-of-segment attitude/velocity parameters, the position projection is based on simplifying assumptions. Based on the approximate Quick-Look projection, the user can accept the trial set of segment parameters, or select a modified set that more closely creates the desired trajectory segment characteristics.

After the trajectory segment parameters are selected and accepted, the End-of-Segment Position Generation option is used to calculate the end-of-segment attitude, velocity and position location with precision. The end-of-segment attitude and velocity are computed exactly from the trajectory segment parameters. The end-of-segment position location is calculated as a velocity integration process using precision strapdown inertial navigation algorithms to arrive at an accurate end-of-segment solution. Initial conditions for the end-of-segment attitude/velocity/position computations are the previous end-of-segment values created by the End-of-Segment Data Generation routines.

The following subsections describe the analytical processes involved in the trajectory shaping operations.

17.1.1 SEGMENT PARAMETER SELECTION

The basic parameters used to characterize a trajectory segment are defined symbolically by T_S, ΔV_S, $\underline{\phi}_S^{\mathcal{V}}$ where:

T_S = Trajectory segment time interval.

ΔV_S = Change across the trajectory segment in the "signed velocity vector magnitude" (to be defined subsequently)

$\mathcal{V}$ = Velocity vector referenced coordinate frame having its X axis parallel to the velocity vector. For an idealized vehicle that maintains its longitudinal axis along the velocity vector direction, the $\mathcal{V}$ Frame would correspond to vehicle fixed reference coordinates (assuming that the vehicle frame has its X axis defined in the general direction of motion).

$\underline{\phi}_S^{\mathcal{V}}$ = Rotation vector describing the rotation of velocity vector referenced coordinate Frame $\mathcal{V}$ across the trajectory segment as projected on $\mathcal{V}$ Frame axes. The $\underline{\phi}_S^{\mathcal{V}}$ rotation vector is defined as a rotation of the $\mathcal{V}$ Frame relative to the locally level navigation N Frame of the wander azimuth type (See Section 4.5 for definition). For $\underline{\phi}_S^{\mathcal{V}}$ equal to zero, the $\mathcal{V}$ Frame maintains a fixed orientation relative to the N Frame.

For the trajectory shaping function, the $\mathcal{V}$ Frame can be considered as equivalent to the strapdown sensor assembly B Frame. We distinguish between the $\mathcal{V}$ and B Frames so that the B Frame attitude can be subsequently separated from the $\mathcal{V}$ Frame (by specification) as part of Section 17.2 Trajectory Regeneration operations.

Based on the previous $\mathcal{V}$ Frame definition, the signed velocity vector magnitude is defined as V where:

V = "Signed velocity vector magnitude" equal to the velocity vector component along the $\mathcal{V}$ Frame X axis.

By definition of the $\mathcal{V}$ Frame, the magnitude of V is equal to the magnitude of the velocity vector, however, it can be positive or negative depending on whether the $\mathcal{V}$ Frame X axis is in the same or opposite direction to the velocity vector. Application of the $\mathcal{V}$ definition proves useful for simulating "backing up" from an original forward motion without requiring a 180 degree rotation of the $\mathcal{V}$ Frame X axis.

Selection of the basic segment parameters is facilitated by aiding routines that enable the user to create specified trajectory characteristics. For example, aiding routines can be designed for selecting trajectory segment parameters that produce the following trajectory segment characteristics:

- A specified end-of-segment attitude and velocity.
- Specified average specific force acceleration characteristics during the segment.
- Rotation vector attitude change defined in the locally level attitude reference L Frame.
- True heading and segment time to reach a desired long range destination.
- Turn to a specified heading.

Aiding routines can be defined to create other effects as well. For those enumerated above, the associated analytics follow.

17.1.1.1 SPECIFIED END-OF-SEGMENT ATTITUDE/VELOCITY

The attitude specification applies to the orientation of the $\mathcal{V}$ Frame. We define the end-of-segment attitude requirement analytically using the Equation (3.2.1-5) chain rule as:

$$C_{\mathcal{V}_{End}}^{\mathcal{V}_{Start}} = \left(C_{\mathcal{V}_{Start}}^{L}\right)^{T} C_{\mathcal{V}_{End}}^{L} \qquad (17.1.1.1\text{-}1)$$

where

Start, End = Designation for the indicated coordinate frame (or parameter) attitude (or value) at the start and end of the trajectory segment.

L = Locally level attitude reference coordinate frame parallel to the locally level navigation N Frame as defined in Equation (4.1.1-2). For the trajectory shaping function, the N frame is of the azimuth wander type (See Section 4.5 for definition).

The $C_{\mathcal{V}_{Start}}^{L}$ matrix in (17.1.1.1-1) is the attitude direction cosine matrix generated at the end of the last trajectory segment using the End-of-Segment Data Generation option. The $C_{\mathcal{V}_{End}}^{L}$ matrix in (17.1.1.1-1) is the desired $\mathcal{V}$ Frame attitude at the end of the current segment. The $C_{\mathcal{V}_{End}}^{L}$ matrix is computed as in Equations (3.2.3.1-2) from user specified roll/pitch/platform heading Euler angles with platform heading calculated from specified true heading corrected for the wander angle (as in the converse of Equation (4.1.2-2)). The wander angle for the previous platform heading calculation is approximated as the wander angle at the start of the trajectory segment defined from the C_{N}^{E} matrix using Equations (4.4.2.1-3).

Once $C_{\mathcal{V}_{End}}^{\mathcal{V}_{Start}}$ is calculated from (17.1.1.1-1), the equivalent $\underline{\phi}_S^{\mathcal{V}}$ rotation vector for the trajectory segment is evaluated using generalized direction cosine matrix to rotation vector conversion Equations (3.2.2.2-10) - (3.2.2.2-12) and (3.2.2.2-15) - (3.2.2.2-19).

The end-of-segment velocity vector is defined in terms of its orientation and magnitude. From the definition of the $\mathcal{V}$ Frame, specification of the $\mathcal{V}$ Frame orientation at the end-of-the segment (by the previous $\underline{\phi}_S^{\mathcal{V}}$ selection process) automatically fixes the line of action of the end-of-segment velocity vector (i.e., along the X axis of the rotated $\mathcal{V}$ Frame). The magnitude of the velocity vector and its direction along the $\mathcal{V}$ Frame X axis (i.e., along the positive or negative X axis direction) is defined by the "signed velocity magnitude" V. The ΔV_S value to achieve a desired end-of-segment V value is then given by:

$$\Delta V_S = V_{End} - V_{Start} \tag{17.1.1.1-2}$$

where

V_{End} = Desired end-of-segment signed velocity vector magnitude.

V_{Start} = Signed velocity vector magnitude at the start of the velocity segment.

The V_{Start} value is the value of V calculated as the sum of ΔV_S values for the previous trajectory segments, or in recursive algorithm form:

$$V_k = V_{k-1} + \Delta V_{S_k} \tag{17.1.1.1-3}$$

where

k = Trajectory segment number since trajectory start (numbered consecutively starting from 1). The V_{Start} value in (17.1.1.1-2) is the k-1 value for V.

17.1.1.2 SPECIFIED $\mathcal{V}$ FRAME AVERAGE SPECIFIC FORCE ACCELERATION

A typical trajectory segment shaping problem is to define the segment time interval T_S that will produce a specified average specific force acceleration (along a particular $\mathcal{V}$ Frame axis). When performing trajectory shaping, the $\mathcal{V}$ Frame is interpreted as simulating classical aircraft axes having the X-axis "forward", the Y-axis along the "right wing", with the Z axis along the negative lift direction (i.e., down during normal level flight). With this interpretation, typical average $\mathcal{V}$ Frame acceleration requirements might be to simulate a coordinated turn (defined as a maneuver with zero average Y axis specific force acceleration), or a maneuver with a specified average lift specific force (a negative Z axis acceleration). A typical problem is to determine the T_S segment time interval to achieve a particular axis average specific force, given that the $\underline{\phi}_S^{\mathcal{V}}$, ΔV_S segment parameters have already been selected. A variation might be to determine T_S and ΔV_S that achieves specified average specific force acceleration conditions along two $\mathcal{V}$ Frame axes given $\underline{\phi}_S^{\mathcal{V}}$. The analytical equations used to achieve such objectives can be derived by neglecting the effect of N Frame and E Frame rotation during the trajectory segment so that a simplified rearranged version of Equation (4.3-18) applies:

$$\underline{a}_{SF}^N \approx \dot{\underline{v}}^N - \underline{g}_P^N \tag{17.1.1.2-1}$$

or transformed to the $\mathcal{V}$ Frame:

$$\underline{a}_{SF}^{\mathcal{V}} = C_N^{\mathcal{V}}\, \underline{a}_{SF}^N \approx C_N^{\mathcal{V}}\, \dot{\underline{v}}^N - C_N^{\mathcal{V}}\, \underline{g}_P^N \tag{17.1.1.2-2}$$

Using generalized Equation (3.4-4) (with B, A replaced by $\mathcal{V}$, N), we find after multiplication by $C_N^{\mathcal{V}}$ that the second term in (17.1.1.2-2) is given by:

$$C_N^{\mathcal{V}}\, \dot{\underline{v}}^N = \dot{\underline{v}}^{\mathcal{V}} + \underline{\omega}_{N\mathcal{V}}^{\mathcal{V}} \times \underline{v}^{\mathcal{V}} \tag{17.1.1.2-3}$$

where

$\underline{\omega}_{N\mathcal{V}}^{\mathcal{V}}$ = Angular velocity of the $\mathcal{V}$ Frame relative to the N Frame as projected on $\mathcal{V}$ Frame axes.

Substitution in (17.1.1.2-2) then yields:

$$\underline{a}_{SF}^{\mathcal{V}} = \dot{\underline{v}}^{\mathcal{V}} + \underline{\omega}_{N\mathcal{V}}^{\mathcal{V}} \times \underline{v}^{\mathcal{V}} - C_N^{\mathcal{V}}\, \underline{g}_P^N \tag{17.1.1.2-4}$$

From the definition of the $\mathcal{V}$ Frame we can write:

$$\underline{v}^{\mathcal{V}} = V\, \underline{u}_{X\mathcal{V}}^{\mathcal{V}} \tag{17.1.1.2-5}$$

and its derivative:

$$\dot{\underline{v}}^{\mathcal{V}} = \dot{V}\, \underline{u}_{X\mathcal{V}}^{\mathcal{V}} \tag{17.1.1.2-6}$$

where

$\underline{u}_{X\mathcal{V}}^{\mathcal{V}}$ = Unit vector along the $\mathcal{V}$ Frame X axis as projected on $\mathcal{V}$ Frame axes.

and V is the signed velocity vector magnitude as defined previously.

The trajectory generator shaping function is based on a linearly increasing velocity (i.e., constant $\dot{V}$) over each trajectory segment, hence:

$$V = V_{Start} + \dot{V}\, t \qquad (17.1.1.2\text{-}7)$$

where

t = Time from the start of the trajectory segment.

For the segment parameter selection process, it is reasonable to approximate plumb-bob gravity ($\underline{g}_P^N$) as a constant g directed along the negative of the N Frame vertical $\underline{u}_{ZN}^N$. Using this approximation and (17.1.1.2-5) - (17.1.1.2-7) in (17.1.1.2-4), then gives after rearrangement:

$$\underline{a}_{SF}^{\mathcal{V}} = \dot{V}\, \underline{u}_{X\mathcal{V}}^{\mathcal{V}} + \left(V_{Start} + \dot{V}\, t\right)\left(\underline{\omega}_{N\mathcal{V}}^{\mathcal{V}} \times \underline{u}_{X\mathcal{V}}^{\mathcal{V}}\right) - \underline{g}_P^{\mathcal{V}}$$
$$\underline{g}_P^{\mathcal{V}} \approx -\, g\, C_N^{\mathcal{V}}\, \underline{u}_{ZN}^N \qquad (17.1.1.2\text{-}8)$$

where

g = Magnitude of plumb-bob gravity $\underline{g}_P$.

We define the average of the $\mathcal{V}$ Frame acceleration terms in (17.1.1.2-8) as:

$$\underline{a}_{SF_{Avg}}^{\mathcal{V}} \equiv \frac{1}{T_S}\int_0^{T_S} \underline{a}_{SF}^{\mathcal{V}}\, dt \qquad (17.1.1.2\text{-}9)$$

$$\underline{g}_{P_{Avg}}^{\mathcal{V}} \equiv -\, \frac{g}{T_S}\int_0^{T_S} C_N^{\mathcal{V}}\, \underline{u}_{ZN}^N\, dt \qquad (17.1.1.2\text{-}10)$$

where

$\underline{a}_{SF_{Avg}}^{\mathcal{V}}$ = Average $\mathcal{V}$ Frame specific force acceleration over the trajectory segment.

$\underline{g}_{P_{Avg}}^{\mathcal{V}}$ = Average $\mathcal{V}$ Frame plumb-bob gravity acceleration over the trajectory segment.

With (17.1.1.2-8) and (17.1.1.2-10), the average specific force acceleration from (17.1.1.2-9) becomes:

$$\underline{a}_{SF_{Avg}}^{\mathcal{V}} = \dot{V}\, \underline{u}_{X\mathcal{V}}^{\mathcal{V}} + \left(V_{Start} + \frac{1}{2}\,\dot{V}\, T_S\right)\left(\underline{\omega}_{N\mathcal{V}}^{\mathcal{V}} \times \underline{u}_{X\mathcal{V}}^{\mathcal{V}}\right) - \underline{g}_{P_{Avg}}^{\mathcal{V}} \qquad (17.1.1.2\text{-}11)$$

From (17.1.1.1-2) and (17.1.1.2-7) it should be obvious that:

$$\Delta V_S = \dot{V}\, T_S \qquad (17.1.1.2\text{-}12)$$

The trajectory generator shaping function is based on constant $\mathcal{V}$ Frame angular rate $\underline{\omega}_{N\mathcal{V}}^{\mathcal{V}}$ over each trajectory segment, hence, as in Equation (7.1.1.1-14), we can also write:

$$\underline{\omega}_{N\mathcal{V}}^{\mathcal{V}} = \frac{1}{T_S}\, \underline{\phi}_S^{\mathcal{V}} \qquad (17.1.1.2\text{-}13)$$

Substituting (17.1.1.2-12) - (17.1.1.2-13) into (17.1.1.2-11) yields the desired equivalent form in terms of the T_S, ΔV_S, $\underline{\phi}_S^{\mathcal{V}}$ basic segment parameters:

$$\underline{a}_{SF_{Avg}}^{\mathcal{V}} = \frac{1}{T_S}\left[\Delta V_S\, \underline{u}_{X\mathcal{V}}^{\mathcal{V}} + \left(V_{Start} + \frac{1}{2}\Delta V_S\right)\left(\underline{\phi}_S^{\mathcal{V}} \times \underline{u}_{X\mathcal{V}}^{\mathcal{V}}\right)\right] - \underline{g}_{P_{Avg}}^{\mathcal{V}} \qquad (17.1.1.2\text{-}14)$$

To make use of Equation (17.1.1.2-14), it remains to derive and equation for $\underline{g}_{P_{Avg}}^{\mathcal{V}}$ in terms of segment parameters. This is achieved by first applying generalized Equation (3.2.2.1-5) with (3.2.2.1-6) to the Equation (3.2.1-5) chain rule for $C_N^{\mathcal{V}}$, at arbitrary time t within the trajectory segment time interval

$$C_N^{\mathcal{V}} = C_{\mathcal{V}_{Start}}^{\mathcal{V}}\, C_L^{\mathcal{V}_{Start}}\, C_N^L = \left[I - \sin\phi\left(\underline{u}_\phi^{\mathcal{V}}\times\right) + (1 - \cos\phi)\left(\underline{u}_\phi^{\mathcal{V}}\times\right)^2\right] C_L^{\mathcal{V}_{Start}}\, C_N^L \qquad (17.1.1.2\text{-}15)$$

in which

$$\underline{u}_\phi^{\mathcal{V}} = \frac{1}{\phi_S}\, \underline{\phi}_S^{\mathcal{V}} \qquad (17.1.1.2\text{-}16)$$

where

ϕ_S = Magnitude of $\underline{\phi}_S^{\mathcal{V}}$.

$\underline{u}_\phi^{\mathcal{V}}$ = Unit vector along $\underline{\phi}_S^{\mathcal{V}}$.

ϕ = Magnitude of the rotation vector associated with $C_{\mathcal{V}_{Start}}^{\mathcal{V}}$ in which $\mathcal{V}$ is interpreted as the $\mathcal{V}$ Frame orientation at some general time within the current trajectory segment. ϕ is generated as a rotation about $\underline{u}_\phi^{\mathcal{V}}$ defined by (17.1.1.2-16).

We then note as in the magnitude of Equation (7.1.1.1-14) that:

$$\phi = \omega_{N\mathcal{V}}\, t \qquad \phi_S = \omega_{N\mathcal{V}}\, T_S \qquad (17.1.1.2\text{-}17)$$

where

$\omega_{N\mathcal{V}}$ = Magnitude of $\underline{\omega}_{N\mathcal{V}}^{\mathcal{V}}$ (assumed constant).

t = Time from the start of the current trajectory segment.

The differential of ϕ in (17.1.1.2-17) gives:

$$dt = \frac{1}{\omega_{N\mathcal{V}}} d\phi \tag{17.1.1.2-18}$$

Using (17.1.1.2-17) - (17.1.1.2-18), the integral of the time varying terms in (17.1.1.2-15) over the trajectory segment are:

$$\int_0^{T_S} \sin\phi \, dt = \frac{1}{\omega_{N\mathcal{V}}} \int_0^{\phi_S} \sin\phi \, d\phi = \frac{1}{\omega_{N\mathcal{V}}} \Big[-\cos\phi \Big]_0^{\phi_S} = \frac{1}{\omega_{N\mathcal{V}}} \left(1 - \cos\phi_S\right)$$
$$= \frac{1}{\phi_S^2} \left(1 - \cos\phi_S\right) \phi_S \, T_S = h_1 \, \phi_S \, T_S \tag{17.1.1.2-19}$$

$$h_1 \equiv \frac{1}{\phi_S^2} \left(1 - \cos\phi_S\right) = \frac{1}{2\,!} - \frac{\phi_S^2}{4\,!} + \frac{\phi_S^4}{6\,!} - \cdots \tag{17.1.1.2-20}$$

$$\int_0^{T_S} (1 - \cos\phi) \, dt = \frac{1}{\omega_{N\mathcal{V}}} \int_0^{\phi_S} (1 - \cos\phi) \, d\phi = \frac{1}{\omega_{N\mathcal{V}}} \Big[\phi - \sin\phi \Big]_0^{\phi_S}$$
$$= \frac{1}{\omega_{N\mathcal{V}}} \left(\phi_S - \sin\phi_S\right) = \left(1 - \frac{\sin\phi_S}{\phi_S}\right) T_S = h_2 \, \phi_S^2 \, T_S \tag{17.1.1.2-21}$$

$$h_2 \equiv \frac{1}{\phi_S^2} \left(1 - \frac{\sin\phi_S}{\phi_S}\right) = \frac{1}{3\,!} - \frac{\phi_S^2}{5\,!} + \frac{\phi_S^4}{7\,!} - \cdots \tag{17.1.1.2-22}$$

where

h_1, h_2 = Defined trigonometric functions of ϕ_S^2.

Finally, we substitute (17.1.1.2-15) into (17.1.1.2-10) with (17.1.1.2-19) and (17.1.1.2-21), and apply (17.1.1.2-16) to obtain the desired expression for $\underline{g}^{\mathcal{V}}_{P_{Avg}}$ as a function of the $\underline{\phi}^{\mathcal{V}}_S$ basic segment parameter:

$$\underline{g}^{\mathcal{V}}_{P_{Avg}} = -\, g \left[I - h_1 \left(\underline{\phi}^{\mathcal{V}}_S \times\right) + h_2 \left(\underline{\phi}^{\mathcal{V}}_S \times\right)^2 \right] C_L^{\mathcal{V}_{Start}} \, C_N^L \, \underline{u}^N_{ZN} \tag{17.1.1.2-23}$$

The h_1, h_2 terms in (17.1.1.2-23) are a function of ϕ_S^2 as calculated with Equations (17.1.1.2-20) and (17.1.1.2-22).

In summary, Equation (17.1.1.2-14) with (17.1.1.2-20), (17.1.1.2-22) and (17.1.1.2-23) defines the average specific force acceleration $\underline{a}_{SF_{Avg}}^{\mathcal{V}}$ as a function of the basic trajectory segment parameters T_S, ΔV_S and $\underline{\phi}_S^{\mathcal{V}}$. Selected inverses of this equation enable particular basic trajectory segment parameters to be calculated that will create specified average specific force acceleration component maneuvers. To best understand the basic procedure involved, let us first write the component form of Equation (17.1.1.2-14):

$$a_{SFAvg_{X\mathcal{V}}} = \frac{1}{T_S} \Delta V_S - g_{PAvg_{X\mathcal{V}}} \tag{17.1.1.2-24}$$

$$a_{SFAvg_{Y\mathcal{V}}} = \frac{1}{T_S} \left(V_{Start} + \frac{1}{2} \Delta V_S \right) \phi_{S_{Z\mathcal{V}}} - g_{PAvg_{Y\mathcal{V}}} \tag{17.1.1.2-25}$$

$$a_{SFAvg_{Z\mathcal{V}}} = - \frac{1}{T_S} \left(V_{Start} + \frac{1}{2} \Delta V_S \right) \phi_{S_{Y\mathcal{V}}} - g_{PAvg_{Z\mathcal{V}}} \tag{17.1.1.2-26}$$

where

$$\phi_{S_{i\mathcal{V}}}, g_{PAvg_{i\mathcal{V}}} = \mathcal{V} \text{ Frame component i of } \underline{\phi}_S^{\mathcal{V}}, \underline{g}_{P_{Avg}}^{\mathcal{V}}.$$

Given previously selected values of ΔV_S and $\underline{\phi}_S^{\mathcal{V}}$, Equation (17.1.1.2-24) can be inverted to find the T_S that achieves a desired average acceleration along $\mathcal{V}$ Frame X axis:

$$T_S = \frac{1}{\left(a_{SFAvg_{X\mathcal{V}}} + g_{PAvg_{X\mathcal{V}}} \right)} \Delta V_S \tag{17.1.1.2-27}$$

Note that $\underline{\phi}_S^{\mathcal{V}}$ is required in the previous expression to calculate $g_{PAvg_{X\mathcal{V}}}$ from (17.1.1.2-23).

Alternatively, given ΔV_S and $\underline{\phi}_S^{\mathcal{V}}$, the T_S segment time can be calculated to achieve a specified average $\mathcal{V}$ Frame Y or Z axis specific force acceleration from (17.1.1.2-25) or (17.1.1.2-26):

$$T_S = \frac{\left(V_{Start} + \frac{1}{2} \Delta V_S \right)}{\left(a_{SFAvg_{Y\mathcal{V}}} + g_{PAvg_{Y\mathcal{V}}} \right)} \phi_{S_{Z\mathcal{V}}} \tag{17.1.1.2-28}$$

or

$$T_S = - \frac{\left(V_{Start} + \frac{1}{2} \Delta V_S \right)}{\left(a_{SFAvg_{Z\mathcal{V}}} + g_{PAvg_{Z\mathcal{V}}} \right)} \phi_{S_{Y\mathcal{V}}} \tag{17.1.1.2-29}$$

As another variation, given $\underline{\phi}_S^{\mathcal{V}}$, we can calculate both T_S and ΔV_S to achieve specified average X and Y axis accelerations by combining/inverting Equations (17.1.1.2-27) and (17.1.1.2-28). The result is:

$$\Delta V_S = \frac{\left(a_{SFAvg_{X\mathcal{V}}} + g_{PAvg_{X\mathcal{V}}}\right) V_{Start}}{\left(a_{SFAvg_{Y\mathcal{V}}} + g_{PAvg_{Y\mathcal{V}}}\right) - \frac{1}{2}\left(a_{SFAvg_{X\mathcal{V}}} + g_{PAvg_{X\mathcal{V}}}\right) \phi_{S_{Z\mathcal{V}}}} \phi_{S_{Z\mathcal{V}}}$$

$$T_S = \frac{V_{Start}}{\left(a_{SFAvg_{Y\mathcal{V}}} + g_{PAvg_{Y\mathcal{V}}}\right) - \frac{1}{2}\left(a_{SFAvg_{X\mathcal{V}}} + g_{PAvg_{X\mathcal{V}}}\right) \phi_{S_{Z\mathcal{V}}}} \phi_{S_{Z\mathcal{V}}} \qquad (17.1.1.2\text{-}30)$$

Combining/inverting Equations (17.1.1.2-27) and (17.1.1.2-29) provides an equivalent result for T_S and ΔV_S to achieve specified average X and Z axis accelerations:

$$\Delta V_S = -\frac{\left(a_{SFAvg_{X\mathcal{V}}} + g_{PAvg_{X\mathcal{V}}}\right) V_{Start}}{\left(a_{SFAvg_{Z\mathcal{V}}} + g_{PAvg_{Z\mathcal{V}}}\right) + \frac{1}{2}\left(a_{SFAvg_{X\mathcal{V}}} + g_{PAvg_{X\mathcal{V}}}\right) \phi_{S_{Y\mathcal{V}}}} \phi_{S_{Y\mathcal{V}}}$$

$$T_S = -\frac{V_{Start}}{\left(a_{SFAvg_{Z\mathcal{V}}} + g_{PAvg_{Z\mathcal{V}}}\right) + \frac{1}{2}\left(a_{SFAvg_{X\mathcal{V}}} + g_{PAvg_{X\mathcal{V}}}\right) \phi_{S_{Y\mathcal{V}}}} \phi_{S_{Y\mathcal{V}}} \qquad (17.1.1.2\text{-}31)$$

Extending the previous approach to include $\underline{\phi}_S^{\mathcal{V}}$ component determination is more involved because of the presence of $\underline{\phi}_S^{\mathcal{V}}$ in the $\underline{g}_{P_{Avg}}^{\mathcal{V}}$ components. In principle, an iterative procedure can be implemented by which any three of the five basic segment parameters (T_S, ΔV_S and the components of $\underline{\phi}_S^{\mathcal{V}}$) can be calculated to generate specified X, Y and Z axis accelerations.

It is important to note that when using the above techniques to calculate basic segment parameters, logic must be incorporated to avoid unrealistic results (e.g., negative T_S or infinity singularities). These can easily be developed based on the form of the equation being applied. For example, if Equations (17.1.1.2-31) are being utilized, we can assure that T_S will remain positive and both T_S and ΔV_S will remain finite if:

$$\left[\left(a_{SFAvg_{Z\mathcal{V}}} + g_{PAvg_{Z\mathcal{V}}}\right) + \frac{1}{2}\left(a_{SFAvg_{X\mathcal{V}}} + g_{PAvg_{X\mathcal{V}}}\right) \phi_{S_{Y\mathcal{V}}}\right] \mathrm{Sign}\left(V_{Start}\, \phi_{S_{Y\mathcal{V}}}\right) < 0 \qquad (17.1.1.2\text{-}32)$$

Equation (17.1.1.2-32) can be used to guide the user in selecting realistic values of X, Z specific force acceleration components for compatibility with the previously selected $\underline{\phi}_S^{\mathcal{V}}$ values. Similar techniques can be used when implementing Equations (17.1.1.2-27) - (17.1.1.2-30).

17.1.1.3 ROTATION VECTOR DEFINED IN THE L FRAME

At times it is advantageous to treat rotation of the $\mathcal{V}$ Frame over a trajectory segment as the analytical equivalent of rotating the locally level L Frame while considering the $\mathcal{V}$ Frame fixed in the rotating L Frame. Then we can define the rotation effect as a rotation vector in the L Frame. An example might be an aircraft maneuver in which $\mathcal{V}$ Frame rotates around the local vertical. Clearly, this maneuver is easily specified as an L Frame rotation vector along the L Frame Z axis. Once the L Frame rotation vector is specified, it must be translated into its equivalent $\mathcal{V}$ Frame form to set the $\underline{\phi}_S^{\mathcal{V}}$ segment parameters. The following development derives the obvious relationship between the L and $\mathcal{V}$ Frame rotation vectors by applying the (3.2.1-5) chain rule, direction-cosine/rotation-vector equivalency Equation (3.2.2.1-8), and cross-product operator transformation Equation (3.2.1-8):

$$
\begin{aligned}
C_{\mathcal{V}_{End}}^{L} &= C_{\mathcal{V}_{Start}}^{L} \, C_{\mathcal{V}_{End}}^{\mathcal{V}_{Start}} = C_{\mathcal{V}_{Start}}^{L} \, C_{\mathcal{V}_{End}}^{\mathcal{V}_{Start}} \, C_{L}^{\mathcal{V}_{Start}} \, C_{\mathcal{V}_{Start}}^{L} \\
&= C_{\mathcal{V}_{Start}}^{L} \left[I + \frac{\sin \phi_S}{\phi_S} \left(\underline{\phi}_S^{\mathcal{V}} \times \right) + \frac{(1 - \cos \phi_S)}{\phi_S^2} \left(\underline{\phi}_S^{\mathcal{V}} \times \right)^2 \right] C_{L}^{\mathcal{V}_{Start}} \, C_{\mathcal{V}_{Start}}^{L} \\
&= \left[I + \frac{\sin \phi_S}{\phi_S} \, C_{\mathcal{V}_{Start}}^{L} \left(\underline{\phi}_S^{\mathcal{V}} \times \right) C_{L}^{\mathcal{V}_{Start}} \right. \\
&\quad \left. + \frac{(1 - \cos \phi_S)}{\phi_S^2} \, C_{\mathcal{V}_{Start}}^{L} \left(\underline{\phi}_S^{\mathcal{V}} \times \right) C_{L}^{\mathcal{V}_{Start}} \, C_{\mathcal{V}_{Start}}^{L} \left(\underline{\phi}_S^{\mathcal{V}} \times \right) C_{L}^{\mathcal{V}_{Start}} \right] C_{\mathcal{V}_{Start}}^{L}
\end{aligned}
\qquad (17.1.1.3\text{-}1)
$$

or

$$
C_{\mathcal{V}_{End}}^{L} = \left[I + \frac{\sin \phi_S}{\phi_S} \left(\underline{\phi}_S^{L} \times \right) + \frac{(1 - \cos \phi_S)}{\phi_S^2} \left(\underline{\phi}_S^{L} \times \right)^2 \right] C_{\mathcal{V}_{Start}}^{L} \qquad (17.1.1.3\text{-}2)
$$

with

$$
\underline{\phi}_S^{L} = C_{\mathcal{V}_{Start}}^{L} \, \underline{\phi}_S^{\mathcal{V}} \qquad (17.1.1.3\text{-}3)
$$

Equations (17.1.1.3-1) - (17.1.1.3-2) show that the identical $C_{\mathcal{V}_{End}}^{L}$ matrix can be created using $\underline{\phi}_S^{\mathcal{V}}$ or the L Frame equivalent $\underline{\phi}_S^{L}$ as defined by (17.1.1.3-3). Moreover, from the form of (17.1.1.3-2) (compared with generalized Equation (3.2.2.1-8) and the (3.2.1-5) chain rule), we see that $\underline{\phi}_S^L$ corresponds to a rotation vector in the L Frame interpreted as rotating the L Frame from the Start to the End attitude, considering the $\mathcal{V}$ Frame fixed. Thus, $\underline{\phi}_S^L$ can be selected based on this interpretation with confidence that its application will produce the equivalent effect on $C_{\mathcal{V}_{End}}^L$ as applying $\underline{\phi}_S^{\mathcal{V}}$. The equivalent $\underline{\phi}_S^{\mathcal{V}}$ corresponding to the selected $\underline{\phi}_S^L$ is then calculated from (17.1.1.3-3) multiplied by $C_L^{\mathcal{V}_{Start}}$:

$$\underline{\phi}_S^{\mathcal{V}} = C_L^{\mathcal{V}_{Start}} \, \underline{\phi}_S^L \qquad (17.1.1.3\text{-}4)$$

17.1.1.4 STARTING HEADING AND SEGMENT TIME TO REACH A SPECIFIED POSITION LOCATION

Consider the case when it is desired to simulate a straight and level constant velocity flight portion to a specified long range location defined in terms of geodetic latitude and longitude. Such a condition can be easily accommodated using a single trajectory segment after first maneuvering to the proper attitude, heading and velocity to start the long range cruise. Section 17.1.1.1 describes how the cruise velocity and a level attitude (i.e., zero roll/pitch) can be created to begin the long range cruise for an arbitrarily assigned heading. Section 17.1.1.5 to follow describes realistic methods for turning to the proper heading from the level attitude flight condition.

For the long range cruise segment, the ΔV_S and $\underline{\phi}_S^{\mathcal{V}}$ segment parameters would be set to zero. Setting ΔV_S to zero assures a constant velocity magnitude along the cruise segment. The rationale for setting $\underline{\phi}_S^{\mathcal{V}}$ to zero is based on the definition for $\underline{\phi}_S^{\mathcal{V}}$ as a rotation angle produced by angular rate relative to the local wander azimuth N Frame, and that the long range cruise trajectory segment will be along a great circle flight path. A great circle between two earth referenced position locations (of the same altitude) is the shortest constant altitude distance path between the two points over the surface of the earth (assuming an approximate spherical earth shape). It is defined as a constant altitude flight path in the earth fixed plane defined by earth's center and the two (start, end) earth referenced position locations. Along the great circle, the angular rate of the local vertical is perpendicular to the earth fixed great circle plane, hence, is horizontal at fixed orientation relative to the earth (and constant in magnitude for the assumed constant velocity). The wander azimuth N Frame is defined to have zero vertical angular rate relative to the earth, with horizontal angular rate equal to the angular rate of the local horizontal relative to the earth (See Section 4.5 for definition of the wander azimuth frame). Along the great circle, therefore, the local wander azimuth N Frame will have its angular rate vector relative to the earth equal to the constant great circle angular rate vector.

Section 3.2.2.1 (Equation (3.2.2.1-6)) shows that when a coordinate frame is rotated from a starting orientation around a fixed axis in the starting frame, that the rotation axis will remain fixed as viewed from the rotating frame. Thus, for N Frame rotation along the great circle, the direction of the great circle rotation rate vector (i.e., the rotation axis) will remain constant in the N Frame, hence, the horizontal N Frame X, Y axes will remain at fixed angular orientation relative to the great circle plane. We also know from its definition that the $\mathcal{V}$ Frame remains at constant attitude relative to the N Frame if $\underline{\phi}_S^{\mathcal{V}}$ is zero. Therefore, if the $\mathcal{V}$ Frame is initially oriented to be horizontal (Z axis down) with its X axis in the great circle plane pointing toward the desired destination, this attitude will be maintained along the great circle trajectory segment if $\underline{\phi}_S^{\mathcal{V}}$ is set to zero. Because the velocity direction is along the $\mathcal{V}$ Frame X axis, we will thereby achieve the objective of creating a horizontal velocity vector toward the desired destination along the great circle flight path.

The desired starting heading along the great circle trajectory segment is calculated from vertical unit vectors at the segment start and end position locations, projected on the start of segment N Frame. The start of segment vertical projected on the start of segment N Frame is simply $\underline{u}_{ZN}^{N}$. The Figure 17.1.1.4-1 Method of Least Work diagram (derived from Figure 4.4.2.1-2) with the Section 2.2 definitions for the E and Geo Frames, is useful for defining the end of segment vertical:

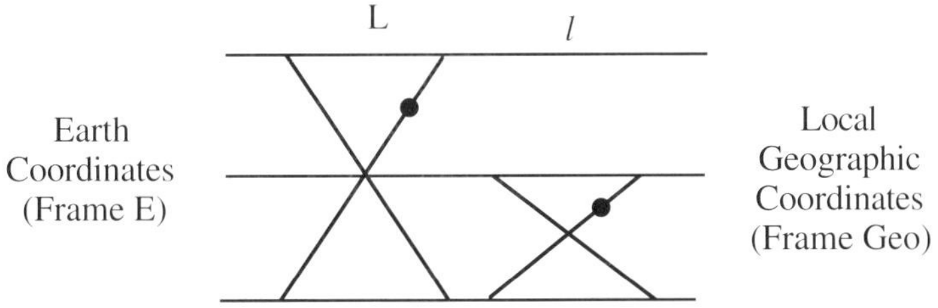

Figure 17.1.1.4-1 Latitude/Longitude Definition

Recall from Section 2.2 that the Z axis of the N and Geo Frames are parallel (i.e., along the upward local vertical). Then using Figure 17.1.1.4-1, we can write the E Frame components for the end of segment position local vertical unit vector $\underline{u}_{ZN}$ (i.e., along the Geo Frame Z axis), and transform the result to the N Frame at the start of the segment:

$$\underline{u}_{ZN_{End}}^{N_{Start}} = \left(C_{N_{Start}}^{E}\right)^T \underline{u}_{ZN_{End}}^{E} = \left(C_{N_{Start}}^{E}\right)^T \underline{u}_{ZGeo_{End}}^{E}$$

$$= \left(C_{N_{Start}}^{E}\right)^T C_{Geo_{End}}^{E} \, \underline{u}_{ZGeo_{End}}^{Geo_{End}} = \left(C_{N_{Start}}^{E}\right)^T \begin{bmatrix} \cos l_{End} \sin L_{End} \\ \sin l_{End} \\ \cos l_{End} \cos L_{End} \end{bmatrix} \qquad (17.1.1.4\text{-}1)$$

where

$\underline{u}_{ZN_{End}}^{N_{Start}}$ = Unit vector along the N Frame Z axis at the end of the trajectory segment, as projected onto the N Frame axes at the start of the trajectory segment.

$\underline{u}_{ZGeo_{End}}$ = Unit vector along the end of segment Geo Frame Z axis (upward).

The cross-product of $\underline{u}_{ZN}^{N}$ with $\underline{u}_{ZN_{End}}^{N_{Start}}$ defines a horizontal vector in the starting location N Frame, that is perpendicular to the great circle plane, and which lies along the angular rate vector for local vertical rotation along the great circle. Then the cross-product of the previous cross-product vector with $\underline{u}_{ZN}^{N}$, describes a vector that lies along the great circle flight path at the start of the trajectory segment:

$$\underline{w}_{GC}^{L_{Start}} = C_N^L \left[\left(\underline{u}_{ZN}^{N} \times \underline{u}_{ZN_{End}}^{N_{Start}} \right) \times \underline{u}_{ZN}^{N} \right] \qquad (17.1.1.4\text{-}2)$$

where

$\underline{w}_{GC}^{L_{Start}}$ = Vector along the great circle flight path at the start of the great circle trajectory segment, projected on the L Frame at the start of the segment.

Given that $\underline{w}_{GC}^{L_{Start}}$ is horizontal (as defined and calculated), the "platform heading" of $\underline{w}_{GC}^{L_{Start}}$ is the angle from the L Frame X axis to $\underline{w}_{GC}^{L_{Start}}$, measured positive for $\underline{w}_{GC}^{L_{Start}}$ having a positive Y component in the L Frame:

$$\psi_{GC/Start_P} = \tan^{-1}\left(\frac{w_{GC_{YLStrt}}}{w_{GC_{XLStrt}}} \right) \qquad (17.1.1.4\text{-}3)$$

where

$\psi_{GC/Start_P}$ = Platform heading of the great circle flight path in the L Frame at the start of the great circle trajectory segment (See Section 4.1.2 for definition).

$w_{GC_{XLStrt}}, w_{GC_{YLStrt}}$ = X, Y components of $\underline{w}_{GC}^{L_{Start}}$.

The equivalent starting great circle true heading is then calculated from (4.1.2-2):

$$\psi_{GC/Start_{True}} = \psi_{GC/Start_P} - \alpha_{Start} \qquad (17.1.1.4\text{-}4)$$

where

$\psi_{GC/Start_{True}}$ = True heading of the great circle flight path at the start of the great circle trajectory segment.

α_{Start} = Wander angle of the N Frame at the start of the great circle trajectory segment (calculated from $C^{E}_{N_{Start}}$ using Equations (4.4.2.1-3)).

To generate a great circle during the great circle trajectory segment, the true heading of $\mathcal{V}$ Frame axis X must be shaped during the previous segment to match $\psi_{GC/Start_{True}}$, and $\underline{\phi}^{\mathcal{V}}_{S}$ must be zero for the great circle segment. The trajectory segment time interval required to move from the start to end position locations on the great circle equals the total angular movement of the local vertical along the great circle (the "range angle") divided by the velocity. We define the range angle to lie in the interval from 0 to 180 degrees, hence, the sine of the range angle is positive. This definition assures that the great circle distance between the selected end points will be the shortest (as opposed to a great circle between the same end points, but starting in the opposite direction). Then, the sine of the range angle is the magnitude of $\underline{u}^{N}_{ZN} \times \underline{u}^{N_{Start}}_{ZN_{End}}$. Since $\underline{u}^{N}_{ZN}$ is a unit vector perpendicular to $\underline{u}^{N}_{ZN} \times \underline{u}^{N_{Start}}_{ZN_{End}}$, the magnitude of $\underline{u}^{N}_{ZN} \times \underline{u}^{N_{Start}}_{ZN_{End}}$ also equals the magnitude of $\left(\underline{u}^{N}_{ZN} \times \underline{u}^{N_{Start}}_{ZN_{End}}\right) \times \underline{u}^{N}_{ZN}$, which from Equation (17.1.1.4-2) is the magnitude of $\underline{w}^{L_{Start}}_{GC}$. Because the magnitude of a vector is the same in any coordinate frame, we therefore can calculate the sine of the range angle as the magnitude $\underline{w}^{L_{Start}}_{GC}$. The cosine of the range angle is simply the dot product between $\underline{u}^{N}_{ZN}$ and $\underline{u}^{N_{Start}}_{ZN_{End}}$. Thus, the range angle can be computed as:

$$\theta_{GC/Range} = \tan^{-1} \frac{\left|\underline{w}^{L_{Start}}_{GC}\right|}{\underline{u}^{N}_{ZN} \cdot \underline{u}^{N_{Start}}_{ZN_{End}}} \qquad (17.1.1.4\text{-}5)$$

where

$\theta_{GC/Range}$ = Great circle trajectory segment range angle.

The segment time interval to cover $\theta_{GC/Range}$ is given by:

$$T_{S/GC} = \frac{\theta_{GC/Range}}{V_{GC}} \qquad (17.1.1.4\text{-}6)$$

where

$T_{S/GC}$ = Great circle trajectory segment time interval.

V_{GC} = Velocity along great circle (established at the end of the previous trajectory segment).

17.1.1.5 TURN TO A SPECIFIED HEADING

In this section we discuss the particular problem of realistically turning from an initial level attitude condition (i.e., from $\mathcal{V}$ Frame Z axis down which corresponds to zero pitch/roll) to a specified heading at level attitude. At first we might try a simple single trajectory segment maneuver for which the X, Y components of $\underline{\phi}_S^{\mathcal{V}}$ are zero, and the Z component of $\underline{\phi}_S^{\mathcal{V}}$ is set equal to the difference between the specified heading and the heading at the end of the last trajectory segment. We would find, however, from Equation (17.1.1.2-25), that such a maneuver requires a corresponding lateral acceleration (i.e., along the $\mathcal{V}$ Frame Y axis). For most manned aircraft, such a maneuver is not comfortable (for the pilot and passengers), and would only be used for small turn angles (over long segment times) to keep the lateral acceleration low. The more common method for executing a heading change is through what is known as a "coordinated turn" in which the aircraft is held at a constant bank angle with zero $\mathcal{V}$ Frame pitch angle, while the aircraft is turned in heading at an angular rate that maintains zero $\mathcal{V}$ Frame Y axis acceleration. The total maneuver required to execute the heading change (e.g., a right turn) would then consist of three trajectory segments: 1. Roll right, 2. Turn right while holding the roll angle, 3. Roll back to wings level at the desired heading.

Let us first address how the roll angle and heading rate can be selected for the turning portion (at constant velocity). During the turn at constant velocity, the following simplified form of (17.1.1.2-8) applies:

$$\underline{a}_{SF}^{\mathcal{V}} = V_{Start}\left(\underline{\omega}_{N\mathcal{V}}^{\mathcal{V}} \times \underline{u}_{X\mathcal{V}}^{\mathcal{V}}\right) + g\,\underline{u}_{ZN}^{\mathcal{V}} \qquad (17.1.1.5\text{-}1)$$

For a constant turning rate about the vertical, $\underline{\omega}_{N\mathcal{V}}^{\mathcal{V}}$ is easily expressed as a function of the platform heading rate around an L Frame vertical:

$$\underline{\omega}_{N\mathcal{V}}^{\mathcal{V}} = \dot{\psi}\,\underline{u}_{ZL}^{\mathcal{V}} \qquad (17.1.1.5\text{-}2)$$

where

$\underline{u}_{ZL}^{\mathcal{V}}$ = Unit vector along the L Frame Z axis (i.e., locally down) as projected on $\mathcal{V}$ Frame axes.

with the platform heading Euler angle ψ as defined in Section 3.2.3. Since the Z axes of the N and L Frames are parallel but opposite in direction (See Section 2.2), we can also write:

$$\underline{u}_{ZN}^{\mathcal{V}} = -\,\underline{u}_{ZL}^{\mathcal{V}} \qquad (17.1.1.5\text{-}3)$$

Substituting (17.1.1.5-2) - (17.1.1.5-3) into (17.1.1.5-1) then gives:

$$\underline{a}_{SF}^{\mathcal{V}} = V_{Start}\,\dot{\psi}\left(\underline{u}_{ZL}^{\mathcal{V}} \times \underline{u}_{X\mathcal{V}}^{\mathcal{V}}\right) - g\,\underline{u}_{ZL}^{\mathcal{V}} \tag{17.1.1.5-4}$$

The dot product of (17.1.1.5-4) with $\underline{u}_{ZL}^{\mathcal{V}}$ and $\left(\underline{u}_{ZL}^{\mathcal{V}} \times \underline{u}_{X\mathcal{V}}^{\mathcal{V}}\right)$ yields:

$$\underline{u}_{ZL}^{\mathcal{V}} \cdot \underline{a}_{SF}^{\mathcal{V}} = -g \qquad \left(\underline{u}_{ZL}^{\mathcal{V}} \times \underline{u}_{X\mathcal{V}}^{\mathcal{V}}\right) \cdot \underline{a}_{SF}^{\mathcal{V}} = V_{Start}\,\dot{\psi} \tag{17.1.1.5-5}$$

Generalized Equation (3.2.1-6) shows that $\underline{u}_{ZL}^{\mathcal{V}}$ is the third column of $C_{L}^{\mathcal{V}}$, hence, its elements correspond to the third row of $C_{\mathcal{V}}^{L}$. Using (3.2.3.1-2) for the general form of $C_{\mathcal{V}}^{L}$, we see then from the third row that during the turn (for which the pitch angle $\theta = 0$):

$$\underline{u}_{ZL}^{\mathcal{V}} = \begin{bmatrix} 0 \\ \sin\phi \\ \cos\phi \end{bmatrix} \tag{17.1.1.5-6}$$

where

ϕ = Roll (or bank) angle during the turn.

Substituting (17.1.1.5-6) into (17.1.1.5-5) then obtains:

$$\begin{aligned} a_{SFY\mathcal{V}}\sin\phi + a_{SFZ\mathcal{V}}\cos\phi &= -g \\ a_{SFY\mathcal{V}}\cos\phi - a_{SFZ\mathcal{V}}\sin\phi &= V_{Start}\,\dot{\psi} \end{aligned} \tag{17.1.1.5-7}$$

where

$a_{SFi\mathcal{V}}$ = Component of $\underline{a}_{SF}^{\mathcal{V}}$ along $\mathcal{V}$ Frame axis i.

For a coordinated turn, $a_{SFY\mathcal{V}} = 0$. Then (17.1.1.5-7) can be solved for the sine and cosine of ϕ in terms of a specified starting velocity, heading rate, and $\mathcal{V}$ Frame Z axis specific force acceleration during the turn (i.e., the negative of a specified lift acceleration):

$$\sin\phi = -\frac{V_{Start}\,\dot{\psi}}{a_{SFZ\mathcal{V}}} \qquad \cos\phi = -\frac{g}{a_{SFZ\mathcal{V}}} \tag{17.1.1.5-8}$$

For a realistic turn, the magnitude of ϕ will be less than 90 degrees. Then $\cos\phi$ will always be positive, hence, $a_{SFZ\mathcal{V}}$ will always be negative (and never smaller than g), and ϕ will always have the same sign as $\dot{\psi}$ (assuming a positive velocity V_{Start}). Under these conditions, we can rearrange and combine Equations (17.1.1.5-8) into the equivalent relationships:

$$\phi = \tan^{-1}\frac{\sin\phi}{\cos\phi} = \tan^{-1}\frac{V_{Start}\,\dot{\psi}}{g} \qquad a_{SF_{Z\mathcal{V}}} = -\frac{g}{\cos\phi} \tag{17.1.1.5-9}$$

Equations (17.1.1.5-9) show how ϕ can be calculated to achieve a specified turning rate $\dot{\psi}$, and the $a_{SF_{Z\mathcal{V}}}$ that would be generated as a result. Alternatively, (17.1.1.5-8) can be rearranged and combined as follows:

$$\phi = \mathrm{Sign}(\dot{\psi})\sec^{-1}\frac{a_{SF_{Z\mathcal{V}}}}{g} \qquad \dot{\psi} = -\frac{a_{SF_{Z\mathcal{V}}}}{V_{Start}}\sin\phi \tag{17.1.1.5-10}$$

Equations (17.1.1.5-10) show how ϕ and $\dot{\psi}$ can be calculated for a specified normal acceleration $a_{SF_{Z\mathcal{V}}}$ and turn rate direction. Note that for very shallow bank angles, $a_{SF_{Z\mathcal{V}}}$ is very close to g (See Equation (17.1.1.5-9)), hence, the inverse arc secant calculation in (17.1.1.5-10) tends to lose accuracy (becomes zero for very small values of ϕ).

After the bank angle ϕ is calculated (from (17.1.1.5-9) or (17.1.1.5-10)), the basic trajectory parameters are shaped for a single trajectory segment to generate ϕ. The ϕ angle so generated will then be the proper initial condition for the $\dot{\psi}$ turning rate to be applied during the next trajectory segment. A common error made in generating ϕ is to set the X component of $\underline{\phi}_S^{\mathcal{V}}$ equal to ϕ with the Y, Z components set to zero. This will indeed produce the desired ϕ bank angle, however, it will also generate an average $\mathcal{V}$ Frame Y axis specific force acceleration in the process from $g_{PAvg_{Y\mathcal{V}}}$ in Equation (17.1.1.2-25). (For an X axis $\underline{\phi}_S^{\mathcal{V}}$ rotation substituted in Equation (17.1.1.2-23), it is readily demonstrated that $g_{PAvg_{Y\mathcal{V}}}$ is non-zero.) To generate a bank angle using a coordinated maneuver (i.e., with zero Y axis lateral specific force), Equation (17.1.1.2-25) shows that $\underline{\phi}_S^{\mathcal{V}}$ must also contain a $\phi_{S_{Z\mathcal{V}}}$ component which when multiplied by $\frac{1}{T_S}\left(V_{Start} + \frac{1}{2}\,\Delta V_S\right)$, balances $g_{PAvg_{Y\mathcal{V}}}$. The following procedure can be used to achieve the desired result of generating the desired bank angle $\phi_{S_{Z\mathcal{V}}}$ with zero average Y axis $\mathcal{V}$ Frame specific force during the maneuver.

We first assume that a small heading change will occur during the roll-to-bank angle ϕ maneuver where:

$\Delta\psi_{Roll}$ = Small platform heading change during the roll-to-bank maneuver.

We then arbitrarily select a reasonable value for $\Delta\psi_{Roll}$ with the same sign as ϕ, and calculate the corresponding $\underline{\phi}_S^{\mathcal{V}}$ rotation vector that will create ϕ and $\Delta\psi_{Roll}$ such that the pitch angle is zero at the end of the maneuver. The corresponding platform heading at completion of roll-to-bank will be:

$$\psi_{End} = \psi_{Start} + \Delta\psi_{Roll} \qquad (17.1.1.5\text{-}11)$$

The Euler angle attitude at completion of the roll-to-bank maneuver should, therefore, correspond to a platform heading of ψ_{End}, a roll angle of ϕ, and zero pitch angle (θ). The corresponding $\underline{\phi}_S^{\mathcal{V}}$ rotation vector is calculated using the procedure of Section 17.1.1.1 to compute $C_{\mathcal{V}_{End}}^{L}$ from ψ_{End}, ϕ, θ, then $C_{\mathcal{V}_{End}}^{\mathcal{V}_{Start}}$ from Equation (17.1.1.1-1), and finally $\underline{\phi}_S^{\mathcal{V}}$ from $C_{\mathcal{V}_{End}}^{\mathcal{V}_{Start}}$ (using the direction cosine to rotation vector conversion formula).

Once $\underline{\phi}_S^{\mathcal{V}}$ is determined, we then find the T_S value for zero $\mathcal{V}$ Frame Y axis specific force using Equation (17.1.1.2-28) with $a_{SFAvgY_{\mathcal{V}}}$ set to zero (and zero for ΔV_S based on no velocity change during the roll-to-bank maneuver). The T_S so calculated should then be checked for reasonableness (e.g., by verifying that the associated roll rate is reasonable as calculated from the X axis component of $\underline{\phi}_S^{\mathcal{V}}$ divided by T_S). If the T_S value is not deemed reasonable (e.g., should be larger by a certain factor), the $\Delta\psi_{Roll}$ value is adjusted (e.g., by the "certain factor") and the process repeated until T_S satisfies the reasonableness criteria.

The previous process lends itself nicely to implementation in a recursive computer algorithm structure. When the process converges, the result will be $\underline{\phi}_S^{\mathcal{V}}$ and T_S values that achieve the desired roll bank angle condition ϕ for a reasonable T_S value, and for which the average $\mathcal{V}$ Frame Y axis specific force is zero during the maneuver. These are the desired conditions to begin the coordinated turn to the desired heading.

The next two trajectory segments are designed to execute the coordinated turn and then return to a "wings level" attitude at the desired heading. The final segment is a roll-to-level maneuver from the ϕ bank angle, which is the inverse of the roll-to-bank angle maneuver discussed in the previous paragraph. The T_S value for this maneuver is identical to the value for the previously discussed roll-to-bank maneuver; the $\underline{\phi}_S^{\mathcal{V}}$ value has the same Y and Z component values but the negative X component value. Applying these $\underline{\phi}_S^{\mathcal{V}}$, T_S values will produce a heading change of $\Delta\psi_{Roll}$ (as in Equation (17.1.1.5-11)) while the roll angle ϕ is being reduced to zero with zero average Y axis specific force, and generating zero pitch angle at segment end. Thus, the total heading change for the initial roll-to-bank and final bank-to-zero-roll maneuvers is $2\ \Delta\psi_{Roll}$.

To achieve a specified heading at the end of the combined maneuver, the heading change during the turning portion of the trajectory segment must, therefore, be:

$$\Delta\psi_{Turn} = \psi_{Desired} - \psi_{Start} - 2\ \Delta\psi_{Roll} \qquad (17.1.1.5\text{-}12)$$

where

$\psi_{Desired}$ = Desired platform heading.

ψ_{Start} = Platform heading at the start of the overall turning maneuver.

$\Delta\psi_{Turn}$ = Platform heading change during the coordinated turn trajectory segment.

The $\underline{\phi}_S^{\mathcal{V}}$ value that will generate $\Delta\psi_{Turn}$ during the turn is easily calculated as in Section 17.1.1.3 by defining the segment rotation vector in the L Frame. To achieve $\Delta\psi_{Turn}$ heading change during the turn (positive around a downward vertical) we set the Z component of $\underline{\phi}_S^{L}$ to $\Delta\psi_{Turn}$, and the X, Y components to zero. The corresponding $\underline{\phi}_S^{\mathcal{V}}$ value is then obtained with Equation (17.1.1.3-4). The segment time T_S to achieve $\Delta\psi_{Turn}$ is simply $\Delta\psi_{Turn}$ divided by the specified turning rate $\dot{\psi}$. During the turning segment (as for the roll-to-bank and bank-to-zero roll segments), ΔV_S would be set to zero corresponding to constant velocity.

17.1.2 QUICK-LOOK PROJECTION

Before a selected set of "trial" basic trajectory parameters ($\underline{\phi}_S^{\mathcal{V}}$, ΔV_S, T_S) can be finalized for a particular trajectory segment, their impact on the trajectory navigation parameters must be assessed. This is achieved through an approximate closed-form analytical estimate ("Quick-Look" projection) of the navigation parameters based on application of the "trial" parameters. The navigation parameters typically of interest are the average $\mathcal{V}$ Frame angular rate and specific force acceleration components over the trajectory segment, the $\mathcal{V}$ Frame specific force acceleration components at the beginning and end of the segment, and the velocity, position and $\mathcal{V}$ Frame attitude at the end of the segment. If, during trajectory regeneration, the simulation program includes angular variations of vehicle coordinate axes (the AC Frame) relative to $\mathcal{V}$ Frame axes (i.e., angle of attack/sideslip effects), then the angle of attack/sideslip, and associated AC Frame attitude, specific force acceleration and angular rate components are also of interest during trajectory segment shaping operations.

17.1.2.1 PROJECTED VELOCITY, ATTITUDE, TIME, ANGULAR RATE, AND SPECIFIC FORCE ACCELERATION

The attitude and signed velocity magnitude (V) at the end of the segment can be calculated from a rearranged form of Equations (17.1.1.1-1) - (17.1.1.1-2):

$$C^{L}_{\mathcal{V}_{End}} = \left(C^{L}_{\mathcal{V}_{Start}}\right) C^{\mathcal{V}_{Start}}_{\mathcal{V}_{End}} \qquad V_{End} = V_{Start} + \Delta V_S \tag{17.1.2.1-1}$$

The initial value for V in (17.1.2.1-1) at the beginning of the trajectory would be an input specification. The $C^{\mathcal{V}_{Start}}_{\mathcal{V}_{End}}$ matrix in (17.1.2.1-1) is calculated using generalized Equation (3.2.2.1-8) with $\underline{\phi}^{\mathcal{V}}_{S}$ for the rotation vector. The initial value for $C^{L}_{\mathcal{V}}$ at the beginning of the trajectory profile would be calculated using Equation (3.2.3.1-2) from input roll, pitch, heading (assuming zero wander angle). Zero wander angle is achieved by appropriate initial setting of the C^{E}_{N} matrix using Equations (4.4.2.1-2) (with latitude/longitude set at input values). The end-of-segment N Frame velocity vector components can be calculated from V_{End} and $C^{L}_{\mathcal{V}_{End}}$ using (17.1.1.2-5) transformed to the N Frame:

$$\underline{v}^{N}_{End} = V_{End}\, C^{N}_{L}\, C^{L}_{\mathcal{V}_{End}}\, \underline{u}^{\mathcal{V}}_{X\mathcal{V}} \tag{17.1.2.1-2}$$

Once $C^{L}_{\mathcal{V}_{End}}$ is calculated from (17.1.2.1-1), the associated end-of-segment roll, pitch, platform heading attitude Euler angles can be extracted using Equations (4.1.2-1). End-of-segment true heading can be determined from platform heading using (4.1.2-2) with the wander angle calculated from the end-of-segment C^{E}_{N} matrix using Equations (4.4.2.1-3). Similarly, the end-of-segment north/east/up velocity components can be calculated from $\underline{v}^{N}_{End}$ using (4.3.1-4) and the wander angle.

The $\mathcal{V}$ Frame angular rate and average specific force acceleration components over the trajectory segment are easily calculated from Equations (17.1.1.2-13) - (17.1.1.2-14) using (17.1.1.2-23) for $\underline{g}^{\mathcal{V}}_{P_{Avg}}$. The $\mathcal{V}$ Frame specific force acceleration components at the start and end of the trajectory segment can be computed with Equations (17.1.1.2-8) and (17.1.1.2-12) - (17.1.1.2-13) using $C^{\mathcal{V}}_{N} = \left(C^{L}_{\mathcal{V}_{Start}}\right)^{T} C^{L}_{N}$ and $t = 0$ for the start-of-segment components, and $C^{\mathcal{V}}_{N} = \left(C^{L}_{\mathcal{V}_{End}}\right)^{T} C^{L}_{N}$ with $t = T_S$ for the end-of-segment components:

$$\underline{a}_{SF_{Start}}^{\mathcal{V}} = \frac{1}{T_S}\left[\Delta V_S\, \underline{u}_{X\mathcal{V}}^{\mathcal{V}} + V_{Start}\left(\underline{\phi}_S^{\mathcal{V}} \times \underline{u}_{X\mathcal{V}}^{\mathcal{V}}\right) + g\, C_L^{\mathcal{V}_{Start}}\, C_N^L\, \underline{u}_{ZN}^N\right]$$
$$\underline{a}_{SF_{End}}^{\mathcal{V}} = \frac{1}{T_S}\left[\Delta V_S\, \underline{u}_{X\mathcal{V}}^{\mathcal{V}} + \left(V_{Start} + \Delta V_S\right)\left(\underline{\phi}_S^{\mathcal{V}} \times \underline{u}_{X\mathcal{V}}^{\mathcal{V}}\right) + g\, C_L^{\mathcal{V}_{End}}\, C_N^L\, \underline{u}_{ZN}^N\right] \qquad (17.1.2.1\text{-}3)$$

The $C_{\mathcal{V}_{End}}^L$ matrix in (17.1.2.1-3) is as calculated in (17.1.2.1-1), and $C_{\mathcal{V}_{Start}}^L$ is the $C_{\mathcal{V}}^L$ value at the end of the previous trajectory segment.

Trivially, the end-of-segment time is calculated as simply:

$$t_{End} = t_{Start} + T_S \qquad (17.1.2.1\text{-}4)$$

17.1.2.2 END-OF-SEGMENT POSITION QUICK-LOOK PROJECTION

The end-of-segment position, in the form of the C_N^E matrix and altitude h, is estimated for quick-look projection using a version of Equations (7.3.1-1), (7.3.1-3), (7.3.1-4), (7.3.1-6), (7.3.1-8) and (7.3.1-9) in which $\underline{\rho}^N$ (or $\underline{\omega}_{EN}^N$ by definition) is calculated from the $\underline{\omega}_{EN}^N$ expression in Equations (12.1.2-6) with ρ_{ZN} set to zero (for the wander azimuth N Frame - See Section 4.5), ∂G_C^N is approximated as $\partial G_{C_{Start}}^N$, r_l is approximated as $r_{l_{Start}}$, and other related calculations from (12.1.2-6) apply:

$$\Delta \underline{R}^N \equiv \int_0^{T_S} \underline{v}^N \, dt \qquad (17.1.2.2\text{-}1)$$

$$h_{End} = h_{Start} + \underline{u}_{ZN}^N \cdot \Delta\underline{R}^N$$
$$C_{N_{End}}^E = C_{N_{Start}}^E\, C_{N_{End}}^{N_{Start}}$$
$$C_{N_{End}}^{N_{Start}} = I + \frac{\sin\xi}{\xi}\left(\underline{\xi}^N \times\right) + \frac{(1 - \cos\xi)}{\xi^2}\left(\underline{\xi}^N \times\right)^2 \qquad (17.1.2.2\text{-}2)$$
$$\underline{\xi}^N \approx \frac{1}{r_{l_{Start}}}\left(I + \partial G_{C_{Start}}^N\right)\left(\underline{u}_{ZN}^N \times \Delta\underline{R}^N\right)$$

(Continued)

$$\underline{u}_{ZN}^{E} = \begin{bmatrix} u_{ZN_{XE}} \\ u_{ZN_{YE}} \\ u_{ZN_{ZE}} \end{bmatrix} = C_N^E \, \underline{u}_{ZN}^{N} = \begin{bmatrix} D_{13} \\ D_{23} \\ D_{33} \end{bmatrix}$$

$$R_{S_{Start}} \approx \left(1 - u_{ZN_{YE/Start}}^{2} \, e\right) R_0$$

$$r_{ls_{Start}} = \left[1 + 2\left(2\, u_{ZN_{YE/Start}}^{2} - 1\right) e\right] R_{S_{Start}} \qquad \text{(17.1.2.2-2) (Continued)}$$

$$r_{l_{Start}} = r_{ls_{Start}} + h_{Start}$$

$$\partial G_{C_{Start}}^{N} = - \frac{2\,e}{\left(1 + h_{Start} / R_S^{'}\right)} \begin{bmatrix} D_{21}^2 & D_{21}\, D_{22} & 0 \\ D_{21}\, D_{22} & D_{22}^2 & 0 \\ 0 & 0 & 0 \end{bmatrix}_{Start}$$

Once $C_{N_{End}}^{E}$ is calculated in (17.1.2.2-2), the end-of-segment latitude, longitude and wander angle can be extracted using Equations (4.4.2.1-3).

A closed-form expression for the $\Delta \underline{R}^{N}$ integrated N Frame velocity in (17.1.2.2-1) can be obtained from the double integral of $\dot{\underline{v}}^{N}$ as defined by Equation (17.1.1.2-3) in the N Frame:

$$\dot{\underline{v}}^{N} = C_{\mathcal{V}}^{N} \left(\dot{\underline{v}}^{\mathcal{V}} + \underline{\omega}_{N\mathcal{V}}^{\mathcal{V}} \times \underline{v}^{\mathcal{V}}\right) \qquad \text{(17.1.2.2-3)}$$

Using (17.1.1.2-5) - (17.1.1.2-7), the bracketed term in (17.1.2.2-3) is:

$$\begin{aligned} \dot{\underline{v}}^{\mathcal{V}} + \underline{\omega}_{N\mathcal{V}}^{\mathcal{V}} \times \underline{v}^{\mathcal{V}} &= \dot{V}\, \underline{u}_{X\mathcal{V}}^{\mathcal{V}} + \left(V_{Start} + \dot{V}\, t\right)\left(\underline{\omega}_{N\mathcal{V}}^{\mathcal{V}} \times \underline{u}_{X\mathcal{V}}^{\mathcal{V}}\right) \\ &= \dot{V}\, \underline{u}_{X\mathcal{V}}^{\mathcal{V}} + V_{Start}\left(\underline{\omega}_{N\mathcal{V}}^{\mathcal{V}} \times \underline{u}_{X\mathcal{V}}^{\mathcal{V}}\right) + \dot{V}\left(\underline{\omega}_{N\mathcal{V}}^{\mathcal{V}} \times \underline{u}_{X\mathcal{V}}^{\mathcal{V}}\right) t \end{aligned} \qquad \text{(17.1.2.2-4)}$$

The $C_{\mathcal{V}}^{N}$ matrix in (17.1.2.2-3) is the transpose of (17.1.1.2-15):

$$C_{\mathcal{V}}^{N} = C_L^N\, C_{\mathcal{V}_{Start}}^{L} \left[I + \sin\phi \left(\underline{u}_{\phi}^{\mathcal{V}} \times\right) + (1 - \cos\phi) \left(\underline{u}_{\phi}^{\mathcal{V}} \times\right)^2 \right] \qquad \text{(17.1.2.2-5)}$$

Substituting (17.1.2.2-4) - (17.1.2.2-5) in (17.1.2.2-3) yields:

$$\begin{aligned} \dot{\underline{v}}^{N} = C_L^N\, C_{\mathcal{V}_{Start}}^{L} \Big\{ & \left[I + \sin\phi \left(\underline{u}_{\phi}^{\mathcal{V}} \times\right) + (1 - \cos\phi) \left(\underline{u}_{\phi}^{\mathcal{V}} \times\right)^2 \right] \left[\dot{V}\, \underline{u}_{X\mathcal{V}}^{\mathcal{V}} \right. \\ & \left. + V_{Start}\left(\underline{\omega}_{N\mathcal{V}}^{\mathcal{V}} \times \underline{u}_{X\mathcal{V}}^{\mathcal{V}}\right) \right] + \left[I\, t + t \sin\phi \left(\underline{u}_{\phi}^{\mathcal{V}} \times\right) + (1 - \cos\phi)\, t \left(\underline{u}_{\phi}^{\mathcal{V}} \times\right)^2 \right] \left[\dot{V}\left(\underline{\omega}_{N\mathcal{V}}^{\mathcal{V}} \times \underline{u}_{X\mathcal{V}}^{\mathcal{V}}\right) \right] \Big\} \end{aligned} \qquad \text{(17.1.2.2-6)}$$

The position change $\Delta \underline{R}^N$ is the double integral of (17.1.2.2-6) over the trajectory time segment T_S including initial conditions on velocity $\underline{v}^N$:

$$\Delta \underline{R}^N = \underline{v}^N_{Start}\, T_S + \int_0^{T_S} \int_0^{\tau} \dot{\underline{v}}^N \, dt\, d\tau \qquad (17.1.2.2\text{-}7)$$

where

$\underline{v}^N_{Start}$ = $\underline{v}^N$ at the start of the trajectory segment (which is $\underline{v}^N$ at the end of the previous segment).

τ = Dummy time integration parameter.

As in Section 17.1.1.2, we apply Equations (17.1.1.2-17) - (17.1.1.2-18) for double integration of the time varying terms to obtain:

$$\int_0^{T_S} \int_0^{\tau} \sin\phi \, dt\, d\tau = \frac{1}{\omega_{N\mathcal{V}}^2} \int_0^{\phi_S} \int_0^{\theta} \sin\phi \, d\phi\, d\theta = \frac{1}{\omega_{N\mathcal{V}}^2} \int_0^{\phi_S} \left(1 - \cos\theta\right) d\theta$$
$$= \frac{1}{\omega_{N\mathcal{V}}^2} \left[\theta - \sin\theta\right]_0^{\phi_S} = \frac{1}{\omega_{N\mathcal{V}}^2} \left(\phi_S - \sin\phi_S\right) = \frac{1}{\phi_S^2}\left(1 - \frac{\sin\phi_S}{\phi_S}\right) \phi_S\, T_S^2 = h_2\, \phi_S\, T_S^2 \qquad (17.1.2.2\text{-}8)$$

$$\int_0^{T_S} \int_0^{\tau} (1 - \cos\phi)\, dt\, d\tau = \frac{1}{\omega_{N\mathcal{V}}^2} \int_0^{\phi_S} \int_0^{\theta} (1 - \cos\phi)\, d\phi\, d\theta = \frac{1}{\omega_{N\mathcal{V}}^2} \int_0^{\phi_S} \left(\theta - \sin\theta\right) d\theta$$
$$= \frac{1}{\omega_{N\mathcal{V}}^2} \left[\frac{1}{2}\theta^2 + \cos\theta\right]_0^{\phi_S} = \frac{1}{\omega_{N\mathcal{V}}^2} \left(\frac{1}{2}\phi_S^2 - 1 + \cos\phi_S\right) \qquad (17.1.2.2\text{-}9)$$
$$= \frac{1}{\phi_S^4}\left(\frac{1}{2}\phi_S^2 - 1 + \cos\phi_S\right) \phi_S^2\, T_S^2 = h_3\, \phi_S^2\, T_S^2$$

$$h_3 \equiv \frac{1}{\phi_S^4}\left(\frac{1}{2}\phi_S^2 - 1 + \cos\phi_S\right) = \frac{1}{4\,!} - \frac{\phi_S^2}{6\,!} + \frac{\phi_S^4}{8\,!} - \cdots \qquad (17.1.2.2\text{-}10)$$

$$\int_0^{T_S}\int_0^{\tau} t \sin\phi \, dt\, d\tau = \frac{1}{\omega_{N\nu}^3}\int_0^{\phi_S}\int_0^{\theta} \phi \sin\phi \, d\phi\, d\theta = \frac{1}{\omega_{N\nu}^3}\int_0^{\phi_S} \left(\sin\theta - \theta\cos\theta\right) d\theta$$

$$= \frac{1}{\omega_{N\nu}^3}\left[- 2\cos\theta - \theta\sin\theta\right]_0^{\phi_S} = \frac{1}{\omega_{N\nu}^3}\left(2 - 2\cos\phi_S - \phi_S\sin\phi_S\right) \qquad (17.1.2.2\text{-}11)$$

$$= \frac{1}{\phi_S^4}\left(2 - 2\cos\phi_S - \phi_S\sin\phi_S\right)\phi_S T_S^3 = h_4\,\phi_S T_S^3$$

$$h_4 \equiv \frac{1}{\phi_S^4}\left(2 - 2\cos\phi_S - \phi_S\sin\phi_S\right) = \left(\frac{1}{3\,!} - \frac{2}{4\,!}\right) - \left(\frac{1}{5\,!} - \frac{2}{6\,!}\right)\phi_S^2 + \left(\frac{1}{7\,!} - \frac{2}{8\,!}\right)\phi_S^4 - \cdots$$

$$= \frac{2}{4\,!} - \frac{4}{6\,!}\phi_S^2 + \frac{6}{8\,!}\phi_S^4 - \cdots \qquad (17.1.2.2\text{-}12)$$

$$\int_0^{T_S}\int_0^{\tau} (1 - \cos\phi)\, t \; dt\, d\tau = \frac{1}{\omega_{N\nu}^3}\int_0^{\phi_S}\int_0^{\theta} \phi\,(1 - \cos\phi)\, d\phi\, d\theta$$

$$= \frac{1}{\omega_{N\nu}^3}\int_0^{\phi_S}\left(\frac{\theta^2}{2} + 1 - \cos\theta - \theta\sin\theta\right) d\theta = \frac{1}{\omega_{N\nu}^3}\left[\frac{\theta^3}{3\,!} + \theta - 2\sin\theta + \theta\cos\theta\right]_0^{\phi_S}$$

$$= \frac{1}{\omega_{N\nu}^3}\left(\frac{\phi_S^3}{3\,!} + \phi_S - 2\sin\phi_S + \phi_S\cos\phi_S\right) \qquad (17.1.2.2\text{-}13)$$

$$= \frac{1}{\phi_S^5}\left(\frac{\phi_S^3}{3\,!} + \phi_S - 2\sin\phi_S + \phi_S\cos\phi_S\right)\phi_S^2 T_S^3 = h_5\,\phi_S^2 T_S^3$$

$$h_5 \equiv \frac{1}{\phi_S^5}\left(\frac{\phi_S^3}{3\,!} + \phi_S - 2\sin\phi_S + \phi_S\cos\phi_S\right)$$

$$= \left(\frac{1}{4\,!} - \frac{2}{5\,!}\right) - \left(\frac{1}{6\,!} - \frac{2}{7\,!}\right)\phi_S^2 + \left(\frac{1}{8\,!} - \frac{2}{9\,!}\right)\phi_S^4 - \cdots = \frac{3}{5\,!} - \frac{5}{7\,!}\phi_S^2 + \frac{7}{9\,!}\phi_S^4 - \cdots \qquad (17.1.2.2\text{-}14)$$

with h_2 as calculated in (17.1.1.2-22) and where

θ = Dummy angle integration parameter.

h_3, h_4, h_5 = Defined trigonometric functions of ϕ_S.

Substituting (17.1.2.2-6) in (17.1.2.2-7) with (17.1.2.2-8) - (17.1.2.2-14), (17.1.1.2-22) and (17.1.1.2-16), then obtains for $\Delta \underline{R}^N$:

$$\Delta \underline{R}^N = \underline{v}^N_{Start} T_S + T_S^2 C_L^N C_{\mathcal{V}Start}^L \left\{ \left[\frac{1}{2} I + h_2 \left(\underline{\phi}_S^{\mathcal{V}} \times \right) + h_3 \left(\underline{\phi}_S^{\mathcal{V}} \times \right)^2 \right] \left[\dot{V} \, \underline{u}_{X\mathcal{V}}^{\mathcal{V}} \right. \right.$$
$$\left. + V_{Start} \left(\underline{\omega}_{N\mathcal{V}}^{\mathcal{V}} \times \underline{u}_{X\mathcal{V}}^{\mathcal{V}} \right) \right] + T_S \left[\frac{1}{6} I + h_4 \left(\underline{\phi}_S^{\mathcal{V}} \times \right) + h_5 \left(\underline{\phi}_S^{\mathcal{V}} \times \right)^2 \right] \left[\dot{V} \left(\underline{\omega}_{N\mathcal{V}}^{\mathcal{V}} \times \underline{u}_{X\mathcal{V}}^{\mathcal{V}} \right) \right] \Bigg\} \qquad (17.1.2.2\text{-}15)$$

With (17.1.1.2-13) for $\underline{\omega}_{N\mathcal{V}}^{\mathcal{V}}$ and the converse of (17.1.1.2-12) for $\dot{V}$, Equation (17.1.2.2-15) for $\Delta \underline{R}^N$ assumes the desired form in terms of the basic trajectory segment parameters:

$$\Delta \underline{R}^N = \underline{v}^N_{Start} T_S + C_L^N C_{\mathcal{V}Start}^L \left\{ \left[\frac{1}{2} I + h_2 \left(\underline{\phi}_S^{\mathcal{V}} \times \right) + h_3 \left(\underline{\phi}_S^{\mathcal{V}} \times \right)^2 \right] \left[\Delta V_S \, \underline{u}_{X\mathcal{V}}^{\mathcal{V}} \right. \right.$$
$$\left. + V_{Start} \left(\underline{\phi}_S^{\mathcal{V}} \times \underline{u}_{X\mathcal{V}}^{\mathcal{V}} \right) \right] + \left[\frac{1}{6} I + h_4 \left(\underline{\phi}_S^{\mathcal{V}} \times \right) + h_5 \left(\underline{\phi}_S^{\mathcal{V}} \times \right)^2 \right] \left[\Delta V_S \left(\underline{\phi}_S^{\mathcal{V}} \times \underline{u}_{X\mathcal{V}}^{\mathcal{V}} \right) \right] \Bigg\} T_S \qquad (17.1.2.2\text{-}16)$$

17.1.2.3 PROJECTED AIRCRAFT AXIS ATTITUDE, SPECIFIC FORCE ACCELERATION AND ANGULAR RATE UNDER ANGLES OF ATTACK AND SIDESLIP

For an aircraft type vehicle, the specific force acceleration calculated by the trajectory generator is created by aerodynamic force and engine thrust. A large portion of the aerodynamic force arises from the angle of attack and angle of sideslip of the aircraft coordinate axes relative to the local airflow velocity vector passing over the aircraft. The aircraft axis orientation to create the angles of attack and sideslip, also defines the direction of the aircraft engine thrust vector in its contribution to the net specific force. To describe the angle of attack/sideslip effects analytically, we must first define the relationship between the local airflow velocity vector $\underline{v}_{Arspd}$, an intermediate relative airspeed coordinate frame ($\mathcal{V}$W) aligned with the $\underline{v}_{Arspd}$ vector, and aircraft coordinate axes (the AC Frame) where:

$\underline{v}_{Arspd}$ = Velocity of the aircraft relative to the local air mass. The $\underline{v}_{Arspd}$ airflow velocity vector is produced by aircraft motion relative to the earth surface minus the average movement of the local air mass (or average wind) relative to the earth's surface.

$\mathcal{V}$W = Coordinate frame whose X axis is aligned with $\underline{v}_{Arspd}$. The relative orientation between the $\mathcal{V}$W and $\mathcal{V}$ Frames is nominally defined to minimize the $\mathcal{V}$-to-$\mathcal{V}$W Frame rotation vector magnitude. Sections 17.1.2.3.1 and 17.1.2.3.2 describe variations from the nominal $\mathcal{V}$W-to-$\mathcal{V}$ Frame orientation for control of the roll angle between the AC and $\mathcal{V}$ Frames.

AC = Coordinate frame parallel to user vehicle ("aircraft" - AC) reference axes. The AC frame for an aircraft is defined to be parallel to the $\mathcal{V}$W Frame under non-maneuvering flight conditions (and parallel to the $\mathcal{V}$ Frame under non-maneuvering flight conditions in still air). A non-parallel angular orientation between the AC Frame X (longitudinal) axis and the $\mathcal{V}$W Frame X axis is produced by the angle of attack (about the $\mathcal{V}$W Frame Y axis) and angle of sideslip (about the $\mathcal{V}$W Frame Z axis). The angular orientation between the AC and $\mathcal{V}$W Frames is defined to include the constraint that the $\mathcal{V}$W-to-AC Frame rotation vector magnitude be minimized.

A simulation of angle of attack/sideslip effects is designed to create an AC Frame attitude (in the form of a $C_{AC}^{\mathcal{V}}$ matrix) that will generate aerodynamic and thrust accelerations corresponding to the trajectory generator specific force acceleration. The $C_{AC}^{\mathcal{V}}$ matrix is computed from the angular relationships between the AC, $\mathcal{V}$W and $\mathcal{V}$ Frames, and the local airflow velocity vector components viewed in these coordinate frames:

$$C_{AC}^{\mathcal{V}} = C_{\mathcal{V}W}^{\mathcal{V}} \, C_{AC}^{\mathcal{V}W} \tag{17.1.2.3-1}$$

$$\underline{v}_{Arspd}^{\mathcal{V}W} = C_{AC}^{\mathcal{V}W} \, \underline{v}_{Arspd}^{AC} \tag{17.1.2.3-2}$$

$$\underline{v}_{Arspd}^{\mathcal{V}} = C_{\mathcal{V}W}^{\mathcal{V}} \, \underline{v}_{Arspd}^{\mathcal{V}W} \tag{17.1.2.3-3}$$

Once determined (as will be described subsequently), the $C_{AC}^{\mathcal{V}}$ matrix can be used to calculate AC Frame attitude relative to the local level L Frame using:

$$C_{AC}^{L} = C_{\mathcal{V}}^{L} \, C_{AC}^{\mathcal{V}} \tag{17.1.2.3-4}$$

The AC Frame specific force acceleration components are computed from $C_{AC}^{\mathcal{V}}$ as:

$$\underline{a}_{SF}^{AC} = \left(C_{AC}^{\mathcal{V}}\right)^T \underline{a}_{SF}^{\mathcal{V}} \tag{17.1.2.3-5}$$

For the Quick Look projection of $\underline{a}_{SF}^{AC}$ at the start and end of the trajectory segment, $\underline{a}_{SF}^{\mathcal{V}}$ in (17.1.2.3-5) is calculated using Equations (17.1.2.1-3).

The AC components of angular rate are approximated for Quick Look projection as the average AC Frame angular rate relative to the L Frame, and are determined from the C_{AC}^{L} matrix calculated from (17.1.2.3-4) at the start and end of the trajectory:

$$C_{AC_{End}}^{AC_{Start}} = \left(C_{AC_{Start}}^{L}\right)^T C_{AC_{End}}^{L} \tag{17.1.2.3-6}$$

where

$C_{AC_{End}}^{AC_{Start}}$ = Direction cosine matrix relating the AC Frame attitude at the start and end of the trajectory segment, relative to the L Frame.

The average AC Frame angular rate is then computed from $C_{AC_{End}}^{AC_{Start}}$ as:

$$\underline{\omega}_{LAC_{Avg}}^{AC} = \frac{1}{T_S}\,\underline{\phi}^{AC} \tag{17.1.2.3-7}$$

where

$\underline{\omega}_{LAC_{Avg}}^{AC}$ = Average angular rate of the AC Frame relative to the L Frame as projected on AC Frame axes.

$\underline{\phi}^{AC}$ = Rotation vector equivalent to $C_{AC_{End}}^{AC_{Start}}$ computed using generalized Equations (3.2.2.2-10) - (3.2.2.2-12) and (3.2.2.2-15) - (3.2.2.2-19).

Equations (17.1.2.3-4) - (17.1.2.3-7) provide useful Quick-look AC Frame attitude, angular rate and specific acceleration data during trajectory shaping. The fundamental input to these equations is $C_{AC}^{\mathcal{V}}$ which must be calculated based on the balance of aerodynamic and thrust forces against the product of vehicle mass with the trajectory generator specific force acceleration. Equations (17.1.2.3-1) - (17.1.2.3-3) are the basic relationships we will now use to solve for $C_{AC}^{\mathcal{V}}$. The derivation procedure is fairly involved consisting of the following steps. First we will calculate $\underline{v}_{Arspd}^{\mathcal{V}}$ from vehicle velocity relative to the earth minus a specified average wind velocity. The $\underline{v}_{Arspd}^{\mathcal{V}W}$ vector will then be defined from the signed magnitude of $\underline{v}_{Arspd}^{\mathcal{V}}$ based on the definition for the $\mathcal{V}$W Frame. Next we will compute $\underline{v}_{Arspd}^{AC}$ in (17.1.2.3-2) based on an <u>assumed</u> angle of attack and sideslip condition. The solutions for $\underline{v}_{Arspd}^{\mathcal{V}}$, $\underline{v}_{Arspd}^{\mathcal{V}W}$ and $\underline{v}_{Arspd}^{AC}$ will be used to calculate $C_{\mathcal{V}W}^{\mathcal{V}}$ and $C_{AC}^{\mathcal{V}W}$ using Equations (17.1.2.3-2) - (17.1.2.3-3) based on the previous definitions for the AC Frame and the nominal $\mathcal{V}$W Frame. The computed $C_{AC}^{\mathcal{V}W}$ and $C_{\mathcal{V}W}^{\mathcal{V}}$ matrices allow $C_{AC}^{\mathcal{V}}$ to be calculated using (17.1.2.3-1). Finally, the angle of attack and sideslip <u>assumed</u> for calculating $\underline{v}_{Arspd}^{AC}$ will be equated to values needed for the vehicle aerodynamic force and thrust components to balance the trajectory generator computed specific force acceleration. Because the results from the previous step are used earlier in calculating $\underline{v}_{Arspd}^{AC}$, an iteration loop is required to compute the angles of attack/sideslip for $C_{AC}^{\mathcal{V}}$ determination that also satisfy the force balance equations. At the conclusion of this section, the overall results obtained will be summarized in order of execution for a simulation program.

We begin with the derivation of the $\underline{v}_{Arspd}^{\mathcal{V}}$ expression. In the $\mathcal{V}$ Frame, the air flow velocity vector can be calculated from its definition as:

$$\underline{v}_{Arspd}^{\mathcal{V}} = V\, \underline{u}_{X\mathcal{V}}^{\mathcal{V}} - \underline{v}_{AvgWnd}^{\mathcal{V}} \qquad (17.1.2.3\text{-}8)$$

where

$\underline{v}_{AvgWnd}^{\mathcal{V}}$ = Local average wind velocity relative to the earth in $\mathcal{V}$ Frame axes. The average wind velocity is defined as the total wind velocity minus high frequency wind gust effects. Wind gust effects will be handled as part of Trajectory Regeneration operations.

Equation (17.1.2.3-8) is based on the underlying assumption for the trajectory generator that the vehicle velocity relative to the earth is along the X axis of the $\mathcal{V}$ Frame with signed magnitude V.

The $\underline{v}_{AvgWnd}^{\mathcal{V}}$ average wind vector components in (17.1.2.3-8) are computed from:

$$\underline{v}_{AvgWnd}^{\mathcal{V}} = \left(C_{\mathcal{V}}^{L}\right)^{T} C_{N}^{L}\, C_{Geo}^{N}\, \underline{v}_{AvgWnd}^{Geo} \qquad (17.1.2.3\text{-}9)$$

where

Geo = Local geographic coordinate frame with Z up and Y north.

The wander azimuth N Frame is rotated from the Geo Frame by the wander angle about the positive Z axis, hence, from Figures 4.4.2.1-2 and 17.1.1.4-1, C_{Geo}^{N} is given by:

$$C_{Geo}^{N} = \begin{bmatrix} \cos \alpha_{Wand} & \sin \alpha_{Wand} & 0 \\ -\sin \alpha_{Wand} & \cos \alpha_{Wand} & 0 \\ 0 & 0 & 1 \end{bmatrix} \qquad (17.1.2.3\text{-}10)$$

where

α_{Wand} = Wander angle computed from the C_{N}^{E} matrix as in Equations (4.4.2.1-3).

The $\underline{v}_{AvgWnd}^{Geo}$ wind vector would be programmed into the simulation as a smooth continuous function of latitude, longitude and altitude. It should be a continuous function to avoid introducing discontinuities in the angles of attack/sideslip calculated from $\underline{v}_{AvgWnd}^{Geo}$.

The magnitude of $\underline{v}_{Arspd}$ will also be useful in subsequent calculations, and can be evaluated from the $\mathcal{V}$ Frame $\underline{v}_{Arspd}$ components using the invariance property of vector dot products between coordinate frames (generalized Equation (3.1.1-29)) by which it is recognized (from (3.1.1-4) - (3.1.1-5)) that the magnitude squared of $\underline{v}_{Arspd}$ is the dot product of $\underline{v}_{Arspd}$ with itself in any coordinate frame:

$$v_{Arspd} = \sqrt{\underline{v}_{Arspd}^{\mathcal{V}} \cdot \underline{v}_{Arspd}^{\mathcal{V}}} \qquad (17.1.2.3\text{-}11)$$

where

v_{Arspd} = Magnitude of $\underline{v}_{Arspd}$.

The $\mathcal{V}$W Frame components of $\underline{v}_{Arspd}$ (i.e., $\underline{v}_{Arspd}^{\mathcal{V}W}$) are easily defined from v_{Arspd} and the definition of the $\mathcal{V}$W Frame having its X axis along $\underline{v}_{Arspd}$:

$$\underline{v}_{Arspd}^{\mathcal{V}W} = v_{Arspd}\, \underline{u}_{X\mathcal{V}W}^{\mathcal{V}W} \qquad (17.1.2.3\text{-}12)$$

where

$\underline{u}_{X\mathcal{V}W}^{\mathcal{V}W}$ = Unit vector along the $\mathcal{V}$W Frame X axis.

The AC Frame components of $\underline{v}_{Arspd}$ (i.e., $\underline{v}_{Arspd}^{AC}$) are calculated as a function of angle of attack, angle of sideslip and v_{Arspd}. The angles about the AC Frame Y, Z axes from $\underline{v}_{Arspd}$ to the AC Frame X axis (i.e., the angles of attack and sideslip) are defined analytically as:

$$\alpha = \tan^{-1} \frac{v_{Arspd\,ZAC}}{v_{Arspd\,XAC}} \qquad \beta = -\tan^{-1} \frac{v_{Arspd\,YAC}}{v_{Arspd\,XAC}} \qquad (17.1.2.3\text{-}13)$$

where

α = Angle of attack.

β = Angle of sideslip.

$v_{Arspd\,iAC}$ = Component of $\underline{v}_{Arspd}^{AC}$ along axis i of the AC Frame.

Given the angles of attack and sideslip, the Y and Z components of $\underline{v}_{Arspd}^{AC}$ can be calculated from the inverse of (17.1.2.3-13):

$$v_{Arspd\,YAC} = -\,v_{Arspd\,XAC} \tan\beta \qquad v_{Arspd\,ZAC} = v_{Arspd\,XAC} \tan\alpha \tag{17.1.2.3-14}$$

The X component of $\underline{v}_{Arspd}^{AC}$ in (17.1.2.3-13) can be computed from the magnitude squared formula:

$$v_{Arspd\,XAC}^2 + v_{Arspd\,YAC}^2 + v_{Arspd\,ZAC}^2 = v_{Arspd}^2 \tag{17.1.2.3-15}$$

Dividing (17.1.2.3-15) by $v_{Arspd\,XAC}^2$, substituting (17.1.2.3-14) for the Y, Z components of $\underline{v}_{Arspd}^{AC}$, rearranging, and taking the positive square root (for a positive $\underline{v}_{Arspd}^{AC}$ forward component), we obtain for $v_{Arspd\,XAC}$:

$$v_{Arspd\,XAC} = v_{Arspd} \sqrt{\frac{1}{1 + \tan^2\alpha + \tan^2\beta}} \tag{17.1.2.3-16}$$

Using (17.1.2.3-16), Equations (17.1.2.3-14) and (17.1.2.3-16) become the normalized set:

$$\begin{aligned} \frac{v_{ArspdXAC}}{v_{Arspd}} &= \left(1 + \tan^2\alpha + \tan^2\beta\right)^{-\frac{1}{2}} \\ \frac{v_{ArspdYAC}}{v_{Arspd}} &= -\tan\beta\left(1 + \tan^2\alpha + \tan^2\beta\right)^{-\frac{1}{2}} \\ \frac{v_{ArspdZAC}}{v_{Arspd}} &= \tan\alpha\left(1 + \tan^2\alpha + \tan^2\beta\right)^{-\frac{1}{2}} \end{aligned} \tag{17.1.2.3-17}$$

The α, β angles in the above expressions create vehicle aerodynamic force and thrust vector direction whose sum equals the trajectory generator computed specific force acceleration. Calculation of α, β values that satisfy the previous condition will be described subsequently.

Once the $\underline{v}_{Arspd}^{\mathcal{V}}$, $\underline{v}_{Arspd}^{\mathcal{V}W}$ and $\underline{v}_{Arspd}^{AC}$ component forms of $\underline{v}_{Arspd}$ are obtained with Equations (17.1.2.3-8) - (17.1.2.3-10), (17.1.2.3-12) and (17.1.2.3-17), the $C_{\mathcal{V}W}^{\mathcal{V}}$ and $C_{AC}^{\mathcal{V}W}$ matrices can be calculated from Equations (17.1.2.3-2) and (17.1.2.3-3). There is no unique solution for $C_{\mathcal{V}W}^{\mathcal{V}}$ and $C_{AC}^{\mathcal{V}W}$ that can be determined from (17.1.2.3-2) and (17.1.2.3-3) without imposing an additional constraint. The constraint we impose is that $C_{\mathcal{V}W}^{\mathcal{V}}$ and $C_{AC}^{\mathcal{V}W}$ satisfy

(17.1.2.3-2) and (17.1.2.3-3) using associated rotation vectors of minimum magnitude. Note that this same constraint was also used in the definition for the AC Frame and for the nominal $\mathcal{V}$W Frame given previously. Then generalized Equations (3.2.1.1-1), (3.2.1.1-22) and (3.2.1.1-26) apply and we can write from (17.1.2.3-2), (17.1.2.3-3) and (17.1.2.3-12):

$$D \equiv \left(\underline{u}_{X\mathcal{V}W}^{\mathcal{V}W}\right)^T \left(\frac{\underline{v}_{Arspd}^{\mathcal{V}}}{v_{Arspd}}\right) \qquad \underline{E} \equiv \left(\underline{u}_{X\mathcal{V}W}^{\mathcal{V}W} \times\right)\left(\frac{\underline{v}_{Arspd}^{\mathcal{V}}}{v_{Arspd}}\right)$$
$$C_{\mathcal{V}W}^{\mathcal{V}} = I + \left(\underline{E}\times\right) + \frac{1}{1+D}\left(\underline{E}\times\right)^2 \tag{17.1.2.3-18}$$

$$G \equiv \left(\frac{\underline{v}_{Arspd}^{AC}}{v_{Arspd}}\right)^T \underline{u}_{X\mathcal{V}W}^{\mathcal{V}W} \qquad \underline{H} \equiv \left[\left(\frac{\underline{v}_{Arspd}^{AC}}{v_{Arspd}}\right)\times\right] \underline{u}_{X\mathcal{V}W}^{\mathcal{V}W}$$
$$C_{AC}^{\mathcal{V}W} = I + \left(\underline{H}\times\right) + \frac{1}{1+G}\left(\underline{H}\times\right)^2 \tag{17.1.2.3-19}$$

It remains to define α, β angles of attack/sideslip to balance the trajectory generator specific force acceleration. We begin by writing expressions for the $\mathcal{V}$ Frame components of force on the vehicle produced by aerodynamic effects and thrust:

$$\underline{F}^{\mathcal{V}} = C_{\mathcal{V}W}^{\mathcal{V}}\left(\underline{F}_{Aero}^{\mathcal{V}W} + \text{Thrst}\; C_{AC}^{\mathcal{V}W}\, \underline{u}_{Thrst}^{AC}\right) \tag{17.1.2.3-20}$$

where

$\underline{F}$ = Total force vector acting on the vehicle.

Thrst = Magnitude of vehicle engine thrust.

$\underline{u}_{Thrst}$ = Unit vector along the vehicle thrust direction.

The $C_{\mathcal{V}W}^{\mathcal{V}}$ matrix in (17.1.2.3-20) is obtained from (17.1.2.3-18) and the $C_{AC}^{\mathcal{V}W}$ matrix is calculated from (17.1.2.3-19).

Assuming that the thrust vector direction is perpendicular to the AC Frame Y (pitch) axis, we can write in general for $\underline{u}_{Thrst}^{AC}$ in (17.1.2.3-20):

$$\underline{u}_{Thrst}^{AC} = \left(\cos\theta_{Thrst},\, 0,\, -\sin\theta_{Thrst}\right)^T \tag{17.1.2.3-21}$$

where

θ_{Thrst} = Angle from the AC Frame X axis to the engine thrust direction axis measured as a positive Euler rotation about the AC Frame Y axis.

The $\mathcal{V}$W Frame components of $\underline{F}_{Aero}^{\mathcal{V}W}$ in (17.1.2.3-20) are traditionally defined as:

$$\underline{F}_{Aero}^{\mathcal{V}W} = \left(-F_{Drag} \;\; F_{Side} \;\; -F_{Lift}\right)^T \tag{17.1.2.3-22}$$

where

$F_{Lift}, F_{Drag}, F_{Side}$ = Aerodynamic lift, drag and side forces produced by airflow over the vehicle at α, β angles of attack, sideslip.

The $F_{Lift}, F_{Drag}, F_{Side}$ force components in (17.1.2.3-22) can be expressed in terms of classical aerodynamic coefficients (e.g., Reference 28, Section 1-2) as:

$$F_{Lift} = C_L \, S \, q \qquad F_{Side} = C_{Side} \, S \, q \qquad F_{Drag} = C_D \, S \, q \tag{17.1.2.3-23}$$

with

$$q = \frac{1}{2} \rho \, v_{Arspd}^2 \tag{17.1.2.3-24}$$

where

C_L, C_D, C_{Side} = Normalized lift, drag, side force aerodynamic coefficients.

S = Vehicle reference area (typically the wing area for an aircraft) which would be an input parameter for the simulation program.

q = Dynamic pressure produced from vehicle velocity relative to the air-mass. If the free air-stream velocity is brought to a stand-still at the aircraft by a loss-less process, the free-stream air pressure would change by q.

ρ = Density of undisturbed air around the vehicle. In a simulation, ρ would be a programmed function of altitude based on "standard" temperature conditions. Variations of ρ due to non-standard temperature can be calculated as the standard ρ multiplied by the standard absolute temperature divided by the simulated non-standard absolute temperature.

The C_L, C_D, C_{Side} coefficients for an aircraft can be approximated for simulation purposes (e.g., using Reference 28 - Sections 2-1 through 2-3 as a guide) by the following functions of angle of attack and sideslip:

$$\begin{aligned} C_L &\approx C_{L\alpha}\left(\alpha + \alpha_0 + \alpha_{Flaps}\right) \\ C_{Side} &\approx C_{Sd\beta} \, \beta \\ C_D &\approx C_{Df} + C_{Dthk} + K_{DInd/L} \, C_L \left(\alpha + \alpha_0 + \alpha_{Flaps}\right) + K_{DInd/Sd} \, C_{Side} \beta \end{aligned} \tag{17.1.2.3-25}$$

where

α_0 = Equivalent angle of attack corresponding to the lift coefficient generated under zero α conditions (a simulation program input parameter).

α_{Flaps} = Equivalent angle of attack bias corresponding to an addition to the lift coefficient generated by extending the vehicle aerodynamic flaps (a simulation program input which can be programmed at different values corresponding to different trajectory times).

$C_{L\alpha}$, $C_{Sd\beta}$ = Lift and side force slope aerodynamic coefficients.

C_{Df}, C_{Dthk} = Friction and thickness drag coefficients.

$K_{DInd/L}$, $K_{DInd/Sd}$ = Lift and side force induced drag coefficients.

Values for the (17.1.2.3-25) aerodynamic coefficients depend on the vehicle aerodynamic shape and Mach number (ratio of true air speed divided by the speed of sound - the speed of sound is proportional to the square root of absolute temperature (Reference 17 - Section 2.8)). Coefficients for different aerodynamic shapes can be found in classical text books such as Reference 1 - Sections 5-4, 5-5, 6-2 and 6-3, and Reference 28 - Chapter 2). In general, the aerodynamic coefficients are approximately constant for low speed subsonic airflow but for higher speed flight, can vary with Mach number. The C_{Dthk} thickness drag coefficient is created by local aerodynamic shock wave induced effects, hence, is zero for subsonic flow.

The angles of attack and sideslip α, β in (17.1.2.3-25) as well as the engine thrust (Thrst) in (17.1.2.3-20) are selected so that the $\underline{F}^{\mathcal{V}}$ components in (17.1.2.3-20) equal the effect on trajectory generator specific force acceleration through Newton's law:

$$\underline{F}^{\mathcal{V}} = M\, \underline{a}_{SF}^{\mathcal{V}} \qquad (17.1.2.3\text{-}26)$$

where

M = Vehicle mass.

For the Quick Look projection, $\underline{a}_{SF}^{\mathcal{V}}$ in (17.1.2.3-26) can be calculated using Equations (17.1.2.1-3). The vehicle mass M in (17.1.2.3-26) can be computed as the starting input mass, minus the mass of jettisoned elements and fuel expended:

$$\begin{aligned} M &= M_0 - \sum \Delta M_{Jtsn} - \Delta M_{Fuel} \\ \Delta M_{Fuel} &= K_{Fuel} \int \mathrm{Thrst}\, dt \end{aligned} \qquad (17.1.2.3\text{-}27)$$

where

M_0 = Vehicle mass at simulation start.

ΔM_{Jtsn} = Mass jettisoned since simulation start.

ΔM_{Fuel} = Fuel mass expended since simulation start.

K_{Fuel} = Fuel mass expenditure rate per unit thrust.

For the Quick-Look projection, ΔM_{Fuel} in (17.1.2.3-27) at a time point within a particular trajectory segment can be determined by the approximate trapezoidal integration algorithm:

$$\Delta M_{Fuel} = \Delta M_{Fuel_{Start}} + K_{Fuel} \frac{1}{2} \left(Thrst_{Start} + Thrst \right) t \qquad (17.1.2.3\text{-}28)$$

where

t = Time from the start of the trajectory segment.

Thrst = Vehicle engine thrust at time t.

The ΔM_{Fuel} value at the end of the trajectory segment (i.e., $\Delta M_{Fuel_{End}}$) is calculated from (17.1.2.3-28) with t set equal to T_S. The $\Delta M_{Fuel_{Start}}$ value in (17.1.2.3-28) is ΔM_{Fuel} at the end of the previous segment.

Equations (17.1.2.3-17) - (17.1.2.3-28) constitute a complete set that can be solved uniquely for α, β and Thrst . Due to the nonlinear character of these equations, it is difficult to combine them into a closed-form solution. The alternative is to use an iteration procedure in which a previous approximate solution is used iteratively to determine a more accurate solution for the next iteration cycle. The iteration cycle continues until the solution error is within prescribed limits.

For the previous system of equations, the iterative solution error can be calculated as the difference between $\underline{F}^{\mathcal{V}}$ calculated with Equations (17.1.2.3-20) and (17.1.2.3-26):

$$\delta \underline{F}_j^{\mathcal{V}} = C_{\mathcal{V}W}^{\mathcal{V}} \left(\underline{F}_{Aero_j}^{\mathcal{V}W} + Thrst_j \, C_{AC_j}^{\mathcal{V}W} \, \underline{u}_{Thrst}^{AC} \right) - M_j \, \underline{a}_{SF}^{\mathcal{V}} \qquad (17.1.2.3\text{-}29)$$

where

j = Iteration procedure cycle index. As a subscript for a parameter, j indicates that the parameter is calculated based on data computed during the jth iteration cycle.

$\delta \underline{F}_j^{\mathcal{V}}$ = The difference for iteration cycle j between $\underline{F}^{\mathcal{V}}$ calculated with Equations (17.1.2.3-20) and (17.1.2.3-26).

We wish to adjust α, β and Thrst in iterative fashion so that $\delta \underline{F}_j^{\mathcal{V}}$ in (17.1.2.3-29) goes to zero. Then Equations (17.1.2.3-20) and (17.1.2.3-26) will balance which corresponds to the correct solution for α, β and Thrst . To streamline the analytical development, we define the sought after parameters α, β and Thrst in a column matrix:

$$\underline{P} \equiv \begin{bmatrix} \text{Thrst} \\ \beta \\ \alpha \end{bmatrix} \qquad (17.1.2.3\text{-}30)$$

where

$\underline{P}$ = Column matrix with components equal to Thrst, β and α.

The $\underline{P}_{j-1}$ value for $\underline{P}$ (i.e., $\underline{P}$ determined during the previous iteration cycle) is used to calculate the j value parameters in Equations (17.1.2.3-29). Let's call the result $\delta \underline{F}_{j-1}^{\mathcal{V}}$ corresponding to $\underline{P}_{j-1}$. Thus, $\delta \underline{F}_{j-1}^{\mathcal{V}}$ from (17.1.2.3-29) is a measure of the error in $\underline{P}_{j-1}$. As the iteration process continues, we adjust $\underline{P}$ iteratively which changes $\delta \underline{F}_j^{\mathcal{V}}$ using (17.1.2.3-29). The change in $\delta \underline{F}_j^{\mathcal{V}}$ for each iteration can be approximated as a linear variation from its last computed (17.1.2.3-29) value:

$$\Delta \delta \underline{F}_j^{\mathcal{V}} \approx \left(\frac{\partial \delta \underline{F}^{\mathcal{V}}}{\partial \underline{P}} \right)_{j-1} \Delta \underline{P}_j \qquad (17.1.2.3\text{-}31)$$

with $\left(\frac{\partial \delta \underline{F}^{\mathcal{V}}}{\partial \underline{P}} \right)_{j-1}$ given by the general form:

$$\frac{\partial (\,)}{\partial \underline{P}} \equiv \begin{bmatrix} \frac{\partial (\,)}{\partial \text{Thrst}} & \frac{\partial (\,)}{\partial \beta} & \frac{\partial (\,)}{\partial \alpha} \end{bmatrix} \qquad (17.1.2.3\text{-}32)$$

where

$\frac{\partial (\,)}{\partial [\,]}$ = Partial derivative of ( ) with respect to variations in the [ ] parameter.

$\frac{\partial (\,)}{\partial \underline{P}}$ = As defined by Equation (17.1.2.3-32) in which ( ) is a vector or scalar quantity.

$\Delta \underline{P}_j$ = Variation in $\underline{P}$ from its j-1 previous iteration cycle value.

$\Delta\delta\underline{F}_j^{\mathcal{V}}$ = Variation in $\delta\underline{F}_{j-1}^{\mathcal{V}}$ from its (17.1.2.3-29) $_{j-1}$ computed value due to $\Delta\underline{P}_j$.

To make the iteration process converge to a zero $\delta\underline{F}_j^{\mathcal{V}}$, we set $\Delta\delta\underline{F}_j^{\mathcal{V}}$ in (17.1.2.3-31) to the negative of $\delta\underline{F}_{j-1}^{\mathcal{V}}$ determined by (17.1.2.3-29) (i.e., $\delta\underline{F}_j^{\mathcal{V}}$ calculated using $\underline{P}_{j-1}$):

$$\Delta\delta\underline{F}_j^{\mathcal{V}} = -\,\delta\underline{F}_{j-1}^{\mathcal{V}} \qquad (17.1.2.3\text{-}33)$$

If $\left(\dfrac{\partial\delta\underline{F}^{\mathcal{V}}}{\partial\underline{P}}\right)_{j-1}$ is known, $\Delta\underline{P}_j$ corresponding to $\Delta\delta\underline{F}_j^{\mathcal{V}}$ can be calculated from the inverse of (17.1.2.3-31) using (17.1.2.3-33) for $\Delta\delta\underline{F}_j^{\mathcal{V}}$:

$$\Delta\underline{P}_j = -\left(\frac{\partial\delta\underline{F}^{\mathcal{V}}}{\partial\underline{P}}\right)_{j-1}^{-1} \delta\underline{F}_{j-1}^{\mathcal{V}} \qquad (17.1.2.3\text{-}34)$$

and $\underline{P}$ can then be corrected from its previous cycle $_{j-1}$ value to its current j cycle value:

$$\underline{P}_j = \underline{P}_{j-1} + \Delta\underline{P}_j \qquad (17.1.2.3\text{-}35)$$

Equations (17.1.2.3-34) and (17.1.2.3-35) constitute the iteration process for determining $\underline{P}$ using (17.1.2.3-29) for $\delta\underline{F}_j^{\mathcal{V}}$. The iteration process is complete when the $\delta\underline{F}_j^{\mathcal{V}}$ components fall within prescribed limits.

It finally remains to determine an analytical expression for $\dfrac{\partial\delta\underline{F}^{\mathcal{V}}}{\partial\underline{P}}$ in (17.1.2.3-34). This is derived by applying (17.1.2.3-32) to (17.1.2.3-29) while recognizing that $\underline{u}_{Thrst}^{AC}$, $C_{\mathcal{V}W}^{\mathcal{V}}$ and $\underline{a}_{SF}^{\mathcal{V}}$ are independent of $\underline{P}$:

$$\frac{\partial\delta\underline{F}^{\mathcal{V}}}{\partial\underline{P}} = C_{\mathcal{V}W}^{\mathcal{V}}\left(\frac{\partial\underline{F}_{Aero}^{\mathcal{V}W}}{\partial\underline{P}} + C_{AC}^{\mathcal{V}W}\,\underline{u}_{Thrst}^{AC}\,\frac{\partial Thrst}{\partial\underline{P}} + Thrst\,\frac{\partial\left(C_{AC}^{\mathcal{V}W}\,\underline{u}_{Thrst}^{AC}\right)}{\partial\underline{P}}\right) - \underline{a}_{SF}^{\mathcal{V}}\,\frac{\partial M}{\partial\underline{P}}$$

(17.1.2.3-36)

From (17.1.2.3-30) and (17.1.2.3-32), the second term on the right in (17.1.2.3-36) is simply:

$$\frac{\partial Thrst}{\partial\underline{P}} = \left[\begin{matrix} 1 & 0 & 0 \end{matrix}\right] \qquad (17.1.2.3\text{-}37)$$

The fourth term on the right in (17.1.2.3-36) is with (17.1.2.3-27) - (17.1.2.3-28):

$$\frac{\partial M}{\partial \underline{P}} = - \frac{\partial \Delta M_{Fuel}}{\partial \underline{P}} = \begin{bmatrix} -\frac{1}{2} K_{Fuel}\, t & 0 & 0 \end{bmatrix} \tag{17.1.2.3-38}$$

The first term on the right of (17.1.2.3-36) can be evaluated from (17.1.2.3-22) for the $\underline{F}_{Aero}^{\mathcal{V}W}$ components. Recognizing from(17.1.2.3-22), (17.1.2.3-23) and (17.1.2.3-25) that $\frac{\partial F_{Side}}{\partial \alpha}$, $\frac{\partial F_{Lift}}{\partial \beta}$, and $\frac{\partial \underline{F}_{Aero}^{\mathcal{V}W}}{\partial Thrst}$ are zero, we can write from the (17.1.2.3-30) and (17.1.2.3-32) definitions:

$$\frac{\partial \underline{F}_{Aero}^{\mathcal{V}W}}{\partial \underline{P}} = \begin{bmatrix} 0 & -\frac{\partial F_{Drag}}{\partial \beta} & -\frac{\partial F_{Drag}}{\partial \alpha} \\ 0 & \frac{\partial F_{Side}}{\partial \beta} & 0 \\ 0 & 0 & -\frac{\partial F_{Lift}}{\partial \alpha} \end{bmatrix} \tag{17.1.2.3-39}$$

The individual terms in (17.1.2.3-39) are evaluated from Equations (17.1.2.3-22), (17.1.2.3-23) and (17.1.2.3-25) combined:

$$\begin{aligned} F_{Drag} &= S\, q \left[C_{Df} + C_{Dthk} + K_{DInd/L}\, C_{L\alpha} \left(\alpha + \alpha_0 + \alpha_{Flaps} \right)^2 + K_{DInd/Sd}\, C_{Sd\beta}\, \beta^2 \right] \\ F_{Side} &= S\, q\, C_{Sd\beta}\, \beta \\ F_{Lift} &= S\, q\, C_{L\alpha} \left(\alpha + \alpha_0 + \alpha_{Flaps} \right) \end{aligned} \tag{17.1.2.3-40}$$

Using (17.1.2.3-40), the components of (17.1.2.3-39) are given by:

$$\begin{aligned} \frac{\partial F_{Drag}}{\partial \beta} &= 2\, S\, q\, K_{DInd/Sd}\, C_{Sd\beta}\, \beta \\ \frac{\partial F_{Drag}}{\partial \alpha} &= 2\, S\, q\, K_{DInd/L}\, C_{L\alpha} \left(\alpha + \alpha_0 + \alpha_{Flaps} \right) \\ \frac{\partial F_{Side}}{\partial \beta} &= S\, q\, C_{Sd\beta} \\ \frac{\partial F_{Lift}}{\partial \alpha} &= S\, q\, C_{L\alpha} \end{aligned} \tag{17.1.2.3-41}$$

The third term on the right in (17.1.2.3-36) is evaluated using (17.1.2.3-19) and generalized Equation (3.1.1-15) as:

$$\left(C_{AC}^{\mathcal{V}W}\, \underline{u}_{Thrst}^{AC}\right) = \underline{u}_{Thrst}^{AC} + \underline{H} \times \underline{u}_{Thrst}^{AC} + \frac{1}{1+G}\, \underline{H} \times \left(\underline{H} \times \underline{u}_{Thrst}^{AC}\right) \qquad (17.1.2.3\text{-}42)$$

Applying (17.1.2.3-32) to (17.1.2.3-42) initially gives:

$$\frac{\partial\left(C_{AC}^{\mathcal{V}W}\, \underline{u}_{Thrst}^{AC}\right)}{\partial \underline{P}} = \frac{\partial\left(\underline{H} \times \underline{u}_{Thrst}^{AC}\right)}{\partial \underline{P}} - \frac{1}{(1+G)^2}\, \underline{H} \times \left(\underline{H} \times \underline{u}_{Thrst}^{AC}\right) \frac{\partial G}{\partial \underline{P}} + \frac{1}{1+G}\, \frac{\partial\left[\underline{H} \times \left(\underline{H} \times \underline{u}_{Thrst}^{AC}\right)\right]}{\partial \underline{P}} \qquad (17.1.2.3\text{-}43)$$

The first and third $\frac{\partial(\;)}{\partial \underline{P}}$ terms on the right in (17.1.2.3-43) are then expanded being careful to maintain the definition for ( ) in $\frac{\partial(\;)}{\partial \underline{P}}$ as being a scalar or vector quantity. For the first term we obtain using general formulas (3.1.1-8) and (3.1.1-13):

$$\frac{\partial\left(\underline{H} \times \underline{u}_{Thrst}^{AC}\right)}{\partial \underline{P}} = -\left(\underline{u}_{Thrst}^{AC} \times\right) \frac{\partial \underline{H}}{\partial \underline{P}} \qquad (17.1.2.3\text{-}44)$$

For the third term we obtain with (17.1.2.3-44) and (3.1.1-15):

$$\begin{aligned}
\frac{\partial\left[\underline{H} \times \left(\underline{H} \times \underline{u}_{Thrst}^{AC}\right)\right]}{\partial \underline{P}} &= \left(\underline{H}\times\right) \frac{\partial\left(\underline{H} \times \underline{u}_{Thrst}^{AC}\right)}{\partial \underline{P}} - \left[\left(\underline{H} \times \underline{u}_{Thrst}^{AC}\right)\times\right] \frac{\partial \underline{H}}{\partial \underline{P}} \\
&= -\left(\underline{H}\times\right)\left(\underline{u}_{Thrst}^{AC} \times \frac{\partial \underline{H}}{\partial \underline{P}}\right) - \left[\left(\underline{H} \times \underline{u}_{Thrst}^{AC}\right)\times\right] \frac{\partial \underline{H}}{\partial \underline{P}} \\
&= -\left\{\left(\underline{H}\times\right)\left(\underline{u}_{Thrst}^{AC}\times\right) + \left[\left(\underline{H} \times \underline{u}_{Thrst}^{AC}\right)\times\right]\right\} \frac{\partial \underline{H}}{\partial \underline{P}}
\end{aligned} \qquad (17.1.2.3\text{-}45)$$

Substituting (17.1.2.3-44) - (17.1.2.3-45) in (17.1.2.3-43) then yields with (3.1.1-15):

$$\frac{\partial\left(C_{AC}^{\mathcal{V}W}\, \underline{u}_{Thrst}^{AC}\right)}{\partial \underline{P}} = -\left(\underline{u}_{Thrst}^{AC}\times\right) \frac{\partial \underline{H}}{\partial \underline{P}} - \frac{1}{1+G}\left\{\left(\underline{H}\times\right)\left(\underline{u}_{Thrst}^{AC}\times\right) + \left[\left(\underline{H} \times \underline{u}_{Thrst}^{AC}\right)\times\right]\right\} \frac{\partial \underline{H}}{\partial \underline{P}} - \frac{1}{(1+G)^2} \left(\underline{H}\times\right)^2 \underline{u}_{Thrst}^{AC} \frac{\partial G}{\partial \underline{P}} \qquad (17.1.2.3\text{-}46)$$

or upon compression:

$$\frac{\partial\left(C_{AC}^{\mathcal{V}W}\,\underline{u}_{Thrst}^{AC}\right)}{\partial \underline{P}} = -\left[\left(\underline{u}_{Thrst}^{AC}\times\right) + \frac{1}{1+G}\left\{\left(\underline{H}\times\right)\left(\underline{u}_{Thrst}^{AC}\times\right) + \left[\left(\underline{H}\times\underline{u}_{Thrst}^{AC}\right)\times\right]\right\}\right]\frac{\partial \underline{H}}{\partial \underline{P}} - \frac{1}{(1+G)^2}\left(\underline{H}\times\right)^2\underline{u}_{Thrst}^{AC}\,\frac{\partial G}{\partial \underline{P}} \qquad (17.1.2.3\text{-}47)$$

From (17.1.2.3-19) and (17.1.2.3-12), using generalized Equations (3.1.1-7), (3.1.1-8), (3.1.1-12) and (3.1.1-13), we see that:

$$G = \frac{\left(\underline{v}_{Arspd}^{\mathcal{V}W}\right)^T}{v_{Arspd}}\,\frac{\underline{v}_{Arspd}^{AC}}{v_{Arspd}} \qquad \underline{H} = -\frac{\left(\underline{v}_{Arspd}^{\mathcal{V}W}\times\right)}{v_{Arspd}}\left(\frac{\underline{v}_{Arspd}^{AC}}{v_{Arspd}}\right) \qquad (17.1.2.3\text{-}48)$$

Recognizing that $\underline{v}_{Arspd}^{\mathcal{V}W}$ and v_{Arspd} are independent of $\underline{P}$, we obtain from (17.1.2.3-48) for $\frac{\partial G}{\partial \underline{P}}$ and $\frac{\partial \underline{H}}{\partial \underline{P}}$:

$$\frac{\partial G}{\partial \underline{P}} = \frac{\left(\underline{v}_{Arspd}^{\mathcal{V}W}\right)^T}{v_{Arspd}}\,\frac{\partial\left(\underline{v}_{Arspd}^{AC} / v_{Arspd}\right)}{\partial \underline{P}} \qquad \frac{\partial \underline{H}}{\partial \underline{P}} = -\frac{\left(\underline{v}_{Arspd}^{\mathcal{V}W}\times\right)}{v_{Arspd}}\,\frac{\partial\left(\underline{v}_{Arspd}^{AC} / v_{Arspd}\right)}{\partial \underline{P}} \qquad (17.1.2.3\text{-}49)$$

Substituting (17.1.2.3-49) into (17.1.2.3-47) then yields $\frac{\partial\left(C_{AC}^{\mathcal{V}W}\,\underline{u}_{Thrst}^{AC}\right)}{\partial \underline{P}}$ as a function of the single parameter $\frac{\partial\left(\underline{v}_{Arspd}^{AC} / v_{Arspd}\right)}{\partial \underline{P}}$:

$$\frac{\partial\left(C_{AC}^{\mathcal{V}W}\,\underline{u}_{Thrst}^{AC}\right)}{\partial \underline{P}} = \left\{\left[\left(\underline{u}_{Thrst}^{AC}\times\right) + \frac{1}{1+G}\left\{\left(\underline{H}\times\right)\left(\underline{u}_{Thrst}^{AC}\times\right) + \left[\left(\underline{H}\times\underline{u}_{Thrst}^{AC}\right)\times\right]\right\}\right]\frac{\left(\underline{v}_{Arspd}^{\mathcal{V}W}\times\right)}{v_{Arspd}} - \frac{1}{(1+G)^2}\left(\underline{H}\times\right)^2\underline{u}_{Thrst}^{AC}\,\frac{\left(\underline{v}_{Arspd}^{\mathcal{V}W}\right)^T}{v_{Arspd}}\right\}\frac{\partial\left(\underline{v}_{Arspd}^{AC} / v_{Arspd}\right)}{\partial \underline{P}} \qquad (17.1.2.3\text{-}50)$$

At this point it is beneficial if we define the components of $\frac{\partial\left(\underline{v}_{Arspd}^{AC} / v_{Arspd}\right)}{\partial \underline{P}}$ in (17.1.2.3-50) based on the (17.1.2.3-30) and (17.1.2.3-32) definitions, while recognizing from (17.1.2.3-17) that $\underline{v}_{Arspd}^{AC} / v_{Arspd}$ is independent of Thrst:

$$\frac{\partial\left(\underline{v}^{AC}_{Arspd} / v_{Arspd}\right)}{\partial \underline{P}} = \begin{bmatrix} 0 & \frac{\partial\left(v_{Arspd_{XAC}} / v_{Arspd}\right)}{\partial\beta} & \frac{\partial\left(v_{Arspd_{XAC}} / v_{Arspd}\right)}{\partial\alpha} \\ 0 & \frac{\partial\left(v_{Arspd_{YAC}} / v_{Arspd}\right)}{\partial\beta} & \frac{\partial\left(v_{Arspd_{YAC}} / v_{Arspd}\right)}{\partial\alpha} \\ 0 & \frac{\partial\left(v_{Arspd_{ZAC}} / v_{Arspd}\right)}{\partial\beta} & \frac{\partial\left(v_{Arspd_{ZAC}} / v_{Arspd}\right)}{\partial\alpha} \end{bmatrix} \qquad (17.1.2.3\text{-}51)$$

The individual elements of (17.1.2.3-51) are easily evaluated from (17.1.2.3-17) as:

$$\frac{\partial\left(v_{Arspd_{XAC}} / v_{Arspd}\right)}{\partial\beta} = -\left(1 + \tan^2\alpha + \tan^2\beta\right)^{-\frac{3}{2}} \tan\beta \sec^2\beta$$

$$\frac{\partial\left(v_{Arspd_{XAC}} / v_{Arspd}\right)}{\partial\alpha} = -\left(1 + \tan^2\alpha + \tan^2\beta\right)^{-\frac{3}{2}} \tan\alpha \sec^2\alpha$$

$$\frac{\partial\left(v_{Arspd_{YAC}} / v_{Arspd}\right)}{\partial\beta} = -\left[1 - \left(1 + \tan^2\alpha + \tan^2\beta\right)^{-1} \tan^2\beta\right] \left(1 + \tan^2\alpha + \tan^2\beta\right)^{-\frac{1}{2}} \sec^2\beta$$

$$\frac{\partial\left(v_{Arspd_{YAC}} / v_{Arspd}\right)}{\partial\alpha} = \left(1 + \tan^2\alpha + \tan^2\beta\right)^{-\frac{3}{2}} \tan\beta \tan\alpha \sec^2\alpha \qquad (17.1.2.3\text{-}52)$$

$$\frac{\partial\left(v_{Arspd_{ZAC}} / v_{Arspd}\right)}{\partial\beta} = -\left(1 + \tan^2\alpha + \tan^2\beta\right)^{-\frac{3}{2}} \tan\alpha \tan\beta \sec^2\beta$$

$$\frac{\partial\left(v_{Arspd_{ZAC}} / v_{Arspd}\right)}{\partial\alpha} = \left[1 - \left(1 + \tan^2\alpha + \tan^2\beta\right)^{-1} \tan^2\alpha\right] \left(1 + \tan^2\alpha + \tan^2\beta\right)^{-\frac{1}{2}} \sec^2\alpha$$

Equations (17.1.2.3-36) with (17.1.2.3-37) - (17.1.2.3-39), (17.1.2.3-41) and (17.1.2.3-50) - (17.1.2.3-52) define $\left(\frac{\partial \delta \underline{F}^{\mathcal{V}}}{\partial \underline{P}}\right)_j$ for input to (17.1.2.3-34) in the iteration routine for evaluating $\underline{P}$.

Let us now summarize the overall results we have obtained for calculating $C^{\mathcal{V}}_{AC}$. The pertinent analytical expressions required are given by Equations (17.1.2.3-1), (17.1.2.3-8) - (17.1.2.3-11), (17.1.2.3-17) - (17.1.2.3-19), (17.1.2.3-21) - (17.1.2.3-25), (17.1.2.3-27) - (17.1.2.3-30), (17.1.2.3-34) - (17.1.2.3-39), (17.1.2.3-41), and (17.1.2.3-50) - (17.1.2.3-52) These equations are repeated below in the order they would be executed in a typical simulation program software structure:

Inputs From Other Routines

α_{Wand}, $C_{\mathcal{V}}^{L}$, $\underline{v}_{AvgWnd}^{Geo}$, ρ, $\underline{a}_{SF}^{\mathcal{V}}$, V, θ_{Thrst}, K_{Fuel}, S, $C_{Sd\beta}$, $C_{L\alpha}$, α_0, α_{Flaps}, C_{Df}, C_{Dthk}, $K_{DInd/L}$, $K_{DInd/Sd}$, $Thrst_{Start}$, M_0, ΔM_{Jtsn}

$\mathcal{V}$ Frame Air Flow Components

$$C_{Geo}^{N} = \begin{bmatrix} \cos\alpha_{Wand} & \sin\alpha_{Wand} & 0 \\ -\sin\alpha_{Wand} & \cos\alpha_{Wand} & 0 \\ 0 & 0 & 1 \end{bmatrix}$$

$$\underline{v}_{AvgWnd}^{\mathcal{V}} = \left(C_{\mathcal{V}}^{L}\right)^{T} C_{N}^{L}\, C_{Geo}^{N}\, \underline{v}_{AvgWnd}^{Geo} \qquad (17.1.2.3\text{-}53)$$

$$\underline{v}_{Arspd}^{\mathcal{V}} = V\, \underline{u}_{X\mathcal{V}}^{\mathcal{V}} - \underline{v}_{AvgWnd}^{\mathcal{V}}$$

$$v_{Arspd} = \sqrt{\underline{v}_{Arspd}^{\mathcal{V}} \cdot \underline{v}_{Arspd}^{\mathcal{V}}}$$

$C_{\mathcal{V}W}^{\mathcal{V}}$ Calculation

$$D \equiv \left(\underline{u}_{X\mathcal{V}W}^{\mathcal{V}W}\right)^{T} \left(\frac{\underline{v}_{Arspd}^{\mathcal{V}}}{v_{Arspd}}\right) \qquad \underline{E} \equiv \left(\underline{u}_{X\mathcal{V}W}^{\mathcal{V}W} \times\right) \left(\frac{\underline{v}_{Arspd}^{\mathcal{V}}}{v_{Arspd}}\right)$$

$$C_{\mathcal{V}W}^{\mathcal{V}} = I + \left(\underline{E}\times\right) + \frac{1}{1 + D}\left(\underline{E}\times\right)^{2} \qquad (17.1.2.3\text{-}54)$$

Engine Thrust Direction

$$\underline{u}_{Thrst}^{AC} = \left(\cos\theta_{Thrst}, 0, -\sin\theta_{Thrst}\right)^{T} \qquad (17.1.2.3\text{-}55)$$

Dynamic Pressure

$$q = \frac{1}{2}\rho\, v_{Arspd}^{2} \qquad (17.1.2.3\text{-}56)$$

$\delta\underline{F}^{\mathcal{V}}$ Variation With P (For Partials That Are Independent Of P)

Thrust Variation

$$\frac{\partial Thrst}{\partial \underline{P}} = \begin{bmatrix} 1 & 0 & 0 \end{bmatrix} \qquad (17.1.2.3\text{-}57)$$

Mass Variation

$$\frac{\partial M}{\partial \underline{P}} = \left[-\frac{1}{2} K_{Fuel}\, t \quad 0 \quad 0 \right] \tag{17.1.2.3-58}$$

Aerodynamic Force Variation

$$\frac{\partial F_{Side}}{\partial \beta} = S\, q\, C_{Sd\beta} \qquad \frac{\partial F_{Lift}}{\partial \alpha} = S\, q\, C_{L\alpha} \tag{17.1.2.3-59}$$

Iteration Loop Initialization

$$\underline{P} = \begin{bmatrix} Thrst \\ \beta \\ \alpha \end{bmatrix} = \begin{bmatrix} 0 \\ 0 \\ 0 \end{bmatrix} \tag{17.1.2.3-60}$$

Iteration Loop

DO UNTIL $\delta \underline{F}^{\mathcal{V}}$ COMPONENTS ARE WITHIN SPECIFIED LIMITS

$\underline{v}_{Arspd}^{AC}$ Calculation

$$\frac{v_{Arspd\,XAC}}{v_{Arspd}} = \left(1 + \tan^2\alpha + \tan^2\beta\right)^{-\frac{1}{2}}$$
$$\frac{v_{Arspd\,YAC}}{v_{Arspd}} = -\tan\beta\left(1 + \tan^2\alpha + \tan^2\beta\right)^{-\frac{1}{2}} \tag{17.1.2.3-61}$$
$$\frac{v_{Arspd\,ZAC}}{v_{Arspd}} = \tan\alpha\left(1 + \tan^2\alpha + \tan^2\beta\right)^{-\frac{1}{2}}$$

$C_{AC}^{\mathcal{V}W}$ Calculation

$$G \equiv \left(\frac{\underline{v}_{Arspd}^{AC}}{v_{Arspd}}\right)^T \underline{u}_{X\mathcal{V}W}^{\mathcal{V}W} \qquad \underline{H} \equiv \left[\left(\frac{\underline{v}_{Arspd}^{AC}}{v_{Arspd}}\right) \times\right] \underline{u}_{X\mathcal{V}W}^{\mathcal{V}W}$$
$$C_{AC}^{\mathcal{V}W} = I + \left(\underline{H}\times\right) + \frac{1}{1+G}\left(\underline{H}\times\right)^2 \tag{17.1.2.3-62}$$

Aerodynamic Forces

$$
\begin{aligned}
C_L &\approx C_{L\alpha}\left(\alpha + \alpha_0 + \alpha_{Flaps}\right) \\
C_{Side} &\approx C_{Sd\beta}\,\beta \\
C_D &\approx C_{Df} + C_{Dthk} + K_{DInd/L}\, C_L\left(\alpha + \alpha_0 + \alpha_{Flaps}\right) + K_{DInd/Sd}\, C_{Side}\,\beta \qquad (17.1.2.3\text{-}63) \\
F_{Lift} &= C_L\, S\, q \qquad F_{Side} = C_{Side}\, S\, q \qquad F_{Drag} = C_D\, S\, q \\
\underline{F}_{Aero}^{\mathcal{V}W} &= \left(-F_{Drag} \quad F_{Side} \quad -F_{Lift} \right)^T
\end{aligned}
$$

Vehicle Mass

$$
\begin{aligned}
\Delta M_{Fuel} &= \Delta M_{Fuel_{Start}} + K_{Fuel}\,\frac{1}{2}\left(Thrst_{Start} + Thrst\right) t \\
M &= M_0 - \sum \Delta M_{Jtsn} - \Delta M_{Fuel}
\end{aligned}
\qquad (17.1.2.3\text{-}64)
$$

Balance Of Applied And Inertial Reaction Forces

$$
\delta\underline{F}^{\mathcal{V}} = C_{\mathcal{V}W}^{\mathcal{V}}\left(\underline{F}_{Aero}^{\mathcal{V}W} + Thrst\, C_{AC}^{\mathcal{V}W}\, \underline{u}_{Thrst}^{AC}\right) - M\, \underline{a}_{SF}^{\mathcal{V}} \qquad (17.1.2.3\text{-}65)
$$

IF ALL $\delta\underline{F}^{\mathcal{V}}$ COMPONENTS ARE WITHIN SPECIFIED LIMITS, EXIT ITERATION LOOP. OTHERWISE, CONTINUE

$\delta\underline{F}^{\mathcal{V}}$ Variation With P (For Partials That Are Functions Of P)

Aerodynamic Force Variation

$$
\begin{aligned}
\frac{\partial F_{Drag}}{\partial \beta} &= 2\, S\, q\, K_{DInd/Sd}\, C_{Sd\beta}\,\beta \\
\frac{\partial F_{Drag}}{\partial \alpha} &= 2\, S\, q\, K_{DInd/L}\, C_{L\alpha}\left(\alpha + \alpha_0 + \alpha_{Flaps}\right) \\
\frac{\partial \underline{F}_{Aero}^{\mathcal{V}W}}{\partial \underline{P}} &= \begin{bmatrix} 0 & -\dfrac{\partial F_{Drag}}{\partial \beta} & -\dfrac{\partial F_{Drag}}{\partial \alpha} \\ 0 & \dfrac{\partial F_{Side}}{\partial \beta} & 0 \\ 0 & 0 & -\dfrac{\partial F_{Lift}}{\partial \alpha} \end{bmatrix}
\end{aligned}
\qquad (17.1.2.3\text{-}66)
$$

$\left(\mathrm{C}_{\mathrm{AC}}^{\mathcal{V}\mathrm{W}}\ \underline{\mathrm{u}}_{\mathrm{Thrst}}^{\mathrm{AC}}\right)$ Variation

$$\frac{\partial\left(\mathrm{v}_{\mathrm{Arspd}_{\mathrm{XAC}}} / \mathrm{v}_{\mathrm{Arspd}}\right)}{\partial\beta} = -\left(1+\tan^2\alpha+\tan^2\beta\right)^{-\frac{3}{2}} \tan\beta \sec^2\beta$$

$$\frac{\partial\left(\mathrm{v}_{\mathrm{Arspd}_{\mathrm{XAC}}} / \mathrm{v}_{\mathrm{Arspd}}\right)}{\partial\alpha} = -\left(1+\tan^2\alpha+\tan^2\beta\right)^{-\frac{3}{2}} \tan\alpha \sec^2\alpha$$

$$\frac{\partial\left(\mathrm{v}_{\mathrm{Arspd}_{\mathrm{YAC}}} / \mathrm{v}_{\mathrm{Arspd}}\right)}{\partial\beta} = -\left[1-\left(1+\tan^2\alpha+\tan^2\beta\right)^{-1}\tan^2\beta\right]\left(1+\tan^2\alpha+\tan^2\beta\right)^{-\frac{1}{2}} \sec^2\beta$$

$$\frac{\partial\left(\mathrm{v}_{\mathrm{Arspd}_{\mathrm{YAC}}} / \mathrm{v}_{\mathrm{Arspd}}\right)}{\partial\alpha} = \left(1+\tan^2\alpha+\tan^2\beta\right)^{-\frac{3}{2}} \tan\beta \tan\alpha \sec^2\alpha$$

$$\frac{\partial\left(\mathrm{v}_{\mathrm{Arspd}_{\mathrm{ZAC}}} / \mathrm{v}_{\mathrm{Arspd}}\right)}{\partial\beta} = -\left(1+\tan^2\alpha+\tan^2\beta\right)^{-\frac{3}{2}} \tan\alpha \tan\beta \sec^2\beta \qquad (17.1.2.3\text{-}67)$$

$$\frac{\partial\left(\mathrm{v}_{\mathrm{Arspd}_{\mathrm{ZAC}}} / \mathrm{v}_{\mathrm{Arspd}}\right)}{\partial\alpha} = \left[1-\left(1+\tan^2\alpha+\tan^2\beta\right)^{-1}\tan^2\alpha\right]\left(1+\tan^2\alpha+\tan^2\beta\right)^{-\frac{1}{2}} \sec^2\alpha$$

$$\frac{\partial\left(\underline{\mathrm{v}}_{\mathrm{Arspd}}^{\mathrm{AC}} / \mathrm{v}_{\mathrm{Arspd}}\right)}{\partial\underline{\mathrm{P}}} = \begin{bmatrix} 0 & \dfrac{\partial\left(\mathrm{v}_{\mathrm{Arspd}_{\mathrm{XAC}}} / \mathrm{v}_{\mathrm{Arspd}}\right)}{\partial\beta} & \dfrac{\partial\left(\mathrm{v}_{\mathrm{Arspd}_{\mathrm{XAC}}} / \mathrm{v}_{\mathrm{Arspd}}\right)}{\partial\alpha} \\ 0 & \dfrac{\partial\left(\mathrm{v}_{\mathrm{Arspd}_{\mathrm{YAC}}} / \mathrm{v}_{\mathrm{Arspd}}\right)}{\partial\beta} & \dfrac{\partial\left(\mathrm{v}_{\mathrm{Arspd}_{\mathrm{YAC}}} / \mathrm{v}_{\mathrm{Arspd}}\right)}{\partial\alpha} \\ 0 & \dfrac{\partial\left(\mathrm{v}_{\mathrm{Arspd}_{\mathrm{ZAC}}} / \mathrm{v}_{\mathrm{Arspd}}\right)}{\partial\beta} & \dfrac{\partial\left(\mathrm{v}_{\mathrm{Arspd}_{\mathrm{ZAC}}} / \mathrm{v}_{\mathrm{Arspd}}\right)}{\partial\alpha} \end{bmatrix}$$

$$\frac{\partial\left(\mathrm{C}_{\mathrm{AC}}^{\mathcal{V}\mathrm{W}}\ \underline{\mathrm{u}}_{\mathrm{Thrst}}^{\mathrm{AC}}\right)}{\partial\underline{\mathrm{P}}} = \left\{\left[\left(\underline{\mathrm{u}}_{\mathrm{Thrst}}^{\mathrm{AC}}\times\right) + \frac{1}{1+\mathrm{G}}\left\{\left(\underline{\mathrm{H}}\times\right)\left(\underline{\mathrm{u}}_{\mathrm{Thrst}}^{\mathrm{AC}}\times\right) + \left[\left(\underline{\mathrm{H}}\times\underline{\mathrm{u}}_{\mathrm{Thrst}}^{\mathrm{AC}}\right)\times\right]\right\}\right] \frac{\left(\underline{\mathrm{v}}_{\mathrm{Arspd}}^{\mathcal{V}\mathrm{W}}\times\right)}{\mathrm{v}_{\mathrm{Arspd}}} - \frac{1}{(1+\mathrm{G})^2}\left(\underline{\mathrm{H}}\times\right)^2 \underline{\mathrm{u}}_{\mathrm{Thrst}}^{\mathrm{AC}} \frac{\left(\underline{\mathrm{v}}_{\mathrm{Arspd}}^{\mathcal{V}\mathrm{W}}\right)^{\mathrm{T}}}{\mathrm{v}_{\mathrm{Arspd}}}\right\} \frac{\partial\left(\underline{\mathrm{v}}_{\mathrm{Arspd}}^{\mathrm{AC}} / \mathrm{v}_{\mathrm{Arspd}}\right)}{\partial\underline{\mathrm{P}}}$$

Combined $\delta\underline{\mathrm{F}}^{\mathcal{V}}$ Variation

$$\frac{\partial\delta\underline{\mathrm{F}}^{\mathcal{V}}}{\partial\underline{\mathrm{P}}} = \mathrm{C}_{\mathcal{V}\mathrm{W}}^{\mathcal{V}}\left(\frac{\partial\underline{\mathrm{F}}_{\mathrm{Aero}}^{\mathcal{V}\mathrm{W}}}{\partial\underline{\mathrm{P}}} + \mathrm{C}_{\mathrm{AC}}^{\mathcal{V}\mathrm{W}}\ \underline{\mathrm{u}}_{\mathrm{Thrst}}^{\mathrm{AC}} \frac{\partial\mathrm{Thrst}}{\partial\underline{\mathrm{P}}} + \mathrm{Thrst}\frac{\partial\left(\mathrm{C}_{\mathrm{AC}}^{\mathcal{V}\mathrm{W}}\ \underline{\mathrm{u}}_{\mathrm{Thrst}}^{\mathrm{AC}}\right)}{\partial\underline{\mathrm{P}}}\right) - \underline{\mathrm{a}}_{\mathrm{SF}}^{\mathcal{V}}\frac{\partial\mathrm{M}}{\partial\underline{\mathrm{P}}} \qquad (17.1.2.3\text{-}68)$$

End $\delta\underline{\mathrm{F}}^{\mathcal{V}}$ Variation With P (For Partials That Are Functions Of P)

Error In Trial P

$$\Delta\underline{P} = -\left(\frac{\partial\delta\underline{F}^{\mathcal{V}}}{\partial\underline{P}}\right)^{-1} \delta\underline{F}^{\mathcal{V}} \qquad (17.1.2.3\text{-}69)$$

P Update

$$\underline{P} = \underline{P} + \Delta\underline{P} \qquad (17.1.2.3\text{-}70)$$

ENDDO

$C_{AC}^{\mathcal{V}}$ Calculation

$$C_{AC}^{\mathcal{V}} = C_{\mathcal{V}W}^{\mathcal{V}} \, C_{AC}^{\mathcal{V}W} \qquad (17.1.2.3\text{-}71)$$

The previous process for calculating $C_{AC}^{\mathcal{V}}$ is based on specifying the $\mathcal{V}$W Frame attitude as a minimum rotation angle movement from the $\mathcal{V}$ Frame. The result of this definition is that to second order in the $C_{AC}^{\mathcal{V}}$ rotation angle, the roll Euler angle between the AC and $\mathcal{V}$ Frames may be non-zero. The following sections discuss refinements to the $C_{AC}^{\mathcal{V}}$ computational process that can be incorporated to directly control the $C_{AC}^{\mathcal{V}}$ roll Euler angle.

17.1.2.3.1 Refinement To $C_{AC}^{\mathcal{V}}$ For Control Of AC Relative To $\mathcal{V}$ Frame Roll Euler Angle

In order to control the $C_{AC}^{\mathcal{V}}$ roll Euler angle, we must introduce a roll control parameter in the $C_{\mathcal{V}W}^{\mathcal{V}}$ matrix that causes it to deviate from the minimum rotation angle configuration. We thereby define:

$$C_{\mathcal{V}W}^{\mathcal{V}} = C_{\mathcal{V}W_0}^{\mathcal{V}} \, C_{\mathcal{V}W}^{\mathcal{V}W_0} \qquad (17.1.2.3.1\text{-}1)$$

where

$\mathcal{V}W_0$ = Coordinate frame with X axis along $\underline{v}_{Arspd}$ and with minimum rotation angle from the $\mathcal{V}$ Frame (i.e., as the $\mathcal{V}$W Frame was defined in Section 17.1.2.3).

$\mathcal{V}W$ = Coordinate frame with X axis along $\underline{v}_{Arspd}$, and oriented relative to the $\mathcal{V}W_0$ Frame by a rotation about the $\mathcal{V}W_0$ Frame X axis.

Based on these definitions, we can write for $C_{\mathcal{V}\,W}^{\mathcal{V}\,W_0}$ using generalized Equation (3.2.2.1-4):

$$C_{\mathcal{V}\,W}^{\mathcal{V}\,W_0} = I + \sin\phi_{Cntrl}\left(\underline{u}_{X\mathcal{V}\,W_0}^{\mathcal{V}\,W_0}\times\right) + (1 - \cos\phi_{Cntrl})\left(\underline{u}_{X\mathcal{V}\,W_0}^{\mathcal{V}\,W_0}\times\right)^2 \qquad (17.1.2.3.1\text{-}2)$$

where

$\underline{u}_{X\mathcal{V}\,W_0}$ = Unit vector along the $\mathcal{V}\,W_0$ Frame X axis.

ϕ_{Cntrl} = Roll control parameter.

and from Equations (17.1.2.3-54):

$$D \equiv \left(\underline{u}_{X\mathcal{V}\,W_0}^{\mathcal{V}\,W_0}\right)^T\left(\frac{\underline{v}_{Arspd}^{\mathcal{V}}}{v_{Arspd}}\right) \qquad \underline{E} \equiv \left(\underline{u}_{X\mathcal{V}\,W_0}^{\mathcal{V}\,W_0}\times\right)\left(\frac{\underline{v}_{Arspd}^{\mathcal{V}}}{v_{Arspd}}\right)$$

$$C_{\mathcal{V}\,W_0}^{\mathcal{V}} = I + \left(\underline{E}\times\right) + \frac{1}{1+D}\left(\underline{E}\times\right)^2 \qquad (17.1.2.3.1\text{-}3)$$

It will prove convenient for the development to rewrite force/inertial balance Equation (17.1.2.3-29) in the following equivalent form:

$$\underline{F}_j^{\mathcal{V}\,W} = \underline{F}_{Aero_j}^{\mathcal{V}\,W} + Thrst_j\, C_{AC_j}^{\mathcal{V}\,W}\, \underline{u}_{Thrst}^{AC}$$

$$\delta\underline{F}_j^{\mathcal{V}} = C_{\mathcal{V}\,W_j}^{\mathcal{V}}\, \underline{F}_j^{\mathcal{V}\,W} - M_j\, \underline{a}_{SF}^{\mathcal{V}} \qquad (17.1.2.3.1\text{-}4)$$

Due to the added roll control parameter, iteration loop Equation (17.1.2.3-31) becomes the expanded form:

$$\Delta\delta\underline{F}_j^{\mathcal{V}} \approx \left(\frac{\partial\delta\underline{F}^{\mathcal{V}}}{\partial\underline{P}}\right)_{j-1}\Delta\underline{P}_j + \left(\frac{\partial\delta\underline{F}^{\mathcal{V}}}{\partial\phi_{Cntrl}}\right)_{j-1}\Delta\phi_{Cntrl_j} \qquad (17.1.2.3.1\text{-}5)$$

where

$\Delta\phi_{Cntrl\,j}$ = Correction to the previous iteration cycle estimate for ϕ_{Cntrl} to achieve the desired control objective (i.e., in this case, control of the $C_{AC}^{\mathcal{V}}$ roll angle).

or with (17.1.2.3-33):

$$\delta\underline{F}_{j-1}^{\mathcal{V}} \approx -\left(\frac{\partial\delta\underline{F}^{\mathcal{V}}}{\partial\underline{P}}\right)_{j-1}\Delta\underline{P}_j - \left(\frac{\partial\delta\underline{F}^{\mathcal{V}}}{\partial\phi_{Cntrl}}\right)_{j-1}\Delta\phi_{Cntrl_j} \qquad (17.1.2.3.1\text{-}6)$$

Substituting (17.1.2.3.1-1) into the (17.1.2.3.1-4) $\delta \underline{F}^{\mathcal{V}}$ expression, recognizing that $\underline{F}^{\mathcal{V}\,W}$ and $C^{\mathcal{V}}_{\mathcal{V}\,W_0}$ are independent of ϕ_{Cntrl}, and taking the partial derivative with respect to ϕ_{Cntrl}, we obtain for $\frac{\partial \delta \underline{F}^{\mathcal{V}}}{\partial \phi_{Cntrl}}$ in (17.1.2.3.1-6):

$$\frac{\partial \delta \underline{F}^{\mathcal{V}}}{\partial \phi_{Cntrl}} = C^{\mathcal{V}}_{\mathcal{V}\,W_0} \frac{\partial C^{\mathcal{V}\,W_0}_{\mathcal{V}\,W}}{\partial \phi_{Cntrl}} \underline{F}^{\mathcal{V}\,W} \tag{17.1.2.3.1-7}$$

The partial in the previous expression is from (17.1.2.3.1-2):

$$\frac{\partial C^{\mathcal{V}\,W_0}_{\mathcal{V}\,W}}{\partial \phi_{Cntrl}} = \left[\cos \phi_{Cntrl} \left(\underline{u}^{\mathcal{V}\,W_0}_{X\mathcal{V}\,W_0} \times\right) + \sin \phi_{Cntrl} \left(\underline{u}^{\mathcal{V}\,W_0}_{X\mathcal{V}\,W_0} \times\right)^2\right] \tag{17.1.2.3.1-8}$$

In order to solve (17.1.2.3.1-6) for $\Delta \underline{P}_j$ and $\Delta \phi_{Cntrl\,j}$, an additional expression is needed defining the roll control requirement. We provide this in the form of the general constraint equation:

$$\vartheta = \underline{u}_{Z\mathcal{V}} \cdot \underline{u}_{YAC} - \phi_{AC/\mathcal{V}} \quad \text{which should be zero} \tag{17.1.2.3.1-9}$$

where

ϑ = General constraint parameter.

$\phi_{AC/\mathcal{V}}$ = Desired roll Euler angle between the AC and $\mathcal{V}$ Frames.

$\underline{u}_{Z\mathcal{V}}, \underline{u}_{YAC}$ = Unit vectors along the $\mathcal{V}$ Frame Z axis and the AC Frame Y axis.

The $\underline{u}_{Z\mathcal{V}} \cdot \underline{u}_{YAC}$ product in (17.1.2.3.1-9) represents the direction cosine between the $\mathcal{V}$ Frame Z axis and the AC Frame Y axis which from generalized Equations (3.2.3.1-2) (the C_{32} term) equals $\sin \phi \cos \theta$ (ϕ. θ are roll, pitch Euler angles). Assuming small roll and pitch angles between the AC and $\mathcal{V}$ Frames, $\sin \phi \cos \theta$ is approximately ϕ. Thus, $\underline{u}_{Z\mathcal{V}} \cdot \underline{u}_{YAC}$ in (17.1.2.3.1-9) measures the computed roll angle between the AC and $\mathcal{V}$ Frames, and ϑ measures the error in the computed roll angle satisfying the $\phi_{AC/\mathcal{V}}$ requirement. The error in the constraint equation is the difference between ϑ and its desired zero value:

$$\delta\vartheta = \vartheta \tag{17.1.2.3.1-10}$$

where

$\delta\vartheta$ = Error in ϑ meeting its zero requirement.

Numerical evaluation of (17.1.2.3.1-9) for ϑ is the following equivalent form using coordinate frame designations for vectors and (3.1.1-12) for the dot product:

$$\vartheta = \left(\underline{u}_{Z\mathcal{V}}^{\mathcal{V}}\right)^{T} C_{AC}^{\mathcal{V}} \, \underline{u}_{YAC}^{AC} - \phi_{AC/\mathcal{V}} \tag{17.1.2.3.1-11}$$

As in (17.1.2.3.1-5), we also define a variation in $\delta\vartheta$ produced by the iteration loop adjustments:

$$\Delta\delta\vartheta_j \approx \left(\frac{\partial\delta\vartheta}{\partial\underline{P}}\right)_{j-1} \Delta\underline{P}_j + \left(\frac{\partial\delta\vartheta}{\partial\phi_{Cntrl}}\right)_{j-1} \Delta\phi_{Cntrl_j} \tag{17.1.2.3.1-12}$$

where

$\Delta\delta\vartheta_j$ = Variation in $\delta\vartheta$ from its (17.1.2.3.1-10) $_{j-1}$ computed value due to $\Delta\underline{P}_j$ and $\Delta\phi_{Cntrl_j}$.

To make the iteration loop converge to a zero $\delta\vartheta$, we set $\Delta\delta\vartheta_j$ equal to the negative of the past value of $\delta\vartheta$ (similar to (17.1.2.3-33)):

$$\Delta\delta\vartheta_j = - \delta\vartheta_{j-1} \tag{17.1.2.3.1-13}$$

with which (17.1.2.3.1-12) becomes:

$$\delta\vartheta_{j-1} \approx - \left(\frac{\partial\delta\vartheta}{\partial\underline{P}}\right)_{j-1} \Delta\underline{P}_j - \left(\frac{\partial\delta\vartheta}{\partial\phi_{Cntrl}}\right)_{j-1} \Delta\phi_{Cntrl\,j} \tag{17.1.2.3.1-14}$$

Equations (17.1.2.3.1-6) and (17.1.2.3.1-14) in combined matrix form are:

$$\begin{bmatrix} \delta\underline{F}^{\mathcal{V}} \\ \delta\vartheta \end{bmatrix}_{j-1} = - \begin{bmatrix} \dfrac{\partial\delta\underline{F}^{\mathcal{V}}}{\partial\underline{P}} & \dfrac{\partial\delta\underline{F}^{\mathcal{V}}}{\partial\phi_{Cntrl}} \\ \dfrac{\partial\delta\vartheta}{\partial\underline{P}} & \dfrac{\partial\delta\vartheta}{\partial\phi_{Cntrl}} \end{bmatrix}_{j-1} \begin{bmatrix} \Delta\underline{P} \\ \Delta\phi_{Cntrl} \end{bmatrix}_{j} \tag{17.1.2.3.1-15}$$

whose solution is the inverse relationship similar to Equation (17.1.2.3-34):

$$\begin{bmatrix} \Delta \underline{P} \\ \Delta \phi_{Cntrl} \end{bmatrix}_j = - \begin{bmatrix} \frac{\partial \delta \underline{F}^{\mathcal{V}}}{\partial \underline{P}} & \frac{\partial \delta \underline{F}^{\mathcal{V}}}{\partial \phi_{Cntrl}} \\ \frac{\partial \delta \vartheta}{\partial \underline{P}} & \frac{\partial \delta \vartheta}{\partial \phi_{Cntrl}} \end{bmatrix}_{j-1}^{-1} \begin{bmatrix} \delta \underline{F}^{\mathcal{V}} \\ \delta \vartheta \end{bmatrix}_{j-1} \tag{17.1.2.3.1-16}$$

which is then used to update our estimates for $\underline{P}$ and ϕ_{Cntrl} similar to Equation (17.1.2.3-35):

$$\begin{bmatrix} \underline{P} \\ \phi_{Cntrl} \end{bmatrix}_j = \begin{bmatrix} \underline{P} \\ \phi_{Cntrl} \end{bmatrix}_{j-1} + \begin{bmatrix} \Delta \underline{P} \\ \Delta \phi_{Cntrl} \end{bmatrix}_j \tag{17.1.2.3.1-17}$$

The $\frac{\partial \delta \underline{F}^{\mathcal{V}}}{\partial \underline{P}}$ partial derivative in (17.1.2.3.1-16) is provided by Equation (17.1.2.3-36) from the previous section. The $\frac{\partial \delta \underline{F}^{\mathcal{V}}}{\partial \phi_{Cntrl}}$ partial derivative is given by Equations (17.1.2.3.1-7) and (17.1.2.3.1-8). The $\frac{\partial \delta \vartheta}{\partial \phi_{Cntrl}}$ partial derivative in (17.1.2.3.1-16) is determined from (17.1.2.3.1-10) and (17.1.2.3.1-11), treating $\phi_{AC/\mathcal{V}}$ as a constant, and using (17.1.2.3-1) with (17.1.2.3.1-1) to expand $C_{AC}^{\mathcal{V}}$:

$$\frac{\partial \delta \vartheta}{\partial \phi_{Cntrl}} = \left(\underline{u}_{Z\mathcal{V}}^{\mathcal{V}}\right)^T C_{\mathcal{V}W_0}^{\mathcal{V}} \frac{\partial C_{\mathcal{V}W}^{\mathcal{V}W_0}}{\partial \phi_{Cntrl}} C_{AC}^{\mathcal{V}W} \, \underline{u}_{YAC}^{AC} \tag{17.1.2.3.1-18}$$

The $\frac{\partial C_{\mathcal{V}W}^{\mathcal{V}W_0}}{\partial \phi_{Cntrl}}$ term in (17.1.2.3.1-18) is provided by (17.1.2.3.1-8).

The $\frac{\partial \delta \vartheta}{\partial \underline{P}}$ partial derivative in (17.1.2.3.1-16) is derived from (17.1.2.3.1-10) - (17.1.2.3.1-11) using (17.1.2.3-1) to expand $C_{AC}^{\mathcal{V}}$:

$$\frac{\partial \delta \vartheta}{\partial \underline{P}} = \left(\underline{u}_{Z\mathcal{V}}^{\mathcal{V}}\right)^T C_{\mathcal{V}W}^{\mathcal{V}} \frac{\partial \left(C_{AC}^{\mathcal{V}W} \, \underline{u}_{YAC}^{AC}\right)}{\partial \underline{P}} \tag{17.1.2.3.1-19}$$

An expression for $\frac{\partial\left(C_{AC}^{\mathcal{V}W}\ \underline{u}_{YAC}^{AC}\right)}{\partial \underline{P}}$ in (17.1.2.3.1-19) is derived using the identical procedure leading to Equation (17.1.2.3-50):

$$\frac{\partial\left(C_{AC}^{\mathcal{V}W}\ \underline{u}_{YAC}^{AC}\right)}{\partial \underline{P}} = \left\{ \left[\left(\underline{u}_{YAC}^{AC}\times\right) + \frac{1}{1+G}\left\{ \left(\underline{H}\times\right)\left(\underline{u}_{YAC}^{AC}\times\right) + \left[\left(\underline{H}\times\underline{u}_{YAC}^{AC}\right)\times \right] \right\} \right] \frac{\left(\underline{v}_{Arspd}^{\mathcal{V}W}\times\right)}{v_{Arspd}} - \frac{1}{(1+G)^2}\left(\underline{H}\times\right)^2 \underline{u}_{YAC}^{AC}\ \frac{\left(\underline{v}_{Arspd}^{\mathcal{V}W}\right)^T}{v_{Arspd}} \right\} \frac{\partial\left(\underline{v}_{Arspd}^{AC} / v_{Arspd}\right)}{\partial \underline{P}} \qquad (17.1.2.3.1\text{-}20)$$

The $\frac{\partial\left(\underline{v}_{Arspd}^{AC} / v_{Arspd}\right)}{\partial \underline{P}}$ term in (17.1.2.3.1-20) is provided by Equations (17.1.2.3-51) - (17.1.2.3-52) of the previous section.

Let us now summarize the previous results in context with variations from the Equations (17.1.2.3-53) - (17.1.2.3-71) summary of the previous section. The pertinent expressions from above are Equations (17.1.2.3.1-1) - (17.1.2.3.1-4), (17.1.2.3.1-7) - (17.1.2.3.1-8), (17.1.2.3.1-10) - (17.1.2.3.1-11) and (17.1.2.3.1-16) - (17.1.2.3.1-20) which when integrated into (17.1.2.3-53) - (17.1.2.3-71), summarize as follows in order of execution in a digital computer program:

<u>Inputs From Other Routines</u>

α_{Wand}, $C_{\mathcal{V}}^{L}$, $\underline{v}_{AvgWnd}^{Geo}$, ρ, $\underline{a}_{SF}^{\mathcal{V}}$, V, θ_{Thrst}, K_{Fuel}, S, $C_{Sd\beta}$, $C_{L\alpha}$, α_0, α_{Flaps}, C_{Df}, C_{Dthk}, $K_{DInd/L}$, $K_{DInd/Sd}$, $Thrst_{Start}$, M_0, ΔM_{Jtsn}

<u>$\mathcal{V}$ Frame Air Flow Components</u>

Equations (17.1.2.3-53)

<u>$C_{\mathcal{V}W_0}^{\mathcal{V}}$ Calculation</u>

$$D \equiv \left(\underline{u}_{X\mathcal{V}W_0}^{\mathcal{V}W_0}\right)^T \left(\frac{\underline{v}_{Arspd}^{\mathcal{V}}}{v_{Arspd}}\right) \qquad \underline{E} \equiv \left(\underline{u}_{X\mathcal{V}W_0}^{\mathcal{V}W_0}\times\right)\left(\frac{\underline{v}_{Arspd}^{\mathcal{V}}}{v_{Arspd}}\right)$$

$$C_{\mathcal{V}W_0}^{\mathcal{V}} = I + \left(\underline{E}\times\right) + \frac{1}{1+D}\left(\underline{E}\times\right)^2 \qquad (17.1.2.3.1\text{-}21)$$

Engine Thrust Direction

Equation (17.1.2.3-55)

Dynamic Pressure

Equation (17.1.2.3-56)

$\delta \underline{F}^{\mathcal{V}}$ Variation With P (For Partials That Are Independent Of P)

Equations (17.1.2.3-57) - (17.1.2.3-59)

Iteration Loop Initialization

$$\underline{P} = \begin{bmatrix} \text{Thrst} \\ \beta \\ \alpha \end{bmatrix} = \begin{bmatrix} 0 \\ 0 \\ 0 \end{bmatrix} \qquad \phi_{Cntrl} = 0 \qquad (17.1.2.3.1\text{-}22)$$

Iteration Loop

DO UNTIL $\delta\vartheta$ AND $\delta \underline{F}^{\mathcal{V}}$ COMPONENTS ARE WITHIN SPECIFIED LIMITS

$C^{\mathcal{V}}_{\mathcal{V}\,W}$ Calculation

$$C^{\mathcal{V}\,W_0}_{\mathcal{V}\,W} = I + \sin\phi_{Cntrl}\left(\underline{u}^{\mathcal{V}\,W_0}_{X\mathcal{V}\,W_0}\times\right) + (1 - \cos\phi_{Cntrl})\left(\underline{u}^{\mathcal{V}\,W_0}_{X\mathcal{V}\,W_0}\times\right)^2$$

$$C^{\mathcal{V}}_{\mathcal{V}\,W} = C^{\mathcal{V}}_{\mathcal{V}\,W_0}\, C^{\mathcal{V}\,W_0}_{\mathcal{V}\,W} \qquad (17.1.2.3.1\text{-}23)$$

$\underline{v}^{AC}_{Arspd}$ Calculation

Equations (17.1.2.3-61)

$C^{\mathcal{V}\,W}_{AC}$ Calculation

Equations (17.1.2.3-62)

$C^{\mathcal{V}}_{AC}$ Calculation

Equation (17.1.2.3-71)

Aerodynamic Forces

Equations (17.1.2.3-63)

Vehicle Mass

Equations (17.1.2.3-64)

Balance Of Applied And Inertial Reaction Forces

$$\underline{F}^{\mathcal{V}\,W} = \underline{F}^{\mathcal{V}\,W}_{Aero} + \text{Thrst}\; C^{\mathcal{V}\,W}_{AC}\; \underline{u}^{AC}_{Thrst}$$
$$\delta\underline{F}^{\mathcal{V}} = C^{\mathcal{V}}_{\mathcal{V}\,W}\; \underline{F}^{\mathcal{V}\,W} - M\; \underline{a}^{\mathcal{V}}_{SF} \qquad (17.1.2.3.1\text{-}24)$$

Error In Meeting AC Frame Roll Angle Constraint

$$\vartheta = \left(\underline{u}^{\mathcal{V}}_{Z\mathcal{V}}\right)^T C^{\mathcal{V}}_{AC}\; \underline{u}^{AC}_{YAC} - \phi_{AC/\mathcal{V}} \qquad \delta\vartheta = \vartheta \qquad (17.1.2.3.1\text{-}25)$$

IF $\delta\vartheta$ AND ALL $\delta\underline{F}^{\mathcal{V}}$ COMPONENTS ARE WITHIN SPECIFIED LIMITS, EXIT ITERATION LOOP. OTHERWISE, CONTINUE

$\delta\underline{F}^{\mathcal{V}}$ Variation With P (For Partials That Are Functions Of P)

Equations (17.1.2.3-66) - (17.1.2.3-68)

$\delta\underline{F}^{\mathcal{V}}$ Variation With ϕ_{Cntrl}

$$\frac{\partial C^{\mathcal{V}\,W_0}_{\mathcal{V}\,W}}{\partial \phi_{Cntrl}} = \left[\cos\phi_{Cntrl}\left(\underline{u}^{\mathcal{V}\,W_0}_{X\mathcal{V}\,W_0}\times\right) + \sin\phi_{Cntrl}\left(\underline{u}^{\mathcal{V}\,W_0}_{X\mathcal{V}\,W_0}\times\right)^2\right]$$
$$\frac{\partial \delta\underline{F}^{\mathcal{V}}}{\partial \phi_{Cntrl}} = C^{\mathcal{V}}_{\mathcal{V}\,W_0}\frac{\partial C^{\mathcal{V}\,W_0}_{\mathcal{V}\,W}}{\partial \phi_{Cntrl}}\underline{F}^{\mathcal{V}\,W} \qquad (17.1.2.3.1\text{-}26)$$

$\delta\vartheta$ Variation

$$\frac{\partial \delta\vartheta}{\partial \phi_{Cntrl}} = \left(\underline{u}^{\mathcal{V}}_{Z\mathcal{V}}\right)^T C^{\mathcal{V}}_{\mathcal{V}\,W_0}\frac{\partial C^{\mathcal{V}\,W_0}_{\mathcal{V}\,W}}{\partial \phi_{Cntrl}} C^{\mathcal{V}\,W}_{AC}\;\underline{u}^{AC}_{YAC} \qquad (17.1.2.3.1\text{-}27)$$
$$\frac{\partial\left(C^{\mathcal{V}\,W}_{AC}\;\underline{u}^{AC}_{YAC}\right)}{\partial \underline{P}} = \left\{\left[\left(\underline{u}^{AC}_{YAC}\times\right) + \frac{1}{1+G}\left\{\left(\underline{H}\times\right)\left(\underline{u}^{AC}_{YAC}\times\right) + \left[\left(\underline{H}\times\underline{u}^{AC}_{YAC}\right)\times\right]\right\}\right]\frac{\left(\underline{v}^{\mathcal{V}\,W}_{Arspd}\times\right)}{v_{Arspd}}\right.$$

(Continued)

$$- \frac{1}{(1+G)^2} \left(\underline{H}\times\right)^2 \underline{u}_{YAC}^{AC} \left(\frac{\left(\underline{v}_{Arspd}^{\mathcal{V}W}\right)^T}{v_{Arspd}} \right) \frac{\partial\left(\underline{v}_{Arspd}^{AC} / v_{Arspd}\right)}{\partial \underline{P}} \qquad (17.1.2.3.1\text{-}27)\ \text{(Continued)}$$

$$\frac{\partial \delta\vartheta}{\partial \underline{P}} = \left(\underline{u}_{Z\mathcal{V}}^{\mathcal{V}}\right)^T C_{\mathcal{V}W}^{\mathcal{V}} \frac{\partial\left(C_{AC}^{\mathcal{V}W}\, \underline{u}_{YAC}^{AC}\right)}{\partial \underline{P}}$$

Error In Trial $\underline{P}$ And ϕ_{Cntrl}

$$\begin{bmatrix} \Delta\underline{P} \\ \Delta\phi_{Cntrl} \end{bmatrix} = - \begin{bmatrix} \frac{\partial \delta\underline{F}^{\mathcal{V}}}{\partial \underline{P}} & \frac{\partial \delta\underline{F}^{\mathcal{V}}}{\partial \phi_{Cntrl}} \\ \frac{\partial \delta\vartheta}{\partial \underline{P}} & \frac{\partial \delta\vartheta}{\partial \phi_{Cntrl}} \end{bmatrix}^{-1} \begin{bmatrix} \delta\underline{F}^{\mathcal{V}} \\ \delta\vartheta \end{bmatrix} \qquad (17.1.2.3.1\text{-}28)$$

$\underline{P}$ And ϕ_{Cntrl} Update

$$\begin{bmatrix} \underline{P} \\ \phi_{Cntrl} \end{bmatrix} = \begin{bmatrix} \underline{P} \\ \phi_{Cntrl} \end{bmatrix} + \begin{bmatrix} \Delta\underline{P} \\ \Delta\phi_{Cntrl} \end{bmatrix} \qquad (17.1.2.3.1\text{-}29)$$

ENDDO

17.1.2.3.2 Utilization Of Constraint Formulation To Calculate $C_{AC}^{\mathcal{V}}$ With Zero ϕ_{Cntrl}

Given that the Section 17.1.2.3.1 formulation is to be used under particular circumstances to calculate $C_{AC}^{\mathcal{V}}$, it may be advantageous to recast the basic Section 17.1.2.3 equations into the same format using the ϑ constraint parameter approach. This is easily achieved by setting ϑ to constrain the control parameter ϕ_{Cntrl} to zero (which is the defining condition in Section 17.1.2.3 for $C_{AC}^{\mathcal{V}}$ determination):

Error In Meeting ϕ_{Cntrl} Control Parameter Constraint

$$\vartheta = \phi_{Cntrl} \qquad \delta\vartheta = \vartheta \qquad (17.1.2.3.2\text{-}1)$$

with the associated partials then becoming:

ϑ Variation

$$\frac{\partial \delta \vartheta}{\partial \underline{P}} = \begin{bmatrix} 0 & 0 & 0 \end{bmatrix} \qquad \frac{\partial \delta \vartheta}{\partial \phi_{Cntrl}} = 1 \tag{17.1.2.3.2-2}$$

Substitution of (17.1.2.3.2-1) - (17.1.2.3.2-2) for (17.1.2.3.1-25) and (17.1.2.3.1-27) in Section 17.1.2.3.1 converts the 17.1.2.3.1 summary equations into a computation for $C_{AC}^{\mathcal{V}}$ that will set the ϕ_{Cntrl} control parameter to zero, which is equivalent to the Section 17.1.2.3 result for $C_{AC}^{\mathcal{V}}$.

17.1.3 END-OF-SEGMENT DATA GENERATION

An accurate end-of-segment attitude, velocity and position (and time) solution is calculated by the End-Of-Segment Data Generation routine. The attitude, velocity, time solution is determined directly from the $\underline{\phi}_S^{\mathcal{V}}$, ΔV_S and T_S segment parameters. The end-of-segment position is calculated as a velocity integration process over the trajectory segment.

End-of-segment attitude is obtained as in Equations (17.1.2.1-1) with the $C_{\mathcal{V}_{End}}^{\mathcal{V}_{Start}}$ matrix computed from $\underline{\phi}_S^{\mathcal{V}}$ using generalized Equation (3.2.2.1-8):

$$C_{\mathcal{V}_{End}}^{L} = \left(C_{\mathcal{V}_{Start}}^{L}\right) C_{\mathcal{V}_{End}}^{\mathcal{V}_{Start}}$$
$$C_{\mathcal{V}_{End}}^{\mathcal{V}_{Start}} = \left[I + \frac{\sin \phi_S}{\phi_S} \left(\underline{\phi}_S^{\mathcal{V}} \times\right) + \frac{(1 - \cos \phi_S)}{\phi_S^2} \left(\underline{\phi}_S^{\mathcal{V}} \times\right)^2 \right] \tag{17.1.3-1}$$

End-of-segment velocity and time are calculated as in Equations (17.1.2.1-1), (17.1.2.1-2) and (17.1.2.1-4):

$$V_{End} = V_{Start} + \Delta V_S \qquad t_{End} = t_{Start} + T_S$$
$$\underline{v}_{End}^{N} = V_{End}\, C_L^N\, C_{\mathcal{V}_{End}}^{L}\, \underline{u}_{X\mathcal{V}}^{\mathcal{V}} \tag{17.1.3-2}$$

The $\underline{v}^N$ integration routines used for the end-of-segment position determination are executed at a selected repetition rate over the trajectory segment. The time interval from the start of the segment to the current integration time within the segment is calculated as:

$$t_m = t_{m-1} + T_m \qquad t_m = 0 \quad \text{At Start Of Trajectory Segment} \tag{17.1.3-3}$$

where

m = Trajectory segment computer integration cycle index. As a subscript, m designates values for parameters at the corresponding cycle time.

T_m = Trajectory segment integration time interval for cycle m.

t_m = Integration time interval since trajectory segment start.

In general, T_m will be constant over the segment integration period (and the same constant for all segments), however, we allow that for the last integration increment in a given segment, T_m will be adjusted to a smaller value so that the last t_m exactly fits T_S.

The velocity $\underline{v}^N$ at t_m is then calculated from (17.1.1.2-7) and (17.1.1.2-12) using a version of (17.1.2.1-1) - (17.1.2.1-2) for the t_m time point:

$$V_m = V_{m-1} + \frac{t_m}{T_S}\,\Delta V_S \qquad \left(C_{\mathcal{V}}^{L}\right)_m = \left(C_{\mathcal{V}_{Start}}^{L}\right)\left(C_{\mathcal{V}}^{\mathcal{V}_{Start}}\right)_m \qquad \underline{v}_m^N = V_m\, C_L^N \left(C_{\mathcal{V}}^{L}\right)_m \underline{u}_{X\mathcal{V}}^{\mathcal{V}} \tag{17.1.3-4}$$

The $C_{\mathcal{V}_m}^{\mathcal{V}_{Start}}$ matrix in (17.1.3-4) is computed from the equivalent rotation vector based on (7.1.1.1-14) and (17.1.1.2-13) for constant $\underline{\omega}_{N\mathcal{V}}^{\mathcal{V}}$ angular rate:

$$\underline{\phi}_m^{\mathcal{V}} = \frac{t_m}{T_S}\,\underline{\phi}_S^{\mathcal{V}} \tag{17.1.3-5}$$

where

$\underline{\phi}_m^{\mathcal{V}}$ = Rotation vector defining the attitude of the $\mathcal{V}$ Frame at time t_m relative to the $\mathcal{V}$ Frame attitude at the start of the trajectory segment.

Then using generalized Equation (3.2.2.1-8), $C_{\mathcal{V}_m}^{\mathcal{V}_{Start}}$ is calculated as:

$$\left(C_{\mathcal{V}}^{\mathcal{V}_{Start}}\right)_m = \left[I + \frac{\sin \phi_m}{\phi_m}\left(\underline{\phi}_m^{\mathcal{V}}\times\right) + \frac{(1 - \cos \phi_m)}{\phi_m^2}\left(\underline{\phi}_m^{\mathcal{V}}\times\right)^2 \right] \tag{17.1.3-6}$$

The position change $\Delta\underline{R}_m^N$ over T_m is defined formally for the position integration algorithm as in (7.3.1-4):

$$\Delta\underline{R}_m^N \equiv \int_{t_{m-1}}^{t_{m-1}+T_m} \underline{v}^N \, dt \tag{17.1.3-7}$$

Using $\underline{v}^N$ at the m-1 time point from (17.1.3-3) - (17.1.3-6), the position change $\Delta\underline{R}_m^N$ over T_m is evaluated as in (17.1.2.2-16) based on constant angular-rate/longitudinal-acceleration over cycle m, using (17.1.1.2-13) for the $\underline{\omega}_{N\mathcal{V}}^{\mathcal{V}}$ angular rate and the converse of (17.1.1.2-12) for the longitudinal acceleration $\dot{V}$:

$$\Delta\underline{R}_m^N = \underline{v}_{m-1}^N T_m + C_L^N \left(C_{\mathcal{V}}^L\right)_{m-1} \left\{ \left[\frac{1}{2} I + h_{2\psi}\left(\underline{\psi}_m^{\mathcal{V}}\times\right) + h_{3\psi}\left(\underline{\psi}_m^{\mathcal{V}}\times\right)^2\right] \left[\Delta V_m \, \underline{u}_{X\mathcal{V}}^{\mathcal{V}} + V_{m-1}\left(\underline{\psi}_m^{\mathcal{V}} \times \underline{u}_{X\mathcal{V}}^{\mathcal{V}}\right)\right] + \left[\frac{1}{6} I + h_{4\psi}\left(\underline{\psi}_m^{\mathcal{V}}\times\right) + h_{5\psi}\left(\underline{\psi}_m^{\mathcal{V}}\times\right)^2\right] \left[\Delta V_m \left(\underline{\psi}_m^{\mathcal{V}} \times \underline{u}_{X\mathcal{V}}^{\mathcal{V}}\right)\right] \right\} T_m \tag{17.1.3-8}$$

with

$$\underline{\psi}_m^{\mathcal{V}} = \frac{T_m}{T_S} \underline{\phi}_S^{\mathcal{V}} \tag{17.1.3-9}$$

$$\Delta V_m = \frac{T_m}{T_S} \Delta V_S \tag{17.1.3-10}$$

where

$\underline{\psi}_m^{\mathcal{V}}$ = Rotation vector defining the relative orientation of Frame $\mathcal{V}$ at the m and m-1 cycle times.

ΔV_m = Integrated signed velocity magnitude over the m-1 to m cycle time interval.

$h_{2\psi}, h_{3\psi}, h_{4\psi}, h_{5\psi}$ = h_2, h_3, h_4, h_5 from Equations (17.1.1.2-22), (17.1.2.2-10), (17.1.2.2-12) and (17.1.2.2-14) using ψ_m (the magnitude of $\underline{\psi}_m^{\mathcal{V}}$) in place of ϕ_S.

Once $\Delta\underline{R}_m^N$ and $\underline{v}_m^N$ are determined as outlined above, the position at cycle m is calculated to precision in the form of the C_N^E matrix and altitude h using a version of Equations (7.3.1-1), (7.3.1-3), (7.3.1-6), (7.3.1-8) and (7.3.1-11) with ρ_{ZN} set to zero (for the wander azimuth N Frame - See Section 4.5):

$$\begin{aligned} h_m &= h_{m-1} + \underline{u}_{ZN}^N \cdot \Delta\underline{R}_m^N \\ \left(C_N^E\right)_m &= \left(C_N^E\right)_{m-1} C_{N_m}^{N_{m-1}} \end{aligned} \tag{17.1.3-11}$$

(Continued)

$$C_{N_m}^{N_{m-1}} = I + \frac{\sin \xi_m}{\xi_m} \left(\underline{\xi}_m^N \times \right) + \frac{(1 - \cos \xi_m)}{\xi_m^2} \left(\underline{\xi}_m^N \times \right)^2$$

(17.1.3-11)
(Continued)

$$\underline{\xi}_m = F_{C_{m-1/2}}^N \left(\underline{u}_{ZN}^N \times \Delta \underline{R}_m^N \right)$$

The F_C^N curvature matrix in (17.1.3-11) (a function of position being computed) is calculated for an ellipsoidal shape earth model with Equation (5.3-18). The $F_{C_{m-1/2}}^N$ evaluation (at the m-1/2 cycle time) is performed by extrapolation of past F_C^N data (as in Equation (7.3.1-12)).

Equations (17.1.3-11) are processed repetitively (a digital integration process) from the start to the end of the trajectory segment. The h and C_N^E position parameters at the start of the segment integration process are initialized at their computed values at the end of the previous trajectory segment. For the first trajectory segment, h would be initialized at a specified input altitude value, and C_N^E set to correspond with input initial latitude, longitude and zero wander angle (using Equations (4.4.2.1-2)).

The end-of-segment latitude, longitude and wander angle would be calculated for output display at the end of the segment integration process from $C_{N_{End}}^E$ using (4.4.2.1-3).

17.2 TRAJECTORY REGENERATION FUNCTION

The trajectory regeneration function creates a navigation data profile at a selected timing rate from the basic trajectory segment parameters (T_S, ΔV_S, $\underline{\phi}_S^{\mathcal{V}}$) that were developed (and saved) during trajectory shaping. The trajectory regeneration function includes smoothing of the basic trajectory parameters to eliminate step changes in angular-rate/acceleration at the segment interfaces, selection of the B Frame attitude relative to the $\mathcal{V}$ Frame, creating integrated B Frame inertial angular rate and specific force acceleration increments at a selected timing rate (simulating the input to stapdown angular rate sensors and accelerometers), and generating attitude, velocity and position corresponding to the B Frame angular-rate/specific-force profile (i.e., the result that would be generated by strapdown inertial navigation integration of the B Frame angular-rate/specific-force data). Additionally, we can add wind gusts and high frequency angular/linear motion effects as an option. The following subsections describe each of these operations.

17.2.1 SEGMENT JUNCTION SMOOTHING

Smoothing of the junctions between trajectory segments is accomplished by filtering the angular rate and acceleration defined from the T_S, ΔV_S, $\underline{\phi}_S^{\mathcal{V}}$ parameters. In particular, we will be regenerating the trajectory data from incremental integrated angular rate and acceleration increments defined during each trajectory segment as:

$$\underline{\psi}_m^{\mathcal{V}} \equiv \int_{t_{m-1}}^{t_m} \underline{\omega}_{N\mathcal{V}}^{\mathcal{V}} \, dt = \frac{T_m}{T_S} \underline{\phi}_S^{\mathcal{V}} \qquad \Delta V_m \equiv \int_{t_{m-1}}^{t_m} \dot{V} \, dt = \frac{T_m}{T_S} \Delta V_S \qquad (17.2.1\text{-}1)$$

where

t = Trajectory regeneration time from the start of the trajectory profile.

m = Trajectory regeneration function computer cycle rate index corresponding to simulation time t_m.

T_m = Trajectory data regeneration time interval $t_m - t_{m-1}$.

$\underline{\psi}_m^{\mathcal{V}}$ = Integrated $\mathcal{V}$ Frame angular rate $\underline{\omega}_{N\mathcal{V}}^{\mathcal{V}}$ from time t_{m-1} to t_m.

ΔV_m = Change in signed velocity magnitude from time t_{m-1} to t_m.

Equations (17.2.1-1) are based on our definition for the trajectory segments as consisting of constant $\dot{V}$ and $\underline{\omega}_{N\mathcal{V}}^{\mathcal{V}}$ over the segment time T_S. Note that the definitions for t and m are now relative to the start of the trajectory profile (in contrast with their definition in subsections of Section 17.1 as being referenced to the start of each trajectory segment).

For the trajectory regeneration function, we generally require T_m to be constant for the total trajectory profile for interface compatibility with simulators receiving the generated trajectory data (i.e., a fixed update time interval). Because the trajectory segment T_S time periods are not generally integer multiples of T_m, Equations (17.2.1-1) must be modified accordingly at the trajectory segment junctions as follows:

Do At Trajectory Segment Junction:

$$\underline{\psi}_m^{\mathcal{V}} = \frac{T_{Fin/Crnt}}{T_{S_{Crnt}}} \underline{\phi}_{S_{Crnt}}^{\mathcal{V}} + \frac{T_m - T_{Fin/Crnt}}{T_{S_{Next}}} \underline{\phi}_{S_{Next}}^{\mathcal{V}} \qquad (17.2.1\text{-}2)$$

$$\Delta V_m = \frac{T_{Fin/Crnt}}{T_{S_{Crnt}}} \Delta V_{S_{Crnt}} + \frac{T_m - T_{Fin/Crnt}}{T_{S_{Next}}} \Delta V_{S_{Next}}$$

where

Crnt, Next = Subscripts designating parameter values for the <u>currently</u> ending trajectory segment and for the <u>next</u> upcoming trajectory segment.

$T_{Fin/Crnt}$ = Last (final) time increment in the current trajectory segment just prior to the next trajectory segment.

The smoothing function can be accomplished using the following general linear filtering operation on $\underline{\psi}_m^{\mathcal{V}}$ and ΔV_m:

$$\begin{aligned} \underline{\psi}_{F_m}^{\mathcal{V}} &= a_0\,\underline{\psi}_m^{\mathcal{V}} + a_1\,\underline{\psi}_{m-1}^{\mathcal{V}} + a_2\,\underline{\psi}_{m-2}^{\mathcal{V}} + \cdots + b_1\,\underline{\psi}_{F_{m-1}}^{\mathcal{V}} + b_2\,\underline{\psi}_{F_{m-2}}^{\mathcal{V}} + \cdots \\ \Delta V_{F_m} &= a_0\,\Delta V_m + a_1\,\Delta V_{m-1} + a_2\,\Delta V_{m-2} + \cdots + b_1\,\Delta V_{F_{m-1}} + b_2\,\Delta V_{F_{m-2}} + \cdots \end{aligned} \tag{17.2.1-3}$$

where

a_i, b_i = Smoothing filter coefficients (constants).

$\underline{\psi}_{F_m}^{\mathcal{V}}, \Delta V_{F_m}$ = Filtered (smoothed) values of $\underline{\psi}_m^{\mathcal{V}}, \Delta V_m$.

We also impose the so-called "exactness" constraint on the filter coefficients which for our case can be stated as the requirement that for extended periods of constant input (i.e., $\underline{\psi}_m^{\mathcal{V}}$ and ΔV_m), the filter output will converge in the steady state to the filter input (assuming a stable filter design). Under extended constant input $\underline{\psi}_m^{\mathcal{V}}$, in the steady state, the $\underline{\psi}_{F_m}^{\mathcal{V}}$ values for successive m cycles will be equal, i.e. $\underline{\psi}_{F_m}^{\mathcal{V}} = \underline{\psi}_{F_{m-1}}^{\mathcal{V}} = \underline{\psi}_{F_{m-2}}^{\mathcal{V}} =$ etc. (The same is true for successive ΔV_{F_m} under extended periods of constant ΔV_m input). Thus, for extended constant $\underline{\psi}_m^{\mathcal{V}}$ we can write for $\underline{\psi}_{F_m}^{\mathcal{V}}$ from (17.2.1-3):

$$\underline{\psi}_{F_m}^{\mathcal{V}} = \left(a_0 + a_1 + a_2 + \cdots\right) \underline{\psi}_m^{\mathcal{V}} + \left(b_1 + b_2 + \cdots\right) \underline{\psi}_{F_m}^{\mathcal{V}}$$

or

$$\underline{\psi}_{F_m}^{\mathcal{V}} \left(1 - b_1 - b_2 - \cdots\right) = \left(a_0 + a_1 + a_2 + \cdots\right) \underline{\psi}_m^{\mathcal{V}}$$

The exactness constraint requires $\underline{\psi}_{F_m}^{\mathcal{V}}$ in the previous expression to equal $\underline{\psi}_m^{\mathcal{V}}$ which is equivalent to the bracketed coefficient groupings being equal, hence:

$$1 - b_1 - b_2 - \cdots = a_0 + a_1 + a_2 + \cdots$$

or

$$a_0 + a_1 + a_2 + \cdots + b_1 + b_2 + \cdots = 1 \tag{17.2.1-4}$$

As we shall see shortly, imposing exactness Equation (17.2.1-4) on the filter coefficients implicitly bounds the integrated filter output to track the integrated filter input within a predictable "following error". This is an important characteristic of the smoothing filter because the integrated filter output represents the attitude/velocity that will result when using the filter outputs (in place of the unfiltered $\underline{\psi}_m^{\mathcal{V}}$, ΔV_m data) during trajectory regeneration. From their definition, the digital integral of $\underline{\psi}_m^{\mathcal{V}}$, ΔV_m (i.e., their summation) over the trajectory profile is the sum of $\underline{\phi}_S^{\mathcal{V}}$, ΔV_S over the profile. To preserve the essential trajectory characteristics of the $\underline{\phi}_S^{\mathcal{V}}$, ΔV_S segment parameters for the entire trajectory, the integral (summation) of $\underline{\psi}_{F_m}^{\mathcal{V}}$ and ΔV_{F_m} must also track the integral of $\underline{\psi}_m^{\mathcal{V}}$ and ΔV_m (within filter transient effects). Before returning to this issue, let us now develop the integration algorithm for generating velocity and attitude from $\underline{\psi}_{F_m}^{\mathcal{V}}$ and ΔV_{F_m}. The velocity algorithm is simply the sum of the ΔV_{F_m}'s:

$$V_{F_m} = V_{F_{m-1}} + \Delta V_{F_m} \qquad (17.2.1\text{-}5)$$

where

V_F = Signed velocity magnitude derived from the filtered ΔV_{F_m} increments.

The velocity vector corresponding to V_F is then obtained as in (17.1.3-4):

$$\underline{v}_m^N = V_{F_m} \, C_L^N \left(C_{\mathcal{V}F}^L \right)_m \underline{u}_{X\mathcal{V}F}^{\mathcal{V}F} \qquad (17.2.1\text{-}6)$$

where

$\mathcal{V}F$ = Coordinate frame whose attitude orientation is generated by integrating the $\underline{\psi}_{F_m}^{\mathcal{V}}$ data. In effect, the $\mathcal{V}F$ Frame represents the filtered $\mathcal{V}$ Frame.

$\underline{u}_{X\mathcal{V}F}^{\mathcal{V}F}$ = Unit vector along the $\mathcal{V}F$ Frame X axis.

The integration algorithm to generate $\left(C_{\mathcal{V}F}^L \right)_m$ from $\underline{\psi}_{F_m}^{\mathcal{V}}$ is designed using the (3.2.1-5) chain rule as an extension of the $\left(C_{\mathcal{V}}^L \right)_m$ integration algorithm:

$$\left(C_{\mathcal{V}}^L \right)_m = \left(C_{\mathcal{V}}^L \right)_{m-1} C_{\mathcal{V}_m}^{\mathcal{V}_{m-1}} \qquad (17.2.1\text{-}7)$$

The $C_{\mathcal{V}_m}^{\mathcal{V}_{m-1}}$ matrix is evaluated from its equivalent rotation vector using generalized Equation (3.2.2.1-8). Assuming constant angular rate over the m-1 to m interval (as in Equation

(7.1.1.1-14)), we equate $\underline{\psi}_m^{\mathcal{V}}$ (the integrated angular rate) to the $C_{\mathcal{V}_m}^{\mathcal{V}_{m-1}}$ equivalent rotation vector. This is generally consistent with $\underline{\psi}_m^{\mathcal{V}}$ being defined as in (17.2.1-1) as the integral of constant trajectory segment angular rate. We note in passing, however, that at the trajectory segment junctions, the $\underline{\omega}_{N\mathcal{V}}^{\mathcal{V}}$ angular rate from m-1 to m is generally not constant because it spans the end and beginning of two different trajectory segments. For the trajectory regeneration function we, never-the-less, still define $\underline{\psi}_m^{\mathcal{V}}$ once calculated (from (17.2.1-1) - (17.2.1-2)) as a rotation vector. The minor error this will introduce will be to generate a $\mathcal{V}$ Frame attitude history during trajectory regeneration that may have very slight deviation from the $\mathcal{V}$ Frame attitude history generated during trajectory shaping. Since the shaped trajectory profile is generally somewhat arbitrary to begin with, the profile created during trajectory regeneration is probably equally valid. Proceeding with $\underline{\psi}_m^{\mathcal{V}}$ calculated from (17.2.1-1) - (17.2.1-2), and treating it as a rotation vector, we then have, using generalized Equation (3.2.2.1-8):

$$C_{\mathcal{V}_m}^{\mathcal{V}_{m-1}} = \left[I + \frac{\sin \psi_m}{\psi_m} \left(\underline{\psi}_m^{\mathcal{V}} \times \right) + \frac{(1 - \cos \psi_m)}{\psi_m^2} \left(\underline{\psi}_m^{\mathcal{V}} \times \right)^2 \right] \tag{17.2.1-8}$$

Having computed $\left(C_{\mathcal{V}}^{L}\right)_m$ by the (17.2.1-7) - (17.2.1-8) integration process, we then calculate the equivalent version based on the filtered $\underline{\psi}_{F_m}^{\mathcal{V}}$ data as:

$$\left(C_{\mathcal{V}F}^{L}\right)_m = \left(C_{\mathcal{V}}^{L}\right)_m C_{\mathcal{V}F_m}^{\mathcal{V}_m} \tag{17.2.1-9}$$

where

$C_{\mathcal{V}F_m}^{\mathcal{V}_m}$ =Direction cosine matrix relating the $\mathcal{V}$F Frame and $\mathcal{V}$ Frame attitudes at cycle time

The $C_{\mathcal{V}F_m}^{\mathcal{V}_m}$ matrix is obtained from the rotation vector between the $\mathcal{V}$F and $\mathcal{V}$ Frames using generalized Equation (3.2.2.1-8):

$$C_{\mathcal{V}F_m}^{\mathcal{V}_m} = \left[I + \frac{\sin \gamma_m}{\gamma_m} \left(\underline{\gamma}_m^{\mathcal{V}} \times \right) + \frac{(1 - \cos \gamma_m)}{\gamma_m^2} \left(\underline{\gamma}_m^{\mathcal{V}} \times \right)^2 \right] \tag{17.2.1-10}$$

where

$\underline{\gamma}_m^{\mathcal{V}}$ = Difference between the integrated (cumulative) $\underline{\psi}_m^{\mathcal{V}}$ and the integrated $\underline{\psi}_{F_m}^{\mathcal{V}}$ increments.

From its definition, $\overset{\mathcal{V}}{\underline{\gamma}}_m$ is given by:

$$\overset{\mathcal{V}}{\underline{\gamma}}_m = \sum_{i=1}^{i=m} \overset{\mathcal{V}}{\underline{\psi}}_{F_i} - \sum_{i=1}^{i=m} \overset{\mathcal{V}}{\underline{\psi}}_i = \sum_{i=1}^{i=m} \left(\overset{\mathcal{V}}{\underline{\psi}}_{F_i} - \overset{\mathcal{V}}{\underline{\psi}}_i \right) \qquad (17.2.1\text{-}11)$$

The recursive algorithm equivalent to (17.2.1-11) for calculating $\overset{\mathcal{V}}{\underline{\gamma}}_m$ in the trajectory generator software is:

$$\overset{\mathcal{V}}{\underline{\gamma}}_m = \overset{\mathcal{V}}{\underline{\gamma}}_{m-1} + \overset{\mathcal{V}}{\underline{\psi}}_{F_m} - \overset{\mathcal{V}}{\underline{\psi}}_m \qquad (17.2.1\text{-}12)$$

Equations (17.2.1-7) - (17.2.1-10) with (17.2.1-12) define a method for calculating $\left(C^{L}_{\mathcal{V}F} \right)_m$ as the integrated effect of $\overset{\mathcal{V}}{\underline{\psi}}_{F_m}$, and which deviates from $\left(C^{L}_{\mathcal{V}} \right)_m$ by $\overset{\mathcal{V}}{\underline{\gamma}}_m$, the difference between the integrated $\overset{\mathcal{V}}{\underline{\psi}}_{F_m}$ and $\overset{\mathcal{V}}{\underline{\psi}}_m$ increments. At first look you might believe it easier to simply calculate $\left(C^{L}_{\mathcal{V}F} \right)_m$ using a direct $\overset{\mathcal{V}}{\underline{\psi}}_{F_m}$ integration algorithm similar to (17.2.1-7) - (17.2.1-8) (i.e., with $C^{\mathcal{V}_{m-1}}_{\mathcal{V}_m}$ replaced by $C^{\mathcal{V}F_{m-1}}_{\mathcal{V}F_m}$ and calculated as in (17.2.1-8), but using $\overset{\mathcal{V}}{\underline{\psi}}_{F_m}$). The problem with this approach is that the $\mathcal{V}$F Frame orientation would not be constrained to remain at a $\overset{\mathcal{V}}{\underline{\gamma}}_m$ rotation vector attitude relative to the $\mathcal{V}$ Frame, and due to second order effects, might gradually drift away from the $\mathcal{V}$ Frame by a value that differed more and more from $\overset{\mathcal{V}}{\underline{\gamma}}_m$. The resulting regenerated trajectory might thereby differ significantly from the trajectory profile created during the trajectory shaping process. By directly constraining the $\mathcal{V}$F Frame to remain at $\overset{\mathcal{V}}{\underline{\gamma}}_m$ relative to the $\mathcal{V}$ Frame, we are limiting the regenerated trajectory to small variations from the shaped profile, created only by the filtering effect itself. As explained below, imposing the (17.2.1-4) exactness constraint on the selected smoothing filter assures that $\overset{\mathcal{V}}{\underline{\gamma}}_m$ will be bounded and will always return to zero during zero angular rate trajectory periods. The same, of course, will be true for the difference between V and V_F (i.e., the difference between integrated ΔV_{F_m} and ΔV_m), because the same smoothing filter is used in (17.2.1-3) to generate ΔV_{F_m} from ΔV_m.

Let us summarize the results obtained for generating smoothed $\mathcal{V}$ Frame attitude and velocity data during trajectory regeneration. We first apply Equations (17.2.1-1) - (17.2.1-2) to calculate the smoothing filter <u>inputs</u> ΔV_m and $\overset{\mathcal{V}}{\underline{\psi}}_m$. Equations (17.2.1-3) are then processed to calculate the smoothing filter <u>outputs</u> ΔV_{F_m} and $\overset{\mathcal{V}}{\underline{\psi}}_{F_m}$. Equation (17.2.1-5) is applied next to compute the

$\mathcal{V}$F Frame signed velocity magnitude V_{F_m} as a ΔV_{F_m} integration process. Then $\underline{\gamma}_m^{\mathcal{V}}$ is determined from (17.2.1-12) as the cumulative difference between $\underline{\psi}_{F_m}^{\mathcal{V}}$ and $\underline{\psi}_m^{\mathcal{V}}$. Finally, the smoothed attitude matrix $\left(C_{\mathcal{V}F}^{L}\right)_m$ is obtained by first performing the (17.2.1-7) - (17.2.1-8) integration process to calculate $\left(C_{\mathcal{V}}^{L}\right)_m$, and then computing $\left(C_{\mathcal{V}F}^{L}\right)_m$ from $\left(C_{\mathcal{V}}^{L}\right)_m$ and $\underline{\gamma}_m^{\mathcal{V}}$ using (17.2.1-9) - (17.2.1-10).

One additional point we might note regards the interpretation of $\underline{\psi}_m^{\mathcal{V}}$ compared to $\underline{\psi}_{F_m}^{\mathcal{V}}$. From its basic definition in (17.2.1-1), $\underline{\psi}_m^{\mathcal{V}}$ represents a rotation vector created by constant angular rate. If a strapdown sensor assembly was somehow constrained to remain parallel to the $\mathcal{V}$ Frame, ideal angular rate sensors would output $\underline{\psi}_m^{\mathcal{V}}$ as integrated angular rate increments (plus the L Frame rotation increment). If, on the other hand, the strapdown sensor assembly was parallel to the $\mathcal{V}$F Frame, we might expect $\underline{\psi}_{F_m}^{\mathcal{V}}$ (plus the L Frame increment) to be the angular rate sensor output. This is not exactly true however (in error to second order), due to the manner in which we have calculated $\left(C_{\mathcal{V}F}^{L}\right)_m$ in (17.2.1-9) rather than from a direct $\left(C_{\mathcal{V}F}^{L}\right)_m$ integration algorithm using $\underline{\psi}_{F_m}^{\mathcal{V}}$ for input (as we did for $\left(C_{\mathcal{V}}^{L}\right)_m$ from $\underline{\psi}_m^{\mathcal{V}}$ in (17.2.1-7) - (17.2.1-8)). To calculate the correct angular rate sensor output for sensors that are parallel to the $\mathcal{V}$F Frame, we must use a rotation vector extraction algorithm based on successive $\left(C_{\mathcal{V}F}^{L}\right)_m$'s (i.e., Equations (3.2.2.2-10) - (3.2.2.2-12) and (3.2.2.2-15) - (3.2.2.2-19)), and on the L Frame attitude history relative to non-rotating inertial space. This will be the approach used during Section 17.2.3 Trajectory Regeneration Operations.

For the remainder of this section we will analyze the (17.2.1-3) - (17.2.1-4) smoothing filter properties in more detail to confirm behavior patterns of the integrated filter output alluded to previously. Using the $\underline{\psi}_{F_m}^{\mathcal{V}}$ filter as representative, we first expand (17.2.1-11) for $\underline{\gamma}_m^{\mathcal{V}}$ using (17.2.1-3) for $\underline{\psi}_{F_m}^{\mathcal{V}}$ to obtain:

$$\underline{\gamma}_m^{\mathcal{V}} = \sum_{i=1}^{i=m} \left[(a_0 - 1)\,\underline{\psi}_i^{\mathcal{V}} + a_1\,\underline{\psi}_{i-1}^{\mathcal{V}} + a_2\,\underline{\psi}_{i-2}^{\mathcal{V}} + \cdots + b_1\,\underline{\psi}_{F_{i-1}}^{\mathcal{V}} + b_2\,\underline{\psi}_{F_{i-2}}^{\mathcal{V}} + \cdots \right] \qquad (17.2.1\text{-}13)$$

The summation of individual i-k terms in (17.2.1-13) can be expanded to the following equivalent form:

$$\sum_{i=1}^{i=m} \underline{\psi}_{i-k}^{\mathcal{V}} = \sum_{i=1}^{i=m} \underline{\psi}_{i}^{\mathcal{V}} - \sum_{i=m-k+1}^{i=m} \underline{\psi}_{i}^{\mathcal{V}} + \sum_{i=1-k}^{i=0} \underline{\psi}_{i}^{\mathcal{V}}$$

$$\sum_{i=1}^{i=m} \underline{\psi}_{F_{i-k}}^{\mathcal{V}} = \sum_{i=1}^{i=m} \underline{\psi}_{F_i}^{\mathcal{V}} - \sum_{i=m-k+1}^{i=m} \underline{\psi}_{F_i}^{\mathcal{V}} + \sum_{i=1-k}^{i=0} \underline{\psi}_{F_i}^{\mathcal{V}} \tag{17.2.1-14}$$

We assume that prior to m = 1 (i.e., prior to the start of the trajectory profile), the $\underline{\psi}_{m}^{\mathcal{V}}$'s will be zero, hence, from (17.2.1-3), the $\underline{\psi}_{F_m}^{\mathcal{V}}$'s will also be zero. Thus, the last terms in (17.2.1-14) are zero and we can write:

$$\sum_{i=1}^{i=m} \underline{\psi}_{i-k}^{\mathcal{V}} = \sum_{i=1}^{i=m} \underline{\psi}_{i}^{\mathcal{V}} - \sum_{i=m-k+1}^{i=m} \underline{\psi}_{i}^{\mathcal{V}}$$

$$\sum_{i=1}^{i=m} \underline{\psi}_{F_{i-k}}^{\mathcal{V}} = \sum_{i=1}^{i=m} \underline{\psi}_{F_i}^{\mathcal{V}} - \sum_{i=m-k+1}^{i=m} \underline{\psi}_{F_i}^{\mathcal{V}} \tag{17.2.1-15}$$

Substituting (17.2.1-15) into (17.2.1-13) then obtains:

$$\begin{aligned} \underline{\gamma}_{m}^{\mathcal{V}} = & \sum_{i=1}^{i=m} \left[\left(a_0 - 1 + a_1 + a_2 + \cdots \right) \underline{\psi}_{i}^{\mathcal{V}} + \left(b_1 + b_2 + \cdots \right) \underline{\psi}_{F_i}^{\mathcal{V}} \right] \\ & - a_1 \, \underline{\psi}_{m}^{\mathcal{V}} - a_2 \left(\underline{\psi}_{m}^{\mathcal{V}} + \underline{\psi}_{m-1}^{\mathcal{V}} \right) - \cdots - b_1 \, \underline{\psi}_{F_m}^{\mathcal{V}} - b_2 \left(\underline{\psi}_{F_m}^{\mathcal{V}} + \underline{\psi}_{F_{m-1}}^{\mathcal{V}} \right) - \cdots \end{aligned} \tag{17.2.1-16}$$

Exactness constraint Equation (17.2.1-4) is equivalently:

$$b_1 + b_2 + \cdots = - \left(a_0 + a_1 + a_2 + \cdots - 1 \right) \tag{17.2.1-17}$$

With (17.2.1-17) and the (17.2.1-11) definition for $\underline{\gamma}_{m}^{\mathcal{V}}$, the summation term in (17.2.1-16) becomes:

$$\begin{aligned} & \sum_{i=1}^{i=m} \left[\left(a_0 + a_1 + a_2 + \cdots - 1 \right) \underline{\psi}_{i}^{\mathcal{V}} + \left(b_1 + b_2 + \cdots \right) \underline{\psi}_{F_i}^{\mathcal{V}} \right] \\ & \quad = - \left(a_0 + a_1 + a_2 + \cdots - 1 \right) \sum_{i=1}^{i=m} \left(\underline{\psi}_{F_i}^{\mathcal{V}} - \underline{\psi}_{i}^{\mathcal{V}} \right) = - \left(a_0 + a_1 + a_2 + \cdots - 1 \right) \underline{\gamma}_{m}^{\mathcal{V}} \end{aligned} \tag{17.2.1-18}$$

Substituting (17.2.1-18) in (17.2.1-16) then yields after rearrangement:

$$\overset{\mathcal{V}}{\underline{\gamma}}_m = - \frac{1}{\left(a_0 + a_1 + a_2 + \cdots\right)} \left[a_1 \overset{\mathcal{V}}{\underline{\psi}}_m + a_2 \left(\overset{\mathcal{V}}{\underline{\psi}}_m + \overset{\mathcal{V}}{\underline{\psi}}_{m-1} \right) + \cdots + b_1 \overset{\mathcal{V}}{\underline{\psi}}_{F_m} + b_2 \left(\overset{\mathcal{V}}{\underline{\psi}}_{F_m} + \overset{\mathcal{V}}{\underline{\psi}}_{F_{m-1}} \right) + \cdots \right] \tag{17.2.1-19}$$

Let us now analyze Equation (17.2.1-19) under sustained constant angular rate conditions. Using the logic that led to the Equation (17.2.1-4) exactness constraint, under sustained constant $\overset{\mathcal{V}}{\underline{\psi}}_m$ conditions, all $\overset{\mathcal{V}}{\underline{\psi}}$ and $\overset{\mathcal{V}}{\underline{\psi}}_F$ terms become equal to the current input $\overset{\mathcal{V}}{\underline{\psi}}_m$, and (17.2.1-19) reduces to:

$$\overset{\mathcal{V}}{\underline{\gamma}}_m = - \frac{\sum_{i=1}^{r_a} i\, a_i + \sum_{i=1}^{r_b} i\, b_i}{\sum_{i=0}^{r_a} a_i} \overset{\mathcal{V}}{\underline{\psi}}_m \tag{17.2.1-20}$$

where

r_a, r_b = Highest numerical subscript for the non-zero a_i, b_i smoothing filter past value coefficients.

Equation (17.2.1-20) also applies for the (17.2.1-3) velocity smoothing filter under steady input:

$$\mu_m = - \frac{\sum_{i=1}^{r_a} i\, a_i + \sum_{i=1}^{r_b} i\, b_i}{\sum_{i=0}^{r_a} a_i} \Delta V_m \tag{17.2.1-21}$$

where

μ_m = Integral of the difference between the ΔV_F filter input and output at cycle m.

Equations (17.2.1-20) - (17.2.1-21) show that under the (17.2.1-4) exactness constraint, the smoothing filter described by Equations (17.2.1-3) has an output under steady input, whose integral follows the integrated input within a fixed value proportional to the input. The constant of proportionality can be expressed as the equivalent to a dynamic response time τ_{Filt} in the filter output relative to the input where:

τ_{Filt} = Filter dynamic response time defined as the time required under constant angular rate for the $\mathcal{V}$ Frame to rotate from the $\mathcal{V}F$ attitude to its current attitude at cycle m (i.e., from $\mathcal{V}F$ through minus $\overset{\mathcal{V}}{\underline{\gamma}}_m$). We also define τ_{Filt} to be positive for a "lag" condition when $\underline{\gamma}_m^{\mathcal{V}}$ is negative (i.e., the sum (or integral) of $\underline{\psi}_{F_m}^{\mathcal{V}}$ is lagging the sum of $\underline{\psi}_m^{\mathcal{V}}$). A negative τ_{Filt} is denoted as a dynamic "lead" condition in which the sum of $\underline{\psi}_{F_m}^{\mathcal{V}}$ leads the sum of $\underline{\psi}_m^{\mathcal{V}}$. To achieve dynamic lead, prediction is implied in the filter coefficients.

From Equation (17.2.1-1) under constant angular rate, the angular rate at cycle m is $\underline{\psi}_m^{\mathcal{V}}$ divided by T_m, and the above τ_{Filt} definition translates analytically to:

$$\underline{\gamma}_m^{\mathcal{V}} = -\frac{1}{T_m}\underline{\psi}_m^{\mathcal{V}}\,\tau_{Filt} \qquad (17.2.1\text{-}22)$$

Combining Equations (17.2.1-20) and (17.2.1-22) we see that:

$$\tau_{Filt} = \frac{\sum_{i=1}^{r_a} i\,a_i + \sum_{i=1}^{r_b} i\,b_i}{\sum_{i=0}^{r_a} a_i}\,T_m \qquad (17.2.1\text{-}23)$$

An example of a simple filter that might be used for the smoothing function is the so-called "walking window" filter which has zero for the b_i coefficients and equal values for the a_i coefficients. For this filter configuration, Equations (17.2.1-3) with (17.2.1-4) simplify to:

$$\underline{\psi}_{F_m}^{\mathcal{V}} = \frac{1}{r_a+1}\sum_{i=0}^{r_a}\underline{\psi}_{m-i}^{\mathcal{V}} \qquad \Delta V_{F_m} = \frac{1}{r_a+1}\sum_{i=0}^{r_a}\Delta V_{m-i} \qquad (17.2.1\text{-}24)$$

and τ_{Filt} from (17.2.1-23) is:

$$\tau_{Filt} = \frac{\sum_{i=1}^{r_a} i}{r_a+1}\,T_m \qquad (17.2.1\text{-}25)$$

The i terms in the previous expression form a simple arithmetic progression whose sum is $r_a(r_a+1)/2$. Then, (17.2.1-25) becomes:

$$\tau_{Filt} = \frac{r_a}{2}\,T_m \qquad (17.2.1\text{-}26)$$

It may be desirable to have the $\mathcal{V}$F Frame attitude and velocity (i.e., the equivalent of the integrated filter output) match the $\mathcal{V}$ Frame attitude/velocity under steady angular-rate/longitudinal-acceleration conditions. Because this is a simulation program, a particularly simple way of accomplishing this is to modify the $\underline{\psi}_m^{\mathcal{V}}$, ΔV_m input data time base to occur τ_{Filt} seconds <u>earlier</u>. Then the ΔV_F, $\underline{\psi}_F^{\mathcal{V}}$ smoothing filter outputs will be brought into time synchronization with the original $\underline{\psi}_m^{\mathcal{V}}$, ΔV_m input data due to the filter dynamic lag. Alternatively, an additional constraint can be placed on the smoothing coefficients by which τ_{Filt} is set to zero. From Equation (17.2.1-23), we see that this occurs for:

$$\sum_{i=1}^{r_a} i\, a_i + \sum_{i=1}^{r_b} i\, b_i = 0 \tag{17.2.1-27}$$

By imposing the (17.2.1-27) constraint, we are in effect, forcing the filter to "catch up" under steady inputs, to time lags developed under previous dynamic conditions. In general, the net effect is a penalty in the smoothing properties of the filter. For example, consider the simple filter configuration (for even values of r_a) defined by:

$$\begin{aligned} & b_i = 0 \qquad\qquad a_0 = a \\ & a_i = \frac{a}{i} \;\; \text{for } i = 1 \text{ to } \frac{r_a}{2} \qquad\qquad a_i = -\frac{a}{i} \;\; \text{for } i = 1 + \frac{r_a}{2} \text{ to } r_a \end{aligned} \tag{17.2.1-28}$$

which satisfies (17.2.1-27). With exactness constraint (17.2.1-4) and the (17.2.1-28) coefficients, Equations (17.2.1-3) become for the $\underline{\psi}_{F_m}^{\mathcal{V}}$ smoothing filter:

$$\begin{aligned} \underline{\psi}_{F_m}^{\mathcal{V}} &= K\left(\underline{\psi}_m^{\mathcal{V}} + \sum_{i=1}^{r_a/2} \frac{1}{i}\,\underline{\psi}_{m-i}^{\mathcal{V}} - \sum_{i=1+r_a/2}^{r_a} \frac{1}{i}\,\underline{\psi}_{m-i}^{\mathcal{V}}\right) \\ K &\equiv \frac{1}{\left(1 + \sum_{i=1}^{r_a/2} \frac{1}{i} - \sum_{i=1+r_a/2}^{r_a} \frac{1}{i}\right)} \end{aligned} \tag{17.2.1-29}$$

For r_a equal to 10 as an example, consider the response of (17.2.1-29) to a step change in $\underline{\psi}_m^{\mathcal{V}}$ (call it $\Delta\underline{\psi}^{\mathcal{V}}$) from a steady state $\underline{\psi}_m^{\mathcal{V}}$ (and $\underline{\psi}_{F_m}^{\mathcal{V}}$) condition. The filter smoothing effectiveness can be measured by the magnitude of the maximum <u>change</u> in the filter output $\underline{\psi}_{F_m}^{\mathcal{V}}$ from cycle to cycle (i.e., $\underline{\psi}_{F_m}^{\mathcal{V}} - \underline{\psi}_{F_{m-1}}^{\mathcal{V}}$) compared to the maximum cycle-to-cycle <u>change</u> in the filter input.

The filter input cycle-to-cycle change profile $\underline{\psi}_m^{V} - \underline{\psi}_{m-1}^{V}$ for the steady input plus the $\Delta\underline{\psi}^{V}$ step is the sequence $\cdots$, 0, 0, $\Delta\underline{\psi}^{V}$, 0, 0, $\cdots$. The (17.2.1-29) filter output cycle-to-cycle change response ($\underline{\psi}_{F_m}^{V} - \underline{\psi}_{F_{m-1}}^{V}$) to the steady input plus the $\Delta\underline{\psi}^{V}$ step is the sequence $\cdots$, 0, 0, $K\,\Delta\underline{\psi}^{V}$, $K\,\Delta\underline{\psi}^{V}$, $\frac{K}{2}\Delta\underline{\psi}^{V}$, $\frac{K}{3}\Delta\underline{\psi}^{V}$, $\frac{K}{4}\Delta\underline{\psi}^{V}$, $\frac{K}{5}\Delta\underline{\psi}^{V}$, $-\frac{K}{6}\Delta\underline{\psi}^{V}$, $-\frac{K}{7}\Delta\underline{\psi}^{V}$, $-\frac{K}{8}\Delta\underline{\psi}^{V}$, $-\frac{K}{9}\Delta\underline{\psi}^{V}$, $-\frac{K}{10}\Delta\underline{\psi}^{V}$, 0, 0, $\cdots$. The sum of the $\underline{\psi}_{F_m}^{V} - \underline{\psi}_{F_{m-1}}^{V}$ changes (with K from (17.2.1-29)) adds exactly to $\Delta\underline{\psi}^{V}$ (as it should due to the exactness constraint). For the previous sequence, the largest change in $\underline{\psi}_{F_m}^{V}$ is $K\,\Delta\underline{\psi}^{V}$ occurring for the first two cycles following the $\Delta\underline{\psi}^{V}$ change. For the 10 stage filter selected, K from (17.2.1-29) is 0.3791. Thus, the filter has the effect of smoothing the input step from a 1.0 $\Delta\underline{\psi}^{V}$ maximum cycle-to-cycle <u>input</u> change to a 0.3791 $\Delta\underline{\psi}^{V}$ maximum filter cycle-to-cycle <u>output</u> change.

The advantage for the (17.2.1-29) filter is that the integral of its output eventually balances the integral of the input following the $\Delta\underline{\psi}^{V}$ step input. For the r_a value of 10 discussed previously, the (17.2.1-29) filter integrated output (i.e., the sum of the $\underline{\psi}_{F_m}^{V}$'s would balance the integrated filter input (the sum of the $\underline{\psi}_m^{V}$'s) after 11 filter cycles following application of a step input (for confirmation, try it as an exercise with $\underline{\psi}_m^{V}$ going from a steady zero condition to a steady $\Delta\underline{\psi}^{V}$ condition). This, of course, is the direct result of applying the Equation (17.2.1-27) constraint in the (17.2.1-29) filter design to achieve zero steady state integral dynamic lag.

Now consider the response of the walking window filter as described by Equations (17.2.1-24). For a ten stage version of this filter (i.e., r_a of 10), the <u>output change</u> (i.e., $\underline{\psi}_{F_m}^{V} - \underline{\psi}_{F_{m-1}}^{V}$) response to the $\Delta\underline{\psi}^{V}$ input step would be $K\,\Delta\underline{\psi}^{V}$ for 11 successive cycles, followed by zero. However, K for this case is $\frac{1}{r_a + 1}$, which equals 0.0909 for the r_a of 10. Thus, the maximum step change in the walking window filter output would be 0.0909 of the input step compared to 0.3791 for the Equation (17.2.1-29) filter. The penalty for the walking window filter is the (17.2.1-26) dynamic lag response time (5 T_m for the 10 stage filter) compared to zero dynamic response time for the Equation (17.2.1-29) filter. For a $\Delta\underline{\psi}^{V}$ step filter input (i.e., $\underline{\psi}_m^{V}$ going from a steady zero condition to a steady $\Delta\underline{\psi}^{V}$ condition), Equation (17.2.1-22) shows that the 5 T_m response time would eventually cause the integrated filter output (i.e., the sum of the $\underline{\psi}_{F_m}^{\mathcal{V}}$'s) to lag the integrated filter input (the sum of the $\underline{\psi}_m^{\mathcal{V}}$'s) by $5 \, \Delta\underline{\psi}^{\mathcal{V}}$. By analysis of the 10 stage walking window filter integrated output response, we would find that the steady $5 \, \Delta\underline{\psi}^{\mathcal{V}}$ lag condition would be reached after 11 filter cycles following application of the step input. Using the simple technique of inputting time advanced data to the walking window (by 5 T_m), the dynamic response time lag can be eliminated while retaining the 0.0909 smoothing factor.

17.2.2 SPECIFYING B FRAME ATTITUDE

The B Frame attitude is created during trajectory regeneration operations by specifying the B Frame orientation in the user vehicle and the orientation of the user vehicle relative to the smoothed $\mathcal{V}$ Frame (i.e., $\mathcal{V}$F) attitude:

$$C_B^{\mathcal{V}F} = C_{AC}^{\mathcal{V}F} \, C_B^{AC} \tag{17.2.2-1}$$

$$C_B^{L} = C_{\mathcal{V}F}^{L} \, C_B^{\mathcal{V}F} \tag{17.2.2-2}$$

The $C_{\mathcal{V}F}^{L}$ matrix in (17.2.2-1) is provided from the smoother integration operations described in Section 17.2.1. The C_B^{AC} matrix would be a specified constant for a "hard-mounted" strapdown INS. In some very high accuracy applications, the strapdown sensor assembly may be mounted on a multi-axis mechanical platform so that its orientation (i.e., the B Frame attitude) relative to AC axes can be controlled in a rotation pattern that cancels error effects. For such an arrangement, a set of Euler angles can be defined to describe the C_B^{AC} attitude of the B Frame relative to the AC frame using Equations similar to (3.2.3.1-2) (depending on the Euler angle sequence represented by the multi-axis platform). The Euler angles for these equations would typically be generated during the trajectory regeneration process by integrating a specified set of Euler angle rate equations. An interesting application of the Euler angle technique arises in spinning sensor assembly applications in which the $\mathcal{V}$F Frame is defined for a hypothetical non-spinning body with the real B Frame considered to be rotating around the $\mathcal{V}$F Frame at a roll Euler angle rate. The spinning vehicle trajectory is shaped for the non-spinning $\mathcal{V}$F Frame, and the B Frame attitude is superimposed on the non-spinning coordinates using a single roll Euler angle rotation. The roll Euler angle rate is then set to an input specified profile.

The $C_{AC}^{\mathcal{V}F}$ matrix in (17.2.2-1) is typically used to handle angle of attack/sideslip effects if desired to be included in the simulation. Otherwise, $C_{AC}^{\mathcal{V}F}$ can be simply set to the identity matrix. To include angle of attack and sideslip effects, the $C_{AC}^{\mathcal{V}F}$ matrix would be calculated using the methods of Section 17.1.2.3 (summarized by Equations (17.1.2.3-53) - (17.1.2.3-71)), Section 17.1.2.3.1 (summarized by Equations (17.1.2.3-53) - (17.1.2.3-71) with the Equation (17.1.2.3.1-21) - (17.1.2.3.1-29) modifications), or Section 17.1.2.3.2 (summarized by Equations (17.1.2.3-53) - (17.1.2.3-71) with the Equation (17.1.2.3.1-21) - (17.1.2.3.1-29) and (17.1.2.3.2-1) - (17.1.2.3.2-2) modifications). The following particulars would also apply:

- All $\mathcal{V}$ and V designations would be replaced by $\mathcal{V}$F and V_F (i.e., smoothed) designations. In particular:

$$V \rightarrow V_F \qquad \underline{(\,)}_{(\,)}^{\mathcal{V}(\,)} \rightarrow \underline{(\,)}_{(\,)}^{\mathcal{V}F(\,)} \qquad \underline{(\,)}_{(\,)\mathcal{V}(\,)}^{(\,)} \rightarrow \underline{(\,)}_{(\,)\mathcal{V}F(\,)}^{(\,)}$$
$$C_{(\,)}^{\mathcal{V}(\,)} \rightarrow C_{(\,)}^{\mathcal{V}F(\,)} \qquad C_{\mathcal{V}(\,)}^{(\,)} \rightarrow C_{\mathcal{V}F(\,)}^{(\,)} \tag{17.2.2-3}$$

  in which ( ) designates other parameters (or no parameter).

- Equation (17.1.2.3-58) and the ΔM_{Fuel} expression in Equation (17.1.2.3-64) (derived from the (17.1.2.3-27) integral expression) would be replaced by the more accurate repetition rate algorithm:

$$\Delta M_{Fuel_m} = \Delta M_{Fuel_{m-1}} + K_{Fuel} \frac{1}{2} \left(Thrst_m + Thrst_{m-1} \right) T_m$$
$$Thrst_{m=0} = 0 \tag{17.2.2-4}$$
$$\frac{\partial M}{\partial \underline{P}} = \left[-\frac{1}{2} K_{Fuel} T_m \quad 0 \quad 0 \right]$$

  where

  m = Computation cycle rate index for the trajectory regeneration function.

  T_m = Time interval for the trajectory regeneration function computation cycle.

- The $\underline{a}_{SF}^{\mathcal{V}} \rightarrow \underline{a}_{SF}^{\mathcal{V}F}$ term in Equations (17.1.2.3-65) and (17.1.2.3.1-24) would be calculated from:

$$\underline{a}_{SF_m}^{\mathcal{V}F} \approx \left(C_{\mathcal{V}F}^{L} \right)_m^T C_N^L \, \underline{a}_{SF_m}^{N} \tag{17.2.2-5}$$

with $\left(C^{L}_{\mathcal{V}F}\right)_m$ computed from smoother Equations (17.2.1-7) - (17.2.1-10) and (17.2.1-12), and $\underline{a}^{N}_{SF_m}$ calculated with the revised form of (7.2-2):

$$\underline{a}^{N}_{SF_m} \approx \frac{1}{2\,T_m}\left(\Delta\underline{v}^{N}_{SF_{m+1}} + \Delta\underline{v}^{N}_{SF_m}\right) \approx \frac{1}{2\,T_m}\left(\underline{v}^{N}_{m+1} - \underline{v}^{N}_{m-1} - 2\,\Delta\underline{v}^{N}_{G/COR_m}\right) \qquad (17.2.2\text{-}6)$$

The $\underline{v}^{N}_{m}$ terms in (17.2.2-6) would be computed with smoothed data as in (17.2.1-6). Note that Equation (17.2.2-6) requires $\underline{v}^{N}_{m}$ data at the m+1 cycle time. This implies that the smoothing algorithms in Section 17.2.1 will be operated one cycle in advance to provide the required $V_{F_{m+1}}, \left(C^{L}_{\mathcal{V}F}\right)_{m+1}$ inputs to (17.2.1-6). Alternatively, a less accurate algorithm can be used such as:

$$\underline{a}^{N}_{SF_m} \approx \frac{1}{T_m}\left(\underline{v}^{N}_{m} - \underline{v}^{N}_{m-1} - \Delta\underline{v}^{N}_{G/COR_m}\right) \qquad (17.2.2\text{-}7)$$

The $\Delta\underline{v}^{N}_{G/COR_m}$ term in (17.2.2-6) and (17.2.2-7) would be calculated as part of trajectory regeneration operations (See Section 17.2.3) based on Equation (7.2.1-1).

- Provisions can be incorporated for a control function to transition into and out of aerodynamic flight during take-off and landing. The control function would allow the above defined $C^{\mathcal{V}F}_{AC}$ matrix to be "faded" in or out from the identity matrix to its full angle of attack/sideslip value during selected time periods as follows:

$\left(C^{\mathcal{V}F}_{AC\,\alpha/\beta}\right)_m$ = Value for $C^{\mathcal{V}F}_{AC}$ calculated as described above that incorporates the full angle of attack/sideslip values.

$\phi_{AC/\mathcal{V}F\alpha/\beta_m}, \theta_{AC/\mathcal{V}F\alpha/\beta_m}, \psi_{AC/\mathcal{V}F\alpha/\beta_m}$ = Euler angle extraction from $\left(C^{\mathcal{V}F}_{AC\alpha/\beta}\right)_m$ using Equations (3.2.3.2-1) - (3.2.3.2-2).

$$\phi_{AC/\mathcal{V}F_m} = \left(1 - f_{\phi Cntrl}\right)\phi_{AC/\mathcal{V}F\alpha/\beta_m}$$
$$\theta_{AC/\mathcal{V}F_m} = \left(1 - f_{\theta Cntrl}\right)\theta_{AC/\mathcal{V}F\alpha/\beta_m} \qquad (17.2.2\text{-}8)$$
$$\psi_{AC/\mathcal{V}F_m} = \left(1 - f_{\psi Cntrl}\right)\psi_{AC/\mathcal{V}F\alpha/\beta_m}$$

$\left(C^{\mathcal{V}F}_{AC}\right)_m$ = Direction cosine matrix reconstruction from $\phi_{AC/\mathcal{V}F_m}$, $\theta_{AC/\mathcal{V}F_m}, \psi_{AC/\mathcal{V}F_m}$ using Equations (3.2.3.1-2).

where

$f_{\phi Cntrl}$, $f_{\theta Cntrl}$, $f_{\psi Cntrl}$ = Roll, pitch, heading control parameter ranging from zero (for no control) to one (for complete control).

$\left(C_{AC}^{\mathcal{V}F}\right)_m$ = Value for $C_{AC}^{\mathcal{V}F}$ used in Equation (17.2.2-1) during trajectory regeneration.

The $f_{\phi Cntrl}$, $f_{\theta Cntrl}$, $f_{\psi Cntrl}$ parameters would be programmed to transition linearly from one to zero during simulated take-off/runway-lift-off, and to linearly transition from zero to one during simulated landing/runway-touchdown. During simulated take-off, the $f_{\phi Cntrl}$, $f_{\psi Cntrl}$ parameters would be programmed to transition from one to zero from the time of vehicle lift-off. To simulate angle of attack build-up prior to liftoff, the $f_{\theta Cntrl}$ parameter would be programmed to transition from one to zero prior to runway liftoff. During simulated landing, the $f_{\phi Cntrl}$, $f_{\psi Cntrl}$ parameters would be programmed to transition from zero to one so that the transition completes at the instant of runway touch-down. During landing, $f_{\theta Cntrl}$ would be programmed to transition to one immediately following runway touchdown.

- Provisions can be incorporated for selecting the desired form of $C_{AC}^{\mathcal{V}F}$ generation algorithm for different trajectory segments (i.e., for direct $C_{AC}^{\mathcal{V}F}$ Euler roll angle control as in Section 17.1.2.3.1, or for minimum $C_{\mathcal{V}WF}^{\mathcal{V}F}$ rotation angle magnitude as in Section 17.1.2.3), and for smoothly transitioning from one form to another. The transition operation can be handled similar to the (17.2.2-8) approach:

$\left(C_{AC\text{-}1}^{\mathcal{V}F}\right)_m$ = Value for $C_{AC}^{\mathcal{V}F}$ based on one computation method (call it Method 1).

$\phi_{AC/\mathcal{V}F\text{-}1_m}, \theta_{AC/\mathcal{V}F\text{-}1_m}, \psi_{AC/\mathcal{V}F\text{-}1_m}$ = Euler angle extraction from $\left(C_{AC\text{-}1}^{\mathcal{V}F}\right)_m$ using Equations (3.2.3.2-1) - (3.2.3.2-2).

$\left(C_{AC\text{-}2}^{\mathcal{V}F}\right)_m$ = Value for $C_{AC}^{\mathcal{V}F}$ based on the other computation method (call it Method 2). (17.2.2-9)

$\phi_{AC/\mathcal{V}F\text{-}2_m}, \theta_{AC/\mathcal{V}F\text{-}2_m}, \psi_{AC/\mathcal{V}F\text{-}2_m}$ = Euler angle extraction from $\left(C_{AC\text{-}2}^{\mathcal{V}F}\right)_m$ using Equations (3.2.3.2-1) - (3.2.3.2-2).

$$\phi_{AC/\mathcal{V}F_m} = \left(1 - f_{1\text{-}2}\right)\phi_{AC/\mathcal{V}F\text{-}1_m} + f_{1\text{-}2}\,\phi_{AC/\mathcal{V}F\text{-}2_m}$$

(Continued)

$$\theta_{AC/\mathcal{V}F_m} = \left(1 - f_{1\text{-}2}\right) \theta_{AC/\mathcal{V}F\text{-}1_m} + f_{1\text{-}2}\, \theta_{AC/\mathcal{V}F\text{-}2_m}$$

$$\psi_{AC/\mathcal{V}F_m} = \left(1 - f_{1\text{-}2}\right) \psi_{AC/\mathcal{V}F\text{-}1_m} + f_{1\text{-}2}\, \psi_{AC/\mathcal{V}F\text{-}2_m} \qquad \text{(17.2.2-9) (Continued)}$$

$\left(C_{AC}^{\mathcal{V}F}\right)_m$ = Direction cosine matrix reconstruction from $\phi_{AC/\mathcal{V}F_m}$, $\theta_{AC/\mathcal{V}F_m}$, $\psi_{AC/\mathcal{V}F_m}$ using Equations (3.2.3.1-2).

where

$f_{1\text{-}2}$ = Control parameter ranging from zero to one that will transition $C_{AC}^{\mathcal{V}F}$ from $\left(C_{AC\text{-}1}^{\mathcal{V}F}\right)_m$ to $\left(C_{AC\text{-}2}^{\mathcal{V}F}\right)_m$.

$\left(C_{AC}^{\mathcal{V}F}\right)_m$ = Value for $C_{AC}^{\mathcal{V}F}$ used in Equation (17.2.2-1) during trajectory regeneration.

The $f_{1\text{-}2}$ control parameter would be programmed to transition linearly from zero to one during the time period that the $C_{AC}^{\mathcal{V}F}$ solution is required to transition from $\left(C_{AC\text{-}1}^{\mathcal{V}F}\right)_m$ to $\left(C_{AC\text{-}2}^{\mathcal{V}F}\right)_m$.

- Provisions can be incorporated for including AC Frame high frequency angular rotation effects (See Section 17.2.3.2.3 for details).

17.2.3 TRAJECTORY REGENERATION

Trajectory regeneration operations take the smoothed data from Section 17.2.1 and the B Frame attitude data from Section 17.2.2 to create a consistent set of position (in the E Frame), N Frame velocity, and B Frame specific-force-acceleration/inertial-angular-rate data at the trajectory generator m cycle times. The B Frame specific-force-acceleration/inertial-angular-rate data are in the form of integrated increments over each m cycle, simulating the output from an error free strapdown inertial sensor assembly. The simulated sensor data is designed such that when operated upon by the high precision strapdown inertial navigation integration algorithms of Chapter 7, the resulting attitude, velocity and position data will be identical to the attitude, velocity position data provided by the trajectory generator. The previous requirement is an important constraint because it goes to proving the validity of the trajectory generator to users during simulation applications. It is reassuring to a user that a documented set of verifiable precision integration algorithms exist that can be applied to the trajectory generator strapdown sensor data outputs and produce attitude, velocity and position time histories that match the equivalent trajectory generator attitude, velocity, position outputs.

The previous operations constitute the basic trajectory regeneration function. Variations can also be incorporated to include the effects of wind gust induced aerodynamic force acceleration, sensor assembly lever arm position displacement in the AC Frame, and additional high frequency linear/angular motion measured by the inertial sensors. The following sections address each of these topics.

17.2.3.1 BASIC TRAJECTORY REGENERATION OPERATIONS

The basic trajectory regeneration operations begin with the Section 17.2.1 and 17.2.2 filtering and B Frame attitude calculations as defined by Equations (17.2.1-1) - (17.2.1-3), (17.2.1-5) - (17.2.1-10), (17.2.1-12) and (17.2.2-1) - (17.2.2-2) repeated below in their order of execution in a typical simulation program software structure. Included are provisions for AC Frame attitude output.

<u>$\mathcal{V}$ Frame Incremental Data</u>

$$\underline{\psi}_m^{\mathcal{V}} = \frac{T_m}{T_S}\,\underline{\phi}_S^{\mathcal{V}} \qquad \Delta V_m = \frac{T_m}{T_S}\,\Delta V_S$$

Do At Trajectory Segment Junctions:

$$\underline{\psi}_m^{\mathcal{V}} = \frac{T_{Fin/Crnt}}{T_{S_{Crnt}}}\,\underline{\phi}_{S_{Crnt}}^{\mathcal{V}} + \frac{T_m - T_{Fin/Crnt}}{T_{S_{Next}}}\,\underline{\phi}_{S_{Next}}^{\mathcal{V}}$$

$$\Delta V_m = \frac{T_{Fin/Crnt}}{T_{S_{Crnt}}}\,\Delta V_{S_{Crnt}} + \frac{T_m - T_{Fin/Crnt}}{T_{S_{Next}}}\,\Delta V_{S_{Next}} \qquad (17.2.3.1\text{-}1)$$

<u>$\mathcal{V}$ Frame Attitude</u>

$$C_{\mathcal{V}_m}^{\mathcal{V}_{m-1}} = \left[I + \frac{\sin\psi_m}{\psi_m}\left(\underline{\psi}_m^{\mathcal{V}}\times\right) + \frac{(1-\cos\psi_m)}{\psi_m^2}\left(\underline{\psi}_m^{\mathcal{V}}\times\right)^2 \right]$$

$$\left(C_{\mathcal{V}}^{L}\right)_m = \left(C_{\mathcal{V}}^{L}\right)_{m-1} C_{\mathcal{V}_m}^{\mathcal{V}_{m-1}}$$

<u>Smoothed Attitude Increments</u>

$$\underline{\psi}_{F_m}^{\mathcal{V}} = a_0\,\underline{\psi}_m^{\mathcal{V}} + a_1\,\underline{\psi}_{m-1}^{\mathcal{V}} + a_2\,\underline{\psi}_{m-2}^{\mathcal{V}} + \cdots + b_1\,\underline{\psi}_{F_{m-1}}^{\mathcal{V}} + b_2\,\underline{\psi}_{F_{m-2}}^{\mathcal{V}} + \cdots$$

(Continued)

$\mathcal{V}$F Frame Attitude

$$\underline{\gamma}_m^{\mathcal{V}} = \underline{\gamma}_{m-1}^{\mathcal{V}} + \underline{\psi}_{F_m}^{\mathcal{V}} - \underline{\psi}_m^{\mathcal{V}}$$

$$C_{\mathcal{V}F_m}^{\mathcal{V}_m} = \left[I + \frac{\sin \gamma_m}{\gamma_m} \left(\underline{\gamma}_m^{\mathcal{V}} \times\right) + \frac{(1 - \cos \gamma_m)}{\gamma_m^2} \left(\underline{\gamma}_m^{\mathcal{V}} \times\right)^2 \right]$$

$$\left(C_{\mathcal{V}F}^{L}\right)_m = \left(C_{\mathcal{V}}^{L}\right)_m C_{\mathcal{V}F_m}^{\mathcal{V}_m}$$

B And AC Frame Attitude

$$\left(C_B^{\mathcal{V}F}\right)_m = \left(C_{AC}^{\mathcal{V}F}\right)_m \left(C_B^{AC}\right)_m$$

$C_{AC}^{\mathcal{V}F}$ and C_B^{AC} calculated as described in Section 17.2.2. (17.2.3.1-1) (Continued)

$$\left(C_B^{L}\right)_m = \left(C_{\mathcal{V}F}^{L}\right)_m \left(C_B^{\mathcal{V}F}\right)_m$$

$$\left(C_{AC}^{L}\right)_m = \left(C_B^{L}\right)_m \left(C_B^{AC}\right)_m^T$$

For Output Information

Smoothed Velocity Increments

$$\Delta V_{F_m} = a_0 \, \Delta V_m + a_1 \, \Delta V_{m-1} + a_2 \, \Delta V_{m-2} + \cdots + b_1 \, \Delta V_{F_{m-1}} + b_2 \, \Delta V_{F_{m-2}} + \cdots$$

N Frame Velocity

$$V_{F_m} = V_{F_{m-1}} + \Delta V_{F_m}$$

$$\underline{v}_m^{N} = V_{F_m} \, C_L^{N} \left(C_{\mathcal{V}F}^{L}\right)_m \underline{u}_{X\mathcal{V}F}^{\mathcal{V}F}$$

The basic outputs from Equations (17.2.3.1-1) are the B Frame attitude C_B^L and N Frame velocity $\underline{v}^N$ at computer cycle m. The E Frame position and B Frame integrated specific-force-acceleration/inertial-angular-rate increments are computed from these parameters using an inverted form of the Chapter 7 stapdown integration algorithms. For the trajectory regeneration function, we place the following restrictions on the Chapter 7 integration algorithms:

- The trajectory generator N Frame is of the azimuth wander type, hence from Section 4.5, $\rho_{ZN} = 0$.

- The trajectory regeneration function will use a single updating rate identified as the m cycle, hence, the Table 7.5-1 algorithms will be set so that n = m.

- The high resolution algorithms will be used for position updating.

- We will assume constant B Frame angular rate and specific force acceleration over an m cycle.

The constant angular-rate/specific-force restriction is incorporated to simplify the trajectory regeneration computations. This assumption should not restrict the generality of the trajectory generator if it is acknowledged that the m cycle rate can be increased to whatever the user feels is required to adequately account for high frequency effects in the simulator generated data. The previous approach is justified because real-time throughput restrictions are generally not an issue for simulation programs. It is important to recognize that while the trajectory generator computations are based on the Chapter 7 algorithms, the m cycle rate selected for the trajectory generator is not related to the execution rate for Chapter 7 algorithms implemented in a real strapdown inertial navigation system, whose execution rate structure is based on performance issues and real-time computer throughput restrictions (see Section 7.4). At the conclusion of this section we will discuss how the trajectory generator can be modified for enhanced accuracy and realism.

Subject to the previous constraints, let us now list the applicable Chapter 7 strapdown integration algorithms we will apply for trajectory regeneration. Using Table 7.5-1 as a guide, but with $\underline{\phi}_m = \underline{\alpha}_m$ in (7.1.1.1-12) and (7.1.1.1-3), using (7.2.2.2.1-5) for $\Delta\underline{v}_{SF_m}^{B_{I(m-1)}}$, with (7.3.3.1-9) and (7.3.3.1-11) for $\Delta\underline{R}_{SF_m}^{B}$ (all based on constant B Frame angular rate and specific force), and eliminating the I and E subscripts in the B_I, L_I and N_E Frame notation for simplicity, the resulting integration algorithms in their order of execution are as follows:

VELOCITY

$$\Delta\underline{v}_{SF_m}^{B_{m-1}} = \left[I + \frac{(1 - \cos \alpha_m)}{\alpha_m^2} (\underline{\alpha}_m \times) + \frac{1}{\alpha_m^2}\left(1 - \frac{\sin \alpha_m}{\alpha_m}\right)(\underline{\alpha}_m \times)^2 \right] \underline{\upsilon}_m \qquad (17.2.3.1\text{-}2)$$

$$\Delta\underline{v}_{SF_m}^{L_{m-1}} = \left(C_B^L\right)_{m-1} \Delta\underline{v}_{SF_m}^{B_{m-1}} \qquad (17.2.3.1\text{-}3)$$

$$\Delta\underline{R}_{v_m}^{N} \approx T_m \left(\frac{3}{2}\,\underline{v}_{m-1}^{N} - \frac{1}{2}\,\underline{v}_{m-2}^{N}\right) \qquad (17.2.3.1\text{-}4)$$

$$F_{C_{m-1}}^{N} = f\left[\left(C_N^E\right)_{m-1}, h_{m-1}\right] \quad \text{Using Equations (5.3-18)} \qquad (17.2.3.1\text{-}5)$$

$$F_{C_{m-1/2}}^{N} = \frac{3}{2}\, F_{C_{m-1}}^{N} - \frac{1}{2}\, F_{C_{m-2}}^{N} \qquad (17.2.3.1\text{-}6)$$

$$\underline{\omega}_{IE_{m-1}}^{N} = \left(C_N^E\right)_{m-1}^{T} \underline{\omega}_{IE}^{E} \qquad (17.2.3.1\text{-}7)$$

$$\underline{\omega}^N_{IE_{m-1/2}} \approx \frac{3}{2}\,\underline{\omega}^N_{IE_{m-1}} - \frac{1}{2}\,\underline{\omega}^N_{IE_{m-2}} \qquad (17.2.3.1\text{-}8)$$

$$\underline{\zeta}_{v_m} \approx C^L_N \left[\underline{\omega}^N_{IE_{m-1/2}}\, T_m + F^N_{C_{m-1/2}} \left(\underline{u}^N_{ZN} \times \Delta\underline{R}^N_{v_m} \right) \right] \qquad (17.2.3.1\text{-}9)$$

$$\mathfrak{C}^{L_m}_{L_{m-1}} \approx I - \left(\underline{\zeta}_{v_m} \times \right) \qquad (17.2.3.1\text{-}10)$$

$$\Delta\underline{v}^{L_m}_{SF_m} = \mathfrak{C}^{L_m}_{L_{m-1}}\, \Delta\underline{v}^{L_{m-1}}_{SF_m} \qquad (17.2.3.1\text{-}11)$$

$$\underline{v}^N_{m-1/2} \approx \frac{3}{2}\,\underline{v}^N_{m-1} - \frac{1}{2}\,\underline{v}^N_{m-2} \qquad (17.2.3.1\text{-}12)$$

$$\underline{g}^N_{P_{m-1/2}} \approx \frac{3}{2}\,\underline{g}^N_{P_{m-1}} - \frac{1}{2}\,\underline{g}^N_{P_{m-2}} \qquad (17.2.3.1\text{-}13)$$

$$\Delta\underline{v}^N_{G/COR_m} \approx \left\{ \underline{g}^N_{P_{m-1/2}} - \left[2\,\underline{\omega}^N_{IE_{m-1/2}} + F^N_{C_{m-1/2}} \left(\underline{u}^N_{ZN} \times \underline{v}^N_{m-1/2} \right) \right] \times \underline{v}^N_{m-1/2} \right\} T_m \qquad (17.2.3.1\text{-}14)$$

$$\underline{v}^N_m = \underline{v}^N_{m-1} + C^N_L\, \Delta\underline{v}^{L_m}_{SF_m} + \Delta\underline{v}^N_{G/COR_m} \qquad (17.2.3.1\text{-}15)$$

POSITION

$$\Delta\underline{R}^B_{SF_m} = \left[\frac{1}{2} I + \frac{1}{\alpha_m^2} \left(1 - \frac{\sin\alpha_m}{\alpha_m} \right) \left(\underline{\alpha}_m \times \right) + \frac{1}{\alpha_m^2} \left(\frac{1}{2} - \frac{(1 - \cos\alpha_m)}{\alpha_m^2} \right) \left(\underline{\alpha}_m^2 \times \right) \right] \underline{\upsilon}_m\, T_m \qquad (17.2.3.1\text{-}16)$$

$$\Delta\underline{R}^L_{SF_m} = -\frac{1}{3}\,\underline{\zeta}_{v_m} \times \Delta\underline{v}^{L_{I(m-1)}}_{SF_m}\, T_m + \left(C^L_B \right)_{m-1} \Delta\underline{R}^B_{SF_m} \qquad (17.2.3.1\text{-}17)$$

$$\Delta\underline{R}^N_m = \left(\underline{v}^N_{m-1} + \frac{1}{2}\,\Delta\underline{v}^N_{G/COR_m} \right) T_m + C^N_L\, \Delta\underline{R}^L_{SF_m} \qquad (17.2.3.1\text{-}18)$$

$$\Delta h_m = \underline{u}^N_{ZN} \cdot \Delta\underline{R}^N_m \qquad (17.2.3.1\text{-}19)$$

$$\underline{\xi}_m \approx F^N_{C_{m-1/2}} \left(\underline{u}^N_{ZN} \times \Delta\underline{R}^N_m \right) \qquad (17.2.3.1\text{-}20)$$

$$C^{N_{m-1}}_{N_m} = I + \frac{\sin\xi_m}{\xi_m} \left(\underline{\xi}_m \times \right) + \frac{(1 - \cos\xi_m)}{\xi_m^2} \left(\underline{\xi}_m \times \right) \left(\underline{\xi}_m \times \right) \qquad (17.2.3.1\text{-}21)$$

$$h_m = h_{m-1} + \Delta h_m \qquad (17.2.3.1\text{-}22)$$

$$\left(C^E_N \right)_m = \left(C^E_N \right)_{m-1} C^{N_{m-1}}_{N_m} \qquad (17.2.3.1\text{-}23)$$

ATTITUDE

$$C_{B_m}^{B_{m-1}} = I + \frac{\sin \alpha_m}{\alpha_m}\left(\underline{\alpha}_m \times\right) + \frac{(1 - \cos \alpha_m)}{\alpha_m^2}\left(\underline{\alpha}_m \times\right)\left(\underline{\alpha}_m \times\right) \qquad (17.2.3.1\text{-}24)$$

$$C_{B_m}^{L_{m-1}} = \left(C_B^L\right)_{m-1} C_{B_m}^{B_{m-1}} \qquad (17.2.3.1\text{-}25)$$

$$\underline{\zeta}_m \approx C_N^L \left[\underline{\omega}_{IE_{m-1/2}}^N T_m + F_{C_{m-1/2}}^N \left(\underline{u}_{ZN}^N \times \Delta\underline{R}_m^N\right)\right] \qquad (17.2.3.1\text{-}26)$$

$$C_{L_{m-1}}^{L_m} = I - \frac{\sin \zeta_m}{\zeta_m}\left(\underline{\zeta}_m \times\right) + \frac{(1 - \cos \zeta_m)}{\zeta_m^2}\left(\underline{\zeta}_m \times\right)\left(\underline{\zeta}_m \times\right) \qquad (17.2.3.1\text{-}27)$$

$$\left(C_B^L\right)_m = C_{L_{m-1}}^{L_m} C_{B_m}^{L_{m-1}} \qquad (17.2.3.1\text{-}28)$$

where

$\underline{\zeta}_{v_m}, \mathfrak{C}_{L_{m-1}}^{L_m}, \Delta\underline{R}_{v_m}^N$ = Estimated values for $\underline{\zeta}_m, C_{L_{m-1}}^{L_m}, \Delta\underline{R}_m^N$ used in the velocity calculations based on extrapolated past value data available at the time the associated equation is processed. The different symbols used for these parameters have been introduced to avoid confusion in subsequent inversion operations.

Given $\left(C_B^L\right)_m$ and $\underline{v}_m^N$ from (17.2.3.1-1), let us now find an inverted form of (17.2.3.1-2) - (17.2.3.1-28) for position in the E Frame (h_m and $\left(C_N^E\right)_m$), and for the integrated B Frame inertial-angular-rate/specific-force increments ($\underline{\alpha}_m$ and $\underline{\upsilon}_m$). The algorithms for calculating the B Frame integrated inertial angular rate increment $\underline{\alpha}_m$ are obtained directly from Equations (17.2.3.1-5) - (17.2.3.1-8) and (17.2.3.1-26) - (17.2.3.1-27) using the combined inverse of (17.2.3.1-25) and (17.2.3.1-28), the inverse of (17.2.3.1-24), and a linear interpolation algorithm for $\Delta\underline{R}_m^N$:

$$F_{C_{m-1}}^N = f\left[\left(C_N^E\right)_{m-1}, h_{m-1}\right] \quad \text{Using Equations (5.3-18)}$$

$$F_{C_{m-1/2}}^N = \frac{3}{2} F_{C_{m-1}}^N - \frac{1}{2} F_{C_{m-2}}^N \qquad (17.2.3.1\text{-}29)$$

$$\underline{\omega}_{IE_{m-1}}^N = \left(C_N^E\right)_{m-1}^T \underline{\omega}_{IE}^E$$

(Continued)

$$\underline{\omega}_{IE_{m-1/2}}^{N} \approx \frac{3}{2}\,\underline{\omega}_{IE_{m-1}}^{N} - \frac{1}{2}\,\underline{\omega}_{IE_{m-2}}^{N}$$

$$\Delta\underline{R}_{Att_m}^{N} \approx \frac{1}{2}\left(\underline{v}_{m}^{N} + \underline{v}_{m-1}^{N}\right) T_m$$

$$\underline{\zeta}_m \approx C_N^L \left[\underline{\omega}_{IE_{m-1/2}}^{N} T_m + F_{C_{m-1/2}}^{N} \left(\underline{u}_{ZN}^{N} \times \Delta\underline{R}_{Att_m}^{N} \right) \right]$$

$$C_{L_{m-1}}^{L_m} = I - \frac{\sin \zeta_m}{\zeta_m} \left(\underline{\zeta}_m \times\right) + \frac{(1 - \cos \zeta_m)}{\zeta_m^{\,2}} \left(\underline{\zeta}_m \times\right)\left(\underline{\zeta}_m \times\right) \qquad \text{(17.2.3.1-29) (Continued)}$$

$$C_{B_m}^{B_{m-1}} = \left(C_B^L\right)_{m-1}^T \left(C_{L_{m-1}}^{L_m}\right)^T \left(C_B^L\right)_m$$

$\underline{\alpha}_m$ = Rotation angle extraction from $C_{B_m}^{B_{m-1}}$ using (3.2.2.2-10) - (3.2.2.2-12) and (3.2.2.2-15) - (3.2.2.2-19).

where

$\Delta\underline{R}_{Att_m}^{N}$ = Value for position change increment $\Delta\underline{R}_{m}^{N}$ used for attitude updating based on trapezoidal integration of velocity.

Note that $\Delta\underline{R}_{Att_m}^{N}$ is used in (17.2.3.1-29) rather than $\Delta\underline{R}_{m}^{N}$ as in (17.2.3.1-26). This is because $\Delta\underline{R}_{m}^{N}$ has not yet been calculated when $\Delta\underline{R}_{m}^{N}$ is needed in Equation (17.2.3.1-29). For enhanced accuracy (in having the (17.2.3.1-29) result being the exact inverse of the (17.2.3.1-24) - (17.2.3.1-28) integration process, Equations (17.2.3.1-29) can be repeated using $\Delta\underline{R}_{m}^{N}$ in place of $\Delta\underline{R}_{Att_m}^{N}$, once $\Delta\underline{R}_{m}^{N}$ is determined. A discussion on continuing this process iteratively for refined accuracy is provided at the conclusion of this section. Note that if $\Delta\underline{R}_{Att_m}^{N}$ as calculated in (17.2.3.1-29) is used for $\Delta\underline{R}_{m}^{N}$ in (17.2.3.1-26), then Equations (17.2.3.1-29) as shown will exactly represent the inverse of the (17.2.3.1-24) - (17.2.3.1-28) integration process without iteration.

Using $\underline{\alpha}_m$, $\underline{\omega}_{IE_{m-1/2}}^{N}$ and $F_{C_{m-1/2}}^{N}$ from (17.2.3.1-29), the algorithms for calculating the B Frame integrated specific force increment $\underline{\upsilon}_m$ are obtained from Equations (17.2.3.1-12) - (17.2.3.1-14) directly, from the inverse of (17.2.3.1-15), from (17.2.3.1-4) and (17.2.3.1-9) - (17.2.3.1-10) directly, from the inverse of (17.2.3.1-11) and (17.2.3.1-3), and from the inverse of (17.2.3.1-2):

$$\underline{v}^{N}_{m-1/2} \approx \frac{3}{2}\,\underline{v}^{N}_{m-1} - \frac{1}{2}\,\underline{v}^{N}_{m-2}$$

$$\underline{g}^{N}_{P_{m-1/2}} \approx \frac{3}{2}\,\underline{g}^{N}_{P_{m-1}} - \frac{1}{2}\,\underline{g}^{N}_{P_{m-2}}$$

$$\Delta\underline{v}^{N}_{G/COR_m} \approx \left\{\underline{g}^{N}_{P_{m-1/2}} - \left[2\,\underline{\omega}^{N}_{IE_{m-1/2}} + F^{N}_{C_{m-1/2}}\left(\underline{u}^{N}_{ZN} \times \underline{v}^{N}_{m-1/2}\right)\right] \times \underline{v}^{N}_{m-1/2}\right\} T_m$$

$$\Delta\underline{v}^{L_m}_{SF_m} = \left(C^{N}_{L}\right)^{T}\left(\underline{v}^{N}_{m} - \underline{v}^{N}_{m-1} - \Delta\underline{v}^{N}_{G/COR_m}\right)$$

$$\Delta\underline{R}^{N}_{v_m} \approx T_m\left(\frac{3}{2}\,\underline{v}^{N}_{m-1} - \frac{1}{2}\,\underline{v}^{N}_{m-2}\right) \qquad (17.2.3.1\text{-}30)$$

$$\underline{\zeta}_{v_m} \approx C^{L}_{N}\left[\underline{\omega}^{N}_{IE_{m-1/2}}\,T_m + F^{N}_{C_{m-1/2}}\left(\underline{u}^{N}_{ZN} \times \Delta\underline{R}^{N}_{v_m}\right)\right]$$

$$\mathfrak{C}^{L_m}_{L_{m-1}} \approx I - \left(\underline{\zeta}_{v_m}\times\right)$$

$$\Delta\underline{v}^{L_{m-1}}_{SF_m} = \left(\mathfrak{C}^{L_m}_{L_{m-1}}\right)^{-1}\Delta\underline{v}^{L_m}_{SF_m}$$

$$\Delta\underline{v}^{B_{m-1}}_{SF_m} = \left(C^{L}_{B}\right)^{T}_{m-1}\Delta\underline{v}^{L_{m-1}}_{SF_m}$$

$$\underline{\upsilon}_m = \left[I + \frac{\left(1 - \cos\alpha_m\right)}{\alpha_m^2}\left(\underline{\alpha}_m\times\right) + \frac{1}{\alpha_m^2}\left(1 - \frac{\sin\alpha_m}{\alpha_m}\right)\left(\underline{\alpha}_m\times\right)^2\right]^{-1}\Delta\underline{v}^{B_{m-1}}_{SF_m}$$

Finally, using $\underline{\alpha}_m$ and $F^{N}_{C_{m-1/2}}$ from (17.2.3.1-29), and $\underline{\upsilon}_m$ and $\Delta\underline{v}^{L_{m-1}}_{SF_m}$ from (17.2.3.1-30), the algorithms for calculating position in the E Frame (h_m and $\left(C^{E}_{N}\right)_m$) are Equations (17.2.3.1-16) - (17.2.3.1-23) directly:

$$\Delta\underline{R}^{B}_{SF_m} = \left[\frac{1}{2}I + \frac{1}{\alpha_m^2}\left(1 - \frac{\sin\alpha_m}{\alpha_m}\right)\left(\underline{\alpha}_m\times\right) + \frac{1}{\alpha_m^2}\left(\frac{1}{2} - \frac{\left(1 - \cos\alpha_m\right)}{\alpha_m^2}\right)\left(\underline{\alpha}_m^2\times\right)\right]\underline{\upsilon}_m\,T_m$$

$$\Delta\underline{R}^{L}_{SF_m} = -\frac{1}{3}\,\underline{\zeta}_{v_m} \times \Delta\underline{v}^{L_{m-1}}_{SF_m}\,T_m + \left(C^{L}_{B}\right)_{m-1}\Delta\underline{R}^{B}_{SF_m} \qquad (17.2.3.1\text{-}31)$$

$$\Delta\underline{R}^{N}_{m} = \left(\underline{v}^{N}_{m-1} + \frac{1}{2}\,\Delta\underline{v}^{N}_{G/COR_m}\right)T_m + C^{N}_{L}\,\Delta\underline{R}^{L}_{SF_m}$$

$$\Delta h_m = \underline{u}^{N}_{ZN} \cdot \Delta\underline{R}^{N}_{m}$$

(Continued)

$$\underline{\xi}_m \approx F_{C_{m-1/2}}^{N}\left(\underline{u}_{ZN}^{N} \times \Delta\underline{R}_m^{N}\right)$$

$$C_{N_m}^{N_{m-1}} = I + \frac{\sin \xi_m}{\xi_m}\left(\underline{\xi}_m \times\right) + \frac{(1 - \cos \xi_m)}{\xi_m^{\,2}}\left(\underline{\xi}_m \times\right)\left(\underline{\xi}_m \times\right) \qquad \text{(17.2.3.1-31) (Continued)}$$

$$h_m = h_{m-1} + \Delta h_m$$

$$\left(C_N^E\right)_m = \left(C_N^E\right)_{m-1} C_{N_m}^{N_{m-1}}$$

Equations (17.2.3.1-1) and (17.2.3.1-29) - (17.2.3.1-31) define the basic computational requirements for the trajectory regeneration function. The output from these equations are B Frame integrated inertial angular rate and specific force increments ($\underline{\alpha}$ and $\underline{\upsilon}$) with a corresponding set of attitude, velocity and position data (C_B^L, $\underline{v}^N$, C_N^E, h). The C_B^L, $\underline{v}^N$ attitude, velocity data correspond to the trajectory shaping results, modified for data smoothing and B Frame attitude selection. The $\underline{\alpha}$ and $\underline{\upsilon}$ data are such that (with one minor exception) if they are processed through high precision strapdown inertial integration algorithms (i.e., Equations (17.2.3.1-2) - (17.2.3.1-28)), the solutions obtained will identically match the C_B^L, $\underline{v}^N$, C_N^E, h trajectory generator outputs. The previous noted exception refers to $\Delta\underline{R}_{Att}^N$ used in (17.2.3.1-29) rather than $\Delta\underline{R}^N$ as in (17.2.3.1-26). As noted under the discussion following (17.2.3.1-29), this is because $\Delta\underline{R}^N$ has not yet been calculated. For enhanced accuracy, an iteration process can be invoked following the first (17.2.3.1-29) - (17.2.3.1-31) solution, in which Equations (17.2.3.1-29) - (17.2.3.1-31) are processed repeatedly using the $\Delta\underline{R}^N$ value from (17.2.3.1-31) in place of $\Delta\underline{R}_{Att}^N$ in (17.2.3.1-29). The iteration process is complete when the $\Delta\underline{R}^N$ solution from (17.2.3.1-31) matches its previous value within prescribed limits. The prescribed limits can be set to the overall regenerated trajectory accuracy requirement divided by the total number of m cycles in the trajectory. Accuracy in this context refers to the accuracy to which we desire the (17.2.3.1-29) - (17.2.3.1-31) solution to match the (17.2.3.1-2) - (17.2.3.1-28) solution when using $\underline{\alpha}$ and $\underline{\upsilon}$ from (17.2.3.1-29) - (17.2.3.1-30) in (17.2.3.1-2) - (17.2.3.1-28). Note, as mentioned under the discussion following (17.2.3.1-29), that if $\Delta\underline{R}_{Att}^N$ as calculated in (17.2.3.1-29) is used for $\Delta\underline{R}^N$ in (17.2.3.1-26), then Equations (17.2.3.1-29) as shown will exactly represent the inverse of the (17.2.3.1-24) - (17.2.3.1-28) integration process without an iteration requirement.

17.2.3.2 VARIATIONS FROM BASIC REGENERATED TRAJECTORY SOLUTION

In this section and its subsections we will discuss methods for generating a consistent set of navigation data parameters for a trajectory that has a specified attitude and position variation from the Section 17.2.3.1 basic regenerated trajectory solution. For convenience, we will identify the Section 17.2.3.1 navigation solution as the "reference trajectory" and the variation navigation solution to be developed as the "variation trajectory". As in Section 17.2.3.1, the equations to be developed will be based on the requirement that integration of the resulting integrated specific-force/inertial-angular-rate increments using the Section 17.2.3.1 precision integration algorithms (Equations (17.2.3.1-2) - (17.2.3.1-28)) will generate the identically same attitude, velocity, position solutions as the equations we will now derive for the variation trajectory attitude, velocity, position. Thus, the equations to follow are based on the inverse of Equations (17.2.3.1-2) - (17.2.3.1-28) with the additional requirement of having a specified attitude and position variation from the reference trajectory.

For a specified difference in the earth referenced position between the reference and variation trajectories, the altitude of the variation solution at computer cycle m will be:

$$h_{Var_m} = h_m + \underline{u}_{ZN}^{N} \cdot \underline{S}_{Var_m}^{N} \qquad (17.2.3.2\text{-}1)$$

where

$\underline{S}_{Var_m}^{N}$ = N Frame components of the difference between earth referenced position locations for the variation trajectory compared to the Section 17.2.3.1 basic regenerated trajectory reference solution.

Var = Subscript identifying that the associated parameter is for the variation trajectory. In Equation (17.2.3.2-1), h_{Var_m} is the altitude for the variation trajectory at computer cycle m.

We can also describe the horizontal position displacement from the reference solution at cycle m by the rotation vector equivalent to $\underline{\xi}$ of Equation (17.2.3.1-20):

$$\underline{\lambda}_m^{N} = \frac{1}{2}\left(F_{C_m}^{N} + F_{CVar_m}^{N}\right)\left(\underline{u}_{ZN}^{N} \times \underline{S}_{Var_m}^{N}\right) \qquad (17.2.3.2\text{-}2)$$

where

$\underline{\lambda}_m^{N}$ = Horizontal angular position displacement over the earth's surface corresponding to $\underline{S}_{Var_m}^{N}$.

$F^N_{CVar_m}$ = F^N_C curvature matrix calculated from h_{Var_m} and $\left(C^E_{NVar}\right)_m$ position data (Note: NVar is defined next).

NVar = N Frame for the variation trajectory position location.

with $F^N_{C_m}$ provided from the reference solution. Since $\left(C^E_{NVar}\right)_m$ is not yet calculated for $F^N_{CVar_m}$ evaluation, past value extrapolation must be used to determine $F^N_{CVar_m}$, such as the linear extrapolation formula:

$$F^N_{CVar_m} \approx F^N_{CVar_{m-1}} + \left[F^N_{CVar_{m-1}} - F^N_{CVar_{m-2}}\right] = 2\, F^N_{CVar_{m-1}} - F^N_{CVar_{m-2}} \tag{17.2.3.2-3}$$

Alternatively, $F^N_{CVar_m}$ can be approximated as $F^N_{C_m}$. If (17.2.3.2-3) is used, the past value F^N_{CVar} terms would be calculated with Equations (5.3-18) using past values of h_{Var} (from Equation (17.2.3.2-1)) and C^E_{NVar} as calculated next.

From generalized Equation (3.2.2.1-8), the $\underline{\lambda}^N$ angular displacement defines the equivalent direction cosine matrix between the reference solution N Frame and the N Frame corresponding to the displaced position:

$$\left(C^N_{NVar}\right)_m = I + \frac{\sin\lambda_m}{\lambda_m}\left(\underline{\lambda}^N_m\times\right) + \frac{(1-\cos\lambda_m)}{\lambda_m^2}\left(\underline{\lambda}^N_m\times\right)^2 \tag{17.2.3.2-4}$$

The NVar Frame orientation relative to the E Frame is calculated at cycle m from the (3.2.1-5) chain rule as:

$$\left(C^E_{NVar}\right)_m = \left(C^E_N\right)_m \left(C^N_{NVar}\right)_m \tag{17.2.3.2-5}$$

Equations (17.2.3.2-1) and (17.2.3.2-5) define the position of the variation trajectory relative to the earth (h_{Var_m} and $\left(C^E_{NVar}\right)_m$) as a specified variation from the reference trajectory. We will now use the inverse of position integration Equations (17.2.3.1-19) - (17.2.3.1-23) to obtain the equivalent to $\Delta\underline{R}^N_m$ for the variation trajectory.

The (17.2.3.2-5) computed value for $\left(C^E_{NVar}\right)_m$ can be applied over successive m cycles to calculate the horizontal angular position movement of the NVar Frame over an m cycle using, by analogy, the inverse of Equation (17.2.3.1-23):

$$C_{NVar_m}^{NVar_{m-1}} = \left(C_{NVar}^{E}\right)_{m-1}^{T} \left(C_{NVar}^{E}\right)_{m} \qquad (17.2.3.2\text{-}6)$$

The equivalent horizontal rotation vector corresponding to $C_{NVar_m}^{NVar_{m-1}}$ is obtained by analogy from the inverse of (17.2.3.1-21):

$$\underline{\xi}_{Var_m} = \begin{matrix}\text{Rotation angle} \\ \text{extraction from}\end{matrix} \; C_{NVar_m}^{NVar_{m-1}} \; \begin{matrix}\text{using (3.2.2.2-10) - (3.2.2.2-12)} \\ \text{and (3.2.2.2-15) - (3.2.2.2-19)}\end{matrix} \qquad (17.2.3.2\text{-}7)$$

By direct analogy with the inverse of (17.2.3.1-19) - (17.2.3.1-20) and (17.2.3.1-22) using h_{Var} and $\underline{\xi}_{Var}$ from (17.2.3.2-1) and (17.2.3.2-7) as input, the NVar Frame components of the variation trajectory position change relative to the earth over an m cycle are then calculated:

$$\begin{aligned} \underline{u}_{ZNVar}^{NVar} \times \Delta\underline{R}_{Var_m}^{NVar} &= \left(F_{CVar_{m-1/2}}^{N}\right)^{-1} \underline{\xi}_{Var_m} \\ \underline{u}_{ZNVar}^{NVar} \cdot \Delta\underline{R}_{Var_m}^{NVar} &= h_{Var_m} - h_{Var_{m-1}} \end{aligned} \qquad (17.2.3.2\text{-}8)$$

where

$\Delta\underline{R}_{Var_m}^{NVar}$ = Position change relative to the earth over an m cycle for the variation trajectory position, expressed in NVar Frame coordinates.

$\underline{u}_{ZNVar}^{NVar}$ = NVar Frame components of a unit vector along the NVar Frame Z axis.

with $F_{CVar_{m-1/2}}^{N}$ calculated by linear extrapolation as in (17.2.3.1-6) from past computed values:

$$F_{CVar_{m-1/2}}^{N} \approx \frac{3}{2} F_{CVar_{m-1}}^{N} - \frac{1}{2} F_{CVar_{m-2}}^{N} \qquad (17.2.3.2\text{-}9)$$

The past value F_{CVar}^{N} terms in (17.2.3.2-9) would be calculated with Equations (5.3-18) using past values of h_{Var} and C_{NVar}^{E} as input.

The $\Delta\underline{R}_{Var_m}^{NVar}$ position change vector is derived from (17.2.3.2-8) using generalized Equation (13.1-9) rearranged with (3.1.1-15):

$$\Delta\underline{R}_{Var_m}^{NVar} = \underline{u}_{ZNVar}^{NVar}\left(\underline{u}_{ZNVar}^{NVar} \cdot \Delta\underline{R}_{Var_m}^{NVar}\right) - \underline{u}_{ZNVar}^{NVar} \times \left(\underline{u}_{ZNVar}^{NVar} \times \Delta\underline{R}_{Var_m}^{NVar}\right) \qquad (17.2.3.2\text{-}10)$$

Substituting (17.2.3.2-8) into (17.2.3.2-10) obtains the equation for the variation trajectory position change increment:

$$\Delta \underline{R}_{Var_m}^{NVar} = \left(h_{Var_m} - h_{Var_{m-1}} \right) \underline{u}_{ZNVar}^{NVar} - \underline{u}_{ZNVar}^{NVar} \times \left[\left(F_{CVar_{m-1/2}}^{N} \right)^{-1} \underline{\xi}_{Var_m} \right] \qquad (17.2.3.2\text{-}11)$$

We can apply $\left(C_{NVar}^{N} \right)_m$ from(17.2.3.2-4) and $\Delta \underline{R}_{Var_m}^{NVar}$ from (17.2.3.2-11) to calculate the body B Frame attitude associated with the variation trajectory. First, we rewrite Equations (17.2.2-1) and (17.2.2-2) based on a modified body (strapdown sensor) coordinate frame definition:

$$\left(C_{BVar}^{\mathcal{V}F} \right)_m = \left(C_{AC}^{\mathcal{V}F} \right)_m \left(C_{BVar}^{AC} \right)_m \qquad \left(C_{BVar}^{L} \right)_m = \left(C_{\mathcal{V}F}^{L} \right)_m \left(C_{BVar}^{\mathcal{V}F} \right)_m \qquad (17.2.3.2\text{-}12)$$

where

BVar = Sensor coordinate frame associated with the variation trajectory.

The $\left(C_{AC}^{\mathcal{V}F} \right)_m$ matrix in (17.2.3.2-12) is provided from the reference trajectory solution. The orientation of the BVar Frame depends on our choice for the BVar Frame attitude relative to the AC Frame as manifested in the $\left(C_{BVar}^{AC} \right)_m$ matrix. The discussion on B Frame selection in Section 17.2.2 also applies for BVar Frame selection. We may also choose to include additional specified angular rotation components associated with the variation trajectory, based on the (3.2.1-5) chain rule:

$$\left(C_{BVar}^{AC} \right)_m = \left(C_{ACVar}^{AC} \right)_m \left(C_{BVar}^{ACVar} \right)_m \qquad (17.2.3.2\text{-}13)$$

where

ACVar = AC Frame associated with the variation trajectory that includes the additional specified variation trajectory angular rotation effects.

The $\left(C_{BVar}^{ACVar} \right)_m$ matrix in Equation (17.2.3.2-13) would be specified based on the Section 17.2.2 discussion on B Frame orientation selection relative to AC, but applied to BVar orientation relative to ACVar. The $\left(C_{ACVar}^{AC} \right)_m$ matrix in (17.2.3.2-13) describes the angular displacement of the ACVar Frame from the reference solution AC Frame and can be defined by a rotation vector using generalized Equation (3.2.2.1-8):

$$\left(C_{ACVar}^{AC} \right)_m = I + \frac{\sin \phi_{Var_m}}{\phi_{Var_m}} \left(\underline{\phi}_{Var_m} \times \right) + \frac{\left(1 - \cos \phi_{Var_m} \right)}{\phi_{Var_m}^2} \left(\underline{\phi}_{Var_m} \times \right)^2 \qquad (17.2.3.2\text{-}14)$$

where

$\underline{\phi}_{Var_m}$ = Rotation vector specified in the AC Frame representing the ACVar Frame angular orientation relative to the AC Frame.

ϕ_{Var_m} = Magnitude of $\underline{\phi}_{Var_m}$.

To determine the BVar Frame orientation relative to the local level of the variation trajectory position, we apply $\left(C_{NVar}^{N}\right)_m$ from (17.2.3.2-4) to $\left(C_{BVar}^{L}\right)_m$ from (17.2.3.2-12):

$$\left(C_{BVar}^{LVar}\right)_m = C_{NVar}^{LVar}\left(C_{NVar}^{N}\right)_m^T C_L^N \left(C_{BVar}^{L}\right)_m = C_N^L \left(C_{NVar}^{N}\right)_m^T C_L^N \left(C_{BVar}^{L}\right)_m \qquad (17.2.3.2\text{-}15)$$

where

LVar = Locally level coordinate frame parallel to the NVar Frame, but with Z axis down and X, Y axes interchanged. The C_N^L matrix also defines the relative LVar to NVar attitude (i.e., $C_{NVar}^{LVar} = C_N^L$).

Summarizing for a moment, we have found a solution for h_{Var_m}, $\left(C_{NVar}^{E}\right)_m$, $\Delta\underline{R}_{Var_m}^N$ and $\left(C_{BVar}^{LVar}\right)_m$ from Equations (17.2.3.2-1) - (17.2.3.2-7), (17.2.3.2-9), and (17.2.3.2-11) - (17.2.3.2-15) repeated below:

$$h_{Var_m} = h_m + \underline{u}_{ZN}^N \cdot \underline{S}_{Var_m}^N$$

$$F_{CVar_{m-1}}^N = f\left[\left(C_{NVar}^{E}\right)_{m-1}, h_{Var_{m-1}}\right] \qquad \text{Using Equations (5.3-18)}$$

$$F_{CVar_m}^N \approx 2\, F_{CVar_{m-1}}^N - F_{CVar_{m-2}}^N$$

$$\underline{\lambda}_m^N = \frac{1}{2}\left(F_{C_m}^N + F_{CVar_m}^N\right)\left(\underline{u}_{ZN}^N \times \underline{S}_{Var_m}^N\right) \qquad (17.2.3.2\text{-}16)$$

$$\left(C_{NVar}^{N}\right)_m = I + \frac{\sin\lambda_m}{\lambda_m}\left(\underline{\lambda}_m^N\times\right) + \frac{(1-\cos\lambda_m)}{\lambda_m^2}\left(\underline{\lambda}_m^N\times\right)^2$$

$$\left(C_{NVar}^{E}\right)_m = \left(C_N^E\right)_m \left(C_{NVar}^{N}\right)_m$$

$$C_{NVar\,m}^{NVar\,m-1} = \left(C_{NVar}^{E}\right)_{m-1}^T \left(C_{NVar}^{E}\right)_m$$

(Continued)

$$\underline{\xi}_{Var_m} = \text{Rotation angle extraction from } C_{NVar_m}^{NVar_{m-1}} \text{ using (3.2.2.2-10) - (3.2.2.2-12) and (3.2.2.2-15) - (3.2.2.2-19)}$$

$$\underline{F}_{CVar_{m-1/2}}^{N} \approx \frac{3}{2}\,\underline{F}_{CVar_{m-1}}^{N} - \frac{1}{2}\,\underline{F}_{CVar_{m-2}}^{N}$$

$$\Delta\underline{R}_{Var_m}^{NVar} = \left(h_{Var_m} - h_{Var_{m-1}}\right)\underline{u}_{ZNVar}^{NVar} - \underline{u}_{ZNVar}^{NVar} \times \left[\left(\underline{F}_{CVar_{m-1/2}}^{N}\right)^{-1} \underline{\xi}_{Var_m}\right]$$

$$\left(C_{ACVar}^{AC}\right)_m = I + \frac{\sin \phi_{Var_m}}{\phi_{Var_m}}\left(\underline{\phi}_{Var_m}\times\right) + \frac{(1 - \cos \phi_{Var_m})}{\phi_{Var_m}^2}\left(\underline{\phi}_{Var_m}\times\right)^2 \qquad \text{(17.2.3.2-16) (Continued)}$$

$$\left(C_{BVar}^{AC}\right)_m = \left(C_{ACVar}^{AC}\right)_m \left(C_{BVar}^{ACVar}\right)_m \qquad C_{BVar}^{ACVar} \text{ from Section 17.2.2, substituting BVar for B and ACVar for AC}$$

$$\left(C_{BVar}^{\mathcal{V}F}\right)_m = \left(C_{AC}^{\mathcal{V}F}\right)_m \left(C_{BVar}^{AC}\right)_m$$

$$\left(C_{BVar}^{L}\right)_m = \left(C_{\mathcal{V}F}^{L}\right)_m \left(C_{BVar}^{\mathcal{V}F}\right)_m$$

$$\left(C_{BVar}^{LVar}\right)_m = C_N^L \left(C_{NVar}^{N}\right)_m^T C_L^N \left(C_{BVar}^{L}\right)_m$$

The $\Delta\underline{R}_{Var_m}^{N}$ and $\left(C_{BVar}^{LVar}\right)_m$ terms from (17.2.3.2-16) can now be used to obtain the integrated specific-force/inertial-angular-rate increments associated with the BVar Frame, and the NVar components of velocity relative to the earth for the variation trajectory. The associated equations to follow are based on the inverse of Equations (17.2.3.1-2) - (17.2.3.1-28).

By direct analogy to Equations (17.2.3.1-29) for the B Frame inertial angular increment, the BVar Frame angular increment is obtained as follows:

$$\underline{\omega}_{IE_{m-1}}^{NVar} = \left(C_{NVar}^{E}\right)_{m-1}^T \underline{\omega}_{IE}^{E}$$

$$\underline{\omega}_{IE_{m-1/2}}^{NVar} \approx \frac{3}{2}\,\underline{\omega}_{IE_{m-1}}^{NVar} - \frac{1}{2}\,\underline{\omega}_{IE_{m-2}}^{NVar}$$

$$\underline{\zeta}_{Var_m} \approx C_N^L \left[\underline{\omega}_{IE_{m-1/2}}^{NVar} T_m + \underline{F}_{CVar_{m-1/2}}^{N}\left(\underline{u}_{ZNVar}^{NVar} \Delta\underline{R}_{Var_m}^{N}\right)\right]$$

$$C_{LVar_{m-1}}^{LVar_m} = I - \frac{\sin \zeta_{Var_m}}{\zeta_{Var_m}}\left(\underline{\zeta}_{Var_m}\times\right) + \frac{(1 - \cos \zeta_{Var_m})}{\zeta_{Var_m}^2}\left(\underline{\zeta}_{Var_m}\times\right)\left(\underline{\zeta}_{Var_m}\times\right) \qquad \text{(17.2.3.2-17)}$$

$$C_{BVar_m}^{BVar_{m-1}} = \left(C_{BVar}^{LVar}\right)_{m-1}^T \left(C_{LVar_{m-1}}^{LVar_m}\right)^T \left(C_{BVar}^{LVar}\right)_m$$

$$\underline{\alpha}_{Var_m} = \text{Rotation angle extraction from } C_{BVar_m}^{BVar_{m-1}} \text{ using (3.2.2.2-10) - (3.2.2.2-12) and (3.2.2.2-15) - (3.2.2.2-19)}$$

where

$$\underline{\alpha}_{Var_m} = \text{Integrated BVar Frame inertial angular rate increment over the m cycle.}$$

The F^N_{CVar} terms in (17.2.3.2-17) are provided from Equations (17.2.3.2-16). Note in Equations (17.2.3.2-17) that $\underline{\zeta}_{Var_m}$ uses $\Delta\underline{R}^N_{Var_m}$, the proper equivalent to $\Delta\underline{R}^N_m$ in Equation (17.2.3.1-26), rather than an approximate form as in Equations (17.2.3.1-29). This is because $\Delta\underline{R}^N_{Var_m}$ is available from Equations (17.2.3.2-16) for the variation trajectory while it had not yet been computed prior to Equations (17.2.3.1-29) for the reference trajectory.

Using $\underline{\alpha}_{Var_m}$ from (17.2.3.2-17) and $\Delta\underline{R}^{NVar}_{Var_m}$ from (17.2.3.2-16), we can now calculate the integrated BVar Frame specific force increment associated with the variation trajectory. This is accomplished in two steps; calculation of the LVar Frame position change due to specific force followed by simultaneous solution of velocity/position updating relationships to determine the BVar Frame specific force increment. Treating velocity/position updating Equations (17.2.3.1-2) - (17.2.3.1-23) as a generalized integration algorithm applied to the lever arm offset position, we have from (17.2.3.1-12) - (17.2.3.1-14) directly and the inverse of (17.2.3.1-18), for the LVar specific force generated position change:

$$\begin{aligned}
\underline{v}^{NVar}_{Var_{m-1/2}} &\approx \frac{3}{2}\,\underline{v}^{NVar}_{Var_{m-1}} - \frac{1}{2}\,\underline{v}^{NVar}_{Var_{m-2}} \\
\underline{g}^{NVar}_{P_{m-1/2}} &\approx \frac{3}{2}\,\underline{g}^{NVar}_{P_{m-1}} - \frac{1}{2}\,\underline{g}^{NVar}_{P_{m-2}} \\
\Delta\underline{v}^{NVar}_{G/COR/Var_m} &\approx \Big\{\underline{g}^{NVar}_{P_{m-1/2}} - \Big[2\,\underline{\omega}^{NVar}_{IE_{m-1/2}} \\
&\qquad + F^N_{CVar_{m-1/2}}\left(\underline{u}^{NVar}_{ZNVar} \times \underline{v}^{NVar}_{Var_{m-1/2}}\right)\Big] \times \underline{v}^{NVar}_{Var_{m-1/2}}\Big\}\,T_m \qquad (17.2.3.2\text{-}18) \\
\Delta\underline{R}^{LVar}_{SFVar_m} &= \left(C^N_L\right)^T \left[\Delta\underline{R}^{NVar}_{Var_m} - \left(\underline{v}^{NVar}_{Var_{m-1}} + \frac{1}{2}\,\Delta\underline{v}^{NVar}_{G/COR/Var_m}\right) T_m\right]
\end{aligned}$$

The BVar Frame integrated specific force increment is then calculated by simultaneous solution of the following direct forms of (17.2.3.1-4) and (17.2.3.1-9), inverted forms of (17.2.3.1-16) - (17.2.3.1-17) and direct variation forms of (17.2.3.1-2) - (17.2.3.1-3):

$$\begin{aligned}
\Delta\underline{R}^{NVar}_{vVar_m} &\approx T_m\left(\frac{3}{2}\,\underline{v}^{NVar}_{Var_{m-1}} - \frac{1}{2}\,\underline{v}^{NVar}_{Var_{m-2}}\right) \\
\underline{\zeta}_{vVar_m} &\approx C^L_N\left[\underline{\omega}^{NVar}_{IE_{m-1/2}}\,T_m + F^N_{CVar_{m-1/2}}\left(\underline{u}^{NVar}_{ZNVar} \times \Delta\underline{R}^{NVar}_{vVar_m}\right)\right]
\end{aligned} \qquad (17.2.3.2\text{-}19)$$

$$\Delta \underline{R}_{SFVar_m}^{BVar} = \left(C_{BVar}^{LVar} \right)_{m-1}^{T} \left(\Delta \underline{R}_{SFVar_m}^{LVar} + \frac{1}{3} \underline{\zeta}_{vVar_m} \times \Delta \underline{v}_{SFVar_m}^{LVar_{m-1}} T_m \right)$$

$$\underline{\upsilon}_{Var_m} = \frac{1}{T_m} \left[\frac{1}{2} I + \frac{1}{\alpha_{Var_m}^2} \left(1 - \frac{\sin \alpha_{Var_m}}{\alpha_{Var_m}} \right) \left(\underline{\alpha}_{Var_m} \times \right) + \frac{1}{\alpha_{Var_m}^2} \left(\frac{1}{2} - \frac{(1 - \cos \alpha_{Var_m})}{\alpha_{Var_m}^2} \right) \left(\underline{\alpha}_{Var_m} \times \right)^2 \right]^{-1} \Delta \underline{R}_{SFVar_m}^{BVar} \tag{17.2.3.2-20}$$

$$\Delta \underline{v}_{SFVar_m}^{BVar_{m-1}} = \left[I + \frac{(1 - \cos \alpha_{Var_m})}{\alpha_{Var_m}^2} \left(\underline{\alpha}_{Var_m} \times \right) + \frac{1}{\alpha_{Var_m}^2} \left(1 - \frac{\sin \alpha_{Var_m}}{\alpha_{Var_m}} \right) \left(\underline{\alpha}_{Var_m} \times \right)^2 \right] \underline{\upsilon}_{Var_m}$$

$$\Delta \underline{v}_{SFVar_m}^{LVar_{m-1}} = \left(C_{BVar}^{LVar} \right)_{m-1} \Delta \underline{v}_{SFVar_m}^{BVar_{m-1}}$$

where

$\underline{\upsilon}_{Var_m}$ = Integrated BVar Frame specific force acceleration increment for the variation trajectory over an m cycle.

The Equations (17.2.3.2-20) simultaneous solution for $\underline{\upsilon}_{Var_m}$ is simplified if we introduce the notation:

$$G_{\alpha R Var_m} \equiv \frac{1}{2} I + \frac{1}{\alpha_{Var_m}^2} \left(1 - \frac{\sin \alpha_{Var_m}}{\alpha_{Var_m}} \right) \left(\underline{\alpha}_{Var_m} \times \right) + \frac{1}{\alpha_{Var_m}^2} \left(\frac{1}{2} - \frac{(1 - \cos \alpha_{Var_m})}{\alpha_{Var_m}^2} \right) \left(\underline{\alpha}_{Var_m} \times \right)^2 \tag{17.2.3.2-21}$$

$$G_{\alpha v Var_m} \equiv I + \frac{(1 - \cos \alpha_{Var_m})}{\alpha_{Var_m}^2} \left(\underline{\alpha}_{Var_m} \times \right) + \frac{1}{\alpha_{Var_m}^2} \left(1 - \frac{\sin \alpha_{Var_m}}{\alpha_{Var_m}} \right) \left(\underline{\alpha}_{Var_m} \times \right)^2$$

where

$G_{\alpha R Var_m}$ = Matrix function of $\underline{\alpha}_{Var_m}$ that translates $\underline{\upsilon}_{Var_m} T_m$ into $\Delta \underline{R}_{SFVar_m}^{BVar}$.

$G_{\alpha v Var_m}$ = Matrix function of $\underline{\alpha}_{Var_m}$ that translates $\underline{\upsilon}_{Var_m}$ into $\Delta \underline{v}_{SFVar_m}^{BVar_{m-1}}$.

Applying (17.2.3.2-21), Equations (17.2.3.2-20) have the simpler form:

$$\Delta \underline{R}_{SFVar_m}^{BVar} = \left(C_{BVar}^{LVar}\right)_{m-1}^{T} \left(\Delta \underline{R}_{SFVar_m}^{LVar} + \frac{1}{3}\, \underline{\zeta}_{vVar_m} \times \Delta \underline{v}_{SFVar_m}^{LVar_{m-1}}\, T_m\right)$$

$$\underline{\upsilon}_{Var_m} = \frac{1}{T_m}\, G_{\alpha RVar_m}^{-1}\, \Delta \underline{R}_{SFVar_m}^{BVar}$$

$$\Delta \underline{v}_{SFVar_m}^{BVar_{m-1}} = G_{\alpha vVar_m}\, \underline{\upsilon}_{Var_m} \qquad (17.2.3.2\text{-}22)$$

$$\Delta \underline{v}_{SFVar_m}^{LVar_{m-1}} = \left(C_{BVar}^{LVar}\right)_{m-1} \Delta \underline{v}_{SFVar_m}^{BVar_{m-1}}$$

By combining the last two of the (17.2.3.2-22) expressions, substituting the result in the first equation, substituting that result in the second equation, rearranging the $\underline{\upsilon}_{Var_m}$ term from the right to the left side of the result, and factoring $\underline{\upsilon}_{Var_m}$ from the result, we obtain the combined form:

$$\left[I - \frac{1}{3}\, G_{\alpha RVar_m}^{-1} \left(C_{BVar}^{LVar}\right)_{m-1}^{T} \left(\underline{\zeta}_{vVar_m} \times\right) \left(C_{BVar}^{LVar}\right)_{m-1} G_{\alpha vVar_m}\right] \underline{\upsilon}_{Var_m}$$

$$= \frac{1}{T_m}\, G_{\alpha RVar_m}^{-1} \left(C_{BVar}^{LVar}\right)_{m-1}^{T} \Delta \underline{R}_{SFVar_m}^{LVar} \qquad (17.2.3.2\text{-}23)$$

The solution for $\underline{\upsilon}_{Var_m}$ is then easily obtained as the inverse of (17.2.3.2-23):

$$\underline{\upsilon}_{Var_m} = \frac{1}{T_m} \Big[I \qquad (17.2.3.2\text{-}24)$$

$$- \frac{1}{3}\, G_{\alpha RVar_m}^{-1} \left(C_{BVar}^{LVar}\right)_{m-1}^{T} \left(\underline{\zeta}_{vVar_m} \times\right) \left(C_{BVar}^{LVar}\right)_{m-1} G_{\alpha vVar_m}\Big]^{-1} G_{\alpha RVar_m}^{-1} \left(C_{BVar}^{LVar}\right)_{m-1}^{T} \Delta \underline{R}_{SFVar_m}^{LVar}$$

In order to calculate the NVar Frame velocity associated with the variation trajectory, we utilize $\underline{\zeta}_{vVar_m}$ from (17.2.3.2-19), $\underline{\upsilon}_{Var_m}$ from (17.2.3.2-24) and $\Delta \underline{v}_{G/COR/Var_m}^{NVar}$ from (17.2.3.2-18) as input to the last two equations in set (17.2.3.2-22) and the variation version of (17.2.3.1-10) - (17.2.3.1-11) and (17.2.3.1-15):

$$\Delta \underline{v}_{SFVar_m}^{BVar_{m-1}} = G_{\alpha vVar_m}\, \underline{\upsilon}_{Var_m}$$

$$\Delta \underline{v}_{SFVar_m}^{LVar_{m-1}} = \left(C_{BVar}^{LVar}\right)_{m-1} \Delta \underline{v}_{SFVar_m}^{BVar_{m-1}} \qquad (17.2.3.2\text{-}25)$$

$$\mathfrak{C}_{LVar_{m-1}}^{LVar_m} \approx I - \left(\underline{\zeta}_{vVar_m} \times\right)$$

(Continued)

$$\Delta \underline{v}_{SFVar_m}^{LVar_m} = \mathfrak{C}_{LVar_{m-1}}^{LVar_m} \Delta \underline{v}_{SFVar_m}^{LVar_{m-1}}$$

$$\underline{v}_{Var_m}^{NVar} = \underline{v}_{Var_{m-1}}^{NVar} + C_L^N \Delta \underline{v}_{SFVar_m}^{LVar_m} + \Delta \underline{v}_{G/COR/Var_m}^{NVar} \qquad (17.2.3.2\text{-}25) \; \text{(Continued)}$$

In summary, the $\underline{\alpha}_{Var}$ BVar Frame integrated inertial angular rate increment is computed with Equations (17.2.3.2-17), the $\underline{\upsilon}_{Var}$ BVar Frame components of the variation trajectory integrated specific force acceleration increment is provided by Equations (17.2.3.2-18) - (17.2.3.2-19), (17.2.3.2-21) and (17.2.3.2-24), and the NVar Frame velocity components $\underline{v}_{Var}^{NVar}$ of the variation trajectory are calculated from Equations (17.2.3.2-25). Equations (17.2.3.2-16) provide BVar Frame attitude orientation relative to the LVar Frame (C_{BVar}^{LVar}) and the variation trajectory position in the E Frame as defined by C_{NVar}^{E} and h_{Var}. The variation trajectory parameters so calculated have been designed to be at a specified N Frame position location $\underline{S}_{Var}^{N}$ relative to the reference position solution, and to have a specified BVar Frame attitude orientation relative to the reference solution AC Frame (C_{BVar}^{AC}) as defined in Equations (17.2.3.2-16) by C_{BVar}^{ACVar} plus a specified angular variation $\underline{\phi}_{Var}$ between the reference solution AC Frame and the ACVar Frame. The $\underline{\alpha}_{Var}$ and $\underline{\upsilon}_{Var}$ BVar Frame integrated inertial angular rate and specific force acceleration increments have been designed so that if processed by the high accuracy Equation (17.2.3.1-2) - (17.2.3.1-28) integration algorithms, they will generate C_{BVar}^{LVar}, $\underline{v}_{Var}^{NVar}$, C_{NVar}^{E} and h_{Var} with identical values as those computed from (17.2.3.2-16) and (17.2.3.2-25).

Equations (17.2.3.2-16) - (17.2.3.2-19), (17.2.3.2-21), (17.2.3.2-24) and (17.2.3.2-25) define how user specified $\underline{\phi}_{Var}$, C_{BVar}^{ACVar}, and $\underline{S}_{Var}^{N}$ parameters can be used to generate a consistent complete set of navigation data for a variation trajectory relative to a reference trajectory. The discussion on B Frame selection in Section 17.2.2 also applies for user selection of the attitude between the ACVar and BVar Frames (as manifested in the C_{BVar}^{ACVar} matrix). In the following subsections, we will discuss how $\underline{\phi}_{Var}$ and $\underline{S}_{Var}^{N}$ can be used to account for three effects: 1. Aerodynamic wind gusts, 2. The trajectory followed by a position location in the vehicle that is displaced from the reference solution by an AC Frame specified lever arm displacement, and 3. High frequency angular and linear motion effects in general.

17.2.3.2.1 Adding Wind Gust Aerodynamic Force Effects

The effect of wind gusts can be added to the trajectory regeneration function using the procedure of Section 17.2.3.2, as a variation in position $\underline{S}^{N}_{Var_m}$ from the Section 17.2.3.1 reference trajectory produced by wind gust induced variations in the aerodynamic specific force. Neglecting rotations of the local level N Frame, we can write from Newton's basic law:

$$\Delta \dot{\underline{v}}^{N}_{Var} = \frac{1}{M} \Delta \underline{F}^{N}_{WndGst} \qquad (17.2.3.2.1\text{-}1)$$

where

$\Delta \underline{v}_{Var}$ = Variation in the Section 17.2.3.1 calculated velocity relative to the earth due to additional wind gust induced aerodynamic forces.

M = Vehicle mass.

$\Delta \underline{F}_{WndGst}$ = Wind gust induced modifications to the Section 17.1.2.3 aerodynamic force.

Using the (3.2.1-5) chain law, the N Frame components of $\Delta \underline{F}_{WndGst}$ can be expressed in terms of components in the relative airspeed coordinate frame used for $\Delta \underline{F}_{WndGst}$ evaluation:

$$\Delta \underline{F}^{N}_{WndGst} = C^{N}_{L}\, C^{L}_{\mathcal{V}F}\, C^{\mathcal{V}F}_{\mathcal{V}WF}\, \Delta \underline{F}^{\mathcal{V}WF}_{WndGst} \qquad (17.2.3.2.1\text{-}2)$$

where

$\mathcal{V}$WF = $\mathcal{V}$W Frame generated using smoothed trajectory segment parameter data.

The aerodynamic force variation is the difference between the aerodynamic force with and without wind gusts:

$$\Delta \underline{F}^{\mathcal{V}WF}_{WndGst} = \underline{F}^{\mathcal{V}WF}_{TotAero} - \underline{F}^{\mathcal{V}WF}_{Aero} \qquad (17.2.3.2.1\text{-}3)$$

where

$\underline{F}^{\mathcal{V}WF}_{Aero}$ = Aerodynamic force exclusive of wind gust effects as calculated with Equations (17.1.2.3-63) in the Section 17.1.2.3 iteration loop (but based on smoothed trajectory profile segment data).

$\underline{F}^{\mathcal{V}WF}_{TotAero}$ = Total aerodynamic force including wind gust effects, but calculated with Equations (17.1.2.3-63) following the Section 17.1.2.3 iteration loop using an addition to Equations (17.1.2.3-53) for $\underline{v}^{\mathcal{V}F}_{Arspd}$ that includes wind gusts effects.

Using the (3.2.1-5) chain law, the previously mentioned addition to the smoothed version of (17.1.2.3-53) for wind gust effects is as follows:

$$\begin{aligned}
\underline{v}^{\mathcal{V}F}_{WndGst} &= \left(C^{L}_{\mathcal{V}F}\right)^{T} C^{L}_{N}\, C^{N}_{Geo}\, \underline{v}^{Geo}_{WndGst} \\
\Delta\underline{v}^{\mathcal{V}F}_{Var} &= \left(C^{L}_{\mathcal{V}F}\right)^{T} \left(C^{N}_{L}\right)^{T} \Delta\underline{v}^{N}_{Var} \\
\underline{v}^{\mathcal{V}F}_{TotArspd} &= V_F\, \underline{u}^{\mathcal{V}F}_{X\mathcal{V}F} + \Delta\underline{v}^{\mathcal{V}F}_{Var} - \underline{v}^{\mathcal{V}F}_{AvgWnd} - \underline{v}^{\mathcal{V}F}_{WndGst} \\
v_{TotArspd} &= \sqrt{\underline{v}^{\mathcal{V}F}_{TotArspd} \cdot \underline{v}^{\mathcal{V}F}_{TotArspd}}
\end{aligned} \qquad (17.2.3.2.1\text{-}4)$$

where

$\mathcal{V}F$ = $\mathcal{V}$ Frame generated in Section 17.1.2.3 using smoothed trajectory segment parameter data.

$\underline{v}_{AvgWnd}$ = Local average wind velocity relative to the earth.

$\underline{v}_{WndGst}$ = Wind gust velocity relative to the earth.

$\underline{v}_{TotArspd}$ = Total vehicle velocity relative to the air mass including wind gust effects.

$v_{TotArspd}$ = Magnitude of $\underline{v}_{TotArspd}$.

The dynamic pressure associated with $\underline{v}_{TotArspd}$ is the smoothed revised form of (17.1.2.3-56):

$$q_{Tot} = \frac{1}{2}\,\rho\, v^{2}_{TotArspd} \qquad (17.2.3.2.1\text{-}5)$$

where

q_{Tot} = Total airspeed dynamic pressure (i.e., including wind gust velocity).

The AC Frame components of $\underline{v}_{TotArspd}$ are given by:

$$\underline{v}^{AC}_{TotArspd} = \left(C^{\mathcal{V}F}_{AC}\right)^{T} \underline{v}^{\mathcal{V}F}_{TotArspd} \qquad (17.2.3.2.1\text{-}6)$$

which generates AC Frame angles of attack and sideslip per Equations (17.1.2.3-13) of:

$$\alpha_{Tot} = \tan^{-1}\frac{v_{TotArspd_{ZAC}}}{v_{TotArspd_{XAC}}} \qquad \beta_{Tot} = -\tan^{-1}\frac{v_{TotArspd_{YAC}}}{v_{TotArspd_{XAC}}} \qquad (17.2.3.2.1\text{-}7)$$

where

$v_{TotArspd_{iAC}}$ = Axis i component of $\underline{v}^{AC}_{TotArspd}$.

$\alpha_{Tot}, \beta_{Tot}$ = Angles of attack and sideslip generated by $\underline{v}^{AC}_{TotArspd}$.

The $\underline{F}^{\mathcal{V}WF}_{TotAero}$ vector in (17.2.3.2.1-3) is calculated with Equations (17.1.2.3-63) using α_{Tot}, β_{Tot} from (17.2.3.2.1-7) for α, β, and q_{Tot} from (17.2.3.2.1-5) for q:

$$\underline{F}^{\mathcal{V}WF}_{TotAero} = f\left(\alpha_{Tot}, \beta_{Tot}, q_{Tot}\right) \quad \text{with Equations (17.1.2.3-63)} \qquad (17.2.3.2.1\text{-}8)$$

The $\underline{v}^{Geo}_{WndGst}$ wind gust vector in (17.2.3.2.1-4) can be modeled as a first order Markov process as in (16.2.3.1-1):

$$\dot{\underline{v}}^{Geo}_{WndGst} = - C_{WndGst} \, \underline{v}^{Geo}_{WndGst} + \underline{n}^{Geo}_{WndGst} \qquad (17.2.3.2.1\text{-}9)$$

where

$\underline{n}^{Geo}_{WndGst}$ = Vector of independent white noise components.

C_{WndGst} = Wind gust correlation frequency (the reciprocal of the correlation time).

For digital simulation purposes, we can approximate $\underline{n}^{Geo}_{WndGst}$ as a finite constant vector over each computer update cycle m that varies randomly from cycle, with a value equal to the integral of $\underline{n}^{Geo}_{WndGst}$ over the m cycle divided by the m cycle time interval. Based on this approximation and using standard methods for solving linear differential equations with constant coefficients (e.g., Reference 38 - Section 18-9), the $\underline{v}^{Geo}_{WndGst}$ components at each m cycle are then provided by the integral solution to the (17.2.3.2.1-9) differential equation over the m cycle, using the $\underline{v}^{Geo}_{WndGst}$ value at the previous cycle as the initial condition (as in (15.1-1), (15.1.1-1), (15.1.1-3), (15.1.1-4), (15.1.1-6) and (15.1.1-12)):

$$v_{WndGst/iGeo_m} = \left(e^{-C_{WndGst} \, T_m}\right) v_{WndGst/iGeo_{m-1}} + \frac{\left(1 - e^{-C_{WndGst} \, T_m}\right)}{C_{WndGst}} \frac{w_{WndGst/iGeo_m}}{T_m} \qquad (17.2.3.2.1\text{-}10)$$

where

$v_{WndGst/iGeo_m}$ = Component i of $\underline{v}^{Geo}_{WndGst}$ at computer cycle m.

$w_{WndGst/iGeo_m}$ = Component i of the integral of $\underline{n}^{Geo}_{WndGst}$ over a computer cycle. In a simulation program, $w_{WndGst/iGeo_m}$ would be programmed as a random number generator with a specified variance.

The variance of $w_{WndGst/iGeo_m}$ for the random number generator can be found from the variance form of (17.2.3.2.1-10). The variance form of (17.2.3.2.1-10) is obtained by squaring the right and left sides, taking the expected value, identifying the expected values of $w_{WndGst/iGeo}^2$ and $v_{WndGst/iGeo}^2$ as their variances, and recognizing from the definitions of $\underline{n}_{WndGst}^{Geo}$ and $w_{WndGst/iGeo_m}$, that $w_{WndGst/iGeo_m}$ is uncorrelated from cycle to cycle with zero expected value for any m cycle. The result is:

$$P_{WndGsti_m} = \left(e^{-2\,C_{WndGst}\,T_m}\right) P_{WndGsti_{m-1}} + \left[\frac{\left(1 - e^{-C_{WndGst}\,T_m}\right)}{C_{WndGst}\,T_m}\right]^2 Q_{WndGsti_m} \qquad (17.2.3.2.1\text{-}11)$$

where

$P_{WndGsti_m}$ = Variance of $v_{WndGst/iGeo_m}$.

$Q_{WndGsti_m}$ = Variance of $w_{WndGst/iGeo_m}$ in (17.2.3.2.1-10) for the simulation program.

In the steady state, $P_{WndGsti_m}$ will equal $P_{WndGsti_{m-1}}$ which allows us to solve (17.2.3.2.1-11) for $Q_{WndGsti_m}$ as a function of the steady state value for $P_{WndGsti}$:

$$Q_{WndGsti_m} = \frac{\left(1 - e^{-2\,C_{WndGst}\,T_m}\right)}{\left[\frac{\left(1 - e^{-C_{WndGst}\,T_m}\right)}{C_{WndGst}\,T_m}\right]^2} \sigma_{WndGsti_m}^2 \qquad (17.2.3.2.1\text{-}12)$$

where

$\sigma_{WndGsti_m}$ = Wind gust velocity standard deviation (i.e., root-mean-square value for $P_{WndGsti}$ in the steady state).

The $Q_{WndGsti_m}$ wind gust noise variance can be calculated from (17.2.3.2.1-12) using $\sigma_{WndGsti_m}$ specified by the user as a function of the variation trajectory latitude, longitude and altitude.

Equation (17.2.3.2.1-1) with (17.2.3.2.1-2) - (17.2.3.2.1-8), (17.2.3.2.1-10) and (17.2.3.2.1-12) defines the rate of change of $\Delta\underline{v}_{Var}^N$ due to wind gusts. The double integral of (17.2.3.2.1-1) generates $\Delta\underline{v}_{Var}^N$ for input to Equation (17.2.3.2.1-4), and $\underline{S}_{Var}^N$ for input to the Section 17.2.3.2 variation trajectory regeneration procedure. We also artificially constrain the double integration routines to assure that $\Delta\underline{v}_{Var}^N$ and $\underline{S}_{Var}^N$ will be bounded and reduce to zero in the absence of wind gusts. Using (17.2.3.2.1-10) as a template with (17.2.3.2.1-1) as the input, we thereby approximate the integration routines in recursive m cycle form as:

$$\Delta \underline{v}^N_{Var_m} = \left(1 - C_{Cnstrnt}\ T_m\right) \Delta \underline{v}^N_{Var_{m-1}} + \frac{1}{2}\left(\Delta \dot{\underline{v}}^N_{Var_m} + \Delta \dot{\underline{v}}^N_{Var_{m-1}}\right) T_m$$

$$\underline{S}^N_{Var_m} = \left(1 - C_{Cnstrnt}\ T_m\right) \underline{S}^N_{Var_{m-1}} + \frac{1}{2}\left(\Delta \underline{v}^N_{Var_m} + \Delta \underline{v}^N_{Var_{m-1}}\right) T_m \qquad (17.2.3.2.1\text{-}13)$$

where

$C_{Cnstrnt}$ = Divergence constraint correlation time. The value for $C_{Cnstrnt}$ should be smaller than C_{WndGst} (say one half C_{WndGst}).

Without the $C_{Cnstrnt}$ terms, Equations (17.2.3.2.1-13) would provide a trapezoidal double digital integration of the $\Delta \dot{\underline{v}}^N_{Var}$ input. The $C_{Cnstrnt}$ terms provide feedback to prevent $\Delta \underline{v}^N_{Var}$ and $\underline{S}^N_{Var}$ build-up to unreasonably large values, and to control $\Delta \underline{v}^N_{Var}$ and $\underline{S}^N_{Var}$ toward zero in the absence of wind gusts (i.e., when the $\Delta \dot{\underline{v}}^N_{Var}$ input is zero).

The $\underline{S}^N_{Var_m}$ position variation obtained from (17.2.3.2.1-13) would be used as input to the Section 17.2.3.2 procedure to generate the variation trajectory that includes wind gust effects. For this application, the $\underline{\phi}_{Var_m}$ parameter for Section 17.2.3.2 input would be equated to zero.

17.2.3.2.2 Adding Sensor Assembly Lever Arm Displacement Effects

Thus far, we have not distinguished between different location points in the vehicle with regard to position/velocity. Obviously, different position locations in the vehicle will have different navigational position solutions in the E Frame, being displaced from one another by the difference in their respective vehicle position locations. The earth referenced velocity of separated vehicle position locations will also differ under vehicle angular rate due to the resulting circular movement of one point about the other. Using the trajectory variation method of Section 17.2.3.2, the trajectory profile for a location point in the vehicle, that is displaced from the reference solution by an AC Frame defined lever arm, is easily determined by setting:

$$\underline{l}^{AC}_m = \underline{l}^{AC}_0 + \Delta \underline{l}^{AC}_{HiF_m} \qquad \underline{S}^N_{Var_m} = C^N_L\, C^L_{AC_m}\, \underline{l}^{AC}_m \qquad (17.2.3.2.2\text{-}1)$$

where

$\underline{l}$ = Lever arm displacement of the strapdown sensor assembly from the Section 17.2.3.1 reference solution position location.

$\underline{l}_0^{AC}$ = Fixed AC Frame components of $\underline{l}$.

$\Delta \underline{l}_{HiF}^{AC}$ = High frequency variation of the $\underline{l}$ lever arm position location in the AC Frame (e.g., due to vehicle linear flexing). If $\Delta \underline{l}_{HiF}^{AC}$ is to be included in the simulation, it would be calculated using the procedure described subsequently in Section 17.2.3.2.3.

High frequency angular motion at the displaced lever arm position location can be accounted for by setting:

$$\underline{\phi}_{Var_m} = \underline{\phi}_{HiFlev_m}^{AC} \qquad (17.2.3.2.2\text{-}2)$$

where

$\underline{\phi}_{HiFlev_m}^{AC}$ = High frequency angular variation between the ACVar Frame and the reference AC Frame at the lever arm location (e.g., due to vehicle angular flexing). If $\underline{\phi}_{HiFlev_m}^{AC}$ is to be included in the simulation, it would be calculated using the procedure described subsequently in Section 17.2.3.2.3.

If the above lever arm effects are to be included in addition to the Section 17.2.3.2.1 wind gust effects, Equations (17.2.3.2.2-1) - (17.2.3.2.2-2) would be applied using the Section 17.2.3.2 method as a variation on the Section 17.2.3.2.1 solution (rather than on the reference Section 17.2.3.1 solution). Alternatively, the $\underline{S}_{Var}^{N}$ wind gust solution from Equation (17.2.3.2.1-13) of the previous section can be added to $\underline{S}_{Var}^{N}$ from Equations (17.2.3.2.2-1), with the result and $\underline{\phi}_{Var}$ from (17.2.3.2.2-2) applied, using the Section 17.2.3.2 method, as a variation on the reference Section 17.2.3.1 solution. The latter method introduces a small second order error which is usually negligible, but which should be verified to be insignificant if lever arm effects are to be simulated for two different vehicle locations in which second order effects can introduce position errors between the two relative lever arm displaced locations.

17.2.3.2.3 Adding High Frequency Effects

High frequency linear and angular rotation effects (e.g., vehicle dynamic bending) can be added to the Section 17.2.3.1 basic reference solution using the variation technique of Section 17.2.3.2 with $\underline{S}_{Var}^{N}$ and $\underline{\phi}_{Var}$ representing the high frequency linear and angular motion. For example, consider that we represent the bending effect as a variational movement of a generalized mass m from the rigid body solution, in which m is connected to the rigid solution by a linear spring of constant k, and experiences damping forces by a linear friction coefficient c acting on the mass motion relative to the rigid solution. Further, consider that the total acceleration of the mass (that makes it generally follow the rigid body solution) is generated from only the spring/damper plus the effect of gravity. Then a simplified one dimensional model for the generalized mass movement can be described from Newton's law as:

$$m\left(\dot{v} + \Delta\dot{v}\right) = m\,g - c\,\Delta v - k\,\Delta x \qquad \Delta\dot{x} = \Delta v \qquad (17.2.3.2.3\text{-}1)$$

where

v = Velocity for the rigid body solution.

Δv = Velocity of the generalized mass relative to the rigid solution.

Δx = Displacement of the generalized mass from the rigid solution.

g = Gravitational acceleration.

We can also write for the rigid solution:

$$\dot{v} = a_{SF} + g \qquad (17.2.3.2.3\text{-}2)$$

where

a_{SF} = Total specific force acceleration for the rigid body solution.

Substituting (17.2.3.2.3-2) into (17.2.3.2.3-1) and rearranging terms then obtains:

$$\Delta\dot{v} = -\,a_{SF} - \frac{c}{m}\,\Delta v - \frac{k}{m}\,\Delta x \qquad \Delta\dot{x} = \Delta v \qquad (17.2.3.2.3\text{-}3)$$

Let's now extend Equations (17.2.3.2.3-3) to three dimensions in the AC Frame including rotational response. In general, the response of a flexible body can be described as the linear sum of the responses of its "natural vibration modes" to externally applied forces (Reference 1 - Chapter 3). Each vibration mode has a characteristic shape (linear and angular in three-dimensions), frequency, and damping characteristic defined by the body's internal structure, material and mass distribution. The amplitude of each mode is determined by the amplitude, distribution and time history of externally applied force acting on the flexible structure. Recognizing (17.2.3.2.3-3) as a classical second order system, we can write a simplified model (as in Section 15.2.1.2, Equations (15.2.1.2-8)) for the generalized response of the lth bending mode as:

$$\begin{aligned}\dot{\mu}_{l\,bnd_i} &= -\,a_{SF_{iAC}} - 2\,\zeta_{l\,bnd}\,\omega_{n_{l\,bnd}}\,\mu_{l\,bnd_i} - \omega_{n_{l\,bnd}}^{2}\,\rho_{l\,bnd_i} \\ \dot{\rho}_{l\,bnd_i} &= \mu_{l\,bnd_i}\end{aligned} \qquad (17.2.3.2.3\text{-}4)$$

$$\Delta R_{l\text{bnd}_{ji}} = K_{R\rho l\text{bnd}_{ji}} \, \rho_{l\text{bnd}l_i} \qquad \Delta\phi_{l\text{bnd}_{ji}} = K_{\phi\rho l\text{bnd}_{ji}} \, \rho_{l\text{bnd}l_i} \tag{17.2.3.2.3-5}$$

where

$l\text{bnd}$ = Designation for the lth bending mode.

$a_{SF_{iAC}}$ = Total rigid body specific force acceleration along AC Frame axis i.

$\zeta_{l\text{bnd}}, \omega_{n_{l\text{bnd}}}$ = Damping ratio and undamped natural frequency for the lth bending mode.

$\mu_{l\text{bnd}_i}, \rho_{l\text{bnd}_i}$ = Damped sinusoidal generalized velocity and position response characteristics of the lth bending mode induced by specific force along the AC Frame i axis.

$\Delta R_{l\text{bnd}_{ji}}$ = Translational position movement along AC Frame axis j of the lth bending mode from the rigid body solution at a particular vehicle location, induced by $a_{SF_{iAC}}$.

$\Delta\phi_{l\text{bnd}_{ji}}$ = Angular displacement around AC Frame axis j of the lth bending mode from the rigid body solution at a particular vehicle location, induced by $a_{SF_{iAC}}$.

$K_{R\rho l\text{bnd}_{ji}}, K_{\phi\rho l\text{bnd}_{ji}}$ = Translational position movement and angular displacement influence coefficients for the lth bending mode that shape the response of $\Delta R_{l\text{bnd}_{ji}}$ and $\Delta\phi_{l\text{bnd}_{ji}}$ as functions of the $\rho_{l\text{bnd}_i}$ generalized position response characteristic. Values for $K_{R\rho l\text{bnd}_{ji}}$ and $K_{\phi\rho l\text{bnd}_{ji}}$ depend on the particular location in the flexing body where bending effects are being evaluated.

The $\underline{S}^N_{\text{Var}}$ and $\underline{\phi}_{\text{Var}}$ parameters for Section 17.2.3.2 are then calculated as the sum of the Equations (17.2.3.2.3-4) - (17.2.3.2.3-5) solutions for the three AC Frame axis specific force acceleration inputs for the number of l bending modes being simulated. Generally, the principal bending effects are produced by wind gusts. Then it makes sense to form $\underline{S}^N_{\text{Var}}$ and $\underline{\phi}_{\text{Var}}$ as the sum of the Section 17.2.3.2.1 solution plus the previous solution:

$$\underline{S}^N_{\text{Var}} = \underline{S}^N_{\text{WndGst}} + C^N_L \, C^L_{AC} \sum_l \Delta\underline{R}^{AC}_{l\text{bnd}}$$

$$\underline{\phi}_{\text{Var}} = \sum_l \Delta\underline{\phi}^{AC}_{l\text{bnd}} \tag{17.2.3.2.3-6}$$

where

$\underline{S}^N_{WndGst}$ = $\underline{S}^N_{Var}$ calculated in Equation (17.2.3.2.1-13).

$\Delta\underline{R}^{AC}_{l\,bnd}$ = Vector in the AC Frame having each of its j components equal to the sum over i (i = 1, 2, 3) of the Equation (17.2.3.2.3-5) $\Delta R_{l\,bnd_{ji}}$'s.

$\Delta\underline{\phi}^{AC}_{l\,bnd}$ = Vector in the AC Frame having each of its j components equal to the sum over i (i = 1, 2, 3) of the Equation (17.2.3.2.3-5) $\Delta\phi_{l\,bnd_{ji}}$'s.

The C^L_{AC} matrix in (17.2.3.2.3-6) would be provided from the reference trajectory (e.g., from Equations (17.2.3.1-1)). The variation trajectory would be built upon the Section 17.2.3.1 reference trajectory using (17.2.3.2.3-6) for $\underline{S}^N_{Var}$ and $\underline{\phi}_{Var}$ with the Section 17.2.3.2 procedure.

The integration of Equations (17.2.3.2.3-4) to obtain $\mu_{l\,bnd_i}$, $\rho_{l\,bnd_i}$ would be handled digitally at the m cycle rate. The digital integration equation can be derived from the general matrix form of (17.2.3.2.3-4)

$$\dot{\underline{x}} = A\,\underline{x} + \underline{y} \tag{17.2.3.2.3-7}$$

with for this case:

$$\underline{x} \equiv \begin{bmatrix} \mu_{l\,bnd_i} \\ \rho_{l\,bnd_i} \end{bmatrix} \qquad A \equiv \begin{bmatrix} -2\,\zeta_{l\,bnd}\,\omega_{n_{l\,bnd}} & -\omega^2_{n_{l\,bnd}} \\ 1 & 0 \end{bmatrix} \qquad \underline{y} \equiv \begin{bmatrix} -a_{SF_{iAC}} \\ 0 \end{bmatrix} \tag{17.2.3.2.3-8}$$

Using the principle of linear superposition, the general solution to (17.2.3.2.3-7) is constructed as:

$$\underline{x} = \underline{x}_{Hmg} + \underline{x}_{Prt} \tag{17.2.3.2.3-9}$$

where

$\underline{x}_{Hmg}$ = Homogeneous solution to (17.2.3.2.3-7) defined as the solution for zero $\underline{y}$ input.

$\underline{x}_{Prt}$ = Particular solution to (17.2.3.2.3-7) that satisfies (17.2.3.2.3-7) for the particular form of $\underline{y}$.

Using Section 15.1.1 as a guide, we construct $\underline{x}_{Hmg}$ as:

$$\underline{x}_{Hmg} = \Phi(t, t_{m-1})\,\underline{C} \tag{17.2.3.2.3-10}$$

where

$\Phi(t, t_{m-1})$ = State transition matrix that propagates $\underline{x}$ from a value at simulation time cycle m-1 to time t following cycle time m-1.

$\underline{C}$ = Constant vector.

Substituting (17.2.3.2.3-10) for $\underline{x}$ in (17.2.3.2.3-7) with $\underline{y}$ set to zero, and using the definition for $\Phi(t, t_{m-1})$ in the form of Equation (15.1.1-6), shows that (17.2.3.2.3-10) satisfies the requirement for the homogeneous solution.

We structure the particular solution based on A being constant, and the approximation that $\underline{y}$ is constant from cycle m-1 to m:

$$\underline{x}_{Prt} = \underline{D} \tag{17.2.3.2.3-11}$$

where

$\underline{D}$ = Constant vector.

Substituting (17.2.3.2.3-11) in (17.2.3.2.3-7) with $\underline{y}$ (and A) constant shows that:

$$\underline{D} = - A^{-1}\, \underline{y} \tag{17.2.3.2.3-12}$$

Substituting (17.2.3.2.3-10) - (17.2.3.2.3-12) in (17.2.3.2.3-9) then gives:

$$\underline{x} = \Phi(t, t_{m-1})\, \underline{C} - A^{-1}\, \underline{y} \tag{17.2.3.2.3-13}$$

The value for $\underline{C}$ is obtained from (17.2.3.2.3-13) by evaluation at m-1 with, from (15.1.1-4), $\Phi(t_{m-1},t_{m-1})$ equal to the identity matrix:

$$\underline{C} = \underline{x}_{m-1} + A^{-1}\, \underline{y} \tag{17.2.3.2.3-14}$$

Finally, we substitute (17.2.3.2.3-14) in (17.2.3.2.3-13) and evaluate $\underline{x}$ at cycle m to obtain the digital integration algorithm equivalent to (17.2.3.2.3-7):

$$\underline{x}_m = \Phi\, \underline{x}_{m-1} + \left(\Phi - I\right) A^{-1}\, \underline{y} \qquad \Phi \equiv \Phi(t_m,t_{m-1}) \tag{17.2.3.2.3-15}$$

The Φ state transition matrix is evaluated as in (15.1.2.1.1.1-10) - (15.1.2.1.1.1-11) as:

$$\Phi = e^{A T_m} \tag{17.2.3.2.3-16}$$

The a_{SF_iAC} components for (17.2.3.2.3-8) can be calculated from:

$$\underline{a}_{SF}^{AC} = \frac{1}{T_m}\left(C_{AC}^{L}\right)^{T} \Delta\underline{v}_{SF}^{L} \qquad (17.2.3.2.3\text{-}17)$$

with C_{AC}^{L} and $\Delta\underline{v}_{SF}^{L}$ provided from the Section 17.2.3.1 reference trajectory (i.e. from Equations (17.2.3.1-1) and (17.2.3.1-30)).

As another example, consider the general problem of accounting for high frequency angular and linear motion (e.g., caused by air turbulence or other high frequency random effects). This general problem can be handled using (17.2.3.2.3-7) as a general equation defining the overall dynamic response $\underline{x}$ of a particular vehicle station to random acceleration inputs in $\underline{y}$. The (17.2.3.2.3-15) - (17.2.3.2.3-16) incremental form would still apply. The general state vector $\underline{x}$ would include all pertinent AC Frame high frequency velocity, position, angular rate and angular components (relative to the rigid body solution). The $\underline{S}_{Var}^{N}$ and $\underline{\phi}_{Var}$ parameters would be calculated as the sum of the position and angular components of $\underline{x}$ similar to Equations (17.2.3.2.3-6). In this case, the $\underline{y}$ acceleration components would be modeled to simulate specified noise/bandwidth characteristics. For example, if we choose to model the acceleration components as having components in a prescribed frequency band, the acceleration can be analytically modeled as the response of a linear band-pass filter to a white noise input. The band-pass filter can be modeled as a high pass filter feeding a low pass filter. The high pass filter can be modeled as its input minus the response of a low pass filter to the same input. Thus, the band-pass filter satisfies the following equations:

$$\begin{aligned}
\dot{z}_1 &= \omega_{Hi}\left(z_{In} - z_1\right) \\
z_{Hi} &= z_{In} - z_1 \qquad (17.2.3.2.3\text{-}18) \\
\dot{z}_{Out} &= \omega_{Lo}\left(z_{Hi} - z_{Out}\right) = \omega_{Lo}\left(z_{In} - z_1 - z_{Out}\right)
\end{aligned}$$

where

z_{In}, z_{Out} = Band-pass filter input and output.

z_{Hi} = Output of high-pass filter stage.

z_1 = Output of low-pass filter used to form the first stage high-pass filter.

ω_{Hi}, ω_{Lo} = Break frequencies for the high-pass and low-pass band-pass filter stages. Note that with this definition, ω_{Lo} is greater than ω_{Hi}.

For this application, we identify:

$$z_{In} = n_{SF_{iRnd}} \qquad a_{SF_{iAC}} = z_{Out} \qquad (17.2.3.2.3\text{-}19)$$

where

$n_{SF_{iRnd}}$ = White acceleration noise along the AC Frame i axis.

$a_{SF_{iAC}}$ = Acceleration input to Equation (17.2.3.2.3-7) (as defined in Equation (17.2.3.2.3-8)) having frequency content from ω_{Hi} to ω_{Lo}.

Equations (17.2.3.2.3-18) and (17.2.3.2.3-19) can be written in the equivalent state vector form:

$$\dot{\underline{z}} = B\,\underline{z} + \underline{\omega}\, n_{SF_{iRnd}} \qquad (17.2.3.2.3\text{-}20)$$

$$\underline{z} \equiv \begin{bmatrix} z_1 \\ a_{SF_{iAC}} \end{bmatrix} \qquad B \equiv - \begin{bmatrix} \omega_{Hi} & 0 \\ \omega_{Lo} & \omega_{Lo} \end{bmatrix} \qquad \underline{\omega} \equiv \begin{bmatrix} \omega_{Hi} \\ \omega_{Lo} \end{bmatrix}$$

The $n_{SF_{iRnd}}$ white noise term can be characterized in terms of its density by the covariance form of (17.2.3.2.3-20) (using (15.1.2.1.1-30) with (15.1-1) as a guide):

$$\dot{P}_Z = B\,P_Z + P_Z\,B^T + \underline{\omega}\left(\underline{\omega}\right)^T Q_{SF_{iRnd}} \qquad (17.2.3.2.3\text{-}21)$$

where

$Q_{SF_{iRnd}}$ = Density for $n_{SF_{iRnd}}$.

P_Z = Covariance of $\underline{z}$.

In the steady state the derivative term in (17.2.3.2.3-21) is zero and the equation reduces to:

$$B\,P_Z + P_Z\,B^T = -\,\underline{\omega}\left(\underline{\omega}\right)^T Q_{SF_{iRnd}} \qquad (17.2.3.2.3\text{-}22)$$

Equation (17.2.3.2.3-22) in component form provides three simultaneous scalar equations for the elements of P_Z in terms of $Q_{SF_{iRnd}}$ (including the fact that because P_Z is symmetrical, its off-diagonal terms are equal). The solution to these scalar equations finds that:

$$Q_{SF_{iRnd}} = \frac{2\left(\omega_{Lo} + \omega_{Hi}\right)}{\omega_{Lo}^2}\, P_{aSF_{iAC}} \qquad (17.2.3.2.3\text{-}23)$$

where

$P_{aSF_{iAC}}$ = Variance of $a_{SF_{iAC}}$, the 2, 2 element of P_Z.

Equation (17.2.3.2.3-23) defines the $Q_{SF_{iRnd}}$ white noise density in terms of the variance of $a_{SF_{iAC}}$ which can be specified by the user. We also note that (17.2.3.2.3-23) can be obtained more rapidly using Table E.2-1 of Reference 26 as will be discussed at the conclusion of this section.

In a digital simulation program, an algorithmic version of (17.2.3.2.3-20) would be incorporated as a digital integration operation to obtain $a_{SF_{iAC}}$. Using Equations (17.2.3.2.3-7), (17.2.3.2.3-15) and (17.2.3.2.3-16) as a template, we can write for the $\underline{z}$ digital integration algorithm (assuming B is constant):

$$\underline{z}_l = \theta\, \underline{z}_{l-1} + (\theta - I)\, B^{-1}\, \underline{\omega}\, w_{SFiRnd_l} \qquad \theta = e^{B\, T_l} \qquad (17.2.3.2.3\text{-}24)$$

where

l = Computer integration rate cycle index for calculating $\underline{z}$.

T_l = Time interval for l cycle.

θ = State transition matrix for Equation (17.2.3.2.3-20).

w_{SFiRnd_l} = Constant over computer cycle l that varies randomly from l cycle to l cycle. Used as an approximation to the integrated effect of $n_{SF_{iRnd}}$.

Equations (17.2.3.2.3-24) are shown at an l cycle rate rather than an m cycle rate to allow for a normally faster l rate to properly account for the frequency width of the band-pass filter (i.e., the magnitude of the highest frequency root ω_{Lo}).

The w_{SFiRnd_l} parameter would be obtained from a random number generator with a variance equivalent to $Q_{SF_{iRnd}}$. The $Q_{SF_{iRnd}}$ equivalence can be defined from the requirement that the P_z variance change produced by the integral of (17.2.3.2.3-21) over T_l match the variance change in $\underline{z}_l$ as computed by Equations (17.2.3.2.3-24). To first order in B T_l, Equations (17.2.3.2.3-24) are given by:

$$\underline{z}_l \approx (I + B\, T_l)\, \underline{z}_{l-1} + \underline{\omega}\, w_{SFiRnd_l}\, T_l \qquad (17.2.3.2.3\text{-}25)$$

where

$\Delta\underline{z}_l$ = Change in $\underline{z}$ over an l cycle ($\Delta\underline{z}_l \equiv \underline{z}_l - \underline{z}_{l-1}$).

The covariance form of (17.2.3.2.3-25) using the (15.1.2.1-4) definition is to first order in B T_l:

$$\Delta P_{z_l} \approx B\, P_{z_{l-1}}\, T_l + P_{z_{l-1}}\, B^T\, T_l + \underline{\omega}\, (\underline{\omega})^T\, P_{wSFiRnd}\, T_l^2 \qquad (17.2.3.2.3\text{-}26)$$

where

ΔP_{z_l} = Change in the $\underline{z}$ covariance over an l cycle.

$$P_{wSFiRnd} = \text{Variance of } w_{SFiRnd_l}.$$

The integral of (17.2.3.2.3-21) over an l cycle is easily found to first order (in B T_l) by treating the terms on the right as constant (at their last computed values) and multiplying by the l cycle time interval:

$$\Delta P_{z_l} \approx B\, P_{z_{l-1}}\, T_l + P_{z_{l-1}}\, B^T\, T_l + \underline{\omega}\left(\underline{\omega}\right)^T Q_{SF_{iRnd}}\, T_l \qquad (17.2.3.2.3\text{-}27)$$

Comparing Equations (17.2.3.2.3-26) and (17.2.3.2.3-27) we see that equivalency is achieved if we set:

$$P_{wSFiRnd} = \frac{1}{T_l}\, Q_{SF_{iRnd}} \qquad (17.2.3.2.3\text{-}28)$$

or with (17.2.3.2.3-23):

$$P_{wSFiRnd} = \frac{1}{T_l}\, \frac{2\left(\omega_{Lo} + \omega_{Hi}\right)}{\omega_{Lo}^2}\, P_{aSF_{iAC}} \qquad (17.2.3.2.3\text{-}29)$$

Equation (17.2.3.2.3-29) provides an analytical method for evaluating $P_{wSFiRnd}$, the variance of w_{SFiRnd_l}. $P_{wSFiRnd}$ would then be applied as the specified variance of the random number generator used in the simulation for generating the w_{SFiRnd_l} random number sequence. If this is the approach taken for selecting $P_{wSFiRnd}$, the w_{SFiRnd_l} sequence created by the random number generator, when applied as the input to Equation (17.2.3.2.3-24), will produce a $\underline{z}_l$ history with an $a_{SF_{iAC}}$ variance (i.e., $P_{aSF_{iAC}}$) that approximately matches the desired $P_{aSF_{iAC}}$ value used in (17.2.3.2.3-29). The match will only be approximate because of the simplifications used in the derivation of (17.2.3.2.3-29). If more accuracy is required, an empirical method can be applied as an alternate, based on directly measuring the response of (17.2.3.2.3-24) to a unity variance random number generator input.

The empirical method for determining $P_{wSFiRnd}$ consists of first running Equation (17.2.3.2.3-24) for many cycles using unity for $P_{wSFiRnd}$, and computing the resulting "normalized $P_{aSF_{iAC}}$" as the mean squared average of the $a_{SF_{iAC}}$ output. The correct $P_{wSFiRnd}$ value is then calculated as the desired specified value for $P_{aSF_{iAC}}$ divided by the computed normalized $P_{aSF_{iAC}}$. Subsequently applying the so determined "correct" $P_{wSFiRnd}$ in (17.2.3.2.3-24) for the w_{SFiRnd_l} random number sequence variance, will then generate an $a_{SF_{iAC}}$ steady state output variance that matches the specified $P_{aSF_{iAC}}$ value. When using this process, it is important that the "normalized $P_{aSF_{iAC}}$" mean-squared average computation begin after Equation (17.2.3.2.3-24) has been run for at least four time constants (i.e., the reciprocal of ω_{Hi}) to assure that covariance transients have had time to decay.

Equations (17.2.3.2.3-25) with (17.2.3.2.3-29) show how $\underline{z}$ can be computed at the l cycle rate to generate a random $a_{SF_{iAC}}$ output (see (17.2.3.2.3-20) for $\underline{z}$ definition) that has a specified variance $P_{aSF_{iAC}}$ and frequency bandwidth ω_{Hi}, ω_{Lo}. These equations would be repeated for each i axis of the AC Frame using a different random number sequence for w_{SFiRnd_l}, but having the same variance. The $a_{SF_{iAC}}$ output would then be input to Equation (17.2.3.2.3-15) in the $\underline{y}$ vector as defined in Equation (17.2.3.2.3-8), for each i axis being simulated (using the broader interpretation of (17.2.3.2.3-15) provided in the paragraph following Equation (17.2.3.2.3-17)). For the general case when the m cycle rate is slower than the l cycle rate, the value for $a_{SF_{iAC}}$ used at the m cycle rate in $\underline{y}$ (in Equations (17.2.3.2.3-8) and (17.2.3.2.3-15)) must be compatible with the $a_{SF_{iAC}}$ value calculated by (17.2.3.2.3-25) at the l cycle rate. The compatibility is based on the integrated effect of $a_{SF_{iAC}}$ used in $\underline{y}$ over an m cycle, matching the integral of $a_{SF_{iAC}}$ from (17.2.3.2.3-25) over an m cycle. Because $a_{SF_{iAC}}$ used in $\underline{y}$ is treated as a constant over an m cycle, the compatibility relation is defined simply by:

$$a_{SF_{iAC\text{-}m}}\, T_m = \sum^{r} a_{SF_{iAC\text{-}l}}\, T_l \qquad (17.2.3.2.3\text{-}30)$$

or

$$a_{SF_{iAC\text{-}m}} = \frac{1}{r} \sum^{r} a_{SF_{iAC\text{-}l}} \qquad (17.2.3.2.3\text{-}31)$$

where

r = Number of l cycles in an m cycle.

$a_{SF_{iAC\text{-}l}}$ = $a_{SF_{iAC}}$ calculated from l cycle rate Equations (17.2.3.2.3-25).

$a_{SF_{iAC\text{-}m}}$ = $a_{SF_{iAC}}$ value used in $\underline{y}$ for Equations (17.2.3.2.3-15) at the m cycle rate.

For the remainder of this section we will address the general problem of calculating the steady variance of a linear system driven by white noise. This is the same problem addressed previously based on Equations (17.2.3.2.3-20) and (17.2.3.2.3-22) leading to Equation (17.2.3.2.3-23) for the input white noise density that will produce a given output variance. The solution method we will use now is based on Table E.2-1 of Reference 26 which is simpler than the approach used previously (given that Table E.2-1 exists). We begin by restating the Chapter 10 definition Equations (10.2.2-9) - (10.2.2-10) for the auto-correlation function of a general stationary random noise process p(t):

$$\varphi_{pp}(t, \tau) \equiv \mathcal{E}\big(p(t)\, p(t+ \tau)\big) \qquad \varphi_{pp}(t, \tau) = \varphi_{pp}(\tau) \qquad (17.2.3.2.3\text{-}32)$$

where

τ = Correlation time for the p(t) ensemble random process.

$\varphi_{pp}(t, \tau)$ = Autocorrelation function for p(t).

Section 10.2.2, Equation (10.2.2-21) defines a power spectral density function associated with $\varphi_{pp}(\tau)$ as:

$$G(\omega) \equiv \frac{2}{\pi} \int_0^{\infty} \varphi_{pp}(\tau) \cos \omega\tau \, d\tau \tag{17.2.3.2.3-33}$$

where

ω = Fourier transform frequency parameter.

$G(\omega)$ = Power spectral density function as defined by Equation (17.2.3.2.3-33).

Equation (4.2-2) of Reference 26 also defines a power spectral density function for $\varphi_{pp}(\tau)$ as:

$$F(S) \equiv \frac{1}{2\pi} \int_{-\infty}^{\infty} \varphi_{pp}(\tau) \, e^{-S\tau} \, d\tau \tag{17.2.3.2.3-34}$$

where

S = Laplace transform parameter.

F(S) = Power spectral density function as defined by Equation (17.2.3.2.3-34).

An equivalency between $G(\omega)$ and F(S) can be established by substituting $S = j\omega$ in (17.2.3.2.3-34) and expanding using (10.2.1-12) while recognizing from Section 10.2.2 that $\varphi_{pp}(\tau)$ is a symmetrical function of τ:

$$\begin{aligned} F(j\omega) &= \frac{1}{2\pi} \int_{-\infty}^{\infty} \varphi_{pp}(\tau) \left(\cos \omega\tau - j \sin \omega\tau\right) d\tau \\ &= \frac{1}{2\pi} \int_{-\infty}^{\infty} \phi_{pp}(\tau) \cos \omega\tau \, d\tau = \frac{1}{\pi} \int_0^{\infty} \phi_{pp}(\tau) \cos \omega\tau \, d\tau \end{aligned} \tag{17.2.3.2.3-35}$$

Comparing Equations (17.2.3.2.3-33) and (17.2.3.2.3-35) we see immediately that:

$$G(\omega) = 2 F(j\omega) \tag{17.2.3.2.3-36}$$

The converse of (17.2.3.2.3-36) is obtained by substituting $\omega = \frac{S}{j} = -jS$ and rearranging:

$$F(S) = \frac{1}{2} G(-jS) \tag{17.2.3.2.3-37}$$

Equation (17.2.3.2.3-37) allows us to apply Equation (4.4-2) of Reference 26 for calculation of the mean squared value of p(t) in which p(t) is defined as the output of a linear system:

$$\mathcal{E}\left(\overline{p_{Out}(t)^2}\right) = \frac{1}{j}\int_{-j\infty}^{j\infty} F_{Out}(S)\, dS \qquad (17.2.3.2.3\text{-}38)$$

where

Out = Designation for linear system output.

$\overline{p_{Out}(t)^2}$ = Mean value of $p_{Out}(t)^2$.

We also know from Equation (4.4-2) of Reference 26 that:

$$F_{Out}(S) = H(S)\, H(-S)\, F_{In}(S) \qquad (17.2.3.2.3\text{-}39)$$

where

In = Designation for linear system input.

H(S) = Linear system input-to-output transfer function as defined in Section 10.2.1 Equation (10.2.1-3).

As in Section 10.2.1, Equation (10.2.1-30), we expand H(S) to:

$$H(S) = \frac{H_{Num}(S)}{H_{Den}(S)} \qquad (17.2.3.2.3\text{-}40)$$

where

$H_{Num}(S)$, $H_{Den}(S)$ = H(S) numerator and denominator polynomials (in powers of S).

Substituting (17.2.3.2.3-39) - (17.2.3.2.3-40) in (17.2.3.2.3-38), and using (17.2.3.2.3-37) for $F_{In}(S)$ obtains:

$$\mathcal{E}\left(\overline{p_{Out}(t)^2}\right) = \frac{1}{2j}\int_{-j\infty}^{-j\infty} \frac{H_{Num}(S)\, H_{Num}(-S)}{H_{Den}(S)\, H_{Den}(-S)}\, G_{In}(-jS)\, dS \qquad (17.2.3.2.3\text{-}41)$$

We now specialize Equation (17.2.3.2.3-41) to the case in which the input to the linear system is white noise. For this situation, we apply Equation (15.1.2.1.1-34) relating the power spectral density of a white noise process to its process noise density so that:

$$G_{In}(\omega) = \frac{1}{\pi} Q_{In} \qquad (17.2.3.2.3\text{-}42)$$

where

Q_{In} = Process noise density of the linear system white noise input.

Based on (17.2.3.2.3-42), the input power spectral density is constant and we can, therefore, write:

$$G_{In}(-jS) = \frac{1}{\pi} Q_{In} \qquad (17.2.3.2.3\text{-}43)$$

We can also define the numerator and denominator H(S) polynomials as:

$$\begin{aligned} H_{Num}(S) &= c_{n-1}\, S^{n-1} + \cdots + c_0 \\ H_{Den}(S) &= d_n\, S^n + \cdots + d_0 \end{aligned} \qquad (17.2.3.2.3\text{-}44)$$

where

c_i, d_i = Coefficients for the H(S) numerator and denominator polynomials in S.

Finally, we substitute (17.2.3.2.3-43) into (17.2.3.2.3-41) to find in factored form:

$$\mathcal{E}\left(\overline{p_{Out}(t)^2}\right) = I_n\, Q_{In} \qquad (17.2.3.2.3\text{-}45)$$

with

$$I_n \equiv \frac{1}{2\,\pi j} \int_{-j\infty}^{-j\infty} \frac{H_{Num}(S)\, H_{Num}(-S)}{H_{Den}(S)\, H_{Den}(-S)}\, dS \qquad (17.2.3.2.3\text{-}46)$$

where

I_n = Function given by Equation (17.2.3.2.3-46) in which n refers to the number of terms in the (17.2.3.2.3-44) polynomials.

Equation (17.2.3.2.3-45) for I_n with (17.2.3.2.3-44) for the H(S) polynomials is now in a form that can be easily evaluated using Table E.2-1 of Reference 26. Table E.2-1 provides values for I_n as a function of the c_i's and d_i's for values of n ranging from 1 to 10.

As an example of the application of the Reference 26 - Table E.2-1 technique, let us return to the linear system described by Equations (17.2.3.2.3-20) to determine the input white noise density that will produce a given steady (average) output variance. Using the Section 10.2.1 methodology, the combined Laplace transform of the previous equation components yields for $a_{SF_{iAC}}$:

$$a_{SF_{iAC}}(S) = \frac{\omega_{Lo}\, S}{S^2 + \left(\omega_{Lo} + \omega_{Hi}\right) S + \omega_{Lo}\,\omega_{Hi}}\, n_{SF_{iRnd}}(S) \qquad (17.2.3.2.3\text{-}47)$$

where

(S) = Indicates Laplace transform of the associated quantity.

From Equation (17.2.3.2.3-47) we see that the $n = 2$ and that the c_i's and d_i's in (17.2.3.2.3-44) are given by:

$$c_0 = 0 \qquad c_1 = \omega_{Lo}$$
$$d_0 = \omega_{Lo}\,\omega_{Hi} \qquad d_1 = \omega_{Lo} + \omega_{Hi} \qquad d_2 = 1 \tag{17.2.3.2.3-48}$$

Using Reference 26 - Table E.2-1, the I_n value for $n = 2$ is:

$$I_2 = \frac{c_1^2\, d_0 + c_0^2\, d_2}{2\, d_0\, d_1\, d_2} \tag{17.2.3.2.3-49}$$

Identifying $\mathcal{E}\left(\overline{p_{Out}(t)^2}\right)$ in (17.2.3.2.3-45) as the $a_{SF_{iAC}}$ steady state variance $P_{aSF_{iAC}}$, and Q_{In} as the $n_{SF_{iRnd}}$ process noise density $Q_{SF_{iRnd}}$, we now substitute (17.2.3.2.3-48) into (17.2.3.2.3-49) and the result in (17.2.3.2.3-45) to find:

$$P_{aSF_{iAC}} = \frac{\omega_{Lo}^2}{2\left(\omega_{Lo} + \omega_{Hi}\right)}\, Q_{SF_{iRnd}} \tag{17.2.3.2.3-50}$$

which shows that:

$$Q_{SF_{iRnd}} = \frac{2\left(\omega_{Lo} + \omega_{Hi}\right)}{\omega_{Lo}^2}\, P_{aSF_{iAC}} \tag{17.2.3.2.3-51}$$

Equation (17.2.3.2.3-51) is identical to the (17.2.3.2.3-23) result obtained previously by a more involved analytical process.

17.3 USING A TRAJECTORY GENERATOR IN AIDED STRAPDOWN INS SIMULATIONS

A trajectory generator has various applications in the simulation of aided inertial navigation systems. It should be noted at the onset, however, that a trajectory generator should not be the primary vehicle used to validate strapdown inertial navigation integration algorithms (i.e., such as developed in Chapter 7). The reason should be obvious. Since trajectory generators are in general formed from strapdown integration algorithms, their use becomes suspect in validating algorithms of the type from which they were constructed. The basic method for strapdown algorithm verification should generally be based on comparisons with known closed-form analytical navigation solutions, rather than solutions derived from a numerical integration process (e.g., Chapter 11). In fact, the validation process for the trajectory generator itself should be based on a Chapter 11 type approach. Never-the-less, an independently created trajectory generator can be used to provide inputs to strapdown inertial navigation algorithms to verify that the algorithm solution matches the trajectory generator within prescribed limits. Of course, a mismatch in such a comparison can be caused by errors in the strapdown algorithms being tested, or in the trajectory generator. If this approach is to be utilized, it is important that the trajectory generator be previously validated accurately including documentation for later substantiation.

Two common uses for trajectory generators are for input to covariance simulation programs, and for input to simulators used in validating Kalman filter operations.

For covariance simulation purposes, the C_B^L attitude matrix, $\underline{v}^N$ velocity vector, C_N^E angular position, h altitude, B Frame integrated inertial-angular-rate/specific-force increments ($\underline{\alpha}$ and $\underline{\upsilon}$), and other parameters computed by the trajectory generator (e.g., gravity, earth rate, and transport rate components) are used in forming the error model matrices (i.e., A, G_P, H_{RW}, G_M, A_{xx}^*, $G_{P_x}^*$, H_x^* and G_M^* in Equations (16.2.3-1) -(16.2.3-6)). Note that for the trajectory generator described in the previous sections, the B Frame inertial angular rate and specific force ($\underline{\omega}_{IB}^B$ and $\underline{a}_{SF}^B$) are approximated as constant over an m cycle, hence, can be calculated for an m cycle as $\underline{\alpha}$ and $\underline{\upsilon}$ divided by T_m. This provides two values for $\underline{\omega}_{IB}^B$ and $\underline{a}_{SF}^B$ at each m cycle junction (for the end of the previous cycle and the start of the current cycle). A simple average of the two can be used to single-value the $\underline{\omega}_{IB}^B$, $\underline{a}_{SF}^B$ components at the m cycle junctions (i.e., the average of the m and m+1 values of $\underline{\alpha}$ and $\underline{\upsilon}$ divided by T_m to obtain the m cycle values for $\underline{\omega}_{IB}^B$ and $\underline{a}_{SF}^B$). However, at cycle m, the m+1 values are not yet available for $\underline{\alpha}$ and $\underline{\upsilon}$. To resolve the difficulty three methods can be used: 1. Approximate $\underline{\omega}_{IB}^B$ and $\underline{a}_{SF}^B$ at cycle m as equal to $\underline{\alpha}$ and $\underline{\upsilon}$ divided by T_m, 2. Use the previous junction averaging technique with extrapolation to obtain $\underline{\omega}_{IB}^B$ and $\underline{a}_{SF}^B$ at m from previously computed m-2 and m-1 values for $\underline{\omega}_{IB}^B$ and $\underline{a}_{SF}^B$, and 3. Use the previous junction averaging technique but run the trajectory generator one cycle faster so that the m+1 values for $\underline{\alpha}$ and $\underline{\upsilon}$ are available one cycle earlier to calculate $\underline{\omega}_{IB}^B$ and $\underline{a}_{SF}^B$.

When using a trajectory generator in an aided strapdown INS time domain simulation, the B Frame integrated inertial angular rate and specific force increments ($\underline{\alpha}$ and $\underline{\upsilon}$) would typically be used for input to a simulation model of a strapdown inertial sensor assembly that includes sensor error characteristics. The output from the strapdown sensor assembly simulator would then be input to a simulation containing the strapdown INS integration algorithms including inertial sensor compensation (e.g., Chapter 7 - Table 7.5-1 and Sections 8.1.2.1 - 8.1.2.2). The output from the simulated strapdown integration algorithms represents a simulated strapdown INS output history including the effect of inertial sensor error and INS sensor compensation error. A similar procedure can be used to simulate the strapdown INS aiding device in which trajectory generator navigation parameters would be used to formulate a simulation of the aiding device and its error characteristics. The simulated strapdown INS and aiding device outputs would then be input to a simulation of a Kalman filter configuration (e.g., for validation purposes as described in Section 15.1.4). The "truth solution" for the simulation would be the trajectory generator output attitude, velocity and position data.

The following sections provide more detail regarding strapdown INS sensor error simulation and an example of a GPS receiver simulation for INS aiding purposes.

17.3.1 SIMULATING STRAPDOWN INS SENSOR ERRORS

In a strapdown INS simulation, sensor errors are accounted for in the sensor assembly simulator (driven by the trajectory generator) and in the simulation of the INS sensor compensation algorithms. The strapdown sensor assembly simulation would implement an integrated increment form of Equations (8.1.1.1-1) and (8.1.1.2-1), e.g.:

$$\begin{aligned}
\underline{\alpha}_{Cnt_m} &= \frac{1}{\Omega_{Wt_0}} \left(I + F_{Scal}\right)\left(F_{Algn}\,\underline{\alpha}_m + \delta\underline{\omega}_{Bias}\,T_m + \Delta\underline{\alpha}_{Quant_m} + \underline{\alpha}_{Rand_m}\right) \\
\underline{\upsilon}_{Cnt_m} &= \frac{1}{A_{Wt_0}} \left(I + G_{Scal}\right)\left(G_{Algn}\,\underline{\upsilon}_m + \delta\underline{a}_{Bias}\,T_m + \Delta\underline{\upsilon}_{Quant_m} + \underline{\upsilon}_{Rand_m}\right)
\end{aligned} \qquad (17.3.1\text{-}1)$$

in which size effect and anisoinertia effects have not been included (for simplicity) and where:

$\underline{\alpha}_{Cnt_m}$, $\underline{\upsilon}_{Cnt_m}$ = Pulse count outputs from the sensor assembly over cycle m.

$\underline{\alpha}_m$, $\underline{\upsilon}_m$ = Integrated cycle m B Frame inertial angular rate and specific force increments from the trajectory generator.

$\Delta\underline{\alpha}_{Quant_m}$, $\Delta\underline{\upsilon}_{Quant_m}$ = Integrated angular rate sensor and accelerometer quantization noise over the m cycle.

$\underline{\alpha}_{Rand_m}$, $\underline{\upsilon}_{Rand_m}$ = Integrated angular rate sensor and accelerometer random noise over the m cycle.

For simulation purposes, $\underline{\alpha}_{Rand_m}$, $\underline{\upsilon}_{Rand_m}$ would be created from a zero mean Gaussian random number generator with variances set to match the angular rate sensor and accelerometer random noise densities multiplied by T_m. The $\Delta\underline{\alpha}_{Quant_m}$, $\Delta\underline{\upsilon}_{Quant_m}$ quantization terms can be elaborately modeled to reflect the quantization mechanism in each inertial component. A simpler approximation for the quantization terms is to model them (similar to the approach taken in Section 12.5) such that their cumulative sum (i.e., digital integral) is a random number of specified variance. Then $\Delta\underline{\alpha}_{Quant_m}$, $\Delta\underline{\upsilon}_{Quant_m}$ would be calculated as:

$$\Delta\underline{\alpha}_{Quant_m} = \underline{\alpha}_{Quant_m} - \underline{\alpha}_{Quant_{m-1}}$$
$$\Delta\underline{\upsilon}_{Quant_m} = \underline{\upsilon}_{Quant_m} - \underline{\upsilon}_{Quant_{m-1}} \qquad (17.3.1\text{-}2)$$

where

$\underline{\alpha}_{Quant_m}$, $\underline{\upsilon}_{Quant_m}$ = Angular rate sensor and accelerometer quantization noise associated with the continuous integrated sensor output.

The $\underline{\alpha}_{Quant_m}$, $\underline{\upsilon}_{Quant_m}$ quantities would be generated from a zero mean random number generator having a uniform statistical distribution over the sensor output pulse size. The associated variance for the random number sequence can be shown to equal the pulse size squared divided by twelve. The pulse size used would generally represent the effective pulse size following quantization compensation (assuming that quantization compensation algorithms will not be explicitly applied in the simulation).

For constant B Frame angular rates and accelerations assumed for the trajectory generator over an m cycle, the $\underline{S}$ terms for the precision position compensation algorithm would be obtained from Equations (8.1.2.1-1) and (8.1.2.2-1) as:

$$\underline{S}_{\alpha Cnt_m} = \frac{1}{2}\,\underline{\alpha}_{Cnt_m}\,T_m \qquad \underline{S}_{\upsilon Cnt_m} = \frac{1}{2}\,\underline{\upsilon}_{Cnt_m}\,T_m \qquad (17.3.1\text{-}3)$$

The strapdown INS sensor compensation simulator would then operate from the (17.3.1-1) - (17.3.1-2) outputs using a simulation of the sensor compensation algorithms (e.g., Equations (8.1.2.1-3) - (8.1.2.1-6) and (8.1.2.2-3) - (8.1.2.2-6) neglecting the quantization, size effect and anisoinertia terms). As mentioned previously, quantization compensation would not be included based on the assumption that the simplified (17.3.1-2) model describes the quantization noise following compensation in the INS computer. Outputs from the sensor compensation simulator would be input to the strapdown integration algorithm simulation (e.g., Chapter 7 Table 7.5-1) with the coning, sculling, and scrolling terms set to zero (i.e., corresponding to constant B Frame angular-rate/specific-force over an m cycle assumed for the trajectory generator).

17.3.2 GPS RECEIVER SIMULATION FOR KALMAN AIDED INS APPLICATION

The simulation of a GPS receiver from a trajectory generator input must calculate the range from the GPS receiver antenna to selected GPS satellites. In the earth E Frame (See Section 2.2 for definition) this can be achieved through:

$$\underline{\rho}_i^E = \underline{R}_{Si}^E - \underline{R}_{GPSAnt}^E \qquad \rho_i = \sqrt{\underline{\rho}_i^E \cdot \underline{\rho}_i^E} \qquad (17.3.2\text{-}1)$$

where

$\underline{R}_{GPSAnt}^E$ = E Frame components of the position vector from earth's center to the strapdown INS GPS receiver antenna.

$\underline{R}_{Si}^E$ = E Frame components of the position vector from earth's center to the ith GPS satellite.

$\underline{\rho}_i^E$ = E Frame components of the range vector from the GPS receiver to the ith GPS satellite.

ρ_i = Range from the GPS receiver to the ith GPS satellite.

The $\underline{R}_{Si}^E$ vector would be computed from a simulation of the GPS satellite orbit in a specified inertial (I) coordinate frame and then transformed from the I to the E Frame (which accounts for the angular rotation of the earth relative to the I Frame). The $\underline{R}_{GPSAnt}^E$ vector would be calculated from trajectory generator input data. For example, the method of Section 17.2.3.2.2 can be used by which a complete navigation solution (with position in terms of C_N^E and h) is calculated for the GPS antenna whose lever arm position location in the vehicle is defined relative to the trajectory generator reference navigation solution. Then $\underline{R}_{GPSAnt}^E$ would be calculated using a variation of generalized Equations (4.4.2.2-2) and (4.4.2.2-5):

$$\underline{R}_{GPSAnt}^E = \underline{R}_{S/GPSAnt}^E + \underline{u}_{ZN/GPSAnt}^E \, h_{GPSAnt}$$

$$\underline{u}_{ZN/GPSAnt}^E = C_{N/GPSAnt}^E \, \underline{u}_{ZN/GPSAnt}^{N/GPSAnt}$$

$$\underline{R}_{S/GPSAnt}^E = R'_{S/GPSAnt} \begin{pmatrix} u_{ZN/GPSAnt\,XE} \\ (1 - e)^2 \, u_{ZN/GPSAnt\,YE} \\ u_{ZN/GPSAnt\,ZE} \end{pmatrix} \qquad (17.3.2\text{-}2)$$

$$R'_{S/GPSAnt} = R_0 / \sqrt{1 + u_{ZN/GPSAnt\,YE}^2 \left[(1 - e)^2 - 1\right]}$$

where

$GPSAnt$ = Reference to GPS antenna navigation solution parameters.

$u_{ZN/GPSAnt\,iE}$ = E Frame i axis component of $\underline{u}^E_{ZN/GPSAnt}$.

The nomenclature in Equations (17.3.2-2) is further defined in Section 4.4.2.2.

The previous method is useful if a complete detailed navigation solution is to be accurately modeled for the INS, the GPS antenna, and other devices in the vehicle. Alternatively, a $\underline{R}^E_{GPSAnt}$ solution can be developed that only accounts for relative position between the INS and GPS antenna using:

$$\underline{R}^E_{GPSAnt} = \underline{R}^E_{INS} + C^E_{N/INS}\, C^N_L\, C^{L/INS}_{AC}\, \underline{l}^{AC}_{INS/GPS} \qquad (17.3.2\text{-}3)$$

where

INS = Reference to INS (inertial navigation system) navigation solution data.

$\underline{R}^E_{INS}$ = E Frame components of the position vector from earth's center to the strapdown INS.

$\underline{l}^{AC}_{INS/GPS}$ = Lever arm position displacement vector from the strapdown INS to the GPS receiver antenna (in AC Frame axes).

For this case, the INS data would be the trajectory generator reference navigation solution, $\underline{R}^E_{INS}$ would be calculated from this data using the equivalent to (17.3.2-2) (for $\underline{R}^E_{INS}$ extraction from $C^E_{N/INS}$ and h_{INS}), and $C^{L/INS}_{AC}$ would be provided from Equations (17.2.3.1-1) (using C^L_B from (17.2.3.1-28) or its variation from Equations (17.2.3.2-16) and the subsections of 17.2.3.2). The $\underline{l}^{AC}_{INS/GPS}$ lever arm can be modeled as a constant or to include bending effects as in Equation (17.2.3.2.2-1).

If GPS receiver derived range rate is to be simulated, the range rate can be calculated from the derivative of the square of the (17.3.2-1) ρ_i expression with rearrangement:

$$\dot{\rho}_i = \frac{1}{\rho_i}\left(\underline{\rho}^E_i \cdot \dot{\underline{\rho}}^E_i\right) \qquad (17.3.2\text{-}4)$$

The $\dot{\underline{\rho}}^E_i$ term in (17.3.2-4) is the derivative of the $\underline{\rho}^E_i$ expression in (17.3.2-1):

$$\dot{\underline{\rho}}_i^E = \dot{\underline{R}}_{Si}^E - \dot{\underline{R}}_{GPSAnt}^E \tag{17.3.2-5}$$

and using (4.3-1) and (4.3-2) as general equations:

$$\dot{\underline{R}}_{GPSAnt}^E = C_{N/GPSAnt}^E \, \underline{v}_{GPSAnt}^{N/GPSAnt} \tag{17.3.2-6}$$

The $\dot{\underline{R}}_{Si}^E$ term in (17.3.2-5) would be obtained from the GPS satellite orbit parameter simulation in the I Frame using generalized Equation (3.4-4) (with $\underline{\omega}_{AB}^A = -\underline{\omega}_{BA}^A$) to compute $\dot{\underline{R}}_{Si}^E$ from $\dot{\underline{R}}_{Si}^I$, $\underline{R}_{Si}^I$ and $\underline{\omega}_{IE}^I$. If the Section 17.2.3.2.2 approach is being used in which a complete navigation solution (with velocity terms) is being calculated for the GPS antenna station, $\underline{v}_{GPSAnt}^{N/GPSAnt}$ in (17.3.2-6) would be the N Frame velocity from this solution. A simplified alternative can be used for cases when the position of the GPS antenna in the AC frame can be approximated as constant (i.e., no bending effects). Then $\dot{\underline{R}}_{GPSAnt}^E$ can be calculated from the derivative of (17.3.2-3) holding $\underline{l}_{INS/GPS}^{AC}$ constant and using (4.3-1) - (4.3-2):

$$\dot{\underline{R}}_{GPSAnt}^E = \dot{\underline{R}}_{INS}^E + \dot{C}_{AC}^E \, \underline{l}_{INS/GPS}^{AC} = C_{N/INS}^E \, \underline{v}^{N/INS} + \dot{C}_{AC}^E \, \underline{l}_{INS/GPS}^{AC} \tag{17.3.2-7}$$

Using generalized Equation (3.3.2-6) for $\dot{C}_{AC}^E$, we then obtain:

$$\begin{aligned} \dot{\underline{R}}_{GPSAnt}^E &= C_{N/INS}^E \, \underline{v}^{N/INS} + C_{AC}^E \left(\underline{\omega}_{EAC}^{AC} \times \underline{l}_{INS/GPS}^{AC} \right) \\ &= C_{N/INS}^E \left[\underline{v}^{N/INS} + C_L^N \, C_{AC}^{L/INS} \left(\underline{\omega}_{EAC}^{AC} \times \underline{l}_{INS/GPS}^{AC} \right) \right] \end{aligned} \tag{17.3.2-8}$$

in which $C_{AC}^{L/INS}$ is obtained as described following Equation (17.3.2-3), and where:

$\underline{\omega}_{EAC}$ = Angular rate of the AC Frame relative to the E Frame.

The $\underline{\omega}_{EAC}^{AC}$ angular rate can be expanded at cycle m as follows, recognizing that angular rates relative to the N Frame and relative to the L Frame are equal because L is fixed relative to N:

$$\begin{aligned} \underline{\omega}_{EAC_m}^{AC} &= \underline{\omega}_{N/INS\text{-}AC_m}^{AC} - \underline{\omega}_{EN/INS_m}^{AC} = \underline{\omega}_{L/INS\text{-}AC_m}^{AC} - \left(C_{N/INS}^{AC} \right)_m \underline{\omega}_{EN/INS_m}^{N/INS} \\ &= \underline{\omega}_{L/INS\text{-}AC_m}^{AC} - \left(C_{B/INS}^{AC} \right)_m \left(C_{B/INS}^{L/INS} \right)_m^T C_N^L \, \underline{\omega}_{EN/INS_m}^{N/INS} \end{aligned} \tag{17.3.2-9}$$

where

$\underline{\omega}_{N/INS\text{-}AC}$, $\underline{\omega}_{L/INS\text{-}AC}$ = Angular rate of the AC Frame relative to the INS N and L Frame navigation solutions.

The $C_{B/INS}^{L/INS}$, $\underline{\omega}_{EN/INS}^{N/INS}$ terms in (17.3.2-9) are normally computed m cycle parameters within the trajectory generator INS navigation solution, and $C_{B/INS}^{AC}$ is determined as in Section 17.2.2. The $\underline{\omega}_{L/INS\text{-}AC}$ term in (17.3.2-9) can be computed from the rotation angle associated with the change in the $C_{AC}^{L/INS}$ matrix over an m cycle:

$$C_{AC_m}^{AC_{m-1}} = \left(C_{AC_{m-1}}^{L/INS}\right)^T C_{AC_m}^{L/INS}$$

$$\underline{\alpha}_{L/INS\text{-}AC_m}^{AC} = \begin{matrix}\text{Rotation angle} \\ \text{extraction from}\end{matrix} \; C_{AC_m}^{AC_{m-1}} \; \begin{matrix}\text{using (3.2.2.2-10) - (3.2.2.2-12)} \\ \text{and (3.2.2.2-15) - (3.2.2.2-19)}\end{matrix} \qquad (17.3.2\text{-}10)$$

Then $\underline{\omega}_{L/INS\text{-}AC}^{AC}$ can be calculated as $\underline{\alpha}_{L/INS\text{-}AC}^{AC}$ divided by T_m, treating the result as the average value midway between m-1 and m:

$$\underline{\omega}_{L/INS\text{-}AC_{m-1/2}}^{AC} \approx \frac{1}{T_m} \underline{\alpha}_{L/INS\text{-}AC_m}^{AC} \qquad (17.3.2\text{-}11)$$

where

m-1/2 = Designation for time point midway between the m-1 and m cycle times.

To obtain $\underline{\omega}_{L/INS\text{-}AC}^{AC}$ at cycle m for Equation (17.3.2-9), the following variation of (17.3.2-11) might be tried:

$$\underline{\omega}_{L/INS\text{-}AC_m}^{AC} \approx \frac{1}{2\,T_m}\left(\underline{\alpha}_{L/INS\text{-}AC_{m+1}}^{AC} + \underline{\alpha}_{L/INS\text{-}AC_m}^{AC}\right) \qquad (17.3.2\text{-}12)$$

Equation (17.3.2-12) requires the m+1 value for $\underline{\alpha}_{L/INS\text{-}AC}^{AC}$ which can be obtained by running the trajectory generator one cycle faster, or by extrapolation from the m and m-1 values:

$$\begin{aligned}\underline{\alpha}_{L/INS\text{-}AC_{m+1}}^{AC} &\approx \underline{\alpha}_{L/INS\text{-}AC_m}^{AC} + \left(\underline{\alpha}_{L/INS\text{-}AC_m}^{AC} - \underline{\alpha}_{L/INS\text{-}AC_{m-1}}^{AC}\right) \\ &= 2\,\underline{\alpha}_{L/INS\text{-}AC_m}^{AC} - \underline{\alpha}_{L/INS\text{-}AC_{m-1}}^{AC} \\ \underline{\omega}_{L/INS\text{-}AC_m}^{AC} &\approx \frac{1}{2\,T_m}\left(\underline{\alpha}_{L/INS\text{-}AC_{m+1}}^{AC} + \underline{\alpha}_{L/INS\text{-}AC_m}^{AC}\right) \\ &\approx \frac{1}{2\,T_m}\left(3\,\underline{\alpha}_{L/INS\text{-}AC_m}^{AC} - \underline{\alpha}_{L/INS\text{-}AC_{m-1}}^{AC}\right)\end{aligned} \qquad (17.3.2\text{-}13)$$

18 Strapdown Inertial System Testing

18.0 OVERVIEW

In this chapter we will describe four system level tests that can be used to estimate inertial sensor errors in strapdown inertial navigation systems: the Schuler Pump test, the Strapdown Drift test, the System Level Angular Rate Sensor Random Noise Estimation test and the Strapdown Rotation test. Each of these tests is primarily based on the measured response of the strapdown analytical platform acceleration output components (specific force acceleration transformed through computed attitude) under controlled angular rotation inputs. Use of the analytic platform eliminates the need for a stable test fixture base because angular base motion is sensed by the strapdown angular rate sensors, hence, integrated into a change in the computed attitude. The registered attitude change in turn, modifies the transformed specific force acceleration to cancel the change in accelerometer output produced by the same angular base motion. Eliminating a stable base requirement for the test fixture allows the tests to be conducted in any quasi-stationary test bed.

Of the four tests described, the Strapdown Drift test, the System Level Angular Rate Sensor Noise Estimation test, and the Strapdown Rotation test are the most useful for reasonably accurate sensor error determination. The Schuler Pump test has been included because of its popularity for use as a means of understanding classical behavior patterns of a strapdown INS under test.

18.1 SCHULER PUMP TEST

The Schuler Pump test is a method for exciting the Schuler oscillation error characteristic of an INS under test (See Section 13.2.2) while the INS is inertially navigating. Schuler oscillations are excited by intentionally rotating the INS about the vertical at an average rate equal to the Schuler frequency (i.e., 84 minutes for a full 360 degree rotation). The induced angular motion excites ("Schuler pumps") the strapdown INS horizontal fixed angular rate and accelerometer errors into measurable horizontal velocity error Schuler oscillations. By properly interpreting the Schuler velocity error response (i.e., sinusoidal or cosinusoidal), several inertial sensor errors can be ascertained as well as initial INS heading error caused by sensor random noise during INS initial self-alignment operations. The Schuler frequency input "pumping" action amplifies fixed horizontal sensor errors in the resulting Schuler response. In contrast, sensor noise induced Schuler errors are not amplified by rotation. As a result of the effective increase in signal to noise ratio (for fixed sensor error compared to sensor noise induced Schuler oscillations), calculation of sensor errors from the velocity measurements is fairly accurate for stable inertial sensors.

Inertial sensor errors determined from the Schuler Pump test are the following:

- Horizontal angular rate sensor bias
- Composite horizontal-accelerometer bias, horizontal-accelerometer vertical misalignment, and horizontal-angular-rate-sensor vertical misalignment.
- Initial heading error caused by horizontal inertial sensor random noise during initial self- alignment.

Composite vertical accelerometer bias, scale factor and scale factor asymmetry error can also be ascertained by analysis of INS vertical acceleration response (if available as an output) during stationary periods. Additionally, vertical angular rate sensor bias error can be estimated from measured heading change during stationary test segments. A more accurate measure of vertical angular rate sensor error can be obtained, however, by repeating the test with the INS at 90 degrees roll angle (assuming the original test was at zero roll angle) so that the previously vertical angular rate sensor becomes horizontal. A repeated test at 90 degrees roll angle also positions the previously vertical accelerometer in a horizontal orientation so that the normal Schuler Pump test horizontal acceleration measurement can be used to ascertain its composite horizontal performance.

In a Schuler Pump test, the strapdown INS is mounted to a test fixture having a vertical rotation axis. The INS is typically positioned on the test fixture with the sensor assembly Z or Y axis vertical (depending on whether the Y or Z axis inertial sensors are to be horizontal for test measurement performance assessment. The X sensor axis is horizontal in either case). Initially, the INS self-alignment mode is engaged (e.g., as described in Chapter 6) to align the INS in vertical and heading using the INS inertial sensors. Although not absolutely necessary, the INS initial alignment orientation is typically chosen to position the horizontal INS inertial sensors along north/east axes so that the normal INS north/east velocity output data can be used for data analysis. To minimize the likelihood of turn-on sensor error transients, the INS is typically allowed to operate with power applied for several minutes prior to engaging the initial alignment mode. After self-alignment is complete, the navigation mode is entered with the vertical control loop engaged to prevent vertical loop divergence, using the test laboratory altitude as a reference input (See Section 4.4.1.2.1). To simplify the test procedure, a square wave (rather than circular) Schuler pumping procedure is then initiated which sequentially positions the INS at 180, then 0 degree heading orientations (relative to the initial heading) at half Schuler cycle (42 minute) time intervals.

Immediately following navigation mode engagement, the INS is rotated plus 180 degrees in heading and allowed to navigate statically for a half Schuler period (42 minutes). At 42 minutes, the INS is then rotated minus 180 degrees in heading to its original alignment heading orientation for the second half Schuler cycle. (A minus rather than a plus 180 degree rotation is used when repositioning the INS, to cancel vertical angular rate sensor scale factor error induced heading error incurred in the original plus 180 degree heading rotation.). At one Schuler period (84 minutes from navigation mode entry) the test can be terminated, or the INS can be again rotated through plus 180 degrees and the test continued. Continued testing would add minus and plus 180 rotations at successive 42 minute intervals for extended error amplification.

Data typically taken during a Schuler Pump test consists of elapsed time from navigation mode entry, north velocity, east velocity and platform heading at quarter Schuler cycle time points (i.e., every 21 minutes). At the half Schuler cycle time points, the heading measurements are taken prior to and after completion of the 180 degree rotations. Vertical acceleration can also be recorded if available for vertical accelerometer error determination. If not available, vertical acceleration can be estimated from the slope of vertical velocity with the vertical control loop disabled. The vertical control loop can be disabled by setting the INS altitude reference input equal to the INS altitude output signal. However, the previous procedure does not disengage integral compensation terms that may be present in the vertical loop (e.g., the e_{vc_3} signal in Section 4.4.1.2.1) that may have been initialized during initial alignment operations to balance accelerometer error effects (See Section 6.3).

At test conclusion, the velocity and heading data are combined to estimate the INS sensor error parameters.

18.1.1 ANALYTICAL BASIS FOR THE SCHULER PUMP TEST

The Schuler Pump test is based on the analytical response of the N Frame horizontal velocity error for a strapdown INS in a static orientation. The effect of the 180 degree rotations at half Schuler cycles is analytically treated as a phase reversal of horizontal angular rate sensor errors feeding the attitude/acceleration-transformation algorithms. Between half Schuler cycle rotations, the velocity error propagates as derived in Section 13.3.2 (Equations (13.3.2-25)), but under zero horizontal velocity ($\underline{v}^N_{H_0}$) and velocity change ($\Delta\underline{v}^N_{SF_H}$) conditions, for which we find in slightly modified notation:

$$
\begin{aligned}
\delta\underline{V}^N_H &= g\,\underline{u}^N_{Up} \times \underline{\gamma}^N_{H_0} \frac{1}{\omega_S} \sin \omega_S t - \gamma_{Up_0}\, R\, \underline{\omega}^N_{IE_H} (1 - \cos \omega_S t) \\
&\quad + \delta\underline{a}^N_{SF_H} \frac{1}{\omega_S} \sin \omega_S t - R\, \underline{u}^N_{Up} \times \delta\underline{\omega}^N_{IB_H} (1 - \cos \omega_S t) \\
&\quad + R\, \delta\omega_{IB_{Up}}\, \underline{\omega}^N_{IE_H} \left(t - \frac{1}{\omega_S} \sin \omega_S t \right)
\end{aligned}
\qquad (18.1.1\text{-}1)
$$

with

$$\delta \underline{a}^{N}_{SF_H} \equiv \left(C^{N}_{B} \, \delta \underline{a}^{B}_{SF} \right)_{H}$$

$$\delta \underline{\omega}^{N}_{IB_H} \equiv \left(C^{N}_{B} \, \delta \underline{\omega}^{B}_{IB} \right)_{H} \qquad \delta \omega_{IB_{Up}} \equiv \left(C^{N}_{B} \, \delta \underline{\omega}^{B}_{IB} \right)_{ZN} \tag{18.1.1-2}$$

$$\underline{u}^{N}_{Up} \equiv \underline{u}^{N}_{ZN} \qquad \gamma_{Up_0} \equiv \gamma_{ZN_0}$$

where

B = Strapdown sensor assembly "body" coordinate frame as defined in Section 2.2.

$\delta \underline{a}^{B}_{SF}$, $\delta \underline{\omega}^{B}_{IB}$ = Composite accelerometer and angular rate sensor B Frame output error vectors (composite effect of fixed bias, scale factor, and misalignment error as defined in Equations (13.3-14)).

N = Navigation coordinate frame (defined for the general case in Section 2.2) of the wander azimuth type (See Section 4.5) that maintains its attitude relative to the earth for an INS that is stationary relative to the earth.

H = Designator for horizontal component.

Up = Designator for upward vertical component.

Equation (18.1.1-1) describes the horizontal velocity error response of an INS as a linear function of initial attitude/heading error ($\underline{\gamma}^{N}_{H_0}$, γ_{Up_0}) and constant N Frame inertial sensor errors ($\delta \underline{a}^{N}_{SF_H}$, $\delta \underline{\omega}^{N}_{IB_H}$, $\delta \omega_{IB_{Up}}$). Because the equation is linear, we can apply the principle of linear superposition to decompose (18.1.1-1) into parts excited by misalignment and sensor errors (that in the Schuler Pump test are constant during half Schuler cycle time segments), and then recombine the parts to form the total solution. To apply this approach, we first describe the inertial sensor errors in a compatible analytical form.

$$\begin{aligned}
\delta \underline{a}^{N}_{SF_H}(t) &= \delta \underline{a}^{N}_{SF_{H1}} \, \mathrm{Step}(t) + \left(\delta \underline{a}^{N}_{SF_{H2}} - \delta \underline{a}^{N}_{SF_{H1}} \right) \mathrm{Step}\left(t - \frac{\pi}{\omega_S} \right) \\
&\quad + \left(\delta \underline{a}^{N}_{SF_{H3}} - \delta \underline{a}^{N}_{SF_{H2}} \right) \mathrm{Step}\left(t - \frac{2\,\pi}{\omega_S} \right) + \cdots \\
\delta \underline{\omega}^{N}_{IB_H}(t) &= \delta \underline{\omega}^{N}_{IB_{H1}} \, \mathrm{Step}(t) + \left(\delta \underline{\omega}^{N}_{IB_{H2}} - \delta \underline{\omega}^{N}_{IB_{H1}} \right) \mathrm{Step}\left(t - \frac{\pi}{\omega_S} \right) \\
&\quad + \left(\delta \underline{\omega}^{N}_{IB_{H3}} - \delta \underline{\omega}^{N}_{IB_{H2}} \right) \mathrm{Step}\left(t - \frac{2\,\pi}{\omega_S} \right) + \cdots
\end{aligned} \tag{18.1.1-3}$$

(Continued)

$$\delta\omega_{IB_{Up}}(t) = \delta\omega_{IB_{Up1}} \, \text{Step}(t) + \left(\delta\omega_{IB_{Up2}} - \delta\omega_{IB_{Up1}}\right) \text{Step}\left(t - \frac{\pi}{\omega_S}\right) + \left(\delta\omega_{IB_{Up3}} - \delta\omega_{IB_{Up2}}\right) \text{Step}\left(t - \frac{2\,\pi}{\omega_S}\right) + \cdots \tag{18.1.1-3 (Continued)}$$

where

$1, 2, 3, \cdots$ = Designation for the indicated parameter values during the first, second, third, $\cdots$ half Schuler cycle time periods.

Step( ) = Unit step function having a value of zero for ( ) < 0 and unity for ( ) ≥ 0.

It should be apparent by inspection of (18.1.1-3) that the $\delta\underline{a}^N_{SF_H}(t)$, $\delta\underline{\omega}^N_{IB_H}(t)$, $\delta\omega_{IB_{Up}}(t)$ functions equal the $1, 2, 3, \cdots$ parameter values during the first, second, third, $\cdots$ half Schuler cycles.

Let us now apply Equations (18.1.1-3) in (18.1.1-1) to develop a solution for the horizontal velocity error using the principle of linear superposition. We first decompose Equation (18.1.1-1) into two parts; a part generated by the $\underline{\gamma}^N_{H_0}$, γ_{Up_0} initial conditions, and a part generated by the sensor error terms $\delta\underline{a}^N_{SF_H}$, $\delta\underline{\omega}^N_{IB_H}$, $\delta\omega_{IB_{Up}}$. The sensor error part is then decomposed into parts excited by each of the unit step function portions of (18.1.1-3) with the results interpreted as starting from the time that the unit steps become unity. We also add additional "initial" tilt and heading error components to account for tilt/heading error introduced during the initial and half Schuler cycle 180 degree heading rotations, generated from horizontal angular rate sensor misalignment (to vertical) and vertical angular rate sensor scale factor error. The total horizontal velocity solution is then obtained as the sum of the decomposed parts. Thus:

$$\delta\underline{V}^N_H(t) = g\,\underline{u}^N_{Up} \times \underline{\gamma}^N_{H_0} \frac{1}{\omega_S} \sin\omega_S t - \gamma_{Up_0} R\,\underline{\omega}^N_{IE_H} (1 - \cos\omega_S t) + \delta\underline{V}^N_{H1} + \delta\underline{V}^N_{H2} + \delta\underline{V}^N_{H3} + \cdots \tag{18.1.1-4}$$

$$\delta\underline{V}^N_{H1} = \left(\delta\underline{a}^N_{SF_{H1}} + g\,\underline{u}^N_{Up} \times \delta\underline{\gamma}^N_{H_1}\right) \frac{1}{\omega_S} \sin\omega_S t$$

(Continued)

$$- R \left(\underline{u}_{Up}^{N} \times \delta\underline{\omega}_{IB_{H1}}^{N} + \delta\gamma_{Up1} \, \underline{\omega}_{IE_H}^{N} \right) (1 - \cos \omega_S t) + R \, \delta\omega_{IB_{Up1}} \, \underline{\omega}_{IE_H}^{N} \left(t - \frac{1}{\omega_S} \sin \omega_S t \right)$$

$$\begin{aligned} \delta\underline{V}_{H2}^{N} = \ & \text{Step}\left(t - \frac{\pi}{\omega_S}\right) \Bigg\{ \left[\left(\delta\underline{a}_{SF_{H2}}^{N} - \delta\underline{a}_{SF_{H1}}^{N} \right) + g \, \underline{u}_{Up}^{N} \times \delta\underline{\gamma}_{H_2}^{N} \right] \frac{1}{\omega_S} \sin \omega_S \left(t - \frac{\pi}{\omega_S} \right) \\ & - R \left[\underline{u}_{Up}^{N} \times \left(\delta\underline{\omega}_{IB_{H2}}^{N} - \delta\underline{\omega}_{IB_{H1}}^{N} \right) + \delta\gamma_{Up2} \, \underline{\omega}_{IE_H}^{N} \right] \left[1 - \cos \omega_S \left(t - \frac{\pi}{\omega_S} \right) \right] \\ & + R \left(\delta\omega_{IB_{Up2}} - \delta\omega_{IB_{Up1}} \right) \underline{\omega}_{IE_H}^{N} \left[t - \frac{\pi}{\omega_S} - \frac{1}{\omega_S} \sin \omega_S \left(t - \frac{\pi}{\omega_S} \right) \right] \Bigg\} \end{aligned} \qquad \begin{matrix} (18.1.1\text{-}4) \\ (\text{Continued}) \end{matrix}$$

$$\begin{aligned} \delta\underline{V}_{H3}^{N} = \ & \text{Step}\left(t - \frac{2\pi}{\omega_S}\right) \Bigg\{ \left[\left(\delta\underline{a}_{SF_{H3}}^{N} - \delta\underline{a}_{SF_{H2}}^{N} \right) + g \, \underline{u}_{Up}^{N} \times \delta\underline{\gamma}_{H_3}^{N} \right] \frac{1}{\omega_S} \sin \omega_S \left(t - \frac{2\pi}{\omega_S} \right) \\ & - R \left[\underline{u}_{Up}^{N} \times \left(\delta\underline{\omega}_{IB_{H3}}^{N} - \delta\underline{\omega}_{IB_{H2}}^{N} \right) + \delta\gamma_{Up3} \, \underline{\omega}_{IE_H}^{N} \right] \left[1 - \cos \omega_S \left(t - \frac{2\pi}{\omega_S} \right) \right] \\ & + R \left(\delta\omega_{IB_{Up3}} - \delta\omega_{IB_{Up2}} \right) \underline{\omega}_{IE_H}^{N} \left[t - \frac{2\pi}{\omega_S} - \frac{1}{\omega_S} \sin \omega_S \left(t - \frac{2\pi}{\omega_S} \right) \right] \Bigg\} \end{aligned}$$

$$\delta\underline{V}_{H4}^{N} = \cdots$$

where

i = Subscript on $\delta\underline{\gamma}_{H_i}^{N}$, $\delta\gamma_{Up_i}$ terms (defined below) referring to the ith 180 degree rotation in the Schuler Pump test ($i = 1, 2, 3, \cdots$).

$\delta\underline{\gamma}_{H_i}^{N}$ = Horizontal tilt due to horizontal angular rate sensor misalignment (into vertical) excited by the 180 deg heading rotation immediately preceding the ith half Schuler cycle time period (equivalent to the $\Delta\underline{\psi}_{TMis}$ term in Equation (13.2.4-16)).

$\delta\gamma_{Up_i}$ = Heading error due to vertical angular rate sensor scale factor (and orthogonality) error excited by the 180 deg heading rotation immediately preceding the ith half Schuler cycle time period (as in Equation (13.4.1.1-17)).

The time offsets in the (18.1.1-4) sinusoidal term arguments are present because of the time offsets in the (18.1.1-3) sensor error inputs that generate the sinusoidal wave forms from the instant their step functions become unity.

Because the INS orientation only changes in heading during the Schuler Pump test, the B Frame specific force acting on the INS inertial components will be the same constant value for all static test orientations, hence, the B Frame accelerometer error $\delta \underline{a}_{SF}^{B}$ will be constant, including scale-factor/misalignment error effects. For the particular B Frame orientation during INS initial self-alignment operations, $\delta \underline{a}_{SF}^{B}$ will generate a particular value for $\delta \underline{a}_{SF_H}^{N}$ in Equations (18.1.1-4), depending on the orientation of the B Frame relative to the N Frame. Because $\delta \underline{a}_{SF}^{B}$ is constant, its horizontal value in the N Frame will be of opposite polarity during the Schuler Pump test when the INS is at 180 degrees around the vertical from its initial alignment orientation. Thus, from the heading orientation profile for the Schuler Pump test described in Section 18.1, we see that the accelerometer error terms in (18.1.1-4) are given by:

$$\delta \underline{a}_{SF_{H1}}^{N} = - \delta \underline{a}_{SF_{H0}}^{N} \qquad \delta \underline{a}_{SF_{H2}}^{N} = \delta \underline{a}_{SF_{H0}}^{N} \qquad \delta \underline{a}_{SF_{H3}}^{N} = - \delta \underline{a}_{SF_{H0}}^{N} \qquad \cdots \qquad (18.1.1\text{-}5)$$

where

$\delta \underline{a}_{SF_{H0}}^{N}$ = Value for $\delta \underline{a}_{SF_H}^{N}$ at the INS initial self-alignment heading orientation.

Unlike the B Frame accelerometer error characteristic (which was constant for all INS heading orientations), the B Frame angular rate sensor error $\delta \underline{\omega}_{IB}^{B}$ will have a different value at the 180 degree heading orientations due to phase reversal (in B Frame coordinates) of the horizontal earth rate component measured by the angular rate sensors. Angular rate sensor scale-factor/misalignment errors multiplying B Frame horizontal earth rate components will thereby be opposite in sign at the 180 degree heading compared to the values at the alignment orientation. Thus, the equivalent to (18.1.1-5) for the angular rate sensor error components in Equations (18.1.1-4) are defined by the more complicated expressions:

$$\begin{aligned} \delta \underline{\omega}_{IB_{H1}}^{N} &= - \delta \underline{\omega}_{IB_{H0}}^{N} + \delta \underline{\omega}_{IB_{H00}}^{N} \\ \delta \underline{\omega}_{IB_{H2}}^{N} &= \delta \underline{\omega}_{IB_{H0}}^{N} + \delta \underline{\omega}_{IB_{H00}}^{N} \\ \delta \underline{\omega}_{IB_{H3}}^{N} &= - \delta \underline{\omega}_{IB_{H0}}^{N} + \delta \underline{\omega}_{IB_{H00}}^{N} \\ &\vdots \end{aligned} \qquad (18.1.1\text{-}6)$$

$$\begin{aligned} \delta \omega_{IB_{Up1}} &= \delta \omega_{IB_{Up0}} - \delta \omega_{IB_{Up00}} \\ \delta \omega_{IB_{Up2}} &= \delta \omega_{IB_{Up0}} + \delta \omega_{IB_{Up00}} \\ \delta \omega_{IB_{Up3}} &= \delta \omega_{IB_{Up0}} - \delta \omega_{IB_{Up00}} \\ &\vdots \end{aligned} \qquad (18.1.1\text{-}7)$$

where

$\delta\underline{\omega}^N_{IB_{H0}}, \delta\omega_{IB_{Up0}}$ = Values for $\delta\underline{\omega}^N_{IB_H}, \delta\omega_{IB_{Up}}$ at the INS initial self-alignment heading orientation, but exclusive of scale-factor/misalignment error effects coupling horizontal B Frame angular rate (i.e., horizontal earth rate).

$\delta\underline{\omega}^N_{IB_{H00}}, \delta\omega_{IB_{Up00}}$ = Values for the portion of $\delta\underline{\omega}^N_{IB_H}, \delta\omega_{IB_{Up}}$ at the INS initial self-alignment heading orientation produced by scale-factor/misalignment error coupling of horizontal B Frame angular rate (i.e., horizontal earth rate).

Due to the equal magnitude but alternating sign of the half Schuler cycle 180 degree rotations (See Section 18.1), the $\delta\underline{\gamma}^N_{H_i}, \delta\gamma_{Up_i}$ terms in (18.1.1-4) will be equal in magnitude but will alternate in sign, hence:

$$
\begin{aligned}
\delta\underline{\gamma}^N_{H_2} &= -\,\delta\underline{\gamma}^N_{H_1} \qquad & \delta\gamma_{Up_2} &= -\,\delta\gamma_{Up_1} \\
\delta\underline{\gamma}^N_{H_3} &= \delta\underline{\gamma}^N_{H_1} \qquad & \delta\gamma_{Up_3} &= \delta\gamma_{Up_1} \\
&\vdots & &\vdots
\end{aligned}
\qquad (18.1.1\text{-}8)
$$

The $\underline{\gamma}^N_{H_0}, \gamma_{Up_0}$ terms in Equation (18.1.1-4) can also be expressed in terms of the sensor errors during alignment by applying the Equation (14.3-38) initial self-alignment solution assuming that the initial position error ($\underline{\varepsilon}^N$) is zero so that at alignment completion, using equivalency Equation (12.2.1-17):

$$
\underline{\gamma}^N = \underline{\psi}^N \qquad (18.1.1\text{-}9)
$$

Using (18.1.1-9) with the previous inertial sensor error definitions in Equation (14.3-38), and applying the (3.1.1-35) mixed vector dot/cross product identity, we find:

$$
\begin{aligned}
\underline{\gamma}^N_{H_0} &\approx \frac{1}{g}\left(\underline{u}^N_{Up} \times \delta\underline{a}^N_{SF_{H0}}\right) \\
\gamma_{Up_0} &= \frac{1}{\omega^2_{IE_H}}\,\underline{u}^N_{Up} \cdot \left[\underline{\omega}^N_{IE_H} \times \left(\delta\underline{\omega}^N_{IB_{H0}} + \delta\underline{\omega}^N_{IB_{H00}}\right)\right] + \delta\gamma_{Up_0} \\
&= -\,\frac{1}{\omega_{IE_H}}\left(\delta\underline{\omega}^N_{IB_{H0}} + \delta\underline{\omega}^N_{IB_{H00}}\right) \cdot \left[\frac{1}{\omega_{IE_H}}\left(\underline{\omega}^N_{IE_H} \times \underline{u}^N_{Up}\right)\right] + \delta\gamma_{Up_0}
\end{aligned}
\qquad (18.1.1\text{-}10)
$$

where

$\delta\gamma_{Up0}^{N}$ = Initial heading error component produced by other than constant angular rate sensor bias error effects during initial alignment.

The γ_{H0}^{N} expression in (18.1.1-10) with generalized Equation (13.1-8) shows that the horizontal tilt term in (18.1.1-4) is given by:

$$g\left(\underline{u}_{Up}^{N} \times \gamma_{H0}^{N}\right) = -\,\delta\underline{a}_{SF_{H0}}^{N} \qquad (18.1.1\text{-}11)$$

The γ_{Up0} expression in (18.1.1-10) can be simplified by recognizing that $\underline{\omega}_{IE_H}^{N}$ is in the direction of north (because north is along the horizontal, pointing toward the positive earth rotation axis), and the cross product between north and upward directed vectors, lies east:

$$\frac{1}{\omega_{IE_H}}\,\underline{\omega}_{IE_H}^{N} = \underline{u}_{North}^{N} \qquad \underline{u}_{North}^{N} \times \underline{u}_{Up}^{N} = \underline{u}_{East}^{N} \qquad (18.1.1\text{-}12)$$

where

$\underline{u}_{North}^{N}$, $\underline{u}_{East}^{N}$ = Horizontal unit vectors pointing north and east.

Substituting (18.1.1-12) and generalized Equation (3.1.1-12) in the (18.1.1-10) γ_{Up0} expression shows the heading error term in (18.1.1-4) to be:

$$\gamma_{Up0}\,\underline{\omega}_{IE_H}^{N} = -\,\underline{u}_{North}^{N}\left(\underline{u}_{East}^{N}\right)^{T}\left(\delta\underline{\omega}_{IB_{H0}}^{N} + \delta\underline{\omega}_{IB_{H00}}^{N}\right) + \delta\gamma_{Up0}\,\underline{\omega}_{IE_H}^{N} \qquad (18.1.1\text{-}13)$$

The following derives an identity that will prove useful based on the cross product between east and north vectors being up:

$$\underline{u}_{Up}^{N} = \underline{u}_{East}^{N} \times \underline{u}_{North}^{N} \qquad (18.1.1\text{-}14)$$

Applying (18.1.1-14) to the cross product of $\underline{u}_{Up}^{N}$ with an arbitrary vector $\underline{W}^{N}$ yields, with generalized Equations (3.1.1-12), (3.1.1-13) and (3.1.1-16):

$$\begin{aligned} \underline{u}_{Up}^{N} \times \underline{W}^{N} &= \left(\underline{u}_{Up}^{N}\times\right)\underline{W}^{N} = \left(\underline{u}_{East}^{N} \times \underline{u}_{North}^{N}\right) \times \underline{W}^{N} \\ &= \underline{u}_{North}^{N}\left(\underline{u}_{East}^{N} \cdot \underline{W}^{N}\right) - \underline{u}_{East}^{N}\left(\underline{u}_{North}^{N} \cdot \underline{W}^{N}\right) \\ &= \left[\underline{u}_{North}^{N}\left(\underline{u}_{East}^{N}\right)^{T} - \underline{u}_{East}^{N}\left(\underline{u}_{North}^{N}\right)^{T}\right]\underline{W}^{N} \end{aligned} \qquad (18.1.1\text{-}15)$$

or, since $\underline{W}^N$ is arbitrary:

$$\left(\underline{u}_{Up}^N \times\right) = \underline{u}_{North}^N \left(\underline{u}_{East}^N\right)^T - \underline{u}_{East}^N \left(\underline{u}_{North}^N\right)^T \tag{18.1.1-16}$$

Thus:

$$\left(\underline{u}_{Up}^N \times\right) - \underline{u}_{North}^N \left(\underline{u}_{East}^N\right)^T = - \underline{u}_{East}^N \left(\underline{u}_{North}^N\right)^T \tag{18.1.1-17}$$

We also know that for the Schuler Pump test, the test measurements are taken at a static condition for which the true horizontal velocity is zero. Hence, the INS output horizontal velocity measurements equal the velocity error $\delta\underline{V}_H^N(t)$. At this point we can also generalize Equations (18.1.1-4) by transformation to any convenient coordinate frame for evaluation (e.g., east, north, up geographic axes). With (18.1.1-5) - (18.1.1-7), (18.1.1-11), (18.1.1-13), (18.1.1-17) and (3.1.1-13), Equations (18.1.1-4) then become:

$$\begin{aligned}
\underline{V}_{INS_H}(t) = & - 2 \left(\delta\underline{a}_{SF_{H0}} - \frac{g}{2}\, \underline{u}_{Up} \times \delta\underline{\gamma}_{H_1}\right) \frac{1}{\omega_S} \sin \omega_S t \\
& + R \left\{ \left[\left(\underline{u}_{Up} \times\right) + \underline{u}_{North}\, \underline{u}_{East}^T \right] \delta\underline{\omega}_{IB_{H0}} - \delta\gamma_{Up_1}\, \underline{\omega}_{IE_H} \right\} (1 - \cos \omega_S t) \\
& - R\, \delta\gamma_{Up_0}\, \underline{\omega}_{IE_H} \left(1 - \cos \omega_S t\right) + R \left(\delta\omega_{IB_{Up0}} - \delta\omega_{IB_{Up00}}\right) \underline{\omega}_{IE_H} \left(t - \frac{1}{\omega_S} \sin \omega_S t\right) \\
& \qquad + R\, \underline{u}_{East}\, \underline{u}_{North}^T\, \delta\underline{\omega}_{IB_{H00}}\ (1 - \cos \omega_S t) \\
& + \mathrm{Step}\left(t - \frac{\pi}{\omega_S}\right) \left\{ 2 \left(\delta\underline{a}_{SF_{H0}} - \frac{g}{2}\, \underline{u}_{Up} \times \delta\underline{\gamma}_{H_1}\right) \frac{1}{\omega_S} \sin \omega_S \left(t - \frac{\pi}{\omega_S}\right) \right. \\
& \quad - R \left(2\, \underline{u}_{Up} \times \delta\underline{\omega}_{IB_{H0}} - \delta\gamma_{Up_1}\, \underline{\omega}_{IE_H}\right) \left[1 - \cos \omega_S \left(t - \frac{\pi}{\omega_S}\right) \right] \\
& \quad \left. + R\, 2\, \delta\omega_{IB_{Up00}}\, \underline{\omega}_{IE_H} \left[t - \frac{\pi}{\omega_S} - \frac{1}{\omega_S} \sin \omega_S \left(t - \frac{\pi}{\omega_S}\right) \right] \right\} \\
& + \mathrm{Step}\left(t - \frac{2\,\pi}{\omega_S}\right) \left\{ - 2 \left(\delta\underline{a}_{SF_{H0}} - \frac{g}{2}\, \underline{u}_{Up} \times \delta\underline{\gamma}_{H_1}\right) \frac{1}{\omega_S} \sin \omega_S \left(t - \frac{2\,\pi}{\omega_S}\right) \right. \\
& \quad + R \left(2\, \underline{u}_{Up} \times \delta\underline{\omega}_{IB_{H0}} - \delta\gamma_{Up_1}\, \underline{\omega}_{IE_H}\right) \left[1 - \cos \omega_S \left(t - \frac{2\,\pi}{\omega_S}\right) \right] \\
& \quad \left. - R\, 2\, \delta\omega_{IB_{Up00}}\, \underline{\omega}_{IE_H} \left[t - \frac{2\,\pi}{\omega_S} - \frac{1}{\omega_S} \sin \omega_S \left(t - \frac{2\,\pi}{\omega_S}\right) \right] \right\} \\
& + \cdots
\end{aligned} \tag{18.1.1-18}$$

where

$\underline{V}_{INS_H}(t)$ = Horizontal velocity output from the INS during the Schuler Pump test at the stationary test orientations.

Note that Equation (18.1.1-18) has the superscript designator removed from the vector quantities, indicating that this is a general vector equation that can be evaluated in any selected coordinate frame. Note also that only the north component of $\delta\underline{\omega}_{IB_{H00}}$ appears in (18.1.1-18). This is because the east component produces an initial heading error in Equation (18.1.1-13) (not included in $\delta\gamma_{Up_0}$) that cancels the effect of east $\delta\underline{\omega}_{IB_{H00}}$ error during navigation.

As an adjunct to Equation (18.1.1-18), static heading measurements can be taken at the start and end of each half Schuler cycle time segment to provide a more direct means for estimating the $\delta\omega_{IB_{Up0}}$, $\delta\omega_{IB_{Up00}}$ vertical inertial sensor error terms. The approach is based in part on Equations (12.2.1-38) which show that for zero pitch angle, the γ_{Up} error (i.e., γ_{ZN}) is equivalent to a platform heading error $\delta\psi_P$. We also know from Equations (13.3.2-1) that during the stationary periods, $\dot{\gamma}_{Up} = - \delta\omega_{IB_{Up}}$, hence:

$$\delta\dot{\psi}_P = - \delta\omega_{IB_{Up}} \qquad (18.1.1\text{-}19)$$

where

$\delta\psi_P$ = Error in INS calculated platform heading.

The $\delta\dot{\psi}_P$ term in (18.1.1-19) can be measured during a stationary half Schuler cycle period as the difference between platform heading measurements taken at the beginning and end of the period divided by the half Schuler cycle time $\frac{\pi}{\omega_S}$. Thus, (18.1.1-19) with (18.1.1-7) yields:

$$\begin{aligned}\delta\omega_{IB_{Up0}} - \delta\omega_{IB_{Up00}} = - \frac{1}{m_{Odd}} \frac{\omega_S}{\pi} \Big[&\left(\psi_{P/INS_{0.5S-}} - \psi_{P/INS_{0.0S+}}\right) \\ &+ \left(\psi_{P/INS_{1.5S-}} - \psi_{P/INS_{1.0S+}}\right) + \cdots \Big]\end{aligned} \qquad (18.1.1\text{-}20)$$

$$\begin{aligned}\delta\omega_{IB_{Up0}} + \delta\omega_{IB_{Up00}} = - \frac{1}{m_{Even}} \frac{\omega_S}{\pi} \Big[&\left(\psi_{P/INS_{1.0S-}} - \psi_{P/INS_{0.5S+}}\right) \\ &+ \left(\psi_{P/INS_{2.0S-}} - \psi_{P/INS_{1.5S+}}\right) + \cdots \Big]\end{aligned} \qquad (18.1.1\text{-}21)$$

where

$\psi_{P/INS}$ = INS computed platform heading.

0.0S+, 0.5S+, 1.0S+, 1.5S+, etc. = Designation for measurement taken immediately following the 180 degree rotation at time zero, the half Schuler cycle time, the full Schuler cycle time, the 1.5 Schuler cycle time, etc.

0.5S-, 1.0S-, 1.5S-, etc. = Designation for measurement taken immediately prior to the 180 degree rotation at the half Schuler cycle time, the full Schuler cycle time, the 1.5 Schuler cycle time, etc.

m_{Odd}, m_{Even} = Number of odd and even half Schuler cycles in the test.

Equations (18.1.1-20) - (18.1.1-21) presume that platform heading ($\psi_{P/INS}$) will be an available INS output. If platform heading is not available, the following development shows how the $\psi_{P/INS}$ differences in (18.1.1-20) - (18.1.1-21) can be computed from INS true heading and longitude outputs. The method is based on Equation (12.2.1-44) at zero pitch angle (θ) combined with the (12.2.3-37) δL expression:

$$\delta\psi_T = \psi_{ZN} + \delta L \sin l \qquad (18.1.1\text{-}22)$$

where

$\delta\psi_T$ = Error in INS calculated true heading.

ψ_{ZN} = Vertical component of $\underline{\psi}^N$.

δL = Error in INS calculated longitude.

l = Latitude.

Equation (12.2.1-42) shows that for zero pitch angle (θ):

$$\psi_{ZN} = \delta\psi_P - \varepsilon_{ZN} \qquad (18.1.1\text{-}23)$$

where

ε_{ZN} = N Frame Z axis (vertical) component of $\underline{\varepsilon}^N$, the rotation angle error associated with the C_N^E matrix.

The heading and longitude error terms in (18.1.1-22) and (18.1.1-23) are defined by the relations:

$$\delta\psi_T = \psi_{T/INS} - \psi_T \qquad \delta\psi_P = \psi_{P/INS} - \psi_P \qquad \delta L = L_{INS} - L \qquad (18.1.1\text{-}24)$$

where

$\psi_{T/INS}$ = INS computed true heading.

L_{INS} = INS computed longitude.

ψ_T, ψ_P, L = True (error free) values for true heading, platform heading and longitude.

Substituting (18.1.1-23) - (18.1.1-24) in (18.1.1-22) with rearrangement then yields:

$$\psi_{P/INS} - \psi_P = \psi_{T/INS} - \psi_T - \left(L_{INS} - L\right) \sin l + \varepsilon_{ZN} \tag{18.1.1-25}$$

Because of the Schuler Pump test environment, ψ_T, L and l in (18.1.1-25) are constant during each stationary Schuler half cycle time period. For the selected wander azimuth N and L Frames (see Section 4.5 for definitions) the vertical transport rate component ρ_{ZN} is set to zero. From Equation (4.4.3-3) we see that for the zero longitude rate laboratory test environment, this equates the wander angle rate $\dot{\alpha}$ to zero, hence the wander angle α to a constant. Therefore, since the true heading ψ_T is constant during stationary half Schuler cycles, Equation (4.1.2-2) shows that platform heading ψ_P will also be constant during half Schuler cycles. Because ρ_{ZN} is zero, the associated error term $\delta\rho_{ZN}$ will also be zero. Because of the stationary test environment (zero horizontal velocity) and the zero ρ_{ZN}, Equation (5.3-17) shows that horizontal transport rate ($\underline{\rho}_H \equiv \underline{\omega}_{EN_H}$) will also be zero. Therefore, Equation (12.3.5-21) shows that $\dot{\varepsilon}_{ZN}$ will be zero and ε_{ZN} will be constant. Thus, we see that during each stationary Schuler half cycle, the ψ_T, L, l, ψ_P and ε_{ZN} parameters in Equation (18.1.1-25) will be constant. Based on this observation, we subtract Equation (18.1.1-25) at the start of a Schuler half cycle from its value at the end of the same Schuler half cycle to obtain:

$$\psi_{P/INS_{End}} - \psi_{P/INS_{Start}} = \psi_{T/INS_{End}} - \psi_{T/INS_{Start}} - \left(L_{INS_{End}} - L_{INS_{Start}}\right) \sin l \tag{18.1.1-26}$$

Equation (18.1.1-26) allows the $\psi_{P/INS}$ differences in (18.1.1-20) - (18.1.1-21) to be calculated from INS true heading and longitude outputs if platform heading outputs are not available.

Equations (18.1.1-18) with adjunct Equations (18.1.1-20)- (18.1.1-21) (and potentially (18.1.1-26)) are the basis for the Schuler Pump test expressing the relationship between strapdown INS measurable horizontal velocity $\underline{V}_{INS_H}(t)$ and platform heading $\psi_{P/INS}$, the inertial sensor errors ($\delta\underline{a}_{SF_{H0}}$, $\delta\underline{\omega}_{IB_{H0}}$, $\delta\omega_{IB_{Up0}}$, $\delta\underline{\omega}_{IB_{H00}}$, $\delta\omega_{IB_{Up00}}$), the residual initial heading error $\delta\gamma_{Up_0}$, and the heading/tilt error $\delta\underline{\gamma}_{H_1}$, $\delta\gamma_{Up_1}$ induced by angular rate sensor misalignment/scale-factor error excitation during the 180 degree heading rotations. During the test, $\underline{V}_{INS_H}(t)$ is measured at selected time points, and $\psi_{P/INS}$ is measured at the half Schuler cycle times (immediately before and after the 180 degree rotations). At completion of the test, the $\underline{V}_{INS_H}(t)$, $\psi_{P/INS}$ measurements in conjunction with (18.1.1-18) and (18.1.1-20) - (18.1.1-21) evaluated at the $\underline{V}_{INS_H}(t)$ time points, define a system of equations that can be solved for $\delta \underline{a}_{SF_{H0}}$, $\delta \underline{\omega}_{IB_{H0}}$, $\delta \omega_{IB_{Up0}}$, $\delta \underline{\omega}_{IB_{H00}}$, $\delta \omega_{IB_{Up00}}$, $\delta \gamma_{Up_0}$, $\delta \underline{\gamma}_{H_1}$ and $\delta \gamma_{Up_1}$. The following section describes classical deterministic approaches for the solution based on $\underline{V}_{INS_H}(t)$ samples taken at quarter Schuler cycle time points.

18.1.2 CLASSIC SCHULER PUMP TEST SOLUTIONS

The Schuler Pump test was originated to provide a simple rapid estimation of strapdown INS composite inertial sensor errors with a minimum of special purpose test equipment and processing software. The classical Schuler Pump test method is to take $\underline{V}_{INS_H}(t)$ measurements at quarter Schuler cycle times and calculate the sensor errors using a deterministic solution approach. If we evaluate Equation (18.1.1-18) at the quarter Schuler cycle time points using (18.1.1-16) when appropriate, we find that:

$$\begin{aligned} \underline{V}_{INS/H_{0.25S}} = & -2\left[\delta \underline{a}_{SF_{H0}} - \frac{g}{2}\left(\underline{u}_{North}\, \underline{u}_{East}^T - \underline{u}_{East}\, \underline{u}_{North}^T\right) \delta \underline{\gamma}_{H_1}\right] \frac{1}{\omega_S} \\ & + R\left[\left(\underline{u}_{Up} \times\right) + \underline{u}_{North}\, \underline{u}_{East}^T\right] \delta \underline{\omega}_{IB_{H0}} - R\, \delta \gamma_{Up_1}\, \underline{\omega}_{IE_H} \\ & - R\, \delta \gamma_{Up_0}\, \underline{\omega}_{IE_H} + R\left(\delta \omega_{IB_{Up0}} - \delta \omega_{IB_{Up00}}\right) \frac{\pi}{\omega_S}\left(\frac{1}{2} - \frac{1}{\pi}\right) \underline{\omega}_{IE_H} \\ & + R\, \underline{u}_{East}\, \underline{u}_{North}^T\, \delta \underline{\omega}_{IB_{H00}} \end{aligned} \tag{18.1.2-1}$$

$$\begin{aligned} \underline{V}_{INS/H_{0.5S}} = & R\left[\left(\underline{u}_{Up} \times\right) + \underline{u}_{North}\, \underline{u}_{East}^T\right] 2\, \delta \underline{\omega}_{IB_{H0}} - 2\, R\, \delta \gamma_{Up_1}\, \underline{\omega}_{IE_H} \\ & - R\, 2\, \delta \gamma_{Up_0}\, \underline{\omega}_{IE_H} + R\left(\delta \omega_{IB_{Up0}} - \delta \omega_{IB_{Up00}}\right) \frac{\pi}{\omega_S}\, \underline{\omega}_{IE_H} \\ & + 2\, R\, \underline{u}_{East}\, \underline{u}_{North}^T\, \delta \underline{\omega}_{IB_{H00}} \end{aligned} \tag{18.1.2-2}$$

$$\begin{aligned} \underline{V}_{INS/H_{0.75S}} = & 4\left[\delta \underline{a}_{SF_{H0}} - \frac{g}{2}\left(\underline{u}_{North}\, \underline{u}_{East}^T - \underline{u}_{East}\, \underline{u}_{North}^T\right) \delta \underline{\gamma}_{H_1}\right] \frac{1}{\omega_S} \\ & + R\, \underline{u}_{East}\, \underline{u}_{North}^T\, \delta \underline{\omega}_{IB_{H0}} \\ & - R\, \delta \gamma_{Up_0}\, \underline{\omega}_{IE_H} + R\, \delta \omega_{IB_{Up0}} \frac{\pi}{\omega_S}\left(\frac{3}{2} + \frac{1}{\pi}\right) \underline{\omega}_{IE_H} \\ & - R\, \delta \omega_{IB_{Up00}} \frac{\pi}{\omega_S}\left(\frac{1}{2} + \frac{3}{\pi}\right) \underline{\omega}_{IE_H} + R\, \underline{u}_{East}\, \underline{u}_{North}^T\, \delta \underline{\omega}_{IB_{H00}} \end{aligned} \tag{18.1.2-3}$$

$$\underline{V}_{INS/H\,1.0S} = -R\,\underline{u}_{Up} \times 4\,\delta\underline{\omega}_{IB_{H0}} + 2\,R\,\delta\gamma_{Up_1}\,\underline{\omega}_{IE_H} + R\,\delta\omega_{IB_{Up0}}\frac{\pi}{\omega_S}\,2\,\underline{\omega}_{IE_H} \qquad (18.1.2\text{-}4)$$

$$\begin{aligned} \underline{V}_{INS/H\,1.25S} = & -6\left[\delta\underline{a}_{SF_{H0}} - \frac{g}{2}\left(\underline{u}_{North}\,\underline{u}_{East}^{T} - \underline{u}_{East}\,\underline{u}_{North}^{T}\right)\delta\underline{\gamma}_{H_1}\right]\frac{1}{\omega_S} \\ & + R\left[\left(\underline{u}_{Up}\times\right) + \underline{u}_{North}\;\underline{u}_{East}^{T}\right]\delta\underline{\omega}_{IB_{H0}} - R\,\delta\gamma_{Up_1}\,\underline{\omega}_{IE_H} \\ & - R\,\delta\gamma_{Up_0}\,\underline{\omega}_{IE_H} + R\,\delta\omega_{IB_{Up0}}\frac{\pi}{\omega_S}\left(\frac{5}{2} - \frac{1}{\pi}\right)\underline{\omega}_{IE_H} \\ & + R\,\underline{u}_{East}\,\underline{u}_{North}^{T}\;\delta\underline{\omega}_{IB_{H00}} - R\,\delta\omega_{IB_{Up00}}\frac{\pi}{\omega_S}\left(\frac{1}{2} - \frac{5}{\pi}\right)\underline{\omega}_{IE_H} \end{aligned} \qquad (18.1.2\text{-}5)$$

$$\begin{aligned} \underline{V}_{INS/H\,1.5S} = & \;2\,R\left[3\left(\underline{u}_{Up}\times\right) + \underline{u}_{North}\;\underline{u}_{East}^{T}\right]\delta\underline{\omega}_{IB_{H0}} - 4\,R\,\delta\gamma_{Up_1}\,\underline{\omega}_{IE_H} \\ & - 2\,R\,\delta\gamma_{Up_0}\,\underline{\omega}_{IE_H} + R\,\delta\omega_{IB_{Up0}}\frac{\pi}{\omega_S}\,3\,\underline{\omega}_{IE_H} \\ & - R\,\delta\omega_{IB_{Up00}}\frac{\pi}{\omega_S}\,\underline{\omega}_{IE_H} + 2\,R\,\underline{u}_{East}\,\underline{u}_{North}^{T}\;\delta\underline{\omega}_{IB_{H00}} \end{aligned} \qquad (18.1.2\text{-}6)$$

where

0.25 S, 0.5 S, 0.75 S, etc. = Designation for parameter values at the quarter, half, three quarter, etc. Schuler cycle times.

Notice in the previous equations, that the horizontal velocity at the half Schuler cycle time points (i.e., 0.5 S, 1.0 S, 1.5 S) is independent of accelerometer and $\delta\underline{\gamma}_{H_1}$ horizontal tilt error. This will form the basis for using these points to determine the angular rate sensor errors. Once the angular rate sensor errors are found, the accelerometer errors can then be determined from the odd quarter Schuler cycle horizontal velocities (i.e., at 0.25 S, 0.75 S and 1.25 S).

Solutions to Equations (18.1.2-1) - (18.1.2-6) are expedited if we note from Equation (18.1.1-16) that:

$$\left(\underline{u}_{Up}\times\right) + \underline{u}_{North}\;\underline{u}_{East}^{T} = 2\,\underline{u}_{North}\;\underline{u}_{East}^{T} - \underline{u}_{East}\,\underline{u}_{North}^{T} \qquad (18.1.2\text{-}7)$$

We also note that the earth rate horizontal component is north (from the definition of north being in the direction of earth's positive polar axis), hence, it has zero easterly component.

Let's first look at a classical solution to the previous equations for a simple case when the test is terminated after half a Schuler cycle. Then only Equations (18.1.2-1) - (18.1.2-2) apply during the half Schuler cycle test period. Let us assume that the available velocity data from the INS is along north/east axes. Applying (18.1.2-7) to Equations(18.1.2-1) - (18.1.2-2) and taking the dot product with $\underline{u}_{North}$ and $\underline{u}_{East}$ then finds for the north/east components:

$$V_{INSNorth\ 0.5S} = 4\,R\,\delta\omega_{IB_{H0East}} - 2\,R\left(\delta\gamma_{Up_0} + \delta\gamma_{Up_1}\right)\omega_{IE_{North}} + R\left(\delta\omega_{IB_{Up0}} - \delta\omega_{IB_{Up00}}\right)\frac{\pi}{\omega_S}\,\omega_{IE_{North}} \qquad (18.1.2\text{-}8)$$

$$V_{INSEast_{0.5S}} = -\,2\,R\left(\delta\omega_{IB_{H0North}} - \delta\omega_{IB_{H00North}}\right)$$

$$V_{INSNorth_{0.25S}} = -\,2\left(\delta a_{SF_{H0North}} - \frac{g}{2}\,\delta\gamma_{H1_{East}}\right)\frac{1}{\omega_S} + 2\,R\,\delta\omega_{IB_{H0East}}$$
$$-\,R\left(\delta\gamma_{Up_0} + \delta\gamma_{Up_1}\right)\omega_{IE_{North}} + R\left(\delta\omega_{IB_{Up0}} - \delta\omega_{IB_{Up00}}\right)\frac{\pi}{\omega_S}\left(\frac{1}{2} - \frac{1}{\pi}\right)\omega_{IE_{North}} \qquad (18.1.2\text{-}9)$$

$$V_{INSEast_{0.25S}} = -\,2\left(\delta a_{SF_{H0East}} + \frac{g}{2}\,\delta\gamma_{H1_{North}}\right)\frac{1}{\omega_S} - R\left(\delta\omega_{IB_{H0North}} - \delta\omega_{IB_{H00North}}\right)$$

where

$V_{INSNorth\ ()}$, $V_{INSEast()}$ = North/east components of $\underline{V}_{INS/H\ ()}$.

$\delta\omega_{IB_{H0North}}$, $\delta\omega_{IB_{H0East}}$, $\delta\omega_{IB_{H00North}}$, $\delta\omega_{IB_{H00East}}$ = North/east components of $\delta\underline{\omega}_{IB_{H0}}$, $\delta\underline{\omega}_{IB_{H00}}$.

$\delta a_{SF_{H0North}}$, $\delta a_{SF_{H0East}}$, $\delta\gamma_{H1_{North}}$, $\delta\gamma_{H1_{East}}$ = North/east components of $\delta\underline{a}_{SF_{H0}}$, $\delta\underline{\gamma}_{H_1}$.

$\omega_{IE_{North}}$ = North component of $\underline{\omega}_{IE_H}$ which is also its magnitude since the east component of $\underline{\omega}_{IE_H}$ is zero.

We also know from (18.1.1-20) for a single half Schuler cycle that:

$$\left(\delta\omega_{IB_{Up0}} - \delta\omega_{IB_{Up00}}\right)\frac{\pi}{\omega_S} = \psi_{P/INS_{0.0S+}} - \psi_{P/INS_{0.5S-}} \qquad (18.1.2\text{-}10)$$

Equations (18.1.2-8) combined with (18.1.2-10) obtains for the angular rate sensor errors:

$$\delta\omega_{IB_{H0North}} - \delta\omega_{IB_{H00North}} = -\frac{1}{2\,R} V_{INSEast_{0.5S}}$$

$$\delta\omega_{IB_{H0East}} - \frac{1}{2}\,\omega_{IE_{North}}\left(\delta\gamma_{Up_0} + \delta\gamma_{Up_1}\right) = \frac{1}{4}\left[\frac{1}{R} V_{INSNorth\ 0.5S} - \left(\psi_{P/INS_{0.0S+}} - \psi_{P/INS_{0.5S-}}\right)\omega_{IE_{North}}\right] \qquad (18.1.2\text{-}11)$$

Equations (18.1.2-11) allow the angular rate sensor errors to be estimated only in the composite form shown. The $\delta\omega_{IB_{H00North}}$, $\frac{1}{2}\left(\delta\gamma_{Up_0} + \delta\gamma_{Up_1}\right)\omega_{IE_{North}}$ terms cannot be separated from $\delta\omega_{IB_{H0North}}$, $\delta\omega_{IB_{H0East}}$ because they have identical horizontal signatures at the quarter and half Schuler cycle time points.

The accelerometer errors can be evaluated by substituting (18.1.2-10) - (18.1.2-11) into (18.1.2-9):

$$\delta a_{SF_{H0North}} - \frac{g}{2}\,\delta\gamma_{H1_{East}} = -\frac{\omega_S}{2}\left[V_{INSNorth_{0.25S}} - \frac{1}{2} V_{INSNorth_{0.5S}} + R\left(\psi_{P/INS_{0.0S+}} - \psi_{P/INS_{0.5S-}}\right)\frac{1}{\pi}\,\omega_{IE_{North}}\right]$$

$$\delta a_{SF_{H0East}} + \frac{g}{2}\,\delta\gamma_{H1_{North}} = -\frac{\omega_S}{2}\left(V_{INSEast_{0.25S}} - \frac{1}{2} V_{INSEast_{0.5S}}\right) \qquad (18.1.2\text{-}12)$$

Equations (18.1.2-12) allow the accelerometer errors to be estimated, but only in the composite form shown. The $\delta\underline{\gamma}_{H_1}$ terms cannot be separated from $\delta\underline{a}_{SF_{H0}}$ because they have identical horizontal signatures at the quarter and half Schuler cycle time points. For the accelerometer, this is a manifestation of the general rule discussed in Section 13.2.4 (following Equation (13.2.4-28)) that only sensor misalignments relative to one another affect strapdown inertial navigation performance. The composite of $\delta\underline{\gamma}_{H_1}$ and $\delta\underline{a}_{SF_{H0}}$ in (18.1.2-12) converts the accelerometer misalignment terms in $\delta\underline{a}_{SF_{H0}}$ to be relative to the angular rate sensor input axes. In this regard, the $\delta\underline{\gamma}_{H_1}$, $\delta\underline{a}_{SF_{H0}}$ composite terms in (18.1.2-12) can be interpreted as accelerometer errors having the misalignment terms referenced to the angular rate sensor input axes.

Equations (18.1.2-11) and (18.1.2-12) are commonly used with the $\psi_{P/INS_{0.0S+}} - \psi_{P/INS_{0.5S-}}$ term neglected, to obtain an approximate estimate of horizontal accelerometer and angular rate sensor error.

Regarding accelerometer error determination, an alternative to Equations (18.1.2-12) is based on the derivative of Equation (18.1.1-18) at t = 0 for which we find upon differentiation that:

$$\dot{\underline{V}}_{INS_H}(t=0) = -2\left(\delta\underline{a}_{SF_{H0}} - \frac{g}{2}\,\underline{u}_{Up} \times \delta\underline{\gamma}_{H_1}\right) \tag{18.1.2-13}$$

hence,

$$\delta\underline{a}_{SF_{H0}} - \frac{g}{2}\,\underline{u}_{Up} \times \delta\underline{\gamma}_{H_1} = -\frac{1}{2}\,\dot{\underline{V}}_{INS_H}(t=0) \tag{18.1.2-14}$$

The $\dot{\underline{V}}_{INS_H}(t=0)$ term in Equation (18.1.2-14) can be evaluated by measuring the change in $\underline{V}_{INS_H}(t)$ over a small time increment immediately following the first 180 degree rotation at self-alignment completion. Then $\dot{\underline{V}}_{INS_H}(t=0)$ is calculated as the change in $\underline{V}_{INS_H}(t)$ divided by the time interval. The time interval should be large enough to attenuate accelerometer pulse quantization uncertainty error in the $\underline{V}_{INS_H}(t)$ measurement, but small compared to the Schuler period so that the computation of $\dot{\underline{V}}_{INS_H}(t=0)$ is valid (e.g., 10 seconds).

It is interesting to note that the Equation (18.1.2-14) method for measuring accelerometer error was the original basis for development of the Strapdown Rotation test described in Section 18.4. The Strapdown Rotation test generalizes the initial rotation (following self-alignment) to a rotation sequence about different sensor assembly axes. The result is a coupling of accelerometer and angular rate sensor errors into measurable horizontal acceleration. By structuring a group of such measurements using different rotation sequences, a set of horizontal acceleration measurements can be obtained, each representing a different linear combination of the sensor errors. Post-processing the horizontal acceleration measurements then allows individual sensor errors to be determined explicitly to high accuracy (i.e., accelerometer misalignment and bias; angular rate sensor misalignment and scale factor error). Vertical acceleration measurements taken during the same test series also allow determination of accelerometer scale factor errors.

Returning to the Schuler Pump test discussion, a more accurate estimate for the sensor errors can be made by allowing the test to continue for two half Schuler cycles. To simplify the analytical form of our deterministic solution, we will not use the first quarter Schuler cycle data so that Equations (18.1.2-2) - (18.1.2-4) then apply. Equations (18.1.2-3) - (18.1.2-4) with (18.1.1-16) finds for the north/east components:

$$\begin{aligned} V_{INSNorth_{1.0S}} &= -4\,R\,\delta\omega_{IB_{H0East}} + 2\,R\,\delta\gamma_{Up_1}\,\omega_{IE_{North}} \\ &\quad + R\,\delta\omega_{IB_{Up0}}\,\frac{\pi}{\omega_S}\,2\,\omega_{IE_{North}} \\ V_{INSEast_{1.0S}} &= 4\,R\,\delta\omega_{IB_{H0North}} \end{aligned} \tag{18.1.2-15}$$

$$V_{INSNorth_{0.75S}} = 4\left(\delta a_{SF_{H0North}} - \frac{g}{2}\,\delta\gamma_{H1_{East}}\right)\frac{1}{\omega_S} - R\,\delta\gamma_{Up0}\,\omega_{IE_{North}}$$

$$+ R\,\delta\omega_{IB_{Up0}}\,\frac{\pi}{\omega_S}\left(\frac{3}{2} + \frac{1}{\pi}\right)\omega_{IE_{North}} - R\,\delta\omega_{IB_{Up00}}\,\frac{\pi}{\omega_S}\left(\frac{1}{2} + \frac{3}{\pi}\right)\omega_{IE_{North}} \qquad (18.1.2\text{-}16)$$

$$V_{INSEast_{0.75S}} = 4\left(\delta a_{SF_{H0East}} + \frac{g}{2}\,\delta\gamma_{H1_{North}}\right)\frac{1}{\omega_S}$$

$$+ R\left(\delta\omega_{IB_{H0North}} + \delta\omega_{IB_{H00North}}\right)$$

We also know from (18.1.1-21) for the second half Schuler cycle that:

$$\left(\delta\omega_{IB_{Up0}} + \delta\omega_{IB_{Up00}}\right)\frac{\pi}{\omega_S} = -\left(\psi_{P/INS_{1.0S-}} - \psi_{P/INS_{0.5+}}\right) \qquad (18.1.2\text{-}17)$$

Summing and differencing Equations (18.1.2-17) and (18.1.2-10) yields:

$$\delta\omega_{IB_{Up0}} = -\frac{\omega_S}{2\,\pi}\left(\psi_{P/INS_{1.0S-}} - \psi_{P/INS_{0.5S+}} + \psi_{P/INS_{0.5S-}} - \psi_{P/INS_{0.0S+}}\right) \qquad (18.1.2\text{-}18)$$

$$\delta\omega_{IB_{Up00}} = -\frac{\omega_S}{2\,\pi}\left(\psi_{P/INS_{1.0S-}} - \psi_{P/INS_{0.5S+}} - \psi_{P/INS_{0.5S-}} + \psi_{P/INS_{0.0S+}}\right) \qquad (18.1.2\text{-}19)$$

With (18.1.2-18), Equations (18.1.2-15) can be solved for the H0 angular rate sensor errors:

$$\delta\omega_{IB_{H0North}} = \frac{1}{4\,R}\,V_{INSEast_{1.0S}}$$

$$\delta\omega_{IB_{H0East}} - \frac{1}{2}\,\delta\gamma_{Up_1}\,\omega_{IE_{North}} = -\frac{1}{4}\left[\frac{1}{R}\,V_{INSNorth_{1.0S}}\right. \qquad (18.1.2\text{-}20)$$

$$\left. + \left(\psi_{P/INS_{1.0S-}} - \psi_{P/INS_{0.5S+}} + \psi_{P/INS_{0.5S-}} - \psi_{P/INS_{0.0S+}}\right)\omega_{IE_{North}}\right]$$

From (18.1.2-20), we see that $\delta\omega_{IB_{H0North}}$ can now be determined explicitly, but that the $\delta\omega_{IB_{H0East}}$ angular rate sensor error appears as a composite with $\frac{1}{2}\,\delta\gamma_{Up_1}\,\omega_{IE_{North}}$ (in contrast with Equations (18.1.2-11) in which $\delta\omega_{IB_{H0North}}$ was combined with $\delta\omega_{IB_{H00North}}$, and $\delta\omega_{IB_{H0East}}$ was combined with $\frac{1}{2}\left(\delta\gamma_{Up_0} + \delta\gamma_{Up_1}\right)$). In general, for a normally calibrated INS, the $\frac{1}{2}\,\delta\gamma_{Up1}\,\omega_{IE_{North}}$ term in (18.1.2-20) is negligible compared to the $\delta\omega_{IB_{H0East}}$ term being evaluated.

The north H00 angular rate sensor error can be determined by summing the east component of Equations (18.1.2-15) with two times the east component of Equations (18.1.2-8):

$$\delta\omega_{IB_{H00North}} = \frac{1}{4\,R}\left(V_{INSEast_{1.0S}} + 2\,V_{INSEast_{0.5S}}\right) \qquad (18.1.2\text{-}21)$$

The initial heading residual $\delta\gamma_{Up0}$ can be calculated by summing the north components of (18.1.2-8) and (18.1.2-15), and applying (18.1.2-18) - (18.1.2-19):

$$\begin{aligned}\delta\gamma_{Up0} = &- \frac{1}{2\,R\,\omega_{IE_{North}}}\left(V_{INSNorth_{0.5S}} + V_{INSNorth_{1.0S}}\right) \\ &- \frac{1}{2}\left[\left(\psi_{P/INS_{1.0S-}} - \psi_{P/INS_{0.5S+}}\right) + 2\left(\psi_{P/INS_{0.5S-}} - \psi_{P/INS_{0.0S+}}\right)\right]\end{aligned} \qquad (18.1.2\text{-}22)$$

Finally, the horizontal composite accelerometer errors are evaluated by combining Equations (18.1.2-8), (18.1.2-15), (18.1.2-16), (18.1.2-18) and (18.1.2-19):

$$\begin{aligned}\delta a_{SF_{H0North}} - \frac{g}{2}\,\delta\gamma_{H1_{East}} = &\frac{\omega_S}{4}\left\{V_{INSNorth_{0.75S}} - \frac{1}{2}\left(V_{INSNorth_{1.0S}} + V_{INSNorth_{0.5S}}\right)\right. \\ &\left. - \frac{R\,\omega_{IE_{North}}}{\pi}\left(\psi_{P/INS_{1.0S-}} - \psi_{P/INS_{0.5S+}} - 2\,\psi_{P/INS_{0.5S-}} + 2\,\psi_{P/INS_{0.0S+}}\right)\right\} \\ \delta a_{SF_{H0East}} + \frac{g}{2}\,\delta\gamma_{H1_{North}} = &\frac{\omega_S}{4}\left[V_{INSEast_{0.75S}} - \frac{1}{2}\left(V_{INSEast_{1.0S}} + V_{INSEast_{0.5S}}\right)\right]\end{aligned} \qquad (18.1.2\text{-}23)$$

Equations (18.1.2-18) - (18.1.2-23) allow determination of $\delta\omega_{IB_{H0North}}$, $\left(\delta\omega_{IB_{H0East}} - \frac{1}{2}\,\delta\gamma_{Up1}\,\omega_{IE_{North}}\right)$, $\delta\omega_{IB_{Up0}}$, $\left(\delta a_{SF_{H0North}} - \frac{g}{2}\,\delta\gamma_{H1_{East}}\right)$, $\left(\delta a_{SF_{H0East}} + \frac{g}{2}\,\delta\gamma_{H1_{North}}\right)$, $\delta\gamma_{Up0}$, $\delta\omega_{IB_{H00North}}$, and $\delta\omega_{IB_{Up00}}$ within the random noise limits of the horizontal velocity data. In general, the horizontal velocity random noise is produced primarily by the integrated effect of accelerometer and angular rate sensor output noise, typically increasing as the square root of time (See Sections 13.6.1 - 13.6.2). For increased accuracy in the Schuler Pump test, later time measurements are generally preferred because their amplification effect on the principal errors builds linearly with time, hence, the error-signal to noise ratio increases as the square root of time. If we extend the test duration for another half Schuler cycle, Equations (18.1.2-5) - (18.1.2-6) would be included for sensor error evaluation. The addition of the North component of Equation (18.1.2-6), in particular, would then allow $\delta\omega_{IB_{H0East}}$ and $\delta\gamma_{Up1}$ to be explicitly evaluated (within the sensor noise limits). This is a minor (if any) improvement since $\delta\gamma_{Up_1}$ is generally small, and its impact on $\delta\omega_{IB_{H0East}}$ determination accuracy is generally minimal.

Comparing Equation (18.1.2-6) at the 1.5 Schuler cycle time with (18.1.2-4) at the full Schuler cycle time, we see (using (18.1.1-16)) that the north and east H0 angular rate sensor errors are amplified by 6 and 8 respectively in (18.1.2-6), compared with 4 in (18.1.2-4). However, Equation (18.1.2-6) contains $\delta\gamma_{Up_0}$ and $\delta\underline{\omega}_{IB_{H00}}$ error terms that are absent in (18.1.2-4). The result is that H0 angular rate error estimates obtained using (18.1.2-6) will not achieve the full benefit of increased amplification because data from the half cycle response (Equation (18.1.2-2)) must be included with (18.1.2-6) for H0 error determination (to eliminate $\delta\gamma_{Up_0}$ and $\delta\underline{\omega}_{IB_{H00}}$ in the solution process). In order to avoid this difficulty, data at the two Schuler cycle time can be used which would be similar in form to (18.1.2-6) (i.e., without the $\delta\gamma_{Up_0}$, $\delta\underline{\omega}_{IB_{H00}}$ terms), but having an H0 angular rate sensor error amplification factor of 8 instead of 4. Then H0 error determination would achieve the full benefit of increased amplification.

Extending the Schuler Pump test duration for increased sensor error amplification has a fundamental limitation in the analytical solutions presented thus far. The limitation arises because of an approximation employed in underlying Equation (18.1.1-1) (a repeat of the velocity error expression in Equation (13.3.2-25)) that neglected earth rate coupling between horizontal tilt and velocity error build-up (i.e., derived in the steps going from (13.3-6) - (13.3-9) to (13.3-15) to (13.3.2-1) to (13.3.2-20)). This approximation enabled the development of (18.1.1-1) in the closed-form indicated, in effect based on the accompanying assumption that Schuler oscillations maintain their planar orientation relative to wander azimuth N Frame coordinates. Without the approximation, the correct Schuler response is for sustained oscillations in <u>free azimuth</u> N Frame coordinates (as discussed in Section 13.2.2). From Section 4.5, we see that the angular difference between free and wander axis coordinate frames (assuming identical initialization of the C_N^E matrix) is a rotation about the vertical equal to the vertical component of earth rate multiplied by navigation time t. Thus, for 45 degrees latitude for which the vertical earth rate is 11 degrees per hour, the angular separation between the free and wander azimuth frames is 0.26 radians at one Schuler cycle (84 minutes), or an average of 0.13 radians over the 84 minute time period. This introduces 13 percent cross-coupling error between the north/east sensor error estimates, which is generally within the ball-park of expected accuracy for the Schuler Pump test solutions. For test periods exceeding one Schuler cycle, the average cross-coupling error effect may be unacceptable.

To eliminate the previous cross-coupling error effect, numerical techniques can be used to compute the horizontal velocity error sensitivity to strapdown sensor error without approximation. The method, of course, is to calculate the state transition matrix Φ that couples

B Frame sensor error to N Frame horizontal velocity error by integration of Φ (as in Equation (15.1.1-13)) from time $t = 0$, using the state dynamic matrix $A(\tau)$ associated with the full wander azimuth error equation set (e.g., Equations (13.3-6)). Equations (18.1.2-1) - (18.1.2-6), etc. would then be replaced by the equivalent to the $\underline{x}(t)$ expression in Equation (15.1.1-13) (based on zero process noise $\underline{w}(t)$), evaluated at the quarter Schuler cycle times. Additional intermediate time points could also be easily included with this approach. The inertial sensor errors can then be estimated by a more sophisticated inversion process, such as weighted least squares (e.g., Reference 24 - Section 6.9) in which the weighting matrix is increased linearly with time (as the variance in horizontal velocity error caused by inertial sensor output noise). Carrying this to a more sophisticated level, we can also consider using a Kalman filter to estimate the sensor errors.

The previous discussion illustrates sophisticated techniques that can be added to enhance Schuler Pump test data analysis. However, the penalty is added complexity which, in a sense, defeats the basic motivation behind the test; i.e., having a simple method for rapidly determining approximate accelerometer and angular rate sensor error, without requiring sophisticated software analysis tools. In fact, if more accurate solutions are required for angular rate sensor and accelerometer sensor evaluation, simpler and more effective techniques are available based on analyzing the performance characteristics of the strapdown "platform" without the attendant navigation solution (e.g., Sections 18.2 and 18.4). Based on these considerations, when used, the Schuler Pump test is usually applied based on the simpler deterministic solution approaches.

18.2 STRAPDOWN DRIFT TEST

The Strapdown Drift test is designed to evaluate angular rate sensor error by processing data generated during extended self-alignment operations. The test is performed on a strapdown analytic platform during an extension of the normal self-alignment initialization mode. The principal measurement of the strapdown drift test is the composite north horizontal angular rate sensor output, determined from the north component of angular rate bias applied to the strapdown analytic platform to render it stationary in tilt around North attitude. Subtracting the known true value of north earth rate from the measurement evaluates the north component of angular rate sensor composite error. East and vertical angular rate sensor errors are ascertained by repeating the test with the previously east and vertical angular rate sensors in the horizontal north orientation.

For situations when the biasing rate to the strapdown analytic platform is not an available INS output, an alternative procedure can be utilized based on INS computed true heading outputs. In this case the east angular rate sensor error is determined from the test based on the heading error it generates at the end of an extended self-alignment run. In order to discriminate east angular rate sensor error from earth rate input, the INS heading output is measured for two individual alignment runs. The second alignment run is performed at a heading orientation that is rotated 180 degrees from the first. The difference between the average heading measurements so obtained cancels the true earth rate input, thereby becoming the measurement for east angular rate sensor error determination. North and vertical angular rate sensor errors are ascertained by repeating the test with the previously north and vertical angular rate sensors in the horizontal east orientation.

In the following subsections we will analyze both Strapdown Drift test methods (based on analytical platform bias measurement, and based on INS heading measurement). The concluding subsection shows how the determined angular rate sensor bias errors can be used to update the INS bias calibration coefficients.

18.2.1 STRAPDOWN DRIFT TEST BASED ON ANALYTICAL PLATFORM REBALANCE BIAS MEASUREMENT

The analytical basis for this version of the Strapdown Drift test can be derived from Equations (6.1.2-2) of Chapter 6. The basic (6.1.2-2) self-alignment equations are processed to initially align the INS (for approximately 5 minutes, depending on strapdown inertial sensor noise content). The Kalman gains are then frozen and the alignment process is extended. Let us write the C_B^L attitude rate equation in (6.1.2-2) as being processed in the INS computer during the extended fine-alignment period, identifying that the computed parameters contain error due to initial condition error and inertial sensor input error:

$$\dot{\hat{C}}_B^L = \hat{C}_B^L \left(\underline{\tilde{\omega}}_{IB}^B \times \right) - \left(\underline{\hat{\omega}}_{IL}^L \times \right) \hat{C}_B^L \qquad (18.2.1\text{-}1)$$

where

$\tilde{}$ = Designation for angular rate sensor output containing angular rate sensor error (as opposed to angular rate sensor signals without the $\tilde{}$ that will, henceforth, signify error free signals).

$\hat{}$ = Designation for computed parameter containing errors (as opposed to quantities without the $\hat{}$ that will, henceforth, signify error free parameters).

L = Locally level attitude referencing coordinate frame as defined in Section 2.2, and as specialized for this section, to be of the wander azimuth type (See Section 4.5).

Based on the (15.2.1-1) $\dot{C}_B^N$ expression compared with $\dot{C}_B^L$ in (18.2.1-1), let's use the $\dot{C}_B^N$ error rate equations from (15.2.1-3) (transformed from the N to the L Frames - See Section 2.2

for definitions) to write the error state dynamic equation for the $\hat{C}_B^L$ angular error rate during alignment. In the process, recall that (15.2.1-3) was derived from (14.2-18) which, in turn, was derived from Equation (14.2-7). From the explanation preceding (14.2-7) we note that $\underline{\gamma}$ is actually $\underline{\gamma}^*$ as defined by Equation (12.5-5), but with the * deleted for notation simplicity (it is in sections like these that one sometimes regrets using such simplifications). The $\underline{\gamma}^*$ parameter is $\underline{\gamma}$ neglecting quantization noise (See Equation (12.5-5)). In this section, it is preferred to restore the original $\underline{\gamma}$ meaning to Equation (15.2.1-3) before applying it as outlined above. We achieve this by changing all $\underline{\gamma}$'s to $\underline{\gamma}^*$'s, then replacing the $\dot{\underline{\gamma}}^*$ terms by the derivative of $\underline{\gamma}^L + C_B^L \, \delta\underline{\alpha}_{Quant}$ per the (12.5-5) $\underline{\gamma}^*$ definition in the L Frame, and finally, moving the $C_B^L \, \delta\underline{\alpha}_{Quant}$ derivative to the right side of the equality with a corresponding sign reversal. The result is the desired error state dynamic equation for the $\hat{C}_B^L$ rotation angle error $\underline{\gamma}$ "without any shenanigans":

$$\dot{\underline{\gamma}}_H^L = -\left(C_B^L \, \delta\underline{\omega}_{IB}^B\right)_H - \gamma^*_{ZL}\left(\underline{\omega}_{IL_H}^L \times \underline{u}_{ZL}^L\right) + \left(\underline{u}_{ZL}^L \times \underline{\gamma}^{*L}_H\right)\omega_{IE_{Up}} + \delta\underline{\omega}_{IL_H}^L$$
$$+ \left(C_B^L\right)_H \left(\underline{\omega}_{Vib}^B \times \delta\underline{\alpha}_{Quant}\right) - \frac{d}{dt}\,\delta\underline{\alpha}_{Quant_H}^L$$
$$\dot{\gamma}_{ZL} = -\,\underline{u}_{ZL}^L \cdot \left(C_B^L \, \delta\underline{\omega}_{IB}^B\right) - \underline{u}_{ZL}^L \cdot \left(\underline{\omega}_{IL_H}^L \times \underline{\gamma}^{*L}_H\right)$$
$$+ \underline{u}_{ZL}^L \cdot \left[C_B^L\left(\underline{\omega}_{Vib}^B \times \delta\underline{\alpha}_{Quant}\right)\right] - \frac{d}{dt}\,\delta\alpha_{Quant_{ZL}} \qquad (18.2.1\text{-}2)$$

with

$$\underline{\gamma}^{*L} = \underline{\gamma}^L + C_B^L \, \delta\underline{\alpha}_{Quant} \qquad (18.2.1\text{-}3)$$

where

$\underline{\gamma}_H^L, \gamma_{ZL}$ = Horizontal and vertical (along the L Frame Z axis) components of the rotation angle error $\underline{\gamma}^L$ in $\hat{C}_B^L$.

$\underline{u}_{ZL}^L$ = Unit vector downward along the L Frame Z axis projected on L Frame axes. Note that $\underline{u}_{ZL}$ is in the opposite direction from $\underline{u}_{ZN}$ in Equations (15.2.1-3) which is upward along the N Frame Z axis.

$\omega_{IE_{Up}}$ = Upward vertical component of earth rate which is $\omega_e \sin l$.

$\underline{\omega}_{Vib}^B$ = B Frame angular vibration rate during alignment.

$\delta\underline{\omega}_{IL_H}^{L}$, $\delta\underline{\omega}_{IB}^{B}$ = Errors in $\underline{\tilde{\omega}}_{IB}^{B}$ and $\underline{\hat{\omega}}_{IL}^{L}$.

$\delta\underline{\alpha}_{Quant}$ = Integrated angular rate sensor output quantization error in the Sensor (B) Frame.

$\delta\underline{\alpha}_{Quant_H}^{L}$, $\delta\alpha_{Quant_{ZL}}$ = Horizontal and vertical (along the L Frame Z axis) components of $C_B^L\,\delta\underline{\alpha}_{Quant}$.

The integral of (18.2.1-2) since the start of the extended Fine Alignment process (showing functional time dependence where needed for clarity), including substitution of (18.2.1-3) in the ZL component result, is:

$$\begin{aligned}
\underline{\gamma}_H^L(t) = {} & \underline{\gamma}_{H\,Start}^L + \int_{t_{Start}}^{t}\Big[-\left(C_B^L\,\delta\underline{\omega}_{IB}^B\right)_H - \gamma^*_{ZL}(\tau)\left(\underline{\omega}_{IL\,H}^L \times \underline{u}_{ZL}^L\right) \\
& + \left(\underline{u}_{ZL}^L \times \underline{\gamma}^{*L}_H\right)\omega_{IE_{Up}} + \delta\underline{\omega}_{IL\,H}^L + \left(C_B^L\right)_H\left(\underline{\omega}_{Vib}^B \times \delta\underline{\alpha}_{Quant}\right)\Big]\,d\tau \\
& - \delta\underline{\alpha}_{Quant_H}^L(t) + \delta\underline{\alpha}_{Quant_H}^L(t_{Start})
\end{aligned} \qquad (18.2.1\text{-}4)$$

$$\begin{aligned}
\gamma^*_{ZL}(t) = {} & \gamma_{ZLStart} + \int_{t_{Start}}^{t}\Big\{-\underline{u}_{ZL}^L \cdot \left(C_B^L\,\delta\underline{\omega}_{IB}^B\right) \\
& - \underline{u}_{ZL}^L \cdot \left(\underline{\omega}_{IL_H}^L \times \underline{\gamma}^{*L}_H(\tau)\right) + \underline{u}_{ZL}^L \cdot \Big[C_B^L\left(\underline{\omega}_{Vib}^B \times \delta\underline{\alpha}_{Quant}\right)\Big]\Big\}\,d\tau \\
& + \delta\alpha_{Quant_{ZL}}(t_{Start})
\end{aligned}$$

where

t_{Start} = Time point at the start of the extended alignment period for initiating the (18.2.1-1) integration process.

Start = Subscript indicating the value for the parameter at time t_{Start}.

τ = Integration time parameter.

Equation (14.3-39) in Section 14.3 shows that in the absence of sensor and measurement noise, $\underline{\gamma}^*_H$ will reach a steady offset value at the end of Fine Alignment determined by horizontal accelerometer error components (Note from our "back-off" discussion preceding Equation (18.2.1-2), that γ_H in (14.3-39) is actually $\underline{\gamma}^*_H$ defined by (12.5-5) and (18.2.1-3)). Chapter 14, Section 14.6 focused on the response of horizontal earth rate estimation error to random noise during Fine Alignment. The procedures in Chapter 14 leading to Equations

(14.6.4.4-1) can also be used to show that the response of horizontal tilt error ($\underline{\gamma}^*_H$) during Fine Alignment caused by random sensor errors and measurement noise, diminishes with time in alignment, becoming negligible at the end of alignment (again, $\underline{\gamma}_H$ in Chapter 14 is actually $\underline{\gamma}^*_H$). We can conclude, therefore, that during extended alignment we can approximate:

$$\underline{\gamma}^{*L}_H(t) \approx \underline{\gamma}^L_{H_{Offset}} \tag{18.2.1-5}$$

where

$\underline{\gamma}^L_{H_{Offset}}$ = Average offset of $\hat{C}^L_B$ from the true C^L_B during extended alignment.

We can write for the components of $C^L_B \, \delta\underline{\omega}^B_{IB}$ in (18.2.1-4):

$$\begin{aligned} \delta\omega_{IB_{ZL}} &= \delta\omega_{ARS/Cnst_{ZL}} + n_{ARS/Rnd_{ZL}} \\ \delta\underline{\omega}^L_{IB_H} &= \delta\underline{\omega}^L_{ARS/Cnst_H} + \underline{n}^L_{ARS/Rnd_H} \end{aligned} \tag{18.2.1-6}$$

where

$\delta\omega_{ARS/Cnst_{ZL}}, \delta\underline{\omega}^L_{ARS/Cnst_H}$ = Vertical and horizontal constant components of $C^L_B \, \delta\underline{\omega}^B_{IB}$.

$n_{ARS/Rnd_{ZL}}, \underline{n}^L_{ARS/Rnd_H}$ = Vertical and horizontal random noise components of $C^L_B \, \delta\underline{\omega}^B_{IB}$.

From (18.2.1-1) compared to the (15.2.1-1) $\dot{C}^N_B$ expression L Frame equivalent, we should also recognize that $\underline{\omega}^L_{IL_H}$ in (18.2.1-2) is horizontal earth rate $\underline{\omega}^L_{IE_H}$, a constant for the wander azimuth L Frame during alignment. Then substituting (18.2.1-5), (18.2.1-6), and $\underline{\omega}^L_{IE_H}$ for $\underline{\omega}^L_{IL_H}$ in the (18.2.1-4) γ^*_{ZL} expression finds:

$$\gamma^*_{ZL}(t) \approx \gamma_{ZL_{Start}} + \int_{t_{Start}}^{t} \Big\{ -\delta\omega_{ARS/Cnst_{ZL}} - n_{ARS/Rnd_{ZL}} - \underline{u}^L_{ZL} \cdot \left(\underline{\omega}^L_{IE_H} \times \underline{\gamma}^L_{H_{Offset}} \right) + \underline{u}^L_{ZL} \cdot \left[C^L_B \left(\underline{\omega}^B_{Vib} \times \delta\underline{\alpha}_{Quant} \right) \right] \Big\} d\tau + \delta\alpha_{Quant_{ZL}}(t_{Start}) \tag{18.2.1-7}$$

$$= \gamma_{ZL_{Start}} - \left[\delta\omega_{ARS/Cnst_{ZL}} + \underline{u}^L_{ZL} \cdot \left(\underline{\omega}^L_{IE_H} \times \underline{\gamma}^L_{H_{Offset}} \right) \right] (t - t_{Start}) - \theta_{ARS/Rnd_{ZL}}(t, t_{Start}) + \theta_{ARS/\omega Vib\text{-}\alpha Quant_{ZL}}(t, t_{Start}) + \delta\alpha_{Quant_{ZL}}(t_{Start})$$

with

$$\theta_{ARS/Rnd_{ZL}}(t, t_{Start}) \equiv \int_{t_{Start}}^{t} n_{ARS/Rnd_{ZL}}\, d\tau$$

$$\theta_{ARS/\omega Vib\text{-}\alpha Quant_{ZL}}(t, t_{Start}) \equiv \int_{t_{Start}}^{t} \underline{u}^L_{ZL} \cdot \left[C^L_B \left(\underline{\omega}^B_{Vib} \times \delta\underline{\alpha}_{Quant} \right) \right] d\tau \tag{18.2.1-8}$$

Substituting (18.2.1-5) - (18.2.1-8) in the (18.2.1-4) $\underline{\gamma}^L_H$ expression obtains at the end of the extended alignment period:

$$\underline{\gamma}^L_{H_{End}} = \underline{\gamma}^L_{H_{Start}} + \int_{t_{Start}}^{t_{End}} \delta\underline{\omega}^L_{IL_H}\, dt - \left[\delta\underline{\omega}^L_{ARS/Cnst_H} - \left(\underline{u}^L_{ZL} \times \underline{\gamma}^L_{H_{Offset}} \right) \omega_{IE_{Up}} \right] T - \underline{\theta}^L_{ARS/Rnd_H} + \underline{\theta}^L_{ARS/\omega Vib\text{-}\alpha Quant_H} - \delta\underline{\alpha}^L_{Quant/H_{End}} + \delta\underline{\alpha}^L_{Quant/H_{Start}}$$
$$- \Big\{ \left(\gamma_{ZL_{Start}} + \delta\alpha_{Quant/ZL_{Start}} \right) T - \frac{1}{2} \left[\delta\omega_{ARS/Cnst_{ZL}} + \underline{u}^L_{ZL} \cdot \left(\underline{\omega}^L_{IE_H} \times \underline{\gamma}^L_{H_{Offset}} \right) \right] T^2 - S\theta_{ARS/Rnd_{ZL}} + S\theta_{ARS/\omega Vib\text{-}\alpha Quant_{ZL}} \Big\} \underline{\omega}^L_{IE_H} \times \underline{u}^L_{ZL} \tag{18.2.1-9}$$

with

$$\underline{\theta}^L_{ARS/Rnd_H} \equiv \int_{t_{Start}}^{t_{End}} \underline{n}^L_{ARS/Rnd_H}\, dt$$

$$\underline{\theta}^L_{ARS/\omega Vib\text{-}\alpha Quant_H} \equiv \int_{t_{Start}}^{t_{End}} \left(C^L_B\right)_H \left(\underline{\omega}^B_{Vib} \times \delta\underline{\alpha}_{Quant}\right) dt$$

$$S\theta_{ARS/Rnd_{ZL}} \equiv \int_{t_{Start}}^{t_{End}} \int_{t_{Start}}^{t} n_{ARS/Rnd_{ZL}}\, d\tau\, dt \qquad (18.2.1\text{-}10)$$

$$S\theta_{ARS/\omega Vib\text{-}\alpha Quant_{ZL}} \equiv \int_{t_{Start}}^{t_{End}} \int_{t_{Start}}^{t} \underline{u}^L_{ZL} \cdot \left[C^L_B \left(\underline{\omega}^B_{Vib} \times \delta\underline{\alpha}_{Quant}\right)\right] d\tau\, dt$$

where

End = Subscript identifying parameter values at the end of extended alignment.

T = Time duration for the extended alignment period (i.e., $t_{End} - t_{Start}$).

Based on the definition for $\delta\underline{\omega}^L_{IL_H}$ and, as noted previously, $\underline{\omega}^L_{IL_H} = \underline{\omega}^L_{IE_H}$ we can write:

$$\delta\underline{\omega}^L_{IL_H} = \hat{\underline{\omega}}^L_{IL_H} - \underline{\omega}^L_{IL_H} = \hat{\underline{\omega}}^L_{IL_H} - \underline{\omega}^L_{IE_H} \qquad (18.2.1\text{-}11)$$

We also define for the integral of the $\hat{\underline{\omega}}^L_{IL_H}$ term in (18.2.1-11):

$$\hat{\underline{\phi}}^L_H \equiv \int_{t_{Start}}^{t_{End}} \hat{\underline{\omega}}^L_{IL_H}\, dt \qquad (18.2.1\text{-}12)$$

where

$\hat{\underline{\phi}}^L_H$ = What will be the measurement for the Strapdown Drift test.

Substituting (18.2.1-11) and (18.2.1-12) in (18.2.1-9), approximating $\underline{\gamma}^L_{H_{Start}}$ as equal to $\underline{\gamma}^L_{H_{End}}$ (i.e., t_{Start} occurs after transients have decayed in the Fine Alignment process), and solving for $\delta\underline{\omega}^L_{ARS/Cnst_H}$:

$$\begin{aligned}\delta\underline{\omega}^L_{ARS/Cnst_H} &= \frac{1}{T}\,\hat{\underline{\phi}}^L_H - \underline{\omega}^L_{IE_H} + \left(\underline{u}^L_{ZL} \times \underline{\gamma}^L_{H_{Offset}}\right)\omega_{IE_{Up}} \\ &- \frac{1}{T}\left(\underline{\theta}^L_{ARS/Rnd_H} - \underline{\theta}^L_{ARS/\omega Vib\text{-}\alpha Quant_H} + \delta\underline{\alpha}^L_{Quant/H_{End}} - \delta\underline{\alpha}^L_{Quant/H_{Start}}\right) \\ - \frac{1}{T}\Bigg\{&\left(\gamma_{ZL_{Start}} + \delta\alpha_{Quant/ZL_{Start}}\right)T - \frac{1}{2}\left[\delta\omega_{ARS/Cnst_{ZL}} + \underline{u}^L_{ZL}\cdot\left(\underline{\omega}^L_{IE_H} \times \underline{\gamma}^L_{H_{Offset}}\right)\right]T^2 \\ &- S\theta_{ARS/Rnd_{ZL}} + S\theta_{ARS/\omega Vib\text{-}\alpha Quant_{ZL}}\Bigg\}\,\underline{\omega}^L_{IE_H} \times \underline{u}^L_{ZL}\end{aligned} \tag{18.2.1-13}$$

A unit vector along $\hat{\underline{\phi}}^L_H$ is:

$$\hat{\underline{u}}^L_{\phi H} = \frac{1}{\hat{\phi}_H}\,\hat{\underline{\phi}}^L_H \tag{18.2.1-14}$$

where

$\hat{\underline{u}}^L_{\phi H}$ = Unit vector along $\hat{\underline{\phi}}^L_H$.

ϕH = Subscript designation for vector component along $\hat{\underline{\phi}}^L_H$.

$\hat{\phi}_H$ = Magnitude of $\hat{\underline{\phi}}^L_H$.

The error in the $\hat{\underline{u}}^L_{\phi H}$ unit vector can be found as follows. First we write the equivalent error free form of (18.2.1-14) as:

$$\underline{u}^L_{\phi H} = \frac{1}{\phi_H}\,\underline{\phi}^L_H \tag{18.2.1-15}$$

Then, we express the equivalent error free version of Equation (18.2.1-13) by setting all sensor error effects to zero and deleting the $\hat{\ }$ notation which finds:

$$\frac{1}{T}\,\underline{\phi}^L_H = \underline{\omega}^L_{IE_H} \tag{18.2.1-16}$$

Thus, as might have been guessed, the nominal $\underline{\phi}^L_H$ lies along the true horizontal earth rate vector (which, of course, is horizontal true north), hence, for $\underline{\omega}^L_{IE_H}$ and the unit vector $\underline{u}^L_{\phi H}$ along $\underline{\phi}^L_H$:

$$\underline{\omega}^L_{IE_H} = \omega_{IE_H}\, \underline{u}^L_{North} \qquad \underline{u}^L_{\phi H} = \underline{u}^L_{North} \tag{18.2.1-17}$$

where

$\underline{u}^L_{North}$ = Unit vector in the direction of horizontal true North.

ω_{IE_H} = Magnitude of $\underline{\omega}^L_{IE_H}$ which is the component of $\underline{\omega}^L_{IE_H}$ along horizontal true North, and equals earth rate magnitude times the cosine of geodetic latitude.

The difference between $\hat{\underline{u}}^L_{\phi H}$ and $\underline{u}^L_{\phi H}$ is the error in $\hat{\underline{u}}^L_{\phi H}$ which we then express analytically as the differential of (18.2.1-14). With (18.2.1-15) this is:

$$\delta\underline{u}^L_{\phi H} = \frac{\phi_H\, \delta\underline{\phi}^L_H - \underline{\phi}^L_H\, \delta\phi_H}{\phi_H^2} = \frac{1}{\phi_H}\, \delta\underline{\phi}^L_H - \frac{\phi_H\, \delta\phi_H}{\phi_H^2}\, \underline{u}^L_{\phi H} \tag{18.2.1-18}$$

where

$\delta\underline{u}^L_{\phi H}$ = Error in $\hat{\underline{u}}^L_{\phi H}$.

$\delta\underline{\phi}^L_H$, $\delta\phi_H$ = The error in $\hat{\underline{\phi}}^L_H$ and its magnitude.

An expression for the $\phi_H\, \delta\phi_H$ term in (18.2.1-18) is determined from the differential of:

$$\hat{\phi}_H^2 = \hat{\underline{\phi}}^L_H \cdot \hat{\underline{\phi}}^L_H \tag{18.2.1-19}$$

which, with (18.2.1-15), gives:

$$\phi_H\, \delta\phi_H = \underline{\phi}^L_H \cdot \delta\underline{\phi}^L_H = \phi_H \left(\underline{u}^L_{\phi H} \cdot \delta\underline{\phi}^L_H\right) \tag{18.2.1-20}$$

Substituting (18.2.1-20) in (18.2.1-18) yields:

$$\delta\underline{u}^L_{\phi H} = \frac{1}{\phi_H}\left[\delta\underline{\phi}^L_H - \left(\underline{u}^L_{\phi H} \cdot \delta\underline{\phi}^L_H\right) \underline{u}^L_{\phi H}\right] \tag{18.2.1-21}$$

Hence, with (18.2.1-17), we see that $\hat{\underline{u}}^L_{\phi H}$ is:

$$\hat{\underline{u}}^L_{\phi H} = \underline{u}^L_{North} + \frac{1}{\phi_H}\left[\delta\underline{\phi}^L_H - \left(\underline{u}^L_{North} \cdot \delta\underline{\phi}^L_H\right) \underline{u}^L_{North}\right] \tag{18.2.1-22}$$

We will subsequently take the dot product of $\underline{\hat{u}}_{\phi H}^{L}$ with Equation (18.2.1-13). Using (18.2.1-22) and (18.2.1-17), we find for the dot product of $\underline{\hat{u}}_{\phi H}^{L}$ with particular terms in (18.2.1-13):

$$\underline{\hat{u}}_{\phi H}^{L} \cdot \underline{\hat{\phi}}_{H}^{L} = \hat{\phi}_{H}$$

$$\underline{\hat{u}}_{\phi H}^{L} \cdot \underline{\omega}_{IE_H}^{L} = \omega_{IE_H} \left(\underline{\hat{u}}_{\phi H}^{L} \cdot \underline{u}_{North}^{L} \right) \approx \omega_{IE_H}$$

$$\begin{aligned} \underline{\hat{u}}_{\phi H}^{L} \cdot \left(\underline{\omega}_{IE_H}^{L} \times \underline{u}_{ZL}^{L} \right) &= \omega_{IE_H} \, \underline{\hat{u}}_{\phi H}^{L} \cdot \left(\underline{u}_{North}^{L} \times \underline{u}_{ZL}^{L} \right) \\ &= - \omega_{IE_H} \left(\underline{\hat{u}}_{\phi H}^{L} \cdot \underline{u}_{East}^{L} \right) = - \frac{\omega_{IE_H}}{\phi_H} \left(\underline{u}_{East}^{L} \cdot \delta \underline{\phi}_{H}^{L} \right) \end{aligned} \qquad (18.2.1\text{-}23)$$

$$\begin{aligned} \underline{\hat{u}}_{\phi H}^{L} \cdot \left(\underline{u}_{ZL}^{L} \times \underline{\gamma}_{H_{Offset}}^{L} \right) &\approx \underline{u}_{\phi H}^{L} \cdot \left(\underline{u}_{ZL}^{L} \times \underline{\gamma}_{H_{Offset}}^{L} \right) \\ &= \underline{u}_{North}^{L} \cdot \left[\underline{u}_{ZL}^{L} \times \left(\gamma_{North_{Offset}} \, \underline{u}_{North}^{L} + \gamma_{East_{Offset}} \, \underline{u}_{East}^{L} \right) \right] \\ &= \underline{u}_{North}^{L} \cdot \left(\gamma_{North_{Offset}} \, \underline{u}_{East}^{L} - \gamma_{East_{Offset}} \, \underline{u}_{North}^{L} \right) = - \gamma_{East_{Offset}} \end{aligned}$$

where

$\underline{u}_{East}^{L}$ = Unit vector in the direction of horizontal East.

$\gamma_{North_{Offset}}, \gamma_{East_{Offset}}$ = North and east components of $\underline{\gamma}_{H_{Offset}}^{L}$.

We now take the dot product of (18.2.1-13) with $\underline{\hat{u}}_{\phi H}^{L}$, substitute (18.2.1-23) for the corresponding terms and, based on (18.2.1-22), identify the dot products with $\underline{\hat{u}}_{\phi H}^{L}$ for the remaining (18.2.1-23) vector elements as being north components. After dropping products of $\delta \underline{\phi}_{H}^{L}$ with error terms as second order, the result is:

$$\begin{aligned} \delta\omega_{ARS/Cnst_{North}} = \frac{1}{T} \hat{\phi}_H - \omega_{IE_H} - \gamma_{East_{Offset}} \, \omega_{IE_{Up}} - \frac{1}{T} \Big(& \theta_{ARS/Rnd_{North}} \\ & - \theta_{ARS/\omega Vib\text{-}\alpha Quant_{North}} + \delta\alpha_{Quant/North_{End}} - \delta\alpha_{Quant/North_{Start}} \Big) \end{aligned} \qquad (18.2.1\text{-}24)$$

where

North = Subscript identifying the North component of the associated vector.

Equation (18.2.1-24) provides the analytical basis for evaluating the constant portion of horizontal north angular rate sensor error ($\delta\omega_{ARS/Cnst_{North}}$) using the Strapdown Drift test. As an aside, if (18.2.1-24) is compared with the Section 18.1 Schuler Pump method (e.g., Equation (18.1.2-11) with (18.1.2-8)) for estimating horizontal angular rate sensor error, it may be noticed that no term comparable to $\gamma_{East_{Offset}} \, \omega_{IE_{Up}}$ is present in the Schuler Pump analyses. This is because the Section 18.1 results are based on Equations (13.3.2-25) which, in turn, are based on the Equation (13.3-8) approximation of neglecting horizontal tilt error products with earth rate compared to heading-error/earth-rate products. The $\gamma_{East_{Offset}} \, \omega_{IE_{Up}}$ term in (18.2.1-24) is the effect of the horizontal-tilt-error/earth-rate product term that was neglected in (13.3-8).

To apply (18.2.1-24), we first specify that T will be of sufficient duration that the noise contributions will be negligible compared to the $\delta\omega_{ARS/Cnst_{North}}$ term being evaluated. We also assume that $\gamma_{East_{Offset}}$ (produced from horizontal accelerometer constant error - See Equation (14.3-39) in Section 14.3) will be negligible due to an accurate previous calibration. Then (18.2.1-24) simplifies to:

$$\delta\omega_{ARS/Cnst_{North}} \approx \frac{1}{T} \hat{\phi}_H - \omega_{IE_H} \qquad (18.2.1\text{-}25)$$

The error in (18.2.1-25) is from the $\gamma_{East_{Offset}}$ and noise terms in (18.2.1-24) which were dropped as negligible. For a previously calibrated INS sensor assembly (e.g., individual accelerometer bias calibration, and use of the Section 18.4 Strapdown Rotation test for accelerometer misalignment calibration), $\gamma_{East_{Offset}}$ produced by accelerometer error should be on the order of 50 μrad (typical for a moderate accuracy INS; i.e., having 1 nmph CEP (i.e., "Circular Error Probable" which is a 50% probability radial horizontal position error growth rate). For $\gamma_{East_{Offset}} = 50$ μrad at 45 degrees latitude (for which $\omega_{IE_{Up}}$ is approximately 15 deg/hr sin 45 deg $\approx$ 11 deg/hr), the $\gamma_{East_{Offset}} \, \omega_{IE_{Up}}$ term in (18.2.1-24) is 0.0005 deg/hr. A typical value of $\delta\omega_{ARS/Cnst_{North}}$ for a moderate accuracy INS is 0.007 deg/hr, for which the 0.0005 deg/hr uncertainty is negligible.

For the $\theta_{ARS/Rnd_{North}}$ angular rate sensor random noise term in (18.2.1-24) (that we have neglected in (18.2.1-25)), we know from the integral of covariance rate Equation (15.1.2.1.1-30), that the variance of the integral of white noise (i.e., for $A(t) = 0$ and $G_P(t) = I$) equals the white noise density multiplied by the integration time. Therefore, for $\theta_{ARS/Rnd_{North}}$ in (18.2.1-24) as defined by (18.2.1-13), the square root of its mean squared value (i.e., the RMS or "root-mean-square" value) is $\sigma_{ARS/Rnd} \sqrt{T}$ where:

$\sigma_{ARS/Rnd}$ = Square root of the angular rate sensor random output noise density.

Then the RMS value for the $\frac{1}{T}\, \theta_{ARS/Rnd_{North}}$ term in (18.2.1-24) is $\sigma_{ARS/Rnd} / \sqrt{T}$. For a typical 1.0 nmph CEP INS, $\sigma_{ARS/Rnd}$ is on the order of 0.002 deg/sqr-rt-hr. For a 1 hour test time T, the error in $\delta\omega_{ARS/Cnst_{North}}$ due to $\theta_{ARS/Rnd_{North}}$, then, is 0.002 deg/sq-rt-hr / $\sqrt{1 \text{ hr}}$ or 0.002 deg/hr. For T = 4 hrs, the error is half this value or 0.001 deg/hr.

The reader can verify using the analytical techniques of Section 16.2.3.1 and Equation (15.1.2.1.1-30) (with $A(t) = 0$ and $G_P(t) = I$ for integrated random noise variance analysis) that the $\theta_{ARS/\omega Vib-\alpha Quant_{North}}$ term in (18.2.1-24) is negligible compared to the $\theta_{ARS/Rnd_{North}}$ effect, particularly under the benign angular rate vibration environment of system testing in the laboratory. For a 1 nmph CEP INS, a typical root-mean-square value for the $\delta\alpha_{Quant/North}$ angular rate sensor quantization noise term in (18.2.1-24) is $1 / \sqrt{12}$ arc sec (i.e., corresponding to a 1 arc sec output pulse size and an associated quantization error that is statistically uniformly distributed over the range of - 0.5 to + 0.5 arc sec - See the discussion in the second and third paragraphs following Equation (15.2.1.2-17)). In one hour this translates into $\left(1 / \sqrt{12}\right)$ arc-sec / 3600 arc-sec per deg / 1 hr = 0.00008 deg/hr for the contribution to the $\delta\omega_{ARS/Cnst_{North}}$ estimation error, clearly negligible compared to the typical $\delta\omega_{ARS/Cnst_{North}}$ accuracy requirement of 0.007 deg/hr.

Equation (18.2.1-25) defines the method for estimating the constant error component in the horizontal north angular rate sensor for the Strapdown Drift test. To estimate the constant error on each angular rate sensor output in a three-axis orthogonal strapdown sensor assembly, the test is performed three times, each with a different sensor oriented north. We should also note that $\delta\omega_{ARS/Cnst_{North}}$ is composed mainly of angular rate bias error components. This is easily verified from Equation (12.4-12) which shows that the remaining constant errors(in $\delta\omega_{ARS/Cnst_{North}}$) are products of angular rate sensor scale factor and misalignment errors with earth rate. For angular rate sensor scale factor and misalignment error residuals (following calibration) on the order of 5 ppm and 50 μrad respectively (typical for an aircraft INS), an analysis similar to the one following Equation (18.2.1-5) would show that their products with earth rate are negligible compared with typical 0.01 deg/hr bias accuracy requirements.

18.2.2 STRAPDOWN DRIFT TEST BASED ON INS HEADING MEASUREMENTS

The Strapdown Drift test as described in Section 18.2.1 implies the existence of special software to execute the associated analytical operations (i.e., extended alignment with integrated $\hat{\underline{\omega}}_{IL_H}^L$ measurements). For situations when such software is not available (internal or external to the INS) and $\hat{\underline{\omega}}_{IL_H}^L$ measurements are not accessible, an alternate procedure can be utilized for a strapdown INS based on true heading output measurements taken at completion of an extended self alignment using Kalman time varying gains during the overall alignment process. This approach differs from the Section 18.2.1 Strapdown Drift Test (based on INS estimated earth rate measurement) which used fixed gains during the extended alignment period. The Strapdown Drift Test based on INS heading measurements is formulated from a single alignment version of Repeated Alignment Test Equation (18.3.1.2-9) (repeated below) using the L Frame (Z axis down) for vertical component (ZL) definition (as opposed to ZN in (18.3.1.2-9) based on Z up in the N Frame):

$$
\begin{aligned}
\theta_{ARS/Rnd_{East}} = & - \omega_{IE_H} \left(\hat{\psi}_T - \delta\alpha_{Quant_{ZL}} \right) T_{Align} \\
& + \left[\omega_{IE_H} \psi_T - \delta\omega_{ARS/Cnst_{East}} + \omega_{IE_{Up}} \gamma_{North\ Offset} \right] T_{Align} \\
& + \frac{1}{2} T_{Align}^2 \left[\delta\omega_{ARS/Cnst_{ZL}} + \omega_{IE_H} \gamma_{East_{Offset}} \right] \omega_{IE_H} \\
& + \delta\omega_{IE/East_{Resid}} T_{Align} + \theta_{ARS/\omega Vib\text{-}\alpha Quant_{East}} \\
& - \left(S\theta_{ARS/Rnd_{ZL}} - \theta_{ARS/Rnd_{ZL}} T_{Align} \right) \omega_{IE_H} \\
& + \left(S\theta_{ARS/\omega Vib\text{-}\alpha Quant_{ZL}} - \theta_{ARS/\omega Vib\text{-}\alpha Quant_{ZL}} T_{Align} \right) \omega_{IE_H}
\end{aligned}
\qquad (18.2.2\text{-}1)
$$

where

T_{Align} = Time from start to end of the overall Fine Alignment process.

ψ_T, $\hat{\psi}_T$, = Actual INS true heading and the INS computed true heading output at the end of alignment.

$\delta\omega_{ARS/Cnst_{East}}$, $\delta\omega_{ARS/Cnst_{ZL}}$ = East and downward vertical components of the constant portion of angular rate sensor error.

$\theta_{ARS/Rnd_{East}}$ = Integral over T_{Align} of the east component of angular rate sensor random noise.

$\delta\alpha_{Quant_{ZL}}$ = Downward vertical component of angular rate sensor output quantization error.

$\theta_{ARS/\omega Vib-\alpha Quant_{East}}$ = Integral over T_{Align} of the east component of angular-vibration/angular-rate-sensor quantization-noise product as defined in Equations (18.3-26).

$\theta_{ARS/Rnd_{ZL}}$, $S\theta_{ARS/Rnd_{ZL}}$ = Single and double integral over T_{Align} of the downward vertical component of angular rate sensor random noise.

$\theta_{ARS/\omega Vib-\alpha Quant_{ZL}}$, $S\theta_{ARS/\omega Vib-\alpha Quant_{ZL}}$ = Single and double integral over T_{Align} of the downward vertical component of angular-vibration/angular-rate-sensor quantization-noise product as defined in Equations (18.3-24) and (18.3-26).

$\delta\omega_{IE/East_{Resid}}$ = East earth rate estimation residual error term that is a function of the time in alignment, the initial values for parameters being estimated during Fine Alignment, noise effects during alignment, and the magnitude of the vibration levels present during the alignment process. Generally speaking, the magnitude of $\delta\omega_{IE/East_{Resid}}$ near the end of Fine Alignment become negligible compared to the other terms in (18.2.2-1)).

$\gamma_{North_{Offset}}$, $\gamma_{East_{Offset}}$ = North and east components of $\underline{\gamma}^{L}_{H_{Offset}}$.

Note in (18.2.2-1) that, in comparison with (18.3.1.2-9), we have added the $\delta\alpha_{Quant_{ZL}}$ quantization error to the $\hat{\psi}_T$ readout. The rationale is the same as in the Section 18.2.1 discussion leading to Equation (18.2.1-2), but applied to the development of (18.3.1.2-9).

Solving (18.2.2-1) for $\delta\omega_{ARS/Cnst_{East}}$ and rearranging with (14.3-39) substituted for the $\underline{\gamma}^{L}_{H_{Offset}}$ components yields:

$$\begin{aligned}
\delta\omega_{ARS/Cnst_{East}} = & - \omega_{IE_H}\left(\hat{\psi}_T - \delta\alpha_{Quant_{ZL}} - \psi_T\right) + \omega_{IE_{Up}} \frac{1}{g}\, \delta a_{SF_{East}} \\
& + \frac{1}{2}\, \omega_{IE_H}\, T_{Align} \left[\delta\omega_{ARS/Cnst_{ZL}} - \omega_{IE_H} \frac{1}{g}\, \delta a_{SF_{North}}\right] \\
& - \frac{1}{T_{Align}}\left(\theta_{ARS/Rnd_{East}} - \theta_{ARS/\omega Vib-\alpha Quant_{East}}\right) \\
& - \omega_{IE_H}\left(\frac{1}{T_{Align}}\, S\theta_{ARS/Rnd_{ZL}} - \theta_{ARS/Rnd_{ZL}}\right) \\
& + \omega_{IE_H}\left(\frac{1}{T_{Align}}\, S\theta_{ARS/\omega Vib-\alpha Quant_{ZL}} - \theta_{ARS/\omega Vib-\alpha Quant_{ZL}}\right) + \delta\omega_{IE/East_{Resid}}
\end{aligned} \tag{18.2.2-2}$$

Equation (18.2.2-2) is of the form:

$$\delta\omega_{ARS/Cnst_{East}} = -\,\omega_{IE_H}\left(\hat{\psi}_T - \psi_T\right) + \omega_{IE_{Up}}\frac{1}{g}\,\delta a_{SF_{East}} + \frac{1}{2}\,\omega_{IE_H}\,T_{Align}\left[\delta\omega_{ARS/Cnst_{ZL}} - \omega_{IE_H}\frac{1}{g}\,\delta a_{SF_{North}}\right] + \text{Random error effects} \qquad (18.2.2\text{-}3)$$

Consider that we perform the extended alignment at two different headings. Equation (18.2.2-3) becomes for each alignment:

$$\begin{aligned} \delta\omega_{ARS/Cnst/East_1} &= -\,\omega_{IE_H}\left(\hat{\psi}_{T_1} - \psi_{T_1}\right) + \omega_{IE_{Up}}\frac{1}{g}\,\delta a_{SF/East_1} \\ &\quad + \frac{1}{2}\,\omega_{IE_H}\,T_{Align}\left[\delta\omega_{ARS/Cnst/ZL_1} - \omega_{IE_H}\frac{1}{g}\,\delta a_{SF/North_1}\right] \\ &\quad + \text{Random error effects 1} \\ \delta\omega_{ARS/Cnst/East_2} &= -\,\omega_{IE_H}\left(\hat{\psi}_{T_2} - \psi_{T_2}\right) + \omega_{IE_{Up}}\frac{1}{g}\,\delta a_{SF/East_2} \\ &\quad + \frac{1}{2}\,\omega_{IE_H}\,T_{Align}\left[\delta\omega_{ARS/Cnst/ZL_2} - \omega_{IE_H}\frac{1}{g}\,\delta a_{SF/North_2}\right] \\ &\quad + \text{Random error effects 2} \end{aligned} \qquad (18.2.2\text{-}4)$$

where

1, 2 = Designation for parameter values for the first and second alignment runs.

We further stipulate that the headings for the two alignments will be approximately 180 degrees apart. Then, since sensor error directions rotate with the sensor assembly axes, we can stipulate that the horizontal (north and east) sensor error terms in (18.2.2-4) will be opposite in sign, while the vertical (ZL) sensor components will be of the same sign:

$$\begin{aligned} &\delta\omega_{ARS/Cnst/East_2} = -\,\delta\omega_{ARS/Cnst/East_1} \\ &\delta a_{SF/East_2} = -\,\delta a_{SF/East_1} \qquad \delta a_{SF/North_2} = -\,\delta a_{SF/North_1} \\ &\delta\omega_{ARS/Cnst/ZL_2} = \delta\omega_{ARS/Cnst/ZL_1} \end{aligned} \qquad (18.2.2\text{-}5)$$

Subtracting the $\delta\omega_{ARS/Cnst/East_2}$ expression in (18.2.2-4) from the $\delta\omega_{ARS/Cnst/East_1}$ expression, substituting (18.2.2-5), and dividing by 2 then obtains:

$$\delta\omega_{ARS/Cnst/East_1} = -\frac{\omega_{IE_H}}{2}\left[\hat{\psi}_{T_1} - \hat{\psi}_{T_2} - \left(\psi_{T_1} - \psi_{T_2}\right)\right]$$
$$+ \omega_{IE_{Up}}\frac{1}{g}\,\delta a_{SF/East_1} - \frac{1}{2}\,\omega_{IE_H}^2\,T_{Align}\,\frac{1}{g}\,\delta a_{SF/North_1} \tag{18.2.2-6}$$
$$+ \frac{1}{2}\,(\text{Random error effects 1 - Random error effects 2})$$

Equation (18.2.2-6) can be approximated as:

$$\delta\omega_{ARS/Cnst/East_1} \approx -\frac{\omega_{IE_H}}{2}\left[\hat{\psi}_{T_1} - \hat{\psi}_{T_2} - \Delta\hat{\psi}_{T_{1\text{-}2}}\right] \tag{18.2.2-7}$$

where

$\Delta\hat{\psi}_{T_{1\text{-}2}}$ = Difference between ψ_{T_1} and ψ_{T_2} taken by an independent measurement. Nominally $\Delta\hat{\psi}_{T_{1\text{-}2}}$ is 180 degrees.

The $\Delta\hat{\psi}_{T_{1\text{-}2}}$ measurement in (18.2.2-7) can be taken using rotation fixture angle output data, or can be obtained using INS heading outputs and the following procedure. At completion of the first alignment run, the INS is transitioned into the navigation mode and rotated to the heading for the second alignment. While in the navigation mode at the new heading, the INS output heading is recorded (call it $\hat{\psi}_{T_2}^*$). $\Delta\hat{\psi}_{T_{1\text{-}2}}$ is then calculated by subtracting $\hat{\psi}_{T_2}^*$ from $\hat{\psi}_{T_1}$.

Equation (18.2.2-7) is the basis for estimating east angular rate sensor error in the Strapdown Drift Test based on INS heading output measurements. From (18.2.2-7) and the previous paragraph for $\Delta\hat{\psi}_{T_{1\text{-}2}}$ measurement, the following defines the test procedure that would then be used for estimating $\delta\omega_{ARS/Cnst/East_1}$.

The INS is initially positioned at an orientation where the angular rate sensor to be evaluated has its input axis East. An extended INS initial alignment is then executed using the normal INS self-alignment mode, and the INS computed true heading is recorded at the end of alignment ($\hat{\psi}_{T_1}$). Following the $\hat{\psi}_{T_1}$ heading measurement, INS operation is transitioned to the free-inertial navigation mode and immediately rotated approximately 180 degrees about the vertical. The true heading of the INS is immediately recorded at the new heading. The $\Delta\hat{\psi}_{T_{1\text{-}2}}$ heading change is calculated by subtracting $\hat{\psi}_{T_1}$ from the true heading reading taken after the 180 degree rotation.

An extended INS initial alignment is then executed at the new heading using the normal INS self-alignment mode, and the INS computed true heading is recorded at the end of alignment ($\hat{\psi}_{T_2}$). The east angular rate sensor constant error component is then calculated from $\hat{\psi}_{T_1}$, $\Delta\hat{\psi}_{T_{1\text{-}2}}$ and $\hat{\psi}_{T_2}$ using Equation (18.2.2-7).

The procedure is repeated for each angular rate sensor to be evaluated.

The accuracy of the above procedure for estimating east angular rate sensor constant error is limited by the error ("uncertainty") in the (18.2.2-7) approximation. The difference between (18.2.2-7) and (18.2.2-6) with (18.2.2-2) for the random error terms can be used to assess the uncertainty in (18.2.2-7):

$$
\begin{aligned}
\delta\delta\omega_{ARS/Cnst/East_1} = {} & \frac{\omega_{IE_H}}{2}\left(\delta\Delta\psi_{T_{1\text{-}2}} - \Delta\psi_{T_2}\right) \\
& - \omega_{IE_{Up}} \frac{1}{g}\,\delta a_{SF/East_1} + \frac{1}{2}\,\omega_{IE_H}^2\, T_{Align}\,\frac{1}{g}\,\delta a_{SF/North_1} \\
& + \frac{1}{2\,T_{Align}}\,\Delta_{1\text{-}2}\left(\theta_{ARS/Rnd_{East}} - \theta_{ARS/\omega Vib\text{-}\alpha Quant_{East}}\right) - \frac{\omega_{IE_H}}{2}\,\Delta_{1\text{-}2}\left(\delta\alpha_{Quant_{ZL}}\right) \\
& + \frac{\omega_{IE_H}}{2}\,\Delta_{1\text{-}2}\left(\frac{1}{T_{Align}}\,S\theta_{ARS/Rnd_{ZL}} - \theta_{ARS/Rnd_{ZL}}\right) \\
& - \frac{\omega_{IE_H}}{2}\,\Delta_{1\text{-}2}\left(\frac{1}{T_{Align}}\,S\theta_{ARS/\omega Vib\text{-}\alpha Quant_{ZL}} - \theta_{ARS/\omega Vib\text{-}\alpha Quant_{ZL}}\right) \\
& - \frac{1}{2}\,\Delta_{1\text{-}2}\left(\delta\omega_{IE/East_{Resid}}\right)
\end{aligned}
\qquad (18.2.2\text{-}8)
$$

where

$\delta\delta\omega_{ARS/Cnst/East_1}$ = Uncertainty in the (18.2.2-7) estimate of $\delta\omega_{ARS/Cnst/East_1}$.

$\delta\Delta\psi_{T_{1\text{-}2}}$ = Uncertainty in the $\Delta\hat{\psi}_{T_{1\text{-}2}}$ measurement.

$\Delta\psi_{T_2}$ = Shift in the actual true heading between the time the $\Delta\hat{\psi}_{T_{1\text{-}2}}$ measurement was made and the second alignment initiated. This additional term is included to account for potential high frequency angular oscillations of the sensor assembly during alignment (e.g., such as produced by mechanically dithered ring laser gyro back reaction torque).

$\Delta_{1\text{-}2}(\,)$ = Functional operator representing the difference between values of the argument ( ) for the two alignment runs.

The dominant contributor to $\delta\delta\omega_{ARS/Cnst/East_1}$ in (18.2.2-8) is east angular rate sensor random noise ($\theta_{ARS/Rnd_{East}}$) which was analyzed at the conclusion of Section 18.2.1 (for the north component) regarding its impact on angular rate sensor error determination. A similar analysis applies here showing that the square root of the $\Delta_{1\text{-}2}\left(\theta_{ARS/Rnd_{East}}\right)$ mean-squared error (i.e., the RMS or "root-mean-square" value) is $\sigma_{ARS/Rnd}\sqrt{2\,T_{Align}}$ where:

$\sigma_{ARS/Rnd}$ = Square root of the angular rate sensor random output noise density.

The $\sqrt{2}$ arises in the previous expression from the two values for $\theta_{ARS/Rnd_{East}}$ in the (18.2.2-8) $\Delta_{1\text{-}2}\left(\theta_{ARS/Rnd_{East}}\right)$ term (one for each alignment) being uncorrelated. Thus, the RMS value for $\frac{1}{2\,T_{Align}}\Delta_{1\text{-}2}\left(\theta_{ARS/Rnd_{East}}\right)$ in (18.2.2-8) is $\sigma_{ARS/Rnd} / \sqrt{2\,T_{Align}}$. For a typical 1.0 nmph CEP INS, $\sigma_{ARS/Rnd}$ is on the order of 0.002 deg/sqr-rt-hr. For a 1 hour alignment time T_{Align}, the error in $\delta\omega_{ARS/Cnst/East_1}$ due to $\theta_{ARS/Rnd_{East}}$, then, is 0.002 deg/sq-rt-hr / $\sqrt{2 \times 1\text{ hr}}$ or 0.0014 deg/hr. For a 4 hour alignment time, the error is half this value or 0.0007 deg/hr. A typical value of $\delta\omega_{ARS/Cnst/East_1}$ for a moderate accuracy INS is 0.007 deg/hr, for which the 0.0007 deg/hr uncertainty is negligible.

To assess the contribution of the $\left(\frac{1}{T_{Align}} S\theta_{ARS/Rnd_{ZL}} - \theta_{ARS/Rnd_{ZL}}\right)$ term in (18.2.2-8), one must account for the fact that the components $S\theta_{ARS/Rnd_{ZL}}$ and $\theta_{ARS/Rnd_{ZL}}$ are created from the same noise source (i.e., vertical angular rate sensor random output noise), hence, are correlated. To find their combined contribution to (18.2.2-8) we can apply the same technique utilized in Section 13.6.1 leading to Equation (13.6.1-4) in which each element of noise is first analyzed independently. For this case, the effect on $\left(\frac{1}{T_{Align}} S\theta_{ARS/Rnd_{ZL}} - \theta_{ARS/Rnd_{ZL}}\right)$ at the end of alignment ($t = T_{Align}$) produced by a vertical angular rate sensor output noise element $\varepsilon_{\omega Rnd/ZL_i}$ occurring over a time interval $\Delta\tau_i$ at time $t = \tau$ is $\left(\frac{T_{Align} - \tau}{T_{Align}} - 1\right)\varepsilon_{\omega Rnd/ZL_i}$ or $-\frac{\tau}{T_{Align}}\varepsilon_{\omega Rnd/ZL_i}$. The $\left(\frac{1}{T_{Align}} S\theta_{ARS/Rnd_{ZL}} - \theta_{ARS/Rnd_{ZL}}\right)$ term in (18.2.2-8) is the sum of all the $\left(-\frac{\tau}{T_{Align}}\varepsilon_{\omega Rnd/ZL_i}\right)$'s from $\tau = 0$ to $\tau = T_{Align}$. Squaring the sum, taking the expected value, setting the expected value of the $\varepsilon_{\omega Rnd/ZL_i}\,\varepsilon_{\omega Rnd/ZL_j}$ products to zero for $i \neq j$ (i.e., the noise element for interval $\Delta\tau_i$ is uncorrelated with the noise element for interval $\Delta\tau_j$), dividing by $\Delta\tau_i$, letting $\Delta\tau_i$ go to zero in the limit so that the summation becomes an integral, identifying the expected value of $\varepsilon^2_{\omega Rnd/ZL_i} / \Delta\tau_i$ in the limit as the angular rate sensor noise density $\sigma^2_{ARS/Rnd}$ (as in (13.6.1-20)), carrying out the integral analytically, and taking the square root, then finds that the $\left(\frac{1}{T_{Align}} S\theta_{ARS/Rnd_{ZL}} - \theta_{ARS/Rnd_{ZL}}\right)$ root-mean-square (RMS) value is $\sigma_{ARS/Rnd} \sqrt{T_{Align} / 3}$. Thus, allowing that this effect contributes twice to $\delta\delta\omega_{ARS/Cnst/East_1}$ in (18.2.2-8) (once for each alignment) and that each contribution is uncorrelated (see $\sqrt{2}$ discussion in previous paragraph), the net RMS contribution of $\frac{\omega_{IE_H}}{2} \Delta_{1\text{-}2}\left(\frac{1}{T_{Align}} S\theta_{ARS/Rnd_{ZL}} - \theta_{ARS/Rnd_{ZL}}\right)$ in (18.2.2-8) is $\omega_{IE_H} \sigma_{ARS/Rnd} \sqrt{T_{Align} / 6}$. At 45 degrees alignment latitude using a 4 hour alignment time and 0.002 deg/sqr-rt-hr for $\sigma_{ARS/Rnd}$, the resulting RMS numerical contribution to $\delta\omega_{ARS/Cnst/East_1}$ in (18.2.2-8) is 15 deg/hr earth rate $\times$ 0.0175 rad/deg $\times$ cosine 45 deg $\times$ 0.002 deg/sqr-rt-hr $\times$ $\sqrt{4 \text{ hr} / 6}$ = 0.0003 deg/hr. This is also clearly negligible compared to the 0.007 deg/hr angular rate sensor error allowance in a 1 nmph CEP accuracy INS.

The reader can verify using the analytical techniques of Section 16.2.3.1 and Equation (15.1.2.1.1-30) (with $A(t) = 0$ and $G_P(t) = I$ for integrated random noise variance analysis) that the effects of $S\theta_{ARS/\omega Vib\text{-}\alpha Quant_{ZL}}$ and $\theta_{ARS/\omega Vib\text{-}\alpha Quant_{ZL}}$ effects on (18.2.2-8) $\delta\omega_{ARS/Cnst/East_1}$ uncertainty are negligible compared to the effect of $\theta_{ARS/Rnd_{East}}$, particularly under the benign angular rate vibration environment of system testing in the laboratory. We assume that the alignment time is sufficiently long that the effect of $\delta\underline{\omega}^N_{IE/H/Resid_j}$ can also be neglected.

The $\frac{1}{g} \delta a_{SF/East_1} \omega_{IE_{Up}}$ term in (18.2.2-8) is equivalent to the $\gamma_{East_{Offset}} \omega_{IE_{Up}}$ term in the Section 18.2.1 north angular rate sensor error Equation (18.2.1-24). The order of magnitude analysis for this term provided at the conclusion of Section 18.2.1 also applies for the (18.2.2-8) term showing a value of 0.0005 deg/hr at 45 degrees latitude (small compared to the 0.007 deg/hr angular rate sensor accuracy in a 1 nmph INS). For a 4 hour alignment time at 45 deg latitude, the effect of $\delta a_{SF/North_1}$ on $\delta\delta\omega_{ARS/Cnst/East_1}$ in (18.2.2-8), is 37% of the $\delta a_{SF/East_1}$ effect.

Assuming that INS navigation mode heading data is used to calculate $\widehat{\Delta\psi_{T_{1\text{-}2}}}$ (as described in the paragraph preceding (18.2.2-8)), the $\delta\Delta\psi_{T_{1\text{-}2}}$ error in (18.2.2-8) will equal 180 degrees times the vertical angular rate sensor scale factor error plus some smaller integrated vertical angular rate sensor bias and random noise effects. For an aircraft quality INS, the $\widehat{\Delta\psi_{T_{1\text{-}2}}}$ measurement error should be small (on the order of 5 ppm scale factor error times π or 15 micro-radians). From Equation (18.2.2-8) we see that an uncertainty is thereby created in $\delta\omega_{ARS/Cnst/East_1}$ equal to $\delta\Delta\psi_{T_{1\text{-}2}}$ multiplied by one half horizontal earth rate (earth rate magnitude times the cosine of geodetic latitude). Thus, at 45 degrees latitude, the $\delta\omega_{ARS/Cnst/East_1}$ uncertainty introduced is 0.5×15 deg/hr $\times$ (cosine 45 deg) $\times 15\ \mu$rad $\times 10^{-6}$ or 0.000080 deg/hr. The error is clearly negligible compared to 0.0007 deg/hr. A similar conclusion applies to the $\Delta\psi_{T_2}$ and $\delta\alpha_{Quant_{ZL}}$ terms in (18.2.2-8). If the $\Delta\psi_{T_2}$ is on the order of 10 arc sec (or 48 μrad), a typical value for a mechanically dithered ring laser gyro sensor assembly, the associated uncertainty introduced in $\delta\omega_{ARS/Cnst/East_1}$ would be 0.00026 deg/hr. For an angular rate sensor pulse size of 1 arc sec (5 μrad), the RMS value for each $\delta\alpha_{Quant_{ZL}}$ component in (18.2.2-8) is $5 / \sqrt{12} = 1.4\ \mu$rad - Forget about it.

The above procedure for estimating east angular rate sensor error was based on Equation (18.3.1.2-9). If Section 18.3.1.2 is reviewed it will be found that (18.3.1.2-9) assumes that Kalman time varying gains will be used throughout the alignment process, and implicitly, that the Kalman filter error state model accurately represents the error characteristics of the INS Fine Alignment process. In particular, if these conditions are not satisfied, the test procedure outlined above may produce additional uncertainties in the angular rate sensor error determination. The basic question that must be addressed in this regard is the behavior of the Fine Alignment process for long alignment times when the Kalman gains become increasingly small. Selection of the maximum allowable alignment time for the test should address this question. If there is doubt regarding selection of a long alignment (to reduce random sensor error effects), the procedure can be altered by executing several shorter alignments at each of the two alignment attitudes, with the results then linearly averaged to find the equivalent for the single long alignment scenario. The penalty for this approach is the potential introduction of additional $\Delta\psi_{T_2}$ type errors for each additional alignment. From the $\Delta\psi_{T_2}$ error analysis provided previously, it would appear that the penalty will not be great if the added alignments are kept to a reasonable minimum. The final alignment time selection for the test should be supported by special validation tests performed with a real INS of the same type to be tested. One of the advantages of the Section 18.2.1 Strapdown Drift Test approach (based on INS estimated earth rate measurements) is that it uses fixed gains during the extended alignment measurement period. As a result, it can be run for lengthy extended alignment times (for increased sensor error estimation accuracy) without incurring an added inaccuracy penalty due to Kalman filter modeling approximations.

18.2.3 CALIBRATION COEFFICIENT UPDATING FROM MEASURED ANGULAR RATE SENSOR BIAS ERRORS

The composite angular rate sensor errors determined from the Strapdown Drift test represent errors in the INS operating with "Sensor Level", and potentially, previous "System Level" calibration coefficients (See Sections 8.1.1.1.1 and 8.1.1.2.1 for definitions). If the INS has been properly calibrated prior to the Drift test (i.e., for accelerometer error and angular rate sensor scale factor error, misalignment and bias error), the composite angular rate sensor errors measured during the Drift test will be dominated by the angular rate sensor bias error. For further elaboration on this point, see the discussion following Equation (18.2.1-25). Thus, the Strapdown Drift test results represent errors in the current INS angular rate sensor bias calibration. In order to update the INS angular rate sensor bias coefficients, we update the $\underline{\kappa}_{SystBias}$ System Level bias coefficients in calibration Equation (8.1.1.1.1-12) (See Table 8.4-1) to equal the values used by the INS during the Strapdown Drift test, plus the corrections found with Equation (18.2.1-25) (or (18.2.2-7)) for each angular rate sensor. The $\delta\underline{\omega}_{SensBias}$ Sensor Level coefficients in (8.1.1.1.1-12) would be kept at the values used during the test. Then the overall $\underline{K}_{Bias}$ compensation coefficients calculated with (8.1.1.1.1-12) would represent the updated angular rate sensor bias coefficient vector.

18.3 SYSTEM LEVEL ANGULAR RATE SENSOR RANDOM NOISE ESTIMATION

In this section we describe two INS System Level test methods that can be utilized to evaluate the random output noise from the strapdown angular rate sensors; the Repeated Alignment test and the Continuous Alignment test. Each method is based on the response of a Kalman filter based INS Fine Alignment process as represented (for example) by Equations (6.1.2-2) using Kalman filter derived gains. In this section we will derive the basic relationship used for random noise estimation; the result will then be applied in the subsections that follow.

To begin, let us write the Chapter 6 quasi-stationary Fine Alignment continuous form process Equations (6.1.2-2) that would be implemented in the INS computer using a wander azimuth N Frame (See Sections 2.2 and 4.5 for definition):

$$\dot{\hat{C}}_B^N = \hat{C}_B^N \left(\tilde{\underline{\omega}}_{IB}^B \times\right) - \left(\hat{\underline{\omega}}_{IN}^N \times\right) \hat{C}_B^N$$

$$\hat{\underline{\omega}}_{IN}^N = \hat{\underline{\omega}}_{IE}^N + \hat{\underline{\omega}}_{Tilt}^N$$

$$\hat{\underline{\omega}}_{Tilt}^N = K_2 \, \underline{u}_{ZN}^N \times \Delta\hat{\underline{R}}_H^N$$

$$\hat{\underline{\omega}}_{IE}^N = \hat{\underline{\omega}}_{IE_H}^N + \underline{u}_{ZN}^N \, \omega_e \sin l \qquad (18.3\text{-}1)$$

$$\dot{\hat{\underline{\omega}}}_{IE_H}^N = K_1 \, \underline{u}_{ZN}^N \times \Delta\hat{\underline{R}}_H^N$$

$$\dot{\hat{\underline{v}}}_H^N = \left(\hat{C}_B^N\right)_H \tilde{\underline{a}}_{SF}^B - K_3 \, \Delta\hat{\underline{R}}_H^N$$

$$\Delta\dot{\hat{\underline{R}}}_H^N = \hat{\underline{v}}_H^N - K_4 \, \Delta\hat{\underline{R}}_H^N$$

where

$\tilde{\ }$ = Designation for angular rate sensor output containing angular rate sensor error.

$\hat{\ }$ = Designation for parameter computed (or estimated) within the INS Fine Alignment process, and containing errors. Terms without this designation are error free.

$\underline{u}_{ZN}^N$ = Unit vector along the navigation N Frame vertical axis (Z), projected on N Frame axes.

K_1, K_2, K_3, K_4 = Fine Alignment process estimation feedback control gains.

$\underline{\omega}_{Tilt}^N$ = Angular rate feedback to correct C_B^N horizontal angular error.

ω_e = Earth rate magnitude.

l = Geodetic latitude (assumed to be available as an error free input).

C_B^N = Direction cosine matrix that transforms vectors from the B Frame (sensor axes) to the wander azimuth N Frame.

$\underline{\omega}_{IB}^B$ = Angular rate of the B Frame relative to I Frame inertial space (measured by the system angular rate sensors).

$\underline{\omega}_{IE}^N$ = Angular rate of the earth fixed E Frame relative to the I Frame.

$\underline{v}_H^N$ = Horizontal velocity relative to the earth in the N Frame.

$\underline{a}_{SF}^{B}$ = Specific force acceleration in the B Frame (measured by the system accelerometers).

$\Delta\underline{R}_{H}^{N}$ = Integrated horizontal velocity in the N Frame ("position" divergence).

Equations (18.3-1) represent the Fine Alignment process using a Kalman filter with error state controls (the K_i terms) to continuously null the estimated error in the (18.3-1) integration parameters. Chapter 15, Section 15.1.5.3.2 shows that an equivalent Kalman filter configuration can be constructed based on pure estimation in which the basic system equations are operated without controls. The equivalent to (18.3-1) for the system equations in this case would be:

Pure Estimator System Computer Equations

$$\dot{\hat{C}}_{B}^{N\star} = \hat{C}_{B}^{N\star}\left(\tilde{\underline{\omega}}_{IB}^{B}\times\right) - \left(\underline{\omega}_{IE_{Exp}}^{N}\times\right)\hat{C}_{B}^{N\star}$$

$$\underline{\omega}_{IE_{Exp}}^{N} = \underline{u}_{ZN}^{N}\,\omega_e \sin l$$

$$\dot{\underline{\omega}}_{IE_{Exp}}^{N} = 0 \qquad (18.3\text{-}2)$$

$$\dot{\hat{\underline{v}}}_{H}^{N\star} = \left(\hat{C}_{B}^{N\star}\right)_{H}\tilde{\underline{a}}_{SF}^{B}$$

$$\Delta\dot{\hat{\underline{R}}}_{H}^{N\star} = \hat{\underline{v}}_{H}^{N\star}$$

where

$\star$ = Designation for parameters calculated in the equivalent uncontrolled alignment process operations.

$\underline{\omega}_{IE_{Exp}}^{N}$ = Expected (or assumed average) value for the earth rate, and treated as a constant. Since the horizontal earth rate components are unknown (i.e., to be determined as a Fine Alignment process output), and since the N Frame can be at any arbitrary heading, we use zero for the horizontal components of $\underline{\omega}_{IE_{Exp}}^{N}$.

If the system computer parameters were error free and the $\underline{\omega}_{IE_{Exp}}^{N}$ value was equal to the actual N Frame earth rate value, the result would be what was actually happening during the quasi-stationary alignment process (the so-called "truth model"):

Pure Estimator Truth Model Equations

$$\dot{C}_B^N = C_B^N \left(\underline{\omega}_{IB}^B \times\right) - \left(\underline{\omega}_{IE}^N \times\right) C_B^N$$

$$\underline{\omega}_{IE}^N = \underline{\omega}_{IE_H}^N + \underline{u}_{ZN}^N \, \omega_e \sin l$$

$$\dot{\underline{\omega}}_{IE}^N = 0 \qquad (18.3\text{-}3)$$

$$\dot{\underline{v}}_H^N = \left(C_B^N\right)_H \underline{a}_{SF}^B$$

$$\Delta\dot{\underline{R}}_H^N = \underline{v}_H^N$$

The difference between (18.3-2) and (18.3-3) defines the errors in the (18.3-2) system computer solution, the basis for the Kalman estimator associated with (18.3-2). From Chapter 15, Equations (15.1.5.3.2-14), (15.1.5.3.2-17), (15.2.1-13) (expanded to include all components of $\underline{\gamma}^N$ and constant sensor errors), (15.2.1-14) and (15.2.1-26), the Kalman estimator for the quasi-stationary Fine Alignment Kalman estimator are given by:

Pure Estimator Kalman Filter Configuration:

$$\dot{\underline{x}}(t) = A(t)\,\underline{x}(t) + G_P(t)\,\underline{n}_P(t) \qquad \text{Error State Dynamics}$$

$$\underline{Z}_{Obs}^{\blacklozenge}(t) = \Delta\widehat{\underline{R}}_H^N \qquad \text{Observation}$$

$$\underline{z}^{\blacklozenge}(t) = \delta\Delta\underline{R}_H^N - \Delta\underline{R}_{Vib}^{N\blacklozenge} = H(t)\,\underline{x}(t) + G_M(t)\,\underline{n}_M^{\blacklozenge}(t) \qquad \text{Measurement} \qquad (18.3\text{-}4)$$

$$\dot{\widehat{\underline{x}}}(t) = \left(A(t) - K^{\blacklozenge}(t)\,H(t)\right)\widehat{\underline{x}}(t) + K^{\blacklozenge}(t)\,\underline{z}^{\blacklozenge}(t) \qquad \text{Estimation}$$

with

$$\underline{x} = \left[\left(\delta\underline{\omega}_{IE_{Exp}}^N\right)^T, \left(\delta\underline{\omega}_{IB_{Cnst}}^B\right)^T, \left(\underline{\gamma}^{N\star}\right)^T, \delta\underline{a}_{SF_{Cnst}}^B, \left(\delta\underline{v}_H^{N\star}\right)^T, \left(\delta\Delta\underline{R}_H^{N\star}\right)^T\right]^T$$

$$\underline{n}_P = \left[\left(\delta\underline{\omega}_{Rand}\right)^T, \left(\delta\underline{\alpha}_{Quant}\right)^T, \left(\delta\underline{a}_{Rand}\right)^T, \left(\delta\underline{\upsilon}_{Quant}\right)^T\right]^T \qquad (18.3\text{-}5)$$

$$\underline{n}_M^{\blacklozenge}(t) = \Delta\underline{R}_{Vib}^{N\blacklozenge}$$

where

$\underline{x}$ = Error state vector.

$\underline{n}_P$ = Process noise vector.

$\underline{n}_M^{\blacklozenge}(t)$ = Continuous form Kalman filter measurement noise.

$\underline{Z}_{Obs}^{\blacklozenge}(t)$ = The continuous form Kalman filter observation vector.

$\underline{z}^{\blacklozenge}(t)$ = The continuous form Kalman filter measurement vector.

$K^{\blacklozenge}(t)$ = The continuous form Kalman filter gain matrix (whose non-zero components are the K_i terms in Equations (18.3-1)).

$A(t)$, $G_P(t)$, $H(t)$, $G_M(t)$ = The continuous form Kalman filter error state dynamic matrix, process noise coupling matrix, measurement matrix and measurement noise coupling matrix.

$\underline{\gamma}^{N\star}$ = Rotation angle error vector associated with the $\hat{C}_B^{N\star}$ matrix.

$\delta\underline{\omega}_{IE_{Exp}}^N$, $\delta\underline{v}_H^{N\star}$, $\delta\Delta\underline{R}_H^{N\star}$ = Errors in $\underline{\omega}_{IE_{Exp}}^N$, $\hat{\underline{v}}_H^{N\star}$, $\Delta\hat{\underline{R}}_H^{N\star}$.

$\delta\underline{\omega}_{IB_{Cnst}}^B$, $\delta\underline{a}_{SF_{Cnst}}^B$ = Constant error portions of $\tilde{\underline{\omega}}_{IB}^B$, $\tilde{\underline{a}}_{SF}^B$.

$\delta\underline{\omega}_{Rand}$, $\delta\underline{\alpha}_{Quant}$ = Random output noise and quantization error portions of $\tilde{\underline{\omega}}_{IB}^B$.

$\delta\underline{a}_{Rand}$, $\delta\underline{\upsilon}_{Quant}$ = Random output noise and quantization error portions of $\tilde{\underline{a}}_{SF}^B$.

$\Delta\underline{R}_{Vib}^{N\blacklozenge}$ = Continuous form horizontal position vibration motion.

The non-zero components of the $K^{\blacklozenge}(t)$ matrix in (18.3-5) are typically calculated in traditional Kalman filter fashion from the covariance form of (18.3-5) based on a simplified reduced state error model that neglects $\delta\underline{\omega}_{IB_{Cnst}}^B$, $\delta\underline{a}_{SF_{Cnst}}^B$ and the vertical component of $\underline{\gamma}^{N\star}$ (see Chapter 15, Section 15.2.1, 15.2.1.1 and Equations (15.2.1.1-21) for the discrete form quasi-stationary Fine Alignment Kalman filter). Chapter 14, Section 14.6 analyzes the covariance characteristics of the simplified reduced state Kalman estimator as summarized for typical alignment times by Equations (14.6.4.4-1) and (14.6.5.2-5). To simplify the analysis, these equations were derived by assuming zero for the initial estimated error state uncertainty for all but the earth rate error state. More extensive analysis would show that for typical alignment times, the equations are also valid for initial uncertainties in all of the error states.

An important part of the Chapter 14 results is that as alignment time t increases, the effect of process and measurement noise on horizontal earth rate estimation error becomes dominated by the angular rate sensor horizontal random output noise (see Equations (14.6.4.4-1) and (14.6.5.2-5)). This effect can be confirmed by numerically analyzing the alignment performance of a simulated or real inertial system. Thus, for typical alignment times, we can assume for analysis purposes, that a reasonable assessment of the estimation process performance can be achieved by neglecting all noise terms but the horizontal angular rate sensor output random noise. Chapter 14, Section 14.6.4.3 (and the Chapter 14 subsections leading to it - Including supporting Sections 15.1.5.4 and 15.1.5.4.1) show that when angular rate sensor random output noise is the only output source, the covariance response of the simplified Kalman alignment estimator is equivalent to that of a hypothetical Kalman estimator with the measurement equal to the derivative of the horizontal attitude angle error state. This result was determined in Chapter 14 for a constant error state dynamic matrix, but a similar derivation would show that in general (using (15.2.1-6) and (18.3-4) as guides for form) for the hypothetical Kalman estimator:

$$\begin{aligned}
&\underline{x}' = \left(\delta\omega_{IE_H}, \text{ other constant error states}\right)^T && \text{Error State Vector} \\
&z^{\blacklozenge\prime}(t) = \dot{\gamma}_H^{\star}(t) = A_\gamma(t)\,\underline{x}' - \delta\omega_{Rand_H}(t) && \text{Measurement} \\
&\dot{\underline{x}}' = 0 \qquad \text{i.e., } A_{x'}(t) = 0 && \text{Error State Dynamics} \\
&\dot{\hat{\underline{x}}}'(t) = -K^{\blacklozenge\prime}(t)\,A_\gamma(t)\,\hat{\underline{x}}'(t) + K^{\blacklozenge\prime}(t)\,z^{\blacklozenge\prime}(t) && \text{Estimation}
\end{aligned} \qquad (18.3\text{-}6)$$

where

$\underline{x}'$ = Error state vector for the hypothetical Kalman estimator.

$A_{x'}(t)$ = Error state dynamic matrix for the $\underline{x}'$ error state vector.

$\delta\omega_{IE_H}$ = Earth rate estimation error along one of the horizontal axes.

$z^{\blacklozenge\prime}(t)$ = Continuous form measurement for the hypothetical Kalman estimator.

$\dot{\gamma}_H^{\star}(t)$ = Component of $\dot{\underline{\gamma}}^{N\star}$ along the $\delta\omega_{IE_H}$ horizontal axis.

$A_\gamma(t)$ = $\dot{\gamma}_H^{\star}(t)$ row of the error state dynamic matrix that couples $\underline{x}'$ into $\dot{\gamma}_H^{\star}(t)$.

$\delta\omega_{Rand_H}(t)$ = Component of $C_B^N\,\delta\underline{\omega}_{Rand}(t)$ along the $\delta\omega_{IE_H}$ horizontal axis.

$K^{\blacklozenge\prime}(t)$ = Continuous form hypothetical Kalman filter gain matrix.

We see from Equation (18.3-6) that the hypothetical Kalman filter treats the $\delta\omega_{Rand_H}(t)$ angular rate sensor process noise as measurement noise in the $z^{\blacklozenge\prime}(t)$ measurement.

The $\delta\omega_{IE_H}$ component of the (18.3-6) estimation equation is:

$$\dot{\hat{\delta\omega}}_{IE_H}(t) = -K_\omega^{\blacklozenge\prime}(t)\,A_\gamma(t)\,\hat{\underline{x}}'(t) + K_\omega^{\blacklozenge\prime}(t)\,z^{\blacklozenge\prime}(t) \qquad (18.3\text{-}7)$$

where

$K_{\omega}^{\blacklozenge\prime}(t) \;=\; \delta\omega_{IE_H}$ element of the $K^{\blacklozenge\prime}(t)$ matrix.

$\widehat{\delta\omega}_{IE_H}(t) \;=\;$ Hypothetical Kalman filter estimate for $\delta\omega_{IE_H}$.

For the Fine Alignment problem, the $\delta\omega_{IE_H}$ element of $A_{\gamma}(t)$ is the dominant term (i.e., the other components of $\underline{x}'$ (call them $\underline{x}''$) have little effect on Fine Alignment estimation accuracy). This, of course, was the justification for neglecting them in the Chapter 14, Section 14.6 analyses. It also follows that if all the terms in $A_{\gamma}(t)$ were included, the ability to estimate $\underline{x}''$ would be nil because of their small effect (compared with $\delta\omega_{IE_H}$) on the measurement $z^{\blacklozenge\prime}(t)$. Based on this rationale, we neglect the $\hat{\underline{x}}''$ components of $\underline{x}'$ in (18.3-7). Recognizing (as in (15.2.1-6)) that the $\delta\omega_{IE_H}$ component of $A_{\gamma}(t)$ is unity, then finds:

$$\dot{\widehat{\delta\omega}}_{IE_H}(t) \;=\; -\; K_{\omega}^{\blacklozenge\prime}(t)\; \widehat{\delta\omega}_{IE_H}(t) + K_{\omega}^{\blacklozenge\prime}(t)\; z^{\blacklozenge\prime}(t) \qquad (18.3\text{-}8)$$

Consistent with the previous approximation, we can use the Chapter 14 results to calculate the $K_{\omega}^{\blacklozenge\prime}(t)$ gain. Using Equation (15.1.5.3.2-14) for $K_{\omega}^{\blacklozenge\prime}(t)$ with (from the (18.3-6) measurement equation) the measurement noise $R^{\blacklozenge}(t)$ set to the $\delta\omega_{Rand_H}(t)$ noise density, $G_M(t)$ set to -1, H(t) set to $A_{\gamma}(t)$, the $\delta\omega_{IE_H}$ component of $A_{\gamma}(t)$ set to unity, and P(t) set to $P_{\Omega_{\omega Rand}}$ in Equations (14.6.4.4-1), we obtain:

$$K_{\omega}^{\blacklozenge\prime}(t) \;=\; \frac{P_{\Omega_0}}{q_{\omega Rand} + P_{\Omega_0}\, t} \qquad (18.3\text{-}9)$$

where

$P_{\Omega_0} \;=\;$ Initial uncertainty in $\widehat{\delta\omega}_{IE_H}(t)$.

$q_{\omega Rand} \;=\; \delta\omega_{Rand_H}(t)$ noise density.

With (18.3-9), Equation (18.3-8) becomes:

$$\dot{\widehat{\delta\omega}}_{IE_H}(t) \;=\; -\left(\frac{P_{\Omega_0}}{q_{\omega Rand} + P_{\Omega_0}\, t}\right) \widehat{\delta\omega}_{IE_H}(t) + \left(\frac{P_{\Omega_0}}{q_{\omega Rand} + P_{\Omega_0}\, t}\right) z^{\blacklozenge\prime}(t) \qquad (18.3\text{-}10)$$

The solution to (18.3-10) follows (as can be verified by substitution in (18.3-10)):

$$\hat{\delta\omega}_{IE_H}(t) = \left(\frac{P_{\Omega_0}}{q_{\omega Rand} + P_{\Omega_0}\, t}\right)\left(\frac{q_{\omega Rand}}{P_{\Omega_0}}\,\hat{\delta\omega}_{IE/H_0} + \int_0^t z^{\blacklozenge\prime}(\tau)\, d\tau\right) \tag{18.3-11}$$

where

$\hat{\delta\omega}_{IE/H_0}$ = Value for $\hat{\delta\omega}_{IE_H}(t)$ at time t = 0.

For large t (i.e., near the end of alignment) with zero for $\hat{\delta\omega}_{IE/H_0}$, and using (18.3-6) for $z^{\blacklozenge\prime}(t)$, Equation (18.3-11) reduces to:

$$\hat{\delta\omega}_{IE_H}(t) \approx \frac{1}{t}\int_0^t \dot{\gamma}_H^{\star}(\tau)\, d\tau \tag{18.3-12}$$

Equation (18.3-12) was derived assuming that angular rate sensor horizontal random noise was the only noise present (i.e., the other noise sources have negligible impact). It also uses a $K_{\omega}^{\blacklozenge\prime}(t)$ value based on $A_{\gamma}(t)$ having only the $\delta\omega_{IE_H}$ element active (Chapter 14, Sections 14.6.4.2 and 14.6.4.3 use only $\delta\omega_{IE_H}$ in $\underline{x}'$ - i.e., look at the ** elements in Equations (14.6.4.2-2)) At this point we make the additional assumption that (18.3-12) can be used in general when all error states and noise sources are present. Then, allowing for residual errors in our approximations and using $\delta\underline{\omega}_{IE_{Exp}}^{N}$ for the earth rate error, the equivalent vector form of (18.3-12) can be written as:

$$\hat{\delta\underline{\omega}}_{IE/Exp_H}^{N}(t) = \frac{1}{t}\int_0^t \dot{\underline{\gamma}}_H^{\star}(\tau)\, d\tau + \delta\underline{\omega}_{IE/H_{Resid}}^{N}(t) \tag{18.3-13}$$

where

$\hat{\delta\underline{\omega}}_{IE/Exp_H}^{N}(t)$ = Kalman filter estimated value for $\delta\underline{\omega}_{IE/Exp_H}^{N}$ defined below.

$\delta\underline{\omega}_{IE/Exp_H}^{N}$ = Horizontal component of $\delta\underline{\omega}_{IE_{Exp}}^{N}$.

$\delta\underline{\omega}_{IE/H_{Resid}}^{N}(t)$ = Residual term that is a function of the time in alignment, the initial values of the error state components in Equations (18.3-1), the magnitude of all noise terms present in Equations (18.3-1), and the magnitude of the vibration levels present during the alignment process. Generally speaking, the magnitude of $\delta \underline{\omega}^{N}_{IE/H_{Resid}}(t)$ becomes negligible compared to the $\frac{1}{t} \int_0^t \underline{\dot{\gamma}}_H^{*}(\tau)\, d\tau$ term near the end of Fine Alignment.

To apply Equation (18.3-13), we need to define $\delta \hat{\underline{\omega}}^{N}_{IE/Exp_H}(t)$ in terms of data that is available in the system computer (i.e., the Equations (18.3-1) data). First, we can define a Kalman filter version of the estimated value for the horizontal earth rate by correcting $\underline{\omega}^{N}_{IE/Exp_H}$ with the filter's estimate of the $\underline{\omega}^{N}_{IE/Exp_H}$ error:

$$\hat{\underline{\omega}}^{N}_{IE_H}{}^{*}(t) = \underline{\omega}^{N}_{IE/Exp_H} - \delta \hat{\underline{\omega}}^{N}_{IE/Exp_H}(t) \qquad (18.3\text{-}14)$$

where

$\hat{\underline{\omega}}^{N}_{IE_H}{}^{*}(t)$ = Estimate for the constant $\underline{\omega}^{N}_{IE_H}$ that would be obtained using the (18.3-4) Kalman estimate for $\delta \hat{\underline{\omega}}^{N}_{IE/Exp_H}(t)$.

But from (18.3-2) we know that:

$$\underline{\omega}^{N}_{IE/Exp_H} = 0 \qquad (18.3\text{-}15)$$

so that (18.3-14) becomes:

$$\hat{\underline{\omega}}^{N}_{IE_H}{}^{*}(t) = - \delta \hat{\underline{\omega}}^{N}_{IE/Exp_H} \qquad (18.3\text{-}16)$$

Next we equate $\hat{\underline{\omega}}^{N}_{IE_H}{}^{*}(t)$ generated using the Equations (18.3-2) - (18.3-5) pure estimator configuration with the equivalent signal generated by the (18.3-1) Kalman controlled configuration:

$$\hat{\underline{\omega}}^{N}_{IE_H} = \hat{\underline{\omega}}^{N}_{IE_H}{}^{*}(t) \qquad (18.3\text{-}17)$$

so that (18.3-16) becomes with rearrangement:

$$\delta \hat{\underline{\omega}}^{N}_{IE/Exp_H} = - \hat{\underline{\omega}}^{N}_{IE_H} \qquad (18.3\text{-}18)$$

Substituting (18.3-18) in (18.3-13) then obtains the important result:

$$\underline{\hat{\omega}}^N_{IE_H}(t) = -\frac{1}{t}\int_0^t \dot{\underline{\gamma}}_H^{\star}(\tau)\, d\tau - \delta\underline{\omega}^N_{IE/H_{Resid}}(t) \qquad (18.3\text{-}19)$$

Equation (18.3-19) is the formula that will form the basis for estimating angular rate sensor random output noise in the Repeated Alignment and Continuous Alignment tests.

Using (15.2.1-3) as a guide, let's now write the complete error state dynamic equation for $\underline{\gamma}^{N\star}$ and $\delta\underline{\omega}^N_{IE/Exp\,H}$ based on (18.3-2) compared with (18.3-3).

$$\begin{aligned}
\dot{\underline{\gamma}}_H^{N\star} &= -\left(C_B^N\, \delta\underline{\omega}^B_{IB}\right)_H - \gamma_{ZN}^{\star}\, \underline{\omega}^N_{IE_H} \times \underline{u}^N_{ZN} - \left(\underline{u}^N_{ZN} \times \underline{\gamma}_H^{N\star}\right) \omega_{IE_{Up}} \\
&\quad + \delta\underline{\omega}^N_{IE/Exp_H} + \left(C_B^N\right)_H \left(\underline{\omega}^B_{Vib} \times \delta\underline{\alpha}_{Quant}\right) \\
\dot{\gamma}_{ZN}^{\star} &= -\underline{u}^N_{ZN} \cdot \left(C_B^N\, \delta\underline{\omega}^B_{IB}\right) - \underline{u}^N_{ZN} \cdot \left(\underline{\omega}^N_{IE_H} \times \underline{\gamma}_H^{N\star}\right) \qquad (18.3\text{-}20) \\
&\quad + \underline{u}^N_{ZN} \cdot \left[C_B^N \left(\underline{\omega}^B_{Vib} \times \delta\underline{\alpha}_{Quant}\right)\right] \\
\delta\dot{\underline{\omega}}^N_{IE/Exp_H} &= 0
\end{aligned}$$

with, as in (18.2.1-6):

$$\begin{aligned}
\delta\omega_{IB_{ZN}} &= \delta\omega_{ARS/Cnst_{ZN}} + n_{ARS/Rnd_{ZN}} \\
\delta\underline{\omega}^N_{IB_H} &= \delta\underline{\omega}^N_{ARS/Cnst_H} + \underline{n}^N_{ARS/Rnd_H}
\end{aligned} \qquad (18.3\text{-}21)$$

where

$\omega_{IE_{Up}}$ = Vertical component of $\underline{\omega}^N_{IE}$ which is $\omega_e \sin l$.

$\underline{\omega}^B_{Vib}$ = B Frame angular vibration rate during alignment.

$\delta\omega_{ARS/Cnst_{ZN}}$, $\delta\underline{\omega}^N_{ARS/Cnst_H}$ = Vertical and horizontal constant components of $C_B^N\, \delta\underline{\omega}^B_{IB}$.

$n_{ARS/Rnd_{ZN}}$, $\underline{n}^N_{ARS/Rnd_H}$ = Vertical and horizontal random noise components of $C_B^N\, \delta\underline{\omega}^B_{IB}$.

The integral of (18.3-20) since the start of Fine Alignment (showing functional time dependence where needed for clarity) is:

$$\int_0^t \dot{\underline{\gamma}}_H^{N\star}(t) = \delta\underline{\omega}_{IE/Exp_H}^N \, t + \int_0^t \Big[-\left(C_B^N \, \delta\underline{\omega}_{IB}^B\right)_H - \gamma_{ZN}^{\star}(\tau)\, \underline{\omega}_{IE_H}^N \times \underline{u}_{ZN}^N$$

$$- \left(\underline{u}_{ZN}^N \times \underline{\gamma}_H^{N\star}(\tau)\right) \omega_{IE_{Up}} + \left(C_B^N\right)_H \left(\underline{\omega}_{Vib}^B \times \delta\underline{\alpha}_{Quant}\right) \Big] \, d\tau$$

$$\gamma_{ZN}^{\star}(t) = \gamma_{ZN\,0}^{\star} + \int_0^t \dot{\gamma}_{ZN}^{\star}(\tau)\, d\tau \qquad (18.3\text{-}22)$$

$$= \int_0^t \Big\{ -\underline{u}_{ZN}^N \cdot \left(C_B^N \, \delta\underline{\omega}_{IB}^B\right) - \underline{u}_{ZN}^N \cdot \left(\underline{\omega}_{IE_H}^N \times \underline{\gamma}_H^N(\tau)^{\star}\right)$$

$$+ \underline{u}_{ZN}^N \cdot \left[C_B^N \left(\underline{\omega}_{Vib}^B \times \delta\underline{\alpha}_{Quant}\right)\right] \Big\} \, d\tau$$

The $\gamma_{ZN\,0}^{\star}$ initial value for $\gamma_{ZN}^{\star}(t)$ in (18.3-22) is zero based on the following rationale. The method for finding initial heading in Section 6.1.2 is based on finding N (and L) Frame heading relative to true north by interpreting N Frame horizontal earth rate components estimated during Fine Alignment. Hence, the heading value selected for the N Frame at the start of alignment is arbitrary, since its value relative to north will be properly found and used in the subsequent navigation mode phase (within the accuracy of the inertial sensors that produce earth rate estimation errors). Stated differently, the initial heading orientation of the N (or L) Frames in the $\hat{C}_B^N$ matrix is error free and, therefore, matches the C_B^N matrix initial heading. This sets the initial value of $\gamma_{ZN}^{\star}(t)$ in (18.3-22) to zero as shown.

To simplify the analytics, we approximate $\underline{\gamma}_H^N(\tau)^{\star}$ in the (18.3-22) $\gamma_{ZN}^{\star}(t)$ equation by its average value over the alignment period. Then, with (18.3-21), $\gamma_{ZN}^{\star}(t)$ in (18.3-22) becomes:

$$\gamma_{ZN}(t)^{\star} \approx \int_0^t \Big\{ -\delta\omega_{ARS/Cnst_{ZN}} - n_{ARS/Rnd_{ZN}}$$

$$- \underline{u}_{ZN}^N \cdot \left(\underline{\omega}_{IE_H}^N \times \underline{\gamma}_{H_{Avg}}^{N\,\star}\right) + \underline{u}_{ZN}^N \cdot \left[C_B^N \left(\underline{\omega}_{Vib}^B \times \delta\underline{\alpha}_{Quant}\right)\right] \Big\} \, d\tau \qquad (18.3\text{-}23)$$

$$= -\left[\delta\omega_{ARS/Cnst_{ZN}} + \underline{u}_{ZN}^N \cdot \left(\underline{\omega}_{IE_H}^N \times \underline{\gamma}_{H_{Avg}}^{N\,\star}\right)\right] t$$

$$- \theta_{ARS/Rnd_{ZN}}(t) + \theta_{ARS/\omega Vib\text{-}\alpha Quant_{ZN}}(t)$$

with

$$\theta_{ARS/Rnd_{ZN}}(t) \equiv \int_0^t n_{ARS/Rnd_{ZN}} \, d\tau$$

$$\theta_{ARS/\omega Vib\text{-}\alpha Quant_{ZN}}(t) \equiv \int_0^t \underline{u}_{ZN}^{N} \cdot \left[C_B^N \left(\underline{\omega}_{Vib}^{B} \times \delta\underline{\alpha}_{Quant} \right) \right] d\tau \qquad (18.3\text{-}24)$$

where

$$\underline{\gamma}_{H_{Avg}}^{N\,\star} = \text{Average value for } \underline{\gamma}_{H}^{N}(t)^{\star} \text{ over the alignment period t.}$$

Now substitute (18.3-23) with (18.3-24) and (18.3-21) in the (18.3-22) $\int \dot{\underline{\gamma}}_{H}^{N\,\star}(t)$ expression to obtain:

$$\begin{aligned}
\int_0^t \dot{\underline{\gamma}}_{H}^{N\,\star}(t) = \; & \delta\underline{\omega}_{IE/Exp_H}^{N}\, t - \left[\delta\underline{\omega}_{ARS/Cnst_H}^{N} + \left(\underline{u}_{ZN}^{N} \times \underline{\gamma}_{H_{Avg}}^{N\,\star} \right) \omega_{IE_{Up}} \right] t \\
& - \underline{\theta}_{ARS/Rnd_H}^{N}(t) + \underline{\theta}_{ARS/\omega Vib\text{-}\alpha Quant_H}^{N}(t) \\
& + \left\{ \frac{1}{2} \left[\delta\omega_{ARS/Cnst_{ZN}} + \underline{u}_{ZN}^{N} \cdot \left(\underline{\omega}_{IE_H}^{N} \times \underline{\gamma}_{H_{Avg}}^{N\,\star} \right) \right] t^2 \right. \\
& \left. + S\theta_{ARS/Rnd_{ZN}}(t) - S\theta_{ARS/\omega Vib\text{-}\alpha Quant_{ZN}}(t) \right\} \underline{\omega}_{IE_H}^{N} \times \underline{u}_{ZN}^{N}
\end{aligned} \qquad (18.3\text{-}25)$$

in which

$$\underline{\theta}_{ARS/Rnd_H}^{N}(t) \equiv \int_0^t \underline{n}_{ARS/Rnd_H}^{N} \, d\tau$$

$$\underline{\theta}_{ARS/\omega Vib\text{-}\alpha Quant_H}^{N}(t) \equiv \int_0^t \left(C_B^N \right)_H \left(\underline{\omega}_{Vib}^{B} \times \delta\underline{\alpha}_{Quant} \right) d\tau$$

$$S\theta_{ARS/Rnd_{ZN}}(t) \equiv \int_0^t \int_0^\tau n_{ARS/Rnd_{ZN}} \, d\tau_1 \, d\tau \qquad (18.3\text{-}26)$$

$$S\theta_{ARS/\omega Vib\text{-}\alpha Quant_{ZN}}(t) \equiv \int_0^t \int_0^\tau \underline{u}_{ZN}^{N} \cdot \left[C_B^N \left(\underline{\omega}_{Vib}^{B} \times \delta\underline{\alpha}_{Quant} \right) \right] d\tau_1 \, d\tau$$

where

$$\underline{\theta}_{ARS/Rnd_H}^{N}(t) = \text{N Frame integral of horizontal angular rate sensor output noise.}$$

$\underline{\theta}^N_{ARS/\omega Vib\text{-}\alpha Quant_H}(t)$ = N Frame integral of horizontal angular-vibration/angular-rate-sensor-quantization-noise product.

$S\theta_{ARS/Rnd_{ZN}}(t)$ = Double integral of the upward vertical component of angular rate sensor random noise.

$S\theta_{ARS/\omega Vib\text{-}\alpha Quant_{ZN}}(t)$ = Double integral of the upward vertical component of angular-vibration/angular-rate-sensor-quantization-noise product.

We also write the definition for the $\delta\underline{\omega}^N_{IE_{Exp}}$ term in (18.3-25) as the difference between the horizontal component of $\underline{\omega}^N_{IE_{Exp}}$ used in (18.3-2) and the true horizontal earth rate value. Then, using (18.3-15), we find:

$$\delta\underline{\omega}^N_{IE/Exp_H} = \underline{\omega}^N_{IE/Exp_H} - \underline{\omega}^N_{IE_H} = - \underline{\omega}^N_{IE_H} \tag{18.3-27}$$

Now in (18.3-25), we substitute (18.3-27) and apply (14.2-33) to the $\underline{\omega}^N_{IE_H} \times \underline{u}^N_{ZN}$ term (as in (14.2-34)). Substituting the result in (18.3-19) with rearrangement finds an expression for the integrated horizontal angular rate sensor random output noise:

$$\begin{aligned}\underline{\theta}^N_{ARS/Rnd_H}(t) = {} & t\,\hat{\underline{\omega}}^N_{IE_H} - t\,\underline{\omega}^N_{IE_H} + t\,\delta\underline{\omega}^N_{IE/H_{Resid}}(t) \\ & - \left[\delta\underline{\omega}^N_{ARS/Cnst_H} + \left(\underline{u}^N_{ZN} \times \underline{\gamma}^N_{H_{Offset}}\right)\omega_{IE_{Up}}\right] t + \underline{\theta}_{ARS/\omega Vib\text{-}\alpha Quant_H}(t) \\ & + \left\{\frac{1}{2}\left[\delta\omega_{ARS/Cnst_{ZN}} + \underline{u}^N_{ZN} \cdot \left(\underline{\omega}^N_{IE_H} \times \underline{\gamma}^N_{H_{Offset}}\right)\right] t^2 \right. \\ & \left. + S\theta_{ARS/Rnd_{ZN}}(t) - S\,\theta_{ARS/\omega Vib\text{-}\alpha Quant_{ZN}}(t) \right\} \omega_{IE_H}\,\underline{u}^N_{East}\end{aligned} \tag{18.3-28}$$

where

$\underline{u}^N_{East}$ = Unit vector in the horizontal East direction.

ω_{IE_H} = Horizontal earth rate magnitude (earth rate times cosine geodetic latitude).

$\underline{\gamma}^N_{H_{Offset}}$ = Portion of $\underline{\gamma}^N_H$ that will remain near the end of Fine Alignment due to horizontal accelerometer error when using the controlled Kalman system configuration of Equations (18.3-1) (See Section 14.3 - Equation (14.3-39) for the relationship between $\underline{\gamma}_H$ and accelerometer error).

Using $\underline{\gamma}^{N}_{H_{Offset}}$ in (18.3-28) (rather than $\underline{\gamma}^{N}_{H_{Avg}}{}^{*}$ in (18.3-25)) is more representative of how the associated terms affect the (18.3-28) result when using the (18.3-1) Kalman controlled system.

Equation (18.3-28) with (18.3-26) defines the integral of horizontal angular rate sensor noise from alignment initiation to time t in alignment. This result will be used in the subsections to follow as the basis for estimating angular rate sensor random noise from measurements taken during Fine Alignment.

18.3.1 REPEATED ALIGNMENT TEST

The Repeated Alignment test method for estimating horizontal angular rate sensor random noise is based on applying Equation (18.3-28) to INS self-alignment process output measurements taken at Fine Alignment completion, for an ensemble of Fine Alignment runs. In the following subsections we describe two versions of the Repeated Alignment test; one based on Kalman filter earth rate estimate measurements; the other based on INS true heading output measurements (for situations when the earth rate estimates are not available outputs).

18.3.1.1 REPEATED ALIGNMENT TEST USING KALMAN EARTH RATE ESTIMATES

The Repeated Alignment test based on Kalman alignment filter earth rate estimates applies Equation (18.3-28) to horizontal earth rate measurements taken at Fine Alignment completion for an ensemble of Fine Alignment runs. With this approach, Equations (18.3-28) and (18.3-26) can be written as:

$$\underline{\theta}^{N}_{ARS/Rnd_H} \equiv \int_0^{T_{Align}} \underline{n}^{N}_{ARS/Rnd_H}\, dt$$

$$\underline{\theta}^{N}_{ARS/\omega Vib\text{-}\alpha Quant_H} \equiv \int_0^{T_{Align}} \left(C^{N}_{B}\right)_H \left(\underline{\omega}^{B}_{Vib} \times \delta\underline{\alpha}_{Quant}\right) dt$$

$$S\theta_{ARS/Rnd_{ZN}} \equiv \int_0^{T_{Align}} \int_0^{t} n_{ARS/Rnd_{ZN}}\, d\tau\, dt \qquad (18.3.1.1\text{-}1)$$

$$S\theta_{ARS/\omega Vib\text{-}\alpha Quant_{ZN}} \equiv \int_0^{T_{Align}} \int_0^{t} \underline{u}^{N}_{ZN} \cdot \left[C^{N}_{B}\left(\underline{\omega}^{B}_{Vib} \times \delta\underline{\alpha}_{Quant}\right)\right] d\tau\, dt$$

$$\begin{aligned}\underline{\theta}^N_{ARS/Rnd/H_j} = \; & \hat{\underline{\omega}}^N_{IE/H_j} T_{Align} - \Big[\underline{\omega}^N_{IE_H} + \delta\underline{\omega}^N_{ARS/Cnst_H} \\ & + \left(\underline{u}^N_{ZN} \times \underline{\gamma}^N_{H_{Offset}}\right) \omega_{IE_{Up}} - \delta\underline{\omega}^N_{IE/H/Resid_j} \Big] T_{Align} \\ & + \frac{1}{2} T^2_{Align} \left[\delta\omega_{ARS/Cnst_{ZN}} + \underline{u}^N_{ZN} \cdot \left(\underline{\omega}^N_{IE_H} \times \underline{\gamma}^N_{H_{Offset}} \right) \right] \omega_{IE_H} \, \underline{u}^N_{East} \\ & + \underline{\theta}^N_{ARS/\omega Vib\text{-}\alpha Quant/H_j} + \left(S\theta_{ARS/Rnd/ZN_j} - S\theta_{ARS/\omega Vib\text{-}\alpha Quant/ZN_j} \right) \omega_{IE_H} \, \underline{u}^N_{East}\end{aligned} \qquad (18.3.1.1\text{-}2)$$

where

T_{Align} = Time from start to end of the Fine Alignment process, assumed the same for each jth Fine Alignment test run.

$\underline{\theta}^N_{ARS/Rnd/H_j}$ = N Frame integral over T_{Align} of horizontal angular rate sensor random noise for the jth Fine Alignment test run.

$\underline{\theta}^N_{ARS/\omega Vib\text{-}\alpha Quant/H_j}$ = N Frame integral over T_{Align} of the horizontal component of angular - vibration / angular - rate - sensor - quantization - noise product for the jth Fine Alignment test run.

$S\theta_{ARS/Rnd/ZN_j}$ = Double integral over T_{Align} of upward vertical angular rate sensor random noise for the jth Fine Alignment test run.

$S\theta_{ARS/\omega Vib\text{-}\alpha Quant/ZN_j}$ = Double integral over T_{Align} of the upward vertical component of angular-vibration/angular-rate-sensor-quantization-noise product for the jth Fine Alignment test run.

$\hat{\underline{\omega}}^N_{IE/H_j}$ = Fine Alignment Kalman filter horizontal earth rate estimate at the end of alignment for the jth Fine Alignment test run.

$\delta\underline{\omega}^N_{IE/H/Resid_j}$ = Value for $\delta\underline{\omega}^N_{IE/H_{Resid}}(t)$ at the end of alignment for the jth Fine Alignment test run.

Equation (18.3.1.1-2) is in a form we can use for determination of angular rate sensor random noise by statistical analysis. First we recognize that since $\underline{\theta}^N_{ARS/Rnd/H_j}$, $\underline{\theta}_{ARS/\omega Vib\text{-}\alpha Quant/H_j}$, $S\theta_{ARS/Rnd/ZN_j}$ and $S\theta_{ARS/\omega Vib\text{-}\alpha Quant/ZN_j}$ are defined to be single and double integrals of random noise, they are also random, hence, their expected values over the ensemble of j alignments is zero. We also assume that the alignment time is of sufficient duration that the initial transient elements in $\delta\underline{\omega}^N_{IE/H/Resid_j}$ are negligible. Therefore,

$\delta\underline{\omega}^N_{IE/H/Resid_j}$ is also random over the ensemble of j alignments. Taking the expected value of (18.3.1.1-2) then yields after rearrangement:

$$\mathcal{E}\left(\hat{\underline{\omega}}^N_{IE_H}\right) T_{Align} = \left[\underline{\omega}^N_{IE_H} + \delta\underline{\omega}^N_{ARS/Cnst_H} + \left(\underline{u}^N_{ZN} \times \underline{\gamma}^N_{H_{Offset}}\right)\omega_{IE_{Up}}\right] T_{Align}$$
$$- \frac{1}{2} T^2_{Align} \left[\delta\omega_{ARS/Cnst_{ZN}} + \underline{u}^N_{ZN} \cdot \left(\underline{\omega}^N_{IE_H} \times \underline{\gamma}^N_{H_{Offset}}\right)\right] \omega_{IE_H}\, \underline{u}^N_{East} \qquad (18.3.1.1\text{-}3)$$

where

$\mathcal{E}(\,)$ = The expected value operator (i.e., average statistical value).

With (18.3.1.1-3), Equation (18.3.1.1-2) reduces to:

$$\underline{\theta}^N_{ARS/Rnd/H_j} = \left[\hat{\underline{\omega}}^N_{IE/H_j} - \mathcal{E}\left(\hat{\underline{\omega}}^N_{IE_H}\right) + \delta\underline{\omega}^N_{IE/H/Resid_j}\right] T_{Align}$$
$$+ \underline{\theta}^N_{ARS/\omega Vib\text{-}\alpha Quant/H_j} + \left(S\theta_{ARS/Rnd/ZN_j} - S\theta_{ARS/\omega Vib\text{-}\alpha Quant/ZN_j}\right)\omega_{IE_H}\,\underline{u}^N_{East} \qquad (18.3.1.1\text{-}4)$$

Assume that the dominant noise term in (18.3.1.1-4) is $\underline{\theta}^N_{ARS/Rnd/H_j}$ (to be proven later). We will also assume that the alignment time is sufficiently long that $\delta\underline{\omega}^N_{IE/H/Resid_j}$ is negligible compared to $\underline{\theta}^N_{ARS/Rnd/H_j}$. Therefore, we can approximate (18.3.1.1-4) as:

$$\underline{\theta}^N_{ARS/Rnd/H_j} \approx \left[\hat{\underline{\omega}}^N_{IE/H_j} - \mathcal{E}\left(\hat{\underline{\omega}}^N_{IE_H}\right)\right] T_{Align} \qquad (18.3.1.1\text{-}5)$$

The variance of $\underline{\theta}^N_{ARS/Rnd/H_j}$ in (18.3.1.1-5) over the ensemble of j alignment runs provides the desired measure of horizontal angular rate sensor random noise. From the integral of covariance rate Equation (15.1.2.1.1-30), we know that the variance of the integral of white noise (i.e., for $A(t) = 0$ and $G_P(t) = I$) equals the white noise density multiplied by the integration time. Thus, for $\underline{\theta}^N_{ARS/Rnd/H_j}$ in (18.3.1.1-5):

$$\mathcal{E}\left(\theta^2_{ARS/Rnd/H\text{-}k_j}\right) = \sigma^2_{ARS/Rnd/H\text{-}k}\, T_{Align} \qquad (18.3.1.1\text{-}6)$$

where

$\theta_{ARS/Rnd/H\text{-}k_j}$ = Component of $\underline{\theta}^N_{ARS/Rnd/H_j}$ along N Frame horizontal axis k.

$\sigma_{ARS/H/Rnd\text{-}k}$ = Square root of the angular rate sensor random process noise density along N Frame horizontal axis k.

$$\mathcal{E}\left(\theta^2_{ARS/Rnd/H\text{-}k_j}\right) = \text{The variance of } \theta_{ARS/Rnd/H\text{-}k_j}.$$

Based on (18.3.1.1-6), we can then write for our estimate of random noise:

$$\sigma_{ARS/Rnd/H\text{-}k} = \sqrt{\frac{\mathcal{E}\left(\theta^2_{ARS/Rnd/H\text{-}k}\right)}{T_{Align}}} \qquad (18.3.1.1\text{-}7)$$

where

$$\mathcal{E}\left(\theta^2_{ARS/Rnd/H\text{-}k}\right) = \text{Variance of } \theta_{ARS/Rnd/H\text{-}k_j} \text{ for the ensemble of j alignments.}$$

The Repeated Alignment test for evaluating $\sigma_{ARS/H/Rnd\text{-}k}$ is based on Equation (18.3.1.1-7) using (18.3.1.1-5) and the $\hat{\underline{\omega}}^N_{IE/H_j}$ measurements for estimating $\mathcal{E}\left(\theta^2_{ARS/Rnd/H\text{-}k}\right)$ in traditional fashion as follows:

$$\mathcal{E}\left(\hat{\underline{\omega}}^N_{IE_H}\right) \approx \frac{1}{n}\sum_{j=1}^{n} \hat{\underline{\omega}}^N_{IE/H_j}$$
$$\mathcal{E}\left(\theta^2_{ARS/Rnd/H\text{-}k}\right) \approx \frac{T^2_{Align}}{n-1}\sum_{j=1}^{n}\left\{\hat{\omega}_{IE/H\text{-}k_j} - \left[\mathcal{E}\left(\hat{\underline{\omega}}^N_{IE_H}\right)\right]_k\right\}^2 \qquad (18.3.1.1\text{-}8)$$

where

n = Number of Fine Alignment run samples used for angular rate sensor random noise estimation.

$\hat{\omega}_{IE/H\text{-}k_j}, \left[\mathcal{E}\left(\hat{\underline{\omega}}^N_{IE_H}\right)\right]_k$ = The k horizontal axis components of $\hat{\underline{\omega}}^N_{IE/H_j}$ and $\mathcal{E}\left(\hat{\underline{\omega}}^N_{IE_H}\right)$.

The overall result for estimating angular rate sensor random noise then, is (18.3.1.1-7) combined with (18.3.1.1-8):

$$\mathcal{E}\left(\hat{\underline{\omega}}^N_{IE_H}\right) \approx \frac{1}{n}\sum_{j=1}^{n} \hat{\underline{\omega}}^N_{IE/H_j}$$
$$\sigma_{ARS/Rnd/H\text{-}k} = \sqrt{\frac{T_{Align}\sum_{j=1}^{n}\left\{\hat{\omega}_{IE/H\text{-}k_j} - \left[\mathcal{E}\left(\hat{\underline{\omega}}^N_{IE_H}\right)\right]_k\right\}^2}{n-1}} \qquad (18.3.1.1\text{-}9)$$

Equations (18.3.1.1-9) allow estimation of angular rate sensor random noise for the two angular rate sensors along the horizontal axes. A random noise estimate for the vertical angular rate sensor can be obtained by repeating the n alignments and the (18.3.1.1-9) analysis with the INS oriented so that the originally vertical angular rate sensor is horizontal.

The previous process is based on a particular self-alignment time T_{Align}. By repeating the process for different alignment times, the random noise coefficients can be estimated from (18.3.1.1-9) as a function of alignment time. If the sensor noise is white (as implied in Equation (18.3.1.1-7)), the calculated $\sigma_{ARS/Rnd/H\text{-}k}$ value should be the same regardless of T_{Align} (within the assumptions used in formulating (18.3.1.1-9)). Variations in $\sigma_{ARS/Rnd/H\text{-}k}$ for different alignment times indicate that the angular rate sensor noise estimate includes other non-white additional sensor noise components (i.e., random terms that were neglected going from (18.3.1.1-4) to (18.3.1.1-5)) whose effect on the earth rate estimates differs from that of white angular rate sensor output noise. An important point to recognize, however, is that for a given T_{Align}, the $\sigma_{ARS/Rnd/H\text{-}k}$ value calculated with the (18.3.1.1-9) Repeated Alignment test method is a direct measurement of the earth rate uncertainty at the end of alignment for this alignment time (exclusive of alignment filter initial transient residuals). As such, it actually represents the composite effect of angular rate sensor output noise on earth rate estimation uncertainty (hence, self-alignment heading determination accuracy), regardless of whether the sensor noise is pure white or a composite of white and other random effects.

To reduce the magnitude of residual random contributors to $\sigma_{ARS/Rnd/H\text{-}k}$, the Kalman filter convergence time should be minimized by proper selection of the initial covariance uncertainties and measurement noise. In this regard, it is beneficial that the Kalman filter measurement noise (i.e., the P_{RVib_H} position disturbance noise in Section 15.2.1, Equation (15.2.1-27) and Section 15.2.1.1, Equation (15.2.1.1-12)) be set to a low value (near zero) that is characteristic of the static test environment. Kalman filter convergence time can be further reduced by initializing the horizontal tilt uncertainty (i.e., P_{γ/H_0} in Equation (15.2.1.2-2)) at a value representative of initial coarse pre-alignment leveling under benign vibration test conditions. Kalman filter convergence time can also be reduced by initializing the horizontal earth rate estimates to values corresponding with the INS heading during each alignment (determined from an initial trial alignment run). The initial estimated earth rate covariance (i.e., $P_{\delta\omega IE/H_0}$ in Equation (15.2.1.2-2)) would then be correspondingly reduced to allow for the smaller initial uncertainty (e.g., by setting it equal to the earth rate uncertainty covariance at the end of the trial alignment run, multiplied by four for a safety factor).

To assure that angular rate sensor turn-on transient errors will not corrupt results obtained, the self-alignment series is best run after the INS has had sufficient running time for transient decay.

For the remainder of this section we will address the uncertainty in the above $\sigma_{ARS/Rnd/H\text{-}k}$ estimation process produced by the terms in the (18.3.1.1-4) $\underline{\theta}^N_{ARS/Rnd/H_j}$ equation that were neglected in the (18.3.1.1-5) approximation; specifically, the $S\theta_{ARS/Rnd/ZN_j}$, $\delta\underline{\omega}^N_{IE/H/Resid_j}$, $\underline{\theta}^N_{ARS/\omega Vib\text{-}\alpha Quant/H_j}$, and $S\theta_{ARS/\omega Vib\text{-}\alpha Quant/ZN_j}$ terms. Note in Equation (18.3.1.1-4) that the $S\theta_{ARS/Rnd/ZN_j}$ and $S\theta_{ARS/\omega Vib\text{-}\alpha Quant/ZN_j}$ terms are only along the east axis, hence, do not affect the north $\underline{\theta}^N_{ARS/Rnd/H_j}$ component (and the corresponding north $\sigma_{ARS/Rnd/H\text{-}k}$ estimate).

Using covariance rate Equation (15.1.2.1.1-30) we can show that the expected value of the square of the $S\theta_{ARS/Rnd/ZN_j}$ term in (18.3.1.1-4) (i.e., doubly integrated random noise) is:

$$\mathcal{E}\left[\left(S\theta_{ARS/Rnd/ZN_j}\right)^2\right] = \frac{1}{3}\,\sigma^2_{ARS/Rnd/ZN}\,T^3_{Align} \qquad (18.3.1.1\text{-}10)$$

where

$\mathcal{E}\left[\left(S\theta_{ARS/Rnd/ZN_j}\right)^2\right]$ = Variance of the integral of $\theta_{ARS/Rnd/ZN_j}$.

$\sigma_{ARS/Rnd/ZN}$ = Square root of the vertical angular rate sensor random process noise density.

Note that the Section 13.6.1 method (as was applied in Section 13.6.2 leading to the $P_{R_H R_H}$ expression in Equations (13.6.2-3)) can also be used to derive (18.3.1.1-10). For this case, we consider each noise pulse (ε_i) to register instantaneously on the output of the first integrator (at time τ_i) and then propagating through the second integrator into $\varepsilon_i\,(t - \tau_i)$ at time t. Taking the sum of the $\varepsilon_i\,(t - \tau_i)$'s for the total effect at time t of previous noise contributions, squaring the sum and taking the expected value with the ε_i's assumed uncorrelated, multiplying and dividing each element in the result by $\Delta\tau_i$, letting $\Delta\tau_i$ go to zero in the limit so that the summation of the ( ) $\Delta\tau_i$ becomes an integral, identifying the $\mathcal{E}\left(\varepsilon_i^2\right) / \Delta\tau_i$ terms in the limit as $\sigma^2_{ARS/Rnd/ZN}$, and analytically integrating the result, then yields (18.3.1.1-10).

Taking the ratio of (18.3.1.1-10) to (18.3.1.1-6) multiplied by the ω_{IE_H} coefficient squared in (18.3.1.1-4) provides a measure of the relative contribution of the variance of $S\theta_{ARS/Rnd/ZN_j}$ in Equation (18.3.1.1-4) compared with the $\underline{\theta}^N_{ARS/Rnd/H_j}$ variance. (We note that the relative variance is the proper performance evaluation factor when statistically summing independent error contributions; e.g., as in taking the root-sum-square in an error budget analysis).

Assuming equal noise densities for the vertical and horizontal angular rate sensors, said variance ratio of vertical over horizontal sensor noise contributions is $\frac{1}{3}\,\omega_{IE_H}^{2}\,T_{Align}^{2}$. For a 10 minute alignment time at 45 degrees latitude, $\frac{1}{3}\,\omega_{IE_H}^{2}\,T_{Align}^{2} = \frac{1}{3} \times$ (15 deg/hr $\times$ 0.0175 rad/deg $\times$ cosine 45 deg $\times$ 10 min $\times$ 0.0167 hrs/min)^2 = 0.00032; i.e., clearly negligible.

The reader can verify using the analytical techniques of Section 16.2.3.1 and Equation (15.1.2.1.1-30) (with $A(t) = 0$ and $G_P(t) = I$ for integrated random noise variance analysis) that the $\underline{\theta}^{N}_{ARS/\omega Vib-\alpha Quant/H_j}\ S\theta_{ARS/\omega Vib-\alpha Quant/ZN_j}$ angular-vibration/angular-rate-sensor-quantization product terms in (18.3.1.1-4) are negligible compared to the $S\theta_{ARS/Rnd/ZN_j}$ term, particularly under the benign angular rate vibration environment of system testing in the laboratory. We assume that the alignment time is sufficiently long that $\delta\underline{\omega}^{N}_{IE/H/Resid_j}$ can also be neglected.

18.3.1.2 REPEATED ALIGNMENT TEST USING INS COMPUTED TRUE HEADING

The Repeated Alignment test described in Section 18.3.1.1 implies the ability to access the horizontal earth rate estimates during INS self-alignment. For situations when such access is not available, an alternate procedure can be utilized for a strapdown INS based on INS true heading output measurements. This approach is based on the Section 18.3.1.1 method of estimating random noise from an ensemble of repeated alignments. To convert the Section 18.3.1.1 results for heading measurement compatibility, we first define the error in the measured true heading as:

$$\delta\psi_T = \hat{\psi}_T - \psi_T \qquad (18.3.1.2\text{-}1)$$

where

ψ_T = Actual true heading.

$\hat{\ }$ = Designation for parameter calculated in the INS computer and containing errors, in contrast with the undesignated same parameter defined to be error free.

Equations (14.2-32) - (14.2-34) and (12.2.1-44) show that for zero pitch angle (with zero position error), the error in true heading produced by horizontal earth rate estimation uncertainty during the fine-alignment process is:

$$\delta\psi_T = \gamma_{ZN} - \frac{1}{\omega_{IE_H}} \delta\underline{\omega}_{IE_H}^N \cdot \underline{u}_{East}^N \qquad (18.3.1.2\text{-}2)$$

with the earth rate estimation error $\delta\underline{\omega}_{IE_H}^N$ defined as:

$$\delta\underline{\omega}_{IE_H}^N = \hat{\underline{\omega}}_{IE_H}^N - \underline{\omega}_{IE_H}^N \qquad (18.3.1.2\text{-}3)$$

where

ω_{IE_H} = Horizontal earth rate magnitude.

$\underline{u}_{East}^N$ = Unit vector along the horizontal easterly direction.

γ_{ZN} = Vertical component of the rotation angle error associated with the $\hat{C}_B^N$ matrix.

For simplicity we define:

$$\hat{\omega}_{IE_{East}} \equiv \hat{\underline{\omega}}_{IE_H}^N \cdot \underline{u}_{East}^N \qquad (18.3.1.2\text{-}4)$$

Note that (18.3.1.2-4) cannot be explicitly evaluated in the INS because $\underline{u}_{East}^N$ (error free, by definition) is unknown in the INS computer. The INS computer version of $\underline{u}_{East}^N$ is by definition, perpendicular to $\hat{\underline{\omega}}_{IE_H}^N$ (i.e., east is perpendicular to north which by definition lies along the horizontal earth rate vector pointing toward earth's positive rotation axis), hence, would generate zero for the equivalent INS computer version of Equation (18.3.1.2-4). Recognizing that the true horizontal earth rate ($\underline{\omega}_{IE_H}^N$) has no east component, we then find from (18.3.1.2-3) with (18.3.1.2-4) that the $\delta\underline{\omega}_{IE_H}^N \cdot \underline{u}_{East}^N$ term in (18.3.1.2-2) is:

$$\delta\underline{\omega}_{IE_H}^N \cdot \underline{u}_{East}^N = \hat{\omega}_{IE_{East}} \qquad (18.3.1.2\text{-}5)$$

The γ_{ZN} term in (18.3.1.2-2) is provided analytically by Section 18.3, Equation (18.3-23) if we recognize that both the theoretical uncontrolled Kalman alignment filter of Section 18.3 (i.e., the ✯ version) and the actual controlled Kalman filter configuration (without the ✯) do not control the vertical component attitude error. Thus, we can equate γ_{ZN} in (18.3.1.2-2) to $\gamma_{ZN}^{✯}$ in (18.3-23). Adopting the Equation (18.3.1.1-2) terminology for the noise terms and substituting $\underline{\gamma}_{H_{Offset}}^N$ for $\underline{\gamma}_{H_{Avg}}^{N✯}$ (based on the rationale following (18.3-28)) then yields at the end of alignment for the jth alignment run:

$$\gamma_{ZN_j} = -\left[\delta\omega_{ARS/Cnst_{ZN}} + \underline{u}^N_{ZN} \cdot \left(\underline{\omega}^N_{IE_H} \times \underline{\gamma}^N_{H_{Offset}}\right)\right] T_{Align} - \theta_{ARS/Rnd/ZN_j} + \theta_{ARS/\omega Vib\text{-}\alpha Quant/ZN_j} \qquad (18.3.1.2\text{-}6)$$

where

j = Designation for parameter value for a particular alignment run j.

γ_{ZN_j} = γ_{ZN} at the end of alignment for the jth alignment run.

As discussed in the paragraph preceding (18.2.1-2), γ_{ZN_j} in (18.3.1.2-6) is actually the upward vertical (ZN) component of $\underline{\gamma}^*$ as defined by Equation (12.5-5), but with the * notation deleted for simplicity. The $\underline{\gamma}^*$ parameter is $\underline{\gamma}$ neglecting quantization noise (See Equation (12.5-5)). For compatibility with (18.3.1.2-2) (which is for the true $\underline{\gamma}$), we substitute the vertical component of $\underline{\gamma}^*$ from (12.5-5) for γ_{ZN_j} in (18.3.1.2-6):

$$\gamma_{ZN_j} = -\left[\delta\omega_{ARS/Cnst_{ZN}} + \underline{u}^N_{ZN} \cdot \left(\underline{\omega}^N_{IE_H} \times \underline{\gamma}^N_{H_{Offset}}\right)\right] T_{Align} - \theta_{ARS/Rnd/ZN_j} + \theta_{ARS/\omega Vib\text{-}\alpha Quant/ZN_j} - \delta\alpha_{Quant/ZN_j} \qquad (18.3.1.2\text{-}7)$$

where

$\delta\alpha_{Quant/ZN_j}$ = Vertical (along the N Frame Z axis) component of $C^N_B \, \delta\underline{\alpha}_{Quant}$ (See Equation (12.5-5) for clarification) at the end of alignment for the jth alignment run..

Combining Equations (18.3.1.2-1), (18.3.1.2-2), (18.3.1.2-5) and (18.3.1.2-7) for the jth alignment run yields with rearrangement:

$$\hat{\omega}_{IE/East_j} = \omega_{IE_H}\left(\gamma_{ZN_j} - \delta\psi_T\right)$$

$$= -\,\omega_{IE_H}\Big\{\hat{\psi}_T - \psi_T + \left[\delta\omega_{ARS/Cnst_{ZN}} + \underline{u}^N_{ZN} \cdot \left(\underline{\omega}^N_{IE_H} \times \underline{\gamma}^N_{H_{Offset}}\right)\right] T_{Align} + \delta\alpha_{Quant/ZN_j} + \theta_{ARS/Rnd/ZN_j} - \theta_{ARS/\omega Vib\text{-}\alpha Quant/ZN_j}\Big\} \qquad (18.3.1.2\text{-}8)$$

where

$\hat{\omega}_{IE/East_j}$ = $\hat{\omega}_{IE_{East}}$ at the end of alignment for the jth alignment run.

We now take the dot product of Section 18.3.1.1 - Equation (18.3.1.1-2) with $\underline{u}^N_{East}$, apply the (18.3.1.2-4) definition, substitute (18.3.1.2-8) for $\hat{\omega}_{IE/East_j}$, equate the east component of $\underline{\omega}^N_{IE_H}$ to zero, apply the (3.1.1-35) mixed vector dot/cross-product identity in the

$\underline{u}_{ZN}^{N} \cdot \left(\underline{\omega}_{IE_H}^{N} \times \underline{\gamma}_{H_{Offset}}^{N}\right)$ and $\underline{u}_{East}^{N} \cdot \left(\underline{u}_{ZN}^{N} \times \underline{\gamma}_{H_{Offset}}^{N}\right)$ terms (recognizing that $\underline{u}_{East}^{N} \times \underline{u}_{ZN}^{N}$ is south with unity magnitude and $\underline{u}_{ZN}^{N} \times \underline{\omega}_{IE_H}^{N}$ is west with ω_{IE_H} magnitude), and re-group. The result is:

$$
\begin{aligned}
\theta_{ARS/Rnd/East_j} = & - \omega_{IE_H} \hat{\psi}_{T_j} T_{Align} \\
& + \left[\omega_{IE_H} \psi_T - \delta\omega_{ARS/Cnst_{East}} + \omega_{IE_{Up}} \gamma_{North_{Offset}}\right] T_{Align} \\
& - \frac{1}{2} T_{Align}^{2} \left[\delta\omega_{ARS/Cnst_{ZN}} - \omega_{IE_H} \gamma_{East_{Offset}}\right] \omega_{IE_H} \\
& + \left(\delta\omega_{IE/East/Resid_j} - \omega_{IE_H} \delta\alpha_{Quant/ZN_j}\right) T_{Align} + \theta_{ARS/\omega Vib\text{-}\alpha Quant/East_j} \\
& + \left(S\theta_{ARS/Rnd/ZN_j} - \theta_{ARS/Rnd/ZN_j} T_{Align}\right) \omega_{IE_H} \\
& - \left(S\theta_{ARS/\omega Vib\text{-}\alpha Quant/ZN_j} - \theta_{ARS/\omega Vib\text{-}\alpha Quant/ZN_j} T_{Align}\right) \omega_{IE_H}
\end{aligned}
\qquad (18.3.1.2\text{-}9)
$$

where

$\hat{\psi}_{T_j}$ = Value for $\hat{\psi}_T$ at the end of alignment for the jth alignment run.

$\theta_{ARS/Rnd/East_j}$, $\delta\omega_{ARS/Cnst_{East}}$, $\delta\omega_{IE/East/Resid_j}$, $\theta_{ARS/\omega Vib\text{-}\alpha Quant/East_j}$ = East components of $\underline{\theta}_{ARS/Rnd/H_j}^{N}$, $\delta\underline{\omega}_{ARS/Cnst_H}^{N}$, $\delta\underline{\omega}_{IE/H/Resid_j}^{N}$, and $\underline{\theta}_{ARS/\omega Vib\text{-}\alpha Quant/H_j}$.

$\gamma_{North_{Offset}}$, $\gamma_{East_{Offset}}$ = North and east components of $\underline{\gamma}_{H_{Offset}}^{N}$.

As in Section 18.3.1.1 - Equations (18.3.1.1-3) and (18.3.1.1-4), we recognize $\delta\omega_{IE/East/Resid_j}$ and the noise terms to be random with zero expected value over the ensemble of j alignments. Hence, lines 2 and 3 in (18.3.1.2-9) equal $\omega_{IE_H} T_{Align}$ times the expected value of $\hat{\psi}_{T_j}$. Dropping the $\delta\omega_{IE/East/Resid_j}$ and noise terms on the right side of the equality as negligible (as in Section 18.3.1.1), Equation (18.3.1.2-9) then becomes:

$$
\theta_{ARS/Rnd/East_j} \approx - \omega_{IE_H} T_{Align} \left[\hat{\psi}_{T_j} - \mathcal{E}\left(\hat{\psi}_T\right)\right] \qquad (18.3.1.2\text{-}10)
$$

The extension of the final Equation (18.3.1.1-9) result in Section 18.3.1.1 to Equation (18.3.1.2-10) should now be obvious:

$$\mathcal{E}\left(\hat{\psi}_T\right) \approx \frac{1}{n} \sum_{j=1}^{n} \hat{\psi}_{T_j}$$

$$\sigma_{ARS/Rnd/East} = \omega_{IE_H} \sqrt{\frac{T_{Align} \sum_{j=1}^{n} \left[\hat{\psi}_{T_j} - \mathcal{E}\left(\hat{\psi}_T\right)\right]^2}{n - 1}} \qquad (18.3.1.2\text{-}11)$$

where

$\sigma_{ARS/Rnd/East}$ = Square root of the horizontal East angular rate sensor random noise density.

Equations (18.3.1.2-11) allow estimation of random noise for the angular rate sensor along the East horizontal axis. Random noise can be estimated for the North angular rate sensor by repeating the alignment and the Equation (18.3.1.2-11) process with the INS rotated 90 degrees in heading from the original orientation (placing the previously North angular rate sensor in the Easterly direction). An estimate for the vertical angular rate sensor noise can be evaluated by repeating the process with the INS at a 90 degree roll orientation that places the previously vertical angular rate sensor in the horizontal Easterly direction. Each of the previous orientations should be at zero pitch angle for the processing equations to be valid.

As for the Section 18.3.1.1 Repeated Alignment test, the $\sigma_{ARS/Rnd/East}$ value calculated with the Equations (18.3.1.2-11) method is a direct measurement of the heading uncertainty at the end of alignment for the selected alignment time. As such, $\sigma_{ARS/Rnd/East}$ calculated with (18.3.1.2-11) represents the composite effect of inertial sensor output noise on heading determination accuracy, regardless of whether the sensor noise is pure white, a composite of white and other random effects, or a mixture of angular rate sensor and accelerometer random noise effects.

18.3.2 CONTINUOUS ALIGNMENT TEST

For situations in which the alignment process Kalman filter error model accurately represents the actual INS sensor error characteristics over extended alignment periods, an alternate approach can be substituted for the Section 18.3.1 Repeated Alignment procedure to estimate strapdown angular rate sensor random noise. The alternate approach is based on data measurements from a continuous single alignment run. In the following subsections we describe two versions of the Continuous Alignment test approach; one based on Kalman filter earth rate estimate measurements and the other based on INS true heading output measurements.

18.3.2.1 CONTINUOUS ALIGNMENT TEST USING KALMAN EARTH RATE ESTIMATES

Included in Equations (18.3-26) is the definition for the integral of horizontal angular rate sensor random noise from alignment initiation to time t in alignment. To estimate the angular rate sensor random noise from a single Fine Alignment run, we will need an ensemble of successive integrated random noise samples for which the integration period is defined to be over successive time intervals during alignment. We define each integrated angular rate sensor noise sample as follows, relating it to $\underline{\theta}^N_{ARS/Rnd_H}(t)$ in Equations (18.3-26):

$$\Delta\underline{\theta}^N_{ARS/Rnd/H_i} \equiv \int_{t_{i-1}}^{t_i} \underline{n}^N_{ARS/Rnd_H}\, dt = \underline{\theta}^N_{ARS/Rnd/H_i} - \underline{\theta}^N_{ARS/Rnd/H_{i-1}} \qquad (18.3.2.1\text{-}1)$$

where

t_i = General time t in alignment for data measurement.

$\Delta\underline{\theta}^N_{ARS/Rnd/H_i}$ = Integral of $\underline{n}^N_{ARS/Rnd_H}$ from t_{i-1} to t_i.

Substituting (18.3-28) for the i and i-1 time points into (18.3.2.1-1) yields after combining terms:

$$\begin{aligned}
\Delta\underline{\theta}^N_{ARS/Rnd/H_i} = \; & t_i\, \hat{\underline{\omega}}^N_{IE/H_i} - t_{i-1}\, \hat{\underline{\omega}}^N_{IE/H_{i-1}} \\
& - \left[\underline{\omega}^N_{IE_H} + \delta\underline{\omega}^N_{ARS/Cnst_H} + \left(\underline{u}^N_{ZN} \times \underline{\gamma}^N_{H_{Offset}}\right) \omega_{IE_{Up}}\right] \Delta T \\
& + \left[\delta\omega_{ARS/Cnst_{ZN}} + \underline{u}^N_{ZN} \cdot \left(\underline{\omega}^N_{IE_H} \times \underline{\gamma}^N_{H_{Offset}}\right)\right] \left(t_i - \frac{1}{2}\Delta T\right) \Delta T\, \omega_{IE_H}\, \underline{u}^N_{East} \\
& + t_i\, \delta\underline{\omega}^N_{IE/H/Resid_i} - t_{i-1}\, \delta\underline{\omega}^N_{IE/H/Resid_{i-1}} \\
& + \underline{\theta}_{ARS/\omega Vib\text{-}\alpha Quant/H_i} - \underline{\theta}_{ARS/\omega Vib\text{-}\alpha Quant/H_{i-1}} \\
& + \left(S\theta_{ARS/Rnd/ZN_i} - S\theta_{ARS/Rnd/ZN_{i-1}}\right) \omega_{IE_H}\, \underline{u}^N_{East} \\
& - \left(S\theta_{ARS/\omega Vib\text{-}\alpha Quant/ZN_i} - S\theta_{ARS/\omega Vib\text{-}\alpha Quant/ZN_{i-1}}\right) \omega_{IE_H}\, \underline{u}^N_{East}
\end{aligned} \qquad (18.3.2.1\text{-}2)$$

where

$\hat{\underline{\omega}}^N_{IE/H_i} = \hat{\underline{\omega}}^N_{IE_H}$ at time $t = t_i$.

$(\;)_i$ = Parameter ( ) at time $t = t_i$.

ΔT = time interval from t_{i-1} to t_i which will be constant for data measurements.

The noise terms in the last three lines of (18.3.2.1-2) are defined for general time t in Equations (18.3-26).

For simplicity, we define the $\hat{\underline{\omega}}^N_{IE/H_i}$ estimation term group in (18.3.2.1-2) as:

$$\hat{\underline{\phi}}^N_{H_i} \equiv t_i \, \hat{\underline{\omega}}^N_{IE/H_i} - t_{i-1} \, \hat{\underline{\omega}}^N_{IE/H_{i-1}} \tag{18.3.2.1-3}$$

where

$\hat{\underline{\phi}}^N_{H_i}$ = What will be the measurement input for horizontal random noise estimation.

Equation (18.3.2.1-2) with (18.3.2.1-3) is in a form we can use for estimating angular rate sensor random noise by statistical analysis. First we recognize that the expected value for the noise terms at time t_i is zero. If random noise estimation during alignment begins after initial error state uncertainty estimation transients have subsided, we can also assume that the $\delta\underline{\omega}^N_{IE/H/Resid_i}$ residual has zero expected value at t_i. Taking the expected value of (18.3.2.1-2) with (18.3.2.1-3) then yields after rearrangement:

$$\begin{gathered} \left[\underline{\omega}^N_{IE_H} + \delta\underline{\omega}^N_{ARS/Cnst_H} + \left(\underline{u}^N_{ZN} \times \underline{\gamma}^N_{H_{Offset}} \right) \omega_{IE_{Up}} \right] \Delta T \\ - \left[\delta\omega_{ARS/Cnst_{ZN}} + \underline{u}^N_{ZN} \cdot \left(\underline{\omega}^N_{IE_H} \times \underline{\gamma}^N_{H_{Offset}} \right) \right] \left(t_i - \frac{1}{2} \Delta T \right) \Delta T \, \omega_{IE_H} \, \underline{u}^N_{East} = \mathcal{E}\left(\hat{\underline{\phi}}^N_{H_i} \right) \end{gathered} \tag{18.3.2.1-4}$$

where

$\mathcal{E}(\;)$ = Expected value operator.

With (18.3.2.1-4), Equation (18.3.2.1-2) with (18.3.2.1-3) reduces to:

$$\begin{aligned} \Delta\underline{\theta}^N_{ARS/Rnd/H_i} = \; & \hat{\underline{\phi}}^N_{H_i} - \mathcal{E}\left(\hat{\underline{\phi}}^N_{H_i} \right) + t_i \, \delta\underline{\omega}^N_{IE/H/Resid_i} - t_{i-1} \, \delta\underline{\omega}^N_{IE/H/Resid_{i-1}} \\ & + \underline{\theta}_{ARS/\omega Vib-\alpha Quant/H_i} - \underline{\theta}_{ARS/\omega Vib-\alpha Quant/H_{i-1}} \\ & + \left(S\theta_{ARS/Rnd/ZN_i} - S\theta_{ARS/Rnd/ZN_{i-1}} \right) \omega_{IE_H} \, \underline{u}^N_{East} \\ & - \left(S\theta_{ARS/\omega Vib-\alpha Quant/ZN_i} - S\theta_{ARS/\omega Vib-\alpha Quant/ZN_{i-1}} \right) \omega_{IE_H} \, \underline{u}^N_{East} \end{aligned} \tag{18.3.2.1-5}$$

It will be shown that the dominant noise term in (18.3.2.1-5) is $\Delta\underline{\theta}^N_{ARS/Rnd/H_i}$. We will also assume that the alignment time is sufficiently long that $\delta\underline{\omega}^N_{IE/H/Resid}$ terms will also be negligible compared to $\Delta\underline{\theta}^N_{ARS/Rnd/H_i}$. Therefore, we can approximate (18.3.2.1-5) as:

$$\Delta\underline{\theta}^N_{ARS/Rnd/H_i} \approx \hat{\underline{\phi}}^N_{H_i} - \mathcal{E}\left(\hat{\underline{\phi}}^N_{H_i}\right) \tag{18.3.2.1-6}$$

The variance of the $\Delta\underline{\theta}^N_{ARS/Rnd/H_i}$ components over the alignment period provides the desired measure of horizontal angular rate sensor random noise. From the integral of covariance rate Equation (15.1.2.1.1-30) we know that for a stationary white noise process (i.e., constant density independent of time), the variance of the integral of white noise (i.e., for $A(t) = 0$ and $G_P(t) = I$) equals the white noise density multiplied by the integration time. Therefore, for $\Delta\underline{\theta}^N_{ARS/Rnd/H_i}$ in (18.3.2.1-6):

$$\mathcal{E}\left(\Delta\theta^2_{ARS/Rnd/H\text{-}k_i}\right) = \sigma^2_{ARS/Rnd/H\text{-}k}\,\Delta T \tag{18.3.2.1-7}$$

where

$\Delta\theta_{ARS/Rnd/H\text{-}k_i}$ = Component of $\Delta\underline{\theta}^N_{ARS/Rnd/H_i}$ along N Frame horizontal axis k.

$\sigma_{ARS/Rnd/H\text{-}k}$ = Square root of angular rate sensor random process noise density along N Frame horizontal axis k (assumed constant independent of time).

$\mathcal{E}\left(\Delta\theta^2_{ARS/Rnd/H\text{-}k_i}\right)$ = Variance of $\Delta\theta_{ARS/Rnd/H\text{-}k_i}$.

Based on (18.3.2.1-7), we can then write for our estimate of random noise:

$$\sigma_{ARS/Rnd/H\text{-}k} = \sqrt{\frac{\mathcal{E}\left(\Delta\theta^2_{ARS/Rnd/H\text{-}k_i}\right)}{\Delta T}} \tag{18.3.2.1-8}$$

The Continuous Alignment test for evaluating $\sigma_{ARS/H/Rnd\text{-}k}$ is based on Equation (18.3.2.1-8) using (18.3.2.1-6) and the $\hat{\underline{\phi}}^N_{H_i}$ measurements for estimating $\mathcal{E}\left(\Delta\theta^2_{ARS/Rnd/H\text{-}k_i}\right)$. In traditional fashion, we assume that the sequence of $\hat{\underline{\phi}}^N_{H_i}$ measurements taken as a function of alignment time during a single alignment run can be treated like a set of independent $\hat{\underline{\phi}}^N_{H_i}$ measurements taken from an ensemble of alignment histories at a particular t_i (the so-called ergodic assumption). Then we calculate $\mathcal{E}\left(\underline{\hat{\phi}}^N_{H_i}\right)$ based on the ensemble of $\underline{\hat{\phi}}^N_{H_i}$ measurements and $\mathcal{E}\left(\Delta\theta^2_{ARS/Rnd/H\text{-}k}\right)$ as the variance of the $\underline{\hat{\phi}}^N_{H_i}$'s about the calculated $\mathcal{E}\left(\underline{\hat{\phi}}^N_{H_i}\right)$. Let us first address the calculation of $\mathcal{E}\left(\underline{\hat{\phi}}^N_{H_i}\right)$.

Equation (18.3.2.1-4) shows that the theoretical value for $\mathcal{E}\left(\underline{\hat{\phi}}^N_{H_i}\right)$ is a constant plus a term proportional to t_i. Thus, to estimate the value for $\mathcal{E}\left(\underline{\hat{\phi}}^N_{H_i}\right)$ from the $\underline{\hat{\phi}}^N_{H_i}$ data, one approach might be to find the best straight line fit to $\underline{\hat{\phi}}^N_{H_i}$ over t_i (e.g., using the method of least squares as in Reference 37 - Chapter 9, Section 11). However, due to the smallness of the term proportional to t_i, we will now show that a simple linear average of $\underline{\hat{\phi}}^N_{H_i}$ over t_i is sufficiently accurate to calculate $\mathcal{E}\left(\underline{\hat{\phi}}^N_{H_i}\right)$ for our purposes.

For the simple linear averaging approach, since the true value of $\mathcal{E}\left(\underline{\hat{\phi}}^N_{H_i}\right)$ is of the form $A + B\,t_i$ (in which A and B are constants), the approximate linear average of $A + B\,t_i$ over the alignment time (T_{Align}) will be $A + \frac{1}{2} B\,T_{Align}$. The associated error is the difference between the true and approximate $\mathcal{E}\left(\underline{\hat{\phi}}^N_{H_i}\right)$ solutions, or $A + B\,t_i - \left(A + \frac{1}{2} B\,T_{Align}\right) = B\left(t_i - \frac{1}{2} T_{Align}\right)$. Using the approximate value for $\mathcal{E}\left(\underline{\hat{\phi}}^N_{H_i}\right)$ in (18.3.2.1-6) will create an error in $\Delta\underline{\theta}^N_{ARS/Rnd/H_i}$ of $B\left(t_i - \frac{1}{2} T_{Align}\right)$. Then, assuming zero ensemble mean at each i time for $\Delta\theta_{ARS/Rnd/H\text{-}k_i}$, $\mathcal{E}\left(\Delta\theta^2_{ARS/Rnd/H\text{-}k}\right)$ to be calculated from the $\Delta\underline{\theta}^N_{ARS/Rnd/H_i}$'s will be:

$$\mathcal{E}\left(\overline{\Delta\theta}^2_{ARS/Rnd/H\text{-}k_i}\right) = \mathcal{E}\left\{\left[\Delta\theta_{ARS/Rnd/H\text{-}k_i} + B\left(t_i - \frac{1}{2} T_{Align}\right)\right]^2\right\}$$

$$= \mathcal{E}\left(\Delta\theta^2_{ARS/Rnd/H\text{-}k_i}\right) + 2\,B\left(t_i - \frac{1}{2} T_{Align}\right)\mathcal{E}\left(\Delta\theta_{ARS/Rnd/H\text{-}k_i}\right) + B^2\left(t_i - \frac{1}{2} T_{Align}\right)^2$$

$$= \mathcal{E}\left(\Delta\theta^2_{ARS/Rnd/H\text{-}k_i}\right) + B^2\left(t_i - \frac{1}{2} T_{Align}\right)^2 \qquad (18.3.2.1\text{-}9)$$

where

$\overline{\Delta\theta}_{ARS/Rnd/H\text{-}k_i}$ = Value calculated for $\Delta\theta_{ARS/Rnd/H\text{-}k_i}$ using the approximate linear average of the $\underline{\hat{\phi}}^N_{H_i}$'s over the alignment time for $\mathcal{E}\left(\underline{\hat{\phi}}^N_{H_i}\right)$.

The $\mathcal{E}\left(\Delta\theta^2_{ARS/Rnd/H\text{-}k_i}\right)$ term in (18.3.2.1-9) is the correct value (based on the correct value for $\mathcal{E}\left(\underline{\hat{\phi}}^N_{H_i}\right)$) whose magnitude is $\sigma^2_{ARS/Rnd/H\text{-}k}\ \Delta T$ as provided by Equation (18.3.2.1-7). The ratio of $B^2\left(t_i - \frac{1}{2}\,T_{Align}\right)^2$ to $\mathcal{E}\left(\Delta\theta^2_{ARS/Rnd/H\text{-}k_i}\right)$ provides a measure of the significance of the $B^2\left(t_i - \frac{1}{2}\,T_{Align}\right)^2$ error in estimating angular rate sensor random noise. The maximum value for $B^2\left(t_i - \frac{1}{2}\,T_{Align}\right)^2$ in (18.3.2.1-9) occurs at time 0 or time T_{Align} for which $B^2\left(t_i - \frac{1}{2}\,T_{Align}\right)^2$ equals $\frac{1}{4}\,B^2\,T^2_{Align}$. The value for B is the $t_i\ \underline{u}^N_{East}$ coefficient in (18.3.2.1-4) or $-\left[\delta\omega_{ARS/Cnst_{ZN}} + \underline{u}^N_{ZN}\cdot\left(\underline{\omega}^N_{IE_H}\times\underline{\gamma}^N_{H_{Offset}}\right)\right]\Delta T\ \omega_{IE_H}$. For most inertial navigation systems, the $\underline{\omega}^N_{IE_H}\times\underline{\gamma}^N_{H_{Offset}}$ term is considerably smaller than the $\delta\omega_{ARS/Cnst_{ZN}}$ term (e.g., ω_{IE_H} times 50 micro radians for $\underline{\omega}^N_{IE_H}\times\underline{\gamma}^N_{H_{Offset}}$ compared to 0.01 deg/hr for $\delta\omega_{ARS/Cnst_{ZN}}$ in an aircraft INS). Therefore, B can be approximated as $-\ \delta\omega_{ARS/Cnst_{ZN}}\ \Delta T\ \omega_{IE_H}$. Then, the maximum value for the $B^2\left(t_i - \frac{1}{2}\,T_{Align}\right)^2$ to $\mathcal{E}\left(\Delta\theta^2_{ARS/Rnd/H\text{-}k_i}\right)$ performance ratio is $\frac{1}{4}\,\frac{\Delta T\ \delta\omega^2_{ARS/Cnst_{ZN}}}{\sigma^2_{ARS/Rnd/H\text{-}k}}\,\omega^2_{IE_H}\,T^2_{Align}$. The angular rate sensor performance figures in a 1 nmph CEP accurate INS are typically 0.01 deg/hr for $\delta\omega_{ARS/Cnst_{ZN}}$ and 0.002 deg/sqr-rt-hr for $\sigma_{ARS/Rnd/H\text{-}k}$. Using these figures with T_{Align} = 2 hrs, ΔT = 2 min and 45 deg for the alignment latitude, the value for the performance ratio is $\frac{1}{4}\times 2\text{ min}\times 0.0167\text{ hrs/min}\times\left(\frac{0.01\text{ deg/hr}}{0.002\text{ deg/sqr-rt-hr}}\right)^2\times(15\text{ deg/hr}\times 0.0175\text{ rad/deg}\times\text{cosine }45\text{ deg}\times 2\text{ hrs})^2 = 0.029$. Thus, approximating $\mathcal{E}\left(\underline{\hat{\phi}}^N_{H_i}\right)$ as a linear average of $\underline{\hat{\phi}}^N_{H_i}$ over the alignment time has negligible impact on estimating horizontal angular rate sensor noise.

Based on the previous analysis, we then write our usual equations for estimating $\mathcal{E}\left(\underline{\hat{\phi}}^N_{H_i}\right)$ in (18.3.2.1-6) and for evaluating $\mathcal{E}\left(\Delta\theta^2_{ARS/Rnd/H\text{-}k_i}\right)$ in (18.3.2.1-8). The combined result for calculating $\sigma_{ARS/Rnd/H\text{-}k}$ with (18.3.2.1-3) is:

$$\underline{\hat{\phi}}^N_{H_i} = t_i \, \underline{\hat{\omega}}^N_{IE/H_i} - t_{i-1} \, \underline{\hat{\omega}}^N_{IE/H_{i-1}}$$

$$\mathcal{E}\left(\underline{\hat{\phi}}^N_{H_i}\right) \approx \frac{1}{n} \sum_{i=1}^{n} \underline{\hat{\phi}}^N_{H_i} \qquad (18.3.2.1\text{-}10)$$

$$\sigma_{ARS/Rnd/H\text{-}k} = \sqrt{\frac{\sum_{i=1}^{n} \left[\hat{\phi}_{H\text{-}k_i} - \mathcal{E}\left(\hat{\phi}_{H\text{-}k_i}\right)\right]^2}{(n - 1)\, \Delta T}}$$

where

n = Total number of $\underline{\hat{\phi}}^N_{H_i}$ measurements.

$\hat{\phi}_{H\text{-}k_i}, \mathcal{E}\left(\hat{\phi}_{H\text{-}k_i}\right)$ = Components of $\underline{\hat{\phi}}^N_{H_i}, \mathcal{E}\left(\underline{\hat{\phi}}^N_{H_i}\right)$ along N Frame horizontal axis k.

Note that the (18.3.2.1-10) summations can legitimately start at i = 1 because we assume that the i-1 value for $\underline{\hat{\omega}}^N_{IE/H_i}$ will be available at i = 1 based on the alignment process being initiated well before random noise estimation begins.

In practice, the following equivalent version of (18.3.2.1-10) can be used that allows $\sigma_{ARS/Rnd/H\text{-}k}$ to be evaluated as a function of the time since the random estimation process was initiated. The equivalent form is easily derived by expanding the term under the summation in the (18.3.2.1-10) $\sigma_{ARS/Rnd/H\text{-}k}$ expression and substituting the equation for $\mathcal{E}\left(\underline{\hat{\phi}}^N_{H_i}\right)$. The result is:

$$\underline{\hat{\phi}}^N_{H_i} = t_i \, \underline{\hat{\omega}}^N_{IE/H_i} - t_{i-1} \, \underline{\hat{\omega}}^N_{IE/H_{i-1}}$$

$$\sigma_{ARS/Rnd/H\text{-}k_n} = \sqrt{\frac{\left(\sum_{i=1}^{n} \hat{\phi}^2_{H\text{-}k_i}\right) - \frac{1}{n}\left(\sum_{i=1}^{n} \hat{\phi}_{H\text{-}k_i}\right)^2}{(n - 1)\, \Delta T}} \qquad (18.3.2.1\text{-}11)$$

where

$\sigma_{ARS/Rnd/H\text{-}k_n}$ = Square root of angular rate sensor random process noise density along N Frame horizontal axis k (assumed constant independent of time) based on data taken up to i = n.

When using (18.3.2.1-11), the $\sum_{i=1}^{n} \hat{\phi}^2_{H\text{-}k_i}$ and $\sum_{i=1}^{n} \hat{\phi}_{H\text{-}k_i}$ summations are updated as each $\underline{\hat{\phi}}^N_{H_i}$ measurement is taken. The current value for $\sigma_{ARS/Rnd/H\text{-}k_n}$ is then calculated at each i time point as an evolving estimate for horizontal random noise during the data taking process. The current value for $\sigma_{ARS/Rnd/H\text{-}k_n}$ would be the most accurate since it was generated with the most data samples.

Equations (18.3.2.1-11) allow estimation of angular rate sensor random noise for the two angular rate sensors along the horizontal axes. A random noise estimate for the vertical angular rate sensor can be evaluated by repeating the alignment and the (18.3.2.1-11) analyses with the INS at a 90 degree roll orientation, which places the previously vertical angular rate sensor in the horizontal attitude.

Equations (18.3.2.1-11) are based on several assumed Kalman filter performance characteristics. In general, the error in these assumptions becomes smaller as the filter converges to the final solution (i.e., after initial filter estimation transients have subsided). Hence, (18.3.2.1-11) should only be applied after the alignment process has been operating for a while (e.g., 5 minutes). The time in alignment when random noise estimation should begin and the time between $\underline{\hat{\phi}}^N_{H_i}$ samples should be selected based on simulation analysis for the class of sensor assemblies to be tested (with their characteristic noise values), and the desired accuracy for the angular rate sensor random noise estimates. It has been the author's very limited experience with a particular sensor assembly configuration that accelerometer quantization noise magnitude was the determining factor for setting the time between samples (i.e., setting it long enough to minimize the impact of accelerometer quantization noise on angular rate sensor random noise estimation). For that case, the accelerometer output pulse size was 0.013 fps and the angular rate sensor random noise density being estimated was on the order of 0.002 deg/sqr-rt-hr. To limit the impact of accelerometer quantization noise on angular rate sensor random noise estimation error to 10% of the 0.002 deg/sqr-rt-hr figure, the time between samples was set to 2 minutes. For a faster 5 second sample time interval, the accelerometer quantization noise caused an uncertainty in angular rate sensor random noise density estimation on the order of 100% of 0.002 deg/sqr-rt-hr.

From a comparison standpoint, even though the Continuous Alignment sensor noise estimation technique presented in this section is procedurally simpler than the Section 18.3.1 Repeated Alignment approach, the Section 18.3.1 method has generally been preferred because it doesn't implicitly rely on assumed internal Kalman filter characteristics, and explicitly measures performance based on its direct impact on earth rate estimation (the determining factor for initial heading accuracy) for a specified alignment time. Hence, the Section 18.3.1 approach will always generate a solution identified as angular rate sensor random noise that is a direct measurement of initial heading determination accuracy. The method summarized by Equations (18.3.2.1-11) has been designed to implicitly measure the same effect, but without direct evaluation for a specified alignment time. Interestingly, the form of Equations (18.3.2.1-11) would allow implementation within the INS as a means for estimating angular rate sensor random noise during each alignment. Such measurements can be useful in assigning a value to the "Quality" of the alignment just executed.

For the remainder of this section we will address the uncertainty in the above $\sigma_{ARS/Rnd/H\text{-}k}$ estimation process produced by the terms in the (18.3.2.1-5) $\Delta\underline{\theta}^N_{ARS/Rnd/H_i}$ equation that were neglected in the (18.3.2.1-6) approximation; specifically, the $\delta\underline{\omega}^N_{IE/H/Resid}$, $\underline{\theta}_{ARS/\omega Vib\text{-}\alpha Quant/H}$, $S\theta_{ARS/\omega Vib\text{-}\alpha Quant/ZN}$, and $S\theta_{ARS/Rnd/ZN}$ terms. From the discussion in the third paragraph following Equation (18.3.2.1-11), the $\delta\underline{\omega}^N_{IE/H/Resid}$ is controlled to be negligible by proper selection of the time in alignment when random noise estimation is initiated and the time between samples for the estimation process. Based on the discussion in the last paragraph of Section 18.3.1.1, we can assume that the $\underline{\theta}_{ARS/\omega Vib\text{-}\alpha Quant/H}$ and $S\theta_{ARS/\omega Vib\text{-}\alpha Quant/ZN}$ terms have negligible impact on $\Delta\underline{\theta}^N_{ARS/Rnd/H_i}$ horizontal noise assessment. Some analysis is required to be assured that the $S\theta_{ARS/Rnd/ZN}$ term is also negligible (Note in (18.3.2.1-5) that $S\theta_{ARS/Rnd/ZN_i}$ only appears along the east axis, hence, has no influence on north horizontal angular rate sensor random noise determination).

The variance of the $\left(S\theta_{ARS/Rnd/ZN_i} - S\theta_{ARS/Rnd/ZN_{i-1}}\right)$ term in (18.3.2.1-5) can be determined from the following development. Using Equations (18.3-26) and (18.3.2.1-1) we can write:

$$\theta_{ARS/Rnd_{ZN}}(t) = \theta_{ARS/Rnd/ZN_{i-1}} + \int_{t_{i-1}}^{t} n_{ARS/Rnd_{ZN}}(\tau)\, d\tau \qquad (18.3.2.1\text{-}12)$$

$$S\theta_{ARS/Rnd/ZN_i} = S\theta_{ARS/Rnd/ZN_{i-1}} + \int_{t_{i-1}}^{t_i} \theta_{ARS/Rnd_{ZN}}(t)\, dt \qquad (18.3.2.1\text{-}13)$$

Substituting (18.3.2.1-12) into (18.3.2.1-13) and analytically integrating the $\theta_{ARS/Rnd/ZN_{i-1}}$ portion yields for $\left(S\theta_{ARS/Rnd/ZN_i} - S\theta_{ARS/Rnd/ZN_{i-1}}\right)$ after rearrangement:

$$S\theta_{ARS/Rnd/ZN_i} - S\theta_{ARS/Rnd/ZN_{i-1}}$$

$$= \theta_{ARS/Rnd/ZN_{i-1}} \Delta T + \int_{t_{i-1}}^{t_i} \int_{t_{i-1}}^{t} n_{ARS/Rnd_{ZN}}(\tau)\, d\tau\, dt \qquad (18.3.2.1\text{-}14)$$

The $\theta_{ARS/Rnd/ZN_{i-1}}$ and double integral terms in (18.3.2.1-14) are over different non-overlapping time periods, hence, because they are both random noise integrals, they are independent of one another. Consequently, the variance of $\left(S\theta_{ARS/Rnd/ZN_i} - S\theta_{ARS/Rnd/ZN_{i-1}}\right)$ from (18.3.2.1-14) can be calculated as the sum of the variance of $\theta_{ARS/Rnd/ZN_{i-1}} \Delta T$ and the variance of the random noise double integral. As discussed in the paragraphs leading to (18.3.1.1-6) and (18.3.1.1-10), the variance of $\theta_{ARS/Rnd/ZN_{i-1}}$ is $\sigma^2_{ARS/Rnd/ZN}\, t_{i-1}$ and the variance of the random noise double integral is $\frac{1}{3} \sigma^2_{ARS/Rnd/ZN}\, \Delta T^3$. Thus, the variance of $\left(S\theta_{ARS/Rnd/ZN_i} - S\theta_{ARS/Rnd/ZN_{i-1}}\right)$ in (18.3.2.1-14) is the sum of $\sigma^2_{ARS/Rnd/ZN}\, t_{i-1}\, \Delta T^2$ and $\frac{1}{3} \sigma^2_{ARS/Rnd/ZN}\, \Delta T^3$ or:

$$\mathcal{E}\left(\Delta S\theta^2_{ARS/Rnd/ZN_i}\right) = \mathcal{E}\left[\left(S\theta_{ARS/Rnd/ZN_i} - S\theta_{ARS/Rnd/ZN_{i-1}}\right)^2\right]$$
$$= \left(t_{i-1} + \frac{1}{3}\Delta T\right) \Delta T^2\, \sigma^2_{ARS/Rnd/ZN} \qquad (18.3.2.1\text{-}15)$$

Taking the ratio of (18.3.2.1-15) to (18.3.2.1-7) multiplied by the ω_{IE_H} coefficient squared in (18.3.2.1-5), provides a measure of the relative contribution of the variance of $\left(S\theta_{ARS/Rnd/ZN_i} - S\theta_{ARS/Rnd/ZN_{i-1}}\right)$ to Equation (18.3.2.1-5). Assuming equal noise densities for the vertical and horizontal angular rate sensors, said variance performance ratio of vertical over horizontal sensor noise contributions is $\left(t_{i-1} + \frac{1}{3}\Delta T\right) \Delta T\, \omega^2_{IE_H}$. For a 2 hour total alignment time with measurements taken at 5 minute intervals (i.e., for $\left(t_{i-1} + \frac{1}{3}\Delta T\right)$ at the end of fine alignment approximately equal to 2 hours), the maximum value for the performance ratio at 45 degrees latitude is 2 hrs $\times$ 5 min $\times$ 0.0167 hrs/min $\times$ (15 deg/hr $\times$ 0.0175 rad/deg $\times$ cosine 45 deg)^2 = 0.0057, i.e., clearly negligible.

18.3.2.2 CONTINUOUS ALIGNMENT TEST USING INS COMPUTED TRUE HEADING

The Continuous Alignment angular rate sensor random noise test described in Section 18.3.2.1 implies the ability to access the horizontal earth rate estimates during INS self-alignment. For situations when such access is not available, an alternate procedure can be utilized for a strapdown INS based on INS true heading output measurements. This approach is based on the Section 18.3.2.1 method of estimating random noise from an ensemble of measurements taken during a single alignment run. To find the equivalent to the Section 18.3.2.1 results for heading measurement compatibility, we can apply the method used in Section 18.3.1.2 that led to Equation (18.3.1.2-9) and its approximate (18.3.1.2-10) form. Alternatively, we can begin with (18.3.1.2-9) and treat the j sample as the latest i sample for the continuous alignment test. Based on the latter approach, the T_{Align} parameter in (18.3.1.2-9) would be treated as t_i, the current time in alignment when the i samples are taken. We then take the difference between current and past Equation (18.3.1.2-9) samples at times t_i and t_{i-1} to form the continuous alignment measurement (as was done in the step leading to Equation (18.3.2.1-2)) which then obtains the equivalent form based on INS true heading measurements. Finally, we drop the noise and residual terms as negligible. The result is the equivalent to (18.3.2.1-3) and (18.3.2.1-6) for continuous alignment angular rate sensor random noise estimation based on INS true heading measurements:

$$
\begin{aligned}
\hat{\phi}_i &\equiv - \omega_{IE_H} \left(t_i \, \hat{\psi}_{T_i} - t_{i-1} \, \hat{\psi}_{T_{i-1}} \right) \\
\Delta\theta_{ARS/Rnd/East_i} &= \hat{\phi}_i - \mathcal{E}\left(\hat{\phi}_i\right)
\end{aligned}
\qquad (18.3.2.2\text{-}1)
$$

where

$\hat{\psi}_{T_i}$ = INS true heading measured at time t_i since the start of alignment.

$\hat{\phi}_i$ = The ith measurement for east angular rate sensor random output noise assessment.

$\Delta\theta_{ARS/Rnd/East_i}$ = Horizontal east component of $\Delta\underline{\theta}^N_{ARS/Rnd/H_i}$, the integrated angular rate sensor horizontal random output noise from time t_{i-1} to t_i (as defined in Equation (18.3.2.1-1)).

The extension of the final Section 18.3.2.1 - Equations (18.3.2.1-10) result to Equations (18.3.2.2-2) should now be obvious:

$$\hat{\phi}_i = -\, \omega_{IE_H} \left(t_i \, \hat{\psi}_{T_i} - t_{i-1} \, \hat{\psi}_{T_{i-1}} \right)$$

$$\mathcal{E}\left(\hat{\phi}_i\right) \approx \frac{1}{n} \sum_{i=1}^{n} \hat{\phi}_i \qquad (18.3.2.2\text{-}2)$$

$$\sigma_{ARS/Rnd/East} = \sqrt{\frac{\sum_{i=1}^{n} \left[\hat{\phi}_i - \mathcal{E}\left(\hat{\phi}_i\right)\right]^2}{(n-1)\,\Delta T}}$$

where

n = Total number of $\hat{\psi}_{T_i}$ measurements.

$\sigma_{ARS/Rnd/East}$ = Square root of the horizontal East angular rate sensor random noise density.

As in Section 18.3.2.1, Equation (18.3.2.1-11), an alternative version of (18.3.2.2-2) can also be derived that allows evaluation of $\sigma_{ARS/Rnd/East}$ at each i time point based on the total number of measurements taken up to t_i:

$$\hat{\phi}_i = -\, \omega_{IE_H} \left(t_i \, \hat{\psi}_{T_i} - t_{i-1} \, \hat{\psi}_{T_{i-1}} \right)$$

$$\sigma_{ARS/Rnd/East_n} = \sqrt{\frac{\left(\sum_{i=1}^{n} \hat{\phi}_i^2\right) - \frac{1}{n}\left(\sum_{i=1}^{n} \hat{\phi}_i\right)^2}{(n-1)\,\Delta T}} \qquad (18.3.2.2\text{-}3)$$

where

$\sigma_{ARS/Rnd/East_n}$ = Square root of the East angular rate sensor random process noise density (assumed constant independent of time) based on data taken up to i = n.

When using (18.3.2.2-3), the $\sum_{i=1}^{n} \hat{\phi}_i^2$ and $\sum_{i=1}^{n} \hat{\phi}_i$ summations are updated as each $\hat{\phi}_i$ measurement is taken. The current value for $\sigma_{ARS/Rnd/East_n}$ is then calculated at each i time point as an evolving estimate for horizontal random noise during the data taking process. The current value for $\sigma_{ARS/Rnd/East_n}$ would be the most accurate since it was generated with the most data samples.

Equations (18.3.2.2-3) allow estimation of random noise for the angular rate sensor along the East horizontal axis. Random noise is estimated for the North angular rate sensor by repeating the alignment and the Equation (18.3.2.2-3) process with the INS rotated in heading by 90 degrees from the original orientation (placing the previously North angular rate sensor in the Easterly direction). An estimate for the vertical angular rate sensor noise is determined by repeating the process with the INS at a 90 degree roll orientation that places the previously vertical angular rate sensor in the horizontal Easterly direction. Each of the previous orientations should be at zero pitch angle for the processing equations to be valid.

A comparison between the Continuous Alignment noise estimation method described in this section with the Section 18.3.1.2 Repeated Alignment method is identical to the discussion given in the fourth paragraph following Equation (18.3.2.1-11) in Section 18.3.2.1.

18.4 STRAPDOWN ROTATION TEST

The basic concept for the Strapdown Rotation Test was originally published in 1977 (Reference 30). Since then, variations of the concept have formed the basis in most strapdown inertial navigation system manufacturing organizations for system level calibration of accelerometer/angular-rate-sensor scale-factors/misalignments and accelerometer biases.

The Strapdown Rotation test consists of a series of rotations of the strapdown sensor assembly using a rotation test fixture for execution. During the test, special software operates on the strapdown angular rate sensor outputs from the sensor assembly to form an analytic platform (L Frame) that nominally maintains a constant orientation relative to the earth. The analytic platform is implemented by processing strapdown attitude-integration/acceleration-transformation algorithms (e.g., as in Chapter 7 - Table 7.5-1 including Chapter 8 - Table 8.4-1 inertial sensor compensation) using a wander azimuth type platform coordinate frame (i.e., Z axis downward along the local geodetic vertical with vertical transport rate ρ_{ZN} set to zero as defined in Section 4.5), assuming zero velocity relative to the earth ($\underline{v}$), and with the L Frame earth rate components ($\underline{\omega}_{IE}^{L}$) set constant. The $\underline{\omega}_{IE}^{L}$ earth rate components are calculated prior to rotation test initiation using special test software that implements strapdown initial alignment algorithms (e.g., Chapter 6 including Chapter 8 sensor compensation). The initial alignment operations also serve to initialize the C_B^L attitude matrix for the test software attitude updating algorithms. Section 18.4.2 describes alternative methods for $\underline{\omega}_{IE}^{L}$ and C_B^L initialization when testing sensor assemblies of lesser accuracy.

The basic measurement taken during the Strapdown Rotation test is the total L Frame acceleration vector computed as the transformed accelerometer specific force (from the Section 2.2 defined sensor axis B Frame to the L Frame) plus gravity. The L Frame acceleration is provided in the form of integrated acceleration increments over successive attitude-update/acceleration-transformation computation cycles based on Equations (7.2-2) and (7.2-4), with the L Frame acceleration calculated by averaging the integrated acceleration increments:

$$\Delta \underline{v}_m^L = \Delta \underline{v}_{SF_m}^L + g_{Tst}\, \underline{u}_{ZL}^L\, T_m \qquad \underline{a}^L = \frac{1}{T_m} \Delta \underline{v}_{Avg}^L \tag{18.4-1}$$

where

m = Test software computation cycle index.

T_m = Time interval between m cycles.

g_{Tst} = Plumb-bob gravity magnitude at the test site.

L = Locally level attitude referencing coordinate frame as defined in Section 2.2, and as specialized for this section, to be of the wander azimuth type (See Section 4.5).

$\underline{u}_{ZL}^L$ = Unit vector along the L Frame Z axis (i.e., downward along the local plumb-bob vertical).

$\Delta \underline{v}_{SF}^L$ = Integrated specific force acceleration increment in the L Frame measured by digital integration of transformed accelerometer data (from the B to the L Frame).

$\Delta \underline{v}_m^L$ = Increment of integrated total acceleration over an m cycle in L Frame coordinates.

$\Delta \underline{v}_{Avg}^L$ = The output from an averaging process performed on the selected succession of $\Delta \underline{v}_m^L$'s used for the particular stationary acceleration measurement. See Equation (18.4.7.3-18) as an example of an algorithm for the "average-of-averages" averaging process.

$\underline{a}^L$ = L Frame total acceleration measurement.

The $\Delta \underline{v}_{SF}^L$ input to the (18.4-1) $\Delta \underline{v}_m^L$ expression would be provided by the appropriate Table 7.5-1 "Velocity Calculations" algorithms.

The fundamental theory behind the Strapdown Rotation test is based on the principle that for a perfectly calibrated sensor assembly (i.e., in effect, error free), following a perfect initial alignment, the computed total acceleration in the L Frame (i.e., $\underline{a}^L$) should be zero at any time the sensor assembly is stationary. Moreover, this should also be the case if the sensor assembly undergoes arbitrary rotations between the time periods that it is set stationary. Therefore, any deviation from zero in $\underline{a}^L$ can be attributed to imperfections in the sensor assembly (i.e., sensor calibration errors) or in the initial alignment process.

The initial alignment process produces tilt error (i.e., attitude error measured from the local horizontal) as well as error in the $\underline{\omega}_{IE}^{L}$ earth rate vector used in the attitude algorithms (primarily in the horizontal components which is the equivalent to an initial heading error). The $\underline{\omega}_{IE}^{L}$ error produces a build-up rate in the attitude tilt which then generates a ramping build-up in $\underline{a}^{L}$ error. The timing for the rotation test can be structured to make the $\underline{\omega}_{IE}^{L}$ error effect negligible (discussed subsequently). The tilt error generates an error in the specific force transformation process that produces a constant horizontal error in $\underline{a}^{L}$ (this will subsequently be demonstrated analytically).

Based on the previous discussion, for a sensor assembly with zero calibrated sensor error, we expect the vertical component of $\underline{a}^{L}$ (call it a_{ZL}) to be zero when the sensor assembly is set stationary, but the horizontal component of $\underline{a}^{L}$ (call it $\underline{a}_{H}^{L}$) may be a small slowly changing constant (the constant portion from the initial horizontal tilt and the slowly changing portion from $\underline{\omega}_{IE}^{L}$ error). It is the characteristic of initial horizontal errors that they will remain constant in the L Frame, regardless of the sensor assembly attitude following initial alignment (to be subsequently shown analytically). Therefore, if we execute an arbitrary rotation sequence between stationary measurement periods, the $\underline{a}_{H}^{L}$ value measured before the rotation should approximately equal the value after the rotation. The approximation error can be minimized by performing the rotation sequence, and the before/after $\underline{a}_{H}^{L}$ measurements, rapidly. Thus, the difference between the before/after $\underline{a}_{H}^{L}$ measurements (call them $\Delta\underline{a}_{H}^{L}$) should be zero for a sensor assembly with zero calibrated sensor error. Deviations from zero for $\Delta\underline{a}_{H}^{L}$ and a_{ZL} measure the effect of strapdown sensor calibration error. By structuring a group of rotation sequences between stationary acceleration measurements, particular sensor error types can be excited and made visible in the $\Delta\underline{a}_{H}^{L}$, a_{ZL} data taken for each rotation sequence. Processing the composite set of $\Delta\underline{a}_{H}^{L}$, a_{ZL} data taken throughout the rotation test allows the sensor calibration errors to be calculated explicitly. Recalibration is then performed by updating the sensor calibration coefficients based on the measured calibration errors.

As an aside, it is to be noted that in the original Reference 30 paper, the measurement for the rotation test was the average acceleration taken at the end of each rotation sequence, with a self-alignment performed before the start of each rotation sequence. The purpose of the realignment was to eliminate attitude error build-up caused by angular rate sensor error during previous rotation sequences. By now taking the measurement as the difference between average accelerations before and after rotation sequence execution (as indicated above), the need for realignment is eliminated. This is because the attitude error at the start of the sequence remains at completion of the sequence (as does is its effect on average horizontal acceleration), hence, its impact on average acceleration measurement is canceled by taking the measurement as the difference between before/after horizontal acceleration readings. The before/after measurement approach was introduced by Downs in Reference 5 for compatibility with an existing Kalman filter used to extract the acceleration measurements. Additional refinements developed by the author for the Reference 30 method include a simple filtering algorithm for calculating the average acceleration measurements (to attenuate acceleration quantization error effects - See Section 18.4.7.3), use of "mean angular rate sensor coordinates" for referencing sensor misalignments (See Section 18.4.3), and the use of L Frame acceleration from a stand-alone analytic platform in determining the average acceleration measurements (i.e., transformed B Frame specific force acceleration, plus gravity). Average acceleration was determined in Reference 30 by operating the sensor assembly using a full set of inertial navigation software; then calculating average acceleration as the change in computed velocity over a specified time interval divided by the time interval (similar to the method described in Section 18.4.7.3.1).

Particular sensor assembly sensor calibration errors that are quickly and accurately determined from the Strapdown Rotation test are:

- Angular Misalignments Between Sensor Input Axes
- Angular Rate Sensor Scale Factor Errors
- Angular Rate Sensor Scale Factor Asymmetries
- Accelerometer Bias Errors
- Accelerometer Scale Factor Errors
- Accelerometer Scale Factor Asymmetries
- Sensor Assembly Misalignment Relative To Its Mount

The principal advantage for this particular method of error determination derives from the combined use of the angular rate sensors and accelerometers to establish a stabilized reference for measuring accelerations. This implicitly forces the angular rate sensors to measure the attitude of the rotation test fixture as the rotations are executed. Consequently, precision rotation test table readout or controls are not required (nor a stable test fixture base), hence, a significant savings can be made in test fixture cost. Inaccuracies in the rotation fixture manifest themselves as second order errors in sensor error determination, which can be made negligibly small if desired through a repeated test sequence. It has been demonstrated, for example, with precision ring laser gyro strapdown inertial navigation systems, that the test method can measure and calibrate gyro misalignments to better than 1 arc sec accuracy with 0.1 deg rotation fixture inaccuracies. In addition, because the orientation of the sensor assembly is being measured by the sensor assembly itself, it is not necessary that the sensor assembly be rigidly connected to the rotation test fixture. This is an important advantage for high accuracy applications in which the sensor assembly is attached to its chassis and mounting bracket through elastomeric isolators of marginal attitude stability.

While most of the sensor calibration errors evaluated by the Strapdown Rotation test can be measured on an individual sensor basis, the rotation test is the only direct method for measuring the relative misalignments between the sensor input axes. It should also be noted that determination of the sensor-assembly-to-mount misalignment (listed above) is not an intrinsic part of the Strapdown Rotation test, however, because the data taken during the rotation test allows for this determination, it can easily be included as part of the test data processing.

18.4.1 ROTATION SEQUENCE DESIGN/SELECTION

Selection of the sensor assembly test rotation sequences should be based on the following general ground-rules:

1. The rotation sequences should excite all sensor calibration errors so that they are made visible as transformed acceleration (horizontal or vertical) measurements.

2. A sufficient number and type of rotation sequences should be executed so that the acceleration measurements taken between rotations have distinctive responses such that the instrument errors can be ascertained by measurement data analysis.

3. The rotation sequences should be designed so that the sensor errors are excited and made visible on the acceleration measurements under sensor orientations (relative to vertical) that are representative of intended usage.

4. The rotation sequences should be designed so that the fewest number of error sources are excited for each rotation sequence (between measurements). Ideally, each sequence should excite only one particular error source. The measurement for this sequence can, therefore, be used alone to uniquely determine this error without being corrupted by other error effects.

5. When more than one sensor error is excited by a rotation sequence, the other rotations/measurements needed to determine these sensor errors should be performed near-together in time to assure that sensor errors do not change during the measurement period.

6. The rotations and measurements should be executed fairly rapidly (30 - 50 deg/sec rotation rates) to assure that sensor outputs are stable over the test period, and that angular rate sensor bias errors (and $\underline{\omega}_{IE}^{L}$ error) are not allowed to develop attitude errors that produce significant acceleration output measurement shifts.

Based on the above considerations, Tables 18.4.1-1 and Table 18.4.1-2 are examples of rotation sequence sets that have proven useful for calibrating strapdown sensor assemblies containing ring laser gyro angular rate sensors. Table 18.4.1-1 defines a set of 16 rotation sequences and Table 18.4.1-2 defines a set of 21 rotation sequences. The sequences in Tables 18.4.1-1 and 18.4.1-2 are designed for implementation on a two-axis rotation fixture with the outer rotation axis horizontal and the inner axis perpendicular to the outer axis.

Note that rotation sequences 1 - 6 (which are identical for the 16 and 21 rotation sequence sets of Tables 18.4.1-1 and 18.4.1-2), have the same attitude at the end of the sequence as at the start of the sequence. This positions the accelerometers at the same start/end attitude so that they produce zero accelerometer error contribution to $\Delta \underline{a}_H^L$ (the difference between the before/after horizontal $\underline{a}^L$ measurements). Thus, the $\Delta \underline{a}_H^L$ measurements in sequences 1 - 6 only respond to the effects of angular rate sensor error, hence, are used to determine angular rate sensor misalignment and scale factor calibration error. This, of course, was the rationale used in the design of rotation sequences 1 - 6.

The advantage of the 21 rotation set over the 16 set is that the last 15 rotations divide into three sets of 5, with each set of 5 designed to determine the calibration coefficient errors for one of the accelerometers (sequences 7 - 11 for accelerometer X, 12 - 16 for Y, and 17 - 21 for Z). Unstable individual accelerometer error effects are thereby more easily isolated, and test repeats can be performed for an individual accelerometer by re-executing only its 5 rotation sequences. In the 16 set, the last 10 rotation sequences are used to calibrate all accelerometers as a group, hence, allows a longer time period (10 sequences) for accelerometer instabilities to corrupt the calibration. Additionally, the ability to isolate individual accelerometer instability effects becomes more difficult due to the inherent coupling in the 10 rotations. Note that sequence 21 of the 21 rotation set (used for Z accelerometer error determination) is a repeat of sequence 12 of the 21 rotation set (used for Y accelerometer error determination). If sequences 17 - 21 are executed immediately before sequences 12 - 16, sequence 12 will occur immediately after sequence 21. For this type of rotation sequence ordering, the sequence 21 output data can be substituted for sequence 12 data, and the need for sequence 12 execution eliminated. However, if the sequences are performed in the Table 18.4.1-2 order, sequences 12 and 21 should both be executed to account for possible performance shifts in the Z accelerometer during the time interval between sequences 12 and 21.

Table 18.4.1-1

16 Set Rotation Test Sequences

SEQUENCE NUMBER	ROTATION SEQUENCE (Degrees, Axis)	STARTING ATTITUDE (+Z Down, Axis Indicated Along Outer Rotation Fixture Axis)
1	+360 Y	+Y
2	+360 X	+X
3	+90 Y, +360 Z, -90 Y	+Y
4	+180 Y, +90 Z, +180 X, -90 Z	+Y
5	+180 X, +90 Z, +180 Y, -90 Z	+X
6	+90 Y, +90 Z, -90 X, - 90 Z	+Y
7	+90 Y	+Y
8	-90 Y	+Y
9	+90 Y, +90 Z	+Y
10	+90 Y, -90 Z	+Y
11	-90 Y, -90 Z	+Y
12	+90 X, +90 Z	+X
13	+90 X, -90 Z	+X
14	+180 Z	+Y
15	+180 Y	+Y
16	+180 X	+X

*Note: Axes Indicated Correspond To Sensor Assembly (B Frame) Axes.

Table 18.4.1-2

21 Set Rotation Test Sequences

SEQUENCE NUMBER	ROTATION SEQUENCE (Degrees, Axis)	STARTING ATTITUDE Axis Down	STARTING ATTITUDE Axis Along Outer Rotation Fixture Axis
1	+360 Y	+Z	+Y
2	+360 X	+Z	+X
3	+90 Y, +360 Z, -90 Y	+Z	+Y
4	+180 Y, +90 Z, +180 X, -90 Z	+Z	+Y
5	+180 X, +90 Z, +180 Y, -90 Z	+Z	+X
6	+90 Y, +90 Z, -90 X, - 90 Z	+Z	+Y
7	+180 Y	+Z	+Y
8	+180 Y	- Z	+Y
9	+90 X	+Z	+X
10	+90 X	+Y	+X
11	+180 Z	+X	+Y
12	+180 X	+Z	+X
13	+180 X	- Z	+X
14	+90 Y	+Z	+Y
15	+90 Y	- X	+Y
16	+180 Z	+Y	+X
17	+180 X	- Y	+X
18	+180 X	+Y	+X
19	+90 Z	- Y	+X
20	+90 Z	- X	+Y
21	+180 X	+Z	+X

*Note: Axes Indicated Correspond To Sensor Assembly (B Frame) Axes.

18.4.2 ROTATION TEST DATA COLLECTION

The data collection procedure for the 16 and 21 rotation sets in Tables 18.4.1-l and 18.4.1-2 is as follows.

A two axis rotation test fixture is employed having its outer (trunion) rotation axis horizontal, its inner rotation axis perpendicular to the outer axis, with a mounting platform located around the inner axis for the sensor assembly test article. The strapdown sensor assembly is mounted on the test fixture mounting platform with the sensor coordinate (B Frame) Z axis along the rotation fixture inner axis. Sensor compensation software (e.g., Chapter 8) is included in the test software for compensating the sensor for previously determined sensor error effects. As a minimum, previously calculated estimates for the angular rate sensor biases should be included in the sensor compensation. It is also helpful if accelerometer and angular rate sensor scale factor estimates are included, however, these can also be estimated using an abbreviated version of the Strapdown Rotation test procedure. Note, that approximate values for the angular rate sensor biases can also be estimated by averaging the sensor assembly angular rate sensor outputs and subtracting the earth rate inputs, based on the known orientation of the sensor assembly during the bias determination process.

Prior to Strapdown Rotation test initiation, the rotation table is positioned so that the sensor assembly B Frame axes are oriented at the "zero starting attitude" defined as Z-axis down with the Y-axis along the rotation fixture horizontal outer-rotation-axis. The initial self-alignment procedure (e.g., Chapter 6) is then executed to initialize the C_B^L attitude matrix in the test software and to calculate $\underline{\omega}_{IE}^L$. The L Frame for the rotation test is defined to be nominally parallel to the B Frame axes during initial alignment, as the Chapter 6 alignment procedure would create. For sensor assemblies for which the angular rate sensors lack sufficient accuracy for an acceptable $\underline{\omega}_{IE}^L$ determination during initial alignment, a reasonable estimate for $\underline{\omega}_{IE}^L$ can be computed directly from the horizontal (north) earth rate component at the test site, transformed around the vertical from north to the "zero starting attitude" heading orientation of the sensor assembly. For this method, the Chapter 6 type initial alignment operation could still be utilized to initialize the C_B^L matrix attitude, but with $\underline{\omega}_{IE}^L$ computed during the alignment process then replaced at alignment completion by the previous described directly computed value. Alternatively, the C_B^L matrix can be initialized directly based on its approximate known angular orientation at the "zero starting attitude". It should also be noted that for sensor assemblies having large angular rate sensor bias instabilities, it may be necessary to periodically reinitialize the C_B^L matrix during the rotation test by repositioning the sensor assembly at the zero starting attitude, and repeating the C_B^L initialization process.

The rotation sequences are then executed as prescribed in Table 18.4.1-1 or Table 18.4.1-2, while the test software is executing its attitude update and acceleration transformation computation routines. Prior to initiation (but after sensor assembly initial attitude positioning shown in the tables), and after completion of each rotation sequence defined in the tables, the transformed acceleration vector components ($\underline{a}^L$) are measured as in Equation (18.4-1) using an appropriate measurement algorithm (e.g., the "average-of-averages" algorithm described in Section 18.4.7.3 Equations (18.4.7.3-18)). The $\Delta\underline{a}_H^L$, a_{ZL} data used to calculate the sensor calibration errors is obtained from the $\underline{a}^L$ before/after measurements. After each rotation sequence data collection, the sensor assembly is rotated to the starting attitude for the next rotation sequence as defined in the tables. In structuring the Table 18.4.1-2 sequences, it is helpful to recognize that the starting attitude for sequences 7, 8, 10, 13, 15, 18, 19 and 20 are the final attitude of the previous sequence.

After completion of the rotation test series, 16 or 21 sets of $\Delta\underline{a}_H^L$, a_{ZL} data should have been taken and stored (depending on whether the Table 18.4.1-1 or Table 18.4.1-2 rotation sequences are used).

18.4.3 MEASUREMENTS IN TERMS OF SENSOR ERRORS

The analytical relationship between the sensor calibration errors and the $\Delta\underline{a}_H^L$, a_{ZL} measurements are described by Equations (18.4.7-13) with Equations (18.4.7.1-4), (18.4.7.1-2) and (18.4.7.1-5) for $\delta\underline{a}_{SF}^B$, Equations (18.4.7.2-23) and (18.4.7.2-11) for $\Delta\underline{\phi}_H^L$, and the C_B^L terms calculated as described in the final paragraph of Section 18.4.7.2.

For the particular rotation sequences of Tables 18.4.1-1 and 18.4.1-2, generalized measurement Equations (18.4.7-13) have been evaluated for $\Delta\underline{a}_H^L$, a_{ZL} in scalar form in terms of particular sensor calibration error components defined by Equations (18.4.7.1-2) and (18.4.7.2-11). The results are summarized in Figure 18.4.3-1 for the 16 rotation set, and in Figure 18.4.3-2 for the 21 rotation set, where the following definitions apply:

Δa_i, Δb_i = L Frame X, Y axis components of $\Delta\underline{a}_H^L$ defined as the difference between $\underline{a}^L$ horizontal measurements (i.e., $\underline{a}_H^L$) taken at the start and end of rotation sequence i.

c_i^1, c_i^2 = L Frame Z axis (vertical) acceleration measurement (i.e., a_{ZL}) taken immediately before (superscript 1) and after (superscript 2) rotation sequence i.

α_i = i axis accelerometer bias calibration error.

λ_{ii} = i axis accelerometer symmetrical scale factor calibration error.

λ_{iii} = i axis accelerometer scale factor asymmetry calibration error.

λ_{ij} = i axis accelerometer misalignment calibration error, coupling B Frame j axis specific force into the i axis.

κ_{ii} = i axis angular rate sensor scale factor calibration error.

κ_{ij} = i axis angular rate sensor misalignment calibration error, coupling B Frame j axis angular rate into the i axis.

Note that ring laser gyro angular rate sensors have no scale factor asymmetry error, hence, the κ_{iii} terms (not shown in Figures 18.4.3-1 and 18.4.3-2) are zero. If the angular rate sensors being evaluated have scale factor asymmetry, three additional rotation sequences would be included in Tables 18.4.1-1 and 18.4.1-2 that parallel rotation sequences 1 - 3, except that the 360 degree rotation portion would be executed in the opposite direction. These additional three sequences produce opposite $\Delta \underline{a}_H^L$ signatures from sequences 1 - 3 for symmetrical scale factor error effects, but the same signatures for asymmetrical scale factor errors. Summing and differencing the results allow the symmetrical and asymmetrical scale factor calibration errors to be individually assessed.

As an aside, let us briefly return to Equation (13.2.4-28) of Section 13.2.4 which shows transformed acceleration as a function of inertial sensor misalignment (and other effects). Except for the $\delta L_{Orth}\, \underline{a}_{SF}^B$ term, all of the misalignment error terms contributing to transformed acceleration in (13.2.4-28) are perpendicular to the $\underline{a}_{SF}$ specific force vector. Furthermore, for the Table 18.4.1-1 or 18.4.1-2 test orientations, $\underline{a}_{SF}$ will be along one of the sensor assembly axes (B Frame) during test measurements. Since the δL_{Orth} matrix is defined in Section 13.2.4 with zero along the diagonal, this assures that $\delta L_{Orth}\, \underline{a}_{SF}^B$ will also be perpendicular to $\underline{a}_{SF}$. We recognize that for the Strapdown Rotation test static acceleration measurements, the $\underline{a}_{SF}$ specific force vector is vertical (i.e., 1 g upward to balance against downward gravity). We conclude that for the Table 18.4.1-1 or 18.4.1-2 test sequences, the effect of sensor misalignment on the test measurements will be horizontal. Thus, we expect no misalignment terms in the c_i^1, c_i^2 expressions of Figures 18.4.3-1 and 18.4.3-2. If misalignment terms did appear in the c_i^1, c_i^2 equations, it would signify an error in their analytical derivation.

$$\Delta a_1 = -2\pi g \kappa_{yy}$$
$$\Delta b_1 = 0$$
$$c_1^1 = c_1^2 = -g(\lambda_{zz} - \lambda_{zzz}) + \alpha_z$$

$$\Delta a_2 = -2\pi g \kappa_{xx}$$
$$\Delta b_2 = 0$$
$$c_2^1 = c_2^2 = -g(\lambda_{zz} - \lambda_{zzz}) + \alpha_z$$

$$\Delta a_3 = 0$$
$$\Delta b_3 = 2\pi g \kappa_{zz}$$
$$c_3^1 = c_3^2 = -g(\lambda_{zz} - \lambda_{zzz}) + \alpha_z$$

$$\Delta a_4 = -\pi g(\kappa_{xx} + \kappa_{yy})$$
$$\Delta b_4 = 2g(\kappa_{zy} + \kappa_{yz} - \kappa_{zx} - \kappa_{xz})$$
$$c_4^1 = c_4^2 = -g(\lambda_{zz} - \lambda_{zzz}) + \alpha_z$$

$$\Delta a_5 = \pi g(\kappa_{yy} - \kappa_{xx})$$
$$\Delta b_5 = 2g(\kappa_{zx} + \kappa_{xz} + \kappa_{zy} + \kappa_{yz})$$
$$c_5^1 = c_5^2 = -g(\lambda_{zz} - \lambda_{zzz}) + \alpha_z$$

$$\Delta a_6 = \frac{\pi}{2} g(\kappa_{xx} - \kappa_{yy})$$
$$\Delta b_6 = g\left(\kappa_{zy} + \kappa_{yz} + \kappa_{yx} + \kappa_{xy} - \kappa_{zx} - \kappa_{xz} + \frac{\pi}{2}\kappa_{zz}\right)$$
$$c_6^1 = c_6^2 = -g(\lambda_{zz} - \lambda_{zzz}) + \alpha_z$$

$$\Delta a_7 = g\left(\lambda_{zx} + \lambda_{xz} - \frac{\pi}{2}\kappa_{yy}\right) - \alpha_x + \alpha_z$$
$$\Delta b_7 = g(\kappa_{xy} + \kappa_{zy} + \lambda_{yx} + \lambda_{yz})$$
$$c_7^1 = -g(\lambda_{zz} - \lambda_{zzz}) + \alpha_z$$
$$c_7^2 = -g(\lambda_{xx} + \lambda_{xxx}) - \alpha_x$$

$$\Delta a_8 = g\left(\lambda_{zx} + \lambda_{xz} + \frac{\pi}{2}\kappa_{yy}\right) - \alpha_x - \alpha_z$$
$$\Delta b_8 = g(\kappa_{zy} - \kappa_{xy} - \lambda_{yx} + \lambda_{yz})$$
$$c_8^1 = -g(\lambda_{zz} - \lambda_{zzz}) + \alpha_z$$
$$c_8^2 = -g(\lambda_{xx} - \lambda_{xxx}) + \alpha_x$$

Figure 18.4.3-1

Acceleration Measurement In Terms Of Sensor Errors For The 16 Sequence Rotation Test

$$\Delta a_9 = -g\left(\kappa_{xz} + \kappa_{yz} + \lambda_{zy} - \lambda_{xz} + \frac{\pi}{2}\kappa_{yy}\right) - \alpha_x + \alpha_z$$

$$\Delta b_9 = g\left(\kappa_{xy} + \kappa_{zy} + \lambda_{yz} - \lambda_{xy} + \frac{\pi}{2}\kappa_{zz}\right) + \alpha_x - \alpha_y$$

$$c_9^1 = -g\left(\lambda_{zz} - \lambda_{zzz}\right) + \alpha_z$$

$$c_9^2 = -g\left(\lambda_{yy} - \lambda_{yyy}\right) + \alpha_y$$

$$\Delta a_{10} = g\left(\kappa_{yz} - \kappa_{xz} + \lambda_{xz} + \lambda_{zy} - \frac{\pi}{2}\kappa_{yy}\right) - \alpha_x + \alpha_z$$

$$\Delta b_{10} = g\left(\kappa_{xy} + \kappa_{zy} + \lambda_{yz} - \lambda_{xy} - \frac{\pi}{2}\kappa_{zz}\right) - \alpha_x - \alpha_y$$

$$c_{10}^1 = -g\left(\lambda_{zz} - \lambda_{zzz}\right) + \alpha_z$$

$$c_{10}^2 = -g\left(\lambda_{yy} + \lambda_{yyy}\right) - \alpha_y$$

$$\Delta a_{11} = g\left(\kappa_{yz} - \kappa_{xz} + \lambda_{zy} + \lambda_{xz} + \frac{\pi}{2}\kappa_{yy}\right) - \alpha_x - \alpha_z$$

$$\Delta b_{11} = g\left(\kappa_{zy} - \kappa_{xy} + \lambda_{yz} + \lambda_{xy} + \frac{\pi}{2}\kappa_{zz}\right) - \alpha_x - \alpha_y$$

$$c_{11}^1 = -g\left(\lambda_{zz} - \lambda_{zzz}\right) + \alpha_z$$

$$c_{11}^2 = -g\left(\lambda_{yy} - \lambda_{yyy}\right) + \alpha_y$$

$$\Delta a_{12} = g\left(\kappa_{yz} - \kappa_{xz} - \lambda_{zx} - \lambda_{yz} - \frac{\pi}{2}\kappa_{xx}\right) + \alpha_y + \alpha_z$$

$$\Delta b_{12} = g\left(\kappa_{zx} - \kappa_{yx} + \lambda_{yx} + \lambda_{xz} + \frac{\pi}{2}\kappa_{zz}\right) - \alpha_x - \alpha_y$$

$$c_{12}^1 = -g\left(\lambda_{zz} - \lambda_{zzz}\right) + \alpha_z$$

$$c_{12}^2 = -g\left(\lambda_{xx} - \lambda_{xxx}\right) + \alpha_x$$

$$\Delta a_{13} = g\left(\kappa_{xz} + \kappa_{yz} + \lambda_{zx} - \lambda_{yz} - \frac{\pi}{2}\kappa_{xx}\right) + \alpha_y + \alpha_z$$

$$\Delta b_{13} = g\left(\kappa_{zx} - \kappa_{yx} + \lambda_{yx} + \lambda_{xz} - \frac{\pi}{2}\kappa_{zz}\right) - \alpha_x + \alpha_y$$

$$c_{13}^1 = -g\left(\lambda_{zz} - \lambda_{zzz}\right) + \alpha_z$$

$$c_{13}^2 = -g\left(\lambda_{xx} + \lambda_{xxx}\right) - \alpha_x$$

$$\Delta a_{14} = -2g\left(\kappa_{xz} - \lambda_{xz}\right) - 2\alpha_x$$

$$\Delta b_{14} = -2g\left(\kappa_{yz} - \lambda_{yz}\right) - 2\alpha_y$$

$$c_{14}^1 = c_{14}^2 = -g\left(\lambda_{zz} - \lambda_{zzz}\right) + \alpha_z$$

$$\Delta a_{15} = -\pi g \kappa_{yy} - 2\alpha_x$$

$$\Delta b_{15} = 2g\left(\kappa_{zy} + \lambda_{yz}\right)$$

$$c_{15}^1 = -g\left(\lambda_{zz} - \lambda_{zzz}\right) + \alpha_z$$

$$c_{15}^2 = -g\left(\lambda_{zz} + \lambda_{zzz}\right) - \alpha_z$$

$$\Delta a_{16} = -\pi g \kappa_{xx} + 2\alpha_y$$

$$\Delta b_{16} = 2g\left(\kappa_{zx} + \lambda_{xz}\right)$$

$$c_{16}^1 = -g\left(\lambda_{zz} - \lambda_{zzz}\right) + \alpha_z$$

$$c_{16}^2 = -g\left(\lambda_{zz} + \lambda_{zzz}\right) - \alpha_z$$

Figure 18.4.3-1 (Continued)

Acceleration Measurement In Terms Of Sensor Errors For The 16 Sequence Rotation Test

$$\Delta a_1 = -2\,\pi\,g\,\kappa_{yy}$$
$$\Delta b_1 = 0$$
$$c_1^1 = c_1^2 = -g\left(\lambda_{zz} - \lambda_{zzz}\right) + \alpha_z$$

$$\Delta a_2 = -2\,\pi\,g\,\kappa_{xx}$$
$$\Delta b_2 = 0$$
$$c_2^1 = c_2^2 = -g\left(\lambda_{zz} - \lambda_{zzz}\right) + \alpha_z$$

$$\Delta a_3 = 0$$
$$\Delta b_3 = 2\,\pi\,g\,\kappa_{zz}$$
$$c_3^1 = c_3^2 = -g\left(\lambda_{zz} - \lambda_{zzz}\right) + \alpha_z$$

$$\Delta a_4 = -\pi\,g\left(\kappa_{xx} + \kappa_{yy}\right)$$
$$\Delta b_4 = 2\,g\left(\kappa_{zy} + \kappa_{yz} - \kappa_{zx} - \kappa_{xz}\right)$$
$$c_4^1 = c_4^2 = -g\left(\lambda_{zz} - \lambda_{zzz}\right) + \alpha_z$$

$$\Delta a_5 = \pi\,g\left(\kappa_{yy} - \kappa_{xx}\right)$$
$$\Delta b_5 = 2\,g\left(\kappa_{zx} + \kappa_{xz} + \kappa_{zy} + \kappa_{yz}\right)$$
$$c_5^1 = c_5^2 = -g\left(\lambda_{zz} - \lambda_{zzz}\right) + \alpha_z$$

$$\Delta a_6 = \frac{\pi}{2}\,g\left(\kappa_{xx} - \kappa_{yy}\right)$$
$$\Delta b_6 = g\left(\kappa_{zy} + \kappa_{yz} + \kappa_{yx} + \kappa_{xy} - \kappa_{zx} - \kappa_{xz} + \frac{\pi}{2}\,\kappa_{zz}\right)$$
$$c_6^1 = c_6^2 = -g\left(\lambda_{zz} - \lambda_{zzz}\right) + \alpha_z$$

$$\Delta a_7 = -\pi\,g\,\kappa_{yy} - 2\,\alpha_x$$
$$\Delta b_7 = 2\,g\left(\kappa_{zy} + \lambda_{yz}\right)$$
$$c_7^1 = -g\left(\lambda_{zz} - \lambda_{zzz}\right) + \alpha_z$$
$$c_7^2 = -g\left(\lambda_{zz} + \lambda_{zzz}\right) - \alpha_z$$

$$\Delta a_8 = -\pi\,g\,\kappa_{yy} + 2\,\alpha_x$$
$$\Delta b_8 = -2\,g\left(\kappa_{zy} + \lambda_{yz}\right)$$
$$c_8^1 = -g\left(\lambda_{zz} + \lambda_{zzz}\right) - \alpha_z$$
$$c_8^2 = -g\left(\lambda_{zz} - \lambda_{zzz}\right) + \alpha_z$$

$$\Delta a_9 = -g\left(\lambda_{zy} + \lambda_{yz} + \frac{\pi}{2}\,\kappa_{xx}\right) + \alpha_y + \alpha_z$$
$$\Delta b_9 = g\left(\kappa_{zx} - \kappa_{yx} + \lambda_{xz} - \lambda_{xy}\right)$$
$$c_9^1 = -g\left(\lambda_{zz} - \lambda_{zzz}\right) + \alpha_z$$
$$c_9^2 = -g\left(\lambda_{yy} - \lambda_{yyy}\right) + \alpha_y$$

$$\Delta a_{10} = g\left(\lambda_{zy} + \lambda_{yz} - \frac{\pi}{2}\,\kappa_{xx}\right) + \alpha_y - \alpha_z$$
$$\Delta b_{10} = g\left(\kappa_{zx} + \kappa_{yx} + \lambda_{xz} + \lambda_{xy}\right)$$
$$c_{10}^1 = -g\left(\lambda_{yy} - \lambda_{yyy}\right) + \alpha_y$$
$$c_{10}^2 = -g\left(\lambda_{zz} + \lambda_{zzz}\right) - \alpha_z$$

Figure 18.4.3-2

Acceleration Measurement In Terms Of Sensor Errors For The 21 Sequence Rotation Test

$$\Delta a_{11} = -2\,g\left(\kappa_{XZ} + \lambda_{ZX}\right)$$
$$\Delta b_{11} = -\pi\,g\,\kappa_{ZZ} - 2\,\alpha_y$$
$$c^1_{11} = -g\left(\lambda_{XX} - \lambda_{XXX}\right) + \alpha_X$$
$$c^2_{11} = -g\left(\lambda_{XX} + \lambda_{XXX}\right) - \alpha_X$$

$$\Delta a_{16} = 2\,g\left(\kappa_{yZ} + \lambda_{Zy}\right)$$
$$\Delta b_{16} = \pi\,g\,\kappa_{ZZ} - 2\,\alpha_X$$
$$c^1_{16} = -g\left(\lambda_{yy} - \lambda_{yyy}\right) + \alpha_y$$
$$c^2_{16} = -g\left(\lambda_{yy} + \lambda_{yyy}\right) - \alpha_y$$

$$\Delta a_{21} = -\pi\,g\,\kappa_{XX} + 2\,\alpha_y$$
$$\Delta b_{21} = 2\,g\left(\kappa_{ZX} + \lambda_{XZ}\right)$$
$$c^1_{21} = -g\left(\lambda_{ZZ} - \lambda_{ZZZ}\right) + \alpha_Z$$
$$c^2_{21} = -g\left(\lambda_{ZZ} + \lambda_{ZZZ}\right) - \alpha_Z$$

$$\Delta a_{12} = -\pi\,g\,\kappa_{XX} + 2\,\alpha_y$$
$$\Delta b_{12} = 2\,g\left(\kappa_{ZX} + \lambda_{XZ}\right)$$
$$c^1_{12} = -g\left(\lambda_{ZZ} - \lambda_{ZZZ}\right) + \alpha_Z$$
$$c^2_{12} = -g\left(\lambda_{ZZ} + \lambda_{ZZZ}\right) - \alpha_Z$$

$$\Delta a_{17} = -\pi\,g\,\kappa_{XX} + 2\,\alpha_Z$$
$$\Delta b_{17} = -2\,g\left(\kappa_{yX} + \lambda_{Xy}\right)$$
$$c^1_{17} = -g\left(\lambda_{yy} + \lambda_{yyy}\right) - \alpha_y$$
$$c^2_{17} = -g\left(\lambda_{yy} - \lambda_{yyy}\right) + \alpha_y$$

$$\Delta a_{13} = -\pi\,g\,\kappa_{XX} - 2\,\alpha_y$$
$$\Delta b_{13} = -2\,g\left(\kappa_{ZX} + \lambda_{XZ}\right)$$
$$c^1_{13} = -g\left(\lambda_{ZZ} + \lambda_{ZZZ}\right) - \alpha_Z$$
$$c^2_{13} = -g\left(\lambda_{ZZ} - \lambda_{ZZZ}\right) + \alpha_Z$$

$$\Delta a_{18} = -\pi\,g\,\kappa_{XX} - 2\,\alpha_Z$$
$$\Delta b_{18} = 2\,g\left(\kappa_{yX} + \lambda_{Xy}\right)$$
$$c^1_{18} = -g\left(\lambda_{yy} - \lambda_{yyy}\right) + \alpha_y$$
$$c^2_{18} = -g\left(\lambda_{yy} + \lambda_{yyy}\right) - \alpha_y$$

$$\Delta a_{14} = g\left(\lambda_{ZX} + \lambda_{XZ} - \frac{\pi}{2}\,\kappa_{yy}\right) - \alpha_X + \alpha_Z$$
$$\Delta b_{14} = g\left(\kappa_{Xy} + \kappa_{Zy} + \lambda_{yX} + \lambda_{yZ}\right)$$
$$c^1_{14} = -g\left(\lambda_{ZZ} - \lambda_{ZZZ}\right) + \alpha_Z$$
$$c^2_{14} = -g\left(\lambda_{XX} + \lambda_{XXX}\right) - \alpha_X$$

$$\Delta a_{19} = g\left(\kappa_{yZ} - \kappa_{XZ} - \lambda_{ZX} + \lambda_{Zy}\right)$$
$$\Delta b_{19} = -g\left(\lambda_{yX} + \lambda_{Xy} + \frac{\pi}{2}\,\kappa_{ZZ}\right) - \alpha_X - \alpha_y$$
$$c^1_{19} = -g\left(\lambda_{yy} + \lambda_{yyy}\right) - \alpha_y$$
$$c^2_{19} = -g\left(\lambda_{XX} + \lambda_{XXX}\right) - \alpha_X$$

$$\Delta a_{15} = -g\left(\lambda_{ZX} + \lambda_{XZ} + \frac{\pi}{2}\,\kappa_{yy}\right) - \alpha_X - \alpha_Z$$
$$\Delta b_{15} = g\left(\kappa_{Zy} - \kappa_{Xy} + \lambda_{yZ} - \lambda_{yX}\right)$$
$$c^1_{15} = -g\left(\lambda_{XX} + \lambda_{XXX}\right) - \alpha_X$$
$$c^2_{15} = -g\left(\lambda_{ZZ} + \lambda_{ZZZ}\right) - \alpha_Z$$

$$\Delta a_{20} = g\left(\kappa_{yZ} + \kappa_{XZ} + \lambda_{ZX} + \lambda_{Zy}\right)$$
$$\Delta b_{20} = g\left(\lambda_{yX} + \lambda_{Xy} - \frac{\pi}{2}\,\kappa_{ZZ}\right) - \alpha_X + \alpha_y$$
$$c^1_{20} = -g\left(\lambda_{XX} + \lambda_{XXX}\right) - \alpha_X$$
$$c^2_{20} = -g\left(\lambda_{yy} - \lambda_{yyy}\right) + \alpha_y$$

Figure 18.4.3-2 (Continued)

Acceleration Measurement In Terms Of Sensor Errors For The 21 Sequence Rotation Test

It is important to note (as explained analytically in Section 13.2.4) that due to the nature of the measurement process, only relative sensor misalignment calibration errors affect the L Frame acceleration measurements during the Strapdown Rotation test (i.e., sensor input axis misalignments relative to one-another). This can be seen directly from Equation (13.2.4-28) which shows that the error in transformed specific force produced by inertial sensor misalignment has a constant component (dependent on C_B^L matrix initialization error and angular rate sensor triad misalignment relative to sensor assembly reference axes), and additional terms dependent on relative misalignment between the accelerometers or between the angular rate sensors (i.e., orthogonality errors). Due to their differential nature, the Strapdown Rotation test $\Delta\underline{a}_{SF_H}^L$ measurements cancel the constant component, and thereby, only measure the effect of relative misalignment between the angular rate sensor and accelerometer triads, and the relative misalignment between the sensors in each triad. The net result is that three scalar misalignment reference constraints (one for each sensor assembly axis) can be arbitrarily introduced into the sensor misalignment calibration error definitions, that in effect, define a particular (not relative) inertial sensor reference frame for the $\Delta\underline{a}_{SF_H}^L$ measurements. For the specific Strapdown Rotation test routines described in this section, we will apply these constraints as the requirement that sensor misalignment calibration errors be evaluated relative to "mean angular rate sensor axes" defined in Figure 18.4.3-3.

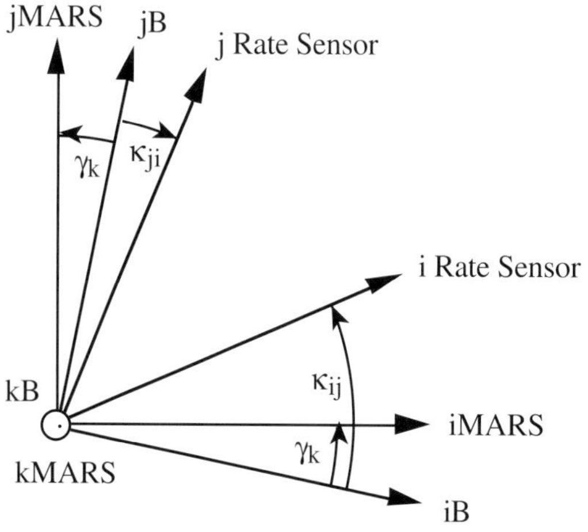

Figure 18.4.3-3 Mean Angular Rate Sensor Axes

where the following definitions apply:

iMARS, jMARS, kMARS = Mutually orthogonal i, j, k Mean Angular Rate Sensor (MARS) coordinate axes.

iB, jB, kB = Mutually orthogonal i, j, k sensor assembly B Frame coordinate axes.

i Rate Sensor, j Rate Sensor = Input axes for the i and j axis angular rate sensors (including compensation correction applied during the rotation test). The Strapdown Rotation test measures the effect of sensor compensation error.

γ_k = Angle measured around the kB axis, of the i, j MARS axes relative to the B Frame i, j axes.

κ_{ij}, κ_{ji} = Misalignment calibration errors (angular errors to first order) for the i, j angular rate sensors, coupling B Frame j, i axis angular rate into the i, j angular rate sensor input axes.

The mean angular rate sensor axes are defined as the orthogonal triad that best fits symmetrically within the actual compensated angular rate sensor input axes. The "best fit" condition is specified as the condition for which the angle between the angular rate sensor i input axis and mean angular rate sensor axis i equals the angle between the j angular rate sensor input and mean sensor axes. From Figure 18.4.3-3 we see that this condition corresponds to:

$$\kappa_{ji} + \gamma_k = \kappa_{ij} - \gamma_k \tag{18.4.3-1}$$

or upon rearrangement for each i, j group:

$$\gamma_x = \frac{1}{2}\left(\kappa_{yz} - \kappa_{zy}\right) \qquad \gamma_y = \frac{1}{2}\left(\kappa_{zx} - \kappa_{xz}\right) \qquad \gamma_z = \frac{1}{2}\left(\kappa_{xy} - \kappa_{yx}\right) \tag{18.4.3-2}$$

Recognizing the B Frame i, j sensor axes in Figure 18.4.3-3 to be approximately perpendicular, allows us to also write for the orthogonality error between the compensated angular rate sensor input axes:

$$\upsilon_{xy} = \kappa_{xy} + \kappa_{yx} \qquad \upsilon_{yz} = \kappa_{yz} + \kappa_{zy} \qquad \upsilon_{zx} = \kappa_{zx} + \kappa_{xz} \tag{18.4.3-3}$$

where

υ_{ij} = Orthogonality compensation error between the i and j angular rate sensor input axes, defined as $\pi/2$ radians minus the angle between the compensated i and j sensor input axes.

Accelerometer misalignment calibration errors relative to mean angular rate sensor axes can now be defined as the difference between misalignment relative to B Frame reference axes (the λ_{ij} coefficients), and misalignment of the mean angular rate sensor axes relative to the B Frame (the γ_k misalignments). Using Figure 18.4.3-3 as a guide (with λ replacing κ) we see that:

$$\mu_{xy} = \lambda_{xy} - \gamma_z \qquad \mu_{yx} = \lambda_{yx} + \gamma_z$$
$$\mu_{yz} = \lambda_{yz} - \gamma_x \qquad \mu_{zy} = \lambda_{zy} + \gamma_x \tag{18.4.3-4}$$
$$\mu_{zx} = \lambda_{zx} - \gamma_y \qquad \mu_{xz} = \lambda_{xz} + \gamma_y$$

where

μ_{ij} = i axis accelerometer misalignment calibration error, coupling specific force from the j axis of the mean angular rate sensor axes into the i axis accelerometer input axis.

Equations (18.4.3-2) through (18.4.3-4) combined provide the equivalent inverse relationships:

$$\kappa_{xy} = \frac{1}{2}\upsilon_{xy} + \gamma_z \qquad \kappa_{yx} = \frac{1}{2}\upsilon_{xy} - \gamma_z$$
$$\kappa_{yz} = \frac{1}{2}\upsilon_{yz} + \gamma_x \qquad \kappa_{zy} = \frac{1}{2}\upsilon_{yz} - \gamma_x$$
$$\kappa_{zx} = \frac{1}{2}\upsilon_{zx} + \gamma_y \qquad \kappa_{xz} = \frac{1}{2}\upsilon_{zx} - \gamma_y$$
$$\tag{18.4.3-5}$$
$$\lambda_{xy} = \mu_{xy} + \gamma_z \qquad \lambda_{yx} = \mu_{yx} - \gamma_z$$
$$\lambda_{yz} = \mu_{yz} + \gamma_x \qquad \lambda_{zy} = \mu_{zy} - \gamma_x$$
$$\lambda_{zx} = \mu_{zx} + \gamma_y \qquad \lambda_{xz} = \mu_{xz} - \gamma_y$$

Equations (18.4.3-5) can be substituted into the 16 and 21 rotation set equations of Figures 18.4.3-1 and 18.4.3-2 to determine the equivalent expressions relating measurement accelerations to sensor calibration errors, with accelerometer misalignments defined relative to mean angular rate sensor axes, and angular rate sensor misalignments defined relative to each other as orthogonality errors. The final results are summarized in Figures 18.4.3-4 and 18.4.3-5. Note, that the γ_k components do not appear in Figures 18.4.3-4 and 18.4.3-5 because they cancel in the (18.4.3-5) substitution process. This is yet another statement that misalignments of the sensor assembly as a whole do not impact the $\Delta \underline{a}_H^L$, a_{ZL} measurements. If γ_k components did appear in the Figure 18.4.3-4 and 18.4.3-5 expressions, it would signify an error in their analytical derivation.

$$\Delta a_1 = -2\pi g \kappa_{yy}$$
$$\Delta b_1 = 0$$
$$c_1^1 = c_1^2 = -g(\lambda_{zz} - \lambda_{zzz}) + \alpha_z$$

$$\Delta a_2 = -2\pi g \kappa_{xx}$$
$$\Delta b_2 = 0$$
$$c_2^1 = c_2^2 = -g(\lambda_{zz} - \lambda_{zzz}) + \alpha_z$$

$$\Delta a_3 = 0$$
$$\Delta b_3 = 2\pi g \kappa_{zz}$$
$$c_3^1 = c_3^2 = -g(\lambda_{zz} - \lambda_{zzz}) + \alpha_z$$

$$\Delta a_4 = -\pi g(\kappa_{xx} + \kappa_{yy})$$
$$\Delta b_4 = 2g(\upsilon_{yz} - \upsilon_{zx})$$
$$c_4^1 = c_4^2 = -g(\lambda_{zz} - \lambda_{zzz}) + \alpha_z$$

$$\Delta a_5 = \pi g(\kappa_{yy} - \kappa_{xx})$$
$$\Delta b_5 = 2g(\upsilon_{zx} + \upsilon_{yz})$$
$$c_5^1 = c_5^2 = -g(\lambda_{zz} - \lambda_{zzz}) + \alpha_z$$

$$\Delta a_6 = \frac{\pi}{2} g(\kappa_{xx} - \kappa_{yy})$$
$$\Delta b_6 = g\left(\upsilon_{yz} + \upsilon_{xy} - \upsilon_{zx} + \frac{\pi}{2}\kappa_{zz}\right)$$
$$c_6^1 = c_6^2 = -g(\lambda_{zz} - \lambda_{zzz}) + \alpha_z$$

$$\Delta a_7 = g\left(\mu_{zx} + \mu_{xz} - \frac{\pi}{2}\kappa_{yy}\right) - \alpha_x + \alpha_z$$
$$\Delta b_7 = g\left(\frac{1}{2}\upsilon_{xy} + \frac{1}{2}\upsilon_{yz} + \mu_{yx} + \mu_{yz}\right)$$
$$c_7^1 = -g(\lambda_{zz} - \lambda_{zzz}) + \alpha_z$$
$$c_7^2 = -g(\lambda_{xx} + \lambda_{xxx}) - \alpha_x$$

$$\Delta a_8 = g\left(\mu_{zx} + \mu_{xz} + \frac{\pi}{2}\kappa_{yy}\right) - \alpha_x - \alpha_z$$
$$\Delta b_8 = g\left(\frac{1}{2}\upsilon_{yz} - \frac{1}{2}\upsilon_{xy} - \mu_{yx} + \mu_{yz}\right)$$
$$c_8^1 = -g(\lambda_{zz} - \lambda_{zzz}) + \alpha_z$$
$$c_8^2 = -g(\lambda_{xx} - \lambda_{xxx}) + \alpha_x$$

Figure 18.4.3-4

Acceleration Measurements In Terms Of Revised Sensor Error Parameters For The 16 Sequence Rotation Test

$$\Delta a_9 = -g\left(\frac{1}{2}\upsilon_{zx} + \frac{1}{2}\upsilon_{yz} + \mu_{zy} - \mu_{xz} + \frac{\pi}{2}\kappa_{yy}\right) - \alpha_x + \alpha_z$$

$$\Delta b_9 = g\left(\frac{1}{2}\upsilon_{xy} + \frac{1}{2}\upsilon_{yz} + \mu_{yz} - \mu_{xy} + \frac{\pi}{2}\kappa_{zz}\right) + \alpha_x - \alpha_y$$

$$c_9^1 = -g\left(\lambda_{zz} - \lambda_{zzz}\right) + \alpha_z$$

$$c_9^2 = -g\left(\lambda_{yy} - \lambda_{yyy}\right) + \alpha_y$$

$$\Delta a_{12} = g\left(\frac{1}{2}\upsilon_{yz} - \frac{1}{2}\upsilon_{zx} - \mu_{zx} - \mu_{yz} - \frac{\pi}{2}\kappa_{xx}\right) + \alpha_y + \alpha_z$$

$$\Delta b_{12} = g\left(\frac{1}{2}\upsilon_{zx} - \frac{1}{2}\upsilon_{xy} + \mu_{yx} + \mu_{xz} + \frac{\pi}{2}\kappa_{zz}\right) - \alpha_x - \alpha_y$$

$$c_{12}^1 = -g\left(\lambda_{zz} - \lambda_{zzz}\right) + \alpha_z$$

$$c_{12}^2 = -g\left(\lambda_{xx} - \lambda_{xxx}\right) + \alpha_x$$

$$\Delta a_{10} = g\left(\frac{1}{2}\upsilon_{yz} - \frac{1}{2}\upsilon_{zx} + \mu_{xz} + \mu_{zy} - \frac{\pi}{2}\kappa_{yy}\right) - \alpha_x + \alpha_z$$

$$\Delta b_{10} = g\left(\frac{1}{2}\upsilon_{xy} + \frac{1}{2}\upsilon_{yz} + \mu_{yz} - \mu_{xy} - \frac{\pi}{2}\kappa_{zz}\right) - \alpha_x - \alpha_y$$

$$c_{10}^1 = -g\left(\lambda_{zz} - \lambda_{zzz}\right) + \alpha_z$$

$$c_{10}^2 = -g\left(\lambda_{yy} + \lambda_{yyy}\right) - \alpha_y$$

$$\Delta a_{13} = g\left(\frac{1}{2}\upsilon_{zx} + \frac{1}{2}\upsilon_{yz} + \mu_{zx} - \mu_{yz} - \frac{\pi}{2}\kappa_{xx}\right) + \alpha_y + \alpha_z$$

$$\Delta b_{13} = g\left(\frac{1}{2}\upsilon_{zx} - \frac{1}{2}\upsilon_{xy} + \mu_{yx} + \mu_{xz} - \frac{\pi}{2}\kappa_{zz}\right) - \alpha_x + \alpha_y$$

$$c_{13}^1 = -g\left(\lambda_{zz} - \lambda_{zzz}\right) + \alpha_z$$

$$c_{13}^2 = -g\left(\lambda_{xx} + \lambda_{xxx}\right) - \alpha_x$$

$$\Delta a_{11} = g\left(\frac{1}{2}\upsilon_{yz} - \frac{1}{2}\upsilon_{zx} + \mu_{zy} + \mu_{xz} + \frac{\pi}{2}\kappa_{yy}\right) - \alpha_x - \alpha_z$$

$$\Delta b_{11} = g\left(\frac{1}{2}\upsilon_{yz} - \frac{1}{2}\upsilon_{xy} + \mu_{yz} + \mu_{xy} + \frac{\pi}{2}\kappa_{zz}\right) - \alpha_x - \alpha_y$$

$$c_{11}^1 = -g\left(\lambda_{zz} - \lambda_{zzz}\right) + \alpha_z$$

$$c_{11}^2 = -g\left(\lambda_{yy} - \lambda_{yyy}\right) + \alpha_y$$

$$\Delta a_{14} = -g\left(\upsilon_{zx} - 2\,\mu_{xz}\right) - 2\,\alpha_x$$

$$\Delta b_{14} = -g\left(\upsilon_{yz} - 2\,\mu_{yz}\right) - 2\,\alpha_y$$

$$c_{14}^1 = c_{14}^2 = -g\left(\lambda_{zz} - \lambda_{zzz}\right) + \alpha_z$$

Figure 18.4.3-4 (Continued)

Acceleration Measurements In Terms Of Revised Sensor Error Parameters For The 16 Sequence Rotation Test

$\Delta a_{15} = - \pi \, g \, \kappa_{yy} - 2 \, \alpha_x$

$\Delta b_{15} = g \left(\upsilon_{yz} + 2 \, \mu_{yz} \right)$

$c^1_{15} = - g \left(\lambda_{zz} - \lambda_{zzz} \right) + \alpha_z$

$c^2_{15} = - g \left(\lambda_{zz} + \lambda_{zzz} \right) - \alpha_z$

$\Delta a_{16} = - \pi \, g \, \kappa_{xx} + 2 \, \alpha_y$

$\Delta b_{16} = g \left(\upsilon_{zx} + 2 \, \mu_{xz} \right)$

$c^1_{16} = - g \left(\lambda_{zz} - \lambda_{zzz} \right) + \alpha_z$

$c^2_{16} = - g \left(\lambda_{zz} + \lambda_{zzz} \right) - \alpha_z$

Figure 18.4.3-4 (Continued)

Acceleration Measurements In Terms Of Revised Sensor Error Parameters For The 16 Sequence Rotation Test

$$\Delta a_1 = -2\pi g \kappa_{yy}$$
$$\Delta b_1 = 0$$
$$c_1^1 = c_1^2 = -g(\lambda_{zz} - \lambda_{zzz}) + \alpha_z$$

$$\Delta a_2 = -2\pi g \kappa_{xx}$$
$$\Delta b_2 = 0$$
$$c_2^1 = c_2^2 = -g(\lambda_{zz} - \lambda_{zzz}) + \alpha_z$$

$$\Delta a_3 = 0$$
$$\Delta b_3 = 2\pi g \kappa_{zz}$$
$$c_3^1 = c_3^2 = -g(\lambda_{zz} - \lambda_{zzz}) + \alpha_z$$

$$\Delta a_4 = -\pi g(\kappa_{xx} + \kappa_{yy})$$
$$\Delta b_4 = 2g(\upsilon_{yz} - \upsilon_{zx})$$
$$c_4^1 = c_4^2 = -g(\lambda_{zz} - \lambda_{zzz}) + \alpha_z$$

$$\Delta a_5 = \pi g(\kappa_{yy} - \kappa_{xx})$$
$$\Delta b_5 = 2g(\upsilon_{zx} + \upsilon_{yz})$$
$$c_5^1 = c_5^2 = -g(\lambda_{zz} - \lambda_{zzz}) + \alpha_z$$

$$\Delta a_6 = \frac{\pi}{2} g(\kappa_{xx} - \kappa_{yy})$$
$$\Delta b_6 = g\left(\upsilon_{yz} + \upsilon_{xy} - \upsilon_{zx} + \frac{\pi}{2}\kappa_{zz}\right)$$
$$c_6^1 = c_6^2 = -g(\lambda_{zz} - \lambda_{zzz}) + \alpha_z$$

$$\Delta a_7 = -\pi g \kappa_{yy} - 2\alpha_x$$
$$\Delta b_7 = g(\upsilon_{yz} + 2\mu_{yz})$$
$$c_7^1 = -g(\lambda_{zz} - \lambda_{zzz}) + \alpha_z$$
$$c_7^2 = -g(\lambda_{zz} + \lambda_{zzz}) - \alpha_z$$

$$\Delta a_8 = -\pi g \kappa_{yy} + 2\alpha_x$$
$$\Delta b_8 = -g(\upsilon_{yz} + 2\mu_{yz})$$
$$c_8^1 = -g(\lambda_{zz} + \lambda_{zzz}) - \alpha_z$$
$$c_8^2 = -g(\lambda_{zz} - \lambda_{zzz}) + \alpha_z$$

$$\Delta a_9 = -g\left(\mu_{zy} + \mu_{yz} + \frac{\pi}{2}\kappa_{xx}\right) + \alpha_y + \alpha_z$$
$$\Delta b_9 = g\left(\frac{1}{2}\upsilon_{zx} - \frac{1}{2}\upsilon_{xy} + \mu_{xz} - \mu_{xy}\right)$$
$$c_9^1 = -g(\lambda_{zz} - \lambda_{zzz}) + \alpha_z$$
$$c_9^2 = -g(\lambda_{yy} - \lambda_{yyy}) + \alpha_y$$

$$\Delta a_{10} = g\left(\mu_{zy} + \mu_{yz} - \frac{\pi}{2}\kappa_{xx}\right) + \alpha_y - \alpha_z$$
$$\Delta b_{10} = g\left(\frac{1}{2}\upsilon_{zx} + \frac{1}{2}\upsilon_{xy} + \mu_{xz} + \mu_{xy}\right)$$
$$c_{10}^1 = -g(\lambda_{yy} - \lambda_{yyy}) + \alpha_y$$
$$c_{10}^2 = -g(\lambda_{zz} + \lambda_{zzz}) - \alpha_z$$

Figure 18.4.3-5

Acceleration Measurements In Terms Of Revised Sensor Error Parameters For The 21 Sequence Rotation Test

$\Delta a_{11} = -g\left(\upsilon_{ZX} + 2\,\mu_{ZX}\right)$

$\Delta b_{11} = -\pi\, g\, \kappa_{ZZ} - 2\,\alpha_y$

$c_{11}^1 = -g\left(\lambda_{XX} - \lambda_{XXX}\right) + \alpha_X$

$c_{11}^2 = -g\left(\lambda_{XX} + \lambda_{XXX}\right) - \alpha_X$

$\Delta a_{16} = g\left(\upsilon_{yZ} + 2\,\mu_{Zy}\right)$

$\Delta b_{16} = \pi\, g\, \kappa_{ZZ} - 2\,\alpha_X$

$c_{16}^1 = -g\left(\lambda_{yy} - \lambda_{yyy}\right) + \alpha_y$

$c_{16}^2 = -g\left(\lambda_{yy} + \lambda_{yyy}\right) - \alpha_y$

$\Delta a_{21} = -\pi\, g\, \kappa_{XX} + 2\,\alpha_y$

$\Delta b_{21} = g\left(\upsilon_{ZX} + 2\,\mu_{XZ}\right)$

$c_{21}^1 = -g\left(\lambda_{ZZ} - \lambda_{ZZZ}\right) + \alpha_Z$

$c_{21}^2 = -g\left(\lambda_{ZZ} + \lambda_{ZZZ}\right) - \alpha_Z$

$\Delta a_{12} = -\pi\, g\, \kappa_{XX} + 2\,\alpha_y$

$\Delta b_{12} = g\left(\upsilon_{ZX} + 2\,\mu_{XZ}\right)$

$c_{12}^1 = -g\left(\lambda_{ZZ} - \lambda_{ZZZ}\right) + \alpha_Z$

$c_{12}^2 = -g\left(\lambda_{ZZ} + \lambda_{ZZZ}\right) - \alpha_Z$

$\Delta a_{13} = -\pi\, g\, \kappa_{XX} - 2\,\alpha_y$

$\Delta b_{13} = -g\left(\upsilon_{ZX} + 2\,\mu_{XZ}\right)$

$c_{13}^1 = -g\left(\lambda_{ZZ} + \lambda_{ZZZ}\right) - \alpha_Z$

$c_{13}^2 = -g\left(\lambda_{ZZ} - \lambda_{ZZZ}\right) + \alpha_Z$

$\Delta a_{14} = g\left(\mu_{ZX} + \mu_{XZ} - \frac{\pi}{2}\kappa_{yy}\right) - \alpha_X + \alpha_Z$

$\Delta b_{14} = g\left(\frac{1}{2}\upsilon_{Xy} + \frac{1}{2}\upsilon_{yZ} + \mu_{yX} + \mu_{yZ}\right)$

$c_{14}^1 = -g\left(\lambda_{ZZ} - \lambda_{ZZZ}\right) + \alpha_Z$

$c_{14}^2 = -g\left(\lambda_{XX} + \lambda_{XXX}\right) - \alpha_X$

$\Delta a_{15} = -g\left(\mu_{ZX} + \mu_{XZ} + \frac{\pi}{2}K_{yy}\right) - \alpha_X - \alpha_Z$

$\Delta b_{15} = g\left(\frac{1}{2}\upsilon_{yZ} - \frac{1}{2}\upsilon_{Xy} + \mu_{yZ} - \mu_{yX}\right)$

$c_{15}^1 = -g\left(\lambda_{XX} + \lambda_{XXX}\right) - \alpha_X$

$c_{15}^2 = -g\left(\lambda_{ZZ} + \lambda_{ZZZ}\right) - \alpha_Z$

$\Delta a_{17} = -\pi\, g\, \kappa_{XX} + 2\,\alpha_Z$

$\Delta b_{17} = -g\left(\upsilon_{Xy} + 2\,\mu_{Xy}\right)$

$c_{17}^1 = -g\left(\lambda_{yy} + \lambda_{yyy}\right) - \alpha_y$

$c_{17}^2 = -g\left(\lambda_{yy} - \lambda_{yyy}\right) + \alpha_y$

$\Delta a_{18} = -\pi\, g\, \kappa_{XX} - 2\,\alpha_Z$

$\Delta b_{18} = g\left(\upsilon_{Xy} + 2\,\mu_{Xy}\right)$

$c_{18}^1 = -g\left(\lambda_{yy} - \lambda_{yyy}\right) + \alpha_y$

$c_{18}^2 = -g\left(\lambda_{yy} + \lambda_{yyy}\right) - \alpha_y$

$\Delta a_{19} = g\left(\frac{1}{2}\upsilon_{yZ} - \frac{1}{2}\upsilon_{ZX} - \mu_{ZX} + \mu_{Zy}\right)$

$\Delta b_{19} = -g\left(\mu_{yX} + \mu_{Xy} + \frac{\pi}{2}\kappa_{ZZ}\right) - \alpha_X - \alpha_y$

$c_{19}^1 = -g\left(\lambda_{yy} + \lambda_{yyy}\right) - \alpha_y$

$c_{19}^2 = -g\left(\lambda_{XX} + \lambda_{XXX}\right) - \alpha_X$

$\Delta a_{20} = g\left(\frac{1}{2}\upsilon_{yZ} + \frac{1}{2}\upsilon_{ZX} + \mu_{ZX} + \mu_{Zy}\right)$

$\Delta b_{20} = g\left(\mu_{yX} + \mu_{Xy} - \frac{\pi}{2}\kappa_{ZZ}\right) - \alpha_X + \alpha_y$

$c_{20}^1 = -g\left(\lambda_{XX} + \lambda_{XXX}\right) - \alpha_X$

$c_{20}^2 = -g\left(\lambda_{yy} - \lambda_{yyy}\right) + \alpha_y$

Figure 18.4.3-5 (Continued)

Acceleration Measurements In Terms Of Revised Sensor Error Parameters For The 21 Sequence Rotation Test

18.4.4 SENSOR CALIBRATION ERRORS COMPUTED FROM MEASUREMENTS

Using Figure 18.4.3-4 for the 16 rotation sequence set, or Figure 18.4.3-5 for the 21 rotation sequence set, analytical relationships are readily derived relating the sensor calibration errors to $\Delta\underline{a}_H^L$, a_{ZL} measurements taken during the tests. In general, the equations in Figures 18.4.3-4 and 18.4.3-5 constitute an over-determined set for computing the sensor calibration errors (i.e., there are more equations than unknowns). The method of Least Squares can be utilized to calculate the sensor error set that best fits the measurements in a least squares error sense (e.g., as in Reference 37 - Chapter 9, Section 11). Alternatively, a particular deterministic set of measurement equations can be selected from the Figure 18.4.3-4 and 18.4.3-5 sets as representing more likely orientations of the sensors in their intended usage. The sensor errors can then be directly calculated from these diminished sets by analytical inversion. Figures 18.4.4-1 and 18.4.4-2 are sensor calibration error solutions to the 16 and 21 rotation sequence equations in Figures 18.4.3-4 and 18.4.3-5 using the latter deterministic solution approach. The Least Squares approach can also be used as a check on the previous results.

If the Least Squares approach is to be incorporated, it is recommended that the measurement data be divided into two separate groups; one for the $\Delta\underline{a}_H^L$ measurements and one for the a_{ZL} measurements. A Least Squares solution can then be performed separately for each group, with the a_{ZL} results used for accelerometer symmetrical/asymmetrical scale factor calibration error determination, and the $\Delta\underline{a}_H^L$ results used for determination of the remaining error terms. The rationale behind this approach is that the a_{ZL} measurement might not be accurate enough for accelerometer bias calibration error determination (when the accelerometers are in the vertical orientation being exposed to 1 g specific force) due to pseudo-bias effects created by scale factor deviations from the analytical model, and micro-heating variations.

ANGULAR RATE SENSOR CALIBRATION ERRORS

Scale Factor Errors

$$\kappa_{xx} = -\frac{1}{2\pi g}\,\Delta a_2$$

$$\kappa_{yy} = -\frac{1}{2\pi g}\,\Delta a_1$$

$$\kappa_{zz} = \frac{1}{2\pi g}\,\Delta b_3$$

Orthogonality Errors

$$\upsilon_{xy} = \frac{1}{g}\left(\Delta b_6 - \frac{1}{2}\Delta b_4 - \frac{1}{4}\Delta b_3\right)$$

$$\upsilon_{yz} = \frac{1}{4g}\left(\Delta b_5 + \Delta b_4\right)$$

$$\upsilon_{zx} = \frac{1}{4g}\left(\Delta b_5 - \Delta b_4\right)$$

ACCELEROMETER CALIBRATION ERRORS

Bias Errors

$$\alpha_x = \frac{1}{4}\Delta a_1 - \frac{1}{2}\Delta a_{15}$$

$$\alpha_y = \frac{1}{2}\Delta a_{16} - \frac{1}{4}\Delta a_2$$

$$\alpha_z = \frac{1}{2}\left(\Delta a_7 - \Delta a_8\right) - \frac{1}{4}\Delta a_1$$

Scale Factor Errors

$$\lambda_{xx} = -\frac{1}{2g}\left(c_{12}^2 + c_{13}^2\right)$$

$$\lambda_{yy} = -\frac{1}{2g}\left(c_9^2 + c_{10}^2\right)$$

$$\lambda_{zz} = -\frac{1}{2g}\left(c_{14}^2 + c_{15}^2\right)$$

Scale Factor Asymmetry

$$\lambda_{xxx} = \frac{1}{2g}\left(c_{12}^2 - c_{13}^2 + \Delta a_{15} - \frac{1}{2}\Delta a_1\right)$$

$$\lambda_{yyy} = \frac{1}{2g}\left(c_9^2 - c_{10}^2 - \Delta a_{16} + \frac{1}{2}\Delta a_2\right)$$

$$\lambda_{zzz} = \frac{1}{2g}\left(c_{14}^2 - c_{15}^2 - \Delta a_7 + \Delta a_8 + \frac{1}{2}\Delta a_1\right)$$

Misalignment Errors Relative To Mean Angular Rate Sensor Axes

$$\mu_{xy} = \frac{1}{2g}\left(\Delta b_{11} - \Delta b_{10} + \Delta b_6 - \frac{3}{4}\Delta b_3 - \frac{1}{2}\Delta b_4\right)$$

$$\mu_{yx} = \frac{1}{2g}\left(\Delta b_7 - \Delta b_8 - \Delta b_6 + \frac{1}{2}\Delta b_4 + \frac{1}{4}\Delta b_3\right)$$

$$\mu_{yz} = \frac{1}{2g}\left(\Delta b_{14} + \Delta a_{16} - \frac{1}{2}\Delta a_2 + \frac{1}{4}\Delta b_4 + \frac{1}{4}\Delta b_5\right)$$

$$\mu_{zy} = \frac{1}{2g}\left(\Delta a_{10} - \Delta a_9 - \frac{1}{4}\Delta b_4 - \frac{1}{4}\Delta b_5\right)$$

$$\mu_{zx} = \frac{1}{2g}\left(\Delta a_{13} - \Delta a_{12} - \frac{1}{4}\Delta b_5 + \frac{1}{4}\Delta b_4\right)$$

$$\mu_{xz} = \frac{1}{2g}\left(\Delta a_{14} - \Delta a_{15} + \frac{1}{2}\Delta a_1 + \frac{1}{4}\Delta b_5 - \frac{1}{4}\Delta b_4\right)$$

Figure 18.4.4-1

Sensor Errors In Terms Of Measurements For The 16 Rotation Sequence Test

ANGULAR RATE SENSOR CALIBRATION ERRORS

Scale Factor Errors

$$\kappa_{xx} = -\frac{1}{2\pi g}\Delta a_2$$

$$\kappa_{yy} = -\frac{1}{2\pi g}\Delta a_1$$

$$\kappa_{zz} = \frac{1}{2\pi g}\Delta b_3$$

Orthogonality Errors

$$\upsilon_{xy} = \frac{1}{g}\left(\Delta b_6 - \frac{1}{2}\Delta b_4 - \frac{1}{4}\Delta b_3\right)$$

$$\upsilon_{yz} = \frac{1}{4g}\left(\Delta b_5 + \Delta b_4\right)$$

$$\upsilon_{zx} = \frac{1}{4g}\left(\Delta b_5 - \Delta b_4\right)$$

ACCELEROMETER CALIBRATION ERRORS

Bias Errors

$$\alpha_x = \frac{1}{4}\left(\Delta a_8 - \Delta a_7\right)$$

$$\alpha_y = \frac{1}{4}\left(\Delta a_{12} - \Delta a_{13}\right)$$

$$\alpha_z = \frac{1}{4}\left(\Delta a_{17} - \Delta a_{18}\right)$$

Scale Factor Errors

$$\lambda_{xx} = -\frac{1}{2g}\left(c_{11}^2 + c_{11}^1\right)$$

$$\lambda_{yy} = -\frac{1}{2g}\left(c_{16}^2 + c_{16}^1\right)$$

$$\lambda_{zz} = -\frac{1}{2g}\left(c_{21}^2 + c_{21}^1\right)$$

Scale Factor Asymmetry

$$\lambda_{xxx} = -\frac{1}{2g}\left(c_{11}^2 - c_{11}^1 - \frac{1}{2}\Delta a_7 + \frac{1}{2}\Delta a_8\right)$$

$$\lambda_{yyy} = -\frac{1}{2g}\left(c_{16}^2 - c_{16}^1 - \frac{1}{2}\Delta a_{13} + \frac{1}{2}\Delta a_{12}\right)$$

$$\lambda_{zzz} = -\frac{1}{2g}\left(c_{21}^2 - c_{21}^1 - \frac{1}{2}\Delta a_{18} + \frac{1}{2}\Delta a_{17}\right)$$

Misalignment Errors Relative To Mean Angular Rate Sensor Axes

$$\mu_{xy} = \frac{1}{2g}\left(\Delta b_{10} - \Delta b_9 - \Delta b_6 + \frac{1}{2}\Delta b_4 + \frac{1}{4}\Delta b_3\right)$$

$$\mu_{yx} = \frac{1}{2g}\left(\Delta b_{14} - \Delta b_{15} - \Delta b_6 + \frac{1}{2}\Delta b_4 + \frac{1}{4}\Delta b_3\right)$$

$$\mu_{yz} = \frac{1}{2g}\left(\Delta b_{15} + \Delta b_{14} - \frac{1}{4}\Delta b_5 - \frac{1}{4}\Delta b_4\right)$$

$$\mu_{zy} = -\frac{1}{2g}\left(\Delta a_{20} + \Delta a_{19} - \frac{1}{4}\Delta b_5 - \frac{1}{4}\Delta b_4\right)$$

$$\mu_{zx} = \frac{1}{2g}\left(-\Delta a_{19} + \Delta a_{20} - \frac{1}{4}\Delta b_5 + \frac{1}{4}\Delta b_4\right)$$

$$\mu_{xz} = \frac{1}{2g}\left(\Delta b_{10} + \Delta b_9 - \frac{1}{4}\Delta b_5 + \frac{1}{4}\Delta b_4\right)$$

Figure 18.4.4-2

Sensor Errors In Terms Of Measurements For The 21 Rotation Sequence Test

18.4.5 SENSOR ASSEMBLY MISALIGNMENT CALIBRATION ERROR DETERMINATION

As a supplemental part of the Strapdown Rotation test data processing routines, misalignment calibration coefficients can also be calculated for the sensor assembly as a unit relative to the rotation fixture test mount. For this particular measurement to be meaningful, the Strapdown Rotation test would have typically been performed with the sensor assembly installed in its INS chassis, and the INS mounted to a standard INS mount attached to the rotation test fixture. Then the calculated misalignment coefficients would correspond to the $\underline{J}$ coefficients in Equation (8.3-2). In determining the $\underline{J}$ coefficients, the formal definition for the B Frame must be specified relative to mean angular rate sensor (MARS) axes (i.e., the $\underline{\gamma}$ rotation angle vector of Section 18.4.3 between the B and MARS Frames).

The procedure for sensor assembly alignment-to-mount determination (derived in Section 18.4.7.4) is to process attitude and acceleration data taken at two selected static sensor assembly attitude orientations during the Strapdown Rotation test. The two data collection attitude orientations must be such that projections of the vertical on the sensor assembly mount for the two orientations are perpendicular to one another. For the 16 Sequence rotation test, the data is taken at the start of rotation sequence 1 and at the end of rotation sequence 7 (see Table 18.4.1-1). For the 21 Sequence rotation test, the data is taken at the start of rotation sequence 1 and at the end of rotation sequence 14 (See Table 18.4.1-2). Other two-orientation measurement attitudes are also possible (as explained in Section 18.4.7.4). At completion of the Strapdown Rotation test, the data taken at the two selected orientations are combined with accelerometer error data computed from the test (i.e. from Figures 18.4.4-1 or 18.4.4-2) to compute the sensor assembly misalignment coefficients. Equations for the sensor assembly alignment calculation for the 16 Sequence and 21 Sequence rotation tests are derived in Section 18.4.7.4 and given by Equations (18.4.7.4-8), (18.4.7.4-9), and (18.4.7.4-18) - (18.4.7.4-21) summarized below.

$$C_M^{MARS} = F^{MARS} \left(F^M\right)^{-1}$$

$$F^{MARS} \equiv \left[-\frac{1}{g_{Tst}} \underline{a}_{SF}^{MARS1} \quad -\frac{1}{g_{Tst}} \underline{a}_{SF}^{MARS2} \quad \frac{1}{g_{Tst}^2} \underline{a}_{SF}^{MARS1} \times \underline{a}_{SF}^{MARS2} \right] \qquad (18.4.5\text{-}1)$$

$$F^M \equiv \left[\underline{u}_{ZL}^{M1} \quad \underline{u}_{ZL}^{M2} \quad \underline{u}_{ZL}^{M1} \times \underline{u}_{ZL}^{M2} \right]$$

(Continued)

$$\underline{a}_{SF}^{MARS1} = \left(I + \lambda_{LinScal} + \mu_{Mis} + \lambda_{Asym}\, A_{SFPulsSign}^{1}\right)^{-1} \left[\left(\hat{C}_{B1}^{L}\right)^{T} \left(\hat{\underline{a}}_{1}^{L} - g_{Tst}\, \underline{u}_{ZL}^{L}\right) - \underline{\lambda}_{Bias}\right]$$

$$\underline{a}_{SF}^{MARS2} = \left(I + \lambda_{LinScal} + \mu_{Mis} + \lambda_{Asym}\, A_{SFPulsSign}^{2}\right)^{-1} \left[\left(\hat{C}_{B2}^{L}\right)^{T} \left(\hat{\underline{a}}_{2}^{L} - g_{Tst}\, \underline{u}_{ZL}^{L}\right) - \underline{\lambda}_{Bias}\right]$$

$$\underline{u}_{ZL}^{M1} = \begin{bmatrix} 0 \\ 0 \\ 1 \end{bmatrix} \qquad \underline{u}_{ZL}^{M2} = \begin{bmatrix} -1 \\ 0 \\ 0 \end{bmatrix}$$

(18.4.5-1)
(Continued)

For B Frame = MARS Frame

$$C_{B}^{M} = \left(C_{M}^{MARS}\right)^{T} \qquad \underline{\gamma} = 0$$

$\underline{J}$ = Rotation Angle Extraction From C_{B}^{M} With Equations (3.2.2.2-10) - (3.2.2.2-12) and (3.2.2.2-15) - (3.2.2.2-17)

For B Frame = M Frame

$$C_{MARS}^{B} = \left(C_{M}^{MARS}\right)^{T} \qquad \underline{J} = 0$$

$\underline{\gamma}$ = Rotation Angle Extraction From C_{MARS}^{B} With Equations (3.2.2.2-10) - (3.2.2.2-12) and (3.2.2.2-15) - (3.2.2.2-17)

where

M = INS standard mount coordinate frame.

MARS = Mean angular rate sensor axis coordinate frame. Nominally the M, MARS and sensor assembly B Frames are parallel.

$\underline{u}_{ZL}^{M}$ = Unit vector along the L Frame Z axis (downward along the plumb-bob vertical) as projected on M Frame axes.

$\hat{\underline{a}}^{L}$ = L Frame total acceleration measurement for the rotation test taken by averaging transformed accelerometer data plus gravity, defined in Equation (18.4-1) as $\underline{a}^{L}$.

g_{Tst} = Plumb-bob gravity magnitude at the test site.

$\underline{u}_{ZL}^{L}$ = Unit vector along the L Frame Z axis (downward along the plumb-bob vertical) as projected on L Frame axes.

$A_{SFPulsSign}$ = Diagonal matrix whose elements are unity in magnitude with sign equal to the sign of the components of $\underline{a}_{SFPuls}^{B}$, the uncompensated accelerometer triad output pulse rate vector.

1, 2 = Designations for the two attitude orientations selected for data collection (Start of Sequence 1 and end of Sequence 7 for the 16 Sequence rotation test; Start of Sequence 1 and end of Sequence 14 for the 21 Sequence rotation test).

M1, MARS1, M2, MARS2 = M, MARS Frames at measurement orientations 1 and 2.

$\underline{\hat{a}}_1^L$, $\underline{\hat{a}}_2^L$, $A^1_{SFPuls_{Sign}}$, $A^2_{SFPuls_{Sign}}$ = Values for $\underline{\hat{a}}^L$ and $A_{SFPuls_{Sign}}$ (definitions above) at test orientations 1 and 2.

$\lambda_{LinScal}$, μ_{Mis}, λ_{Asym}, $\underline{\lambda}_{Bias}$ = Accelerometer calibration error terms calculated from the Strapdown Rotation test data as in Figures 18.4.4-1 or 18.4.4-2.

18.4.6 CALIBRATION COEFFICIENT UPDATING FROM MEASURED CALIBRATION ERRORS

Once the calibration errors have been determined from the Strapdown Rotation test in the form of κ_{ii}, κ_{iii}, υ_{ij}, α_i, λ_{ii}, λ_{iii}, μ_{ij}, $\underline{\gamma}$ and $\underline{J}$ (e.g., using Figures 18.4.4-1 or 18.4.4-2, and Equations (18.4.5-1)), they are used to update the calibration coefficients in the strapdown sensor compensation and INS attitude output algorithms. The individual inertial sensor misalignment calibration algorithms are referenced to B Frame sensor assembly axes while the rotation test misalignment calibration error terms (υ_{ij}, μ_{ij}) have been calculated relative to mean angular rate sensor axes. Before the sensor calibration coefficients can be updated, the misalignment terms must be converted to their equivalent form referenced to B Frame axes. Equations (18.4.3-5) can be used to calculate κ_{ij}, λ_{ij} from υ_{ij}, μ_{ij}, given a specified B Frame orientation relative to mean angular rate sensor axes (the γ_k terms in Equations (18.4.3-5)). Generally, two approaches can be used for B Frame specification; 1. B aligned with mean angular rate sensor (MARS) axes and 2. B aligned with INS mount (M Frame) axes.

For the first B Frame specification approach (B = MARS), $\underline{\gamma}$ is zero by definition which tends to minimize the magnitude of the κ_{ij}, λ_{ij} compensation error terms generated from Equations (18.4.3-5). The calculated κ_{ij}, λ_{ij} compensation error terms are used to update the sensor misalignment coefficients. If the same procedure was used for previous sensor misalignment determinations, the misalignment coefficients being updated by κ_{ij}, λ_{ij} will also be minimized. The result is that the magnitudes of the individual sensor misalignment calibration coefficients will be minimized, thereby reducing second order error effects that have been neglected in the Strapdown Rotation test error determination equations (i.e., Figures 18.4.4-1, 18.4.4-2 and Equations (18.4.3-5)). Use of this method for B Frame selection treats the B Frame orientation as unique to a particular sensor assembly (and its misalignments), rather than having it represent a single entity across sensor assemblies.

For the second B Frame specification method (i.e., B = M), $\underline{J}$ is zero by definition, which specifies a common fixed B Frame orientation relative to the INS mount. If the mount is standardized to be compatible with all INS's, this fixes the B Frame attitude relative to the user vehicle for any INS installed on the mount. Then, from Equation (18.4.5-1), we see that $\underline{\gamma}$ (from C_{MARS}^{B}) would be equated to the rotation angle of C_{MARS}^{M}. The κ_{ij}, λ_{ij} misalignment calibration errors generated with Equations (18.4.3-5) would then be larger in magnitude, generating larger errors in the sensor misalignment coefficients when using κ_{ij}, λ_{ij} for coefficient updating, thereby amplifying second order error effects in the sensor compensation algorithms. These added error effects may be significant, considering that C_{MARS}^{M} rotation angle vector components tend to be larger than υ_{ij}, μ_{ij} due to the less stable nature of the sensor assembly attachment to the INS chassis/mount (compared with the generally stable sensor-to-sensor alignments on a rigid mount within the sensor assembly).

Once κ_{ij}, λ_{ij} are calculated using the previous procedure, these terms together with the κ_{ii}, κ_{iii}, α_i, λ_{ii}, λ_{iii} calibration error terms determined from the Strapdown Rotation test are used to update the sensor calibration coefficients. The method for coefficient updating is based on the form of the sensor compensation algorithms utilized in the strapdown inertial computation software. As an example, let us assume that the angular rate sensor compensation algorithms are based on Equations (8.1.1.1-8) in which non-linear scale factor correction (i.e., κ_{iii}) is not required (as with ring laser gyros). For the accelerometers, let us assume that compensation Equations (8.1.1.3-19) for $\underline{a}'_{SF}$ and (8.1.1.2-8) for $\underline{a}_{SF}$ is being used in which scale factor asymmetry (λ_{iii}) is included for error correction. Then Table 8.4-1 applies, for which the compensation coefficients are calculated from Equations (8.1.1.1.1-13), (8.1.1.1.1-14), (8.1.1.1.1-16), (8.1.1.1-5), (8.1.1.1-6) and (8.1.1.1.1-12) for the angular rate sensors, and (8.1.1.3-20), (8.1.1.3-23), (8.1.1.2.1-16), (8.1.1.2-6) and (8.1.1.2.1-17) for the accelerometers. The κ_{ii}, κ_{ij}, α_i, λ_{ii}, λ_{iii}, λ_{ij} calibration error terms would be used to update the System Level sensor calibration coefficients in these equations by first formatting them into the following matrix/vector format:

$$\kappa_{ii}, \kappa_{ij} \rightarrow \Delta\kappa_{SystScal/Mis}$$

$$\alpha_i \rightarrow \Delta\underline{\lambda}_{SystBias} \qquad \lambda_{ii}, \lambda_{ij} \rightarrow \Delta\lambda_{SystScalLin/Mis} \tag{18.4.6-1}$$

$$\lambda_{iii} \rightarrow \Delta\lambda_{SystScalAsym}$$

where

Δ = Change required in indicated quantity.

The $\kappa_{SystScal/Mis}$, $\underline{\lambda}_{SystBias}$, $\lambda_{SystScalLin/Mis}$ and $\lambda_{SystScalAsym}$ coefficients would then be updated by adding the (18.4.6-1) corrections to the values used for these parameters during the Strapdown Rotation test. The remaining sensor calibration coefficients (e.g., Sensor Level) would not be changed. In the case of the $\underline{J}$ sensor assembly triad misalignment coefficients, the values determined from Equations (18.4.5-1) would be used to replace the old $\underline{J}$ values.

18.4.7 ANALYTICAL BASIS FOR THE STRAPDOWN ROTATION TEST

The analytical basis for the Strapdown Rotation test is derived from the continuous form of Equation (18.4-1) for the rotation test measurement:

$$\underline{a}^L = \underline{a}_{SF}^L + g_{Tst}\, \underline{u}_{ZL}^L \qquad (18.4.7\text{-}1)$$

where

$\underline{a}^L$ = L Frame total acceleration measurement.

$\underline{a}_{SF}^L$ = Specific force acceleration in the L Frame.

For a strapdown sensor assembly containing errors, the $\underline{a}^L$ measured can be represented by:

$$\hat{\underline{a}}^L = \hat{\underline{a}}_{SF}^L + g_{Tst}\, \underline{u}_{ZL}^L \qquad (18.4.7\text{-}2)$$

where

$\hat{\ }$ = Designation for the parameter containing sensor or test system errors. Without the $\hat{\ }$ designation, the parameter is assumed to be error free (as in Equation (18.4.7-1)).

We can also define:

$$\hat{\underline{a}}_{SF}^L = \underline{a}_{SF}^L + \delta\underline{a}_{SF}^L \qquad (18.4.7\text{-}3)$$

where

$\delta\underline{a}_{SF}^L$ = Error in $\hat{\underline{a}}_{SF}^L$.

For the Strapdown Rotation test measurement of $\underline{a}^L$, the sensor assembly is stationary, hence $\underline{a}^L$ is zero, and Equation (18.4.7-3) reduces to:

$$\hat{\underline{a}}^L = \delta \underline{a}_{SF}^L \tag{18.4.7-4}$$

Equation (18.4.7-4) represents the fundamental principle underlying the Strapdown Rotation test: Non-zero acceleration measurements in the L Frame are produced by inertial sensor errors or by system test errors.

Equation (18.4.7-4) can be expressed in terms of particular error sources from the transformation equation:

$$\underline{a}_{SF}^L = C_B^L \, \underline{a}_{SF}^B \tag{18.4.7-5}$$

The differential of (18.4.7-5) is $\delta \underline{a}_{SF}^L$ in (18.4.7-4):

$$\delta \underline{a}_{SF}^L = \delta C_B^L \, \underline{a}_{SF}^B + C_B^L \, \delta \underline{a}_{SF}^B \tag{18.4.7-6}$$

The δC_B^L term in (18.4.7-6) is from generalized Equation (3.5.2-27):

$$\delta C_B^L = - \left(\underline{\phi}^L \times \right) C_B^L \tag{18.4.7-7}$$

where

$\underline{\phi}^L$ = Rotation angle error vector associated with C_B^L considering the L Frame to be misaligned.

From (18.4.7-5) we can also write:

$$\underline{a}_{SF}^B = \left(C_B^L \right)^T \underline{a}_{SF}^L \tag{18.4.7-8}$$

The $\underline{a}_{SF}^L$ term in (18.4.7-8) is obtained from (18.4.7-1) with, as noted previously, $\underline{a}^L$ equal to zero for the Strapdown Rotation test:

$$\underline{a}_{SF}^L = - g_{Tst} \, \underline{u}_{ZL}^L \tag{18.4.7-9}$$

With (18.4.7-9), Equation (18.4.7-8) becomes:

$$\underline{a}_{SF}^B = - g_{Tst} \left(C_B^L \right)^T \underline{u}_{ZL}^L \tag{18.4.7-10}$$

We then substitute (18.4.7-10) and (18.4.7-7) into (18.4.7-6), and the result into (18.4.7-4) to obtain a more explicit expression for the $\hat{\underline{a}}^L$ Strapdown Rotation test measurement:

$$\hat{\underline{a}}^L = - g_{Tst} \left(\underline{u}_{ZL}^L \times \underline{\phi}^L \right) + C_B^L \, \delta\underline{a}_{SF}^B \tag{18.4.7-11}$$

The horizontal and vertical components of (18.4.7-11) are obtained by noting that $\underline{u}_{ZL}^L$ is vertical, hence, $\underline{u}_{ZL}^L \times \underline{\phi}^L$ has no vertical component, and that the vertical component of $\underline{\phi}^L$ has no contribution to $\underline{u}_{ZL}^L \times \underline{\phi}^L$:

$$\begin{aligned} \hat{a}_{ZL} &= \left(C_B^L \, \delta\underline{a}_{SF}^B \right)_Z \\ \hat{\underline{a}}_H^L &= - g_{Tst} \left(\underline{u}_{ZL}^L \times \underline{\phi}_H^L \right) + \left(C_B^L \, \delta\underline{a}_{SF}^B \right)_H \end{aligned} \tag{18.4.7-12}$$

where

$(\,)_H, (\,)_Z$ = Designation for the horizontal (H) and Z axis (vertical) component of the ( ) L Frame parameters.

$\hat{\underline{a}}_H^L, \hat{a}_{ZL}$ = Horizontal and vertical components of $\hat{\underline{a}}^L$.

The specific measurements for each rotation sequence in the Strapdown Rotation test are the values for $\hat{a}_{ZL}$ at the start and end of the rotation sequence, and the difference between $\hat{\underline{a}}_H^L$ measurements taken at the start and end of the rotation sequence. From Equations (18.4.7-12) we see that these are given:

$$\begin{aligned} \hat{a}_{ZL}^1 &= \left(C_{B1}^L \, \delta\underline{a}_{SF}^{B1} \right)_Z \qquad \hat{a}_{ZL}^2 = \left(C_{B2}^L \, \delta\underline{a}_{SF}^{B2} \right)_Z \\ \Delta\underline{a}_H^L &= - g_{Tst} \left(\underline{u}_{ZL}^L \times \Delta\underline{\phi}_H^L \right) + \left(C_{B2}^L \, \delta\underline{a}_{SF}^{B2} \right)_H - \left(C_{B1}^L \, \delta\underline{a}_{SF}^{B1} \right)_H \\ \Delta\underline{a}_H^L &\equiv \hat{\underline{a}}_{H_2}^L - \hat{\underline{a}}_{H_1}^L \end{aligned} \tag{18.4.7-13}$$

and

$$\Delta\underline{\phi}_H^L \equiv \underline{\phi}_{H_2}^L - \underline{\phi}_{H_1}^L \tag{18.4.7-14}$$

where

$\hat{a}_{ZL}^1, \hat{a}_{ZL}^2, \hat{\underline{a}}_{H_1}^L, \hat{\underline{a}}_{H_2}^L$ = Values for $\hat{a}_{ZL}, \hat{\underline{a}}_H^L$ measured at the start (1) and end (2) of the rotation sequence.

B1, B2 = B Frame (sensor assembly) orientations at the start (1) and end (2) of the rotation sequence.

$\Delta \underline{a}_{H}^{L}$ = Difference between the $\hat{\underline{a}}_{H_1}^{L}$ and $\hat{\underline{a}}_{H_2}^{L}$ measurements.

$\underline{\phi}_{H_1}^{L}, \underline{\phi}_{H_2}^{L}$ = $\underline{\phi}_{H}^{L}$ error in C_B^L at the start and end of the rotation sequence.

$\Delta \underline{\phi}_{H}^{L}$ = Change in the C_B^L matrix $\underline{\phi}_{H}^{L}$ error over the rotation sequence.

The $\delta \underline{a}_{SF}^{B}$ terms in (18.4.7-13) are caused by accelerometer error. The $\Delta \underline{\phi}_{H}^{L}$ term in (18.4.7-13) is produced by the integrated effect of angular rate sensor error over the time period for the rotation sequence (plus a small negligible error due to earth rate operating on $\underline{\phi}_{H}^{L}$ over the sequence period - To be discussed subsequently). Once analytical forms are available for $\delta \underline{a}_{SF}^{B}$ and $\Delta \underline{\phi}_{H}^{L}$ in terms of particular sensor error effects, the sensor errors can be determined by inversion of (18.4.7-13). This, then, is the essence of the Strapdown Rotation test.

Sections 18.4.7.1 and 18.4.7.2 define $\delta \underline{a}_{SF}^{B}$ and $\Delta \underline{\phi}_{H}^{L}$ in terms of strapdown sensor error characteristics. Section 18.4.7.2 defines the C_B^L terms in Equations (18.4.7-13) for the executable class of rotations selected for the Strapdown Rotation test. Section 18.4.7.3 describes a computer algorithm for making the Equation (18.4.7-13) $\hat{a}_{ZL}^{1}$, $\hat{a}_{ZL}^{2}$, $\hat{\underline{a}}_{H_1}^{L}$, $\hat{\underline{a}}_{H_2}^{L}$ Strapdown Rotation test acceleration measurements that minimizes quantization error effects. Section 18.4.7.4 describes how data taken during the Strapdown Rotation test can be used to calculate calibration errors in sensor assembly misalignment relative to the rotation test fixture mount.

18.4.7.1 ACCELEROMETER ERROR MODEL FOR THE STRAPDOWN ROTATION TEST

The analytical model used for accelerometer error determination in the Strapdown Rotation test (the $\delta \underline{a}_{SF}^{B}$ terms in Equations (18.4.7-13)) is the following linearized version of Equations (8.1.1.2.1-5) and (8.1.1.3-2) for $\lambda_{SystScal/Mis}$, and (8.1.1.3-15):

$$\delta \underline{a}_{SF}^{B} = \left(\lambda_{LinScal/Mis} + \lambda_{Asym} \, A_{SFSign}\right) \underline{a}_{SF}^{B} + \underline{\lambda}_{Bias} \qquad (18.4.7.1\text{-}1)$$

with

$$\lambda_{LinScal/Mis} \equiv \begin{bmatrix} \lambda_{11} & \lambda_{12} & \lambda_{13} \\ \lambda_{21} & \lambda_{22} & \lambda_{23} \\ \lambda_{31} & \lambda_{32} & \lambda_{33} \end{bmatrix} \qquad \lambda_{Asym} \equiv \begin{bmatrix} \lambda_{111} & 0 & 0 \\ 0 & \lambda_{222} & 0 \\ 0 & 0 & \lambda_{333} \end{bmatrix}$$
$$\underline{\lambda}_{Bias} \equiv \begin{bmatrix} \alpha_1 \\ \alpha_2 \\ \alpha_3 \end{bmatrix} \tag{18.4.7.1-2}$$

and

$$A_{SFSign} \equiv \begin{bmatrix} \mathrm{Sign}\left(a_{SF_{XB}}\right) & 0 & 0 \\ 0 & \mathrm{Sign}\left(a_{SF_{YB}}\right) & 0 \\ 0 & 0 & \mathrm{Sign}\left(a_{SF_{ZB}}\right) \end{bmatrix} \tag{18.4.7.1-3}$$

where

$\mathrm{Sign}(\,) = 1$ for $(\,) \geq 0$ and -1 for $(\,) < 0$.

$\lambda_{LinScal/Mis}$ = Accelerometer triad linear scale factor/misalignment error matrix.

λ_{Asym} = Accelerometer triad scale factor asymmetry error matrix.

$\underline{\lambda}_{Bias}$ = Accelerometer triad bias error vector.

$a_{SF_{iB}}$ = B Frame component i of $\underline{a}_{SF}^{B}$.

Equation (18.4.7-10) shows that the sign of the $\underline{a}_{SF}^{B}$ components equals the negative sign of the $\underline{u}_{ZL}$ B Frame components. Then, using (18.4.7-10) and (18.4.7.1-3), Equation (18.4.7.1-1) becomes the following for the Strapdown Rotation test acceleration measurements:

$$\delta\underline{a}_{SF}^{B} = -\, g_{Tst}\left(\lambda_{LinScal/Mis} - \lambda_{Asym}\, U_{ZLBSign}\right)\underline{u}_{ZL}^{B} + \underline{\lambda}_{Bias} \tag{18.4.7.1-4}$$

with

$$\underline{u}_{ZL}^{B} = \left(C_B^L\right)^T \underline{u}_{ZL}^{L}$$
$$U_{ZLBSign} \equiv \begin{bmatrix} \mathrm{Sign}\left(u_{ZL_{XB}}\right) & 0 & 0 \\ 0 & \mathrm{Sign}\left(u_{ZL_{YB}}\right) & 0 \\ 0 & 0 & \mathrm{Sign}\left(u_{ZL_{ZB}}\right) \end{bmatrix} \tag{18.4.7.1-5}$$

where

$u_{ZL_{iB}}$ = B Frame component i of $\underline{u}_{ZL}^{B}$.

Equation (18.4.7.1-4) with (18.4.7.1-2) and (18.4.7.1-5) is used to analytically define the $\delta\underline{a}_{SF}^{B}$ terms in Equations (18.4.7-13) as a function of the $\lambda_{LinScal/Mis}$, λ_{Asym} and $\underline{\lambda}_{Bias}$ accelerometer error parameters.

18.4.7.2 ATTITUDE ERROR AS FUNCTION OF ANGULAR RATE SENSOR ERRORS FOR THE STRAPDOWN ROTATION TEST

The analytical model used for attitude error in the Strapdown Rotation test (the $\Delta\underline{\phi}_{H}^{L}$ term in Equations (18.4.7-13)) as a function of angular rate sensor error parameters, is derived from the following equivalent version of Equation (4.1-1):

$$\dot{C}_{B}^{L} = C_{B}^{L}\left(\underline{\omega}_{IB}^{B}\times\right) - \left(\underline{\omega}_{IE}^{L}\times\right)C_{B}^{L} \qquad (18.4.7.2\text{-}1)$$

in which $\underline{\omega}_{IE}^{L}$ is used in place of $\underline{\omega}_{IL}^{L}$ based on the L Frame being of the wander azimuth type having zero angular rate relative to the earth fixed E Frame for the Strapdown Rotation test laboratory conditions.

Assuming that the $\underline{\omega}_{IE}^{L}$ earth rate term is known without error, the following L Frame version of Equation (12.3.4-10) applies for C_{B}^{L} error propagation:

$$\dot{\underline{\phi}}^{L} = - C_{B}^{L}\,\delta\underline{\omega}_{IB}^{B} - \underline{\omega}_{IE}^{L} \times \underline{\phi}^{L} \qquad (18.4.7.2\text{-}2)$$

where

$\underline{\phi}^{L}$ = Rotation angle error vector associated with C_{B}^{L} considering the L Frame to be misaligned.

The $\underline{\phi}^{L}$ error can be defined during one of the rotation sequences in the Strapdown Rotation test as:

$$\underline{\phi}^{L} = \underline{\phi}_{1}^{L} + \Delta\underline{\phi}^{L} \qquad (18.4.7.2\text{-}3)$$

where

$\underline{\phi}_1^L$ = Value for $\underline{\phi}^L$ at the start of the rotation sequence.

$\Delta\underline{\phi}^L$ = Change in $\underline{\phi}^L$ since the start of the rotation sequence.

The derivative of (18.4.7.2-3) gives:

$$\dot{\underline{\phi}}^L = \Delta\dot{\underline{\phi}}^L \qquad (18.4.7.2\text{-}4)$$

Substituting (18.4.7.2-4) into (18.4.7.2-2) then obtains:

$$\Delta\dot{\underline{\phi}}^L = - C_B^L \, \delta\underline{\omega}_{IB}^B - \underline{\omega}_{IE}^L \times \underline{\phi}^L \qquad (18.4.7.2\text{-}5)$$

We now make the assumption for the Strapdown Rotation test that $\underline{\phi}^L$ will be controlled to remain relatively small (e.g., by periodically reinitializing C_B^L if needed to cancel error build-up caused by large angular rate sensor errors), and that each rotation sequence will be executed fairly rapidly so that the integral of $\underline{\omega}_{IE}^L \times \underline{\phi}^L$ over the rotation sequence will be small. These assumptions allow us to neglect the $\underline{\omega}_{IE}^L \times \underline{\phi}^L$ term in (18.4.7.2-5) resulting in the following simplified form:

$$\Delta\dot{\underline{\phi}}^L \approx - C_B^L \, \delta\underline{\omega}_{IB}^B \qquad (18.4.7.2\text{-}6)$$

The horizontal component of (18.4.7.2-6) is the derivative of $\Delta\underline{\phi}_H^L$ in Equations (18.4.7-13) required for Strapdown Rotation test data analysis:

$$\Delta\dot{\underline{\phi}}_H^L = - \left(C_B^L \, \delta\underline{\omega}_{IB}^B \right)_H \qquad (18.4.7.2\text{-}7)$$

Equation (18.4.7.2-7) is a linear differential equation whose integral solution $\Delta\underline{\phi}_H^L$ satisfies the principle of linear superposition. Thus, $\Delta\underline{\phi}_H^L$ can be defined as the sum of angle errors generated during each rotation in a particular rotation sequence:

$$\Delta\underline{\phi}_H^L = \sum_i \Delta\underline{\phi}_{H_i}^L \qquad (18.4.7.2\text{-}8)$$

where

$\Delta\underline{\phi}_{H_i}^L$ = The portion of $\Delta\underline{\phi}_H^L$ created by angular rate sensor error in the ith rotation during one of the rotation sequences in the Strapdown Rotation test.

Equation (18.4.7.2-7) applies for any segment of the rotation sequence, hence:

$$\dot{\Delta\underline{\phi}}^{L}_{H_i} = -\left(C^{L}_{B}\,\delta\underline{\omega}^{B}_{IB_i}\right)_H \qquad (18.4.7.2\text{-}9)$$

where

$\delta\underline{\omega}^{B}_{IB_i}$ = Angular rate sensor error during the ith rotation in the particular rotation sequence of the Strapdown Rotation test

The integral of Equation (18.4.7.2-9) for each ith rotation in a given rotation sequence, when substituted in (18.4.7.2-8), provides $\Delta\underline{\phi}^{L}_{H}$ for Equations (18.4.7-13) as a function of angular rate sensor error $\delta\underline{\omega}^{B}_{IB}$. Expressing $\delta\underline{\omega}^{B}_{IB}$ in terms of characteristic angular rate sensor error parameters will then provide an expression for $\Delta\underline{\phi}^{L}_{H}$ as a function of these parameters.

The analytical model used for angular rate sensor error determination in the Strapdown Rotation test is the following linearized version of Equations (8.1.1.1.1-5) and (8.1.1.3-7) for $\kappa_{SystScal/Mis}$, the equivalent to (8.1.1.3-15) for angular rate sensor asymmetry error, and in which the bias error term has been neglected under the assumption that each rotation sequence will be executed rapidly to prohibit bias error induced build-up in $\Delta\underline{\phi}^{L}_{H}$:

$$\delta\underline{\omega}^{B}_{IB} = \left(\kappa_{LinScal} + \kappa_{Mis} + \kappa_{Asym}\,\Omega_{IBSign}\right)\underline{\omega}^{B}_{IB} \qquad (18.4.7.2\text{-}10)$$

with

$$\kappa_{LinScal} \equiv \begin{bmatrix} \kappa_{11} & 0 & 0 \\ 0 & \kappa_{22} & 0 \\ 0 & 0 & \kappa_{33} \end{bmatrix} \qquad \kappa_{Mis} \equiv \begin{bmatrix} 0 & \kappa_{12} & \kappa_{13} \\ \kappa_{21} & 0 & \kappa_{23} \\ \kappa_{31} & \kappa_{32} & 0 \end{bmatrix}$$

$$\kappa_{Asym} \equiv \begin{bmatrix} \kappa_{111} & 0 & 0 \\ 0 & \kappa_{222} & 0 \\ 0 & 0 & \kappa_{333} \end{bmatrix} \qquad (18.4.7.2\text{-}11)$$

$$\Omega_{IBSign} \equiv \begin{bmatrix} \mathrm{Sign}\left(\omega_{IB_{XB}}\right) & 0 & 0 \\ 0 & \mathrm{Sign}\left(\omega_{IB_{YB}}\right) & 0 \\ 0 & 0 & \mathrm{Sign}\left(\omega_{IB_{ZB}}\right) \end{bmatrix} \qquad (18.4.7.2\text{-}12)$$

where

$\kappa_{LinScal}$ = Angular rate sensor triad linear scale factor error matrix.

κ_{Mis} = Angular rate sensor triad misalignment error matrix.

κ_{Asym} = Angular rate sensor triad scale factor asymmetry error matrix.

$\omega_{IB_{iB}}$ = B Frame component i of $\underline{\omega}_{IB}^{B}$.

At this point in the development we now restrict the angular rotations in each rotation sequence of the Strapdown Rotation test to be composed of a sequence of rotations about individual B Frame (sensor assembly) axes (i.e., a sequence of Euler rotations). Under this restriction, the angular rate vector for the jth rotation in a given rotation sequence will be:

$$\underline{\omega}_{IB_i}^{B} = \dot{\beta}_i \, \underline{u}_{ij}^{B} \qquad (18.4.7.2\text{-}13)$$

where

$\underline{\omega}_{IB_i}^{B}$ = Angular rate vector for the ith rotation in the rotation sequence.

$\underline{u}_{ij}^{B}$ = Unit vector along $\underline{\omega}_{IB_i}^{B}$ which is now specialized to lie along a particular B Frame axis (j = 1, 2 or 3, for $\underline{\omega}_{IB_i}^{B}$ along B Frame axis X, Y or Z) for the ith Euler rotation in the sequence.

$\dot{\beta}_i$ = Signed magnitude of $\underline{\omega}_{IB_i}^{B}$ defined as the projection of $\underline{\omega}_{IB_i}^{B}$ along $\underline{u}_{ij}^{B}$.

Substituting (18.4.7.2-13) in Equation (18.4.7.2-10) provides:

$$\delta\underline{\omega}_{IB_i}^{B} = \dot{\beta}_i \, \kappa_i \, \underline{u}_{ij}^{B} \qquad (18.4.7.2\text{-}14)$$

with

$$\kappa_i \equiv \kappa_{LinScal} + \kappa_{Mis} + \kappa_{Asym} \, \mathrm{Sign}\left(\dot{\beta}_i\right) \qquad (18.4.7.2\text{-}15)$$

The C_B^L term in Equation (18.4.7.2-9) during the ith Euler rotation can be written as:

$$C_B^L = C_{B_i}^L \, C_B^{B_i} \qquad (18.4.7.2\text{-}16)$$

where

$C_{B_i}^L$ = Value for C_B^L at the start of the ith Euler rotation.

$C_B^{B_i}$ = Direction cosine matrix relating B Frame axes during the ith rotation to B Frame axes at the start of the ith Euler rotation.

Generalized Equations (3.2.3-2) show that for an Euler rotation, $C_B^{B_i}$ in (18.4.7.2-16) is given by:

$$C_B^{B_i} = I + \sin\beta_i \left(\underline{u}_{ij}^B \times\right) + (1 - \cos\beta_i)\left(\underline{u}_{ij}^B \times\right)^2 \qquad (18.4.7.2\text{-}17)$$

Substituting (18.4.7.2-14), (18.4.7.2-16), and (18.4.7.2-17) into (18.4.7.2-9) then yields:

$$\Delta\dot{\underline{\phi}}_{H_i}^L = -\left\{C_{B_i}^L\left[\kappa_i\,\underline{u}_{ij}^B + \underline{u}_{ij}^B \times \left(\kappa_i\,\underline{u}_{ij}^B\right)\sin\beta_i + \left(\underline{u}_{ij}^B \times\right)^2\left(\kappa_i\,\underline{u}_{ij}^B\right)\left(1 - \cos\beta_i\right)\right]\right\}_H \dot{\beta}_i \qquad (18.4.7.2\text{-}18)$$

Recognizing that $\dot{\beta}_i\,dt = d\beta_i$, Equation (18.4.7.2-18) can be integrated over the range of $\beta_i = 0$ to $\beta_i = \theta_i$ where:

θ_i = Total angle traversal for the ith Euler rotation.

Equating the result to $\Delta\underline{\phi}_{H_i}^L$ for Equation (18.4.7.2-8) obtains the generalized expression:

$$\begin{aligned}\Delta\underline{\phi}_{H_i}^L = -\Big(C_{B_i}^L\Big\{&\left[I + \left(\underline{u}_{ij}^B \times\right)^2\right]\kappa_i\,\underline{u}_{ij}^B\,\theta_i \\ &- \left(\underline{u}_{ij}^B \times\right)^2\kappa_i\,\underline{u}_{ij}^B\sin\theta_i + \left(\underline{u}_{ij}^B \times\right)\left(\kappa_i\,\underline{u}_{ij}^B\right)\left(1 - \cos\theta_i\right)\Big\}\Big)_H\end{aligned} \qquad (18.4.7.2\text{-}19)$$

Equation (18.4.7.2-19) can be simplified when the properties of κ_i as defined in (18.4.7.2-15) are taken into account. Because the ith Euler rotation is about a single B Frame axis (i.e., $\underline{u}_{ij}^B$ is along B Frame X, Y, or Z depending on j = 1, 2 or 3), Equations (18.4.7.2-11) show that the $\kappa_{LinScal}$ and κ_{Asym} terms in $\kappa_i\,\underline{u}_{ij}^B$ of (18.4.7.2-19) simplify to:

$$\begin{aligned}\kappa_{LinScal}\underline{u}_{ij}^B &= \kappa_{jj}\,\underline{u}_{ij}^B \\ \mathrm{Sign}\left(\dot{\beta}_i\right)\kappa_{Asym}\,\underline{u}_{ij}^B &= \mathrm{Sign}\left(\dot{\beta}_i\right)\kappa_{jjj}\,\underline{u}_{ij}^B\end{aligned} \qquad (18.4.7.2\text{-}20)$$

Since each of these terms is directed along $\underline{u}_{ij}^B$, their products with the $\left(\underline{u}_{ij}^B \times\right)$ and $\left(\underline{u}_{ij}^B \times\right)^2$ terms in (18.4.7.2-19) are zero. Because the ith Euler rotation is about a single B Frame axis, the form for κ_{Mis} in (18.4.7.2-11) shows that in (18.4.7.2-19), its product with $\underline{u}_{ij}^B$ (in $\kappa_i\,\underline{u}_{ij}^B$) is perpendicular to $\underline{u}_{ij}^{B}$. Applying generalized Equation (13.1-11) then shows that for this situation:

$$\left(\underline{u}_{ij}^{B} \times\right)^2 \kappa_{Mis}\, \underline{u}_{ij}^{B} = - \kappa_{Mis}\, \underline{u}_{ij}^{B} \tag{18.4.7.2-21}$$

Substituting (18.4.7.2-20) and (18.4.7.2-21) with (18.4.7.2-15) into (18.4.7.2-19), then obtains the simplified form for $\Delta\underline{\phi}_{H_i}^{L}$:

$$\begin{aligned}\Delta\underline{\phi}_{H_i}^{L} = -\Big(C_{B_i}^{L} \Big\{ &\left[\kappa_{LinScal} + \mathrm{Sign}\left(\dot{\beta}_i\right) \kappa_{Asym} \right] \underline{u}_{ij}^{B}\, \theta_i \\ &+ \left[I \sin\theta_i + \left(1 - \cos\theta_i\right)\left(\underline{u}_{ij}^{B} \times\right) \right] \kappa_{Mis}\, \underline{u}_{ij}^{B} \Big\} \Big)_H\end{aligned} \tag{18.4.7.2-22}$$

Substitution of (18.4.7.2-22) into (18.4.7.2-8) provides the desired expression for $\Delta\underline{\phi}_{H}^{L}$ as a function of the individual test sequence rotations and angular rate sensor errors:

$$\begin{aligned}\Delta\underline{\phi}_{H}^{L} = - \sum_i \Big(C_{B_i}^{L} \Big\{ &\left[\kappa_{LinScal} + \mathrm{Sign}\left(\dot{\beta}_i\right) \kappa_{Asym} \right] \underline{u}_{ij}^{B}\, \theta_i \\ &+ \left[I \sin\theta_i + \left(1 - \cos\theta_i\right)\left(\underline{u}_{ij}^{B} \times\right) \right] \kappa_{Mis}\, \underline{u}_{ij}^{B} \Big\} \Big)_H\end{aligned} \tag{18.4.7.2-23}$$

Equation (18.4.7.2-23) with (18.4.7.2-11) defines the relationship between the $\kappa_{LinScal}$, κ_{Mis}, κ_{Asym} angular rate sensor error parameters and the $\Delta\underline{\phi}_{H}^{L}$ term in measurement Equations (18.4.7-13). The C_{B2}^{L} term in (18.4.7-13) is obtained by successive application of Equation (18.4.7.2-16) with (18.4.7.2-17) for each β_i, $\underline{u}_{ij}^{B}$ in a given rotation sequence, using C_{B1}^{L} (the C_{B}^{L} matrix at the start of the rotation sequence) as the initial value for $C_{B_i}^{L}$.

18.4.7.3 MAKING THE STRAPDOWN ROTATION TEST ACCELERATION MEASUREMENT

The acceleration measurements taken during the Strapdown Rotation test are based on an averaging process for the $\Delta\underline{v}_{m}^{L}$'s in Equation (18.4-1) repeated below:

$$\Delta\underline{v}_{m}^{L} = \Delta\underline{v}_{SF_m}^{L} + g_{Tst}\, \underline{u}_{ZL}^{L}\, T_m \qquad \underline{a}^{L} = \frac{1}{T_m} \Delta\underline{v}_{Avg}^{L} \tag{18.4.7.3-1}$$

In this section we will develop an algorithm for calculating $\Delta \underline{v}^L_{Avg}$ in (18.4.7.3-1) based on processing direct $\Delta \underline{v}^L_{SF}$ L Frame integrated specific force acceleration increment data (i.e., transformed integrated B Frame specific force increments). For situations when direct access to $\Delta \underline{v}^L_{SF}$ is not available, Section 18.4.7.3.1 describes a simpler (though less accurate) method for calculating $\underline{a}^L$ in (18.4.7.3-1) using INS output N Frame velocity measurements.

One method for calculating $\Delta \underline{v}^L_{Avg}$ in (18.4.7.3-1) is by a direct linear averaging of a sequence of Equation (18.4.7.3-1) $\Delta \underline{v}^L_m$ L Frame integrated total acceleration increments for each $\Delta \underline{v}^L_{Avg}$ component:

$$\overline{\Delta v_j} = \frac{1}{r} \sum_{m=j}^{j+r-1} \Delta v_m \tag{18.4.7.3-2}$$

where

Δv_m = One of the components of $\Delta \underline{v}^L_m$.

j = m cycle number for the first $\Delta \underline{v}^L_m$ sample used in the average computation.

$\overline{\Delta v_j}$ = Simple linear average of one of the components of $\Delta \underline{v}^L_m$, averaged from m cycle j to m cycle j + r - 1.

r = Number of successive values of $\Delta \underline{v}^L_m$ used to calculate $\overline{\Delta v_j}$.

The accuracy of the $\overline{\Delta v_j}$ average calculated in (18.4.7.3-2) is limited by accelerometer quantization noise in the $\Delta \underline{v}^L_{SF}$ data. In order to reduce the quantization noise in $\overline{\Delta v_j}$, a succession of $\overline{\Delta v_j}$ averages can be taken (for successive values of j), and the results then averaged to obtain the so-called "average of averages":

$$\Delta v_{Avg} = \frac{1}{s} \sum_{j=k}^{k+s-1} \overline{\Delta v_j} \tag{18.4.7.3-3}$$

where

Δv_{Avg} = Refined average for one of the components of $\Delta \underline{v}^L_m$ (the average of averages) with reduced accelerometer quantization noise compared to each of the $\overline{\Delta v_j}$'s.

k = m cycle number for the first $\Delta \underline{v}_m^L$ sample used in the first $\overline{\Delta v_j}$ for the Δv_{Avg} computation.

s = Number of successive $\overline{\Delta v_j}$'s used in calculating Δv_{Avg}.

The vector form of (18.4.7.3-2) - (18.4.7.3-3) is:

$$\overline{\Delta \underline{v}_j^L} = \frac{1}{r} \sum_{m=j}^{j+r-1} \Delta \underline{v}_m^L \qquad \Delta \underline{v}_{Avg}^L = \frac{1}{s} \sum_{j=k}^{k+s-1} \overline{\Delta \underline{v}_j^L} \tag{18.4.7.3-4}$$

From Equation (18.4.7.3-4), the total number of m cycles spanning the $\Delta \underline{v}_{Avg}^L$ computation is:

$$w = r + s - 1 \tag{18.4.7.3-5}$$

where

w = Total number of successive m cycles required to calculate $\Delta \underline{v}_{Avg}^L$.

The need for the -1 term in (18.4.7.3-5) might not be readily apparent. You can verify its correctness by a simple numerical test (e.g., for the trivial case of $s = 1$, w is obviously r which agrees with (18.4.7.3-5)).

In order to minimize the quantization noise induced error in $\Delta \underline{v}_{Avg}^L$ for a specified w value, we must appropriately balance r and s in Equation (18.4.7.3-5). The balance is based on minimizing the mean squared error in $\Delta \underline{v}_{Avg}^L$ produced by quantization noise.

An analytical model for quantization noise can be developed by considering how the Δv_m summation in (18.4.7.3-2) is formed as a summing of pulses (admittedly transformed) from accelerometers. If each accelerometer output pulse was transformed and summed over the $m = j$ to $m = j + r - 1$ time span, the result would be identical to the Δv_m summation in (18.4.7.3-2) over the same time period (allowing for the gravity term in (18.4-1)), including the effect of pulse quantization error. Imagine a situation in which the pulse count from the $m = j$ cycle time happened to begin <u>instantaneously after</u> a pulse was emitted (call it the "start" pulse). Then the pulse count from this time forward will be a true indication of integrated acceleration (i.e., velocity change) at any instant that a pulse has been received and counted. Now consider that the $m = j$ cycle time pulse count is initiated a small time interval <u>before</u> the "start pulse" in which the small time interval is less than the local time interval between pulses. The first pulse that is counted (i.e., the "start" pulse) will be in error (a quantization error) because it is registered as a full pulse when in fact the time period for the count was less than a full pulse period. The maximum error under this condition occurs when a pulse is received (and counted) instantaneously after the $m = j$ time instant, and will equal one pulse. Thus, the quantization error introduced at $m = j$ can range in magnitude from zero to one pulse with a mean value of half a pulse (for a uniform statistical quantization error distribution). The sign of the quantization error will be the sign of the instantaneous pulse rate (positive or negative). The pulse count from this point forward will add no additional quantization error until the $m = j + r - 1$ cycle is reached to halt the count.

If the count is halted at a finite time interval following a pulse occurrence, the pulse count at $m = j + r - 1$ will experience an additional quantization error because the integrated acceleration since the last pulse has not been registered in the count. The error will be maximum at one pulse magnitude if the count is halted at the instant prior to receipt of the next pulse. Thus, the added quantization error at $m = j + r - 1$ will be in the range of zero to minus one pulse with a mean value of minus half a pulse. The sign of the error in this case will be the negative of the instantaneous pulse rate at $m = j + r - 1$.

The previous discussion is the basis for the following quantization error model for the Δv_m pulse count:

$$\delta\left(\sum_{m\,=\,j}^{j+r-1} \Delta v_m\right) = \zeta_j - \zeta_{j+r-1} \qquad\qquad \zeta_j = \frac{\varepsilon_a}{2}\,\mathrm{Sign}(\Delta v_j) + \chi_j \qquad (18.4.7.3\text{-}6)$$

where

ζ_j = Quantization error at computer cycle j.

ε_a = Accelerometer output pulse size.

χ_j = Random accelerometer quantization error at cycle j having a value between $-\frac{\varepsilon_a}{2}$ and $+\frac{\varepsilon_a}{2}$ with equal probability (i.e., uniformly distributed).

We assume that Δv_m will have the same sign at $m = j$ and $m = j + r - 1$ for the Strapdown Rotation test so that (18.4.7.3-6) when combined yields:

$$\delta\left(\sum_{m\,=\,j}^{j+r-1} \Delta v_m\right) = \chi_j - \chi_{j+r-1} \qquad (18.4.7.3\text{-}7)$$

With (18.4.7.3-7), the error in (18.4.7.3-2) and (18.4.7.3-3) is:

$$\delta\Delta v_{Avg} = \frac{1}{r\,s}\sum_{j\,=\,k}^{k+s-1}\left(\chi_j - \chi_{j+r-1}\right)$$

We now take the expected value of the square of (18.4.7.3-7) based on χ_j and χ_{j+r-1} being uncorrelated over the χ_j range of j = k to j = k + s - 1, and the χ_{j+r-1} range from j = k + r -1 to j = k + r + s - 2 (which will be verified later):

$$\sigma^2_{\Delta v_{Avg}} \equiv \mathcal{E}\left(\delta \Delta v^2_{Avg}\right) = \frac{1}{r^2 s^2} \sum_{j=k}^{k+s-1} \left[\mathcal{E}\left(\chi_j^2\right) + \mathcal{E}\left(\chi^2_{j+r-1}\right)\right] = \frac{2\, s\, \sigma^2_\chi}{r^2 s^2} = \frac{2\, \sigma^2_\chi}{r^2 s} \qquad (18.4.7.3\text{-}8)$$

$$\sigma^2_\chi \equiv \mathcal{E}\left(\chi_j^2\right) = \mathcal{E}\left(\chi^2_{j+r-1}\right)$$

where

$\sigma_{\Delta v_{Avg}}$ = Root-mean-square value of $\delta \Delta v_{Avg}$.

σ_χ = Root-mean-square value of the χ_j quantization error.

Based on the previous χ_j definition, the mean value of χ_j is zero, and from the discussion in the second and third paragraphs following (15.2.1.2-17), its probability density is $1 / \varepsilon_a$ over the χ_j range from $-\frac{\varepsilon_a}{2}$ to $+\frac{\varepsilon_a}{2}$. Then the χ_j variance about its mean (i.e., the formal definition for σ^2_χ) is:

$$\sigma^2_\chi = \int_{-\varepsilon_a/2}^{+\varepsilon_a/2} \frac{1}{\varepsilon_a} \chi^2 \, d\chi = \frac{\varepsilon_a^2}{12} \qquad (18.4.7.3\text{-}9)$$

With (18.4.7.3-5) and (18.4.7.3-9), Equation (18.4.7.3-8) is equivalently:

$$\sigma^2_{\Delta v_{Avg}} = \frac{\varepsilon_a^2}{6\, r^2 (w - r + 1)} \qquad (18.4.7.3\text{-}10)$$

For a specified w, the condition for minimum $\sigma_{\Delta v_{Avg}}$ is found by setting the derivative of (18.4.7.3-10) to zero, and solving for r (and s using (18.4.7.3-5)). The result is:

$$r = \frac{2}{3}(w + 1) \qquad s = \frac{1}{3}(w + 1) \qquad (18.4.7.3\text{-}11)$$

As is easily verified by substitution, for the (18.4.7.3-11) solution, the χ_j and χ_{j+r-1} ranges (defined following (18.4.7.3-7)) do not overlap. Thus, (18.4.7.3-11) is compatible with an earlier assumption in (18.4.7.3-8) that χ_j and χ_{j+r-1} are uncorrelated.

Applying (18.4.7.3-11) in (18.4.7.3-10) then finds $\sigma_{\Delta v_{Avg}}$ for the optimal r, s settings:

$$\sigma_{\Delta v_{Avg}} = \sqrt{\frac{9}{8\,(w+1)}\,\frac{w}{(w+1)}\,\frac{\varepsilon_a}{w}} \qquad (18.4.7.3\text{-}12)$$

For comparison, the same treatment can be performed for the simple average $\overline{\Delta v}$ in Equation (18.4.7.3-2) (setting r = w) for which we would find:

$$\sigma_{\overline{\Delta v}} = \sqrt{\frac{1}{6}\,\frac{\varepsilon_a}{w}} \qquad (18.4.7.3\text{-}13)$$

where

$\sigma_{\overline{\Delta v}}$ = Root-mean-square value for $\overline{\Delta v}_j$.

The benefit of the average-of-averages approach in this application becomes significant as the averaging time (the length of w) is increased. As and example, consider an m cycle rate of 50 Hz and a 20 second $\Delta \underline{v}^L_{Avg}$ calculation time allowance, for which w = 50 × 20 = 1000. To maintain r and s at integer values (from (18.4.7.3-11)), let's set w = 998. Substituting w = 998 in (18.4.7.3-12) and (18.4.7.3-13) shows that $\sigma_{\Delta v_{Avg}}$ is one twelfth of $\sigma_{\overline{\Delta v}}$ under this condition.

With (18.4.7.3-11), $\Delta \underline{v}^L_{Avg}$ Equations (18.4.7.3-4) become:

$$\overline{\Delta \underline{v}}_j^L = \frac{3}{2\,(w+1)} \sum_{m=j}^{j-1+2(w+1)/3} \Delta \underline{v}^L_m \qquad \Delta \underline{v}^L_{Avg} = \frac{3}{w+1} \sum_{j=k}^{k-1+(w+1)/3} \overline{\Delta \underline{v}}^L_j \qquad (18.4.7.3\text{-}14)$$

Use of (18.4.7.3-14) for $\Delta \underline{v}^L_{Avg}$ in Equation (18.4.7.3-1) provides the $\underline{a}^L$ measurements for the Strapdown Rotation test.

A recursive algorithm at the computer m cycle execution rate can be constructed for $\Delta \underline{v}^L_{Avg}$ from (18.4.7.3-14) by first defining w and r as a function of s using (18.4.7.3-11):

$$r = 2\,s \qquad w = 3\,s - 1 \qquad (18.4.7.3\text{-}15)$$

Substituting (18.4.7.3-15) into (18.4.7.3-14) obtains the simpler form:

$$\overline{\Delta\underline{v}}_j^L = \frac{1}{2\,s} \sum_{m=j}^{j+2s-1} \Delta\underline{v}_m^L \qquad \Delta\underline{v}_{Avg}^L = \frac{1}{s} \sum_{j=k}^{k+s-1} \overline{\Delta\underline{v}}_j^L \qquad (18.4.7.3\text{-}16)$$

The algorithmic form of (18.4.7.3-15) is easily derived from a numerical example illustrating the summation operations. Consider the case when s = 4 (for which from (18.4.7.3-15), r = 8 and w = 11), and the following diagrammatic representation:

Δv_1	Δv_2	Δv_3	Δv_4	Δv_5	Δv_6	Δv_7	Δv_8	0	0	0
0	Δv_2	Δv_3	Δv_4	Δv_5	Δv_6	Δv_7	Δv_8	Δv_9	0	0
0	0	Δv_3	Δv_4	Δv_5	Δv_6	Δv_7	Δv_8	Δv_9	Δv_{10}	0
0	0	0	Δv_4	Δv_5	Δv_6	Δv_7	Δv_8	Δv_9	Δv_{10}	Δv_{11}

Figure 18.4.7.3-1 Diagrammatic Representation Of $\Delta\underline{v}_{Avg}^L$ Summing Operations

Each row in Figure 18.4.7.3-1 represents the elements of $\overline{\Delta\underline{v}}_j^L$ used in the (18.4.7.3-16) $m = j$ to $j + 2\,s - 1$ summing operation. The sum of the row sums is the $j = k$ to $k + s - 1$ summation operation for $\Delta\underline{v}_{Avg}^L$. The sum of all the elements in the figure is the combined double summation in (18.4.7.3-16). Notice that the figure can be divided into three sections: an upper triangle at the left for the first three columns (i.e., the first s - 1 columns), a rectangle in the center for columns 4 through 8 (i.e., columns s to 2 s), and a lower triangle at the right for columns 9 through 11 (i.e., columns 2 s + 1 to 3 s -1). Based on this observation we can write the total sum as the sum of the column sums in each section:

$$\sum_{1-3} = 1 \times \Delta v_1 + 2 \times \Delta v_2 + 3 \times \Delta v_3 = \sum_{i=1}^{s-1} i\,\Delta v_i$$

$$\sum_{4-8} = 4 \times \sum_{i=4}^{8} \Delta v_i = \sum_{i=s}^{2s} s\,\Delta v_i$$

$$\sum_{9-11} = 3 \times \Delta v_9 + 2 \times \Delta v_{10} + 1 \times \Delta v_{11} = \sum_{i=2s+1}^{3s-1} (3\,s - i)\,\Delta v_i \qquad (18.4.7.3\text{-}17)$$

$$\sum_{Total} = \sum_{1-3} + \sum_{4-8} + \sum_{9-11}$$

where

$\sum_{1-3}, \sum_{4-8}, \sum_{9-11}$ = Sum of terms in columns 1 - 3, 4 - 8, and 9 - 11.

$\sum_{\text{Total}}$ = Sum of all terms in Figure 18.4.7.3-1 representing the double summation operation in Equation (18.4.7.3-16).

The m cycle computation algorithm for (18.4.7.3-16) is then easily constructed from (18.4.7.3-17) including evaluation of $\underline{a}^L$ in (18.4.7.3-1) for the Strapdown Rotation test measurement:

Initialization: $\underline{\text{Sum}} = 0, \quad i = 0$

Execute For 3 s - 1 cycles in the m loop
If (i < 3 s) Then
  $i = i + 1$
  If (i < s) Then
    $\underline{\text{Sum}} = \underline{\text{Sum}} + i\, \Delta\underline{v}_i^L$
  Else If ($s \leq i \leq 2\,s$) Then
    $\underline{\text{Sum}} = \underline{\text{Sum}} + s\, \Delta\underline{v}_i^L$ (18.4.7.3-18)
  Else If ($2\,s < i < 3\,s$) Then
    $\underline{\text{Sum}} = \underline{\text{Sum}} + (3\,s - i)\, \Delta\underline{v}_i^L$
    If (i = 3 s -1) Then
      $\Delta\underline{v}_{Avg}^L = \dfrac{1}{2\,s^2}\, \underline{\text{Sum}} \qquad \underline{a}^L = \dfrac{1}{T_m}\, \Delta\underline{v}_{Avg}^L$
    End If
  End If
End If

18.4.7.3.1 Alternative Strapdown Rotation Test Measurement Approach

The Strapdown Rotation test acceleration measurement as described in Section 18.4.7.3 implies the existence of special software to execute the associated analytical operations. For situations when such software is not available (internal or external to the INS), an alternate procedure can be utilized for a strapdown INS based on INS free-inertial navigation mode velocity output measurements. The alternate procedure uses the change in INS output N Frame horizontal velocity over a specified averaging time T_{Avg} (e.g., 10 seconds), transformed to the L Frame and divided by T_{Avg}, as the measure of $\underline{a}^L$ in Equation (18.4.7.3-1):

$$\underline{a}^L = \frac{1}{T_{Avg}} C_N^L \left(\underline{v}_{End}^N - \underline{v}_{Start}^N \right) \tag{18.4.7.3.1-1}$$

where

$\underline{v}_{Start}^N, \underline{v}_{End}^N$ = $\underline{v}^N$ measurements taken at the start and end of the averaging period.

The $\underline{v}_{Start}^N, \underline{v}_{End}^N$ difference in (18.4.7.3.1-1) is the sum of changes in $\underline{v}^N$ over the averaging period, or using the beginning of Section 18.4.7.3 nomenclature:

$$\underline{a}^L = \frac{1}{r\, T_m} \sum_{m=j}^{j+r-1} \Delta\underline{v}_m^L \tag{18.4.7.3.1-2}$$

If Equation (18.4.7.3.1-2) is compared with the simple linear averaging technique of Section 18.4.7.3 (i.e., Equation (18.4.7.3-2) with (18.4.7.3-1) for $\underline{a}^L$), it should be clear that they are equivalent. Thus, the alternate (18.4.7.3.1-1) approach for calculating $\underline{a}^L$ contains the higher acceleration quantization error present in (18.4.7.3-2), hence, is not as accurate as the (18.4.7.3-18) algorithm averaging technique.

18.4.7.4 SENSOR ASSEMBLY MISALIGNMENT CALIBRATION RELATIVE TO TEST FIXTURE MOUNT

As an adjunct to the Strapdown Rotation test, it is convenient to also calculate the alignment of the sensor assembly relative to the test fixture mount. For the test results to be meaningful, the data is usually taken in rotation tests for which the sensor assembly has been installed in the INS, and the INS is the test article on the rotation test fixture mount. The INS mount would then be a standard mount for INS installation in a user vehicle. The result of this calculation can be used to update the associated misalignment calibration coefficients for the strapdown INS attitude output function (e.g., the $\underline{J}$ vector in Equation (8.3-2)). As in Equation (8.3-2), the alignment calibration matrix is C_B^M which can be equated to the computed attitude data in the INS by:

$$C_B^M = C_{Geo}^M \, C_L^{Geo} \, C_B^L \tag{18.4.7.4-1}$$

where

M = Test fixture mount coordinate frame.

Geo = Local geographic coordinate frame with Y axis north and Z axis up along the local plumb-bob vertical.

Ideally, C_B^M can be calculated from Equation (18.4.7.4-1) for a particular INS based on static measurements, using C_B^L and C_L^{Geo} data calculated by the INS and a measured value for C_{Geo}^M at the particular mount orientation for which the INS data was evaluated. In principle, the same method can be applied to the Strapdown Rotation test, using the computed value for C_B^L at a particular measurement position, and C_L^{Geo} (a Z axis wander angle Euler transformation matrix) based on a wander angle calculated from L Frame components of horizontal earth rate (e.g., Equations (6.2.1-6) - (6.2.1-7)). With this approach, the horizontal earth rate components can be calculated using the Chapter 6, Section 6.1.2 Fine Alignment procedure. For improved accuracy, the more time consuming Strapdown Drift test method of Section 18.2.1 might be used to reduce angular rate sensor random output noise and systematic bias induced error in the computed horizontal earth rates. However, the latter earth rate determination process is still limited in accuracy to the stability of the angular rate sensor bias over the earth rate component determination period. To achieve reasonable accuracy in the C_{Geo}^M measurement, elaborate optical techniques can be brought to bear in which the heading (azimuth) orientation of the mount is determined relative to true north using stellar sightings as the fundamental reference (e.g., through the traditional hole in the ceiling precision sighting of the North Star). Fortunately, the vertical orientation of C_{Geo}^M is more easily determined using bubble level devices.

Sounds complicated doesn't it? And it really isn't necessary if the following simpler formulation is used based on:

$$C_M^{MARS} \, \underline{a}_{SF}^M = \underline{a}_{SF}^{MARS} \tag{18.4.7.4-2}$$

$$C_M^B = C_{MARS}^B \, C_M^{MARS} \tag{18.4.7.4-3}$$

where

$\underline{a}_{SF}$ = Specific force acceleration during static measurements caused by reaction force against local plumb-bob gravity.

MARS = Mean angular rate sensor axes as described in Section 18.4.3.

The C_{MARS}^B matrix in (18.4.7.4-3) is defined by the selected B Frame alignment definition angle $\underline{\gamma}$ of Section 18.4.3. The $\underline{a}_{SF}^M$ term in (18.4.7.4-2) can be calculated based on:

$$\underline{a}_{SF}^{M} = C_{L}^{M} \underline{a}_{SF}^{L} \tag{18.4.7.4-4}$$

which, with Equation (18.4.7-10) for the Strapdown Rotation test static measurement condition, is:

$$\underline{a}_{SF}^{M} = - g_{Tst} \underline{u}_{ZL}^{M} \tag{18.4.7.4-5}$$

where

$\underline{u}_{ZL}^{M}$ = M Frame components of a unit vector along the L Frame Z axis (i.e., downward along the local plumb-bob vertical).

g_{Tst} = Plumb-bob gravity magnitude at the Strapdown Rotation test fixture location.

Substituting (18.4.7.4-5) into (18.4.7.4-2) obtains:

$$C_{M}^{MARS} \underline{u}_{ZL}^{M} = - \frac{1}{g_{Tst}} \underline{a}_{SF}^{MARS} \tag{18.4.7.4-6}$$

Section 3.2.1.1 shows how Equation (18.4.7.4-6) can be solved for C_{M}^{MARS} if we evaluate it for two distinct orientations of the M and MARS Frames. Because the M and MARS Frames are fixed relative to one another, the C_{M}^{MARS} matrix is identical for any M or MARS Frame attitude, and we can write (18.4.7.4-6) at the two distinct orientations as:

$$\begin{aligned} C_{M1}^{MARS1} \underline{u}_{ZL}^{M1} &= C_{M}^{MARS} \underline{u}_{ZL}^{M1} = - \frac{1}{g_{Tst}} \underline{a}_{SF}^{MARS1} \\ C_{M2}^{MARS2} \underline{u}_{ZL}^{M2} &= C_{M}^{MARS} \underline{u}_{ZL}^{M2} = - \frac{1}{g_{Tst}} \underline{a}_{SF}^{MARS2} \end{aligned} \tag{18.4.7.4-7}$$

where

M1, MARS1, M2, MARS2 = M, MARS Frames at measurement orientations 1 and 2.

The solution for C_{M}^{MARS} in (18.4.7.4-7) is provided by Equation (3.2.1.1-8) based on (3.2.1.1-1) - (3.2.1.1-2) with (3.2.1.1-3), (3.2.1.1-5) and (3.2.1.1-6):

$$C_{M}^{MARS} = F^{MARS} \left(F^{M}\right)^{-1} \tag{18.4.7.4-8}$$

with

$$F^{MARS} \equiv \begin{bmatrix} -\frac{1}{g_{Tst}} \underline{a}_{SF}^{MARS1} & -\frac{1}{g_{Tst}} \underline{a}_{SF}^{MARS2} & \frac{1}{g_{Tst}^2} \underline{a}_{SF}^{MARS1} \times \underline{a}_{SF}^{MARS2} \end{bmatrix}$$

$$F^{M} \equiv \begin{bmatrix} \underline{u}_{ZL}^{M1} & \underline{u}_{ZL}^{M2} & \underline{u}_{ZL}^{M1} \times \underline{u}_{ZL}^{M2} \end{bmatrix} \qquad (18.4.7.4\text{-}9)$$

As will be explained subsequently, the $\underline{a}_{SF}^{MARS}$ terms in (18.4.7.4-9) can be computed from static acceleration measurements taken at the two selected sensor assembly orientations during the Strapdown Rotation test. The $\underline{u}_{ZL}^{M}$ terms in (18.4.7.4-9) are readily defined by the orientation of the sensor mount (M Frame) relative to vertical for the two selected measurement orientations. As discussed in Section 3.2.1.1, the F^{M} inverse in (18.4.7.4-8) will be non-singular if the determinant is non-singular, and the determinant will be non-singular if $\underline{u}_{ZL}^{M1}$ is not parallel to $\underline{u}_{ZL}^{M2}$. Ideally, the 1, 2 orientations should be selected so that $\underline{u}_{ZL}^{M1}$ and $\underline{u}_{ZL}^{M2}$ are perpendicular to each other. It remains to define the $\underline{a}_{SF}^{MARS}$ terms as a function of sensor assembly output derived data.

Equations (18.4.7.4-9) are based on idealized error free values for the $\underline{a}_{SF}^{MARS}$ terms. The error free values can be calculated based on Strapdown Rotation test acceleration measurements, corrected for rotation test determined errors as defined in (18.4.7.1-1):

$$\hat{\underline{a}}_{SF}^{B} = \underline{a}_{SF}^{B} + \delta\underline{a}_{SF}^{B} = \left(I + \lambda_{LinScal} + \lambda_{Mis} + \lambda_{Asym}\, A_{SFSign}\right) \underline{a}_{SF}^{B} + \underline{\lambda}_{Bias} \qquad (18.4.7.4\text{-}10)$$

where

$\hat{\underline{a}}_{SF}^{B}$ = Value for $\underline{a}_{SF}^{B}$ calculated in the Strapdown Rotation test software that contains sensor errors. The transformed components of $\hat{\underline{a}}_{SF}^{B}$ (to the L Frame) are what is used to form the $\hat{\underline{a}}_{SF}^{L}$ measurements taken for the Strapdown Rotation test.

$\lambda_{LinScal}$, λ_{Mis} = Scale factor and misalignment calibration error components of $\lambda_{LinScal/Mis}$.

A_{SFSign} = Diagonal matrix whose elements are unity in magnitude with sign equal to the sign of the components of $\underline{a}_{SF}^{B}$.

$\underline{\lambda}_{Bias}$ = Accelerometer triad bias error vector.

The $\underline{a}_{SF}^{B}$ term in (18.4.7.4-10) can be expressed in terms of the desired $\underline{a}_{SF}^{MARS}$ term as:

$$\underline{a}_{SF}^{B} = C_{MARS}^{B} \, \underline{a}_{SF}^{MARS} \tag{18.4.7.4-11}$$

The C_{MARS}^{B} matrix in (18.4.7.4-11) can be defined to first order by the rotation angle vector $\underline{\gamma}$ of Section 18.4.3 using generalized Equation (3.2.2.1-8):

$$C_{MARS}^{B} \approx I + (\underline{\gamma} \times) \tag{18.4.7.4-12}$$

The λ_{Mis} matrix in (18.4.7.4-10) can be expressed relative to mean angular rate sensor axes using the matrix form of the accelerometer misalignment expressions in Equations (18.4.3-5):

$$\lambda_{Mis} = \mu_{Mis} - (\underline{\gamma} \times) \tag{18.4.7.4-13}$$

We also make the approximation as in (8.1.1.3-16) that:

$$A_{SFSign} \approx A_{SFPuls_{Sign}} \tag{18.4.7.4-14}$$

where

$A_{SFPuls_{Sign}}$ = Diagonal matrix whose elements are unity in magnitude with sign equal to the sign of the components of $\underline{a}_{SFPuls}^{B}$, the uncompensated accelerometer triad output pulse rate vector.

Substituting (18.4.7.4-11) - (18.4.7.4-14) in (18.4.7.4-10) and dropping $\underline{\gamma}$ squared and products of $\underline{\gamma}$ with error terms as second order, then gives:

$$\hat{\underline{a}}_{SF}^{B} \approx \left(I + \lambda_{LinScal} + \mu_{Mis} + \lambda_{Asym} \, A_{SFPuls_{Sign}}\right) \underline{a}_{SF}^{MARS} + \underline{\lambda}_{Bias} \tag{18.4.7.4-15}$$

The $\hat{\underline{a}}_{SF}^{B}$ vector in (18.4.7.4-15) can also be expressed in terms of the Equation (18.4.7-2) Strapdown Rotation test acceleration measurement vector components as:

$$\hat{\underline{a}}_{SF}^{L} = \hat{\underline{a}}^{L} - g_{Tst} \, \underline{u}_{ZL}^{L} \qquad \hat{\underline{a}}_{SF}^{B} = \left(\hat{C}_{B}^{L}\right)^{T} \hat{\underline{a}}_{SF}^{L} \tag{18.4.7.4-16}$$

where

$\hat{\underline{a}}^{L}$ = L Frame total acceleration measurement for the rotation test taken by averaging transformed accelerometer data plus gravity, defined as $\underline{a}^{L}$ in Equation (18.4-1).

g_{Tst} = Plumb-bob gravity magnitude at the test site.

$\underline{u}_{ZL}^{L}$ = Unit vector along the L Frame Z axis (downward along the plumb-bob vertical) as projected on L Frame axes.

$\hat{C}_B^L$ = Value for C_B^L computed by the Strapdown Rotation test attitude computation algorithm as an integration of Equation (18.4.7.2-1) using inertial sensor data that is compensated with previously determined values for the compensation coefficients.

Taking the inverse of (18.4.7.4-15) with (18.4.7.4-16) for $\hat{\underline{a}}_{SF}^B$ yields the desired expression for $\underline{a}_{SF}^{MARS}$:

$$\underline{a}_{SF}^{MARS} = \left(I + \lambda_{LinScal} + \mu_{Mis} + \lambda_{Asym}\, A_{SF_{PulsSign}}\right)^{-1} \left[\left(\hat{C}_B^L\right)^T \left(\hat{\underline{a}}^L - g_{Tst}\, \underline{u}_{ZL}^L\right) - \underline{\lambda}_{Bias}\right] \tag{18.4.7.4-17}$$

With (18.4.7.4-17), the $\underline{a}_{SF}^{MARS}$ terms in (18.4.7.4-9) become for each of the two selected measurement orientations:

$$\underline{a}_{SF}^{MARS1} = \left(I + \lambda_{LinScal} + \mu_{Mis} + \lambda_{Asym}\, A^1_{SF_{PulsSign}}\right)^{-1} \left[\left(\hat{C}_{B1}^L\right)^T \left(\hat{\underline{a}}_1^L - g_{Tst}\, \underline{u}_{ZL}^L\right) - \underline{\lambda}_{Bias}\right] \tag{18.4.7.4-18}$$

$$\underline{a}_{SF}^{MARS2} = \left(I + \lambda_{LinScal} + \mu_{Mis} + \lambda_{Asym}\, A^2_{SF_{PulsSign}}\right)^{-1} \left[\left(\hat{C}_{B2}^L\right)^T \left(\hat{\underline{a}}_2^L - g_{Tst}\, \underline{u}_{ZL}^L\right) - \underline{\lambda}_{Bias}\right]$$

where

$\hat{\underline{a}}_1^L$, $\hat{\underline{a}}_2^L$, $A^1_{SF_{PulsSign}}$, $A^2_{SF_{PulsSign}}$ = Values for $\hat{\underline{a}}^L$ and $A_{SF_{PulsSign}}$ at test orientations 1 and 2.

$\lambda_{LinScal}$, μ_{Mis}, λ_{Asym}, $\underline{\lambda}_{Bias}$ Accelerometer calibration error terms calculated from Strapdown Rotation test data as in Figures 18.4.4-1 or 18.4.4-2.

Equations (18.4.7.4-8), (18.4.7.4-9) and (18.4.7.4-18) constitute a complete set for computing C_M^{MARS}. Note that no north referenced data is required in these equations.

Once C_M^{MARS} is calculated, C_M^B can be computed with (18.4.7.4-3) based on the definition for the B Frame relative to the MARS Frame (i.e., C_{MARS}^B which is the direction cosine equivalent of the $\underline{\gamma}$ rotation vector). The $\underline{J}$ calibration coefficients in (8.3-2) can be extracted from C_M^B transpose using direction cosine to rotation vector inversion Equations (3.2.2.2-10) - (3.2.2.2-12) and (3.2.2.2-15) - (3.2.2.2-17).

In the above development, we have treated C_{MARS}^{B} (and its associated $\underline{\gamma}$ vector) as an independent parameter. In practice, $\underline{\gamma}$ is generally selected so that the B Frame is defined to be either the MARS Frame or the Sensor assembly mount M Frame. For the former case (i.e., B = MARS), C_{MARS}^{B} is identity and Equation (18.4.7.4-3) reduces to:

For B Frame = MARS Frame

$$C_{B}^{M} = \left(C_{M}^{MARS}\right)^{T} \qquad \underline{\gamma} = 0 \tag{18.4.7.4-19}$$

$$\underline{J} = \begin{matrix}\text{Rotation Angle}\\ \text{Extraction From}\end{matrix}\; C_{B}^{M} \; \begin{matrix}\text{With Equations (3.2.2.2-10) - (3.2.2.2-12)}\\ \text{and (3.2.2.2-15) - (3.2.2.2-17)}\end{matrix}$$

For the latter case (i.e., B = M), the C_{M}^{B} matrix is identity corresponding to zero for the $\underline{J}$ coefficients in Equation (8.3-2). Then the above process must be inverted to solve for the C_{MARS}^{B} (and the corresponding $\underline{\gamma}$ rotation vector) using identity for C_{M}^{B} in Equation (18.4.7.4-3):

For B Frame = M Frame

$$C_{MARS}^{B} = \left(C_{M}^{MARS}\right)^{T} \qquad \underline{J} = 0 \tag{18.4.7.4-20}$$

$$\underline{\gamma} = \begin{matrix}\text{Rotation Angle}\\ \text{Extraction From}\end{matrix}\; C_{MARS}^{B} \; \begin{matrix}\text{With Equations (3.2.2.2-10) - (3.2.2.2-12)}\\ \text{and (3.2.2.2-15) - (3.2.2.2-17)}\end{matrix}$$

As an example of the above procedure, we might set one of the attitudes for C_{M}^{MARS} determination having the M Frame Z axis down (i.e., nominally level), with the second attitude having the M Frame X axis up (i.e., 90 degree pitch up). For these orientations, it is clear that $\underline{u}_{ZL}^{M1}$ and $\underline{u}_{ZL}^{M2}$ will be along the M Frame positive Z and negative X axes. Thus, $\underline{u}_{ZL}^{M1}$ and $\underline{u}_{ZL}^{M2}$ will be perpendicular to each other, and the determinant of F^{M} in (18.4.7.4-9) will be maximized for an optimized inverse calculation in (18.4.7.4-8). For the previous orientations, the $\underline{u}_{ZL}^{M1}$, $\underline{u}_{ZL}^{M2}$ vectors in (18.4.7.4-9) are given by:

$$\underline{u}_{ZL}^{M1} = \begin{bmatrix} 0 \\ 0 \\ 1 \end{bmatrix} \qquad \underline{u}_{ZL}^{M2} = \begin{bmatrix} -1 \\ 0 \\ 0 \end{bmatrix} \tag{18.4.7.4-21}$$

For the 16 and 21 sequence Strapdown Rotation tests, we see from Tables 18.4.1-1 and 18.4.1-2 that the previous conditions apply for the 16 Sequence test at the start of Rotation 1 and completion of Rotation 7, and for the 21 Sequence test, at the start of Rotation 1 and completion of Rotation 14 . Other rotation test positions also satisfy (18.4.7.4-21) (e.g., for the 16

Sequence test, the start of any of the test sequence rotations for $\underline{u}_{ZL}^{M1}$, and completion of Rotation 13 for $\underline{u}_{ZL}^{M2}$). Perpendicular $\underline{u}_{ZL}^{M1}$, $\underline{u}_{ZL}^{M2}$ conditions other than (18.4.7.4-21) could also have been selected from the multitude of available measurements in the rotation tests. In practice, it is advantageous to select a particular $\underline{u}_{ZL}^{M1}$, $\underline{u}_{ZL}^{M2}$ orientation pair so that the sensor assembly mount can be installed on the rotation fixture at these orientations at the exact $\underline{u}_{ZL}^{M1}$, $\underline{u}_{ZL}^{M2}$ prescribed orientation relative to the vertical (e.g., by shimming and use of a bubble level aligned to precision machined perpendicular reference flats affixed to the sensor assembly mount to define the M Frame). Then the $\underline{u}_{ZL}^{M1}$, $\underline{u}_{ZL}^{M2}$ terms in (18.4.7.4-21) will accurately represent the orientation of the M Frame when the rotation test fixture is positioned at these attitudes.

References

1. Bisplinghoff, R. L., Ashley, H. & Halfman, R. L., *Aeroelasticity*, Addison-Wesley, Reading Mass., 1956.

2. Bortz J. E., "A New Mathematical Formulation for Strapdown Inertial Navigation", *IEEE Transactions on Aerospace and Electronic Systems*, Volume AES-7, No. 1, January 1971, pp. 61-66.

3. Britting, K. R., *Inertial Navigation System Analysis*, John Wiley and Sons, New York, 1971.

4. "Department Of Defense World Geodetic System 1984", NIMA TR8350.2, Third Edition, 4 July 1997.

5. Downs, Harry B., "A Lab Test To Find The Major Error Sources In A Laser Strapdown Inertial Navigator", 38th Annual Meeting of the ION, Colorodo Springs, CO, June 15-17, 1982.

6. Gelb, A., *Applied Optimal Estimation*, The MIT Press, Cambridge Mass., London, England, 1978.

7. Gille, J. C., Pelegrin, M. J., & Decaulne, P., *Feedback Control Systems Analysis, Synthesis, And Design*, McGraw-Hill, New York, Toronto, London, 1959.

8. Halfman, R. L., *Dynamics: Particles, Rigid Bodies, and Systems, Volume I*, Addison-Wesley, Reading Mass., Palo Alto, London, 1962.

9. Hills, F. B., "A Study Of Coordinate-Conversion Errors In Strapped-Down Navigation", MIT Electronics Systems Laboratory, E SL-4-244, Cambridge, MA, August 1965.

10. Holbrook, J. G., *Laplace Transforms For Electronics Engineers*, Pergamon Press, New York, London, Paris, Los Angeles, 1959.

11. Ignagni, M. B., "Optimal Strapdown Attitude Integration Algorithms", *AIAA Journal Of Guidance, Control, And Dynamics*, Vol. 13, No. 2, March-April 1990, pp. 363-369.

12. Ignagni, M. B., "Efficient Class Of Optimized Coning Compensation Algorithms", *AIAA Journal Of Guidance, Control, And Dynamics*, Vol. 19, No. 2, March-April 1996, pp. 424-429.

13. Ignagni, M. B., "Duality of Optimal Strapdown Sculling and Coning Compensation Algorithms", *Journal of the ION*, Vol. 45, No. 2, Summer 1998.

14. Jordan, J. W., "An Accurate Strapdown Direction Cosine Algorithm", NASA TN-D-5384, Sept.ember 1969.

15. Kachickas, G. A., "Error Analysis For Cruise Systems", *Inertial Guidance*, edited by Pitman, G. R., Jr., John Wiley & Sons, New York, London, 1962.

16. Lawrence, Anthony, *Modern Inertial Technology*, Springer-Verlag New York, Inc., 1993.

17. Liepmann, H. W. & Roschko, A., *Elements of Gasdynamics*, John Wiley & Sons, New York, and Chapman & Hall, London, 1957.

18. Litmanovich, Y. A., Lesyuchevsky, V. M. & Gusinsky, V. Z., "Two New Classes of Strapdown Navigation Algorithms", *AIAA Journal Of Guidance, Control, And Dynamics*, Vol. 23, No. 1, January- February 2000.

19. Mark, J.G. & Tazartes, D.A., "On Sculling Algorithms", 3rd St. Petersburg International Conference On Integrated Navigation Systems, St. Petersburg, Russia, May 1996.

20. Mckern, R. A., "A Study of Transformation Algorithms For Use In A Digital Computer", Massachusetts Institute of Technology, Master's Thesis, Department of Aeronautics and Astronautics, Cambridge, MA, January 1968.

21. Merhav, Schmuel, *Aerospace Sensor Systems and Applications*, Springer-Verlag New York, Inc., 1996.

22. Miller, R., "A New Strapdown Attitude Algorithm", *AIAA Journal Of Guidance, Control, And Dynamics*, Vol. 6, No. 4, July-August 1983, pp. 287-291.

23. Minor, J. W., "Low-Cost Strapdown-Down Inertial Systems", AIAA/ION Guidance and Control Conference, August 16-18, 1965.

24. Morrisson, N., *Introduction to Sequential Smoothing and Prediction*, Mcgraw-Hill, New York, St. Louis, San Francisco, London, Sydney, Toronto, Mexico, Panama, 1969.

25. Morse, P. M. and Feshbach, H., *Methods of Theoretical Physics, Part 1*, McGraw-Hill, New York, Toronto, London, 1953.

26. Newton, G. C., Gould, L. A. & Kaiser, J. F., *Analytical Design of Linear Feedback Controls*, John Wiley & Sons, New York, and Chapman & Hall, London, 1957.

27. Parkinson, B. W. & Spilker, Jr., J. J., *Global Positioning System: Theory and Applications, Volume I*, American Institute of Aeronautics & Astronautics, Washington DC, 1996.

28. Perkins, C. D. & Hage, R. E., *Airplane Performance Stability and Control*, John Wiley & Sons, New York, and Chapman & Hall, London, 1957.

29. Savage, P. G., "A New Second-Order Solution for Strapped-Down Attitude Computation", AIAA/JACC Guidance & Control Conference, Seattle, Washington, August 15-17, 1966.

30. Savage, P. G., "Calibration Procedures For Laser Gyro Strapdown Inertial Navigation Systems", 9th Annual Electro-Optics / Laser Conference and Exhibition, Anaheim, California, October 25 - 27, 1977.

31. Savage, P. G., "Strapdown Sensors", *Strapdown Inertial Systems - Theory And Applications*, NATO AGARD Lecture Series No. 95, June 1978, Section 2.

32. Savage, P. G., "Advances In Strapdown Sensors", *Advances In Strapdown Inertial Systems*, NATO AGARD Lecture Series No. 133, May 1984, Section 2.

33. Savage, P. G., "Strapdown System Algorithms", *Advances In Strapdown Inertial Systems*, NATO AGARD Lecture Series No. 133, May 1984, Section 3.

34. Savage, P. G., "Strapdown Inertial Navigation System Integration Algorithm Design Part 1 - Attitude Algorithms", *AIAA Journal Of Guidance, Control, And Dynamics*, Vol. 21, No. 1, January-February 1998, pp. 19-28.

35. Savage, P. G., "Strapdown Inertial Navigation System Integration Algorithm Design Part 2 - Velocity and Position Algorithms", *AIAA Journal Of Guidance, Control, And Dynamics*, Vol. 21, No. 2, March-April 1998, pp. 208-221.

36. Shepperd, S. W., "Quaternion From Rotation Matrix", *AIAA Journal Of Guidance, Control, And Dynamics*, Vol. 1, No. 3, May- June 1978.

37. Sikolnikoff, I. S. & Redheffer, R. M., *Mathematics of Physics and Modern Engineering, McGraw-Hill*, New York, Toronto, London, 1958.

38. Thomas, Jr., G. B., *Calculus And Analytic Geometry*, Addison-Wesley, Cambridge Mass., 1955.

39. Turley, A. R., "A Solution For The Problems Of The No-Gimbal Inertial Navigator Concept", Air Force Avionics Laboratory, AFAL-TR-64-307, Wright Patterson AFB, OH, January 1965.

40. United Aircraft Corporation, "A Study of Critical Computational Problems Associated with Strapdown Inertial Navigation Systems", NASA Report CR-968, April 1968.

Subject Index

(Subject Located By Section Number Unless Otherwise Indicated)

Coordinate Frame Index

Parameter Index

$A^*(t)$ (15.1.5.4.1-3)
$\underline{\hat{a}}_1^L$ (18.4.5-1)
$\underline{\hat{a}}_1^L$ (18.4.7.4-18)
$\underline{\hat{a}}_2^L$ (18.4.5-1)
$\underline{\hat{a}}_2^L$ (18.4.7.4-18)
A_a (15.2.1.1-3)
$\underline{a}_{Accl}$ (8.1.4.1-8)
$\underline{a}_{Accl}(t)$ (10.1.4.2-2)
$a_{Accl0_{Inpt}}$ (10.1.4.1-2)
$a_{Accl0_{Pend}}$ (10.1.4.1-2)
a_{Accl0_x} (10.3-14)
a_{Accl0_y} (10.1.4.2-2)
a_{Accl0_y} (10.3-14)
$a_{Accl_{Inpt}}(t)$ (10.1.4.1-2)
$a_{Accl_{Pend}}(t)$ (10.1.4.1-2)
A_b (15.2.1.1-3)
$A_{\delta K_{Bias}}$ (13.5-21)
a_F (10.5.1-12)
$A_F(S)$ (10.5.1-12)
$A_\gamma(t)$ (18.3-6)
$\underline{\hat{a}}_{H_1}^L$ (18.4.7-14)
$\underline{\hat{a}}_{H_2}^L$ (18.4.7-14)
$\underline{\hat{a}}_H^L$ (18.4.7-12)
$\underline{A}_H^N$ (13.1-1)
A_i (8.2.2.1-14)
a_i (10.2.2-1)
a_i, b_i, etc. (14.6.2-20)
a_{IA_X} (12.6-1)
$\underline{a}^L$ (18.4-1)
$\underline{a}^L$ (18.4.7-1)
$\underline{\hat{a}}^L$ (18.4.5-1)
$\underline{\hat{a}}^L$ (18.4.7.4-16)
Algo (10.1.3.2.2-3)
$(\,)_{Algo}$ (10.1.1.2.2-1)
$(\,)_{Algo}$ (10.1.2.2.2-1)
Align (14.5-10)
$\underline{\alpha}$ (3.5.2-10)
α (4.1.2-2)
α (4.3.1-1)
α (4.4.2.1-1)
α (5.3-8)
$\underline{\alpha}$ (8.1.4.1-10)
α (11.2.1.1-5)
$\underline{\alpha}$ (11.2.2.2-1)
α (12.2.1-39)
α (12.2.2-21)
α (12.2.3-33)
α (17.1.2.3-13)
α Fig. 13.4.2-1
$\underline{\alpha}(t)$ (7.1.1.1-9)
$\alpha(t)$ (7.2.2.2-7)
$\underline{\alpha}(t)$ (10.1.1.2-1)
$\underline{\alpha}(t)$ (10.1.2.2-1)
$\alpha(\tau)$ (7.3.3.1-2)
α_0 (13.4.2-22)
α_0 (17.1.2.3-25)
α_1 (6.1.3-8)
$\underline{\alpha}_{A_1 \text{ to } A}^{A}$ (3.5.3-34)
$\underline{\alpha}_{A_2 \text{ to } A_1}^{A_1}$ (3.5.3-34)
$\underline{\alpha}_{A \text{ to } B}^{B}$ (3.5.2-37)
$\underline{\alpha}_{B \text{ to } A_2}^{A_2}$ (3.5.3-34)
$\underline{\alpha}_{B \text{ to } A}^{A}$ (3.5.2-26)
$\underline{\alpha}_{B \text{ to } A}^{A}$ (3.5.3-34)

$\underline{a}_{SF}$ (8.2.2-2)
$\underline{a}_{SF}$ (8.2.3-1)
$\underline{a}_{SF}$ (9.1-1)
$\underline{a}_{SF}$ (11.2.3.1-7)
a_{SF} (17.2.3.2.3-2)
$\underline{a}_{SF}$ (18.4.7.4-3)
a''_{SF_i} (8.1.1.3-11)
$\underline{a}_{SF}(t)$ (10.1.2-3)
$\underline{a}_{SF}(t)$ (10.3-3)
$\underline{a}_{SF}(t)$ (10.1.3-1)
$\underline{a}^{*}_{SF}$ (8.1.1.2.1-2)
a_{SF_0} (7.4.1-1)
a_{SF_0} (10.1.3-1)
a_{SF0_y} (10.1.2-3)
$\underline{a}^{\mathcal{V}}_{SF_{Avg}}$ (17.1.1.2-10)
$\underline{a}^{B}_{SF}$ (4.2-4)
$\underline{a}^{B}_{SF}$ (7.2-1)
$\underline{a}^{B}_{SF}$ (11.2.2.1-1)
$\underline{a}^{B}_{SF}$ (11.2.4.3.2.2-1)
$\underline{a}^{B}_{SF}$ (12.1-12)
$\underline{a}^{B}_{SF}$ (12.1.4-10)
$\underline{a}^{B}_{SF}$ (18.3-1)
$\underline{a}^{\hat{B}}_{SF}$ (18.4.7.4-10)
$\underline{a}^{I}_{SF}$ (12.1.4-7)
a_{SF_i} (8.1.1.3-15)
a_{SF_i} (11.2.3.2-1)
a_{SF_i} (16.2.3.2-9)
$a_{SF_{iAC}}$ (17.2.3.2.3-19)
$a_{SF_{iAC}}$ (17.2.3.2.3-5)
$a_{SF_{iAC\text{-}l}}$ (17.2.3.2.3-31)
$a_{SF_{iAC\text{-}m}}$ (17.2.3.2.3-31)
$\underline{a}^{B}_{SF_i}$ (11.2.3.1-9)
$a_{SF_{iB}}$ (18.4.7.1-3)
$a_{SF_{i\mathcal{V}}}$ (17.1.1.5-7)
a_{SF_k} (8.1.4.1-7)
$\underline{a}^{I}_{SF_k}$ (8.1.4.1-3)
$\underline{a}^{L}_{SF}$ (4.2-4)
$\underline{a}^{L}_{SF}$ (12.1-12)
$\underline{a}^{L}_{SF}$ (18.4.7-1)
$\underline{a}^{B}_{SF_{Mean}}$ (12.6-1)
$a_{SF\text{-}Puls\,i}$ (8.1.1.3-20)
$\underline{a}^{N}_{SF}$ (12.1-12)
$\underline{a}_{SF_{+Puls}}$ (8.1.1.3-20)
$\underline{a}_{SF_{-Puls}}$ (8.1.1.3-20)
$a_{SF+Puls\,i}$ (8.1.1.3-20)
$\underline{a}_{SF_{Puls}}$ (8.1.1.2-1)
$\underline{a}_{SF_{Puls}}$ (8.1.2.2-1)
$\underline{a}_{SF_{Puls}}$ (8.2.2.1-5)
$\underline{a}_{SF_{Puls}}$ (9.3-1)
$A^{1}_{SFPuls_{Sign}}$ (18.4.5-1)
$A^{1}_{SFPuls_{Sign}}$ (18.4.7.4-18)
$A^{2}_{SFPuls_{Sign}}$ (18.4.5-1)
$A^{2}_{SFPuls_{Sign}}$ (18.4.7.4-18)
a_{SFPuls_i} (8.1.1.3-12)
$A_{SFPuls_{Sign}}$ (18.4.5-1)
$A_{SFPuls_{Sign}}$ (18.4.7.4-14)
a_{SFPuls_X} (8.2.2.1-25)
a_{SFPuls_Y} (8.2.2.1-25)
a_{SFPuls_Z} (8.2.2.1-25)
$\underline{a}_{SF_{Ref}}$ (8.1.4.1-7)

$d\underline{\alpha}$	(7.1.1.1.1-18)
$d\alpha'_i$	(8.1.4.1.1.1-6)
$d\underline{\alpha}_{+Cnt}$	(8.1.2.1-10)
$d\underline{\alpha}_{-Cnt}$	(8.1.2.1-10)
$d\underline{\alpha}_{Cnt}$	(8.1.2.1-1)
$d\underline{\alpha}_{Cnt}$	(8.2.1.1-14)
$d\underline{\alpha}_{Cnt}$	(8.2.2.1-34)
$d\underline{\alpha}_{Cnt}$	(8.2.3.1-2)
$d\alpha_{iCnt}$	(8.1.4.1.4-1)
$d\alpha_{i-Cnt}$	(8.1.4.1.4-7)
$d\alpha_{i+Cnt}$	(8.1.4.1.4-7)
$d\underline{\alpha}_{-Cnt}$	(8.2.1.1-22)
$d\underline{\alpha}_{+Cnt}$	(8.2.1.1-22)
$\delta\underline{\alpha}_{QuantC_{l:m}}$	(8.1.4.1.4-10)
db_a	(8.1.3.3-6)
db_ω	(8.1.3.3-6)
$d\left(\delta\underline{g}_{Mdl}\right)$	(16.2.3.3-1)
$d\Delta\underline{R}_{ScrlCnt}$	(8.2.3.1-2)
$\partial\underline{a}_{Quant}$	(12.4-15)
$\partial\underline{a}_{QuantC}$	(12.4-15)
$\Delta\hat{\underline{R}}^E_{REF_n}$	(15.2.2.1-40)
∂G^N_C	(12.1.1-13)
$\partial g_{P_{North}}$	(12.1.1-12)
$\partial g_{P_{Up}}$	(12.1.1-12)
∂h	(4.4.1.2.1-3)
∂h	(12.1-12)
∂l	(12.1-28)
∂l	Fig. 5.2-1
$\partial\underline{\omega}_{Quant}$	(12.4-2)
$\partial\underline{\omega}_{Quant}$	(12.4-5)
$\partial\underline{\omega}_{QuantC}$	(12.4-5)

$\partial P'_{jk_{nTrans}}$	(16.2.4-15)
δ	(15.2.4-3)
Δ	(18.4.6-1)
$\delta(\ )$	(4.4.1.2.1-4)
$\delta(\ )$	(12.3.1-5)
$\delta(\ )$	(14.2-1)
$\delta(\ )$	(15.2.1-3)
$\delta(\tau_\alpha - \tau_\beta)$	(15.1.2.1.1-27)
$\Delta_{1-2}(\ )$	(18.2.2-8)
$\delta\underline{a}'_{Size}$	(8.1.4.1.1-8)
$\delta a'_{Size/X_k}$	(8.1.4.1.1.2-2)
$\delta a'_{Size/Z_k}$	(8.1.4.1.1.2-2)
$\delta\underline{a}'_{Size_k}$	(8.1.4.1.1-8)
$\delta a'_{Size_X}$	(8.1.4.1.1.2-2)
$\delta a'_{Size_Y}$	(8.1.4.1.1.1-2)
$\delta a'_{Size_Z}$	(8.1.4.1.1.2-2)
$\delta\underline{a}^*_{Aniso}$	(8.1.1.2.1-2)
$\delta\underline{a}^*_{Quant}$	(8.1.1.2.1-2)
$\delta\underline{a}^*_{Size}$	(8.1.1.2.1-2)
$\underline{\delta a_{Accl_{Aniso}}}(t)$	(10.1.4.1-2)
$\delta a_{Accl_{Aniso}}$	(10.1.4.1-9)
$\underline{\delta a_{Accl_{G2}}}(t)$	(10.1.4.1-2)
$\delta a_{Accl_{G2}}$	(10.1.4.1-9)
$\Delta\underline{a}^N_{SF}$	(13.2.2-18)
$\delta\underline{a}_{Aniso}$	(8.1.1.2-1)
$\delta\underline{a}_{Aniso}$	(8.2.2.1-5)
δa_{Aniso_k}	(8.1.4.2-1)
$\delta\underline{a}_{Bias}$	(8.1.1.2-1)
$\Delta\underline{a}^L_H$	(18.4.7-14)
Δa_i	Sect. No. 18.4.3

$\delta\underline{a}^{B}_{SF_{Scal/Mis}}$	(13.2.4-6)
$\Delta\underline{a}^{B}_{JTR}$	(9.1-12)
$\Delta\underline{a}^{N}_{JTR}$	(9.1-12)
$\delta\alpha$	(12.2.1-40)
$\delta\alpha$	(12.2.2-24)
$\delta\alpha$	(14.2-20)
$\delta\underline{a}''_{Size}$	(8.1.4.1.2-5)
$\Delta\alpha'_{i_l}$	(8.1.4.1.1.1-6)
$\Delta\alpha'_{il}$	(8.2.1.1-23)
$\Delta\alpha'_{i_m}$	(8.1.4.1.1.1-6)
$\Delta\alpha'_{i_{m+}}$	(8.1.4.1.1.1-6)
$\Delta\alpha'_{i_{m-1}}$	(8.1.4.1.1.1-6)
$\Delta\alpha'_{i_{(m-1)+}}$	(8.1.4.1.1.1-6)
$\Delta\underline{\alpha}'_{m}$	(8.1.3.3-9)
$\Delta\underline{\alpha}'_{(l:m)-1}$	(8.1.3.3-9)
$\Delta\alpha'^{Q}_{i_m}$	(8.1.4.1.4-10)
$\Delta\underline{\alpha}_{Cnt(l:m)-1}$	(8.1.3.3-6)
$\Delta\underline{\alpha}_{Cnt_l}$	(8.1.3.3-6)
$\Delta\underline{\alpha}_{Cnt_l}$	(8.2.2.1-37)
$\Delta\underline{\alpha}_{Cnt_m}$	(8.1.3.3-6)
$\Delta\alpha_{CntX_l}$	(8.2.2.1-37)
$\Delta\alpha_{CntY_l}$	(8.2.2.1-37)
$\Delta\alpha_{CntZ_l}$	(8.2.2.1-37)
$\Delta\alpha_{iCnt_l}$	(8.2.1.1-23)
$\Delta\underline{\alpha}_{l}$	(11.2.3.2-10)
$\delta\underline{\alpha}_{\psi Quant}$	(16.2.3.1-6)
$\delta\underline{\alpha}_{\psi VQuant}$	(16.2.3.1-6)
$\delta\underline{\alpha}_{Quant}$	(12.5-1)
$\delta\underline{\alpha}_{Quant}$	(18.2.1-3)
$\delta\underline{\alpha}_{Quant}$	(18.3-5)
$\delta\alpha_{Quant/ZN_j}$	(18.3.1.2-7)
$\delta\underline{\alpha}_{QuantC_m}$	(8.1.2.1-6)
$\delta\alpha_{Quant_H}$	(14.6.1-8)
$\delta\underline{\alpha}^{L}_{Quant_H}$	(18.2.1-3)
$\delta\alpha_{Quant_i}$	(16.2.3.1-13)
$\Delta\underline{\alpha}_{Quant_m}$	(17.3.1-1)
$\delta\alpha_{Quant_{Ni}}$	(15.2.1.1-3)
$\delta\alpha_{Quant_{ZL}}$	(18.2.1-3)
$\delta\alpha_{Quant_{ZL}}$	(18.2.2-1)
$\delta\underline{a}_{Quant}$	(8.1.1.2-1)
$\delta\underline{a}_{Quant}$	(12.4-15)
$\delta\underline{a}_{Rand}$	(8.1.1.2-1)
$\delta\underline{a}_{Rand}$	(12.4-15)
$\delta\underline{a}_{Rand}$	(18.3-5)
$\delta a_{Rand_{Ni}}$	(15.2.1.1-3)
$\delta\underline{a}_{SensBias}$	(8.1.1.2.1-2)
δa_{SFAsym_X}	(12.6-1)
$\delta\underline{a}^{B}_{SF}$	(18.1.1-2)
$\delta\underline{a}^{B}_{SF}{}^{*}$	(12.5-13)
$\delta\underline{a}^{B}_{SF_{Cnst}}$	(18.3-5)
$\delta a_{SF_{H0East}}$	(18.1.2-9)
$\delta\underline{a}^{N}_{SF_{H0}}$	(18.1.1-5)
$\delta a_{SF_{H0North}}$	(18.1.2-9)
$\delta\underline{a}^{N}_{SF_H}$	(13.3.1-3)
$\delta\underline{a}^{L}_{SF}$	(18.4.7-3)
$\Delta\underline{a}^{N}_{SF}$	(12.3.5-23)
$\delta\underline{a}^{N}_{SF}$	(13.3.1-2)
$\delta\underline{a}^{B}_{SF_{Other}}$	(13.2.4-24)
$\delta a_{SF_{ZN}}$	(4.4.1.2.1-4)
$\delta a_{SF_{ZN}}$	(13.2.1-16)

$\delta a_{SF_{ZN}}$ (13.3.1-3)
$\delta \dot{A}_{SF_{ZN}}(S)$ (4.4.1.2.1-13)
$\delta \underline{a}_{Size}$ (8.1.1.2-1)
$\delta \underline{a}_{Size}$ (8.1.4.1-9)
$\delta \underline{a}_{Size}$ (8.2.2.1-5)
δa_{Size_k} (9.3-2)
$\delta \underline{a}_{Size_{Lo-f/\omega^2}}$ (8.1.4.1.2-5)
$\delta \dot{C}^A_{B_{Comp}}$ (3.5.1-20)
$\delta \dot{\beta}_{Algo-m_z}$ (10.3-20)
$\delta \underline{\beta}_{Algo_m}$ (10.1.1.2.2-6)
$\delta \dot{\underline{\beta}}_{Algo_m}$ (10.1.1.2.2-28)
Δb_i Sect. No. 18.4.3
δc (10.5.1-4)
δC^A_B (3.5.2-1)
δC^E_B (12.2.1-4)
$\delta \dot{C}^L_{B_{Comp}}$ (7.1.1.3-14)
$\Delta \underline{\chi}$ (16.1.1.3-13)
$\Delta \underline{\chi}$ (16.1.2.1-12)
$\delta \underline{C}_i$ (7.1.1.3-2)
ΔCnt_l (8.1.3.1-2)
$\Delta Cnt_{l=(m-1)k-1}$ (8.1.3.2-8)
$\Delta Cnt_{l=mk}$ (8.1.3.2-8)
ΔCnt_m (8.1.3.2-8)
ΔCnt_{m-1} (8.1.3.2-8)
$\delta ConAlg_i$ (16.2.3.2-1)
$\delta ConAlg_{Norm}$ (16.2.3.2-6)
δCon_i (16.2.3.2-1)
$\Delta \dot{v}_{Scul-m_z}$ (10.3-9)
$\Delta \delta \underline{a}_{Size}$ (8.1.4.1.1-10)
$\Delta \delta \underline{F}^{\nu}_j$ (17.1.2.3-32)
$\delta \delta \omega_{ARS/Cnst/East_1}$ (18.2.2-8)
$\delta \Delta \psi_{T_{1-2}}$ (18.2.2-8)
$\delta \Delta R_H$ (14.6.1-8)
$\delta \Delta \underline{R}^{N*}_H$ (18.3-5)
$\delta \Delta R_{Ni}$ (15.2.1.1-3)
$\Delta \delta \underline{\upsilon}_{SizeC_m}$ (8.1.4.1.1-12)
$\Delta \delta \vartheta_j$ (17.1.2.3.1-12)
$\delta \Delta \upsilon_{SizeCX_m}$ (8.1.4.1.1.2-5)
$\delta \Delta \upsilon_{SizeCY_m}$ (8.1.4.1.1.2-5)
$\delta \Delta \upsilon_{SizeCZ_m}$ (8.1.4.1.1.2-5)
$\delta \Delta \underline{R}^N_{Ref_H}$ (15.2.1-7)
$\delta \Delta \dot{v}_{Scul/Algo-m_z}$ (10.3-20)
$\delta \Delta \dot{\underline{v}}_{Scul/Algo_m}$ (10.1.2.2.2-22)
$\delta \Delta \underline{v}_{Scul/Algo_m}$ (10.1.2.2.2-7)
$\Delta \eta_{ij_l}$ (8.1.4.1.1.1-11)
$\delta \underline{F}^{\nu}_j$ (17.1.2.3-29)
δf_{Osc} (15.2.4-15)
$\delta f_{Osc/Mark}$ (15.2.4-16)
$\delta f_{Osc/RndCnst}$ (15.2.4-16)
$\Delta \underline{F}_{WndGst}$ (17.2.3.2.1-1)
$\delta \gamma_{H1_{East}}$ (18.1.2-9)
$\delta \gamma_{H1_{North}}$ (18.1.2-9)
$\Delta \underline{\gamma}^N_{H_i}$ (13.6.1-4)
$\delta \underline{\gamma}^N_{H_i}$ (18.1.1-4)
$\Delta \underline{\gamma}^{N+}$ (14.2-37)
$\delta \gamma^N_{Up_0}$ (18.1.1-10)
$\delta \gamma_{Up_i}$ (18.1.1-4)
$\Delta \gamma_{ZN_i}$ (13.6.1-4)
$\delta \underline{g}^E_{Mdl}$ (12.2.4-3)

$\delta \underline{g}^{I}_{Mdl}$ (12.2.4-8)
$\delta \underline{g}^{N}_{Mdl}$ (12.2.4-1)
$\delta \underline{g}^{E}_{P}$ (12.2.4-3)
$\delta \underline{g}^{I}_{P}$ (12.2.4-8)
$\delta \underline{g}^{N}_{P}$ (12.2.4-1)
δh (4.4.1.2.1-4)
δh (12.2.3-5)
$\delta H(S)$ (4.4.1.2.1-13)
Δh_{Osc} (11.2.4.1.1-10)
δh_{Prsr} (4.4.1.2.1-4)
δh_{Prsr} (12.3.2-23)
$\delta H_{Prsr}(S)$ (4.4.1.2.1-13)
δ_{i} (16.2.3.2-8)
$\Delta_{i} \delta \underline{V}^{N}_{H}(t)$ (13.6.1-5)
$\Delta Integ_{l}$ (8.1.3.1-2)
$\Delta I \underline{\omega}^{B}$ (11.2.1.2-3)
$\Delta I \omega_{iR}$ (11.2.1.1-6)
$\Delta I \underline{\omega}^{R}$ (11.2.1.2-3)
$\delta \underline{J}_{c}$ (15.1.2.3-2)
δk (10.5.1-4)
$\delta \underline{K}_{0Bias}$ (12.5.6-2)
$\delta \underline{K}_{Bias}$ (12.4-10)
$\delta \underline{K}^{G0}_{Bias_F}$ (13.5-32)
$\delta K_{Bias_{ZG0}}$ (13.5-32)
$\delta K_{Scal/Mis}$ (13.2.4-1)
δK_{G2Bias_X} (12.6-1)
δK_{ii} (13.2.4-1)
δK_{ij} (13.2.4-1)
$\delta \underline{K}_{MBias}$ (12.5.6-3)
δK_{Mis} (12.4-10)
δK_{Orth} (13.2.4-5)
δK_{Scal} (12.4-10)
δK_{Scal} (13.2.4-5)
$\delta K_{Scal/Mis}$ (12.4-13)
$\delta K_{Scal/Orth}$ (13.2.4-2)
δK_{TMis} (13.2.4-2)
$\delta \underline{K}_{TMis}$ (13.2.4-4)
$\left(\delta \underline{K}_{TMis} \times\right)$ (13.2.4-4)
δL (18.1.1-22)
δl (10.5.1-3)
$\delta \underline{L}_{Bias}$ (12.4-15)
$\delta \underline{l}^{E}_{H_E}$ (15.1.2.2-13)
$\Delta \underline{l}^{AC}_{HiF}$ (17.2.3.2.2-1)
δL_{Mis} (12.4-15)
δL_{Orth} (13.2.4-5)
δL_{Scal} (12.4-15)
δL_{Scal} (13.2.4-5)
$\delta L_{Scal/Mis}$ (12.4-15)
$\delta L_{Scal/Mis}$ (13.2.4-1)
$\delta L_{Scal/Orth}$ (13.2.4-2)
$\delta \underline{l}^{VRF}_{Stat}$ (15.1.2.2-25)
$\delta \underline{l}^{VRF}_{Stat_c}$ (15.1.2.3-2)
δL_{TMis} (13.2.4-2)
$\delta \underline{L}_{TMis}$ (13.2.4-4)
$\left(\delta \underline{L}_{TMis} \times\right)$ (13.2.4-4)
$\delta \underline{l}^{VRF}_{Vib}$ (15.1.2.2-25)
$\delta \underline{l}^{VRF}_{VRF}$ (15.1.2.2-15)
$\delta \underline{M}^{E}_{0}$ (15.2.2.1-41)
ΔM_{Fuel} (17.1.2.3-27)
ΔM_{Jtsn} (17.1.2.3-27)
δ_{Norm_i} (3.5.1-13)

$\Delta\omega$ (10.2.2-1)

$\delta\underline{\omega}$ (12.4-6)

$\delta\underline{\omega}^{*}_{Quant}$ (8.1.1.1.1-2)

$\delta\omega_{ARS/Cnst_{East}}$ (18.2.2-1)

$\delta\omega_{ARS/Cnst_{East}}$ (18.3.1.2-9)

$\delta\underline{\omega}^{L}_{ARS/Cnst_{H}}$ (18.2.1-6)

$\delta\underline{\omega}^{N}_{ARS/Cnst_{H}}$ (18.3-21)

$\delta\omega_{ARS/Cnst_{ZL}}$ (18.2.1-6)

$\delta\omega_{ARS/Cnst_{ZL}}$ (18.2.2-1)

$\delta\omega_{ARS/Cnst_{ZN}}$ (18.3-21)

$\delta\underline{\omega}_{Bias}$ (8.1.1.1-1)

$\delta\underline{\omega}^{E}_{EN_{H}}$ (12.3.6.2-13)

$\delta\omega_{G2Bias_{X}}$ (12.6-1)

$\delta\underline{\omega}^{A}_{IA}$ (3.5.1-31)

$\delta\omega_{IB}$ (13.2.4-10)

$\delta\omega_{IB}$ (13.4.1.1-4)

$\delta\underline{\omega}^{B}_{IB}$ (3.5.1-31)

$\delta\underline{\omega}^{B}_{IB}$ (18.1.1-2)

$\delta\underline{\omega}^{B}_{IB}$ (18.2.1-3)

$\delta\underline{\omega}^{B}_{IB}{}^{*}$ (12.5-6)

$\delta\underline{\omega}^{B}_{IB_{Cnst}}$ (18.3-5)

$\delta\omega_{IB_{H00East}}$ (18.1.2-9)

$\delta\underline{\omega}^{N}_{IB_{H00}}$ (18.1.1-7)

$\delta\omega_{IB_{H00North}}$ (18.1.2-9)

$\delta\omega_{IB_{H0East}}$ (18.1.2-9)

$\delta\underline{\omega}^{N}_{IB_{H0}}$ (18.1.1-7)

$\delta\omega_{IB_{H0North}}$ (18.1.2-9)

$\delta\underline{\omega}^{N}_{IB_{H}}$ (13.3.1-3)

$\delta\underline{\omega}^{B}_{IB_{i}}$ (18.4.7.2-9)

$\delta\underline{\omega}^{N}_{IB}$ (13.3.1-2)

$\delta\omega_{IB_{Scal/Mis}}$ (13.2.4-11)

$\delta\underline{\omega}^{B}_{IB_{Scal/Mis}}$ (13.2.4-6)

$\delta\omega_{IB_{Up0}}$ (18.1.1-7)

$\delta\omega_{IB_{Up00}}$ (18.1.1-7)

$\delta\omega_{IB_{ZN}}$ (13.3.1-3)

$\delta\omega_{IE/East/Resid_{j}}$ (18.3.1.2-9)

$\delta\omega_{IE/East\,Resid}$ (18.2.2-1)

$\delta\underline{\omega}^{N}_{IE/Exp_{H}}$ (18.3-13)

$\delta\hat{\underline{\omega}}^{N}_{IE/Exp_{H}}(t)$ (18.3-13)

$\delta\underline{\omega}^{N}_{IE/H/Resid_{j}}$ (18.3.1.1-2)

$\delta\hat{\omega}_{IE/H_{0}}$ (18.3-11)

$\delta\underline{\omega}^{N}_{IE/H_{Resid}}(t)$ (18.3-13)

$\delta\underline{\omega}^{N}_{IE_{Exp}}$ (18.3-5)

$\delta\omega_{IE_{H}}$ (14.6.1-8)

$\delta\omega_{IE_{H}}$ (18.3-6)

$\delta\hat{\omega}_{IE_{H}}(t)$ (18.3-7)

$\delta\omega_{IE_{Ni}}$ (15.2.1.1-3)

$\delta\underline{\omega}^{L}_{IL_{H}}$ (18.2.1-3)

$\Delta\underline{\omega}^{B}_{JTR}$ (9.1-27)

$\delta\underline{\omega}_{Quant}$ (8.1.1.1-1)

$\delta\underline{\omega}_{Rand}$ (8.1.1.1-1)

$\delta\underline{\omega}_{Rand}$ (12.4-2)

$\delta\underline{\omega}_{Rand}$ (18.3-5)

$\delta\omega_{Rand_{H}}$ (14.6.1-8)

$\delta\omega_{Rand_{H}}(t)$ (18.3-6)

$\delta\omega_{Rand_{Ni}}$ (15.2.1.1-3)

δR_{HmgTot}	(13.2.1-11)
$\Delta \underline{R}_{H}^{N}$	(18.3-1)
$\delta \underline{R}_{H}^{N}$	(13.2-7)
$\delta \underline{R}_{H_{LngTrm}}^{N}$	(13.2.3-4)
$\delta \underline{r}^{I}$	(12.2.3-27)
$\delta \underline{R}_{INS/H_c}^{N}$	(15.1.2.3-2)
$\Delta \underline{R}_{INS}^{E}(t)$	(15.2.2.1-5)
$\Delta \underline{R}_{JTR}^{N}$	(9.1-5)
$\Delta \underline{R}_{l\,bnd}^{AC}$	(17.2.3.2.3-6)
$\Delta R_{l\,bnd_{ji}}$	(17.2.3.2.3-5)
$\Delta \underline{R}^{N}$	(6.1.2-1)
$\delta \underline{R}^{N}$	(12.2.3-2)
ΔR_{NX}	(15.2.1.1-17)
ΔR_{NY}	(15.2.1.1-17)
$\delta \underline{R}_{OTH_H}^{E}$	(15.1.2.2-14)
$\Delta \underline{R}_{REF}^{E}(t)$	(15.2.2.1-5)
$\Delta \underline{R}_{Ref_H}^{N}$	(14.1-1)
$\Delta \underline{R}_{Ref_H}^{N}$	(15.2.1-2)
$\delta \Delta \underline{R}_{Ref_H}^{N}$	(14.2-1)
$\Delta \underline{R}_{Rot\,m}$	(7.3.3-11)
$\Delta \underline{R}_{Rot\,m}$	(8.2.3-1)
$\delta \underline{R}_{ScrlA\,l}$	(7.3.3.2-9)
$\delta \underline{R}_{ScrlB_l}$	(7.3.3.2-9)
$\Delta \underline{R}_{ScrlCnt}$	(8.2.3.1-2)
$\Delta \underline{R}_{Scrl_m}$	(7.3.3-11)
$\Delta \underline{R}_{Scrl_m}$	(8.2.3-1)
$\delta R_{SF/Algo}(t)$	(10.4.2-10)
$\delta R_{SF/Algo\,i}(t)$	(10.4.2-10)
$\delta \underline{R}_{SF/Algo\,i}(t)$	(10.4.2-9)
$\delta \underline{R}_{SF/Algo_M}$	(10.1.3.2.4-1)
$\Delta \underline{R}_{SF/Typ_m}^{B}$	(7.4-2)
$\Delta \underline{R}_{SF_m}$	(8.2.3-1)
$\Delta \underline{R}_{SF_m}$	(10.1.3.1-3)
$\Delta \underline{R}_{SF_m}^{L}$	(7.3.3-2)
$\Delta \underline{R}_{True_H}^{N}$	(15.2.1-8)
$\Delta \underline{R}_{True_H}^{N}$	(14.2-2)
$\Delta \underline{R}_{Var_m}^{NVar}$	(17.2.3.2-8)
$\Delta \underline{R}_{Vib_H}^{N\,\blacklozenge}$	(14.2-3)
Δr_{Vib_H}	(14.6.1-12)
$\Delta \underline{R}_{Vib_H}^{N}$	(15.2.1-9)
$\Delta \underline{R}_{Vib}^{N\blacklozenge}$	(18.3-5)
$\Delta r_{Vib_{NX}}$	(15.2.1.1-9)
$\Delta r_{Vib_{NY}}$	(15.2.1.1-9)
$\Delta \underline{R}_{v_m}^{N}$	(17.2.3.1-28)
δR_{XGeo}	(12.2.2-29)
δR_{XN}	(12.2.1-45)
δR_{XN}	(12.2.3-38)
δR_{XN}	(12.2.4-28)
δR_{YGeo}	(12.2.2-29)
δR_{YN}	(12.2.1-45)
δR_{YN}	(12.2.3-38)
δR_{YN}	(12.2.4-28)
δR_{ZN}	(12.2.3-40)
$\delta SculAlg_{i}$	(16.2.3.2-1)
$\delta SculAlg_{ij}$	(16.2.3.2-3)
$\delta SculAlg_{ik}$	(16.2.3.2-3)
$\delta SculAlg_{Norm}$	(16.2.3.2-6)
$\delta Scul_{i}$	(16.2.3.2-1)

$\delta L_{SFAsymBias_X}$ (12.6-1)
$\Delta \underline{a}^{N}_{SF}$ (14.5-2)
$\Delta \underline{S}_{\upsilon/Algo_l}$ (10.1.3.2.2-3)
ΔT (18.3.2.1-2)
$\Delta \tau_i$ (13.6.1-19)
$\delta \tau_{Sen}$ (15.2.2.1-21)
δt_f (15.2.4-15)
$\delta \vartheta$ (17.1.2.3.1-10)
$\delta \theta$ (12.2.1-28)
$\Delta \theta_{ARS/Rnd/East_i}$ (18.3.2.2-1)
$\Delta \theta_{ARS/Rnd/H-k_i}$ (18.3.2.1-7)
$\Delta \bar{\theta}_{ARS/Rnd/H-k_i}$ (18.3.2.1-9)
$\Delta \underline{\theta}^{N}_{ARS/Rnd/H_i}$ (18.3.2.1-1)
$\Delta \underline{\theta}^{B}_{JTR}$ (9.1-5)
$\delta \underline{u}^{B}_{\omega}$ (13.4.1.1-8)
$\delta \underline{u}^{L}_{\phi H}$ (18.2.1-18)
$\delta \underline{\upsilon}$ (12.2.2-9)
$\delta \underline{\upsilon}''_{SizeC}$ (8.1.4.1.2-5)
$\Delta \upsilon'_{il}$ (8.2.2.1-46)
$\delta \underline{\upsilon}'_{SizeC_m}$ (8.1.4.1.1-12)
$\delta \upsilon'_{SizeCX_m}$ (8.1.4.1.1.1-15)
$\delta \upsilon'_{SizeCY_m}$ (8.1.4.1.1.1-15)
$\delta \upsilon'_{SizeCZ_m}$ (8.1.4.1.1.1-15)
$\delta \upsilon'_{SizeCX/k_m}$ (8.1.4.1.1.2-5)
$\delta \upsilon'_{SizeCY/k_m}$ (8.1.4.1.1.2-5)
$\delta \upsilon'_{SizeCZ/k_m}$ (8.1.4.1.1.2-5)
$\delta \underline{\upsilon}'_{SizeC/k_m}$ (8.1.4.1.1.2-5)
$\delta \upsilon_{AnisoCk_m}$ (8.1.4.2-2)
$\delta \underline{\upsilon}_{AnisoC_m}$ (8.1.2.2-6)
$\Delta \underline{\upsilon}_{Cnt_l}$ (8.1.3.3-6)
$\Delta \underline{\upsilon}_{Cnt_l}$ (8.2.2.1-37)
$\Delta \underline{\upsilon}_{Cnt_m}$ (8.1.3.3-6)
$\Delta \upsilon_{CntX_l}$ (8.2.2.1-37)
$\Delta \upsilon_{CntY_l}$ (8.2.2.1-37)
$\Delta \upsilon_{CntZ_l}$ (8.2.2.1-37)
$\delta \dot{\upsilon}_{Force}$ (13.2.1-3)
$\delta \dot{\underline{\upsilon}}^{N}_{H_{Force}}$ (13.2.2-3)
$\delta \underline{\upsilon}^{N}_{H}$ (13.2-7)
$\delta \underline{\upsilon}^{I}$ (12.2.2-9)
$\Delta \upsilon_{iCnt_l}$ (8.2.2.1-46)
$\Delta \upsilon_{i_l}$ (11.2.3.2-4)
$\Delta \underline{\upsilon}_l$ (10.1.3.2.2-4)
$\delta \underline{\upsilon}^{N*}$ (12.5-26)
$\delta \underline{\upsilon}_{Quant}$ (12.5-1)
$\delta \underline{\upsilon}_{Quant}$ (18.3-5)
$\delta \underline{\upsilon}_{QuantC_m}$ (8.1.2.2-6)
$\Delta \underline{\upsilon}_{Quant_m}$ (17.3.1-1)
$\delta \upsilon_{Quant_{Ni}}$ (15.2.1.1-3)
$\delta \upsilon_R$ (13.2-7)
$\delta \underline{\upsilon}_{SizeC}$ (8.1.4.1-12)
$\delta \upsilon_{SizeCi_m}$ (8.1.4.1.3-7)
$\delta \underline{\upsilon}_{SizeC_m}$ (8.1.2.2-6)
$\delta \underline{\upsilon}_{SizeC_{Lo-f/\omega^2}}$ (8.1.4.1.2-5)
$\delta \underline{\upsilon}^{I}_{Snsr_0}$ (13.4.1.2-20)
$\delta \dot{\underline{\upsilon}}^{I}_{Snsr}$ (13.4.1.2-3)
$\delta \underline{\upsilon}_{VQuant}$ (16.2.3.1-11)
$\delta \underline{\upsilon}_{VRQuant}$ (16.2.3.1-11)
$\delta \upsilon_{Quant_H}$ (14.6.1-8)
$\delta \underline{u}^{E}_{ZN}$ (12.2.3-8)
Δv (17.2.3.2.3-1)

$\delta \underline{V}$	(12.2.2-2)
$\delta \underline{\mathcal{V}}$	(12.2.2-17)
$\delta \underline{v}$	(12.2.2-2)
$\delta \underline{V}_{A}^{A}$	(3.5.4-3)
Δv_{Avg}	(18.4.7.3-3)
$\Delta \underline{v}_{Avg}^{L}$	(18.4-1)
$\delta \underline{V}_{B}^{A}$	(3.5.4-1)
$\Delta \underline{v}_{SF_n}^{N}$	(15.2.2.3-4)
$\delta \underline{V}^{E}$	(12.2.2-2)
ΔV_{F_m}	(17.2.1-3)
$\delta \underline{\mathcal{V}}^{Geo}$	(12.2.2-17)
δv_{H}	(14.6.1-8)
$\delta \underline{V}_{H_0}^{N}$	(13.4.3-28)
$\delta \dot{\underline{V}}_{H_0}^{N}$	(13.4.3-28)
$\delta \underline{v}_{H}^{E}$	(12.3.6.1-41)
$\delta \underline{V}_{H_{Hmg}}^{N}$	(13.3.2-10)
$\Delta \delta \underline{V}_{H_i}^{N}$	(13.6.1-4)
$\delta \underline{v}_{H}^{N\star}$	(18.3-5)
$\delta \underline{V}_{H_{Prt}}^{N}$	(13.3.2-11)
$\delta \underline{\underline{V}}_{H_{Prt}}^{N}$	(13.4.3-22)
Δv_{j}	(18.4.7.3-2)
$\Delta \underline{v}_{JTR}^{N}$	(9.1-5)
ΔV_{m}	(17.1.3-10)
ΔV_{m}	(17.2.1-1)
Δv_{m}	(18.4.7.3-2)
$\Delta \underline{v}_{m}^{L}$	(18.4-1)
$\delta \underline{V}^{N}$	(12.2.2-3)
$\delta \underline{v}^{N}$	(12.2.2-2)
$\delta \underline{v}^{N}$	(12.2.4-9)
$\delta \underline{V}^{N*}$	(12.5-26)
$\delta \underline{v}^{N*}$	(12.5-13)
δv_{Ni}	(15.2.1.1-3)
$\delta \underline{V}_{Ref/Other}^{E}$	(15.2.2.1-20)
$\delta \underline{V}_{Ref/Sen}^{E}$	(15.2.2.1-20)
$\delta \underline{v}_{Rot/Scul\text{-}SizeC_m}$	(8.1.4.1-20)
$\delta v_{Rot/Scul\text{-}SizeCX_m}$	(8.1.4.1.3-7)
$\delta v_{Rot/Scul\text{-}SizeCY_m}$	(8.1.4.1.3-7)
$\delta v_{Rot/Scul\text{-}SizeCZ_m}$	(8.1.4.1.3-7)
$\Delta \underline{v}_{Rot/Scul_m}$	(7.2.2.2-27)
$\Delta \underline{v}_{Rot/Scul_m}$	(8.2.2-2)
$\Delta \underline{v}_{Rot_m}$	(7.2.2.2-25)
$\Delta \underline{v}_{Rot_m}$	(8.2.2-2)
ΔV_{S}	Sect. No. 17.1.1
$\Delta \underline{v}_{Scul}(t)$	(7.3.3-4)
$\Delta \underline{v}_{Scul}(t)$	(8.2.3-1)
$\delta \underline{v}_{Scul\text{-}SizeC_m}$	(8.1.4.1-15)
$\delta v_{Scul\text{-}SizeCX_m}$	(8.1.4.1.2-13)
$\delta v_{Scul\text{-}SizeCY_m}$	(8.1.4.1.2-13)
$\delta v_{Scul\text{-}SizeCZ_m}$	(8.1.4.1.2-13)
$\Delta \dot{v}_{Scul/Algo\text{-}m_z}$	(10.3-20)
$\Delta \dot{\underline{v}}_{Scul/Algo_m}$	(10.1.2.2.2-22)
$\Delta \underline{v}_{SculCnt}$	(8.2.3.1-2)
$\Delta \dot{\underline{v}}_{SculCnt}(t)$	(8.2.2.1-24)
$\Delta \underline{v}_{Scul_i}$	(10.1.2.2-3)
$\Delta \underline{v}_{Scul_m}$	(7.2.2.2-25)
$\Delta \underline{v}_{Scul_m}$	(8.2.2-2)
$\Delta \dot{\underline{v}}_{Scul_m}$	(10.1.2.2.1-11)
$\delta \dot{v}_{SF/Algo\text{-}m_x}$	(10.6.1-12)
$\delta \dot{v}_{SF/Algo\text{-}m_z}$	(10.3-20)

$\delta\dot{\underline{v}}_{SF/Algo_m}$ (10.1.2.2.2-22)

$\Delta\underline{v}_{SF/Algo_i}$ (10.1.3.2.2-3)

$\delta\underline{v}_{SF/Algo_m}$ (10.1.2.2.2-6)

$\delta\dot{v}_{SF/Scul/SnsDyn_x}$ (10.6.3-20)

$\delta\dot{v}_{SF/Scul/SnsDyn_z}$ (10.3-14)

$\delta\dot{\underline{v}}_{SF/Scul_{SnsDyn}}$ (10.1.4.2-5)

$\Delta\underline{v}_{SF_m}^{B_{m-1}}$ (11.2.4.3.2.2-4)

$\Delta\underline{v}_{SF_H}^{N}$ (13.3-13)

$\Delta\underline{v}_{SF_m}^{I}$ (11.2.4.3.2.2-1)

$\Delta\underline{v}_{SF_m}^{I}$ (11.2.4.3.2.1-1)

$\Delta\underline{v}_{SF}^{L}$ (18.4-1)

$\Delta\underline{v}_{SF_m}$ (8.2.2-2)

$\Delta\underline{v}_{Var}$ (17.2.3.2.1-1)

δV_{XN} (13.6.1-10)

δV_{YN} (13.6.1-10)

δv_{ZN} (4.4.1.2.1-4)

$\delta V_{ZN}(S)$ (4.4.1.2.1-13)

$\Delta\ddot{\underline{v}}'_{Scul}(t)$ (8.2.2.1-13)

$\Delta\underline{x}$ (14.6.1-4)

$\Delta\underline{x}$ (15.1.2.1-1)

$\Delta\underline{x}$ (16.1.1-13)

Δx (17.2.3.2.3-1)

$\Delta\underline{x}^*(t)$ (15.1.5.4.1-5)

$\Delta\underline{x}^{**}$ (14.6.4.2-5)

Δx_1 (15.1.2.1-5)

Δx_2 (15.1.2.1-5)

Δx_3 (15.1.2.1-5)

$\Delta\underline{x}_{\mathcal{M}}(t)$ (15.1.5.4.1-5)

$\Delta\underline{x}_{Opt}$ (16.1.1.3-21)

$\Delta\underline{y}$ (16.1.1-26)

$\Delta\underline{y}_{Opt}$ (16.1.1.3-21)

$\delta\zeta$ (3.5.3-20)

$\delta\zeta_1$ (3.5.3-25)

$\delta\zeta_2$ (3.5.3-25)

$\delta\zeta_3$ (3.5.3-25)

$\Delta\underline{z}_l$ (17.2.3.2.3-25)

$\partial\underline{x}'_{Tr_{nTrans}}$ (16.2.4-13)

$\Delta\underline{x}^*$ (14.6.4-5)

D_i (10.2.1-7)

d_i (17.2.3.2.3-44)

Diag( ) (8.1.1.1.1-14)

$(\ )^{\blacklozenge}$ (14.6.1-1)

DL (9.2-1)

$d\underline{R}_H^E$ (5.2.4-13)

$d\underline{R}_{HE}^{Geo}$ (5.2.4-29)

$dR_{SE_{East}}$ (5.2.4-22)

$d\underline{R}_{SE}^{Geo}$ (5.2.4-21)

$dR_{SE_{North}}$ (5.2.4-23)

ds (16.2.3.3-1)

$d\vartheta_{EN_{East}}$ (5.2.4-23)

$d\underline{\vartheta}_{EN_H}^{E}$ (5.2.4-4)

$d\vartheta_{EN_{North}}$ (5.2.4-22)

Dtr ( ) (15.1.5.4.1-4)

Dtr(t) (14.6.2-16)

Dtr(t) (14.6.3-6)

Dtr(t) (14.6.4.1-7)

Dtr(t) (14.6.4.2-17)

Dtr(t) (14.6.5.1-15)

$d\underline{\upsilon}$ (7.2.2.2.2-15)

$d\underline{\upsilon}_{+Cnt}$ (8.1.2.2-10)

$d\underline{\upsilon}_{-Cnt}$ (8.1.2.2-10)

$d\underline{\upsilon}_{Cnt}$ (8.1.2.2-1)

$d\underline{\upsilon}_{Cnt}$ (8.2.2.1-34)
$d\underline{\upsilon}_{Cnt}$ (8.2.3.1-2)
$d\underline{u}^{E}_{Up}$ (5.2.4-4)
$d\underline{u}^{Geo}_{UpE}$ (5.2.4-29)
$\mathcal{E}(\,)$ (14.6.1-5)
$\mathcal{E}(\,)$ (14.6.4-5)
$\mathcal{E}(\,)$ (10.2.2-6)
$\mathcal{E}(\,)$ (10.4.1-6)
$\mathcal{E}(\,)$ (10.4.2-11)
E (3.5.1-1)
E (8.1.1.1.1-21)
$\underline{E}$ (8.2.2.1-18)
E (11.2.1.4-1)
e (3.2.4-1)
e (3.2.4-7)
e (4.4.2.2-5)
e (5.1-1)
e (5.3-18)
e (11.2.4.1.2-3)
e (12.1-28)
E' (3.5.1-16)
E'_{SYM} (3.5.1-17)
E'_{SYM} (7.1.1.3-11)
$\mathcal{E}(\,)$ (13.6.1-9)
$\mathcal{E}(\,)$ (15.1.2.1-4)
$\mathcal{E}(\,)$ (15.2.1.1-5)
$\mathcal{E}(\,)$ (18.3.1.1-3)
$\mathcal{E}(\,)$ (18.3.2.1-4)
$\mathcal{E}\left(\overline{a_{SF}(t)^2}\right)$ (16.2.3.1-22)
$\mathcal{E}\left(\overline{a_{Vib}(t)^2}\right)$ (16.2.3.1-22)
$\mathcal{E}\left(\overline{a_{Vib}(t)^2}\right)$ (16.2.3.2-7)
$\mathcal{E}\left(\overline{a_{Vib}(t)^2}\right)$ (10.6.1-25)
$\mathcal{E}\left(\delta\dot{\Phi}_{Algo\text{-}m_z}\right)$ (16.2.3.2-7)
$\mathcal{E}\left(\Delta\theta^2_{ARS/Rnd/H\text{-}k_i}\right)$ (18.3.2.1-7)
$\mathcal{E}\left(\delta\dot{v}_{SF/Algo\text{-}m_z}\right)$ (16.2.3.2-7)
$\mathcal{E}\left(\varepsilon^2_{aRnd_i}\right)$ (13.6.1-15)
$\mathcal{E}\left(\varepsilon^2_{\omega Rnd_i}\right)$ (13.6.1-15)
$\mathcal{E}\left(\overline{\omega(t)^2}\right)$ (16.2.3.1-23)
$\left[\mathcal{E}\left(\underline{\hat{\omega}}^{N}_{IE_H}\right)\right]_k$ (18.3.1.1-8)
$\mathcal{E}\left(\dot{\Phi}_{Con_z}\right)$ (16.2.3.2-7)
$\mathcal{E}\left(\hat{\phi}_{H\text{-}k_i}\right)$ (18.3.2.1-10)
$\mathcal{E}\left(\overline{\theta(t)^2}\right)$ (16.2.3.1-22)
$\mathcal{E}\left(\theta^2_{ARS/Rnd/H\text{-}k}\right)$ (18.3.1.1-7)
$\mathcal{E}\left(\theta^2_{ARS/Rnd/H\text{-}k_j}\right)$ (18.3.1.1-6)
$\mathcal{E}\left(\dot{v}_{SF/Scul_z}\right)$ (16.2.3.2-7)
E_i (10.2.1-7)
e^{M}_{j} (15.1.2.1.1.3-26)
End (17.1.1.1-1)
End (18.2.1-10)
ε_a (18.4.7.3-6)
ε_α (15.2.1.2-18)
$\underline{\varepsilon}_{aRnd_i}$ (13.6.1-4)
ε_c (10.5.1-18)
$\underline{\varepsilon}^{N}_{c}$ (15.1.2.3-11)
ε_{c_x} (10.6.1-15)
ε_{c_y} (10.6.1-15)
$\underline{\varepsilon}^{Geo}$ (12.2.2-24)
$\underline{\varepsilon}^{E}_{H}$ (12.3.6.2-1)
$\underline{\varepsilon}^{Geo}_{H}$ (12.2.2-26)
ε_{ii} (7.1.1.3-2)

$f_{Att/Att}$ (16.2.3.1-18)
$f_{Att/Vel}$ (15.2.1.2-18)
$f_{Att/Vel}$ (16.2.3.1-21)
F_{CS}^{Geo} (5.2.4-28)
F_{C}^{Geo} (5.2.4-34)
F_{C}^{N} (4.1.1-6)
F_{C}^{N} (5.3-18)
F_{C}^{N} (5.3-7)
F_{C}^{N} (7.3.1-10)
F_{C}^{N} (12.1-28)
$F_{CVar_m}^{N}$ (17.2.3.2-2)
F_{Drag} (17.1.2.3-22)
f_e (12.1-28)
f_{eh} (12.1-28)
f_h (12.1-28)
F_{ij} (3.5.1-5)
F_{Lift} (17.1.2.3-22)
$F_{LP}(\ )$ (9.2-1)
f_{Osc_0} (15.2.4-12)
$\tilde{f}_{Osc}$ (15.2.4-14)
$f_{\phi Cntrl}$ (17.2.2-8)
$f_{\psi Cntrl}$ (17.2.2-8)
F_{Scal} (8.1.1.1-1)
$F_{ScalLin}$ (8.1.1.3-8)
$F_{ScalNonLin}$ (8.1.1.3-8)
$F_{SensAlgn}$ (8.1.1.1.1-2)
$F_{SensScal}$ (8.1.1.1.1-2)
$F_{SensScalAsym}$ (8.1.1.3-22)
$F_{SensScalLin}$ (8.1.1.3-8)
$F_{SensScalNonLin}$ (8.1.1.3-8)
F_{Side} (17.1.2.3-22)
f_{Size} (8.1.4.1.1.1-6)
Fst (15.1.5.3.1-6)
$f_{\theta Cntrl}$ (17.2.2-8)
$\underline{F}_{TotAero}^{\mathcal{V}WF}$ (17.2.3.2.1-3)
$f_{Vel/Pos}$ (15.2.1.2-18)
g (3.2.4-7)
g (12.2.4-2)
g (12.3.5-22)
g (14.2-16)
g (14.5-2)
g (17.1.1.2-8)
g (17.2.3.2.3-1)
$\underline{g}$ (5.4.1-1)
$g(\ )$ (10.4.2-14)
$G(\omega)$ (10.2.2-21)
$G(\omega)$ (15.1.2.1.1-32)
$G(\omega)$ (17.2.3.2.3-33)
$G_{a/a}$ (16.2.3.1-26)
$g_{Aid}(\ )$ (15.1-5)
G_{Algn} (8.1.1.2-1)
$\underline{G}_{Algn_k}^{T}$ (8.1.4.1-7)
$G_{Algn_{Off\text{-}Diag}}$ (8.1.1.2.1-14)
G_{Align} (8.2.2.1-5)
$G_{\alpha RVar_m}$ (17.2.3.2-21)
$G_{\alpha vVar_m}$ (17.2.3.2-21)
γ (10.1.3.2.2-25)
γ (11.2.3.1-10)
$\underline{\gamma}_{+}^{N+}$ (14.2-35)
$\underline{\gamma}_{c}^{N}$ (15.1.2.3-11)
$\gamma_{East_{Offset}}$ (18.2.1-23)
$\gamma_{East_{Offset}}$ (18.2.2-1)
$\gamma_{East_{Offset}}$ (18.3.1.2-9)

G_P (14.6.1-1)
g_P (6.1.1-1)
g_P (6.1.3-3)
$\underline{g}_P$ (5.4.1-1)
$\underline{g}_P$ (9.1-1)
$G_P(t)$ (15.1-1)
$G_P(t)$ (15.1.5.3.1-23)
$G_P(t)$ (18.3-5)
$G_{P*}(t)$ (15.1.5.4.1-3)
G_{P_a} (15.2.1.1-3)
$G_{P_a}\, Q_{P_{Dens/a}}\, G_{P_a}^T$ (15.2.1.1-8)
G_{P_b} (15.2.1.1-3)
$G_{P_b}\, Q_{P_{Dens/b}}\, G_{P_b}^T$ (15.2.1.1-8)
g_ϕ (5.4-1)
g_ϕ (12.1-28)
$\left(\frac{g_\phi}{\sin\phi}\right)_S$ (5.4-2)
$\left(\frac{g_\phi}{\sin\phi}\right)_S$ (11.2.4.3.2.1-2)
$\left(\frac{g_\phi}{\sin\phi}\right)_S$ (12.1-28)
g_ϕ (11.2.4.3.2.1-2)
$\underline{g}_P^I$ (4.3-16)
$\underline{g}_P^I$ (12.1.4-8)
$G_{P_{\mathcal{M}}}(t)$ (15.1.5.4.1-3)
$\underline{g}_P^N$ (7.2-1)
$\underline{g}_P^N$ (12.1-12)
$g_{P_{North}}$ (5.4.1-9)
$g_{P_{North}}$ (6.1.3-8)
$g_{P_{North}}$ (12.1-28)
$G_{P_{\psi V}}$ (16.2.3.1-8)
GPSAnt (17.3.2-2)
$g_{P_{Up}}$ (5.4.1-9)
$g_{P_{Up}}$ (6.1.3-8)
$g_{P_{Up}}$ (12.1-28)
$G_{P_{Vib}}(\omega)$ (10.4.1-8)
$G_{P_{VR}}$ (16.2.3.1-11)
g_r (5.4-1)
g_r (11.2.4.3.2.1-2)
g_r (12.1-28)
$\underline{g}_{Ref}^I$ (8.1.4.1-3)
g_{r_S} (5.4-2)
g_{r_S} (12.1-28)
g_{r_S} (11.2.4.3.2.1-2)
$G_{S/DL}(\,)$ (9.4-1)
G_{Scal} (8.1.1.2-1)
$G_{ScalAsym}$ (8.1.1.3-11)
$G_{ScalLin}$ (8.1.1.3-1)
$G_{ScalNonLin}$ (8.1.1.3-1)
$G_{SDL}(\,)$ (9.4-1)
$G_{SensAlgn}$ (8.1.1.2.1-2)
$G_{SensScal}$ (8.1.1.2.1-2)
$G_{SensScalAsym}$ (8.1.1.3-14)
$G_{SensScalLin}$ (8.1.1.3-2)
$G_{SensScalNonLin}$ (8.1.1.3-2)
g_θ (5.4-1)
g_θ (11.2.4.3.2.1-2)
g_{Tst} (18.4-1)
g_{Tst} (18.4.5-1)
g_{Tst} (18.4.7.4-16)
g_{Tst} (18.4.7.4-5)
g_{Up} (5.4-3)
g_{Up} (5.4.1-9)
g_{Up} (11.2.4.3.2.1-8)

κ_{ji}	Fig. 18.4.3-3
$\kappa_{LinScal}$	(18.4.7.2-12)
κ_{Mis}	(18.4.7.2-12)
$\underline{\kappa}_{SystBias}$	(8.1.1.1.1-5)
$\kappa_{SystScal/Mis}$	(8.1.1.1.1-5)
$\kappa_{SystScalAsym}$	(8.1.1.3-22)
$\kappa_{SystScalLin/Mis}$	(8.1.1.3-8)
$\kappa_{SystScalNonLin}$	(8.1.1.3-8)
kB	Fig. 18.4.3-3
$\underline{K}_{Bias}$	(8.1.1.1-7)
$\underline{K}_{Bias}$	(8.2.1.1-2)
$\underline{K}_{Bias}$	(8.2.2.1-5)
$\underline{K}_{Bias}$	(12.4-2)
$\underline{K}_{Bias}$	(12.4-5)
$\underline{K}_{BiasC}$	(12.4-5)
$K_{\delta\Delta R}$	(15.2.1.1-19)
$K_{\delta\omega IE}$	(15.2.1.1-19)
$K_{\delta v}$	(15.2.1.1-19)
$K^{\blacklozenge}$	(14.6.1-1)
$K^{\blacklozenge}$	(14.6.1-2)
$K^{\blacklozenge\prime}(t)$	(18.3-6)
$K^{\blacklozenge}(t)$	(18.3-5)
$K^{\blacklozenge}_{m}$	(15.1.5.3.2-10)
$K^{\blacklozenge}_{Rev}(t)$	(15.1.5.4.1-26)
$K_{DInd/L}$	(17.1.2.3-25)
$K_{DInd/Sd}$	(17.1.2.3-25)
K_{Fst}	(15.1.5.3.1-1)
K_{Fuel}	(17.1.2.3-27)
K_{γ}	(15.2.1.1-19)
K_{jk_i}	(16.2.3.2-8)
kMARS	Fig. 18.4.3-3
K_{Mis}	(8.1.1.1-7)
K_{Mis}	(8.2.1.1-2)
K_{Mis}	(8.2.2.1-5)
K_{Mis}	(12.4-2)
K_{Mis}	(12.4-5)
K_{MisC}	(12.4-5)
$K_{Mis_{ij}}$	(8.1.4.1.1.2-1)
$K_{Mis_{ij}}$	(8.2.1.1-9)
$K_{Mis_{ij}}$	(8.2.2.1-31)
K_{n}	(15.1.2-13)
$K^{\blacklozenge\prime}_{\omega}(t)$	(18.3-7)
K_{Opt}	(16.1.1.3-10)
K_{Opt}	(16.1.2.1-9)
$K_{\phi\rho/bnd_{ji}}$	(17.2.3.2.3-5)
K_{PsCon_i}	(16.2.3.2-1)
K_{ResCon_i}	(16.2.3.2-6)
$K_{R\rho/bnd_{ji}}$	(17.2.3.2.3-5)
L	(3.2.2-10)
L	(4.4.2.1-1)
L	(10.5.1-18)
L	(11.2.4.1.1-4)
L	(11.2.4.2-1)
L	(12.2.3-33)
L	(18.1.1-24)
l	(4.4.2.1-1)
l	(6.1.2-2)
l	(6.1.3-8)
l	(7.1.1.1.1-3)
l	(7.3.3.2-1)
l	(8.1.4.1.1.1-6)
l	(8.1.4.1.2-3)
l	(10.1.3.2.2-3)
l	(10.5.1-3)
l	(11.2.1.1-6)
$\underline{l}$	(11.2.3.1-1)

l	(11.2.4.1.1-1)
l	(11.2.4.3.2.1-8)
l	(12.2.3-33)
l	(13.4.2-34)
l	(14.1-2)
l	(14.5-3)
$\underline{l}$	(15.1.2.2-3)
$\underline{l}$	(17.2.3.2.2-1)
l	(17.2.3.2.3-24)
l	(18.1.1-22)
l	(18.3-1)
$l$bnd	(17.2.3.2.3-5)
l	Fig. 5.2-1
$\underline{l}_k$	(9.3-2)
$\underline{L}_0$	(3.2.2-13)
L_0	(11.2.4.1.1-1)
l_0	(11.2.4.1.1-1)
l_0	Fig. 13.5-1
$\underline{l}_0^{AC}$	(17.2.3.2.2-1)
$\underline{l}_0^{B}$	(11.2.3.1-9)
$\underline{L}_1$	(3.2.2-13)
L_1	(11.2.4.1.1-1)
$\underline{L}_2$	(3.2.2-13)
λ	(3.2.2-10)
λ	(4.4.1.2.1-7)
λ	(13.2.1-6)
$\underline{\lambda}$	(14.6.2-1)
$\underline{\lambda}(0)$	(15.1.5.4-8)
$\underline{\lambda}(t)$	(15.1.5.4-2)
$\underline{\lambda}^*(t)$	(15.1.5.4.1-13)
λ_{Asym}	(18.4.5-1)
λ_{Asym}	(18.4.7.1-3)
λ_{Asym}	(18.4.7.4-18)
$\underline{\lambda}_{Bias}$	(18.4.5-1)
$\underline{\lambda}_{Bias}$	(18.4.7.1-3)
$\underline{\lambda}_{Bias}$	(18.4.7.4-10)
$\underline{\lambda}_{Bias}$	(18.4.7.4-18)
Λ_i	(14.6.2-6)
Λ_i	(14.6.5.1-2)
λ_i	(14.6.2-2)
λ_{i_0}	(14.6.2-6)
λ_{ii}	Sect. No. 18.4.3
λ_{iii}	Sect. No. 18.4.3
λ_{ij}	Sect. No. 18.4.3
$\lambda_{LinScal}$	(18.4.5-1)
$\lambda_{LinScal}$	(18.4.7.4-10)
$\lambda_{LinScal}$	(18.4.7.4-18)
$\lambda_{LinScal/Mis}$	(18.4.7.1-3)
$\underline{\lambda}_{\mathcal{M}}(t)$	(15.1.5.4.1-13)
λ_{Mis}	(18.4.7.4-10)
$\underline{\lambda}_m^N$	(17.2.3.2-2)
$\underline{\lambda}_{SystBias}$	(8.1.1.2.1-5)
$\lambda_{SystScal/Mis}$	(8.1.1.2.1-5)
$\lambda_{SystScalAsym}$	(8.1.1.3-15)
$\lambda_{SystScalLin/Mis}$	(8.1.1.3-2)
$\lambda_{SystScalNonLin}$	(8.1.1.3-2)
L_{Amp}	(11.2.4.1.1-1)
l_{Amp}	(11.2.4.1.1-1)
L_{Aniso}	(10.1.4.1-2)
$\underline{L}_{Bias}$	(8.1.1.2-7)
$\underline{L}_{Bias}$	(8.2.2.1-5)
$\underline{L}_{Bias}$	(12.4-15)
$\underline{L}_{Bias}^*$	(8.2.2.1-5)
$\underline{L}_{BiasC}$	(12.4-15)
$\underline{l}_{Cnst}^{B}$	(15.2.2.1-14)

$\underline{\omega}_{BI}^{P}$ (13.4.2-2)
ω_c Fig. 11.2.1.1-1
ω_e (4.1.1-4)
$\underline{\omega}_e$ (5.4.1-1)
ω_e (5.4.1-4)
ω_e (6.1.2-2)
ω_e (11.2.4.1.1-3)
ω_e (12.1-12)
ω_e (12.1.4-9)
ω_e (12.2.4-21)
ω_e (14.5-3)
ω_e (18.3-1)
ω_e Fig. 13.5-1
$\underline{\omega}_{EAC}$ (17.3.2-8)
ω_{EB} (13.4.3-2)
$\underline{\omega}_{EB}^{B}$ (13.4.3-2)
$\underline{\omega}_{e}^{I}$ (11.2.4.2-2)
$\underline{\omega}_{EN}$ (9.1-1)
$\underline{\omega}_{EN}^{E}$ (5.2.4-1)
$\underline{\omega}_{EN_H}^{E}$ (12.3.6.2-11)
$\underline{\omega}_{EN_H}^{E}$ (5.2.4-3)
$\underline{\omega}_{EN}^{N}$ (4.1.1-1)
$\underline{\omega}_{EN}^{N}$ (7.2-1)
$\underline{\omega}_{EN}^{N}$ (12.1-12)
$\underline{\omega}_{EN}^{N}$ (12.3.3-3)
$\underline{\omega}_{EN}^{N}$ (14.5-2)
$\underline{\omega}_{EN_H}^{N}$ (12.3.5-19)
ω_{h1} (11.2.4.1.1-2)
ω_{h2} (11.2.4.1.1-2)
ω_{Hi} (17.2.3.2.3-18)
ω_i' (10.4.2-9)
$\underline{\omega}_{IA}^{A}$ (3.3.2-10)
$\underline{\omega}_{IA}^{A}$ (3.5.1-19)
$\underline{\omega}_{IA}^{\wedge A}$ (3.5.1-20)
$\left(\underline{\omega}_{IA}^{A}\times\right)$ (3.3.2-10)
$\underline{\omega}_{IB}$ (9.1-1)
ω_{IB} (13.2.4-9)
ω_{IB} (13.4.1.1-1)
ω_{IB} (13.4.1.2-6)
$\underline{\omega}_{IB}(t)$ (10.1.1-2)
$\underline{\omega}_{IB}(t)$ (10.1.2-3)
$\underline{\omega}_{IB}(t)$ (10.1.3-1)
$\omega IB0_{Accl/Inpt}$ (10.1.4.1-2)
$\omega IB0_{Accl/Pend}$ (10.1.4.1-2)
$\omega_{IB_{Accl/Inpt}}(t)$ (10.1.4.1-2)
$\omega_{IB_{Accl/Pend}}(t)$ (10.1.4.1-2)
$\underline{\omega}_{IB}^{B}$ (3.3.2-10)
$\underline{\omega}_{IB}^{B}$ (3.5.1-19)
$\underline{\omega}_{IB}^{\wedge B}$ (3.5.1-20)
$\underline{\omega}_{IB}^{B}$ (4.1-2)
$\underline{\omega}_{IB}^{B}$ (4.1-2)
$\underline{\omega}_{IB}^{B}$ (7.1.1.1-4)
$\underline{\omega}_{IB}^{B}$ (11.2.3.1-4)
$\underline{\omega}_{IB}^{B}$ (12.1-12)
$\underline{\omega}_{IB}^{B}$ (12.1.4-1)
$\underline{\omega}_{IB}^{B}$ (18.3-1)
$\left(\underline{\omega}_{IB}^{B}\times\right)$ (3.3.2-10)

ω_X	(10.5.1-18)
ω_{XGeo}	Fig. 4.4.3-1
Ω_y	(10.1.1-2)
Ω_y	(10.1.2-3)
ω_y	(10.6.1-6)
ω_{YGeo}	Fig. 4.4.3-1
$\underline{\omega}_\zeta$	(11.2.2.1-5)
ω_ζ	(11.2.2.1-6)
ω_{ZGeo}	Fig. 4.4.3-1
$\underline{\omega}'^*$	(8.1.1.1.1-2)
ω'_X	(8.2.2.1-23)
ω'_Y	(8.2.2.1-23)
ω'_Z	(8.2.2.1-23)
Out	(17.2.3.2.3-38)
P	(14.6.1-5)
P	(15.1.2.1-4)
$\underline{P}$	(17.1.2.3-30)
P'	(14.6.2-13)
P'	(16.1.1.1-1)
P'_0	(14.6.2-13)
P'_{ii}	(16.2.4-11)
P'_{jk_0}	(16.2.4-8)
P'_{Meas}	(16.2.5-16)
P'_{Meas/ii_m}	(16.2.4-9)
P'_{Proc}	(16.2.5-16)
P'_{Proc/ii_l}	(16.2.4-9)
P(+)	(15.1.5.3.1-1)
P(-)	(15.1.5.2-1)
P(-)	(15.1.5.3.1-1)
P(0)	(15.1.5.4-9)
$P(\psi_{pVib_i})$	(10.4.2-13)
$P(p_{Vib_{0/i}})$	(10.4.2-13)
$P(p_{Vib_{0/i}}, \psi_{pVib_i})$	(10.4.2-13)
P(t)	(15.1.2.1.1-16)
p(t)	(10.2.2-1)
$\overline{p(t)^2}$	(10.2.2-2)
$P(t_1)$	(15.1.2.1.1-16)
P*	(14.6.4-5)
P*	(14.6.4.2-5)
P*	(15.2.1.1-13)
P*	(16.1.1-37)
P*(t)	(15.1.5.4.1-12)
P**	(14.6.4.2-5)
P^{***}_0	(14.6.4.3-8)
P^{**}_0	(14.6.4.2-7)
P^*_0	(14.6.4-7)
P^*_0	(15.2.1.2-2)
P_{aOut}	(16.2.5-4)
$P_{aOut/ii}$	(16.2.5-21)
$P_{aOut/Meas}$	(16.2.5-16)
$P_{aOut/Meas/ii_m}$	(16.2.5-21)
$P_{aOut/Meas_m}$	(16.2.5-18)
$P_{aOut/Proc}$	(16.2.5-16)
$P_{aOut/Proc/ii_l}$	(16.2.5-21)
$P_{aOut/Proc_l}$	(16.2.5-18)
$P_{aSF_{iAC}}$	(17.2.3.2.3-23)
$P_{\delta\Delta R/H_0}$	(15.2.1.2-2)
$P_{\delta\Delta R\delta v/H_0}$	(15.2.1.2-2)
$P_{\delta g}$	(16.2.3.3-3)
$P_{\delta g_{StdSt}}$	(16.2.3.3-6)
$P_{\delta\omega IE/H_0}$	(15.2.1.2-2)
$P_{\delta v/H_0}$	(15.2.1.2-2)
$P_{\delta v\delta\Delta R/H_0}$	(15.2.1.2-2)
P_{γ/H_0}	(15.2.1.2-2)
Φ	(15.1.5.3.1-1)
ϕ	(3.1-3)

$\underline{\phi}^{AC}_{HiFlev_m}$	(17.2.3.2.2-2)
$\hat{\underline{\phi}}^{L}_{H}$	(18.2.1-12)
$\hat{\underline{\phi}}^{N}_{H_i}$	(18.3.2.1-3)
φ_{h_x}	(10.1.4.1-4)
ϕ_{h_x}	(10.3-20)
φ_{h_y}	(10.1.4.1-4)
ϕ_{h_y}	(10.3-20)
$\underline{\varphi}^{I}$	(12.2.1-18)
$\hat{\phi}_i$	(18.3.2.2-1)
$\hat{\phi}_{H-k_i}$	(18.3.2.1-10)
$\underline{\phi}^{L}$	(18.4.7-7)
$\underline{\phi}^{L}$	(18.4.7.2-2)
$\Phi_{\lambda\lambda}(t)$	(14.6.4.1-4)
$\Phi_{\lambda\lambda}(t)$	(15.1.2.1.1.3-12)
$\Phi_{\lambda\lambda}(t)$	(15.1.5.4-8)
$\Phi^{j}_{\lambda\lambda}$	(15.1.2.1.1.3-28)
$\Phi_{\lambda\lambda_m}$	(15.1.2.1.1.3-7)
$\Phi_{\lambda y}(t)$	(14.6.4.1-4)
$\Phi_{\lambda y}(t)$	(15.1.2.1.1.3-12)
$\Phi_{\lambda y}(t)$	(15.1.5.4-8)
$\Phi^{j}_{\lambda y}$	(15.1.2.1.1.3-28)
$\Phi_{\lambda y_m}$	(15.1.2.1.1.3-7)
$\underline{\Phi}_m$	(10.1.1.2-1)
Φ_m	(15.1.2.1.1.3-28)
ϕ_m	(7.1.1.1-3)
$\underline{\phi}_m$	(7.1.1.1-3)
ϕ_m	(7.1.2.1-3)
$\underline{\phi}_m$	(7.1.2.1-3)
$\underline{\phi}_m$	(8.2.1-1)
$\underline{\phi}^{\nu}_m$	(17.1.3-5)
$\Phi_{n,n-1+c}$	(15.1.2.4-11)
$\Phi_{n-1+c,n-1}$	(15.1.2.4-11)
$\phi_{Num/\vartheta 1_x}(\Omega)$	(10.6.1-15)
$\phi_{Num/\vartheta 1_z}(\Omega)$	(10.6.1-15)
$\phi_{Num_{A1}}(\omega)$	(10.5.1-24)
$\phi_{Num_i}(\omega)$	(10.2.1-32)
$\phi_{Num_{\vartheta 1}}(\omega)$	(10.5.1-24)
$\varphi_{\omega ARS_x}$	(10.1.4.2-2)
$\varphi_{\omega ARS_x}$	(10.3-14)
$\phi_{\omega ARS_x}$	(10.3-16)
$\varphi_{\omega ARS_y}$	(10.3-14)
$\phi_{\omega ARS_y}$	(10.3-16)
$\phi_{\omega IB_{Accl/Inp}}$	(10.3-20)
$\varphi_{\omega IB_{Accl/Inp}}$	(10.1.4.1-2)
$\varphi_{\omega IB_{Accl/Pend}}$	(10.1.4.1-2)
$\phi_{\omega IB_{Accl/Pend}}$	(10.3-20)
$\varphi_{pp}(t, \tau)$	(10.2.2-9)
$\varphi_{pp}(t, \tau)$	(17.2.3.2.3-32)
ϕ_S	(17.1.1.2-16)
$\phi_{S_{i\nu}}$	(17.1.1.2-26)
$\underline{\phi}^{\nu}_S$	Sect. No. 17.1.1
$\phi_{\vartheta}(\omega)$	(10.5.1-25)
φ_{θ_x}	(10.1.2-3)
φ_{θ_x}	(10.1.1-2)
ϕ_{θ_x}	(10.3-3)
$\phi_{\theta_{x/i}}$	(10.4.1-1)
$\varphi_{a_{SF_y}}$	(10.1.2-3)
φ_{θ_y}	(10.1.1-2)
ϕ_{θ_y}	(10.3-3)

$\phi\theta_{y/i}$ (10.4.1-1)
$\phi\theta_z$ (10.3-3)
$\phi\theta_z$ (10.6.1-13)
$\phi\theta_{z/i}$ (10.4.1-1)
ϕVar_m (17.2.3.2-14)
ϕVar_m (17.2.3.2-14)
$\Phi_{V_H\,k}(t,\tau_i)$ (13.6.1-7)
ϕ_X (3.2.2.2-17)
ϕ_x (3.2.4-18)
Φ_{xx} (16.1.1-5)
Φ_{xy} (16.1.1-5)
ϕ_Y (3.2.2.2-17)
ϕ_y (3.2.4-18)
$\Phi_{y\lambda}(t)$ (14.6.4.1-4)
$\Phi_{y\lambda}(t)$ (15.1.2.1.1.3-12)
$\Phi_{y\lambda}(t)$ (15.1.5.4-8)
Φ_{yx} (16.1.1-5)
Φ_{yy} (16.1.1-5)
$\Phi_{yy}(t)$ (14.6.4.1-4)
$\Phi_{yy}(t)$ (15.1.2.1.1.3-12)
$\Phi_{yy}(t)$ (15.1.5.4-8)
Φ_{yy_m} (15.1.2.1.1.3-7)
ϕ_Z (3.2.2.2-17)
ϕ_z (3.2.4-18)
$p_i(t)$ (10.2.2-1)
P_{ii} (15.1.2.1.1.4-1)
$P_{ii_{Max}}$ (15.1.2.1.1.4-4)
$P_{ii_{Min}}$ (15.1.2.1.1.4-1)
P_{ij} (15.1.2.1.1.4-2)
$\underline{\phi}^{AC}$ (17.1.2.3-7)
p_{jk} (10.1.1.2.2-14)
p_{jk} (10.1.2.2.2-14)
P_Ω (14.6.2-16)
P_{Ω_0} (14.6.2-13)
P_{Ω_0} (18.3-9)
$P_{\Omega\,\alpha Quant/aRand}$ (14.6.4.2-20)
$P_{\Omega\,\omega Rand}$ (14.6.4.3-11)
$P_{\Omega\,RVib}$ (14.6.3-9)
$P_{\Omega\,Simult}$ (14.6.5.2-1)
$P_{\Omega_{Sum}}$ (14.6.5.2-3)
$P_{\Omega\,\upsilon Quant}$ (14.6.4.1-10)
P_{Opt} (16.1.1.3-16)
$\underline{P_{Opt}}$ (16.1.2.1-15)
$p_{Out}(t)^2$ (17.2.3.2.3-38)
$P_{R_{HLng}R_{HLng}}$ (13.6.1-30)
P_{RVib_H} (14.6.1-12)
P_{RVib_H} (15.2.1-27)
P_{RVib_H} (15.2.1.1-9)
$P_{Sen/Mrk}$ (15.2.2.1-38)
$P_{Sen/Mrk/Std}$ (15.2.2.1-38)
ψ (3.2.3-1)
ψ (3.5.3-26)
ψ (10.2.1-4)
ψ_0 (11.2.2.1-18)
$\underline{\psi}_0^A$ (13.2.4-20)
$\underline{\psi}_0^I$ (13.4.1.2-12)
$\underline{\psi}_0^I$ (13.5-7)
$\underline{\psi}_0^N$ (14.5-5)
$\dot{\psi}_{Avg_i}$ (13.4.2-41)
$\dot{\psi}_{Bias_i}$ (13.4.2-34)
ψ_{BL} (11.2.1.3-2)
$\psi_{Desired}$ (17.1.1.5-12)
$\underline{\psi}^E$ (12.2.1-4)

q_{aRand} (14.6.1-12)
q_{aRand} (15.2.1-17)
$q_{aRnd}(\tau)$ (13.6.1-20)
$q_{B(t)}^{B_{I(m-1)}}$ (7.1.2.1-2)
$q_{B_{I(m)}}^{B_{I(m-1)}}$ (7.1.2-1)
$q_{B_{I(m)}}^{L_{I(n)}}$ (7.1.2-1)
$q_{B_{I(m-1)}}^{L_{I(n-1)}}$ (7.1.2-1)
q_{B}^{L} (4.1-2)
$Q_{\delta gDens}$ (16.2.3.3-3)
Q_{ij} (15.1.2.1.1.4-4)
Q_{In} (17.2.3.2.3-42)
$q_{L_{I(n-1)}}^{L(t)}$ (7.1.2.2-2)
$q_{L_{I(n-1)}}^{L_{I(n)}}$ (7.1.2-1)
Q_{m} (15.1.2.1.1.3-28)
Q_{n} (15.1.2.1.1-10)
$q_{\omega Rand}$ (14.6.1-12)
$q_{\omega Rand}$ (15.2.1-17)
$q_{\omega Rand}$ (18.3-9)
$q_{\omega Rnd}(\tau)$ (13.6.1-20)
$Q_{P^{*}{}_{Dens}}$ (15.2.1.1-11)
$Q_{P^{*}{}_{Dens}}(t)$ (15.1.5.4.1-3)
q_{PDens} (15.1.2.1.1-33)
$Q_{P_{Dens}}$ (14.6.1-6)
$Q_{P_{Dens}}(t)$ (15.1.5.3.1-23)
$Q_{PDens}(\tau_{\beta})$ (15.1.2.1.1-28)
$Q_{P_{Dens/a}}$ (15.2.1.1-7)
$Q_{P_{Dens/b}}$ (15.2.1.1-7)
$Q_{P_{Dens/Assoc}}(t)$ (15.1.2.1.1.3-32)
$q_{PDens_{i}}(\tau_{\beta})$ (15.1.2.1.1-27)
$q_{PDens_{l}}$ (16.2.4-9)
$Q_{P_{Dens}}^{N}$ (15.2.1.1-6)
$Q_{P_{\mathcal{M}\ Dens}}(t)$ (15.1.5.4.1-3)
$q_{\psi\alpha VibQuant_{ij}}$ (16.2.3.1-18)
$Q_{\psi V\alpha Quant}$ (16.2.3.1-9)
$q_{\psi V\alpha Quant}$ (16.2.3.1-5)
$q_{VR\upsilon Quant}$ (16.2.3.1-5)
Q_{RW} (16.1.1.3-16)
$Q_{Sen/Mrk}$ (15.2.2.1-38)
$Q_{SF_{iRnd}}$ (17.2.3.2.3-21)
q_{Tot} (17.2.3.2.1-5)
$q_{\upsilon Quant}$ (14.6.1-12)
$q_{\upsilon Quant}$ (15.2.1-17)
$q_{V\alpha VibQuant_{ij}}$ (16.2.3.1-21)
$q_{Vib/Dens}$ (15.2.1.2-10)
$q_{V\upsilon VibQuant_{ij}}$ (16.2.3.1-21)
$Q_{WndGsti_{m}}$ (17.2.3.2.1-11)
R (4.1.1-5)
$\underline{R}$ (4.4.2.2-1)
R (5.2.1-2)
R (5.4-1)
$\underline{R}$ (5.4.1-1)
$\underline{R}$ (9.1-1)
R (9.4-3)
R (11.2.4.1.2-4)
R (11.2.4.3.2.1-2)
R (12.1-28)
R (12.2.1-45)
R (12.2.3-6)
R (15.1.5.2-1)
$\underline{R}$ Fig. 5.2-1
$\underline{r}$ (3.2.4-25)
r (7.2.1-3)
r (7.2.2.1-5)

$\sum_{1-3}$ (18.4.7.3-17)
$\sum_{4-8}$ (18.4.7.3-17)
$\sum_{9-11}$ (18.4.7.3-17)
$\sigma_{a_i Vib}$ (16.2.3.1-21)
$\sigma_{a_j Vib}$ (16.2.3.2-12)
$\sigma_{a_k Vib}$ (16.2.3.2-12)
σ_{aOut/Tr_i} (16.2.5-26)
σ_{aOut_i} (16.2.5-21)
$\sigma_{ARS/H/Rnd\text{-}k}$ (18.3.1.1-6)
$\sigma_{ARS/Rnd}$ (18.2.1-25)
$\sigma_{ARS/Rnd}$ (18.2.2-8)
$\sigma_{ARS/Rnd/East}$ (18.3.1.2-11)
$\sigma_{ARS/Rnd/East}$ (18.3.2.2-2)
$\sigma_{ARS/Rnd/East_n}$ (18.3.2.2-3)
$\sigma_{ARS/Rnd/H\text{-}k}$ (18.3.2.1-7)
$\sigma_{ARS/Rnd/H\text{-}k_n}$ (18.3.2.1-11)
$\sigma_{ARS/Rnd/ZN}$ (18.3.1.1-10)
σ_{aVib} (16.2.3.1-21)
σ_{aVib} (16.2.3.2-13)
σ_{χ} (18.4.7.3-8)
$\sigma_{\Delta v_{Avg}}$ (18.4.7.3-8)
$\overline{\sigma_{\Delta v}}$ (18.4.7.3-13)
σ_{Meas_m} (16.2.4-9)
σ_{Misc_i} (16.2.4-11)
$\sigma_{\omega_i Vib}$ (16.2.3.1-15)
$\sigma_{\omega Vib}$ (16.2.3.1-16)
σ_{PDens_i} (16.2.4-9)
$\sum_{Total}$ (18.4.7.3-17)
$\sigma_{WndGsti_m}$ (17.2.3.2.1-12)
$\sigma_{x'_i}$ (16.2.4-6)
$\sigma_{x'_{j0}}$ (16.2.4-8)
$\sigma_{x'_{j\,nTrans}}$ (16.2.4-15)
$\sigma_{x'Tr_i}$ (16.2.4-14)
S_{ij_n} (16.2.4-1)
sinh (13.2.1-11)
$\underline{S}_{j_n}$ (16.2.4-1)
✯ (18.3-2)
* (14.6.4-2)
** (14.6.4.2-2)
Start (17.1.1.1-1)
Start (18.2.1-4)
$S\theta_{ARS/\omega Vib\text{-}\alpha Quant/ZN_j}$ (18.3.1.1-2)
$S\theta_{ARS/\omega Vib\text{-}\alpha Quant_{ZL}}$ (18.2.2-1)
$S\theta_{ARS/\omega Vib\text{-}\alpha Quant_{ZN}}(t)$ (18.3-26)
$S\theta_{ARS/Rnd/ZN_j}$ (18.3.1.1-2)
$S\theta_{ARS/Rnd_{ZL}}$ (18.2.2-1)
$S\theta_{ARS/Rnd_{ZN}}(t)$ (18.3-26)
S_{Tr} (16.2.4-4)
S_{Tr_i} (16.2.4-13)
$S_{Tr_{ij}}$ (16.2.4-15)
$\underline{S}_{\upsilon}$ (7.3.3-5)
$\underline{S}_{\upsilon Cnt}$ (8.2.3.1-2)
$\underline{S}_{\upsilon Cnt_m}$ (8.1.2.2-1)
$\underline{S}_{\upsilon_m}$ (8.2.3-1)
$\underline{S}_{\upsilon_m}$ (10.1.3.1-8)
$\underline{S}^{N}_{Var_m}$ (17.2.3.2-1)
$S^{E}_{\hat{v}REF}(t_{REF})$ (15.2.2.1-40)
$\underline{S}^{N}_{WndGst}$ (17.2.3.2.3-6)
T (10.1.4.1-5)
T (10.2.2-1)
T (10.4-2)

T (11.2.2.2-1)
T (18.2.1-10)
t (11.2.1.1-5)
t (11.2.2.1-1)
t (11.2.4.1.1-1)
t (13.2.1-12)
t (14.3-11)
t (15.2.2.1-40)
t (17.1.1.2-17)
t (17.1.1.2-7)
t (17.1.2.3-28)
t (17.2.1-1)
t Fig. 13.5-1
T_0 (11.2.4.1.1-7)
t_0 (10.1.1.1-2)
t_0 (10.1.2.1-2)
t_0 (10.1.4.1-5)
T_{Align} (7.4.1-4)
T_{Align} (18.2.2-1)
T_{Align} (18.3.1.1-2)
τ (4.4.1.2.1-10)
τ (7.1.1.1-5)
τ (10.1.1.1-3)
τ (10.2.2-7)
τ (11.2.2.1-1)
τ (13.2.1-12)
τ (17.1.2.2-7)
τ (17.2.3.2.3-32)
τ (18.2.1-4)
τ_α (15.1.2.1.1-22)
τ_β (15.1.2.1.1-22)
τ_{Filt} (17.2.1-21)
τ_i (13.6.1-5)
$T_{Fin/Crnt}$ (17.2.1-2)
t_{GPSi} (15.2.4-2)
ϑ (17.1.2.3.1-9)
θ (3.2.3-1)
θ (3.5.3-26)
θ (10.5.1-3)
θ (12.2.1-28)
θ (17.1.2.2-14)
θ (17.2.3.2.3-24)
$\vartheta(S)$ (10.5.1-11)
$\underline{\theta}(t)$ (10.1.1-2)
$\underline{\theta}(t)$ (10.3-3)
θ_0 (11.2.2.1-18)
θ_{0_x} (10.1.1-2)
θ_{0_x} (10.1.2-3)
θ_{0_y} (10.1.1-2)
$\theta_{ARS/\omega Vib\text{-}\alpha Quant/East_j}$ (18.3.1.2-9)
$\underline{\theta}^N_{ARS/\omega Vib\text{-}\alpha Quant/H_j}$ (18.3.1.1-2)
$\theta_{ARS/\omega Vib\text{-}\alpha Quant_{East}}$ (18.2.2-1)
$\underline{\theta}^N_{ARS/\omega Vib\text{-}\alpha Quant_H}(t)$ (18.3-26)
$\theta_{ARS/\omega Vib\text{-}\alpha\ Quant_{ZL}}$ (18.2.2-1)
$\theta_{ARS/Rnd/East_j}$ (18.3.1.2-9)
$\theta_{ARS/Rnd/H\text{-}k_j}$ (18.3.1.1-6)
$\underline{\theta}^N_{ARS/Rnd/H_j}$ (18.3.1.1-2)
$\theta_{ARS/Rnd_{East}}$ (18.2.2-1)
$\underline{\theta}^N_{ARS/Rnd_H}(t)$ (18.3-26)
$\theta_{ARS/Rnd_{ZL}}$ (18.2.2-1)
θ_{BL} (11.2.1.3-2)
$\vartheta_F(S)$ (10.5.2-12)
θ_F (10.5.2-3)
$\theta_{GC/Range}$ (17.1.1.4-5)
$\underline{\vartheta}^{Geo}$ (12.2.2-19)

$\underline{u}_{\zeta E}^{E}$ (12.2.3-33)
$\underline{u}_{\zeta L_1}^{L_1}$ (12.2.1-28)
$\underline{u}_{\zeta L_2}^{L_2}$ (12.2.1-28)
$\underline{u}_{\zeta L}^{L}$ (12.2.1-28)
$\underline{u}_{\zeta}^{L}$ (11.2.2.1-4)
$\underline{u}_{ZGeo_{End}}$ (17.1.1.4-1)
$\underline{u}_{ZGeo}^{Geo}$ (12.2.2-24)
$\underline{u}_{ZL}$ (6.1.1-1)
$\underline{u}_{ZL}^{B}$ (6.1.1-3)
$u_{ZL_{iB}}$ (18.4.7.1-5)
$\underline{u}_{ZL}^{L}$ (18.2.1-3)
$\underline{u}_{ZL}^{L}$ (18.4-1)
$\underline{u}_{ZL}^{L}$ (18.4.5-1)
$\underline{u}_{ZL}^{L}$ (18.4.7.4-16)
$\underline{u}_{ZL}^{M}$ (18.4.5-1)
$\underline{u}_{ZL}^{M}$ (18.4.7.4-5)
$\underline{u}_{ZL}^{\mathcal{V}}$ (17.1.1.5-2)
$\underline{u}_{ZN}$ (4.4.2.2-2)
$u_{ZN/GPSAnt_{iE}}$ (17.3.2-2)
$\underline{u}_{ZN_1}^{N_1}$ (6.1.3-9)
$\underline{u}_{ZN}^{B}$ (13.4.1.2-14)
$\underline{u}_{ZN}^{E}$ (4.4.1.1-5)
$\underline{u}_{ZN}^{E}$ (12.1-13)
$\underline{u}_{ZN}^{E}$ (12.2.3-6)
$\underline{u}_{ZN}^{E}$ Sect. No. 15.1.2.2
$\underline{u}_{ZN_{End}}^{N_{Start}}$ (17.1.1.4-1)
$\underline{u}_{ZN_{INS}}^{E}$ (15.1.2.2-1)
$\underline{u}_{ZN}^{N}$ (4.1.1-6)
$\underline{u}_{ZN}^{N}$ (4.4.1.2-1)
$\underline{u}_{ZN}^{N}$ (5.3-18)
$\underline{u}_{ZN}^{N}$ (6.1.2-2)
$\underline{u}_{ZN}^{N}$ (9.4-2)
$\underline{u}_{ZN}^{N}$ (12.1-12)
$\underline{u}_{ZN}^{N}$ (12.2.3-9)
$\underline{u}_{ZN}^{N}$ (13.1-1)
$\underline{u}_{ZN}^{N}$ (13.3-1)
$\underline{u}_{ZN}^{N}$ (14.5-2)
$\underline{u}_{ZN}^{N}$ (18.3-1)
$\underline{u}_{ZN_{OTH}}^{E}$ (15.1.2.2-1)
$\underline{u}_{ZNVar}^{NVar}$ (17.2.3.2-8)
$u_{ZN_{XE}}$ (4.4.2.2-4)
$u_{ZN_{XE}}$ (12.1-36)
$u_{ZN_{XI}}$ (12.1.4-14)
$u_{ZN_{YE}}$ (4.4.2.2-4)
$u_{ZN_{YE}}$ (12.1-28)
$u_{ZN_{YE}}$ (12.1-36)
$u_{ZN_{YI}}$ (12.1.4-11)
$u_{ZN_{YI}}$ (12.1.4-14)
$u_{ZN_{ZE}}$ (4.4.2.2-4)
$u_{ZN_{ZE}}$ (12.1-36)
$u_{ZN_{ZI}}$ (12.1.4-14)
$\underline{u}_{ZREF}$ (4.4.2.2-2)
$u_{ZREF_{XE}}$ (4.4.2.2-4)
$u_{ZREF_{YE}}$ (4.4.2.2-4)
$u_{ZREF_{ZE}}$ (4.4.2.2-4)
$\underline{u}_{Z\mathcal{V}}$ (17.1.2.3.1-9)
$\underline{u}_{ZE}^{N}$ (6.2.1-3)

$V_{INSEast()}$ (18.1.2-9)
$\underline{V}_{INS_H}(t)$ (18.1.1-18)
$V_{INSNorth()}$ (18.1.2-9)
$\underline{v}_M(t)$ (15.1.5.3.1-25)
$\underline{v}^N$ (4.1.1-5)
$\underline{v}^N$ (5.3-18)
$\underline{v}^N$ (6.1.2-1)
$\underline{v}^N$ (7.2-1)
$\underline{v}^N$ (11.2.3.3-5)
$\underline{v}^N$ (12.1-12)
$\hat{\underline{v}}^N$ (15.2.2.1-7)
V_{North} (11.2.4.1.1-6)
v_{North} (11.2.4.2-2)
$V_{P_{Dens/Assoc}}(t)$ (15.1.2.1.1.3-32)
$\underline{V}_{\perp}$ (3.1.1-44)
V_{ϕ} (3.2.1.1-13)
$\underline{v}^E_{REF}$ (15.2.2.1-5)
$\underline{v}_{SF}(t)$ (10.1.2.1-2)
$\underline{v}_{SF}(t)$ (10.1.3.1-1)
$\dot{v}_{SF/Scul_x}$ (10.6.1-9)
$\dot{v}_{SF/Scul_z}$ (10.3-20)
$\dot{\underline{v}}_{SF/Scul_{SnsDyn}}$ (10.1.4.2-4)
$\underline{v}^L_{SF}$ (11.2.2.1-1)
$\underline{v}_{SF_m}$ (10.1.2.2-1)
$\underline{v}_{SF_{m-1}}$ (10.1.3.1-6)
$\dot{\underline{v}}_{SF_{Scul}}$ (10.1.2.1-13)
V_{Start} (17.1.1.1-2)
$\underline{v}^N_{Start}$ (17.1.2.2-7)
$\underline{v}^N_{Start}$ (18.4.7.3.1-1)
$v_{TotArspd}$ (17.2.3.2.1-4)
$\underline{v}_{TotArspd}$ (17.2.3.2.1-4)
$v_{TotArspd_{iAC}}$ (17.2.3.2.1-7)
v_{Up} (11.2.4.2-2)
v_{Vib_i} (15.2.1.2-6)
$\underline{v}_{WndGst}$ (17.2.3.2.1-4)
$v_{WndGst/iGeo_m}$ (17.2.3.2.1-10)
v_X (11.2.4.1.1-4)
V_{XA} (3.1-1)
V_{XA} (3.1.1-1)
V_{XA} (3.2.1-8)
V_{XB} (3.1-2)
V_{XB} (3.2.1-8)
v_{XN} (4.4.3-7)
v_Y (11.2.4.1.1-4)
V_{YA} (3.1-1)
V_{YA} (3.1.1-1)
V_{YA} (3.2.1-8)
V_{YB} (3.1-2)
V_{YB} (3.2.1-8)
v_{YN} (4.4.3-7)
V_{ZA} (3.1-1)
V_{ZA} (3.1.1-1)
V_{ZA} (3.2.1-8)
V_{ZB} (3.1-2)
V_{ZB} (3.2.1-8)
v_{ZN} (12.3.5-8)
W (3.1-3)
$\underline{W}$ (3.1.1-1)
w (18.4.7.3-5)
$\underline{w}'$ (16.1.1-28)
$\underline{w}(t,t_1)$ (15.1.1-13)
$\underline{w}^{L_{Start}}_{GC}$ (17.1.1.4-2)
$w_{GC_{XLStrt}}$ (17.1.1.4-3)
$w_{GC_{YLStrt}}$ (17.1.1.4-3)
w_{SFiRnd_l} (17.2.3.2.3-24)

Wt	(8.1.3.1-1)	$\underline{x}_b$	(15.2.1.1-3)
^{w}WndGst/iGeo^o_m	(17.2.3.2.1-10)	$\underline{\tilde{x}}_{c_n}(+_e)$	(15.1.2.3-1)
$\underline{w}_x$	(16.1.1-5)	x_F	(10.5.1-3)
W_{XA}	(3.1.1-1)	$X_F(S)$	(10.5.1-11)
$\underline{w}_y$	(16.1.1-5)	$\underline{x}_{Hmg}$	(17.2.3.2.3-9)
W_{YA}	(3.1.1-1)	$\underline{x}_{Hmg}(t,t_1)$	(15.1.1-1)
W_{ZA}	(3.1.1-1)	$\underline{x}_{Hmg}(t_i)$	(15.1.1-7)
x	(3.2.2-14)	ξ	(5.1-1)
x	(10.1.4.1-4)	ξ_{Aid}	(15.1-3)
x	(10.5.1-3)	$\xi_{Aid/Out_n}(+_c)$	(15.1.2-28)
$\underline{x}$	(14.6.1-2)	ξ_{INS}	(15.1-3)
$\underline{x}$	(16.1.1-5)	$\xi_{INS/Out_n}(+_c)$	(15.1.2-28)
$\underline{x}$	(18.3-5)	$\xi^N_{JTR_{SDL}}$	(9.4-3)
$\underline{x}'$	(14.6.1-8)	ξ_n	(7.3.1-8)
x'	(16.1.1-28)	$\underline{\xi}_n$	(7.3.1-8)
$\underline{x}'$	(18.3-6)	$\underline{x}_M(t)$	(15.1.5.4.1-3)
$\underline{x}'_0$	(16.2.4-5)	$\underline{x}_{Prt}$	(17.2.3.2.3-9)
x'_i	(16.2.4-5)	$\underline{x}_{Prt}(t,t_1)$	(15.1.1-1)
x'_{Meas_i}	(16.2.4-5)	$\underline{x}_{RW}$	(16.2.3-3)
$\underline{x}'_n(+_e)$	(16.1.1-30)	$\underline{\tilde{x}}$	(14.6.1-2)
$\underline{x}'_n(+_e)$	(16.1.2-25)	$\underline{\tilde{x}}_{Cntrld}$	(15.2.2.1-34)
x'_{Proc_i}	(16.2.4-5)	y	(10.1.4.1-4)
x'_{Tr_i}	(16.2.4-13)	$\underline{y}$	(14.6.2-1)
X(S)	(10.2.1-3)	$\underline{y}$	(16.1.1-5)
X(S)	(10.5.1-11)	$\underline{y}(0)$	(15.1.5.4-8)
x(t)	(10.2.1-1)	Y(S)	(10.2.1-3)
$\underline{x}(t)$	(15.1-1)	y(t)	(10.2.1-1)
$\underline{x}^*$	(16.2.3-6)	$\underline{y}(t)$	(15.1.5.4-2)
$\underline{x}^*$	(14.6.1-9)	$\underline{y}^*(t)$	(15.1.5.4.1-13)
$\underline{x}^*(t)$	(15.1.5.4.1-3)	y_F	(10.5.2-3)
x_1	(10.5.1-3)	Y_i	(14.6.2-6)
x_2	(10.5.1-3)	Y_i	(14.6.5.1-2)
$\underline{x}_a$	(15.2.1.1-3)	y_i	(14.6.2-2)

NOTES

NOTES

NOTES